Parameter Estimation in Stochastic Volatility Models

Jaya P. N. Bishwal

Parameter Estimation in Stochastic Volatility Models

Springer

Jaya P. N. Bishwal
Department of Mathematics and Statistics
University of North Carolina at Charlotte
Charlotte, NC, USA

ISBN 978-3-031-03860-0 ISBN 978-3-031-03861-7 (eBook)
https://doi.org/10.1007/978-3-031-03861-7

Mathematics Subject Classification: 60F05, 60F25, 60E05, 60G15, 60G18, 60G22, 60H05, 60H10, 60H35, 60J60, 62F12, 62M09, 62P05, 65C30, 91G20, 91G30

This Springer imprint is published by the registered company Springer Nature Switzerland AG
The registered company address is: Gewerbestrasse 11, 6330 Cham, Switzerland

To the memory of my late grand parents

and to my parents and brothers for their love and affection

Contents

Pages

Chapter 12

Chapter 13

Chapter 14

Basic Notations

$(\Omega, \mathcal{F}, P)$	Probability space
$(\Omega, \mathcal{F}, \{\mathcal{F}_t\}_{t\geq 0}, P)$	Stochastic basis
$\mathbb{R}$	Real line
$\mathbb{C}$	Complex plane
I_A	Indicator function of a set A
$\stackrel{\mathcal{D}[P]}{\to}$	Convergence in distribution under the measure P
$\stackrel{P}{\to}$	Convergence in probability P
a.s. [P]	Almost surely under the measure P
P-a.s.	Almost surely under the measure P
$a_n = o(b_n)$	$\frac{a_n}{b_n} \to 0$
$a_n = O(b_n)$	$\frac{a_n}{b_n}$ is bounded
$X_n = o_P(b_n)$	$\frac{X_n}{b_n} \stackrel{P}{\to} 0$
$X_n = O_P(b_n)$	$\frac{X_n}{b_n}$ is stochastically bounded, i.e., $\lim_{A\to\infty} \sup_n P\{\lvert\frac{X_n}{b_n}\rvert > A\} = 0$
$\square$	End of a proof
$A := B$	A is defined by B
$A =: B$	B is defined by A
$\equiv$	Identically equal
$\ll$	Absolute continuity of two measures
i.i.d.	Independent and identically distributed
$\mathcal{N}(a, b)$	Normal distribution with mean a and variance b
$\Phi(\cdot)$	Standard normal distribution function
$\mathcal{C}(\cdot)$	Standard Cauchy distribution function
$\mathcal{R}(\cdot)$	Standard Rosenblatt distribution function
$X \sim F$	X has the distribution F
$a \bigvee b$	Maximum of a and b
$a \bigwedge b$	Minimum of a and b

Preface

In this book, we study stochastic volatility models and methods of pricing, hedging, and estimation. Among models, we will study models with heavy tails and long memory or long range dependence. In methods, we will pay special attention to higher order asymptotics. In Chap. 1, we study Methods of Pricing, Hedging and Estimation in Stochastic Volatility Models. Section 1.1 presents Introduction. Section 1.2 discusses Stochastic Derivative and Malliavin Calculus for Stochastic Volatility Models. Section 1.3 discusses Pricing European Options. Section 1.4 discusses Hedging and Greek Estimation. Section 1.5 discusses Bootstrap Method for Volatility Estimation. Section 1.6 discusses Stochastic Gradient Descent Algorithm for American Options. Section 1.7 discusses Peacock Process and Indian Option. Section 1.8 discusses Brownian Excursion and Volatility Estimation in Limit Order Book. Section 1.9 discusses Rate of Cauchy Convergence of Brownian Winding and Asian Options. Section 1.10 discusses Bond Pricing for the Fractional Vasicek Model. Section 1.11 discusses Extreme Value Theory in Finance. Section 1.12 discusses Epsilon-Markov Processes.

In Chap. 2, we study Sequential Monte Carlo Methods. Section 2.1 presents Introduction. Section 2.2 discusses Volatility Models and SMC Methods. Section 2.3 discusses Spline Method in Volatility Estimation. Section 2.4 discusses Multilevel Monte Carlo Methods. Section 2.5 discusses Third Order Composition Method.

In Chap. 3, we study Parameter Estimation in the Heston Model. Section 3.1 presents Introduction. Section 3.2 discusses Continuous Observation. Section 3.3 discusses Discrete Observations. Section 3.4 discusses Supercritical CIR Process. Section 3.5 discusses Sequential Estimation in CIR Process. Section 3.6 discusses Berry-Esseen Bound for Heston Model. Section 3.7 discusses Method of Moments Estimation in Heston Model.

In Chap. 4, we study Fractional Ornstein-Uhlenbeck Processes, Levy-Ornstein-Uhlenbeck Processes and Fractional Levy-Ornstein-Uhlenbeck Processes. Section 4.1 presents Introduction. Section 4.2 discusses Continuous Sampling. Section 4.3 discusses Discrete Sampling. Section 4.4 discusses Ornstein-Uhlenbeck-Gamma Process. Section 4.5 discusses Fractional Levy Ornstein-Uhlenbeck Process. Section 4.6 discusses FIECOGARCH Process. Section 4.7 discusses Fractional Gamma-Ornstein-Uhlenbeck Process.

In Chap. 5, we study Inference for General Semimartingales and Self-similar Processes. Section 5.1 presents Introduction. Section 5.2 discusses Continuous Sampling of Semimartingales. Section 5.3 discusses Discrete Sampling of Semimartingales. Section 5.4 discusses Asymptotics of the Log-likelihood

Function. Section 5.5 discusses Random Observation Period: Sequential Inference. Section 5.6 discusses Fractional and Sub-Fractional Levy O-U Process.

In Chap. 6, we study Estimation in Gamma-Ornstein-Uhlenbeck Stochastic Volatility Model. Section 6.1 presents Introduction. Section 6.2 discusses Gamma-Ornstein-Uhlenbeck Model. Section 6.3 discusses Method of Moments Estimators. Section 6.4 discusses Fractional MS-OU and OU-MS Processes. Section 6.5 discusses Fractional Hermite OU Processes.

In Chap. 7, we study Berry-Esseen Inequalities for the Functional Ornstein-Uhlenbeck-Inverse-Gaussian Processes. Section 7.1 Presents Introduction. Section 7.2 discusses Approximate Maximum Likelihood Estimators. Section 7.3 discusses Berry-Esseen Bounds for AMLE1. Section 7.4 discusses Berry-Esseen Bounds for AMLE2. Section 7.5 discusses Berry-Esseen Bounds for AMCEs. Section 7.6 discusses Geometric Mean-Reversion Process: Black-Karasinski Model.

In Chap. 8, we study Maximum Quasi-likelihood Estimation in Fractional Levy Stochastic Volatility Model. Section 8.1 presents Introduction. Section 8.2 discusses Fractional Levy Process. Section 8.3 discusses Quasi-Maximum Likelihood Estimator. Section 8.4 presents Conclusion.

In Chap. 9, we study Estimation in Barndorff-Nielsen-Shephard Ornstein-Uhlenbeck Stochastic Volatility Model. Section 9.1 presents Introduction. Section 9.2 discusses Modified Tempered Stable Models. Section 9.3 discusses Method of Moments Estimators. Section 9.4 discusses Robust Estimators. Section 9.5 discusses Quasi Maximum Likelihood Estimation in Stable-OU Process. Section 9.6 discusses Empirical Characteristic Function Estimator.

In Chap. 10, we study Parameter Estimation in Student Ornstein-Uhlen beck Model. Section 10.1 presents Introduction. Section 10.2 discusses Student O-U Process. Section 10.3 discusses Speed of Convergence in Cauchy Approximation. Section 10.4 presents Conclusion.

In Chap. 11, we study Berry-Esseen Asymptotics for Pearson Diffusions. Section 11.1 presents Introduction. Section 11.2 discusses Pearson Diffusions. Section 11.3 discusses Fractional Student Ornstein-Uhlenbeck Process. Section 11.4 discusses several Estimators.

In Chap. 12, we study Bayesian Maximum Likelihood Estimation in Fractional Stochastic Volatility Models. Section 12.1 presents Introduction. Section 12.2 discusses Fractional Stochastic Volatility Model. Section 12.3 discusses Properties of Bayesian Maximum Likelihood Estimator. Section 12.4 discusses Fractional Heston Model.

In Chap. 13, we study Berry-Esseen-Stein-Malliavin Theory for Fractional Ornstein-Uhlenbeck Process. Section 13.1 presents Introduction. Section 13.2 discusses Exact Berry-Esseen Bounds. Section 13.3 discusses Continuous Sampling. Section 13.4 discusses Discrete Sampling. Section 13.5 discusses Approximate Minimum Contrast Estimator. Section 13.6 discusses Higher Order Estimators. Section 13.7 discusses Fractional Discrete Sampling.

In Chap. 14 we study Approximate Maximum Likelihood Estimation for Sub-fractional Hybrid Stochastic Volatility Model. Section 14.1 presents Introduction. Section 14.2 discusses Term Structure Models and Derivative Pricing. Section 14.3 discusses Newton-Cotes Distribution and Drift Estimation. Section 14.4 discusses Test Function Estimator of Elasticity of Volatility and Stochastic Elasticity Model. Section 14.5 presents Conclusion.

I would like to take this opportunity to pay my sincere thanks and gratitude to Robert Adler, Yacine Ait-Sahalia, Vlad Bally, Erhan Bayraktar, James Berger, Lloyd Blenman, Arup Bose, Amarjit Budhiraja, Zongwu Cai, Rene Carmona, Peter Carr, Siddhartha Chib, Dan Crisan, Arnaud Doucet, Alan Dow, Darrel Duffie, Robert Elliott, Raya Feldman, Jean-Pierre Fouque, Eric Ghysel, Emmanuel Gobet, Geoffrey Grimmet, Chris Hennessy, Steve Heston, David Hinkley, S.R. Jammalamadaka, Robert Kimmel, Yuri Kutoyants, Roger Lee, Terry Lyons, Mike Ludkovski, Dilip Madan, Joanna Mitro, Stanislav Molchanov, Marek Musiela, James Norris, George Papanicolaou, Edsel Pena, Philp Protter, Zari Rachev, C.R. Rao, Eric Renault, Gareth Roberts, Chris Rogers, Jose Scheinkman, David Siegmund, Chris Sims, Ronnie Sircar, Michael Sørensen, Osnat Stramer, S.R.S. Varadhan, and Ruth Williams, from whom I acquired a lot of knowledge and inspiration during my discussions over the years.

Charlotte, NC, USA Jaya P. N. Bishwal
March 3, 2022

Introduction

Stochastic volatility models are important both for theory and practice. We study methods of pricing, hedging, and estimation. We study models with heavy tails and long memory or long range dependence. Higher order asymptotics methods are paid special attention in this book.

Abergel et al. (2016) studied mathematical modeling of limit order books and empirical properties of order-driven markets. They also studied numerical simulation. Ait-Sahalia and Hansen (2009, 2010) studied financial econometrics including pricing and inference. Ait-Sahalia and Jacod (2014) studied financial econometrics with high frequency. Applebaum (2009) studied stochastic calculus for Levy processes. Arato (1982) studied parameter estimation in linear stochastic systems with constant coefficients. Arnold (1974) studied SDEs with Brownian motion. Bain and Crisan (2009) studied stochastic filtering and particle filters. Bally et al. (2016) studied functional Itô calculus and stochastic integration by parts. Barndorff-Nielsen and Shiryayev (2015) studied change of time and change of measure in finance. Basawa and Scott (1983) studied asymptotic optimal inference for non-ergodic models. Basawa and Prakasa Rao (1980) studied statistical inference for discrete and continuous time stochastic models. Beran (1994) studied statistical for long-memory processes. Bergstrom (1976) studied statistical inference for continuous time economic models. Bertoin (1996) studied Levy processes. Billingsley (1961) studied statistical inference for Markov processes. Bilecki and Rutkowski (2004) studied valuation and hedging in credit risk models. Bjork (1998) studied arbitrage theory in continuous time. Cappe et al. (2005) studied inference in hidden Markov models. Carmona and Tehranchi (2006) studied an infinite dimensional perspective to interest rate models. Cartea et al. (2015) studied algorithmic and high frequency trading. Chen et al. (2011) studied normal approximation by Stein's method. Chow and Teicher (1997) studied discrete time martingales theory. Cohen and Elliott (2015) stochastic calculus for finance with general semimartingales theory. Cont and Tankov (2004) studied financial modeling with jump processes. Crepey (2013) studied a BSDE approach to financial modeling. Curtain and Prichard (1978) studied infinite dimensional linear systems theory. Dacunha-Castelle and Duflo (1986) studied probability and statistics for dependent models. Del Moral (2004) studied Feynman-Kac formulae. Delbaen and Schachermayer (2006) studied arbitrage theory of pricing. Dembo and Zeitouni (2010) studied large deviations techniques and applications. Doob (1953) studied stochastic processes and introduced martingales. Doucet et al. (2001) studied SMC method. Duffie (1995) studied dynamic asset pricing models. Elliot (1982) studied stochastic calculus with respect to martingales. Embrechts and Maejima (2002) studied self-similar processes. Embrechts et al. (1997) studied modeling extremal events in insurance and finance. Filipovic (2009) studied term structure models. Wilmott et al. (1995) studied mathematics of financial derivatives.

Friz et al. (2015) studied large deviations and asymptotic methods in finance. Gallant and Long (1997) studied efficient estimation in SDE by minimum chi-square method. Gallardo and Yor (2006) gave a chaotic representation of multidimensional Dunkl process. Garbaczewski and Olkiewicz (2000) studied Ornstein-Uhlenbeck-Cauchy process. Numerical analysis of SDE was studied in Gard (1988). Gaver and Lewis (1980) studied autoregressive gamma process. Gourieroux and Jasiak (2001) studied estimation in continuous and discrete time models in finance. Gouriéroux et al. (1993) and Gouriéroux and Monfort (1994) studied indirect inference method. Grigelionis (2012) studied Student t-processes. Heyde (1998) studied quasi-likelihood method and optimal estimation for stochastic processes. Harvey et al. (1994) studied multivariate stochastic volatility models. Hopfner (2014) studied inference for stochastic process using the local asymptotic approach. Iacus (2008) studied simulation method in inference for stochastic process using R program. Iacus and Yoshida (2018) studied simulation method in inference for stochastic process using YUIMA program. For a good introduction to Malliavin calculus and Bessel processes, see Ikeda and Watanabe (1989). Itô and McKean (1974) studied sample path properties of diffusion processes. Itô (1984) studied SDEs in infinite dimensional spaces. Ivenenko et al. (2015) studied uniform LAN property of locally stable Levy process observed at high frequency. Jacod and Protter (2012) studied discretization and asymptotics of general SDEs. Jacod (1993) studied inference with random sampling in estimation problems of continuous Gaussian processes with independent increments. Janicki and Weron (1994) studied simulation of stable stochastic processes. Jasra et al. (2019) studied Bayesian inference for stable Levy driven SDEs with high frequency. Jeanblanc et al. (2009) studied option pricing in financial models with jumps. Jondeau et al. (2007) studied financial modeling under non-Gaussian distributions. Kallianpur and Karandikar (2000) studied option pricing for semimartingale type models. Kallianpur and Xiong (1995) studied existence and uniqueness theory of infinite dimensional SDEs. For a basic stochastic processes in continuous and discrete time, see Karlin and Taylor (1975). For a detailed treatment of diffusion processes, see Karlin and Taylor (1981). Kawai and Masuda (2011a) studied simulation of tempered stable random variables. Kawai and Masuda (2011b) studied local asymptotic behavior of Miexner Levy processes under high frequency sampling. Kawai and Masuda (2012) studied infinite variation tempered stable O-U process with discrete observation. Kawai and Masuda (2013) studied LAN for NIG Levy process with high frequency sampling. Kim et al. (2008) studied modified tempered stable distributions with volatility clustering. Kim et al. (2009) studied option pricing for modified tempered stable distributions for GARCH models. Kitagawa algorithm for non-gaussian state space models was introduced in Kitagawa (1987). Quantile regression was studied in Koenker (2005). FBM was introduced by Kolmogorov (1940). Krylov (1995) studied theory of infinite dimensional diffusions. Küchler and Tappe (2008) studied application of bilateral Gamma processes in finance. Küchler and Tappe (2013) studied tempered stable processes. Kunita (1990) studied stochastic flows for SDEs. Kusuoka (1982) studied nonlinear transformation of Gaussian measure on Banach apace and its applications. Lai and Xing (2008) studied discrete statistical models for financial markets. Larsen and Sørensen (2007) studied diffusion models for exchange rate. Laurini and Hotta (2013) studied indirect inference method for fractional Vasicek model. Le Cam (1953) studied asymptotic properties of maximum likelihood and Bayes estimators and equivalence. Le Cam (1986) studied asymptotic methods in statistical decision theory. Le Cam and Yang (1990) studied asymptotics in LAMN type models. Levanony (1994) studied parameter estimation in diffusions using martingale LLN and conditional tail probabilities. Levanony et al. (1993) studied equicontinuity of normalized parameter dependent stochastic integrals with asymptotics as the parameter grows to infinity. Levanony et al. (1994) studied parameter estimation in SDE with unbounded parameter space. Levy process was introduced in Levy (1948). Linetsky (2004) studied applications of Bessel processes and non-affine models to finance. Ling (2005) studied self-weighted least absolute deviation estimation for infinite variation autoregressive models. Linkov (1993) studied asymptotic methods for inference for SDE models. Liu (1988) studied bootstrap procedure under non-i.i.d. models. Liu and Chen (1995b) studied SMC method for dynamical systems. Mandl (1968) studied analytical treatment of one-dimensional Markov processes. Malliavin and Thalmaier (2006) studied Mallian calculus method for finance. Metivier and Pellaumail (1980) studied stochastic integration with respect to martingales. McShane (1974) introduced some numerical methods for SDEs. McNeil et al. (2005) techniques and tools for quantitative risk management. Merton (1990) studied option pricing in continuous-time models. Milshtein (1995) studied numerical integration of SDEs. Stochastic calculus for fBM was studied in Mishura (2008). Maximum of fractional Brownian motion was studied in Molchan (2000). Musiela and Rutkowski (1997) studied martingale methods in financial modeling. Nielsen (1999) studied pricing and hedging of derivative Markets. Nourdin and Peccati (2012) studied Malliavin calculus and Stein's method for normal approximation. Nunno et al. (2009) studied Malliavin calculus for Levy processes with applications to finance. Nualart (1995) studied Malliavin calculus. Peccati and Taqqu (2011) studied Wiener chaos theory. Pardoux and Veretennikov (2001, 2003) studied Poisson equation and diffusion approximation. Platen and Bruti-Liberati (2010) studied numerical solution of SDEs with jumps in finance. Prigent (2003) studied weak convergence of discrete models to continuous models in finance. For a detailed treatment of stochastic interest rate models and bond pricing, see Privault (2012). For option pricing with real Market examples, see Privault (2014). Profeta et al. (2010) have a new look at generalized Black-Scholes model by viewing option prices as probabilities. Protter (1990) studied SDEs with respect to martingales. Rachev and Mittnik (2000) studied stable Paretian models in finance. Rachev et al.

(2005) studied heavy-tailed and skewed asset return distributions in finance. Rebonato (2003) studied interest rate option models. Resnick (2007) studied probabilistic and statistical modeling with heavy tail distributions. Resnick (2008) studied extreme values, regular variation and point processes. Revuz and Yor (1991) studied continuous martingales and Brownian motion. Rogers and Williams (1987) studied Itô calculus, Markov processes and Martingales. Roussas (1972) studied contiguity of probability measures. Ruppert (2004) connected statistical and financial models, studied fitting from real data using SAS program and simulation was done by MATLAB program. Ruppert (2011) studied statistical modeling and data analysis for financial engineering using R program. Sato (1999) studied Levy processes with infinitely divisible distributions. Samorodnitsky and Taqqu (1994) studied nan-gaussian processes with infinite variance. Asymptotic theory of mathematical statistics is studied in Strasser (1985). Stroock and Varadhan (1979) studied multidimensional diffusion processes. Taniguchi and Kakizawa (2000) studied asymptotic theory of statistical inference for time series. Thorisson (2000) studied coupling, stationary, and regeneration of stochastic processes. Tudor (2013) studied analysis of variations for self-similar processes using a stochastic calculus approach. Van der Vaart (2000) studied asymptotic theory of statistics. Van Kampen (1981) studied stochastic processes model in Physics and chemistry. Yor (2001) studied exponential functionals of Brownian motion. Walsh (1986) gave an introduction to SPDEs. Zagst (2002) studied interest rate management theory.

Ait-Sahalia (1996) gave nonparametric test for diffusion models of spot interest rates. Ait-Sahalia (1999) studied transition densities of interest rates and other nonlinear diffusions. Ait-Sahalia (2002) studied maximum likelihood estimation of discretely sampled diffusions based on a closed form Hermie expansion of the likelihood. Ait-Sahalia et al. (2005) studied sampling frequency of continuous time process in the presence of microstructure noise. Ait-Sahalia et al. (2011) studied ultrahigh frequency volatility estimation with dependent microstructure noise. Barndorff-Nielsen and Schmiegel (2009) studied Brownian semistationary processes with its connection to volatility and intermittency. Barndorff-Nielsen and Shephard (2002) studied normal modified stable processes. Barndorff-Nielsen and Shephard (2001) non-Gaussian Ornstein-Uhlenbeck models of volatility. Barndorff-Nielsen and Shephard (2002a,b,c), Barndorff-Nielsen and Shephard (2002d) used realized volatility in estimating stochastic volatility model. Barndorff-Nielsen and Shephard (2003) studied realized power variation in estimating stochastic volatility model. Barndorff-Nielsen and Shephard (2004a,b) studied realized covariation and bipower variation. Barndorff-Nielsen and Shephard (2005b) studied power variation and time change. Barndorff-Nielsen and Shephard (2005c, 2006a) studied multipower variation. Barndorff-Nielsen and Shephard (2006a) studied testing for jumps using bipower variation. Barndorff-Nielsen and Shephard (2006e) studied time-deformed Levy process. Barndorff-Nielsen and Shephard (2007) studied variations, jumps, market frictions and high frequency data in financial econometrics. Barndorff-Nielsen et al. (2005) obtained a central limit theorem for realized power and bipower variations of continuous semimartingales. Barndorff-Nielsen et al. (2004) studied gave a review of power variations and stochastic volatility. Barndorff-Nielsen et al. (2006) gave limit theorems for bipower variation. Bender and Marquardt (2007) studied stochastic calculus for convoluted Levy processes. Bentkus et al. (2003) gave optimal bounds for Cauchy approximations of windings distributions of planar Brownian motion. Bertin et al. (2011) studied drift parameter estimation in fractional diffusions perturbed by perturbed random walks. Berger (1985) studied statistical decision theory and Bayesian analysis. Blumenthal and Getoor (1961) introduced activity index for a jump diffusion process. Brockwell et al. (2007) studied estimation in Levy driven O-U process. Buchmann and Kluppelberg (2005) studied maxima of stochastic processes driven by fractional Brownian motion. Buchmann and Kluppelberg (2006) studied fractional integral equations and state space transform. Buchmann and Chan (2007) studied asymptotic theory of least squares estimators for nearly unstable processes under strong dependence. Carr et al. (2002) introduced the CGMY model. Carr et al. (2003) studied the Levy type stochastic volatility models. Chen and Glasserman (2007) studied Malliavin Greeks without Malliavin calculus. Chib et al. (2006) studied estimation in high dimensional multivariate stochastic volatility models. Comte et al. (2012) studied option pricing in affine fractional stochastic volatility model. Comte and Renault (1996) studied estimation in fractional Vasicek model. Comte and Renault (1996) studied fractional stochastic volatility model. Conley et al. (1997) studied short term interest rate as a subordinated diffusion. Duffie and Glynn (2004) studied estimation of continuous time Markov processes observed at random time intervals. Duffie and Kan (1996) introduced affine models for interest rates. Duffie et al. (2000) studied asset pricing for affine jump- diffusions. Duffie and Singleton (1997) studied term structure models of interest rate swap yields. Fan et al. (2003) studied nonhomogeneous diffusion processes for interest rates. Fearnhead (2002, 2005) studied MCMC and particle filters. Friedman (1975) studied stochastic differential equations with respect to Brownian motion. Gallant and Long (1997) studied efficient estimation in SDEs by minimum chi-squared method. Gallardo and Yor (2006) studied chaotic representation property of the multidimensional Dunkl process. Gelb (1974) studied applied optimal estimation. Genon-Catalot (1987) studied parameter estimation in diffusions based on discrete observations. Genon-Catalot (1990) studied maximum contrast estimation in diffusions based on discrete observations. Genon-Catalot and Jacod (1993, 1994) studied estimation of the diffusion coefficient based on discrete observations. Genon-Catalot et al. (1998a) studied parameter estimation in discretely observed stochastic volatility model. Genon-Catalot et al. (1998b) studied limit theorems for discretely observed stochastic volatility model. Ghosal (1994) studied asymptotics of posterior distribution and Bayes estimators in some nonregular problems. Ghosh et al. (1994), Ghosal et al. (1995) studied the convergence of the posterior distributions in some nonregular cases. Gikhman and Skorohod (1972) studied theory of stochastic differential equations. Goncalves and Meddahi (2009) studied bootstrap for the realized volatility model. Grambsch (1983) studied sequential estimation where the stopping time is defined based on observed Fisher information. Gulisashvili (2012) studied a class of analytically tractable stock price models. Guttorp and Kulperger (1984) statistical inference for some Volterra population processes in a random environment. Hall (1992) studied bootstrap and Edgeworth expansion methods. Hall and Heyde (1980) studied limit theorems for martingales. Hansen and Lunde (2006) studied realized variance and market microstructure. Hansen and Scheinkman (2009) studied an operator theoretic approach to long-term risk. Harison (1992, 1996) studied drift estimation for discretely observed linear diffusion processes. Harvey et al. (1994) studied multivariate stochastic volatility model. Haug and Czado (2006a) studied quasi maximum likelihood estimation in compound Poisson ECOGARCH model. Haug and Czado (2006b) studied fractionally integrated ECOGARCH model. Haug and Czado (2007) studied exponential COGARCH process. Haug et al. (2007) studied method of moments estimation in COGARCH model. Yen and Yor (2013) studied local times and excursion theory for Brownian motion. Hol (2003a,b) studied volatility forecasting in stochastic volatility model. Hu and Long (2010) studied parameter estimation in SDEs driven by stable processes from discrete observations. Hull (2015) studied options, futures, and other derivatives. Hutton and Nelson (1984) studied interchanging the order of differentiation and stochastic integration. Jeganathan (1982, 1995) studied LAMN and LAQ models for time series, respectively. Jennrich and Bright (1976) studied fitting linear systems of SDEs. Jensen and Poulsen (1999) studied approximate maximum likelihood estimation for diffusions. Jensen and Poulsen (2002) studied approximate maximum likelihood estimation based on transition density for diffusions and made numerical comparison. Jones (1984) studied fitting of multivariate models to unequally spaced data. Kallianpur and Selukar (1993) studied estimation of Hilbert space valued parameter by the method of sieves. Karandikar (1983) studied interchanging the order of stochastic integration and ordinary differentiation. Kasonga (1988) studied the conditional least squares estimator for nonlinear diffusion processes. Kasonga (1990) studied minimum distance type estimation in diffusions. Kessler (1997) studied method of moments estimation of an ergodic diffusion from discrete observations. Kessler (2000) studied estimating function for discretely observed diffusions. Kessler and Sørensen (1999) estimation equation based on the eigenfunction of a discretely observed diffusions. Khasminskii et al. (1999) studied estimation of the parameter of linear SDEs. Kim (1996) studied estimation in an infinite dimensional SDE. Adler et al. studied stochastic modeling in physical oceanography. Aihara (1992) studied maximum likelihood estimation of an infinite dimensional parameter in a parabolic SPDE. Aihara (1994) studied maximum likelihood estimation of a discontinuous parameter in a hyperbolic SPDE. Aihara (1998) studied identification of a discontinuous parameter in a parabolic SPDE. Aihara and Bagchi (1988) studied parameter estimation for a diffusion process with unknown boundary conditions. Aihara and Bagchi (1989) studied infinite dimensional parameter identification for stochastic parabolic stochastic systems. Aihara and Bagchi (1991) studied parameter identification for hyperbolic stochastic systems. Aldous and Eagleson (1978) studied mixing and stable convergence in law. Yor (1997) studied speed of convergence towards Cauchy distribution for the asymptotics windings of Planar Brownian motion.

Al Gerbi et al. (2018) studied asymptotics of normalized error of the Ninomiya-Victoir scheme. Anscombe (1952) studied large sample theory of sequential estimation. Arato (1978) studied statistical inference for continuous time Markov processes. Arato et al. (1962) studied estimation in a complex stationary Gaussian process. Bagchi and Borkar (1984) studied parameter identification in infinite dimensional linear systems. Bahadur et al. (1980) studied large deviation for tests and estimates. Bandi and Russel (2008) studied microstructure noise, realized variance and optimal sampling of diffusions. Barczy et al. (2018) studied MLE in a CIR model with jumps. Basu (1983a) studied MLE in a nonlinear SDE for the multiparameter case. Basu (1983b) studied bayes estimators for nonlinear SDEs. Bauer (1980) studied drift estimation for a diffusion process. Bellach (1980, 1983) studied MLE for linear SDE. Bergstrom (1988) gave a review of continuous time economic models. Bertoin (2019) studied ergodic properties of O-U process driven by Levy processes. Bibby and Sørensen (1995a,b) studied martingale estimation function for diffusion processes. Borkar and Bagchi (1982) studied consistency of MLE in a diffusion type process. Borkovec and Kluppelberg (2007) studied extremal behavior of diffusion models in finance. Borup and Jakobsen (2019) studied volatility persistence in a EGARCH model. Borwanker et al. (1971) studied Bernstein-von Mises theorem and Bayes estimators for Markov processes. Bose (1985) studied rate of convergence of the MLE in the O-U process. Bose (1986a) studied Berry-Esseen bound for the MLE in the O-U process. Bose (1986b) studied asymptotic theory of estimators in diffusion processes and ARMA processes. Bossy and Olivero (2018) studied strong convergence of symmetric Milstein scheme for CEV models. Brandt and Santa-Clara (2002) studied simulated MLE for diffusions with application to exchange rate dynamics. Brassesco and Garcia-Pire (2014) studied the density of winding number of planar Brownian motion. Brown and Hewitt (1975a) studied MLE for linear diffusion processes. Brown and Hewitt (1975b) sequential inference for diffusion branching process. Broze et al. (1998) studied quasi indirect inference for diffusion processes. Campbell et al. (1997) studied econometrics of financial markets. Caravenna and Corbetta (2018) studied asymptotic smile of a multiscaling stochastic volatility model. Carr et al. (2011) studied unspanned stochastic volatility and interest rate options. Chang and Chen (2011) studied approximate maximum likelihood estimation of diffusion processes. Chronopoulou and Spiliopoulos (2018) studied sequential Monte Carlo methods for stochastic volatility model. Clement (1993) studied inference for discretely observed diffusions. Clement (1995, 1997a) studied estimation of diffusions by the simulated method of moments. Clement (1997b) studied bias correction for the estimation of discretized diffusion processes from an approximate likelihood. Clement and Gloter (2019) studied estimation function for SDEs driven by stable Levy processes. Clement et al. (2018) studied small time asymptotics for the density of SDEs driven by stable Levy processes. Clinet and Yoshida (2017) studied statistical inference fore ergodic point processes and applied to limit order book. Dereich et al. (2012) studied a Euler type method for strong approximation of CIR model. Dieterich et al. (1980) studied diffusion models for supersonic conductors. Dietz (1989) studied mMLE for diffusion type processes. Dietz (1992) studied a non-Markovian relative to the O-U process. Dietz and Kutoyants (1997) studied minimum distance estimators for diffusion processes. Dohnal (1987) studied estimation of the volatility parameter of a diffusion process based on discrete observations. Dorogovcev (1976) studied consistency of the CLSE of a nonlinear SDE. Dunkl (1992) studied Hankel transforms associated with fine reflection groups. Dupuis and Kushner (1989) studied stochastic approximations and large deviations. Durbin and Koopman (1997) studied Monte Carlo MLE for non-Gaussian state space models. Durbin and Koopman (2000) studied estimation in nao-Gaussian state space models. Duvernet et al. (2010) studied testing for Itâgainst multifractal diffusions. Elerian (1998) studied closed form conditional transition density for diffusions based on Milstein scheme. Elerian et al. (2001) studied likelihood inference for discretely observed nonlinear diffusions based on importance sampling. Es-Sebaiy and Viens (2019) studied optimal rates for parameter estimation of a stationary Gaussian process. Feigin (1979) studied unit root type problems for diffusion processes. Filipovic et al. (2019) studied unspanned stochastic volatility for the multi-factor CIR model. Fisk (1963) studied stratonovich integral with respect to quasimartingales. Florens-Zmirou (1989) studied approximate discrete-time schemes for statistics of diffusion processes. Florens-Zmirou (1991) studied statistics on the crossings of discretized diffusions and local time. Florens-Zmirou (1993) studied estimation of the diffusion coefficient from discrete observations. Fournier (2011) studied simulation and approximation of Levy driven SDes. Giesecke and Smelov (2013) studied exact sampling for jump diffusions. Greenwood and Shiryayev (1992) asymptotic minimaxity of a sequential estimator of a first order autoregressive model. Gushchin (1995) studied asymptotic optimality of estimators under LAQ condition. Gushchin and Küchler (1999) studied asymptotic inference in linear SDE with delay. Hefter and Jentzen (2019) studied rate of convergence of strong numerical approximations of CIR and squared Bessel model. Hoffmann et al. (2018) studied nonparametric inference in SDE with jumps. Huang et al. (2011) studied estimation of discretely observed continuous state branching process with immigration. Huebner (1993) studied parameter estimation in SPDE. Huebner and Rozovskii (1995) studied symptotic properties of MLE for parabolic SPDEs. Huebner et al. (1992) studied parameter estimation in SPDEs by giving two different examples. Ivanov (1976) studied berry-Esseen inequality for the distribution of the least squares estimate. Jacobsen (1998) studied estimation functions for discretely observed diffusions. Jacobsen and Sørensen (2019) studied estimation functions for jump diffusions. Jacod (2018) studied limit of random measures associated with the increments of a Brownian semimartingale. Jacod (2010) studied statistics inference and high frequency data. Jankunas and Khasminskii (1997) studied estimation in a linear SDE. Jazaerli and Saporito (2017) studied functional Itô calculus, path dependence and computation of Greeks. Jiao et al. (2017) studied stable CIR model with branching processes in modeling sovereign interest rate. Khasminskii et al. (1999) studied estimation in a linear SDE with fractional Brownian motion. Klebaner (2012) studied stochastic calculus with applications. Kloeden and Platen (1995) studied numerical solution of SDEs. Kloeden et al. (1994) numerical solution of SDEs through computer experiments. Kloeden et al. (1996) studied the effects of discretization on the estimators of diffusion processes. Koopman and Uspensky (2002) studied the stochastic volatility mean model. Koski and Loges (1985) studied asymptotic inference in a stochastic heat equation. Koski and Loges (1986) studied minimum contrast estimation in Hilbert space valued SDEs. Kruhner and Schnurr (2018) studied time change equations for Levy processes. Küchler and Sørensen (1994a) studied exponential families of stochastic processes and Levy processes. Küchler and Sørensen (1994b) studied exponential families of stochastic processes with time continuous likelihood functions. Küchler and Sørensen (1997) studied exponential family of stochastic processes. Kulinich (1975) studied estimation of the drift of a diffusion process. Laredo (1990) studied asymptotic sufficiency of incomplete observations of a diffusion process. Lebovits and Levy Vehel (2014) studied white noise based stochastic calculus with respect to multifractional Brownian motion. Lee and Kozin (1977) studied asymptotic likelihood theory for diffusion processes. Leonenko et al. (2020) studied approximation of heavy-tailed fractional Pearson diffusion. Leonenko and Suvak (2010a) studied statistical inference for a student diffusion process. Leonenko and Suvak (2010b) studied statistical inference for a reciprocal gamma process. Linkov (1990) studied local asymptotic mixed normality for diffusion type processes. Liptser and Shiryayev (1982) studied rate of convergence in the central limit theorem for semimartingales. Liu and Chen (1995a) studied blind deconvolutions via sequential imputations. Liu and Chen (1995b) studied sequential Monte Carlo methods for dynamical systems. Liu et al. (2019) studied LAN property of SDE with fractional noise. Liu (2017) studied jump robust estimation of volatility with the presence of microstructure noise and multiple observations. Loges (1984) studied Girsanov theorem and parameter estimation in a Hilber-space valued SDE. Long et al. (2017) studied least squares estimation of SDEs driven by Levy noise. Ma (2015) studied Lamperti transformation for continuous state branching process. Masuda (2005) studied method of moments for partially and discretely observed models driven by Levy process. Meddahi (2002) studied a theoretical comparison between integrated and realized volatility. Mider et al. (2019) studied simulation of bridges using confluent diffusions. Oblakova

(1989) studied rate of convergence in the central limit theorem for stochastic integrals. Nualart and Xu (2019) studied asymptotic behavior of additive functional of two independent self-similar Gaussian processes. El Omari et al. (2020) studied nonparametric estimation of fractional diffusions with random effects. Overbeck (1998) studied estimation of continuous branching diffusion processes. Overbeck and Ryden (1997) studied estimation in CIR model based on discrete observations. Pedersen (1994) studied quasi likelihood inference for discretely observed diffusion. Pedersen (1995a,b) MLE for SDE based on discrete observations. Renyi (1963) introduced stable convergence in law. Sabzikar and Surgailis (2018) studied invariance principles for tempered fractionally integrated processes. Sandmann and Koopman (1998) studied estimation in stochastic volatility model via Monte Carlo maximum likelihood. Schwartz (1965) studies Bayes procedures. Shephard and Bos (2006) studied inference for stochastic volatility model using conditional Gaussian state space form. Shephard et al. (2005) studied parallel computation in econometrics. Shephard (2006) studied stochastic volatility. Shephard and Pitt (1997) studied likelihood analysis of non-Gaussian processes. Shevchenko and Tudor (2020) studied parameter estimation for the Rosenblatt O-U process with periodic mean. Shimizu (2006) studied M-estimation for discretely observed diffusions with jumps. Shimizu and Yoshida (2006) studied parameter estimation for diffusion processes with jumps from discrete observations. Shoji (1997) studied estimators derived by Euler method for diffusion processes. Singer (1993) studied estimation in continuous time dynamical models with discrete data. Spokoiny (1992) studied lower bounds for minimax risk in estimation problems without LAN condition. Stockmarr and Jacobsen (1994) studied weak convergence of autoregressive processes to Gaussian diffusions and statistical inference. Strauch (2015) studied sharp adaptive drift estimation for ergodic diffusions. Su and Wang (2019) studied parameter estimation for fractional diffusions based on discrete observations. Vetter (2015) studied estimation of integrated volatility of volatility with application to goodness of fit. Yoshida (1990) studied asymptotic behavior of M-estimator and related random fields for diffusion processes. Yoshida (1992) studied estimation of diffusions processes from discrete observations by a Euler type approximation. Uehara (2019) studied statistical inference for misspecified ergodic Levy driven SDEs. Tsitovich (1977) studied the estimation of the drift of a diffusion process from observations in a domain. Trofimov (1982) studied a stochastic differential equation for the maximum likelihood estimator of the drift parameter of a diffusion process. Tikhov (1978) studied optimal sequential estimation for non-quadratic loss functions. Tikhov (1980) studied optimal sequential estimation for non-quadratic loss functions. Touti (1993) studied the rate of convergence in the central limit theorem for stochastic integrals. Taraskin (1974, 1980) studied drift estimation in multidimensional diffusion processes. Sweeting and Adekola (1987) studied asymptotic posterior normality for stochastic processes. Surace and Pfister (2016) studied maximum likelihood estimation in partially observed diffusions. Strauch (2015) studied sharp adaptive drift estimation for ergodic diffusions in the multivariate case. Strasser (1975) studied asymptotic equivalence of the Bayes and maximum likelihood estimation. Strasser (1976) studied asymptotic properties of the posterior distributions. Strasser (1977) obtained improved bounds on the Bayes and maximum likelihood estimators. Stoyanov (1984) studied estimation problems in continuous discrete stochastic models. Stoyanov and Vladeva (1982) studied estimation of unknown parameter of continuous time stochastic processes observed at random moments. Stockmarr and Jacobsen (1994) studied weak convergence of autoregressive processes to Gaussian diffusions and statistical inference. Sørensen (1999) studied asymptotics of estimating functions. Sørensen (1983) studied maximum likelihood estimation in a randomly stopped diffusion type process. Sørensen (2001) gave a review of discretely observed diffusions based on approximation of the continuous time score function. Sørensen (1992) studied continuous record asymptotics in systems of SDEs. Robinson (1977) studied estimation of the Ornstein-Uhlenbeck model based on discrete observations. Rootzen (1980) studied limit distributions of the error in approximation of stochastic integrals. Piterbarg and Rozovskii (1996) studied maximum likelihood estimation in SPDEs. Penev (1985) studied parametric statistical inference for multivariate diffusions based on discrete observations. Pages and Panloup (2012) studied rate of convergence in the ergodic approximation of the distribution of a stationary diffusion. Novikov (1972) studied sequential estimation of the parameters of a diffusion process. Musiela (1976) studied risk of maximum likelihood estimator of the parameters of a diffusion process. Musiela (1984) studied inference for diffusion processes. Mohapl (1994, 1997) studied estimation in infinite dimensional diffusions. McLeish and Kolkiewicz (1997) studied fitting of diffusion models to finance. Mao (1995) studied stochastic differential equations and applications. Luschgy (1992) studied local asymptotic mixed normality for semimartingales. Luschgy (1993) studied second order behavior of Neyman-Pearson tests for diffusions. Luschgy (1994a) studied asymptotic expansions for error probabilities for tests. Luschgy (1994b) studied asymptotic behavior of Neyman-Pearson tests for autoregressive processes. Luschgy (1994c) studied asymptotic inference for semimartingale models with singular parameter points. Luschgy (1995) studied local asymptotic quadracity for stochastic process based on stopping times. Lo (1988) studied maximum likelihood estimation of generalized Itô processes with discretely sampled data. Le Cam (1953) studied asymptotics of maximum likelihood estimators and Bayes estimators. Le Cam (1973) studied convergence of estimates under dimensionality restrictions. Le Cam and Yang (1990) studied asymptotic methods in statistical decision theory. Ibragimov and Khasminskii (1981) studied asymptotic theory of statistical inference based on local asymptotic approach. Van der Meulen and Schauer (2018) studied Bayesian estimation of incompletely observed diffusions. Woerner (2005) studied estimation of integrated volatility in stochastic volatility model. Woerner (2007) studied inference in Levy-type stochastic volatility models. Zhang (2006) studied efficient estimation of stochastic volatility model with noisy observations using a multiscale approach. Zhang et al (2005) two scale approach to estimation of integrated volatility. Zhang et al. (2011) studied Edgeworth expansion of realized volatility. Zhou (1996) studied Itô conditional moments generator and short rate process.

Bishwal (2022a) studied mixingale estimation function for SPDEs with random sampling. Bishwal (2022b) studied Berry-Esseen bounds of the quasi maximum likelihood estimators for the discretely observed diffusions. Bishwal (2022c) studied strong consistency of the Stratonovich estimator for the Ito diffusion. Bishwal (2022d) studied parameter estimation in Levy driven stochastic volatility models. Bishwal (2021a) studied a new algorithm for approximate maximum likelihood estimation in sub-fractional Chan-Karolyi-Longstaff-Sanders model. Bishwal (2021b) studied parameter estimation and hypothesis testing for nonlinear SPDEs. Bishwal (2021c) studied Berry-Esseen bounds of approximate Bayes estimators for the discretely observed Ornstein-Uhlenbeck process. Bishwal (2018a) studied Bernstein-von Mises theorem and asymptotic properties of Bayes estimators for parabolic SPDEs with small noise. Bishwal (2018b) studied sequential maximum likelihood estimation in nonlinear non-Markov diffusion type processes. Bishwal (2017a) studied conditional least squares estimation for nonergodic diffusion processes based on discrete observations. Bishwal (2017b) studied valuation of real options when the shocks are persistent. Bishwal (2017c) studied hypothesis testing for fractional SPDEs and applied to neurophysiology and finance. Lee et al. (2012) studied sequential maximum likelihood estimation for reflected O-U process. Bishwal (2011a) studied Berry-Esseen inequalities for the discretely observed Ornstein-Uhlenbeck-Gamma process. Bishwal (2011b) studied minimum contrast estimation in fractional O-U process based on continuous and discrete sampling. Bishwal (2011c) studied sieve estimator for fractional SPDE. Bishwal (2011d) studied Milstein approximation for posterior density of diffusions and related Bayes estimators. Bishwal (2011e) studied sufficiency and Rao-Blackwellization of Vasicek model. Bishwal (2011f) studied some new estimators of integrated volatility. Bishwal (2011g) studied estimation in interacting diffusions based on continuous and discrete sampling. Bishwal (2011h) studied stochastic moment problems and hedging of Black-Scholes options. Bishwal (2011i) studied maximum quasi-likelihood estimation in fractional Levy stochastic volatility model. Bishwal (2011j) studied extreme value theory in finance. Bishwal (2010b) studied maximum likelihood estimation in Skorohod anticipative stochastic differential equations. Bishwal (2010c) studied sequential Monte Carlo methods for stochastic volatility models. Bishwal (2010d) studied conditional least squares estimation in diffusion process based on random sampling of Poisson type. Bishwal (2009a) studied Berry-Esseen inequalities for discretely observed diffusions. Bishwal (2009b) studied M-estimation for discretely sampled diffusions. Bishwal (2008b) studied large deviations in testing fractional O-U models. Bishwal (2007b) studied large deviations and Berry-Esseen inequalities for estimators of nonlinear nonhomogeneous diffusions. Bishwal (2006b) studied sequential maximum likelihood estimation in semimartingales. Masuda (2005) studied method of moments for partially observed Levy type ergodic models based on discrete observations. Masuda (2009a) studied joint estimation discretely observed stable Levy processes with symmetric Levy density. Masuda (2009b)studied statistical aspects in calibrating geometric skewed stable asset price model. Masuda (2010a) studied approximate quadratic estimating function for discretely observed Levy driven SDEs with application to a noise normality test. Masuda (2010b) studied asymptotic mixed normality in the self-weighted LAD estimation of Levy OU process. Masuda (2011) studied quasi-likelihood analysis for stochastic differential equations with jumps. Masuda (2013a) studied convergence of Gaussian quasi-likelihood random fields for ergodic Levy driven SDE observed art high frequency. Masuda (2013b) studied optimal estimation of stable Ornstein-Uhlenbeck processes. Masuda (2013c) studied asymptotics for functionals of self-normalized residuals of discretely observed stochastic processes. Masuda (2013d) studied multi-step estimation procedure for stable Ornstein-Uhlenbeck processes. Masuda (2015) studied parametric estimation of Levy processes. Masuda (2019) studied non-Gaussian quasi-likelihood estimation of SDE driven by locally stable Levy process. Masuda and Uehara (2017) studied two-step estimation of ergodic Levy driven SDE. Masuda and Uehara (2018) studied estimation of diffusion with compound Poisson jumps based on self-normalized residuals. Masuda and Yoshida (2005) studied asymptotic expansion for Barndorff-Nielsen and Shephard's stochastic volatility models. Masuda and Yoshida (2013) studied Edgeworth expansion for the integrated Levy driven Ornstein-Uhlenbeck process. Kawai and Masuda (2011a) studied simulation of tempered stable random variables. Kawai and Masuda (2011b) studied local asymptotic behavior of the likelihood function for Meixner Levy processes under high-frequency sampling. Kawai and Masuda (2012) studied Infinite variation tempered stable Ornstein-Uhlenbeck processes with discrete observations. Kawai and Masuda (2013) studied local asymptotic normality for normal inverse Gaussian Levy processes with high-frequency sampling. Mishra (1989) studied Bernstein-von Mises theorem for a class of non-homogeneous diffusion processes. Mishra and Bishwal (1995) studied approximate maximum likelihood estimation for diffusion processes from discrete observations. Mishra and Prakasa Rao (1985a) studied the Berry-Esseen theorem for maximum likelihood estimator for linear homogeneous diffusion processes. Mishra and Prakasa Rao (1985b) studied asymptotic theory of maximum likelihood estimation for nonhomogeneous diffusion processes. Mishra and Prakasa Rao (1987) studied rate of convergence in the Bernstein-von Mises theorem for a class of diffusion processes. Mishra and Prakasa Rao (1991) studied bounds on the equivalence of Bayes and maximum likelihood estimators for a class of diffusion processes. Prakasa Rao (1973) studied the rate of convergence of estimators for Markov Processes. Prakasa Rao (1978) studied the rate of convergence of the Bernstein-von Mises approximation for Markov processes. Prakasa Rao (1980) studied Bernstein-von Mises theorem for a class of diffusion processes. Prakasa Rao (1981) studied Bayes estimation for diffusion fields. Prakasa Rao (1982) studied maximum probability estimation for diffusion processes. Prakasa Rao (1983) studied asymptotic theory for non-linear least squares estimator for diffusion processes. Prakasa Rao (1984) studied the exponential rate of convergence of the least squares estimator in the nonlinear regression model with Gaussian errors. Prakasa Rao (1985) studied estimation of the drift for diffusion processes. Prakasa Rao (1986a) studied weak convergence of the least squares random process in the smooth case. Prakasa Rao (1986b) studied the weak convergence of the least squares random fields and its applications. Prakasa Rao (1988a) studied law of the iterated logarithm for fluctuation of posterior distributions for a class of diffusion processes and a sequential test of power one. Prakasa Rao (1988b) studied statistical inference from sampled data for stochastic processes. Prakasa Rao (1999a) studied statistical inference for diffusion type processes. Prakasa Rao (1999b) studied semimartingales and their Statistical Inference. Prakasa Rao (2010) studied statistical inference for fractional diffusion processes. Prakasa Rao and Rubin (1981) studied asymptotic theory of estimation in nonlinear stochastic differential equations. Prakasa Rao and Rubin (1998) studied uniform approximation for families of stochastic integrals.

Kutoyants (1977) studied estimation of the trend parameter of a diffusion process in the smooth case. Kutoyants (1978) studied estimation of a parameter of a diffusion process. Kutoyants (1984b) studied parameter estimation for diffusion type processes of observations. Kutoyants (1994b) studied nonconsistent estimation by diffusion type observations. Kutoyants (2004) studied statistical inference for ergodic diffusion processes. Kutoyants (1984a) studied Parameter Estimation for Stochastic Processes. Kutoyants and Pohlman (1994) studied parameter estimation for Kalman-Bucy filter with small noise. Kutoyants (1994a) studied identification of dynamical systems with small noise. Le Breton (1976) studied continuous and discrete sampling for parameter estimation in diffusion type processes.

Le Breton (1977) studied parameter estimation in a linear stochastic differential equation. Le Breton (1998a) studied filtering and parameter estimation in a simple linear system driven by fractional Brownian motion. Le Breton (1998b) studied a Girsanov type approach to filtering in a simple linear system driven by fractional Brownian motion. Le Breton and Musiela (1984) studied one dimensional bilinear differential model for stochastic processes.

Chapter 1

Stochastic Volatility Models: Methods of Pricing, Hedging and Estimation

1.1 Introduction

Stochastic volatility models are partially observed diffusions and are hidden Markov models when the driving noises are Brownian motions. The chapter is concerned with the study of statistics, econometrics, and financial engineering of high-frequency financial data. The development of increasingly complex financial products requires the use of advanced statistical methods. The purpose of the chapter is to present generalized bootstrap methods for estimation, calibration, and Malliavin calculus methods for pricing, hedging of derivative products (on equities, interest rate, credit risk), and portfolio optimization. Special attention will be paid to models in high dimension, models with jumps, models with long memory in stochastic volatility models.

More specifically, high-frequency financial data, Malliavin calculus, accuracy of normal approximation, Edgeworth expansion, Berry–Esseen theorems, large deviations, fractional Brownian motion, stochastic volatility, realized volatility, integrated volatility, range-based estimator, option pricing, fractional Black-Scholes model, fractional Hull-White model, volatility persistence, fractional volatility, volatility testing, bias reduction, empirical likelihood, hypothesis testing, local-linear smoothing, local time, nonparametric testing, nonparametric estimation, variance stabilization, resampling, bootstrap, weighted bootstrap, wild bootstrap, paired bootstrap, generalized bootstrap, jackknife, and bootstrap specification test will be given special emphasis in this chapter.

Since Black and Scholes (1973) established the theory of option pricing which won Scholes a Nobel prize, volatility has played an important role not only in derivative pricing but also in portfolio selection and risk management. Despite the assumption of constant volatility in Black and Scholes (1973), it is widely recognized that volatility changes over time. The *stylized facts* of volatility are: (1) volatility changes in time, (2) volatility is random, (3) volatility has heavy tails, (4) volatility clusters on high level. Starting with Engle (1982)'s autoregressive conditional heteroskedasticity (ARCH) model which won him a Nobel prize, various stochastic volatility models have been proposed. On the other hand, volatility is often modeled as a parametrized diffusion coefficient of a continuous time diffusion process and then the pa-

J. P. N. Bishwal, *Parameter Estimation in Stochastic Volatility Models*,
https://doi.org/10.1007/978-3-031-03861-7_1

rameters are estimated by various methods like filtered maximum likelihood or method of moments. One should also focus on nonparametric estimation of volatility process. In principle, the more data we can use, the more accurate the estimate will be. However, one will have technological restriction on the amount of data. Recently this kind of restriction have been removed by development of computer power and data recording system. Those kind of data are called *high-frequency data.* Such high-frequency data lend the validity of the method based on quadratic variation formula, that is called the realized volatility. Barndorff-Nielsen and Shephard (2001–2007) have suggested modeling the volatility as a Levy driven Ornstein–Uhlenbeck process.

In the stochastic volatility model, the log-price $y^* = \log S$ with S being the asset price follows

$$dy^*(t) = \alpha(t)dt + \sigma(t)dW(t),$$
$$d\sigma^2(t) = \tilde{\alpha}(t)dt + \tilde{\sigma}(t)d\tilde{W}(t),$$

where σ and α are assumed to be independent of the standard Brownian motion W. The process σ is called the *instantaneous volatility* or *spot volatility* and α is called the *mean process* and W and $\tilde{W}$ are uncorrelated. A simple example of this is

$$\alpha(t) = \mu t + \beta\sigma^{2*}(t) \quad \text{where} \quad \sigma^{2*}(t) = \int_0^t \sigma^2(u)du,$$

in which case β is called the *risk premium* and σ^{2*} is called the *integrated variance.*

Over an interval of time length $h > 0$, returns are defined as

$$y_i := y^*(hi) - y^*((i-1)h), \quad i = 1, 2, \cdots, T,$$

which implies that

$$y_i|\alpha_i, \sigma_i^2 \sim \mathcal{N}(\alpha_i, \sigma_i^2),$$

where

$$\alpha_i := \alpha(ih) - \alpha((i-1))h)$$

and

$$\sigma_i^2 := \sigma^{2*}(ih) - \sigma^{2*}((i-1))h) = \int_{(i-1)h}^{ih} \sigma^2(u)du.$$

Here σ_i^2 is called the *actual variance* and α_i is called the *actual mean.* Suppose one is interested in estimating the actual volatility σ_i using M intra-h observations. A natural candidate is the *realized volatility* given by

$$\sqrt{[y_M^*]} := \sqrt{\sum_{i=1}^{M} y_{j,i}^2},$$

where

$$y_{j,i} := y^*\left((i-1)h + \frac{jh}{M}\right) - y^*\left((i-1)h + \frac{(j-1)h}{M}\right), \quad j = 1, 2, \cdots, M.$$

When $h \to 0$, realized volatility converges in L_2 to the integrated volatility. We consider the fixed h case. One could also obtain large deviations and moderate deviations results.

First one should obtain Berry–Esseen type bound on

$$\sup_{x\in\mathbb{R}} \left| P\left(\frac{\sqrt{\frac{M}{h}}([y_M^*] - \sigma_i^2)}{\sqrt{2\sigma_i^{[4]}}} \le x \right) - \Phi(x) \right|,$$

where

$$\sigma_i^{[4]} := \int_{(i-1)h}^{ih} \sigma^4(u)du$$

is the *actual quarticity* and σ^4 is the *spot quarticity*.

One could use the decomposition technique (Michel and Pfanzagl 1971 developed for the classical i.i.d. case): Decompose the rate of convergence of the numerator to normal distribution and rate of convergence of the denominator to a constant. Note that

$$y_{j,i} = \alpha_j + \epsilon_j,$$

where by Itô formula

$$\alpha_j := \int_{(i-1)h+\frac{(j-1)h}{M}}^{(i-1)h+\frac{jh}{M}} \alpha(u)du, \quad \epsilon_j := \int_{(i-1)h+\frac{(j-1)h}{M}}^{(i-1)h+\frac{jh}{M}} \sigma(u)dW(u).$$

Hence

$$\begin{aligned} y_{j,i}^2 &= \alpha_j^2 + 2\alpha_j\epsilon_j + \epsilon_j^2 \\ &= \sigma_j^2 + \alpha_j^2 + 2\alpha_j\epsilon_j + (\epsilon_j^2 - \sigma_j^2) \\ &= \sigma_j^2 + e_j, \end{aligned}$$

where

$$e_j := \alpha_j^2 + 2\alpha_j\epsilon_j + 2\int_{(i-1)h+\frac{(j-1)h}{M}}^{(i-1)h+\frac{jh}{M}} \left(\int_{(i-1)h+\frac{(j-1)h}{M}}^{u} \sigma(s)dW_s \right) \sigma(u)dW_u.$$

Thus

$$[y_M^*] - \sigma_i^2 = \sum_{j=1}^{M} e_j := e_M^*(h).$$

The term e_j is called the *realized volatility error.* Since $E(e_j|\sigma_j^2) = 0$, realized volatility is an unbiased estimator of actual volatility. When $M \to \infty$, $y_M^* \to \sigma_i^2$ almost surely, so it is strongly consistent estimator.

$$\lim_{h\to 0} \frac{E(e_M^*(h))}{\sqrt{h}} = 0$$

since

$$e_M^*(h) = \sum_{j=1}^{M} e_j = \sum_{j=1}^{M} \sigma_j^2(v_j^2 - 1),$$

where $v_j \sim^{iid} N(0,1)$ and independent of σ_j^2. It is clear that e_j is a weak white noise sequence which is uncorrelated with the actual volatility process σ_j^2. Thus

$$\frac{\sqrt{\frac{M}{h}}([y_M^*] - \sigma_i^2)}{\sqrt{2\sigma_i^{[4]}}} = \frac{\sqrt{\frac{M}{h}} \sum_{j=1}^{M} \sigma_j^2(v_j^2 - 1)}{\sqrt{2\sigma_i^{[4]}}}.$$

One can apply the characteristic function technique followed by Esseen's lemma for the numerator. Then one can apply the splitting technique of Michel and Pfanzagl (1971) or the more precise squeezing technique of Pfanzagl (1971). Then one should refine the result using the generalized bootstrap techniques.

We will emphasize on model calibration and inference. The implied volatility smile phenomenon has led to the appearance of a large variety of extensions of the Black-Scholes model: the local volatility models, diffusions with stochastic volatility, jump diffusions, and long-memory models. The model calibration is the reconstruction of model parameters from the prices of traded options. It is an inverse problem to that of option pricing and as such, typically ill-posed.

We will also emphasize on interest rate models. The calibration problem is yet more complex in the interest rate markets since in this case the empirical data that can be used includes a wider variety of financial products from standard obligations to swaptions. Efficient computational algorithms are thus most needed.

A stochastic volatility model models both the underlier's value and its volatility as stochastic processes. As with a jump-diffusion model, this has the effect of giving the underlier's value a leptokurtic distribution. Heston's (1993) model is a popular stochastic volatility model.

A jump-diffusion model adds random jumps to the geometric Brownian motion that Black-Scholes (1973) assumes for the underlier. Among other things, this has the effect of giving the underlier's value a leptokurtic distribution.

It is well accepted that estimating financial volatility is a central issue in financial theory and practice. To estimate volatility, researchers have given attention to high-frequency data. Unfortunately, high-frequency data is often contaminated my movements dictated by institutional factors of the market. Removing these institutionally driven factors, termed *microstructure effects*, is vital to accurately determine the price volatility. Thus high frequency data is contaminated by market microstructure noise. The available market microstructure noise models have used i.i.d. noise with constant variance to describe market microstructure effects. In continuous time, i.i.d. process is a process with extreme volatility and because of this, the observed price process is dominated by the noise process and carries no information on the volatility of the efficient price process. To overcome this problem, we propose fractional Brownian motion driving noise for market microstructure noise which carries long memory of volatility.

The main difficulty in modeling and forecasting financial volatility is the fact that the volatility is latent and not directly observable, like a *hidden Markov model.* Usually researchers have used realized volatility as an observable proxy for latent volatility. Realized volatility is defined as the summation of squared intraday returns. It is a good estimate to estimate the object of interest, the integrated volatility of an asset, so is optimal to use all the available data. However, it is a common practice in measuring realized volatility to avoid the data at highest frequency available, the tick-by-tick data. Instead, researchers tend to sample intra daily returns at somewhat moderate frequencies such as 5, 10, or 30 minutes. This is due to the suspicion that the market microstructure effects become more severe as sampling frequency increases.

In Sect. 1.2, we study some aspects of the method of Malliavin calculus, also known as the stochastic calculus of variations, for the Monte Carlo estimation of the sensitivity parameters of financial models. We also study bootstrap method for estimation and testing in continuous time stochastic volatility models based on discrete observations. We put special emphasis on jumps and long memory in the volatility process.

1.2 Stochastic Derivative and Malliavin Calculus for Stochastic Volatility Models

The idea of stochastic (or Malliavin) derivative is to define the notion of differentiability within the family of random variables that are equal to or can be approximated by functions of independent increments of Brownian motion. Under suitable assumptions, this family is wide enough to contain solution of stochastic differential equations.

Let $(\Omega, \mathcal{F}, P)$ denote the Wiener space, i.e., $\Omega = C_0[0,T]$ is the Wiener space of continuous functions on $[0,T]$ with initial value 0 which is equipped with the supremum norm $\| \,.\, \|, \mathcal{F}$ the Borel σ-algebra of subsets of Ω, and P is the standard Wiener measure on $(\Omega, \mathcal{F})$. We will represent by $W(\omega,t) = \omega(t), 0 \leq t \leq T,\ \omega \in \Omega$ the Wiener process on the canonical probability space $(\Omega, \mathcal{F}, P)$.

Wiener–Itô Chaos Expansion

The stochastic Sobolev space $\mathbb{D}_{1,2}$ consists of all F_T measurable random variables $F \in L^2(P)$ with chaos expansion

$$F = \sum_{n=0}^{\infty} I_n(f_n), \quad f_n \in \tilde{L}^2((\lambda \times \nu)^n)$$

satisfying the convergence criterion

$$\|F\|_{\mathbb{D}_{1,2}} := \sum_{n=1}^{\infty} nn!\|f\|_{L^2((\lambda\times\nu)^n)}.$$

$F \in \mathbb{D}_{1,2} \subset L^2(P)$. The operator D is defined as

$$D_{t,z}F = \sum_{n=1}^{\infty} nI_{n-1}((f_n(\cdot,t,z)).$$

On $(\Omega, \mathcal{F}, P)$ consider the homogeneous nonlinear Skorohod stochastic differential equation

$$\begin{aligned} dX(t) &= f(\theta, X(t))dt + g(\sigma, X(t))dW(t),\ 0 \le t \le T \\ X(0) &= \psi(X(T)) \end{aligned} \tag{2.1}$$

ψ is a known function, f is a known real valued function defined on $\mathbb{R}$ satisfying the existence and uniqueness of solution to (2.1), see Nualart (1995), $\theta \in \Theta \subset \mathbb{R}$ and $\sigma \in \Sigma \subset \mathbb{R}^+$ are the unknown parameters which are to be estimated on the basis of n- independent copies $X_1(t), X_2(t), \ldots, X_n(t)$ of $\{X(t), 0 \le t \le T\}$. Let θ_0 and σ_0 be the true values of the unknown parameters.

Next we introduce some basic tools from Malliavin calculus. Let $C_b^\infty(\mathbb{R}^n)$ be the set of C^∞ functions $g : \mathbb{R}^n \to \mathbb{R}$ which are bounded and have bounded derivatives of all orders. The class of all real random variables of the form $F(\omega) = g(W(t_1), W(t_2), \ldots, W(t_n)),\ g \in C_b^\infty(\mathbb{R}^n)$, called Wiener functionals, is denoted by $\mathcal{S}$. The space $\mathbb{D}^{1,p}$ designates the Banach space which is the completion of $\mathcal{S}$ with respect to the norm

$$\|F\|_{1,p} = \{E|F|^p\}^{1/p} + \left(E\left[\left\{\int_0^T |D_sF|^2 ds\right\}^{p/2}\right]\right)^{1/p},$$

where

$$D_sF := \sum_{i=1}^{n} \frac{\partial g}{\partial x_i}(W(t_1), W(t_2), \ldots, W(t_n))I_{[0,t_i]}(t).$$

More generally, the k-th order derivative of F is defined as the k-parameter process given by $D^{(k)}_{s_1\ldots s_k}(F) = D_{s_1}D_{s_2}\ldots D_{s_k}F$.

The space $\mathbb{D}^{m,p}$ is defined analogously and its associated norm is denoted by $\|.\|_{m,p}$. That is,

$$\|F\|_{m,p} = \{E|F|^p\}^{1/p} + \left(E\left[\left\{\int_0^T \cdots \int_0^T |D^{(m)}_{s_1\ldots s_m}F|^2 ds_1\ldots ds_m\right\}^{p/2}\right]\right)^{1/p}.$$

The space $\mathcal{S}$ is dense in $(\mathbb{D}^{1,p}, \|.\|_{1,p})$, $p \geq 1$. Set $\mathbb{D}^{1,\infty} = \cap_{p\geq 1}\mathbb{D}^{1,p}$. The adjoint of the closed unbounded operator $D : L^2(\Omega) \to L^2([0,T] \times \Omega)$ is usually denoted by δ and is called the *Skorohod integral*. The domain of δ is the class of processes $u \in L^2([0,T] \times \Omega$ such that

$$\left| E\left(\int_0^T D_t F u_t dt\right)\right| \leq C\|F\|_{L^2(\Omega)},$$

for all $F \in \mathcal{S}$. Here C is a constant which may depend on u. In the case that u is in the domain of δ, then $\delta(u)$ is the square integrable random variable defined by the *duality relation*

$$E(\delta(u)F) = E\left(\int_0^T D_t F u_t dt\right)$$

for all $F \in \mathbb{D}^{1,2}$. The Skorohod integral $\delta(u)$ turns out to be an extension of the classical Itô integral and it allows one to integrate processes u not necessarily adapted. For this reason one also writes $\delta(u) = \int_0^T u(s)dW(s)$.

For any real number $p \geq 1$ and any integer $n \geq 1$ we set $\mathbb{L}^{n,p} = \mathbb{L}^p([0,T]; \mathbb{D}^{n,p})$ and $\mathbb{L}^{1,\infty} = \cap_{p\geq 1}\mathbb{L}^{1,p}$. The processes u of the space $\mathbb{L}^{1,2}$ verify $uI_{[0,t]} \in$ D*om* δ for each $t \in [0,T]$, and for those processes one can define the *indefinite Skorohod integral* $Y_t = \delta(uI_{[0,T]})$ denoted also by $\int_0^t u(s)dW(s)$. This process has continuous modification provided $u \in \mathbb{L}^{1,2}$ and $E(\int_0^T(\int_0^T |D_s u_r|^2 ds)^p dr)$ $< \infty$ for some $p > 2$. The trajectory of this process is continuous but highly irregular. Imkeller (1994) showed that one can represent the Skorohod integral as the composition of a Gaussian semimartingale depending on an infinite dimensional parameter with a Gaussian vector.

The following proposition is a trivial consequence of the fact that Itô integral and thus also iterated Itô integrals have zero expectation.

Proposition 2.1 For any $u \in Dom(\delta)$, the Skorohod integral has zero expectation, that is,

$$E(\delta(u)) = 0.$$

Duality Relation
For every $X \in \mathcal{S}$ and $U \in Dom(\delta)$,

$$E\left(\int_0^T (D_t X)U_t dt\right) = E\left(X\int_0^T U_t \diamond dW_t\right).$$

Example 2.1 Let $X = W_t$. Then

$$D_s X = \begin{cases} 1 & : \quad s \leq t \\ 0 & : \quad s > t \end{cases}$$

Example 2 Let $X = \int_0^t u_r dW_r$ where u is a deterministic function. Then

$$D_s X = \begin{cases} u_s & : \quad s \leq t \\ 0 & : \quad s > t \end{cases}$$

That is,

$$D_t \int_0^T s^2 dW_s = t^2.$$

Chain Rule Let $\varphi \in C_b^1$ and let $X \in \mathbb{D}^{1,2}$. Then

$$D\varphi(X) = \varphi'(X) DX.$$

$$D_s W_t^2 = 2W_t I_{[0,t}(s).$$

Let $X = \int_0^t u_r dW_r$. Then

$$D_s \int_0^t u_r dW_r = u_s + \int_s^t D_s u_r dW_r.$$

The Fundamental Theorem of Calculus

Let $\{u_t, 0 \leq t \leq T\}$ be a stochastic process such that

$$E(\int_0^T u_s^2 ds) < \infty$$

and assume that for all $s \in [0, T]$, $u_s \in \mathbb{D}_{1,2}$ and that for all $t \in [0, T]$, $D_t u \in Dom(\Delta)$. Assume also that

$$E[\int_0^T (\delta(D_t u))^2 dt] < \infty.$$

Then

$$\int_0^T u_s dW_s \in \mathbb{D}_{1,2}$$

and

$$D_t \left(\int_0^T u_s dW_s \right) = \int_0^T D_t u_s dW_s + u_t.$$

Clark–Haussman–Ocone Formula

The Clark–Haussman–Ocone formula expresses the martingale representation theorem in terms of the Malliavin derivative.

Recall that the martingale representation theorem asserts that every square integrable martingale can be represented as a stochastic integral with respect to Brownian motion, i.e., for every square integrable martingale X, there exists an square integrable process u such that

$$X = E(X) + \int_0^T u_s dW_s.$$

If X is Malliavin differentiable, we have

$$D_t X = u_t + \int_t^T D_t u_s dW_s.$$

Thus

$$E(D_t X|\mathcal{F}_t^W) = u_t.$$

Hence if $X \in \mathbb{D}^{1,2}$, then

$$X = E(X) + \int_0^T E(D_t X|\mathcal{F}_t^W) dW_t.$$

An interesting fact of the Clark–Ocone formula is that if $X \in \mathbb{D}^{1,2}$ and $DX = 0$, then X is a.s. constant.

The method based on Clack-Ocone formula has the advantage that it does not depend on a Markovian setup.

Clark–Ocone Formula Under Change of Measure

Let

$$\widetilde{W}_t = \int_0^t u_s ds + W_t,$$

where u_s is an adapted stochastic process satisfying the Novikov condition

$$E(\exp(\int_0^T u_s^2 ds)) < \infty.$$

By Girsanov theorem $Q(d\omega) = Z_T P(d\omega)$ where

$$Z_t = \exp\left(-\int_0^t u_s dW_s - \frac{1}{2}\int_0^t u_s^2 ds\right), 0 \le t \le T.$$

Let $X \in \mathbb{D}_{1,2}$ is F_T measurable and satisfies

$$E_Q \|X\|^2 < \infty,$$

$$E_Q \int_0^T \|D_t X\|^2 dt < \infty,$$

$$E_Q \left[|X| \int_0^T \left(\int_0^T u_s dW_s + \frac{1}{2}\int_0^t (u_s D_t u_s) ds \right)^2 \right] < \infty.$$

Then

$$X = E_Q(X) + \int_0^T E_Q[(D_t X - X \int_0^T D_t u_s d\widetilde{W}_s)|\mathcal{F}_t^W]d\widetilde{W}_t.$$

Bayes Rule

Let μ and ν be two probability measures on a measurable space $(\Omega, \mathcal{F})$ such that $\nu(d\omega) = f(\omega)\nu(d\omega)$ for some $f \in L^1(\mu)$. Let X be a random variable on $(\Omega, \mathcal{F})$ such that $X \in L^1(\nu)$. Let $\mathcal{G} \subset \mathcal{F}$ be a σ-algebra. Then

$$E_\mu(fX|\mathcal{G}) = E_\mu(f|\mathcal{G})E_\nu(X|\mathcal{G}).$$

Let

$$Q(d\omega) = Z_T(\omega)P(d\omega),$$

where

$$Z_t = \exp\left\{-\int_0^t u_s dW_s - \frac{1}{2}\int_0^t u_s^2 ds\right\}.$$

Corollary to Bayes Rule Suppose $G \in L^1(Q)$. Then

$$E_Q[G|\mathcal{F}_t] = \frac{E[Z_T G|\mathcal{F}_t]}{Z_t}$$

Lemma (Using Clark–Ocone Formula)

$$D_t(Z_T F) = Z_T\left[D_t F - F\left(u_t + \int_t^T D_t u_s d\tilde{W}_s\right)\right]$$

Black-Scholes Delta Consider the Black-Scholes model

$$\begin{aligned} dS_t &= \mu S_t dt + \sigma S_t dW_t \\ &= (\mu - \sigma u)S_t dt + \sigma S_t d\widetilde{W}_t \\ &= rS_t dt + \sigma S_t d\widetilde{W}_t \end{aligned}$$

since $\frac{\mu - r}{\sigma} = u$ is the risk premium and $D_t u = 0$. Under the risk-neutral measure

$$dS_t = rS_t dt + \sigma S_t d\widetilde{W}_t$$

which gives

$$S_t = S_0 \exp\{(r - \frac{1}{2}\sigma^2)t + \sigma\widetilde{W}_t\}$$

$$\Delta_t = e^{-r(t-T)}\sigma^{-1}S_t^{-1}E_Q[D_t(S_T - K)^+|\mathcal{F}_t.]$$

But

$$D_t(S_T - K)^+ = I_{[K,\infty)}(S_T)S_T\sigma.$$

By Markov property of S_t, this is the same as

$$\Delta_t = e^{r(t-T)}S_t^{-1}E_Q^y[S_{T-t}I_{[K,\infty)}(S_{T-t})]|_{y=S_t},$$

where E_Q^y is the expectation when $S_0 = y$.

Hence

$$\Delta_t = e^{r(t-T)}S_t^{-1}E^y[Y_{T-t}I_{[K,\infty)}(Y_{T-t})]_{|y=S_t},$$

where

$$Y_t = S_0 \exp\{(r - \frac{1}{2}\sigma^2)t + \sigma W_t\}$$

Since the distribution of W_t is known, Δ_t can be explicitly expressed in terms of quantities involving S_t and the normal distribution function.

Δ_t represents the number of units we must invest in the risky investment at time t in order to be guaranteed to get the payoff $(S_T - K)^+$ at time T.

V_0^Δ represents the corresponding initial fortune needed to achieve this. Thus V_0^Δ is the unique initial fortune, which makes it possible to establish a self-financing portfolio with the same payoff at time T as the option gives

$$V_0^\Delta = E_Q[e^{-rT}(S_T - K)^+] = E[e^{-rT}(Y_T - K)^+]$$

which again can be expressed explicitly by the normal distribution.

Financial Interpretation of the Clark–Ocone Formula

Let X be the payoff of an European option on an asset S. The dynamics of the discounted price under the equivalent martingale measure (EMM) is given by

$$d\tilde{S}_t = \sigma_t \tilde{S}_t dW_t.$$

If (α, β) is a replication strategy of the option, we have

$$\tilde{X} = E[\tilde{X}] + \int_0^T \alpha_t d\tilde{S}_t = E[\tilde{X}] + \int_0^T \alpha_t \sigma_t \tilde{S}_t dW_t.$$

On the other hand, by Clark–Ocone Formula,

$$\tilde{X} = E[\tilde{X}] + \int_0^T E[D_t\tilde{X}|\mathcal{F}_t^W]dW_t.$$

Hence the replication strategy is given by

$$\alpha_t = \frac{E[D_t\tilde{X}|\mathcal{F}_t^W]}{\sigma_t\tilde{S}_t}, \quad t \in [0, T].$$

Stochastic Integration by Parts (SIBP)

The technique based on Malliavin calculus can be effective also when poor regularity conditions are assumed on the payoff function F, i.e., direct application of the Monte Carlo methods gives unsatisfactory results, even if the underlying asset follows geometric Brownian motion.

The stochastic integration by parts allows removing the derivative of the payoff function, thus improving the numerical approximation. More precisely, let us suppose that we want to determine $\partial_\alpha E[F(S_T)Y]$ where S_T denotes the final price of the underlying asset depending on a parameter α, e.g., α is S_T in the case of Delta, α is volatility in the case of Vega and Y is some random variable, e.g., a discount factor. The idea is to try to express $\partial_\alpha E[F(S_T)Y]$ in the form $\int_0^T D_s F(S_T) Y U_s ds$ for some adapted integrable process U. By using the duality relation formally we obtain

$$\partial_\alpha E[F(S_T)Y] = E[F(S_T)D^*(YU)].$$

Let Y be some random variable, e.g., a discount factor. Let $F \in C_b^1$ and let $X \in \mathbb{D}^{1,2}$. Then the following integration by parts holds:

$$E[F'(X)Y] = E\left[F(X)\int_0^T \frac{u_t Y}{\int_0^T u_s D_s X ds} \diamond dW_t\right].$$

In general it is not allowed to take out a random variable from Itô integral. This can be made more precise in the case of anticipative stochastic integral.

Let $X \in \mathbb{D}^{1,2}$ and let U be a second order Skorohod integrable process. Then

$$\int_0^T XU_t \diamond dW_t = X\int_0^T U_t \diamond dW_t - \int_0^T (D_t X)U_t dt.$$

The typical case when U is adapted the above formula becomes

$$\int_0^T XU_t \diamond dW_t = X\int_0^T U_t dW_t - \int_0^T (D_t X)U_t dt$$

and so it is possible to express Skorohod integral as the sum of an Itô integral and a Lebesgue integral.

Recall that by Itô formula

$$\int_0^T W_t dW_t = \frac{1}{2}(W_T^2 - T)$$

but by Integration by Parts (IBP) formula in Malliavin calculus, that is, a direct application of the above formula

$$\int_0^T W_T \diamond dW_t = W_T^2 - T.$$

This shows the difference between adapted and nonadapted integrands in stochastic calculus.

Thus even if the integrand does not depend on T, we have

$$\int_0^T W_T \diamond dW_t \neq W_T \int_0^T dW_t.$$

Bismut-Elworthy Formula

Let the diffusion be given by

$$S_t = x + \int_0^t b(s, S_s)ds + \int_0^t \sigma(s, S_s)dW_s,$$

which gives the stochastic flow

$$D_s S_t = \sigma(s, X_s) + \int_s^t \partial_x b(r, X_r) D_s X_r dr + \int_s^t \partial_x \sigma(r, X_r) D_s X_r dr.$$

Then

$$E[\partial_x F(S_T)G] = \frac{1}{T} E\left[F(S_T)\left(G\int_0^T \frac{\partial_x S_t}{\sigma(t, S_t)} dW_t - \int_0^T D_t G \frac{\partial_x S_t}{\sigma(t, S_t)} dt\right)\right]$$

for every $G \in \mathbb{D}^{1,\infty}$.

Stochastic Flow of Diffeomorphisms and Ocone-Karatzas Hedging
Let

$$S_t = x + \int_0^t b(s, S_s)ds + \int_0^t \sigma(s, S_s)dW_s.$$

The stochastic flow (see Kunita (1990)) is given by

$$D_s S_t = \sigma(s, X_s) + \int_s^t \partial_x b(r, S_r) D_s S_r dr + \int_s^t \partial_x \sigma(r, S_r) D_s S_r dr.$$

Ocone-Karatzas Hedging
Assume that the payoff function is F. Then

$$F(S_T) - E(F(S_T)) = \int_0^T \beta(t)dW_t$$

$$\beta(t) = E\left[D_t F(S_T) | \mathcal{F}_t^W\right],$$

where

$$\beta(t) = E\left[\int_0^T \zeta(t)dF(S_t)\right], \quad \zeta(t) = J_t b(S_t)$$

and J_t is Jacobian of the stochastic flow.

Anticipative Girsanov Theorem

The following is the nonadapted (anticipative) extension of the Girsanov theorem proved by Kusuoka (1982, Theorem 6.4). See also Theorem 4.1.2 in Nualart (1995).

Theorem 2.1 *Let $V : \Omega \to \Omega$ be a mapping of the form*

$$V(t,\omega) = \omega(t) + \int_0^t U(s,\omega)ds,$$

where U is a measurable mapping from Ω in to $H = L^2(0,T)$ and suppose that the following conditions are satisfied:
(i) V is bijective.
(ii) For all $\omega \in \Omega$, there exists a Hilbert–Schmidt operator $DU(\omega)$ from H into itself such that:
(a)

$$\|U(\omega + \int_0^\bullet h_s ds) - U(\omega) - DU(\omega)h\|_H = o(\|h\|_H)$$

for all $\omega \in \Omega$ as $\|h\|_H \to 0$,
(b) $h \to DU(\omega + \int_0^\bullet h_s ds)$ is continuous from H into $L^2([0,T]^2)$ the space of Hilbert–Schmidt operators for all ω,
(c) $I + DU(\omega) : H \to H$ is invertible.
Then if Q is the measure on $\Omega, \mathcal{F}$ such that $F = QV^{-1}, Q$ is absolutely continuous with respect to P and

$$\frac{dQ}{dP} = |d_c(-DU)| \exp\left(-\int_0^T U(t)dW_t - \frac{1}{2}\int_0^T U_t^2 dt\right),$$

where $d_c(-DU)$ denotes the Carleman-Fredholm determinant of the Hilbert–Schmidt operator $-DU$ and $\int_0^T U(t)dW(t)$ is the Skorohod integral.

We recall that the Carleman-Fredholm determinant of a Hilbert–Schmidt operator B from $L^2[0,T]$ into itself is defined by the product expansion

$$d_c(B) = \prod_j (1-\lambda_j)\exp(\lambda_j),$$

where λ_j s are the nonzero eigenvalues of B counted as many times as its multiplicities, see Simon (1979). In particular, if the operator B is nuclear, then

$$d_c(B) = \det(I - B)\exp\{\text{ trace } B\}.$$

Thus if DU is nuclear, then

$$d_c(-DU) = \det(I + DU)\exp\{\text{trace } (-DU)\}.$$

1.3. Pricing European Options

Let $\{W_t, t \geq 0\}$ be a standard Brownian motion. For the Itô SDE

$$dX_t = a(t, X_t)dt + b(t, X_t)dW_t,$$

with the coefficients a and b satisfying the existence and uniqueness of the solution, for a smooth function $f(t, x) \in \mathbb{C}^{1,2}$, the Itô formula is given by

$$\begin{aligned} df(t, X_t) &= [f_t(t, X_t) + f_x(t, X_t)a(t, X_t) + \frac{1}{2} f_{xx}(t, X_t)b^2(t, X_t)]dt \\ &\quad + f_x(t, X_t)b(t, X_t)dW_t \\ &= \mathcal{L}^0 f(t, X_t)dt + \mathcal{L}^1 f(t, X_t)dW_t, \end{aligned}$$

where the Itô operators are given by

$$\mathcal{L}^0 = \frac{\partial}{\partial t} + a\frac{\partial}{\partial x} + \frac{1}{2}b^2\frac{\partial^2}{\partial x^2}, \quad \mathcal{L}^1 = b\frac{\partial}{\partial x}.$$

Consider the Black-Scholes model for an asset price

$$dS_t = \mu S_t dt + \sigma S_t dW_t,$$

where μ is the mean rate of return and σ is the volatility of return of the price S_t. Using Itô formula the solution is given by

$$S_t = S_0 \exp\{(\mu - \frac{1}{2}\sigma^2)t + \sigma W_t\},$$

which is called the geometric Brownian motion.

The Black-Scholes option price formulae for European call and put options are defined, respectively, as $E[e^{-rT}(S_T - K)^+]$ and $E[e^{-rT}(K - S_T)^+]$ (the expectations are under the risk-neutral measure or martingale measure which makes the model arbitrage free) and given, respectively, by:

$$C = S\Phi(d_1) - Ke^{-r(T-t)}\Phi(d_2)$$

$$P = Ke^{-r(T-t)}\Phi(-d_2) - S\Phi(-d_1),$$

where

$$d_1 := \frac{\log\left(\frac{S}{K}\right) + (r + \frac{\sigma^2}{2})(T-t)}{\sigma\sqrt{T-t}},$$

$$d_2 := \frac{\log\left(\frac{S}{K}\right) + (r - \frac{\sigma^2}{2})(T-t)}{\sigma\sqrt{T-t}} = d_1 - \sigma\sqrt{T-t},$$

where T is option maturity, K is the strike price.

For the Vasicek model for the interest rate

$$dr_t = a(b - r_t)dt + \sigma dW_t, \quad a > 0$$

the integrated process $I_T = \int_0^T r_t dt$ is normally distributed since the process r_t is stationary and normally distributed and the price of a zero coupon bond which pays \$1 at time T is given by

$$P(t,T) = E(e^{-\int_t^T r_u du}|r_t) = A(t,T)e^{-B(t,T)r(t)},$$

where

$$B(t,T) = \frac{1 - e^{-a(T-t)}}{a}$$

and

$$A(t,T) = \exp\left(\frac{(B(t,T) - T + t)(a^2b - \sigma^2/2)}{a^2} - \frac{\sigma^2 B(t,T)^2}{4a}\right).$$

The call option price on the bond is defined as $E(e^{-\int_0^T r_u du}(S_T - K)^+)$ (the expectation is under the risk-neutral measure) and is given by

$$C = LP(0,s)\Phi(h) - KP(0,T)\Phi(h - \sigma_p)$$

$$h = \frac{1}{\sigma_P}\ln\frac{LP(0,s)}{P(0,T)K} + \frac{\sigma_P}{2}$$

$$\sigma_P = \frac{\sigma}{a}(1 - e^{-a(s-T)})\sqrt{\frac{1 - e^{-2aT}}{2a}},$$

where L is the bond principal, s is the bond maturity, T is option maturity, K is the strike price.

For the Cox–Ingersoll–Ross (CIR) model for interest rate the bond price and call option price on the bond are, respectively, given by formula (3.76) on page 129 of Glasserman (2004) and formula (32) on page 396 of Cox–Ingersoll–Ross (1985).

For the CIR model

$$dX_t = a(b - X_t)dt + \sigma\sqrt{X_t}dW_t$$

the price at time t of a zero coupon bond that pays \$1 at time T is given by

$$P(t,T) = E_Q\left(e^{-\int_t^T X_t dt}\right) = A(t,T)e^{-B(t,T)X_t},$$

where Q is the risk-neutral measure and

$$B(t,T) = \frac{2(e^{\gamma(T-t)-1})}{(\gamma + a)(e^{\gamma(T-t)-1}) + 2\gamma}$$

$$A(t,T) = \left(\frac{2\gamma e^{(a+\gamma)(T-t)/2}}{(\gamma+a)(e^{\gamma(T-t)-1}) + 2\gamma} \right)^{2ab/\sigma^2},$$

where $\gamma = \sqrt{a^2 + 2\sigma^2}$ and Q is the risk-neutral measure.

The European call option on the zero coupon bond is given by

$$C_t = P(t,s)\chi^2 \left(2\frac{\log(A(t,s)/K)}{B(T,s)}[\phi + \psi + B(T,s)]; \frac{4ab}{\sigma^2}, \frac{2\phi^2 X_t e^{\gamma(T-t)}}{\phi + \psi + B(T,s)} \right)$$

$$-KP(t,T)\chi^2 \left(2\frac{\log(A(t,s)/K)}{B(T,s)}[\phi + \psi]; \frac{4ab}{\sigma^2}, \frac{2\phi^2 X_t e^{\gamma(T-t)}}{\phi + \psi} \right),$$

where

$$\phi =: \frac{2\gamma}{\sigma^2(e^{\gamma(T-t)-1})}, \quad \psi =: \frac{a+\gamma}{\sigma^2}$$

and $\chi^2(x; d, \lambda)$ is the noncentral chi-square distribution with d degrees of freedom and noncentrality parameter λ.

Note that $B(t,T)$ can be written as

$$B(t,T) = \frac{2(e^{\gamma(T-t)-1})}{(\gamma+a)(e^{\gamma(T-t)-1}) + 2\gamma}.$$

The integrated volatility is given by

$$I_T := \int_0^T X_t dt.$$

The following characteristic function of I_T is closely associated with *Levy's stochastic area formula* and is well known from Brownian motion literature and also from the work of Cox et al. (1985).

Let $\phi_T(u) := E\exp(iuI_T), u \in \mathbb{R}$. Then

$$E\exp(iuI_T) = \exp\left(\frac{2iu}{\beta + \gamma\coth\frac{\gamma T}{2}} \right) \left[\cosh\frac{\gamma T}{2} + \frac{\beta}{\gamma}\sinh\frac{\gamma T}{2} \right]^{-1},$$

where $\gamma := (\beta^2 - 2iu)^{1/2}$.

Further

$$\lim_{T\to\infty} E\left(e^{-uI_T}\right) = \lim_{T\to\infty} \exp\left(-\frac{ab}{2\sigma}T\left(\sqrt{1 + \frac{4u\sigma}{b^2}} - 1 \right) \right).$$

$$\begin{aligned} & E\exp(-uX_T - vI_T) \\ = & \left(\frac{2\rho e^{(b-\rho)T/2}}{2\sigma u(1-e^{-\rho T}) + (\rho - b)e^{-\rho T} + (\rho + b)} \right)^{a/\sigma} \\ & \times \exp\left(\frac{u((\rho+b)e^{-\rho T} + (\rho - b) + 2v(1-e^{-\rho T})}{2\sigma u(1-e^{-\rho T}) + (\rho - b)e^{-\rho T} + (\rho + b)} \right), \end{aligned}$$

where $\rho =: \sqrt{b^2 + 4\sigma v}$.

Kummer's confluent hypergeometric function is given by

$$ {}_1F_1(r,s,z) = \frac{\Gamma(s)}{\Gamma(r)\Gamma(s-r)} z^{1-s} \int_0^z e^u u^{r-1}(z-u)^{s-r-1} du. $$

Whittaker function of first kind is given by

$$ M_{s,r}(z) = z^{r+\frac{1}{2}} e^{-z/2} \ {}_1F_1(r-s+\frac{1}{2}, 2r+1, z). $$

Let

$$ J_T = \int_0^T X_t^{-1} dt. $$

Then we have the Laplace transform of J_T

$$ \begin{aligned} & E\exp(-uJ_T) \\ = \ & \frac{\Gamma(k+\frac{\nu}{2}+\frac{1}{2})}{\Gamma(\nu+1)} (\frac{x}{\alpha})^{-k} \beta^{\frac{\nu}{2}+\frac{1}{2}} \exp(\frac{b}{2\sigma}[at - \frac{2x}{e^{bt}-1}]) \\ & \times {}_1F_1(k+\frac{\nu}{2}+\frac{1}{2}, \nu+1, \beta), \end{aligned} $$

where

$$ k =: \frac{a}{2\sigma}, \quad \alpha =: \frac{be^{bt}}{\sigma(e^{bt}-1)}, \quad \nu =: \frac{1}{\sigma}\sqrt{(\alpha-\sigma)^2 + 4u\sigma}, $$

where ${}_1F_1$ is Kummer's confluent hypergeometric function.

The joint Laplace transform of I_T and J_T is given by

$$ \begin{aligned} & E\exp(-uI_T - vJ_T) \\ = \ & \frac{\Gamma(k+\frac{\nu}{2}+\frac{1}{2})}{\Gamma(\nu+1)} \left(\frac{\sqrt{\sigma u}x \coth(\sqrt{\sigma u}t)}{\sigma}\right)^{-k} \exp\left(\frac{\sqrt{\sigma u}x}{\sigma}\coth(\sqrt{\sigma u}t)\right) \\ & \times \exp\left(\frac{\sqrt{\sigma u}x}{2\sigma \sinh(\sqrt{\sigma u}t)\cosh\sqrt{\sigma u}t}\right) M_{-k,\frac{\nu}{2}}\left(\frac{\sqrt{\sigma u}x}{\sigma \sinh(\sqrt{\sigma u}t)\cosh\sqrt{\sigma u}t}\right), \end{aligned} $$

where $M_{s,r}$ is the Whittaker function of first kind.
Consider the Heston stochastic volatility model

$$ dS_t = \mu \ S_t \ dt + \sqrt{X_t} \ S_t \ \ dW_t $$

$$ dX_t = (1 + 2\theta \ X_t) \ dt + \ 2\sqrt{X_t} \ \ dZ_t, $$

where $\{W_t\}$ is a standard Brownian motion independent of another standard Brownian motion $\{Z_t\}$ and $\theta < 0$.

Note that the volatility X_t which is given by the CIR process is not observed.

Integrated volatility I_T has to be estimated on the basis of discrete observations of the price process $\{S_t\}$ at times $0 = t_0 < t_1 < \cdots < t_n = T$ with $t_i - t_{i-1} = \frac{T}{n}, i = 1, 2 \cdots, n$. Let $s_t = \log S_t$ be the log-price process. Then

$$ds_t = (\mu - \frac{1}{2}X_t)dt + \sqrt{X_t}\ dW_t.$$

Thus the drift term depends on the volatility.

Denote

$$\Delta s^2_{t_{i-1}} := (s_{t_i} - s_{t_{i-1}})^2.$$

The realized volatility is defined as

$$R_{n,T} := \sum_{i=1}^{n} \Delta s^2_{t_{i-1}}.$$

It is well known that

$$\underset{n\to\infty}{\text{P-lim}}\, R_{n,T} = I_T.$$

Consider the Heston model with correlated noises:

$$dS_t = \mu S_t dt + \sqrt{S_t} dW_t^1$$

$$dV_t = \kappa(\theta - V_t)dt + \sigma\sqrt{V_t}dW_t^2$$

$$dW_t^1 dW_t^2 = \rho dt.$$

Under the risk-neutral measure,

$$dS_t = rS_t dt + \sqrt{S_t} d\tilde{W}_t^1$$

$$dV_t = \kappa^*(\theta^* - V_t)dt + \sigma^*\sqrt{V_t}d\tilde{W}_t^2$$

$$d\tilde{W}_t^1 d\tilde{W}_t^2 = \rho dt,$$

where

$$\kappa^* = \kappa + \lambda, \quad \theta^* = \frac{\kappa\theta}{\kappa + \lambda}$$

$$d\tilde{W}_t^1 = dW_t^1 + \theta_t dt, \quad d\tilde{W}_t^2 = dW_t^2 + k\sqrt{V_t}dt$$

$$\frac{dQ}{dP} = \exp\left\{-\frac{1}{2}\int_0^t (\theta_s^2 + k^2 V_s)ds - \int_0^t \theta_s dW_s^1 - \int_0^t k^2 V_s dW_s^2\right\}$$

$$\theta_t = \frac{\mu - r}{\sqrt{V_t}}.$$

The characteristic function of s_T under Q is given by

$$\psi_T(u) = E_Q(e^{is_T u}) = A(u,T)e^{B(u,T)},$$

where

$$A(u,T) \quad = \quad e^{iu(s_0+rT)}$$

and

$$\begin{aligned} B(u,T) &= \frac{-(u^2+iu)(1-e^{\gamma T})V_0}{2\gamma-(\gamma-(\kappa-\rho\sigma ui)(1-e^{\gamma T})} \\ &- \frac{\kappa\theta}{\sigma^2}\left[2\log\left(\frac{2\gamma-(\gamma-\kappa-\rho\sigma ui)(1-e^{\gamma T})}{2\gamma}\right)\right. \\ &\left. +(\gamma-\kappa-\rho\sigma ui)T\right] \end{aligned}$$

with

$$\gamma =: \sqrt{(\kappa-\rho\sigma ui)^2+(u^2+iu)}.$$

Option Price The European option price with the payoff $H(T)$ at time T is defined as

$$E_t^T[e^{r(T-t)}H(T)]$$

and is given by

$$C(S_t,V_t,t,T) = S_t P_1 - Ke^{-r(T-t)}P_2,$$

where for $j=1,2$,

$$P_j(x,V_t,T,K) = \frac{1}{2}+\frac{1}{\pi}\int_0^\infty Re\left(\frac{e^{-iu\ln K}f_j(x,V_t,T,u)}{iu}\right)du$$

$$x = \ln S_t$$

$$f_j(x,V_t,T,u) = \exp(C(T-t,u)+D(T-t,u)V_t+iux)$$

$$C(T-t,u) = ruir+\frac{a}{\sigma^2}\left[(b_j-\rho\sigma ui+d)\tau-2\ln\left(\frac{1-ge^{dr}}{1-g}\right)\right]$$

$$D(T-t,u) = \frac{b_j-\rho\sigma ui+d}{\sigma^2}\left(\frac{1-e^{dr}}{1-ge^{dr}}\right)$$

$$g = \frac{b_j-\rho\sigma ui+d}{b_j-\rho\sigma ui-d}$$

$$d = \sqrt{(\rho\sigma ui-b_j)^2-\sigma^2(2\nu_j ui-u^2)}$$

for $j=1,2$ where

$$\nu_1=\frac{1}{2},\nu_2=-\frac{1}{2},a=\kappa\theta,b_1=\kappa+\lambda-\rho\sigma,b_2=\kappa+\lambda.$$

Hybrid Model

A hybrid model under the risk-neutral measure is given by

$$dS_t = r_t S_t dt + \sqrt{\sigma_t} S_t dW_t^1$$

$$dr_t = \alpha(\beta - r_t)dt + \gamma\sqrt{r_t}dW_t^2$$

$$d\sigma_t = \alpha_1(\beta_1 - \sigma_t)dt + \gamma_1\sqrt{\sigma_t}dW_t^2,$$

where r_t and σ_t are stochastic CIR interest rate and stochastic CIR volatility processes with $(\alpha, \beta, \gamma, \alpha_1, \beta_1, \gamma_1)$ being positive parameters and W_t^1, W_t^2, W_t^3 are three correlated Brownian motions specifying the leverage effects.

Option pricing for such models are challenging due to additional stochasticity and unobserved stochastic volatility.

1.4. Hedging and Greek Estimation

Due to recent (2008–2011) financial crisis in the USA, both modeling and valuation of financial products face challenges. Mathematical finance deals with pricing derivatives (that derives its value from the performance of an underlying asset, e.g., a stock, a mutual fund, or an interest rate) products, e.g., options on stocks and bonds. For the derivation of option price formula, one calculates mathematical expectation (the population mean) of a discounted payoff function. Monte Carlo methods (based on the philosophy of law of large numbers, which basically says that for large sample, the sample mean converges in an appropriate sense to the population mean) in mathematical finance and financial engineering are alternative methods of pricing derivatives when analytical pricing formula is not available and when it is available it is too difficult to use, see Glasserman (2004). European options are contracts that are signed between two parties (usually a bank and a costumer) that allows to obtain certain monetary benefits if the price of certain asset falls above (known as call option) or falls below (known as put option) a certain fixed value, the strike price, at a certain fixed date, the expiration date. Whereas European options can be exercised only at the expiration date, American options are contracts which give the holder the right to exercise at any time on or before the exercise date. A Greek is the derivative (mathematical differentiation as in calculus) of the option price with respect to a parameter, also known as the *sensitivity parameter*. Greeks or sensitivity parameters are frequently used for hedging (mitigating risk by taking investment position to offset potential losses/gains suffered by an individual or an organization), market risk management and profit and loss attribution. Greeks measure the stability of a portfolio with respect to change in parameters. For instance, one would be interested to know how sensitive is the change in the option price for a corresponding change in the stock price. It is important to carry out these calculations with high accuracy to prevent future instabilities in the position of a company holding these options. If the Greek does not have a closed form formula, then one may think of performing Monte Carlo simulations in order to approximate it. There are 13 sensitivity parameters of a derivative product: *Delta, Gamma, Vega, Theta, Rho, Phi, Speed, Volga, Vanna, Veta, Charm, Color, and Zomma.* In an idealized setting of continuous trading in a complete financial market, the payoff of a contingent claim can be manufactured or hedged through trading in underlying assets. The risk in a short position in an option, for example, is offset, by a delta-hedging strategy of holding delta units of each underlying asset, where delta is simply the partial derivative of the option price with respect to the current price of that underlying asset. Implementation of the strategy requires the knowledge of these price sensitivities. Sensitivities with respect to other parameters, for example, derivative of the option price with respect to the volatility, called the *Vega*, and the second derivative of the option price with respect to the asset price called the *gamma* are also widely used to measure and manage risk. *Theta* is the first derivative with respect to time to maturity. *Rho* is the first derivative with respect to the interest rate. *Phi* is the first derivative with respect to dividend yield. Speed is the third derivative of the option price with respect to the stock price. *Volga*, also called *Vomma*, is the second partial derivative of the option with respect to volatility. *Vanna* is the second partial derivative once with respect to stock and once with respect to volatility. *Veta* is the second partial derivative once with respect to maturity and once with respect to maturity. *Charm* is the second partial derivative of option price, once with respect to stock and once with respect to time to maturity. *Color* is the third partial derivative, twice with respect to stock and once with respect to time to maturity. *Zomma* is the third partial derivative of the option twice with respect to stock and once with respect to volatility.

One of the fundamental problems in finance is that the asset pricing models are continuous but the data are discrete, see Bishwal (2008a). Whereas the prices themselves can often be observed in the market, their sensitivities cannot, so accurate calculation of sensitivities is even more important than calculation of prices. Thus derivative estimation presents both theoretical and practical challenges to Monte Carlo simulation.

Traditionally, the methods for estimating sensitivities are broadly of two types: methods that involve simulating at two or more values of the parameter of differentiation and methods that do not. The first category- the finite difference approximation methods are easier to implement but produce biased estimates. Methods of the second category produce unbiased estimates. Second category methods are of two types: the pathwise method and the likelihood ratio method. Pathwise method differentiates each simulated outcome with respect to the parameter of interest. The likelihood ratio method differentiates a probability density rather than an outcome. When applicable, pathwise method provides best estimates of the sensitivities. Compared with finite difference methods, pathwise estimates require less computing time. Compared with likelihood ratio method, pathwise estimates usually have smaller variance. The application of the pathwise method requires interchanging the order of differentiation and integration. Pathwise method yields unbiased estimates of the derivative of an option price if the option's discounted payoff is almost surely continuous in the parameter of differentiation. This excluded many options, for example, digital and barrier options. Finite difference methods are easy to implement, but they have large mean square errors. The likelihood ratio method does not require any smoothness in the discounted payoff because it is based on probability density instead. But its application is limited, because it requires the explicit knowledge of the relevant probability densities and often has large variance. Estimating second derivative is even more difficult than estimating first derivative regardless of the method used. The pathwise method is generally inapplicable to second derivative of option prices since a kink in an option payoff becomes a discontinuity in the derivative of the payoff. Combination of pathwise method and likelihood ratio method generally produces better gamma estimates than likelihood ratio method alone. We will study a new method for Greek estimation known as the sequential Monte Carlo method combined with Malliavin calculus method. In Malliavin calculus, also known as the stochastic calculus of variations, the Malliavin derivative differentiates a random variable with respect to its underlying noise generating process, see Nualart (1995). The Malliavin calculus method for the estimation of Greeks developed in Fournie et al. (1999, 2001), Malliavin and Thalmaier (2006) first differentiates, but implementation requires the simulation of a discrete time approximation, thus discretizes next. The question remains as to discretize the underlying asset price process first and then differentiate or differentiate first and then discretize the asset price process. The first route leads to *Malliavin estimators.* Chen and Glasserman (2007) interchange the order and produce the same Malliavin estimators, that is, they discretize first and then differentiate. For the discretization, Chen and Glasserman (2007) used Euler approximation which is a first order approximation. In this project, our first step is to use Milstein approximation which a second order approximation to improve the discretization error and have faster rate of convergence. In Bishwal (2011a,b), We have developed many higher order approximate schemes for approximation of stochastic integrals. In the second step, one can employ some of these schemes, especially the stochastic Boole's rule, for Greek estimation in order to further improve the accuracy of approximation. Finally numerical experiments should be performed. We suppose that the underlying model dynamics, e.g., price of a stock, are given by the stochastic differential equation

$$dX_t = f(X_t)dt + g(X_t)dW_t, X_0 = x,$$

where W_t is a standard Brownian motion and f and g are smooth functions satisfying the existence and uniqueness of the solution of the equation. With a discounted payoff function h and maturity T, option price is

$$u(x) = E(h(X_T))$$

and option delta is

$$u'(x) = E(h'(X_T))\frac{dX_T}{dx},$$

where

$$\frac{dX_t}{dx} = Y_t$$

is the pathwise derivative of X_t with respect to the initial state x. Thus

$$\xi_1 := h'(X_T)\frac{dX_T}{dx} = h'(X_T)Y_T$$

is an unbiased estimator of $u'(x)$. The dynamics of the pathwise derivative is given by the stochastic differential equation

$$dY_t = f'(X_t)Y_t dt + g'(X_t)Y_t dW_t, \quad Y_0 = 1,$$

that is

$$Y_t = \exp\left\{(f'(X_s) - \frac{1}{2}(g'(X_s))^2]ds + \int_0^t g'(X_s)dW_s\right\}.$$

It can be shown that

$$u'(x) = E^x\left[h(X_T)\int_0^T a(t)\frac{Y_t}{g(X_t)}dW_t\right],$$

where $\int_0^T a(t)dt = 1$.

The stochastic integral $\int_0^T a(t)\frac{Y_t}{g(X_t)}dW_t$ is called Malliavin weight. With $a(t) = \frac{1}{T}, \quad 0 \le t \le T$, the Malliavin estimator is given by the stochastic integral

$$\xi_3 := h(X_T)\frac{1}{T}\int_0^T \frac{Y_t}{g(X_t)}dW_t,$$

see Fournie et al. (1999). For the Black-Scholes model

$$dX_t = \mu X_t dt + \sigma X_t dW_t,$$

$$u'(x) = E^x[h(X_T)\frac{W_T}{x\sigma_0 T}].$$

Thus the Malliavin estimator of Δ is

$$h(X_T)\frac{W_T}{x\sigma T}.$$

With $p(x, X_T)$ being the transition density of X_T given x, the likelihood ratio method (LRM) estimator is given by

$$\xi_2 := h(X_T)\frac{d}{dx}\log p(x, X_T).$$

The LRM B-S call delta estimator is

$$e^{-rT}(X_T - K)^+\frac{\log(S_t/x) - (r - \frac{1}{2}\sigma^2)T}{x\sigma T},$$

which can be simulated as

$$e^{-rT}(X_T - K)^+\frac{Z}{x\sigma\sqrt{T}}.$$

Based on Euler discretization of the price SDE, this estimator can be seen as a limit of the average of the combinations of pathwise and LRM estimators. One could study higher order efficient Monte Carlo simulation methods for the derivation of the derivative (differentiation) of the derivative price (option price), that is estimate the Greeks or sensitivities without using Malliavin derivative but using elementary techniques. One could study the convergence of the Greek estimators when the discretization of the SDE is done using the Milstein scheme and other higher order schemes to the corresponding Malliavin estimators. Unbiasedness and the rate of convergence of the new estimators will be studied. Based on discrete observations, higher order approximation of the stochastic integral, studied in Bishwal (2008a,b, 2011a,b), could be used to obtain better Greek estimators. The performance of the new estimators could be numerically tested using computer programs, e.g., MATLAB.

Delta, Vega, and Gamma of European Options

Weak regularity properties are assumed for the payoff F. The stochastic integration by parts formula allows removing the derivative of the payoff function, thus improving the numerical approximation. It is possible to prove the validity of the integration by parts formula for weakly differentiable (or even differentiable in a distributional sense) functions.

Examples:

Consider the Black-Scholes model for the underlying asset under the physical measure

$$dS_t = \mu S_t dt + \sigma S_t dW_t.$$

The Black-Scholes dynamics for the underlying asset of an option under the equivalent martingale measure is

$$S_T = x\exp(\sigma W_T + (r - \frac{1}{2}\sigma^2)T)$$

Delta:
$D_sS_T = \sigma S_T$ and $\partial_x S_T = \frac{S_T}{x}$.
In the SIBP if $u = 1$ and $Y = \partial_\alpha X$, we obtain

$$E[\partial_\alpha F(X)] = E\left[F(X)\int_0^T \frac{\partial_\alpha X}{\int_0^T D_sXds} \diamond dW_t\right].$$

Using this we obtain

$$\Delta = e^{-rT}\partial_x E[F(S_T)]$$

$$= e^{-rT}E\left[F(S_T)\int_0^T \frac{\partial_x S_T}{\int_0^T D_sS_Tds} \diamond dW_t\right] = e^{-rT}E\left[F(S_T)\int_0^T \frac{1}{\sigma Tx}dW_t\right]$$

$$= \frac{e^{-rT}}{\sigma Tx}E[F(S_T)W_T]$$

Vega:

$$\partial_\sigma S_T = (W_T - 2\sigma T)S_T, \quad D_sS_T = \sigma S_T.$$

$$\mathcal{V} = e^{-rT}\partial_\sigma E[F(S_T)] = e^{-rT}E\left[F(S_T)\left(\frac{W_T - \sigma T}{\sigma T}W_T - \frac{1}{\sigma}\right)\right]$$

Gamma:

$$\begin{aligned}
\Gamma = e^{-rT}\partial_{xx}E[F(S_T)] &= \frac{e^{-rT}}{\sigma T}E[\partial_x \frac{F(S_T)}{x}W_T] \\
&= \frac{e^{-rT}}{\sigma Tx^2}E[F(S_T)W_T] + \frac{e^{-rT}}{\sigma Tx}E[\partial_x F(S_T)W_T] \\
&= \frac{e^{-rT}}{\sigma Tx^2}E[F(S_T)W_T] + \frac{e^{-rT}}{\sigma Tx}E[F'(S_T)\partial_x S_T W_T] \\
&= \frac{e^{-rT}}{\sigma Tx^2}E[F(S_T)W_T] + \frac{e^{-rT}}{\sigma Tx}E[F(S_T)\int_0^T \frac{W_T}{\sigma Tx} \diamond dW_t] \\
&= \frac{e^{-rT}}{\sigma Tx^2}E[F(S_T)W_T] + \frac{e^{-rT}}{\sigma Tx}\frac{1}{\sigma Tx}E[F(S_T)(W_T^2 - T)] \\
&= \frac{e^{-rT}}{\sigma Tx^2}E[F(S_T)(\frac{W_T^2 - T}{\sigma T} - W_T).
\end{aligned}$$

Delta of Asian Option

Let the arithmetic average be defined as

$$X = \frac{1}{T}\int_0^T S_t dt.$$

Observe that

$$\partial_x X = \frac{X}{x}.$$

$$\int_0^T D_s X ds = \int_0^T \int_0^T D_s S_t dt ds = \sigma \int_0^T \int_0^t S_t ds dt = \sigma \int_0^T t S_t dt$$

The Delta of the Asian option is given by

$$\Delta = e^{-rT} \partial_x E[F(X)]$$

$$= \frac{e^{-rT}}{x} E[F'(X)X] = \frac{e^{-rT}}{\sigma x} E\left[F(X) \int_0^T \frac{\int_0^T S_s ds}{\int_0^T s S_s ds} \diamond dW_t \right]$$

$$= \frac{e^{-rT}}{x} E\left[F(X) \left(\frac{1}{I_1} \left(\frac{W_T}{\sigma} + \frac{I_2}{I_1} \right) - 1 \right) \right],$$

where

$$I_j = \frac{\int_0^T t^j S_t dt}{\int_0^T S_t dt}, \quad j = 1, 2.$$

Sensitivity with Respect to Correlation in Heston Model

Consider the Heston model

$$dS_t = \sqrt{V_t} S_t \left\{ \sqrt{1 - \rho^2} dW_t^1 + \rho dW_t^2 \right\}$$

$$dV_t = \alpha(\beta - V_t) dt + \sigma \sqrt{V_t} dW_t^2,$$

where W_t^1 and W_t^2 are two independent Brownian motions and the leverage $\rho \in (-1, 1)$. Observe that

$$S_T = S_0 + \int_0^T \sqrt{V_t} dW_t^1 + \int_0^T \sqrt{V_t} dW_t^2 - \frac{1}{2} \int_0^T v_t dt$$

and

$$\partial_\rho S_T = S_T G,$$

where

$$G := -\frac{\rho}{\sqrt{1 - \rho^2}} \int_0^T \sqrt{V_t} dW_t^1 + \int_0^T \sqrt{V_t} dW_t^2.$$

$$\partial_\rho E[F(S_T)] = \frac{1}{T\sqrt{1 - \rho^2}} E\left[F(S_T) \left(G \int_0^T \frac{1}{\sqrt{V_t}} dW_t^1 + \frac{\rho T}{\sqrt{1 - \rho^2}} \right) \right].$$

1.5 Bootstrap Method for Volatility Estimation

In a series of papers, given in the reference, by Barndorff-Nielsen and Shephard (2001–2007) (henceforth BN-S) have developed an asymptotic theory for *realized volatility-like measures.* In particular, for a general stochastic volatility model, BN-S establish a central limit theorem (CLT) for realized volatility over a fixed interval of time, for instance, a day, as the number of intraday returns increases to infinity. They have also showed that a CLT applies to empirical measures based on powers of intraday returns (realized power variation) and products of powers of absolute returns (e.g., bipower variation). BN-S also provided a joint asymptotic distribution theory for the realized volatility and realized bipower variation and showed how to use this distribution to test for the presence of jumps in asset prices.

For the first time, Goncalves and Meddahi (2009) proposed bootstrap methods for evaluating high-frequency data such as realized volatility to improve upon the first order theory of asymptotic mixed normal approximations of BN-S. They studied two bootstrap methods for realized volatility: an i.i.d. (uniform) bootstrap and wild bootstrap. The i.i.d. bootstrap (cf. Efron (1979)) generates bootstrap pseudo intraday returns by resampling with replacement the original set of intraday returns. The wild bootstrap observations are generated by multiplying each original intraday return by an i.i.d. draw from a distribution that is completely independent of the original data. The wild bootstrap was introduced by Wu (1986), and further studied by Liu (1988) and Mammen (1993), in the context of cross-section linear regression models subject to unconditional heteroskedasticity in the error term. Zhang et al (2005) and Zhang (2006) considered an application of subsampling method to realized volatility under stochastic volatility. In particular, they use subsampling plus averaging the bias to correct the realized volatility measure when microstructure noise is present. The generalized bootstrap techniques could be used to estimate the distribution (as opposed to bias) of realized volatility.

A popular bootstrap for serially dependent data is block bootstrap. In our context, intraday returns are (conditionally on the volatility path) independent, and this implies that blocking is not necessary for asymptotic refinements of the bootstrap. The issue here is heteroskedasticity and not serial correlation.

Malliavin calculus is a powerful tool to study the asymptotic expansion of distribution of estimators in diffusion type models, see Yoshida (1997). The nondegeneracy of the Malliavin covariance plays the same role in diffusion case as the Cramer condition in the i.i.d. case. One should use Malliavin calculus for obtaining Edgeworth expansion of generalized bootstrap statistics.

Goncalves and Meddahi (2009) used Monte Carlo simulations and formal Edgeworth expansions to compare the accuracy of the bootstrap and normal approximations. They showed that i.i.d. provides an asymptotic refinement when volatility is constant. The absolute magnitude of the coefficients describing the i.i.d. bootstrap error is smaller than the coefficients entering in the first term Edgeworth expansion of the original statistic. The Edgeworth expansion for wild bootstrap shows that it provides an asymptotic refinement when the volatility is heterogeneous if one chooses the external random variable used to construct the wild bootstrap observations appropriately.

One could use generalized bootstrap schemes for higher order accuracy in estimation and testing in classical stochastic volatility models driven by standard Brownian motion. Then one could also consider stochastic volatility models with jumps (for instance, Poisson type jumps) and stochastic volatility models with long memory driven by fractional Brownian motion.

Hall and Presnell (1999) introduced a class of *weighted bootstrap techniques*, called *biased-bootstrap* or *b-bootstrap* methods. It is motivated by the need to adjust more conventional, uniform-bootstrap methods in a surgical way, so as to alter some of their features while leaving others unchanged. In the b-bootstrap, resampling probabilities are chosen to minimize the distance of the weighted bootstrap distribution from the usual, uniform-bootstrap distribution, conditional on the data and subject to constraints that are designed to improve the statistical performance. Empirical likelihood methods may be viewed as a particular case of the b-bootstrap as also many techniques suggested for refining the generalized method of moments. The b-bootstrap has connection to tilting methods. It has also application to dependent data in the context of empirical likelihood. Methods related to both empirical likelihood and the b-bootstrap include the bootstrap likelihood technology of implied likelihood method for making likelihood calculations from confidence intervals. Depending on the nature of the adjustment, the biased bootstrap can be used to reduce bias, or reduce variance, or render some characteristic equal to a predetermined quantity. More specifically, applications of b-bootstrap methods include hypothesis testing (b-bootstrap enables simulation under the null hypothesis even when the hypothesis is false), b-bootstrap competitor to Tibshirani's variance stabilization method, both density estimation and nonparametric regression under constraints, robustification of general statistical procedures, sensitivity analysis, generalized method of moments, outlier trimming, skewness and kurtosis reduction, shrinkage, and many more.

Bootstrap confidence intervals could also be studied, together with error reduction techniques. Many ideas of linear regression problems could be extended to stochastic volatility models.

Chatterjee and Bose (2005) introduced a generalized bootstrap technique for estimators obtained by solving estimating equations. Special cases are the classical bootstrap of Efron, the delete-d jackknife, and variations of the Bayesian bootstrap. Under fairly general conditions they investigate to establish (a) asymptotic normality of estimator and consistency of bootstrap when model dimension is fixed or increasing with data size; (b) asymptotic representation of resampling variance estimator and (c) higher order accuracy of the new generalized bootstrap estimator for the bias corrected, studentized estimator.

In the classical setting, for instance, consider estimating the variance of the least squares estimate in linear regressions. There are several resampling schemes available in the literature. By establishing representation results, Liu and Singh (1992) classified these into two groups: Those that are *efficient* but not consistent under heteroscedasticity and those that are consistent under heteroscedasticity (*robust*) but not efficient.

Classes of generalized bootstrap are introduced and in some sense all of the above schemes are special cases of these bootstraps. By establishing higher order expansions, one can distinguish between the estimators within the robust and the efficient class. First order representation results are also established for high dimensional regression models where the number of parameters increases with the sample size.

For the related problem of estimating the entire distribution of the realized volatility, one should study consistency of the generalized bootstrap. It is known from the existing works that the paired bootstrap (which is robust) is not second order accurate for the ordinary least squares estimator. One could show that with proper bias correction and studentization, a (smooth) generalization of the paired bootstrap may be second order accurate.

One could then extend these ideas to estimates obtained by solving martingale estimating equations. One could establish representation results for the bootstrap estimator and obtain some first and second order distribution results. Representation results could also be obtained for the bootstrap variance estimator.

One should also think to investigate how these ideas can be implemented in estimating the distribution of M estimators. Generalized bootstrap schemes in *martingale estimation function* should be introduced and their first and second order behavior should be studied. New as well as unifying results on resampling plans in a large class of M estimators could be obtained which includes all standard resampling schemes and a large class of common estimators used in statistics. The performance of the bootstrap for *nonregular models* when the standard conditions often assumed are violated could be also investigated. The following jackknife and bootstrap techniques could be studied for stochastic volatility models:

(1) Delete-1 jackknife (Quenoville 1949), (2) Weighted Jackknife (Hinkley (1977)), (3) Weighted Jackknife (Wu (1986)), (4) Weighted Jackknife (Liu and Singh (1992)), (5) Residual bootstrap (Efron (1979)), (6) External bootstrap (Wu (1986)), (7) Weighted bootstrap (Liu 1988), (8) Uncorrelated weights bootstrap (Chatterjee (1997)).

Goncalves and Meddahi (2009) compared wild bootstrap and i.i.d. bootstrap for realized volatility. One could study generalized bootstrap methods for realized volatility. One could also study other estimation methods. One should also study bootstrapping the Fourier estimator and the Wavelet estimator of the realized volatility. One should employ several different generating mechanisms for the instantaneous volatility process, Ornstein–Uhlenbeck, long memory and jump processes. Market microstructure contamination could also be entertained using a model with bid-ask bounce. Fourier estimators are known to be superior to realized volatility and wavelet estimators when considering bid-ask bounce. One should also consider bootstrapping the weighted realized volatility. This class nests several estimators.

One should also study bootstrapping the realized range-based variance, a statistics that replaces every squared return of realized variance with a normalized squared range. Realized range-based variance is asymptotically mixed normal and *five times more efficient* than the realized variance. This is not surprising. In contrast to the estimator based on daily returns, the range utilizes the information from the entire sample path and one would expect it to be superior.

One should also study bootstrapping continuous time GARCH model which is called COGARCH model. In this context one should consider heavy-tailed distributions (in the sense that all moments are not finite) of the volatility.

Recently in Bishwal (2011f) obtained the rate of weak convergence of the realized volatility thereby refining the results of Barndorff-Nielsen and Shephard (2002). He showed that the normalized realized volatility converges to the standard normal distribution with an error rate $O(M^{-1/2})$ where M is the number of intra-h observations and h is the fixed lag.

Beyond the first order theory of asymptotic mixed normal distribution of the realized volatility, one should obtain the Berry–Esseen bounds. Then one should consider the properties of generalized bootstrap estimators of realized volatility and study their Edgeworth expansion.

One should also study the problem of testing of the parametric form of the volatility in a stochastic differential equation driven by fractional Brownian motion with Hurst parameter $H > \frac{1}{2}$. This class of models is important as it captures the long-memory behavior of the log-share prices. One should then extend to fractional stochastic volatility model.

Testing about volatility is an important problem in derivative pricing. A misspecification of the volatility function could lead to misspecified derivative prices. When the price process follows the classical diffusion, Ait-Sahalia (1996) studied the volatility testing problem using nonparametric density matching. His test is based on the comparison of parametric marginal density with its nonparametric estimator. However, the classical diffusion does not take into account the long-memory behavior of the price process. One could take the long-memory behavior of the price into account using the fractional diffusion and study the testing problem for volatility. One should study the asymptotic behavior of the test statistic.

A normalized fractional Brownian motion $\{W_t^H, t \geq 0\}$ with Hurst parameter $H \in (0,1)$ is a centered Gaussian process with continuous sample paths whose covariance kernel is given by

$$E(W_t^H W_s^H) = \frac{1}{2}(s^{2H} + t^{2H} - |t-s|^{2H}), \quad s,t \geq 0.$$

The process is self-similar (scale invariant) and it can be represented as a stochastic integral with respect to standard Brownian motion. For $H = \frac{1}{2}$, the process is a standard Brownian motion. For $H \neq \frac{1}{2}$, the fBm is not a semimartingale and not a Markov process, but a Dirichlet process. The increments of the fBm are negatively correlated for $H < \frac{1}{2}$ (anti-persistent) and positively correlated for $H > \frac{1}{2}$ and in this case they display long-range dependence (persistence). The parameter H, which is also called the self similarity parameter, measures the intensity of the long-range dependence.

We assume that the price process of an underlying is a one dimensional fractional diffusion satisfying the fractional Itô SDE

$$dX_t = \mu(X_t)dt + \sigma(X_t)dW_t^H, \ 0 < t \leq T,$$

where μ and σ are smooth functions such that a unique solution of the above SDE exists and $\{W_t^H, t \in [0,T]\}$ is a fractional Brownian motion with Hurst coefficient $H > 1/2$. Assume the σ^2 has continuous derivatives up to second order.

For the pricing of contingent claims with a payoff function that depends on the evolution of the underlying process X during the time period $[0,T]$, the derivative prices are calculated under the risk-neutral quasi-martingale measure, which is not influenced by the drift function, but the diffusion function σ captures the volatility of the underlying.

In the field of interest rate modeling, the specification of μ is also important. However, the first step in evaluating a particular parametric interest rate model could be test of parametric form of its volatility coefficient.

It is often assumed that the diffusion coefficient belongs to a set of parametric functions, i.e., there exists an unknown parameter $\theta_0 \in \Theta$ such that $\sigma(x) = \sigma(\theta_0, x)$. The hypotheses to be tested are the null hypothesis
H_0: There exists $\theta_0 \in \Theta$: for every $t \in [0,T]$: $\sigma(X_t) = \sigma(\theta_0, X_t)$ P-a.s.
versus the alternative hypothesis
H_1: For all $\theta \in \Theta$: for every $t \in [0,T]$: $|\sigma(X_t) - \sigma(\theta, X_t)| \geq c_n \, \Delta_n(X_t)$ P-a.s.

Here Δ_n is the local shift in the alternative, a sequence of bounded functions and c_n is the order of difference between H_0 and H_1.
Assumptions:
(A1) The following holds for σ^2:

$$|\sigma^2(\theta, x) - \sigma^2(\theta_0, x)| \leq D(x)|\theta - \theta_0| \ \ \forall x \in I_x,$$

where $D(x)$ is a constant depending on x and the set I_x is defined by

$$I_x := \{x : L_T(x) \geq \epsilon > 0\}$$

with an arbitrary number ϵ, where $L_T(x)$ denotes the local time of X at time T.
(A2) $\hat{\theta}$ is a $1-H$ root consistent parametric estimator of θ within the family of the parametric model, i.e., $|\hat{\theta}_T - \theta| = O_P(n^{-(1-H)})$.

We introduce the following test statistics

$$\tau_n(x) = \{nh_n L_n(x_i)\}^{1-H} \left(\frac{S_n(x)}{\tilde{\sigma}^2(\hat{\theta}, x)} - 1 \right).$$

The proposed test statistic is asymptotically equivalent to the L_2 distance between $S_n(\cdot)$ and $\tilde{\sigma}^2(\hat{\theta}, \cdot)$.

One should study the asymptotic behavior of the above test statistic. Corradi and Swanson (2005) studied bootstrap specification test using an i.i.d. bootstrap. One should study bootstrap specification test using the generalized bootstrap methods.

Jumps and Long Memory

The following processes, Fractional Poisson process, Fractional Levy process, and Sub-fractional Levy processes, could be used for noise driving the stochastic volatility process.

Fractional Poisson process

A fractional Poisson process $\{W_H(t), t > 0\}$ with Hurst parameter $H \in (1/2, 1)$ is defined as

$$W_H(t) = \frac{1}{\Gamma(H - \frac{1}{2})} \int_0^t u^{\frac{1}{2}-H} \left(\int_0^u \tau^{H-\frac{1}{2}} (\tau - u)^{H-\frac{3}{2}} d\tau \right) dQ(u),$$

where

$$Q(u) = \frac{N(u)}{\sqrt{\lambda}} - \sqrt{\lambda} u,$$

where $N(u)$ is a Poisson process.

The process is self-similar in the wide sense, has wide sense stationary increments, has fat-tailed non-Gaussian distribution, and exhibits long range dependence. The process converges to fractional Brownian motion in distribution. The process is self-similar in the asymptotic sense.

Strict sense, wide sense and asymptotic sense self-similarity are equivalent for fractional Brownian motion. Stock returns are far from being self-similar in strict sense. The stochastic volatility model we consider is

$$dS_t = \sqrt{V}_t dB(t)$$

$$dV_t = \mu V_t dt + \sigma V_t dW_H(t).$$

The advantage of fractional Poisson process is that whereas fractional Brownian motion allows for arbitrage when used as noise in the stock price process, the shot noise process itself can be chosen arbitrage free.

Fractional Levy Process
Recently long-memory processes, i.e., processes with slowly decaying autocorrelation and processes with jumps, have received attention in finance, engineering, and physics. The simplest continuous time long memory process is the fractional Brownian motion discovered by Kolmogorov (1940) and later on studied by Levy (1948) and Mandelbrot and Van Ness (1968). Continuous time long-memory jump process is fractional Levy process. Hence fractional Levy process can also be called the *Kolmogorov–Levy process.*

A normalized fractional Brownian motion $\{W_t^H, t \geq 0\}$ with Hurst parameter $H \in (0,1)$ is a centered Gaussian process with continuous sample paths whose covariance kernel is given by

$$E(W_t^H W_s^H) = \frac{1}{2}(s^{2H} + t^{2H} - |t-s|^{2H}), \quad s,t \geq 0.$$

The process is self-similar (scale invariant) and it can be represented as a stochastic integral with respect to standard Brownian motion. For $H = \frac{1}{2}$, the process is a standard Brownian motion. For $H \neq \frac{1}{2}$, the fBm is not a semimartingale and not a Markov process, but a Dirichlet process. The increments of the fBm are negatively correlated for $H < \frac{1}{2}$ and positively correlated for $H > \frac{1}{2}$ and in this case they display long-range dependence. The parameter H, which is also called the self-similarity parameter, measures the intensity of the long-range dependence. The ARIMA(p,d,q) with autoregressive part of order p, moving average part of order q and fractional difference parameter $d \in (0,\ 0.5)$ process converge in Donsker sense to fBm.

Levy driven processes of Ornstein–Uhlenbeck type have been extensively studied over the last few years and widely used in finance, see Barndorff-Nielsen and Shephard (2001). FLOU process generalizes fOU process to include jumps.

The fractional Levy Ornstein–Uhlenbeck (fOU) process is an extension of fractional Ornstein–Uhlenbeck process with fractional Levy motion (fLM) driving term. In finance, it could be useful as a generalization of fractional Vasicek model, as one-factor short-term interest rate model which could take into account the long-memory effect and jump of the interest rate. This process was introduced by Marquardt (2006a,b) who also introduced fractional Levy process in Marquardt (2006a,b). Using suitable transformation of the process, one can obtain a nonlinear stationary process satisfying a fractional SDE, see Buchmann and Kluppelberg (2005). Benth (2003) used the process as temperature and obtained weather derivative arbitrage free pricing formulas for European and average type options. Cheridito et al. obtained the fOU process as a Lamperti transformation of the fBM. The model parameter is usually unknown and must be estimated from data.

Fractional Levy Process is defined as

$$M_{H,t} = \frac{1}{\Gamma(H+\frac{1}{2})} \int_{\mathbb{R}} [(t-s)_+^{H-1/2} - (-s)_+^{H-1/2}] dM_s, \quad t \in \mathbb{R},$$

where $M_t, t \in \mathbb{R}$ is a Levy process on $\mathbb{R}$ with $E(M_1) = 0$, $E(M_1^2) < \infty$ and without Brownian component. FLP has the following properties.
(1) The covariance of the process is given by

$$cov(M_{H,t}, M_{H,s}) = \frac{E(L(1)^2}{2\Gamma(2H+1)\sin(\pi H)}[|t|^{2H} + |s|^{2H} - |t-s|^{2H}].$$

(2) M_H is not a martingale. For a large class of Levy processes, M_d is neither a semimartingale.
(3)M_H is Hölder continuous of any order β less than $H - \frac{1}{2}$.
(4) M_H has stationary increments.
(5) M_H is symmetric.
(6) M_H is self-similar.
(7) M_H has infinite total variation on compacts.
(8) The FIMA process is defined as

$$Y_H(t) = \int_{-\infty}^{t} g_H(t-u)L(du),\ t \in \mathbb{R},$$

where

$$g_H(t) = \frac{1}{\Gamma(H-\frac{1}{2}} \int_0^t g(t-s)s^{H-\frac{3}{2}}ds,\ t \in \mathbb{R}$$

and the kernel g is the kernel of a short memory moving average process.

The process $Y_H(t)$ can be written as

$$Y_H(t) = \int_{-\infty}^{t} g(t-u)dM_{H,u},\ t \in \mathbb{R}.$$

Assuming that the kernel $g : \mathbb{R} \to \mathbb{R}$ satisfies $g(t) = 0$ for all $t < 0$ (causality) and $|g(t)| \leq Ce^{-ct}$ for some constants $C > 0$ and $c > 0$ (has short memory), the FIMA (fractionally integrated moving average) process is stationary and is infinite divisible.

Consider the kernel

$$g(t-s) = \sigma e^{\theta(t-s)} I_{(0,\infty)}(t-s)$$

then

$$g_H(t) = \frac{\sigma}{\Gamma(H-\frac{1}{2})} \int_0^{\infty} e^{\theta(t-s)} I_{(0,\infty)}(t-s)s^{H-\frac{3}{2}}ds,\ t \in \mathbb{R}$$

Note that

$$U_t^{H,\theta,\sigma} = \int_{\mathbb{R}} g_H(t-u)L(du),\ t \in \mathbb{R}$$

is the fractional Levy Ornstein–Uhlenbeck (FLOU) process satisfying the fractional Langevin equation

$$dU_t = \theta U_t dt + \sigma dM_{H,t},\ t \in \mathbb{R}.$$

The process has long memory. Consider the Ornstein–Uhlenbeck process X_t satisfying the Itô stochastic differential equation

$$dX_t = \theta X_t dt + dM_t^H, \quad t \geq 0,$$

where $\{M_t^H\}$ is a fractional Levy motion with $H > 1/2$ with the filtration $\{\mathcal{F}_t\}_{t\geq 0}$ and $\theta > 0$ is the unknown parameter to be estimated on the basis of continuous observation of the process $\{X_t\}$ on the time interval $[0, T]$.

$$X_t = \int_{-\infty}^{t} e^{\theta(t-s)} dM_s.$$

This process is stationary and is a process with long memory.

It can be shown that X_{t_i} is a stationary discrete time AR(1) process with autoregression coefficient $\phi \in (0, 1)$ with the following representation

$$X_{t_i} = \phi X_{t_{i-1}} + \epsilon_{t_{i-1}},$$

where

$$\phi = e^{-\theta h}, \quad \epsilon_{t_{i-1}} = \int_{t_{i-1}}^{t_i} e^{-\theta(t_i - u)} dM_u.$$

For example, M is a gamma process or an inverse-Gaussian process. The fractional Levy process (FLP) is a generalization of FBM. The FLP is not a martingale and not even a semimartingale in general. FLP is a natural counterpart to FBM. It is Holder continuous. It has stationary increments. It is symmetric. It is self-similar. Fractional stable process is a special case of FLP.

The fractional Levy Ornstein–Uhlenbeck (FLOU) process X was introduced in Marquardt (2006a,b). FLOU process is a LOU process which includes long memory.

We introduce the ratio estimator of the drift. The ratio estimator of θ is defined as

$$\hat{\theta}_n := -\log\left[\min_{1\leq i\leq n} \frac{X_{i\Delta}}{X_{(i-1)\Delta}}\right].$$

This estimator is motivated by the extreme value theory for the correlation parameter of an AR(1) process whose innovation distribution is positive. See Davis and McCormick (1989). In the case of exponential AR(1) process, it coincides with the maximum likelihood estimator. See Nielsen and Shephard (2003).

The weak consistency of the ratio estimator in the LOU process was studied in Jongbloed et al. (2005). The strong consistency and asymptotic Weibullness was studied in Brockwell et al. (2007) in the case of Gamma innovations. One can obtain strong consistency and asymptotic Weibullness of the ratio estimator for the fLOU process.

Sub-fractional Levy Process

Recently, sub-fractional Brownian (sub-FBM) motion which is a centered Gaussian process with covariance function

$$C_H(s,t) = s^{2H} + t^{2H} - \frac{1}{2}\left[(s+t)^{2H} + |s-t|^{2H}\right], \quad s,t > 0$$

for $0 < H < 1$ introduced by Bojdecki et al. (2004) has received some attention recently. The interesting feature of this process is that this process has some of the main properties of FBM, but the increments of the process are nonstationary, more weakly correlated on non-overlapping time intervals than that of FBM, and its covariance decays polynomially at a higher rate as the distance between the intervals tends to infinity. It would be interesting to see extension of this section to sub-FBM case. We generalize sub-FBM to sub-FLP.

Sub-fractional Levy process is defined as

$$S_{H,t} = \frac{1}{\Gamma(H+\frac{1}{2})}\int_{\mathbb{R}}[(t-s)_+^{H-1/2} - (-s)_+^{H-1/2}]dM_s, \quad t \in \mathbb{R},$$

where $M_t, t \in \mathbb{R}$ is a Levy process on $\mathbb{R}$ with $E(M_1) = 0,\ E(M_1^2) < \infty$ and without Brownian component. SFLP has the following properties:
(1) The covariance of the process is given by

$$cov(S_{H,t}, S_{H,s}) = s^{2H} + t^{2H} + \frac{E(L(1)^2}{2\Gamma(2H+1)\sin(\pi H)}[|t|^{2H} + |s|^{2H} - |t-s|^{2H}].$$

(2) S_H is not a martingale. For a large class of Levy processes, S_H is neither a semimartingale.
(3) S_H is Hölder continuous of any order β less than $H - \frac{1}{2}$.
(4) S_H has nonstationary increments.
(5) S_H is symmetric.
(6) S_H is self-similar.
(7) S_H has infinite total variation on compacts.

1.6 Stochastic Gradient Descent Algorithm for American Option

1.6.1 Stochastic Gradient Descent in Continuous Time

Consider the ordinary SDE

$$dX_t = f(\theta, X_t, t)dt + dW_t, \tag{6.1}$$

where W is a standard Brownian motion.

In standard discrete time version of stochastic gradient descent, data is usually considered to be i.i.d. at every step. Thus it is natural to ask if one can discretize (6.1), for example, by Euler–Maruyama method and apply traditional stochastic gradient descent. This can result in loss of accuracy or may not even converge. For example, there is no guarantee that using a higher order discretization scheme, for example, the second order Milstein scheme, to discretize the dynamics of the SDE (6.1) and then applying the traditional stochastic gradient descent will produce a statistical learning scheme which is higher order accurate in time. Hence it makes sense to first develop the continuous-time statistical learning equation and then apply higher order accurate numerical scheme.

Stochastic gradient descent in continuous time (SGDCT) provides a computationally efficient method for the statistical learning of continuous-time models. SGDCT algorithm follows a descent direction along a continuous stream of data. The parameter updates occur in continuous time and satisfy a stochastic differential equation (SDE). We analyze the asymptotic convergence rate by proving a central limit theorem. An L^p convergence rate is also proven.

Statistical estimation in SDEs have been studied using entire observed path of X, i.e., batch optimization, see Bishwal (2008a). The vast majority of statistical learning, machine learning, and stochastic gradient descent literature address discrete time algorithm. This paper analyzes statistical learning in continuous time (SGDCT). SGDCT can estimate unknown parameters and functions in SDE models. It is related to online maximum likelihood filtering and identification.

SGDCT can be used to solve continuous-time optimization problem such as American options. The value function is approximated by a parametric function and the parameter is estimated by SGDCT algorithm. Recall that Longstaff-Schwartz estimated the parameter by least squares method. One could discretize the dynamics and then use the Q-learning algorithm. The Q-learning algorithm is biased while SGDCT algorithm is unbiased.

Consider the SDE

$$dX_t = f^*(X_t)dt + \sigma dW_t, \tag{6.2}$$

where $f^*(x)$ is an unknown function. The goal is to estimate $f(x,\theta)$ from continuous observations of $(X_t)_{t\geq 0}$. The function may be convex or non-convex.

The SGD update in continuous time for the parameter $\theta \in \mathbb{R}$ satisfies the SDE

$$d\theta_t = \frac{\alpha_t}{\sigma^2} f'(X_t,\theta_t)dX_t - \frac{\alpha_t}{\sigma^2} f'(X_t,\theta_t)f(X_t,\theta_t)dt, \tag{6.3}$$

where α_t is the *learning rate*. For example, α_t could be $\frac{C_\alpha}{C_0+t}$. We assume that θ_0 is initialized according to some distribution with compact support.

The parameter update can be used both for statistical estimation given previously observed data as well as online learning, i.e., statistical estimation in real time as data becomes available.

Define the objective function

$$g(x,\theta) = \frac{1}{2}\|f(x,\theta) - f^*(x)\|^2_{\sigma^2} = \frac{1}{2\sigma^2}\langle (f(x,\theta) - f^*(x))\rangle^2 \tag{6.4}$$

which measures the distance between the model $f(x,\theta)$ and true dynamics $f^*(x)$ for a specific x. This is a minimum distance type estimator.

We assume that X_t is ergodic and it has some well behaved $\pi(dx)$ as its unique invariant distribution. Let the average over $\pi(dx)$ be denoted by

$$\bar{g}(\theta) = \int g(x,\theta)\pi(dx), \tag{6.5}$$

where $\pi(dx)$ is the invariant measure of X_t when it is ergodic, which is the natural objective function for our analysis of the asymptotic behavior of the algorithm θ_t. This is a weighted average of the distance between $f(x,\theta)$ and $f^*(x)$ where the weights are given by $\pi(dx)$, which is the distribution that X_t tends to when t become large. The distance $g(x,\theta)$ is decreased by moving θ in the descent direction $-g'(x,\theta)$ which motivates the algorithm

$$\begin{aligned} d\theta_t &= -\alpha_t g'(X_t,\theta_t)dt \\ &= \tfrac{\alpha_t}{\sigma^2} f'(X_t,\theta_t)(f^*(X_t) - f(X_t,\theta))dt \\ &= -\alpha_t g'(x,\theta_t)dt + \tfrac{\alpha_t}{\sigma^2} f'(X_t,\theta_t)f(X_t,\theta_t))\sigma dW_t \\ &= -\alpha_t \bar{g}'(x,\theta_t)dt - \alpha_t(g'(x,\theta_t) - \bar{g}'(x,\theta_t))dt \\ &\quad + \tfrac{\alpha_t}{\sigma^2} f'(X_t,\theta_t)f(X_t,\theta_t))\sigma dW_t \\ &=: I_1 + I_2 + I_3, \end{aligned} \tag{6.6}$$

where I_1= Descent term, I_2 = fluctuation term, I_3= Noise term.

If α_t decays with time, e.g., $\alpha_t = \frac{C_\alpha}{C_0+t}$, the descent term I_1 will dominate the fluctuation term and the noise term for large t. Then θ_t will converge to a local minimum of $\bar{g}(\theta)$. Sirignano and Spiliopoulos (2017) proved that θ_t converges to a critical point of the objective function $\bar{g}(\theta)$: $|\bar{g}'(\theta_t)| \to 0$ almost surely as $t \to \infty$.

Since $\lim_{t\to\infty} \alpha_t = 0$, the descent term $\alpha_t \bar{g}'(\theta_t) \to 0$ as $t \to \infty$. Descent term converges to zero as $t \to \infty$. Sirignano and Spiliopoulos (2018) proved the rate at which θ_t converges to zero. They obtained a central limit theorem (CLT) and an L^p convergence rate.

When $\bar{g}(\theta)$ has a single critical point θ^*, let

$$J(\theta^*) := C_\alpha^2 \int_0^\infty e^{-2s(C_\alpha \bar{g}(\theta^*)-1)}\bar{h}(\theta^*)ds. \tag{6.7}$$

Define

$$\Psi_{t,s}^{(p)} := \exp(-pC \int_s^t \alpha_u du), \; p \geq 1 \tag{6.8}$$

and let $\Phi_{t,s}^*$ be the solution to the ODE

$$d\Phi_{t,s}^* = -\alpha_t \bar{g}(\theta^*)\Phi_{t,s}^* dt, \;\; \Phi_{s,s}^* = 1. \tag{6.9}$$

We assume that the general learning rate α_t satisfies the following conditions:

(B1) $\int_0^\infty \alpha_t dt = \infty$,

(B2) $\int_0^\infty \alpha_t^2 dt < \infty$,

(B3) $\int_0^\infty |\alpha_t'|dt < \infty$,
(B4) There exists a $p > 0$ such that $\lim_{t\to\infty} \alpha_t^2 t^p = 0$,
(B5) $\int_0^t \alpha_s^{5/2} \Psi_{t,s}^{(2)} ds \le o(\alpha_t)$,
(B6) $\int_0^t \alpha_s^2 {\Phi^*}_{t,s}^2 ds = O(\alpha_t)$,
(B7) $\int_0^t \alpha_s^2 \Psi_{t,s}^{(1)} ds \le o(\alpha_t^{1/2})$,
(B8) For all $p \ge 2$, $\int_0^t (\alpha_s^2 + |\alpha_s'|)\Psi_{t,s}^{(p)} \alpha_s^{p/2-1} ds \le o(\alpha_t^{p/2})$,
(B9) For all $p \ge 2$, $\Psi_{t,0}^{(p)} \le O(\alpha_t^{p/2})$
(B10) $\Psi_{t,0}^{(1)} \le o(\alpha_t^{1/2})$.

The L^p convergence rate is given by

$$E|\theta_t - \theta^*|^p \le K\alpha_t^{p/2} \tag{6.10}$$

for $p \ge 1$. The CLT is given by

$$\alpha_t^{-1/2}(\theta_t - \theta^*) \to^D \mathcal{N}(0, J(\theta^*)). \tag{6.11}$$

A standard choice of the learning rate α_t which satisfies (B1)–(B10) is $\alpha_t = C_\alpha(C_0 + t)^{-1}$. Hence the L_p convergence rate is given by

$$E[|\theta_t - \theta^*|^p] \le \frac{K}{(C_0 + t)^{p/2}} \tag{6.12}$$

for $p \ge 1$ and the CLT for θ_t is given by

$$\sqrt{t}(\theta_t - \theta^*) \to^D \mathcal{N}(0, J(\theta^*)) \quad \text{as} \quad t \to \infty. \tag{6.13}$$

Sirignano and Spiliopoulos (2020) derived a stochastic integral to represent the $\sqrt{t}(\theta_t - \theta^*)$ using Duhamel's principle and the fundamental solution of the random ODE $d\Psi_{t,s} = -\alpha_t \bar{g}(\tilde{\theta}_t)\Psi_{t,s}dt$ where $\tilde{\theta}_t$ lies on a line connecting θ^* and θ_t. The integrand of this stochastic integral includes the fluctuation term and the noise term as well as Ψ_t and is *anticipative.* Hence standard approach such as Itô isometry cannot be applied directly. Also since $f(x, \theta)$ is allowed to grow with θ, hence the fluctuations as well as other terms can grow with θ. Hence they prove an a priori stability estimate for $|\theta_t|$. Proving central limit theorem for non-convex $\bar{g}(\theta)$ is not straightforward since the convergence speed of θ_t can arbitrarily be slow in certain regions, and the gradient can even point away from the global minimum θ^*. To address this, we consider the stochastic integral after the time τ_δ which is defined as the final time θ_t enters a neighborhood of θ^*. However, τ_δ is anticipative, i.e., is not a stopping time; therefore, careful analysis is required to study the behavior of the stochastic integral.

Let $Y_t := \theta_t - \theta^*$ be the error term. It satisfies

$$dY_t = -\alpha_t \Delta\bar{g}(\theta_t^1)Y_t dt + \alpha_t(\bar{g}'(\theta_t) - g'(X_t, \theta_t))dt + \alpha_t f'(X_t, \theta_t)dW_t. \tag{6.14}$$

$$E|Y_t|^2 \leq Kt^{-1}, \quad E|Y_t|^p \leq Kt^{-p/2}. \tag{6.15}$$

The stochastic integral $\sqrt{t}\int_1^t \alpha_s^2 \Phi_{t,s}\zeta(X_s, \theta_s)dW_s \to 0$ in probability as $t \to \infty$ where $\zeta(x, \theta)$ is a function that can grow at most polynomially in x and θ.

For the analysis of fluctuation term, the proofs use Poisson PDE for ergodicity, given below in Proposition 6.1. The central limit theorem for non-convex $\bar{g}(\theta)$ is challenging since the convergence speed of θ_t can become arbitrarily slow in certain regions and the gradient can even point away from the global minimum θ^*. *Interalia,* they prove convergence to zero of multidimensional stochastic integrals. The proof requires the analysis of stochastic integral with anticipative integrands, which is challenging since standard approaches like Itô isometry cannot be directly applied.

Sirignano and Spiliopoulos (2020) remark that $t^{-1/2}$ is the fastest possible convergence rate given that the noise is Brownian motion. This is due to the quadratic variation of Brownian motion growing linearly in t. With other noises whose variances grow sublinearly in time, one could allow for faster rate of convergence than $t^{-1/2}$. An example of a stochastic process whose variance grows sublinearly in time is fractional Brownian motion with appropriately chosen Hurst parameter.

Proposition 6.1 *(A Poisson Equation)*
Let $\mathcal{L}_x$ be the infinitesimal generator of the X process. Let $G(x, \theta) \in C^{\alpha,2}(\mathcal{X}, \mathbb{R}^n)$ which satisfies

$$\int_{\mathcal{X}} G(x, \theta)\pi(dx) = 0$$

and for some positive constants M and q, and

$$|G(x, \theta)| + \left|\frac{\partial}{\partial \theta} G(x, \theta)\right| + \left|\frac{\partial^2}{\partial \theta^2} G(x, \theta)\right| \leq M(1 + |x|^q).$$

Then the Poisson equation

$$\mathcal{L}_x u(x, \theta) = G(x, \theta), \quad \int_{\mathcal{X}} u(x, \theta)\pi(dx) = 0$$

has a unique solution that satisfies $u(x, \cdot) \in C^2$ for every $x \in \mathcal{X}, \partial_\theta^2 u \in C(\mathcal{X} \times \mathbb{R}^n)$ and there exist positive constants K and p such that

$$|u(x, \theta)| + \left|\frac{\partial}{\partial \theta} u(x, \theta)\right| + \left|\frac{\partial^2}{\partial \theta^2} u(x, \theta)\right| \leq K(1 + |x|^p).$$

1.6.2 Stochastic Gradient Descent Algorithm for American Option

We study the SGDCT algorithm for American option. We compare it with Longstaff-Schwartz method. Longstaff-Schwartz developed an algorithm for the solution of a discrete time version of the a class of free boundary. Their algorithm, commonly called Longstaff-Schwartz Regression based method, uses dynamic programming and approximates the solution using a separate function approximator at each discrete time, typically a linear combination of basis functions.

Given a continuous stream of data, stochastic gradient descent in continuous time (SGDCT) can estimate unknown parameters or functions in stochastic differential equation (SDE) models for stocks, bonds, interest rates, and financial derivatives. High dimensional American option has been a long standing computational challenge in finance with traditional methods like the finite difference. SGDCT can accurately solve American options even in 100 dimensions.

Batch optimization for statistical estimation of continuous-time models can be impractical for large data sets where observations occur over a long time period. Batch optimization takes a sequence of descent steps for the model error for the entire observed path.

SGDCT provides a computationally efficient method for statistical learning over long time periods and for complex models. SGDCT continuously follows a descent direction along the path of the observation. Parameters are updated in continuous time, with the parameter updates θ_t satisfying an SDE.

Numerical analysis of SGDCT in model estimation of the drift and volatility functions of two common financial models like the O-U process and CIR process is studied. One has to simulate using Euler scheme, a single path of X_t for given θ^* and simultaneously solve for the path of θ_t.

Sirignano and Spiliopoulos (2018) studied deep learning algorithm or "Deep Galerkin Method" (DGM) which is Galerkin method with neural network. Neural network is trained on the batches of randomly sampled time and space points. Deep Galerkin method uses a deep neural network instead of basis functions. The deep neural network is trained to satisfy the differential operator, initial condition, and the boundary conditions using stochastic gradient descent at randomly sampled spatial points. By randomly sampling spatial points, we avoid the need to form a mesh (which is infeasible in higher dimensions) and instead convert the PDE problem to a machine learning problem. DGM is natural merger of Galerkin methods and machine learning.

Sirignano and Spiliopoulos (2017) obtained central limit theorem for the SGDCT estimator.

An American option is a financial derivative which the owner can choose to exercise at any time $t \in [0, T]$. If the owner exercises the option, they receive the payoff $g(X_t)$ where X_t is the prices of the underlying stocks. If the owner does not choose to exercise the option, they receive the payoff $g(X_T)$ at the final time T. The value (or price) of the American option at time t is $u(t, X_t)$ which satisfies a free boundary PDE:

$$\frac{\partial u}{\partial t}(t,x)+\mu(x)\frac{\partial u}{\partial x}(t,x)+\frac{1}{2}\sum_{i,j=1}^{n}\rho_{i,j}\sigma(x_j)\frac{\partial^2 u}{\partial x_i \partial x_j}(t,x)-ru(t,x)=0, \quad (6.16)$$

$$\begin{gathered}\forall\{(t,x):u(t,x)>g(x)\},\\ u(t,x)\geq g(x)\ \ \forall(t,x),\\ u(t,x)\in C^1(\mathbb{R}_+\times\mathbb{R}^d),\ \ \forall\{(t,x):u(t,x)=g(x)\},\\ u(T,x)=g(x),\ \ \forall x.\end{gathered}$$

The free boundary set is $F = \{(t,x) : u(t,x) = g(x)\}$. The value function $u(t,x)$ satisfies a PDE "above" the free boundary set F and $u(t,x)$ equals the function $g(x)$ "below" the free boundary set F. The free boundary set F is approximated using the current parameter estimate θ_n.

The SGDCT Algorithm

First, we recall the Q-learning algorithm: The Q-learning algorithm uses stochastic gradient descent to minimize an approximation to the discrete time HJB equation. Consider the Q-learning algorithm to estimate the value function

$$V(x) = E\left[\int_0^\infty e^{-\gamma t} r(X_t)dt \ \Big|\ X_0 = x\right], \quad X_t = x + W_t, \tag{6.17}$$

where $\gamma > 0$ is a discount factor and $r(x)$ is a reward function. The function $Q(x,\theta)$ is an approximation for the value function $V(x)$. The traditional approach is to discretize the dynamics of $V(x)$ and apply a stochastic gradient descent update to the objective function:

$$E\left[\left(r(X_t)\Delta + e^{-\gamma\Delta}E[Q(X_{t+\Delta};\theta)|X_t] - Q(X_t;\theta)\right)^2\right]. \tag{6.18}$$

The result is the stochastic gradient descent algorithm:

$$\begin{aligned}\theta_{t+\Delta} = \theta_t - \frac{\alpha_t}{\Delta}&\left(e^{-\gamma\Delta}E[Q_\theta(X_{t+\Delta};\theta_t)|X_t] - Q_\theta(X_t;\theta_t)\right)\big)\\ &\times\left(r(X_t)\Delta + e^{-\gamma\Delta}E[Q(X_{t+\Delta};\theta_t)|X_t] - Q(X_t;\theta_t)\right).\end{aligned} \tag{6.19}$$

The learning rate is Δ^{-1}. The Q-learning algorithm has a major computational issue. The expectation $E[Q_\theta(X_{t+\Delta};\theta_t)|X_t]$ is challenging to calculate if the process X_t is high dimensional. To circumvent this situation, Q-learning algorithm ignores the inner expectation leading to

$$\begin{aligned}\theta_{t+\Delta} = \theta_t - \tfrac{\alpha_t}{\Delta}&(e^{-\gamma\Delta}Q_\theta(X_{t+\Delta};\theta_t) - Q_\theta(X_t;\theta_t)\\ &\times(r(X_t)\Delta + e^{-\gamma\Delta}Q(X_{t+\Delta};\theta_t) - Q(X_t;\theta_t)).\end{aligned} \tag{6.20}$$

Although computationally efficient, the Q-learning algorithm is biased. The SGDCT algorithm can be directly derived by letting $\Delta \to 0$ and using Itô formula:

$$d\theta_t = -\alpha_t(\frac{1}{2}Q_{\theta xx}(X_t;\theta_t) - \gamma Q_\theta(X_t;\theta_t))(r(X_t) + \frac{1}{2}Q_{xx}(X_t;\theta_t) - \gamma Q(X_t;\theta_t))dt. \tag{6.21}$$

Furthermore, when $\Delta \to 0$, the Q-learning algorithm blows up.

SGDCT Algorithm for American Option

Let $X_t \in \mathbb{R}^d$ be the prices of d stocks. The maturity time is T and the payoff function is $g(x) : \mathbb{R}^d \to \mathbb{R}$. The stock price dynamics and the value functions are given by

$$dX_t^i = \mu(X_t^i)dt + \sigma(X_t^i)dW_t^i, \quad i = 1, 2, \ldots, d \tag{6.22}$$

$$V_{t,x} = \sup_{\tau \geq t} E[e^{-r(\tau \wedge T)} g(X_{\tau \wedge T})|X_t = x], \tag{6.23}$$

where $W_t \in \mathbb{R}^d$ is a Brownian motion. The distribution of W_t is specified by $Var(W_t^i) = t, \ i = 1, 2, \ldots, d$ and $Corr(W_t^i, W_t^j) = \rho_{i,j}dt$ for $i \neq j$. The price of the American option is $V_{0,x}$.

SGDCT for American option is given by

$$\theta_{t\wedge T}^{n+1} = \theta_0^n -$$

$$- \int_0^{\tau\wedge T} \alpha_t^{n+1} \left(\frac{\partial}{\partial t} Q_\theta(t, X_t; \theta_t^{n+1}) + \mathcal{L}_x Q_\theta(t, X_t; \theta_t^{n+1}) - rQ_\theta(t, X_t; \theta_n^{n+1}) \right)$$

$$\times \left(\frac{\partial}{\partial t} Q(t, X_t; \theta_t^{n+1}) + \mathcal{L}_x Q(t, X_t; \theta_t^{n+1}) - rQ(t, X_t; \theta_t^{n+1}) \right) dt$$

$$+\alpha_{\tau\wedge T}^{n+1} \, Q_\theta \left(\tau \wedge T, X_{\tau\wedge T}; \theta_{\tau\wedge T}^{n+1}\right) \left(g(X_{\tau\wedge T}) - Q(\tau \wedge T, X_{\tau\wedge T}; \theta_{\tau\wedge T}^{n+1})\right), \tag{6.24}$$

$$\tau := \inf\{t \geq 0 : Q(t, X_t; \theta_t^{n+1}) < g(X_t)\}, \quad X_0 \sim \nu(dx). \tag{6.25}$$

The function $Q(x, \theta)$ is an approximation of the value function. The parameter θ must be estimated. Here $\mathcal{L}_x$ is the infinitesimal generator of the X process. The algorithm is run for many iterations $n = 0, 1, 2, \ldots$ until convergence.

The Longstaff-Schwarz algorithm works well in low dimension, in high dimension the convergence is slow. In high dimension, SGD algorithm works very well. Sirignano and Spiliopoulos (2017) implemented the American option in 100 dimensions and showed the accuracy of the SGD algorithm for Bachelier model and Black-Scholes model.

1.6.3 Berry-Esseen Inequality of Stochastic Gradient Descent Algorithm for American Option

We study the Berry–Esseen inequality for SGDCT algorithm for American option. We compare it with Longstaff-Schwartz method. We will use anticipative stochastic integral, Duhamel's principle for the stochastic gradient descent algorithm.

Stochastic gradient descent in continuous time (SGDCT) provides a computationally efficient method for the statistical learning of continuous-time models. SGDCT can estimate unknown parameters and functions in SDE models. SGDCT algorithm follows a descent direction along a continuous stream of data. The parameter updates occur in continuous time and satisfy a stochastic differential equation (SDE). The authors analyze the asymptotic convergence rate by proving a central limit theorem. An L^p convergence rate is also proven.

The vast majority of statistical learning, machine learning, and stochastic gradient descent literature address discrete time algorithm. This section analyzes statistical learning in continuous time.

Statistical estimation in SDEs has been studied using entire observed path of X, i.e., batch optimization. MLE can be calculated via batch optimization. Maximum likelihood based on the entire observation path of X has been extensively studied, see Bishwal (2008a).

SGDCT can be used to solve continuous-time optimization problem such as American options. The value function is approximated by a parametric function and the parameter is estimated by SGDCT algorithm. Recall that Longstaff-Schwartz estimated the parameter by least squares method. One could discretize the dynamics and then use the Q-learning algorithm. The Q-learning algorithm is biased while SGDCT algorithm is unbiased.

Consider the SDE

$$dX_t = f^*(X_t)dt + \sigma dW_t,$$

where $f^*(x)$ is an unknown function. The goal is to estimate $f(x,\theta)$ from continuous observations of $(X_t)_{t\geq 0}$. The function may be convex or non-convex.

The SGD update satisfies

$$d\theta_t = \frac{\alpha_t}{\sigma^2} f'(X_t,\theta)dX_t - \frac{\alpha_t}{\sigma^2} f'(X_t,\theta_t) f(X_t,\theta_t)dt,$$

where α_t is the *learning rate*. For example, α_t could be $\frac{C_\alpha}{C_0+t}$. Assume that θ_0 is initialized according to some distribution with compact support.

The parameter update can be used both for statistical estimation given previously observed data as well as online learning, i.e., statistical estimation in real time as data becomes available.

Define the objective function

$$g(x,\theta) = \frac{1}{2}\|f(x,\theta) - f^*(x)\|^2 = \frac{1}{2\sigma^2}\langle(f(x,\theta) - f^*(x))\rangle^2$$

which measures the distance between the model $f(x,\theta)$ and true dynamics $f^*(x)$ for a specific x. This is a minimum distance type estimator.

We assume that X_t is ergodic and it has some well behaved $\pi(dx)$ as its unique invariant distribution. Let the average be denoted by

$$\bar{g}(\theta) = \int g(x,\theta)\pi(dx),$$

where $\pi(dx)$ is the invariant measure of X_t when it is ergodic which is the natural objective function. This is a weighted average of the distance between $f(x,\theta)$ and $f^*(x)$ where the weights are given by $\pi(dx)$, which is the distribution that X_t tends to when t become large. The distance $g(x,\theta)$ is decreased by moving θ in the descent direction $-g'(x,\theta)$ which motivates the algorithm

$$\begin{aligned} d\theta_t &= -\alpha_t g'(X_t,\theta_t)dt \\ &= \tfrac{\alpha_t}{\sigma^2} f'(X_t,\theta_t)(f^*(X_t) - f(X_t,\theta))dt \\ &= -\alpha_t g'(x,\theta_t)dt + \tfrac{\alpha_t}{\sigma^2} f'(X_t,\theta_t) f(X_t,\theta_t))\sigma dW_t \\ &= -\alpha_t \bar{g}'(x,\theta_t)dt - \alpha_t(g'(x,\theta_t) - \bar{g}'(x,\theta_t))dt \\ &\quad + \tfrac{\alpha_t}{\sigma^2} f'(X_t,\theta_t) f(X_t,\theta_t))\sigma dW_t \\ &=: I_1 + I_2 + I_3, \end{aligned}$$

where I_1= Descent term, I_2 = fluctuation term, I_3= Noise term.

If α_t decays with time, e.g., $\alpha_t = \frac{C_\alpha}{C_0+t}$, the descent term I_1 will dominate the fluctuation term and the noise term for large t. Then θ_t will converge to a local minimum of $\bar{g}(\theta)$. Sirignano and Spiliopoulos (2017) proved that θ_t converges to a critical point of the objective function $\bar{g}(\theta)$: $|\bar{g}'(\theta_t)| \to 0$ almost surely as $t \to \infty$.

Since $\lim_{t\to\infty} \alpha_t = 0$, the descent term $\alpha_t \bar{g}'(\theta_t) \to 0$. Descent term converges to zero as $t \to \infty$. Sirignano and Spiliopoulos (2018) proved the rate at which θ_t converges to zero. They obtained a central limit theorem (CLT) and an L^p convergence rate.

Sirignano and Spiliopoulos (2018) derived a stochastic integral to represent the $\sqrt{t}(\theta_t - \theta^*)$ using Duhamel's principle and the fundamental solution of the random ODE $d\Psi_{t,s} = -\alpha_t \bar{g}'(\theta_t)\Psi_{t,s}dt$ where $\tilde{\theta}_t$ lies on a line connecting θ^* and θ_t. The integrand of this stochastic integral includes the fluctuation term and the noise term as well as Ψ_t and is *anticipative*. Also since $f(x,\theta)$ is allowed to grow with θ, hence the fluctuations as well as other terms can grow with θ. Hence they prove an a priori stability estimate for $|\theta_t|$. Proving central limit theorem for non-convex $\bar{g}(\theta)$ since the convergence speed of θ_t can arbitrarily slow in certain regions, and the gradient can even point away from the global minimum θ^*. To address this, we consider the stochastic integral after the time τ_δ which is defined as the final time θ_t enters a neighborhood of θ^*. However, τ_δ is anticipative, i.e., is not a stopping time; therefore, careful analysis is required to study the behavior of the stochastic integral.

Let $Y_t := \theta_t - \theta^*$ be the error term. It satisfies

$$dY_t = -\alpha_t \Delta \bar{g}(\theta_t^1) Y_t dt + \alpha_t(\bar{g}'(\theta_t) - g'(X_t,\theta_t))dt + \alpha_t f'(X_t,\theta_t)dW_t.$$

$$E|Y_t|^2 \le Kt^{-1}, \quad E|Y_t|^p \le Kt^{-p/2}.$$

The stochastic integral

$$\sqrt{t}\int_1^t \alpha_s^2 \Phi_{t,s}\zeta(X_s,\theta_s)dW_s \to 0$$

in probability as $t \to \infty$ where $\zeta(x, \theta)$ is a function that can grow at most polynomially in x and θ.

For the analysis of fluctuation term, the proofs use Poisson PDE for ergodicity, see Proposition 6.1. The central limit theorem for non-convex $\bar{g}(\theta)$ is challenging since the convergence speed of θ_t can become arbitrarily slow in certain regions and the gradient can even point away from the global minimum θ^*. *Interalia,* Sirignano and Spiliopoulos (2018) proved convergence to zero of multidimensional stochastic integrals. The proof requires the analysis of stochastic integral with anticipative integrands, which is challenging since standard approaches like Itô isometry cannot be directly applied. It is related to online maximum likelihood filtering and identification.

We remark that $t^{-1/2}$ is the fastest possible convergence rate given that the noise is Brownian motion. This is due to the quadratic variation of Brownian motion growing linearly in t. With other noises whose variances grow sublinearly in time, one could allow for faster rate of convergence than $t^{-1/2}$. An example of a stochastic process whose variance grows sublinearly in time is fractional Brownian motion with appropriately chosen Hurst parameter discussed in Sect. 1.

In this section we investigate the rate of weak convergence to normality of the update θ_t.

We assume the following conditions:

(A1) The diffusion is nondegenerate and $\lim_{|x|\to\infty} f^*(x) \cdot x = -\infty$.

(A2) $g'(x, \cdot) \in C^2(\mathbb{R})$ for all x.

(A3) The function $f^*(x) \in C^{2+\alpha}(\mathcal{X})$, that is, it has two derivatives in x, with all partial derivatives being Hölder continuous, with exponent α, with respect to x.

(A4) SGD-SDE equation is well-posed.

(A5) There exists a constant $R < \infty$ and almost everywhere positive function $\kappa(x)$ such that
$\langle -g'(x, \theta), \theta/|\theta| \rangle \le -\kappa(x)|\theta|$ for $|\theta| \ge R$.

(A6) Consider the function $\tau(x, \theta) = \langle f'^2(x, \theta), \theta/|\theta| \rangle$. Then there exists a function $\lambda(x)$ growing not faster than polynomially in $|x|$ such that for any $x, \theta_1, \theta_2 \in \mathbb{R}$, $|\tau(x, \theta_1) - \tau(x, \theta_2)| \le |\lambda(x)|\rho(|\theta_1 - \theta_2|)$ where $\rho(u)$ is an increasing function on $[0, \infty)$ with $\rho(0) = 0$ and $\int_{u>0} \rho^{-2}(u)du = \infty$.

(A7) The learning rate is $\frac{C_\alpha}{C_0 + t}$ where $C_\alpha > 0$ and C_0 are constants.

(A8) $f_\theta^{(i)}(x, \theta) \le K(1 + |x|^q + |\theta|^{(2-i)\vee 0})$, $i = 0, 1, 2$ for some finite constants $K, q < \infty$.

(A9) $\bar{g}(\theta)$ is strongly convex with constant C.

(A10) $CC_\alpha > 1$.

(A11) $\bar{g}(\theta) \in C^3$ and $|\bar{g}^{(i)}(\theta)| \le K(1 + |\theta|^{4-i})$ for $i = 0, 1, 2, 3$ and some finite constant $K < \infty$.

The following theorem gives the rate of convergence to normal distribution of the SGDCT estimator:

Theorem 6.1 *Under (A1)–(A11) and (B1)–(B10), we have as $t \to \infty$*

$$\sup_{x\in\mathbb{R}} \left| P\left(\sqrt{\frac{t}{J(\theta^*)}}(\theta_t - \theta^*) \le x\right) - \Phi(x)\right| \le Ct^{-1/2},$$

where

$$J(\theta^*) = C_\alpha^2 \int_0^\infty e^{-2s(C_\alpha g'(\theta^*)-1)}\bar{h}(\theta^*)ds, \ \ \bar{h}(\theta) = \int h(x,\theta)\pi(dx),$$

$$h(x,\theta) = \left(\frac{1}{\sigma^2}f'(x,\theta) - \dot{v}(x,\theta)\right)^2 \sigma^2$$

and $v(x,\theta)$ is the solution to the Poisson equation with

$$H(x,\theta) = g'(x,\theta) - \bar{g}'(\theta).$$

Proof Using second order Taylor expansion

$$\bar{g}'(\theta_t) = \bar{g}'(\theta_t) + \bar{g}''(\theta^*)(\theta_t - \theta^*) + \frac{1}{2}\frac{\partial^3}{\partial\theta^3}\bar{g}(\theta_t^1)(\theta_t - \theta^*)^2.$$

The error term satisfies the SDE

$$d(\theta_t - \theta^*) = -\alpha_t\bar{g}'(\theta_t^1)dt - \frac{\alpha_t}{2}\frac{\partial^3}{\partial\theta^3}\bar{g}(\theta_t^1)dt + \alpha_t(\bar{g}'(\theta_t) - g'(X_t,\theta_t))dt$$
$$+\alpha_t f'(X_t,\theta_t)dW_t.$$

Let $Y_t := \theta_t - \theta^*$. Then Y_t satisfies the SDE

$$dY_t = -\alpha_t\bar{g}'(\theta_t^1)Y_t dt - \frac{\alpha_t}{2}\frac{\partial^3}{\partial\theta^3}\bar{g}(\theta_t^1)Y_t^2 dt + \alpha_t(\bar{g}'(\theta_t) - g'(X_t,\theta_t))dt$$
$$+\alpha_t f'(X_t,\theta_t)dW_t.$$

Let $\Phi_{t,s}$ be the fundamental solution satisfying

$$d\Phi_{t,s} = -\alpha_t\bar{g}'(\theta^*)\Phi_{t,s}dt, \Phi_{s,s} = 1.$$

The process Y_t can be written in terms of $\Phi_{t,s}$:

$$Y_t = \Phi_{t,1}Y_1 - \frac{1}{2}\int_1^t \Phi_{t,s}\alpha_s\bar{g}^{(3)}(\theta_s)Y_s^2 ds + \int_1^t \Phi_{t,s}(\bar{g}'(\theta_s) - g'(X_s,\theta_s))ds$$
$$+ \int_1^t \Phi_{t,s}\alpha_s f'(X_s,\theta_s)dW_s =: \Gamma_t^1 + \Gamma_t^2 + \Gamma_t^3 + \Gamma_t^4.$$

The problem is the weak convergence of *anticipative stochastic integral* which has not been studied much in the literature since standard approach such as Itô isometry cannot be directly applied. We use Malliavin calculus approach as in Bishwal (2010b).

We show the rate at which the stochastic integral converges to the normal distribution $\mathcal{N}$,

$$\sqrt{t}\int_1^t \alpha_s \Phi_{t,s}\zeta(X_s,\theta_s)dW_s \to \mathcal{N}$$

in distribution as $t\to\infty$. By using large deviations, we show the rate at which $\sqrt{t}(\Gamma_t^1+\Gamma_t^2+\Gamma_t^3)\to 0$. By using large deviations, we show the rate at which

$$t\int_1^t \alpha_s^2\Phi_{t,s}^2\zeta^2(X_s,\theta_s)ds \to J(\theta^*)$$

in probability as $t\to\infty$. Combining all these in Y_t and the using squeezing technique in Chapter-1 in Bishwal (2008a), we obtain the result.

Remark Bishwal (2011c) studied parameter estimation in interacting diffusions based on continuous and discrete sampling. The idea was used in Giesecke et al. (2020) for inference in large financial systems.

Next we focus on Monte Carlo method. Let $\hat{\theta}_{n,t}$ be the Monte Carlo estimate of θ_t based on n independent replications of the sample path, i.e.,

$$\hat{\theta}_{n,t} = \frac{1}{n}\sum_{i=1}^{n}\theta_{i,t}.$$

Theorem 6.2 *Under (A1)–(A11) and (B1)–(B10), we have as* $n\to\infty$

$$\sup_{x\in\mathbb{R}}\left|P\left(\sqrt{\frac{n}{J(\theta^*)}}(\hat{\theta}_{n,t}-\theta^*)\le x\right)-\Phi(x)\right|\le Cn^{-1/2},$$

where

$$J(\theta^*) = C_\alpha^2\int_0^\infty e^{-2s(C_\alpha g'(\theta^*)-1)}\bar{h}(\theta^*)ds,\quad \bar{h}(\theta)=\int h(x,\theta)\pi(dx),$$

$$h(x,\theta)=\left(\frac{1}{\sigma^2}f'(x,\theta)-\dot{v}(x,\theta)\right)^2\sigma^2$$

and $v(x,\theta)$ *is the solution to the Poisson equation with*

$$H(x,\theta)=g'(x,\theta)-\bar{g}'(\theta).$$

Proof Berry–Esseen theorem for independent random variables along with anticipative Girsanov theorem (Theorem 2.1) gives the result. Details are omitted.

1.7 Peacock Process and Indian Option

We study peacocks, lyrebirds, and Australian option. We introduce a new option based on harmonic mean which we call Indian option.

We start with some basic definitions.

Definition 1: A stochastic process (X_t) is said to have the *marginals of a martingale* iff there exists a martingale (M_t) such that for each $t \geq 0$, $X_t \sim M_t$ (equal in distribution).

Definition 2: A process X_t is called increasing in convex order or *peacock* if and only if $E(|X_t|) < \infty$ for all t and every convex function $f(\cdot)$ and $0 \leq s \leq t$ such that $E(f(X_s))$ and $E(f(X_t))$ are finite, we have $E(f(X_s)) \leq E(f(X_t))$.

A stochastic process is a peacock iff it has the *marginals of a martingale.*

The name peacock is derived from the French word increasing in convex order-Processes Croissant por l'Ordre Convexe, in short, PCOC- which is a homephone of the word peacock. Also peacock is an Asian bird.

A generalization of peacock is lyrebird.

Definition 3: A process X_t is called a *lyrebird* if and only if $E(|X_t|) < \infty$ for all t and every *increasing* convex function $h(\cdot)$ and $0 \leq s \leq t$ such that $E(h(X_s))$ and $E(h(X_t))$ are finite, we have $E(h(X_s)) \leq E(h(X_t))$.

A stochastic process is a lyrebird iff it has the *marginals of a submartingale.*

Definition 4: A process X_t is called a *parrot* if and only if $E(|X_t|) < \infty$ for all t and every increasing convex function $h(\cdot)$ and $0 \leq s \leq t$ such that $E(h(X_s))$ and $E(h(X_t))$ are finite, we have $E(h(X_s)) \geq E(h(X_t))$.

A stochastic process is a parrot iff it has the *marginals of a supermartingale.*

Let the stock price follow the geometric Brownian motion

$$S_t = S_0 \exp\{(\mu - \frac{1}{2}\sigma^2)t + \sigma W_t\}$$

which is the solution of the Black-Scholes model

$$dS_t = \mu S_t dt + \sigma S_t dW_t.$$

Consider the European call option on this stock $C = E(e^{-rt}(S_t - K)^+)$. The European option has the payoff $(S_t - K)^+$ and the Asian option has the payoff

$$(A_t - K)^+ = \left(\frac{1}{t}\int_0^t S_u du - K\right)^+.$$

The distribution of A_t is *Hartman-Watson distribution.* Australian option has the payoff

$$X_t = \frac{A_t}{S_t} = \frac{\frac{1}{t}\int_0^t S_u du}{S_t} = \frac{1}{t}\int_0^t \frac{S_u}{S_t} du.$$

This Australian option is Asian/European type option.

If $\mu = 0$, S_t is a martingale and $A_t = \frac{1}{t}\int_0^t S_u du$ is a peacock. If $\mu = \sigma^2$, then X_t is a peacock. If $\mu < \sigma^2$, then X_t is a lyrebird. If $\mu > \sigma^2$, then X_t is a parrot.

The prices of Australian option and Asian Option for $\mu \le \sigma^2$, respectively, $\mu \ge 0$ are increasing with maturity length when the premium is paid at maturity.

Indian Option
Let us define the payoff

$$H_t := \frac{t}{\int_0^t S_u^{-1} du}$$

which is related to harmonic mean. We call it *Indian option.* The call option price is given by

$$C = E\left(e^{-rT}\left(\frac{T}{\int_0^T S_t^{-1} dt} - K\right)^+\right).$$

The put option price is given by

$$P = E\left(e^{-rT}\left(K - \frac{T}{\int_0^T S_t^{-1} dt}\right)^+\right).$$

The *Indian-Australian option* has the payoff

$$I_t = \frac{H_t}{S_t} = \frac{tS_t}{\int_0^t S_u^{-1} du}.$$

The *American–Indian–Australian option* has the payoff

$$I_\tau = \left(K - \frac{\tau S_\tau}{\int_0^\tau S_u^{-1} du}\right)^+.$$

The put price of the American–Indian (AI) option is given by

$$P = \sup_{\tau \in \mathcal{T}} E\left(e^{-r\tau}\left(K - \frac{\tau}{\int_0^\tau S_\tau^{-1} dt}\right)^+\right),$$

where $\mathcal{T}$ is the set of stopping times.

The put price of the American–Indian–Australian (AIA) option is given by

$$P = \sup_{\tau \in \mathcal{T}} E\left(e^{-r\tau}\left(K - \frac{\tau S_\tau}{\int_0^\tau S_\tau^{-1} dt}\right)^+\right),$$

where $\mathcal{T}$ is the set of stopping times. One should study the properties of the new options.

Remark This option is Asian type option but based on harmonic mean instead of arithmetic mean. Recall that Arithmetic Mean $\geq$ Harmonic Mean. We will extend this model to fractional Brownian motion and implement it in Matlab. All expectations for options are defined with respect to risk-neutral measure.

1.8 Brownian Excursion and Volatility Estimation in Limit Order Book

High-frequency estimation of the integrated squared volatility of a semimartingale efficient price process which forms a stochastic boundary based on observations of intraday order book quotes with microstructure noise is studied.

We will use Brownian excursion, limit order book, Poisson point process (PPP) model, bid-ask prices, continuous semimartingale, stochastic boundary, high-frequency data, sufficient statistic, regression, nonparametric estimation, Feynman-Kac formula.

Consider observations (Y_i) above a stochastic boundary $(X_t, t \in [0,1])$ which is formed by the graph of a continuous semimartingale. The objective is to optimally recover the driving characteristic $\langle X, X \rangle_t$ of the boundary X_t given the observations (Y_i).

Stochastic boundary problem stems from limit order books. From a microeconomic point of view, *ask prices* typically lie above the efficient market price whereas *bid prices* are symmetric and independent and lie below the market price. Here the underlying latent efficient log-price of a stock $(X_t, t \in [0,1])$ observed over a trading period like a day serves as the boundary, whereas the ask prices form the observations (Y_i). Bid prices can be handled symmetrically and independently, which can be used to validate the model.

Consider the stochastic regression model

$$Y_i = X_{t_i} + \varepsilon_i, \quad i = 0, 1, 2, \ldots, n,$$

where

$$X_t = X_0 + \int_0^t a_s ds + \int_0^t \sigma_s dW_s, \quad t \in [0,1]$$

with possibly stochastic drift and volatility a_s and σ_s defined on a filtered probability space $(\Omega, \mathcal{F}, (\mathcal{F}_t), P)$ with a standard $(\mathcal{F}_t)$-Brownian motion W.

The error distribution function F_λ satisfies $F_\lambda = \lambda x(1 + o(1))$ as $x \downarrow 0$. One natural parametric specification is $\varepsilon_i \sim Exp(\lambda), \quad \lambda > 0$. The noise is assumed to be independent of the signal part X. The quantity

$$\langle X, X\rangle := \int_0^1 \sigma_s^2 ds$$

is called the *integrated squared volatility.*

Conditional on $(X_t, t \in [0,1])$, we observe a Poisson Point Process (PPP) on $[0,1] \times \mathbb{R}$ with intensity measure

$$\Lambda(A) = \int_0^1 \int_{\mathbb{R}} I_A(t,y)\lambda_{t,y} dtdy,$$

where

$$\lambda_{t,y} = n\lambda I(y \geq X_t).$$

We denote by $(T_j, \mathcal{Y}_j)$ the observations of that point process which are homogeneously dispersed over the graph of $(X_t,\ t \in [0,1])$.

Let

$$\mathcal{T}_n^k := [kh, (k+1)h), \quad k = 0, 1, \ldots, h^{-1} - 1$$

and

$$\mathcal{I}_n^k := [khn, khn+1, \ldots, (k+1)hn - 1), \quad k = 0, 1, \ldots, h^{-1} - 1.$$

Define the local minima

$$m_{n,k} := \min_{i \in \mathcal{I}_k^n} Y_i$$

which is a *sufficient statistic* in the regression model and the local minima

$$m_{n,k} := \min_{T_j \in \mathcal{T}_k^n} \mathcal{Y}_j$$

which is a *sufficient statistic* in the PPP model.

We have from Bibinger et al. (2015) the following theorem:

Theorem 8.1

$$P(m_{n,k} > x) = E\left[\exp\left(-\int_{kh}^{(k+1)h} (X_t + x)_+ dt\right)\right], x \in \mathbb{R},$$

where $A_+ = \max(A, 0)$.

For $X_t = W_t$, Brownian excursion area is given by

$$\exp\left(-\int_{kh}^{(k+1)h} (W_t)_+ dt\right).$$

It is a *Brownian bridge process conditioned to be positive.* No explicit form of the expectation of the above expression is available.

We estimate the latent volatility based on observations from a *limit order book.* We assume one sided noise. The optimal convergence rate for the Gaussian noise is $n^{-1/4}$. In our case the optimal rate of convergence is $n^{-1/3}$.

Recall that

$$X_t = X_{(k-1)h} + \int_{(k-1)h}^{t} dW_s.$$

The natural statistic to assess the quadratic variation of the boundary process X is $(m_{n,2k} - m_{n,2k-1})^2$, the squared differences between the consecutive local minima.

Define

$$\Psi(\sigma^2) := h^{-1}E[(m_{n,2k} - m_{n,2k-1})^2]$$

$$\sum_k [(m_{n,2k} - m_{n,2k-1})^2] \to \int \Psi(\sigma_t^2)dt.$$

For the PPP model,

$$\widetilde{IV}_{h,r,n} = \sum_{l=1}^{h^{-1}r} \Psi^{-1}\left(\sum_{k=(l-1)r^{-1}/2+1}^{lr^{-1}/2} (m_{n,2k} - m_{n,2k-1})^2 2h^{-1}r \right) hr^{-1}.$$

$$\Psi_n(\sigma^2) = E[(m_{n,2k} - m_{n,2k-1})^2].$$

$$\Psi_n \to \Psi \quad as \quad n \to \infty.$$

$$\Psi_n^{-1}\left(\sum_{k=(l-1)r^{-1}/2+1}^{lr^{-1}/2} (m_{n,2k} - m_{n,2k-1})^2 2h^{-1}r \right) \approxeq \sigma^2_{lr^{-1}h}.$$

For the regression model

$$\widehat{IV}_{h,r,n} = \sum_{l=1}^{h^{-1}r} \Psi_n^{-1}\left(\sum_{k=(l-1)r^{-1}/2+1}^{lr^{-1}/2} (m_{n,2k} - m_{n,2k-1})^2 2h^{-1}r \right) hr^{-1}.$$

For a parametric estimator of $\sigma_t = \sigma$, one can apply the *moment type estimator* $\widehat{IV}$.

Let $R_{n,k} := m_{n,k} - X_{kh}$ and $L_{n,k} := m_{n,k-1} - X_{kh}$. Thus $m_{n,k} - m_{n,k-1} = R_{n,k} - L_{n,k}$.

The distribution of $R_{n,k}$ is given by the convolution between exponential distribution and the law of the minimum of Brownian motion on the discrete grid $\mathcal{I}_k^n$.

$$\lim_{n\to\infty} P(h^{-1/2}R_{n,k} > x\sigma) = E\left[\exp\left(-k\sigma \int_0^1 (x + W_t)_+ dt\right)\right].$$

$$Var(R_t) = \frac{1}{2}\Psi(\sigma_t^2).$$

Theorem 8.2

$$E\left[\int_0^\infty \exp\left(-st - \sqrt{2}\theta\ \sigma \int_0^t (x + W_t)_+ dt\right)\right] = \theta^{-2/3}\zeta_s(x,\theta),$$

where

$$\zeta_s(x,\theta) = \zeta_{s,-}(x,\theta)I_{(-\infty,0)}(x) + \zeta_{s,+}(x,\theta)I_{(0,\infty)}(x).$$

ζ_s is defined in terms of the Airy function and the Scorer functions, respectively, given by

$$Ai(x) = \pi^{-1}\int_0^\infty cos(t^3/3 + xt)dt,$$

$$Gi(x) = \pi^{-1}\int_0^\infty sin(t^3/3 + xt)dt,$$

$$AI(x) = \int_x^\infty Ai(y)dy,$$

$$\zeta_{s,+} = \frac{\pi(\theta^{1/3}Gi'(\theta^{-2/3}s) - \sqrt{s}\theta^{1/3}Gi(\theta^{-2/3}s)) + \theta^{2/3}s^{-1/2}}{\sqrt{s}Ai(\theta^{-2/3}s) - \theta^{1/3}Ai'(\theta^{-2/3}s)}$$
$$\times Ai(\sqrt{2}\theta^{1/3}x + \theta^{-2/3}s) + \pi Gi(\sqrt{2}\theta^{1/3}x + \theta^{-2/3}s),$$

$$\zeta_{s,-} = \left(\frac{\theta^{2/3}s^{-1/2}Ai(\theta^{-2/3}s) + \theta^{1/3}AI(\theta^{-2/3}s)}{\sqrt{s}Ai(\theta^{-2/3}s) - \theta^{1/3}Ai'(\theta^{-2/3}s)} - \theta^{2/3}s^{-1/2}\right)$$
$$\times \exp(\sqrt{2}sx) + s^{-1}\theta^{2/3}.$$

This result generalizes the Laplace transform of exponential integrated positive part of a Brownian motion studied by Perman and Wellner (1996). By inversion of the Laplace transform one obtains the distribution function.

Theorem 8.3 (Rate of Convergence) We consider $h \to 0, n \to \infty, nh \to \infty$. Choose $r = kn^{-1/3}$ for some $k > 0$. We have

$$(a)\quad \widetilde{IV}_{h,r,n} - IV = O_P(n^{-1/3}).$$

$$(b)\quad \widehat{IV}_{h,r,n} - IV = O_P(n^{-1/3}).$$

The rate $n^{-1/3}$ is optimal.

The rate $n^{-1/4}$ is optimal in Gaussian case.

Jump Processes
Theorem 8.4 Consider the stochastic volatility model with jumps in X and σ:

$$X_t = X_0 + \int_0^t a_s ds + \int_0^t \sigma_s dW_s + \int_0^t \int_{\mathbb{R}} x d\mu^X(ds, dx)$$

$$\sigma_t = \sigma_0 + \int_0^t \tilde{a}_s ds + \int_0^t \tilde{\sigma}_s dW_s + \int_0^t \tilde{\eta}_s dW_s^{\perp} + \int_0^t \int_{\mathbb{R}} x d\mu^\sigma(ds, dx)$$

with finite random measures μ^X, μ^σ.
Then

$$(a) \quad \widehat{IV}_{h,r,n} - IV = O_P(n^{-1/3}).$$

$$(b) \quad \widetilde{IV}_{h,r,n,\tau} - IV = O_P(n^{-1/3}).$$

Theorem 8.5 (Local Asymptotic Minimax Bound) *For any sequence of estimators $\hat{\sigma}_n^2$ of $\sigma^2 \in (0, \infty)$ from the parametric PPP model, for each $\sigma_0^2 > 0$, there exists $\delta > 0$ such that*

$$\liminf_{n\to\infty} \inf_{\hat{\sigma}_n^2} \max_{\sigma^2 \in \{\sigma_0^2, \sigma_0^2 + \delta n^{-1/3}\}} P_{\sigma^2}(n^{1/3}|\hat{\sigma}_n^2 - \sigma^2| \geq \delta) > 0,$$

where the infimum extends over all estimators $\hat{\sigma}_n^2$ based on PPP-model with $\lambda = 1$ and $X_t = \sigma W_t$. The law of the later is denoted by P_{σ^2}.

The times T_j^s are given by a Poisson sampling of intensity $(2/3)nb$ on $[0, 1]$.

Fractional Brownian Motion

We extend the results to fractional Brownian motion with Hurst parameter $H > 0.5$. Molchan (2000) studied logarithmic asymptotics of the distribution of the maximum of a fBm.

Let

$$M_T^H := \max_{s \in [0,T]} W_s^H.$$

Then we have
Theorem 8.6 As $T \to \infty$

$$(a) \quad E\left[\left(\int_0^T \exp(W_s^H) ds\right)\right]^{-1} = HT^{(1-H)}\left(E \max_{s\in[0,T]} W_s^H + o(1)\right).$$

$$(b) \quad \log P\left(M_T^H < 1\right) = (1-H)\log T^{-1}(1 + O(\log T)^{-1/2})$$

1.9 Rate of Cauchy Convergence of Brownian Winding and Asian Options

The accuracy of Cauchy approximation for the winding of Brownian motion and Asian option is studied in this section.

Consider a two-dimensional Brownian motion $(W_t, t \geq 0)$ starting from $Z_0 = (1,0)$ (for simplicity), and let $(\theta_t, t \geq 0)$ denote the continuous determination of the argument of $(W_u, u \leq t)$ around $(0,0)$ as t evolves in $(0,\infty)$ such that $\theta_0 = 0$.

Since $W_0 \neq 0$, $(W_t, t \geq 0)$ does not visit the point 0 but keeps winding around it infinitely often. Hence its continuous winding process

$$\theta_t^W := Im\left(\int_0^t \frac{1}{|W_s|} dW_s\right), \quad t \geq 0$$

is well defined. The skew product representation is given by

$$\log|W_t| + i\theta_t = \int_0^t \frac{1}{|W_s|} dW_s = (\beta_u + i\gamma_u)|_{u=H_t=\int_0^t \frac{ds}{|W_s|^2}},$$

where $\beta_u + i\gamma_u$ is another planar Brownian motion starting from $\log x_0 + i0$.

Spitzer (1958) proved that

$$\lim_{t\to\infty} P\left(\frac{2\theta_t}{\log t} \leq x\right) = \mathcal{C}(x), x \in \mathbb{R},$$

where $\mathcal{C}$ denotes the standard Cauchy distribution

$$\mathcal{C}(x) = \frac{1}{2} + \frac{1}{\pi} \arctan x, \ x \in \mathbb{R}.$$

Pap and Yor (2000) proved that

$$\sup_{x\in\mathbb{R}} \left| P\left(\frac{2\theta_t}{\log t} \leq x\right) - \mathcal{C}(x)\right| = O(\log t)^{-1}.$$

Asian Option

Recall that the payoff of an Asian call option is given by

$$E\left(\frac{1}{t}\int_0^t \exp(B_s + \nu s) ds - K\right)^+,$$

where $x^+ = \max(x, 0)$, B_s is a real Brownian motion, $\nu \in \mathbb{R}$ and $K \in \mathbb{R}^+$ is the strike price. This follows by studying

$$E\left(\int_0^t \exp(B_s + \nu s)ds\right).$$

Let

$$A_t^\nu := \int_0^t \exp(B_s + \nu s)ds.$$

The random variable $2A_t^\nu$ taken up to a random time T_λ which follows the exponential distribution with parameter $\lambda > 0$ has the law

$$\frac{Q_{1,a}}{2G_b} = \frac{1 - U^{1/a}}{2G_b},$$

where $Q_{1,a} \sim Beta(1, a)$, $G_b \sim Gamma(b)$, $U \sim Unif[0,1]$, $a = (\nu/2) + (1/2)\sqrt{2\lambda + \nu^2}$, $b = a - \nu$ and the random variables are independent. Let us suppose that $\nu = 0$. Consider exponential functional of planar Brownian motion taken up to a random time

$$\int_0^{T_c^\gamma} \exp(2B_s)ds \quad \text{where} \quad T_c^\gamma = \inf\{u \geq 0 : \gamma_u = c\}, c > 0$$

and $(\gamma_u, u \geq 0)$ is another real Brownian motion independent of B.

The pdf of T_1^γ is inverse Gaussian given by

$$g(x) = \frac{1}{\sqrt{2\pi x^3}} e^{-1/(2x)}, \quad x > 0.$$

Bougerol's Identity:

$$\sinh(B_t) \stackrel{D}{=} \widehat{B}_{A_t^W(B)},$$

where $\widehat{B}$ is a Brownian motion independent of B.

$$T_c^{\theta^W} = \inf\{u \geq 0 : \theta_u^W = c\} = A_{T_c^\gamma}^W$$

and similarly

$$T_c^{|\theta^W|} = \inf\{u \geq 0 : |\theta_u^W| = c\} = A_{T_c^{|\gamma|}}^W$$

$$T_c^{\theta^W} = A_{T_c^\gamma}^W(B) \stackrel{D}{=} A_u^W(B)|_{u=c^2 T_1^\gamma}.$$

$T_c^{\theta^W}$ is a subordinator. Thus

$$\frac{1}{c^2} T_c^{\theta^W} \stackrel{D}{=} \int_0^{T_1^\gamma} \exp(2cB_s)ds.$$

$$\frac{1}{t} \log A_{t^2}^W \stackrel{D}{\to} 2|B|_{T_1^\gamma} = 2|C|_1.$$

Let β, B, and δ be three independent Brownian motions:

$$\sinh(Y_t^{(\nu,\mu)}) \stackrel{D}{=} \int_0^t \exp(\beta_s + \nu s) d(B_s + \mu s) \stackrel{D}{=} \delta_{\int_0^t \exp(2(\beta_s+\nu s))ds},$$

where $Y_t^{(\nu,\mu)}$ is a diffusion with infinitesimal generator

$$\frac{1}{2}\frac{d^2}{dy^2} + \left(\nu \tanh y + \frac{\mu}{\cosh y}\right)\frac{d}{dy}$$

starting from $y = \arg\sinh(x)$. In particular, $\theta_{T_c^\gamma} \stackrel{D}{=} \mathcal{C}_{a(c)}$ where $a(c) = \arg \sinh c = \log(c + \sqrt{1+c^2})$, $c \in \mathbb{R}$ and $\mathcal{C}_1 = \gamma_{T_1^\beta}$.

Cauchy Approximation

We start with the following lemma:

Lemma 9.1 (Esseen's Smoothing Lemma:) Let F be a non-decreasing function and H be a differentiable function of bounded variation on the real line with $F(\pm\infty) = G(\pm\infty)$. Denote the corresponding Fourier-Stieltjes transforms by $\widehat{F}$ and $\widehat{G}$, respectively. Then for all $\Lambda > 0$,

$$\sup_{x\in\mathbb{R}} |F(x) - G(x)| \leq \frac{1}{\pi}\int_{-\Lambda}^{\Lambda} \frac{\left|\widehat{F}(\lambda) - \widehat{G}(\lambda)\right|}{|\lambda|} d\lambda + \frac{24}{\pi\Lambda}\sup_{x\in\mathbb{R}} |G'(x)|.$$

Proof See Petrov (1975) or Feller (1971).

Theorem 9.2

$$\sup_{x\in\mathbb{R}} \left| P\left(\frac{1}{t}\log A_{t^2}^W \leq x\right) - 2|C|_1(x)\right| = O(t^{-1}).$$

Proof Note that

$$A_{t^2}^W = \int_0^{t^2} \exp(W_s) ds.$$

Applying Esseen's smoothing lemma by choosing $\Lambda = t$, we have

$$\sup_{x\in\mathbb{R}} \left| P\left(\frac{1}{t}\log A_{t^2}^W \leq x\right) - 2|C|_1(x)\right|$$

$$\leq \frac{1}{\pi}\int_{-t}^{t} \frac{\left|E\exp\left(iu\frac{1}{2t}\log A_{t^2}^W\right) - e^{-|u|/2}\right|}{|u|} du + \frac{24}{\pi t}\sup_{x\in\mathbb{R}}\left|\frac{d}{dx}e^{-|x|/2}\right|$$

$$=: I + J \leq Ct^{-1},$$

where

$$I := \int_{-t}^{t} \frac{\left|E \exp\left(iu\frac{1}{2t}\log A_{t^2}^W\right) - e^{-|u|/2}\right|}{|u|} du$$

and

$$J := \frac{24}{\pi t} \sup_{x\in\mathbb{R}} \left|\frac{d}{dx} e^{-|x|/2}\right|.$$

Clearly,

$$\sup_{x\in\mathbb{R}} \left|\frac{d}{du} e^{-|x|/2}\right| < \infty.$$

Hence

$$J = O(t^{-1}).$$

Further

$$I \leq I_1 + I_2.$$

We consider two possibilities:

(i) $|u| \leq t$:

$$I_1 = \int_{|u|\leq t} \frac{\left|E \exp\left(iu\frac{1}{2t}\log A_{t^2}^W\right) - e^{-|u|/2}\right|}{|u|} du$$

$$\leq \frac{C}{t} \int_{|u|\leq t} |u|^{-1/2} e^{\sqrt{-|u|/2}} du \leq \frac{C}{t} \int_{|u|\leq t} |u|^{-1/2} e^{\sqrt{-|u|/2}} du = O(t^{-1}).$$

(ii) $|u| > t$:

$$I_2 = \int_{|u|\geq t} \frac{\left|E \exp\left(iu\frac{1}{2t}\log A_t^W\right) - e^{-|u|/2}\right|}{|u|} du \leq \int_{|u|\geq t} |u|^{-1} e^{\sqrt{-|u|/2}} du$$

$$\leq Ct^{-1} \int_{|u|\geq t} e^{\sqrt{-|u|/2}} du = O(t^{-1}).$$

This completes the proof of the theorem. □

Let the *clock process* be defined as

$$H_t^W := \int_0^t |W_s|^{-2} ds.$$

Theorem 9.3

$$\sup_{x\in\mathbb{R}} \left|P\left(\frac{4}{(\log t)^2} H_t^W \leq x\right) - 2T_1^\gamma(x)\right| = O\left((\log t)^{-1}\right).$$

Proof Applying Esseen's smoothing lemma by choosing $\Lambda = (\log t)^2$, we have

$$\sup_{x\in\mathbb{R}}\left|P\left(\frac{4}{(\log t)^2}H_t^W \leq x\right) - 2T_1^\gamma(x)\right|$$

$$\leq \frac{1}{\pi}\int_{-(\log t)^2}^{(\log t)^2} \frac{\left|E\exp\left(iu\frac{4}{(\log t)^2}H_t^W\right) - e^{-\sqrt{-2iu}}\right|}{|u|}du + \frac{24}{\pi(\log t)^2}\sup_{x\in\mathbb{R}}|\frac{d}{dx}e^{-\sqrt{2ix}}|$$

$$=: I + J,$$

where

$$\sqrt{-2iu} = ((1 - sign(u))\sqrt{|u|}.$$

Clearly,

$$\sup_{x\in\mathbb{R}}|\frac{d}{du}e^{-\sqrt{2iu}}| < \infty.$$

Hence

$$J = O(\log t)^{-2}).$$

$$I \leq I_1 + I_2.$$

We consider two possibilities:

(i) $|u| \leq (\log t)^2$:

$$I_1 = \int_{|u|\leq(\log t)^2} \frac{\left|E\exp\left(iu\frac{4}{(\log t)^2}H_t^W\right) - e^{-\sqrt{-2iu}}\right|}{|u|}du$$

$$\leq \frac{C}{\log t}\int_{|u|\leq(\log t)^2} |u|^{-1/2}e^{\sqrt{-|u|/2}}du \leq \frac{C}{\log t}\int_{|u|\leq(\log t)^2} |u|^{-1/2}e^{\sqrt{-|u|/2}}du$$

$$= O(\log t)^{-1}.$$

(ii) $|u| > (\log t)^2$:

$$I_2 = \int_{|u|\geq(\log t)^2} \frac{\left|E\exp\left(iu\frac{4}{(\log t)^2}H_t^W\right) - e^{-\sqrt{-2iu}}\right|}{|u|}du$$

$$\leq \int_{|u|\geq(\log t)^2} |u|^{-1}e^{\sqrt{-|u|/2}}du \leq C(\log t)^{-2}\int_{|u|\geq(\log t)^2} e^{\sqrt{-|u|/2}}du$$

$$= O(\log t)^{-1}).$$

Hence

$$I + J = O(\log t)^{-1}).$$

This completes the proof of the theorem. □

Fractional Brownian Winding

Consider the fractional Brownian motion W_t^H with Hurst parameter $H > 1/2$. The winding process is defined as

$$\theta_t^{W^H} := Im\left(\int_0^t \frac{1}{|W_s^H|} dW_s^H\right), \quad t \geq 0.$$

The *clock process* be defined as

$$H_t^{W^H} := \int_0^t |W_s^H|^{-2} ds.$$

We have

Theorem 9.4

$$\sup_{x\in\mathbb{R}} \left| P\left(\frac{1}{t}\log A_{t^2}^W \leq x\right) - 2|C|_1(x)\right| = O(t^{-2(1-H)}).$$

Theorem 9.5

$$\sup_{x\in\mathbb{R}} \left| P\left(\frac{4}{(\log t)^2} H_t^{W^H} \leq x\right) - 2T_1^\gamma(x)\right| = O\left((\log t)^{-2(1-H)}\right).$$

Let the maximum of a fBm be defined as

$$M_T^H := \max_{s\in[0,T]} W_s^H.$$

Molchan (2000) studied logarithmic asymptotics of the distribution of the maximum of a fBm M_T^H as $T \to \infty$.:

$$E\left[\left(\int_0^T \exp(W_s^H) ds\right)\right]^{-1} = HT^{(1-H)}\left(E \max_{s\in[0,T]} W_s^H + o(1)\right),$$

$$\log P\left(\max_{s\in[0,T]} W_s^H < 1\right) = (1-H)\log T^{-1}\left(1 + O(\log T)^{-1/2}\right).$$

Equivalently,

$$E\left[\left(\int_0^T \exp(W_s^H) ds\right)\right]^{-1} = HT^{(1-H)}\left(E(M_T^H) + o(1)\right),$$

$$\log P\left(M_T^H < 1\right) = (1-H)\log T^{-1}\left(1 + O(\log T)^{-1/2}\right)$$

1.10 Bond Pricing for the Fractional Vasicek Model

We obtain the bond price formula for the fractional Vasicek model.

Let $(\Omega, \mathcal{F}, \{\mathcal{F}_t\}_{t\geq 0}, P)$ be a stochastic basis on which is defined the Ornstein-Uhlenbeck process X_t satisfying the Itô stochastic differential equation

$$dX_t = \theta X_t dt + dW_t^H, \quad t \geq 0, \quad X_0 = \xi, \tag{10.1}$$

where $\{W_t^H\}$ is a fractional Brownian motion with $H > 1/2$ with the filtration $\{\mathcal{F}_t\}_{t\geq 0}$ and $\theta < 0$ is the unknown parameter.

Define

$$\begin{aligned}
\kappa_H &:= 2H\Gamma(3/2 - H)\Gamma(H + 1/2), \\
k_H(t,s) &:= \kappa_H^{-1}(s(t-s))^{\frac{1}{2}-H} \\
\lambda_H &:= \frac{2H\Gamma(3-2H)\Gamma(H+\frac{1}{2})}{\Gamma(3/2-H)} \\
v_t \equiv v_t^H &:= \lambda_H^{-1} t^{2-2H}, \quad M_t^H := \int_0^t k_H(t,s) dW_s^H.
\end{aligned}$$

From Norros et al. (1999) it is well known that M_t^H is a Gaussian martingale, called the *fundamental martingale* whose variance function $\langle M^H \rangle_t$ is v_t^H. The natural filtration of the martingale M^H coincides with the natural filtration of the fBm W^H since

$$W_t^H := \int_0^t K(t,s) dM_s^H$$

holds for $H \in (1/2, 1)$ where

$$K_H(t,s) := H(2H-1)\int_s^t r^{h-\frac{1}{2}}(r-s)^{H-\frac{3}{2}}, \quad 0 \leq s \leq t$$

and for $H = 1/2$, the convention $K_{1/2} \equiv 1$ is used.

Define

$$Q_t := \frac{d}{dv_t}\int_0^t k_H(t,s) X_s ds.$$

It is easy to see that

$$Q_t = \frac{\lambda_H}{2(2-2H)}\left\{ t^{2H-1} Z_t + \int_0^t r^{2H-1} dZ_s \right\}.$$

X admits the representation

$$X_t = \int_0^t K_H(t,s) dZ_s.$$

The natural filtration generated by the fundamental semimartingale process

$$Z_t = \theta \int_0^t Q_s dv_s + M_t^H \tag{10.2}$$

and the process X coincide, see Kleptsyna and Le Breton (2002). The available information for X and Z are strictly equivalent.

Let the realization $\{X_t, 0 \leq t \leq T\}$ or equivalently $\{Z_t, 0 \leq t \leq T\}$ be denoted by Z_0^T. Let P_θ^T be the measure generated on the space (C_T, B_T) of continuous functions on $[0, T]$ with the associated Borel σ-algebra B_T generated under the supremum norm by the process X_0^T and P_0^T be the standard Wiener measure. Applying fractional Girsanov formula, when θ is the true value of the parameter, P_θ^T is absolutely continuous with respect to P_0^T and the Radon–Nikodym derivative (likelihood) of P_θ^T with respect to P_0^T based on Z_0^T is given by

$$L_T(\theta) := \frac{dP_\theta^T}{dP_0^T}(Z_0^T) = \exp\left\{\theta \int_0^T Q_t dZ_t - \frac{\theta^2}{2}\int_0^T Q_t^2 dv_t\right\}. \tag{10.3}$$

Consider the score function, the derivative of the log-likelihood function, which is given by

$$Y_T(\theta) := \int_0^T Q_t dZ_t - \theta \int_0^T Q_t^2 dv_t. \tag{1.4}$$

Under the hypothesis $\mathcal{H}_0$, the log-likelihood ratio process admits the representation

$$\Lambda_T := \log \frac{dP_{\theta_1}^T}{dP_{\theta_0}^T}(Z_0^T) = \left\{(\theta_1 - \theta_0)\int_0^T Q_t dM_t - \frac{(\theta_1-\theta_0)^2}{2}\int_0^T Q_t^2 dv_t\right\}. \tag{10.5}$$

The Hellinger integral of order $\varepsilon \in (-\infty, \infty)$ is defined as

$$h_T(\varepsilon) = E_{\theta_0}[\exp(\varepsilon \Lambda_t)].$$

Note that

$$h_T(\varepsilon) := h_T(\varepsilon; P_{\theta_1}^T, P_{\theta_0}^T) = h_T(1-\varepsilon; P_{\theta_0}^T, P_{\theta_1}^T).$$

We have

$$\int_0^T Q_t dM_t = \frac{1}{2}\left(U_T + 2\theta V_T - T\right), \tag{10.6}$$

where

$$U_T := \frac{\lambda_H}{2-2H} Z_T \int_0^T t^{2H-1} dZ_s \tag{10.7}$$

and

$$V_T := \int_0^T Q_s^2 dv_s. \tag{10.8}$$

Let $\phi_T^H(u) := E\exp(-uV_T),\ u > 0$. Then $\phi_T^H(u)$ is given by

$$\phi_T^H(u) = E\exp(-uV_T) = \left\{ \frac{4\sin\pi H\sqrt{\theta^2+2u}e^{-\theta T}}{\pi T D_T^H(\theta;\sqrt{\theta^2+2u})} \right\}^{1/2}, \tag{10.9}$$

where

$$\begin{aligned} D_T^H(\theta;\alpha) \quad := \quad & [\alpha\cosh(\frac{\alpha}{2}T) - \theta\sinh(\frac{\alpha}{2}T)]^2 J_{-H}(\frac{\alpha}{2}T)J_{H-1}(\frac{\alpha}{2}T) \\ & -[\alpha\sinh(\frac{\alpha}{2}T) - \theta\cosh(\frac{\alpha}{2}T)]^2 J_{1-H}(\frac{\alpha}{2}T)J_H(\frac{\alpha}{2}T) \end{aligned}$$

for $\alpha > 0$ and J_ν is the modified Bessel function of first kind of order ν. (See Kleptsyna and Le Breton 2002.)
We generalize the above Cameron–Martin type formula to obtain joint generating function of U_T and V_T.

Lemma 10.1 Let $\Psi_T^H(z_1,z_2) := E\exp(z_1V_T + z_2U_T), z_1, z_2 \in \mathbb{C}$. Then $\Psi_T^H(z_1,z_2)$ exists for $|z_i| \le \delta$, $1 = 1{,}2$ for some $\delta > 0$ and is given by

$$\Psi_T^H(z_1,z_2) = \exp\left(\frac{-\theta T}{2}\right)\left[\frac{4(\sin\pi H)\gamma}{\pi T\mathcal{D}_T^H(\gamma;z_2)}\right]^{1/2}, \tag{10.10}$$

where $\gamma := (\theta^2 + 2z_1)^{1/2}$ and we choose the principal branch of the square root and

$$\begin{aligned} \mathcal{D}_T^H(\gamma;z_2) \quad := \quad & [\gamma\cosh(\frac{\gamma}{2}T) - (\theta - 2z_2)\sinh(\frac{\gamma}{2}T)]^2 J_{-H}(\frac{\gamma}{2}T)J_{H-1}(\frac{\gamma}{2}T) \\ & -[\gamma\sinh(\frac{\gamma}{2}T) - (\theta - 2z_2)\cosh(\frac{\gamma}{2}T)]^2 J_{1-H}(\frac{\gamma}{2}T)J_H(\frac{\gamma}{2}T) \end{aligned}$$

and J_ν is the modified Bessel function of first kind of order ν.

Proof We use analytic continuation, Fractional Itô formula (Lemma D.2.3) along with change of measure technique as in Liptser and Shiryayev (1978) and Kleptsyna and Le Breton (2002).

First assume $z_i = a_i \in \mathbb{R}, i = 1, 2$ and a_i are sufficiently small. Define $\gamma := (\theta^2 + 2z_1)^{1/2}$ and $dX_t^\gamma = \gamma X_t^\gamma + dW_t^H, X_0^\gamma = 0$. Also recall that $dX_t^\theta = \theta X_t^\theta + dW_t^H$.

Then

$$\frac{dP_\theta^T}{dP_\gamma^T}(Z_0^T) = \left\{(\theta-\gamma)\int_0^T Q_t^\gamma dZ_t^\gamma - \frac{(\theta^2-\gamma^2)}{2}\int_0^T (Q_t^\gamma)^2 dv_t\right\}. \tag{10.11}$$

Note that

$$\Psi_T^H(a_1,a_2) := E\exp(a_1 V_T^\theta + a_2 U_T^\theta).$$

If we change the measure to that generated by Z_t^γ, then by (10.11),

$$\begin{aligned}\Psi_T^H(a_1,a_2) &= E\left[\exp(a_1\int_0^T (Q_t^\gamma)^2 dv_t + a_2(Z_T^\gamma)^2)\right.\\ &\quad\left. -(\gamma-\theta)\int_0^T Q_t^\gamma dZ_t^\gamma - a_1\int_0^T (Q_t^\gamma)^2 dv_t\right]\\ &= E\exp\left[a_2 U_T^\gamma - (\gamma-\theta)\int_0^T Q_t^\gamma dZ_t^\gamma\right] \qquad (10.12)\end{aligned}$$

By fractional Itô formula (Lemma D.2.3)

$$\int_0^T Q_s^\gamma dZ_s^\gamma = \frac{1}{2}\left[\frac{\lambda_H}{2-2H} Z_T^\gamma \int_0^T t^{2H-1} dZ_s^\gamma - T\right]. \tag{10.13}$$

Using this in (10.12), we have

$$\Psi_T^H(a_1,a_2) = E\exp\left[\left(a_2 - \frac{\gamma-\theta}{2}\right)\frac{\lambda_H}{2-2H} Z_T^\gamma \int_0^T t^{2H-1} dZ_s^\gamma - \left(\frac{\gamma-\theta}{2}\right)T\right]. \tag{10.14}$$

Let $\lambda_H^* := \frac{\lambda_H}{2-2H}$.

Note that $(Z_t^\gamma, \int_0^T t^{2H-1} dZ_s^\gamma)'$ is a Gaussian pair with covariance matrix denoted by $R^\gamma(T)$.

Thus

$$\Psi_T^H(a_1,a_2) = \exp\left(\frac{(\gamma-\theta)T}{2}\right)\left\{\det[I + \frac{(\gamma-\theta)\lambda_H^*}{2} J R^\gamma(T)]\right\}^{-1/2}, \tag{10.15}$$

where the 2×2 matrices I and J are defined by

$$I = \begin{pmatrix}1 & 0\\ 0 & 1\end{pmatrix}$$

and

$$J = \begin{pmatrix} 0 & 1 \\ 1 & 0 \end{pmatrix}.$$

$R^\gamma(T)$ satisfies the Lyapunov differential equation

$$\dot{R}^\gamma(T) = \frac{\gamma}{2}\{A_H(T)R^\gamma(T) + R^\gamma(T)A'_H(T)\} + \frac{1}{\lambda_H^*}B_H(T);\ T \geq 0;\ R^\gamma(0) = 0, \tag{10.16}$$

where

$$A_H(s) = \begin{pmatrix} 1 & s^{1-2H} \\ s^{2H-1} & 1 \end{pmatrix}$$

and

$$B_H(s) = \begin{pmatrix} s^{1-2H} & 1 \\ 1 & s^{2H-1} \end{pmatrix}.$$

Now solving for $R^\gamma(T)$ as in Kleptsyna and Le Breton (2002)

$$\begin{aligned} &\det[I + \frac{(\gamma-\theta)\lambda_H^*}{2} JR^\gamma(T)] \\ = &\frac{\gamma\pi T}{4\pi \sin \pi H}\left\{\left[1 + \frac{\gamma-\theta}{\gamma}e^{\frac{\gamma T}{2}} \sinh\frac{\gamma T}{2}\right]^2 J_{-H}(\frac{\gamma}{2}T)J_{H-1}(\frac{\gamma}{2}T)\right. \\ &\left. - \left[1 - \frac{\gamma-\theta}{\gamma}e^{\frac{\gamma T}{2}} \cosh\frac{\gamma T}{2}\right]^2 J_{1-H}(\frac{\gamma}{2}T)J_H(\frac{\gamma}{2}T)\right\}, \end{aligned}$$

where J_ν is the modified Bessel function of first kind of order ν.

Substituting this into (10.15) yields (10.10) for z_i's real, around a neighborhood of zero. Thus there is no problem of existence of $\Psi_T^H(z_1, z_2)$ around zero in $\mathbb{C}$ and since we have shown that the moment generating function exists, $\Psi_T^H(z_1, z_2)$. On the other hand, (1.10) defines an analytic function in the relevant domain and agrees with $\Psi_T^H(z_1, z_2)$ for z_1, z_2 real. □

1.11 Extreme Value Theory in Finance

Extreme risk in price fluctuations is determined by financial models. The section studies extreme value theory in finance. It discusses extreme value theory in interest rate models, stochastic volatility models, and long-memory models.

1.11.1 Basic Extremes

Extreme Value Distributions

The basic extreme value distributions are given by

Frechet: $\Phi_\alpha(x) = \exp(-x^{-\alpha}),\ x > 0,$

Gumbel: $\Lambda(x) = \exp(-e^{-x}),\ x \in \mathbb{R},$

Weibull: $\Psi_\alpha(x) = \exp(-(-x)^\alpha),\ x \leq 0.$

See Resnick (2008).

Regular Variation

A positive measurable function f on $(0, \infty)$ is *regularly varying* at ∞ with index α if

$$\lim_{t\to\infty} \frac{f(tx)}{f(t)} = x^\alpha,\ x > 0.$$

We use the notation $f \in \mathcal{R}(\alpha)$ for a regularly varying function with index α. The function f is said to be *slowly varying* if $\alpha = 0$, *rapidly varying* if the above limit is 0 for $x > 1$ and ∞ for $0 < x < 1$.

Denote $\overline{F}(u) = 1 - F(u) = P(X > u)$.

We denote$F \in \mathcal{L}(\gamma), \gamma \geq 0$ if for every $y \in \mathbb{R}$,

$$\lim_{x\to\infty} \frac{\overline{F}(x-y)}{\overline{F}(x)} = e^{\gamma y}.$$

Convolution Equivalent Distributions

Let X have d.f. F. Then $F \in \mathcal{S}(\gamma), \gamma \geq 0$ if $F \in \mathcal{L}(\gamma)$ and

$$\lim_{x\to\infty} \frac{\overline{F^{*2}}(x)}{\overline{F}(x)} = 2\hat{f}(\gamma),$$

where $\hat{f}(\gamma) = Ee^{\gamma X}$ is the moment generating function of X at γ. The class $\mathcal{S} := \mathcal{S}(0)$ is the class of *subexponential distributions.*

Subexponential distributions are heavy-tailed in the sense that no exponential moments exist. $\mathcal{S}$ contains all df.f.s F with regularly varying tails and is a much larger class. Distribution functions in $\mathcal{S}(\gamma)$ for some $\gamma > 0$ have exponential tails, hence are lighter tailed than subexponential distributions.

Theorem 11.1 (Fisher-Tippett Theorem) For an i.i.d. sequence, the limit distribution of the maxima is one of the three extreme value distributions: Frechet, Gumbel, and Weibull.

1.11.2 Diffusion Extremes

Consider the Itô stochastic differential equation

$$dX_t = \mu(X_t)dt + \sigma(X_t)dW_t, \ t > 0, \ X_0 = x,$$

where X: diffusion process, W: standard Brownian motion, $\mu \in \mathbb{R}$: *drift coefficient*, $\sigma > 0$: *diffusion coefficient* or *volatility*.
Running maxima is given by

$$M_t = \max_{0 \le s \le t} X_s, \ \ t > 0.$$

Scale function is given by

$$s(x) = \int_z^x \exp\left(-2\int_z^y \frac{\mu(t)}{\sigma^2(t)}dt\right) dy, \ x \in (l, r),$$

where z is any interior point in (l, r).
Speed measure m has Lebesgue density

$$m'(x) = \frac{2}{\sigma^2(x)s'(x)}, \ \ x \in (l, r)$$

and total mass $|m| = m((l, r))$. s' is the Lebesgue density of s. X_t is ergodic and its stationary distribution is absolutely continuous with Lebesgue density

$$h(x) = \frac{m'(x)}{|m|}, x \in (l, r).$$

X_t satisfies the *usual conditions* which guarantees that X is ergodic with stationary density:

$$s(r) = -s(l) = \infty, \quad |m| < \infty.$$

Lemma 11.1 (Davis 1982) Let $(X_t)_{t \ge 0}$ satisfy the usual conditions. Then for any initial value $X_0 = y \in (l, r)$ and any $u_t \uparrow r$,

$$\lim_{t \to \infty} |P^y(M_t \le u_t) - F^t(u_t)| = 0,$$

where F is a df defined for any $z \in (l, r)$ by

$$F(x) = \exp\left(-\frac{1}{|m|s(x)}\right), \ \ x \in (z, r).$$

Outline of Proof A diffusion process can be represented as an Ornstein–Uhlenbeck process after a random time-change. Then standard theory of extremes of Gaussian processes apply.

Corollary 11.1

$$\overline{F}(x) \sim \left(|m| \int_z^x s'(y)dy \right)^{-1} \sim (|m|s(x))^{-1}, \quad x \uparrow r.$$

F is in the maximum domain of attraction of $G : F \in MDA(G)$

$$\frac{M_t - b_t}{a_t} \to^{\mathcal{D}} G, \quad t \to \infty.$$

$G \in \{\Phi_\alpha, \Lambda\}, \alpha > 0$ where Φ_α is Frechet distribution, Λ is Gumbel distribution.

Theorem 11.2 If μ and σ are differentiable in some left neighborhood of r such that

$$\lim_{x \to r} \frac{d}{dx} \frac{\sigma^2(x)}{\mu(x)} = 0$$

$$\lim_{x \to r} \frac{\sigma^2(x)}{\mu(x)} \exp\left(-2 \int_z^x \frac{\mu(t)}{\sigma^2(t)} dt \right) = -\infty,$$

then

$$\overline{F}(x) \sim |\mu(x)| h(x), \quad x \uparrow r.$$

1.11.3 Interest Rate Extremes

Borkovec and Kluppelberg (2007) studied extreme value theory in interest rate models.

Vasicek Model The most popular model for short-term interest rate is given by

$$dX_t = (a - bX_t)dt + \sigma dW_t, \quad X_0 = x, \quad a \in \mathbb{R}, b > 0$$

$$X_t = \frac{a}{b} + (x - \frac{a}{b})e^{-bt} + \sigma \int_0^t e^{-b(t-s)} dW_s.$$

$$E(X_t) = \frac{a}{b} + (x - \frac{a}{b})e^{-bt} \to \frac{a}{b}, \quad t \to \infty.$$

$$Var(X_t) = \frac{\sigma^2}{2b}(1 - e^{-2bt}) \to \frac{\sigma^2}{2b}, \quad t \to \infty.$$

X_t has a normal stationary distribution $N(\frac{a}{b}, \frac{\sigma^2}{2b})$.
Condition of Theorem 11.2 holds which gives

$$\overline{F}(x) \sim \frac{2b^2}{\sigma^2}(x - \frac{a}{b})^2 \overline{H}(x), \quad x \to \infty,$$

where $\overline{H}(x)$ is the tail of a stationary normal distribution, hence F has heavier tail than H.

$F \in MDA(\Lambda)$ with norming constants

$$a_t = \frac{\sigma}{a\sqrt{b\log t}}, \quad b_t = \frac{\sigma}{\sqrt{b}}\sqrt{\log t} + \frac{a}{b} + \frac{\sigma}{4\sqrt{b}} \frac{\log\log t + \log(\sigma^2 d/2\pi)}{\sqrt{\log t}}.$$

Cox–Ingersoll–Ross (CIR) Model Since Vasicek model has normal stationary distribution, interest rate could be negative. In order to avoid this problem, Cox–Ingersoll–Ross introduced the Feller square-root process as a model for short-term interest rate:

$$dX_t = (a - bX_t)dt + \sigma\sqrt{X_t}dW_t$$

$$E(X_t) = \frac{a}{b} + (x - \frac{a}{b})e^{-bt} \to \frac{a}{b}, \ t \to \infty.$$

$$V(X_t) = \frac{a\sigma^2}{2b^2}(1 - (1 + (x - \frac{a}{b})\frac{2b}{a})e^{-2bt} + (x - \frac{a}{b})\frac{2b}{a}e^{-3bt}) \to \frac{a\sigma^2}{2b^2}, \ t \to \infty.$$

X_t has a Gamma stationary distribution $\Gamma(\frac{2a}{\sigma^2}, \frac{2b}{\sigma^2})$

Condition of the Theorem 11.2 holds which gives

$$\overline{F}(x) \sim \frac{2ab}{\sigma^2}\overline{G}(x) \sim Ax\overline{H}(x), \ x \to \infty,$$

where $\overline{G}(x)$ is the tail of a stationary gamma distribution $\Gamma(\frac{2a}{\sigma^2}+1, \frac{2b}{\sigma^2})$ hence F has heavier tail than H.

$F \in MDA(\Lambda)$ with norming constants

$$a_t = \frac{\sigma^2}{2b}, \quad b_t = \frac{\sigma}{2b}(\log t + \frac{2a}{\sigma^2}\log\log t + \log(\frac{b}{\Gamma(2a/\sigma^2)})).$$

Chan-Karolyi-Longstaff-Sanders (CKLS) Model In CIR process, the elasticity of volatility is 0.5. Empirically CKLS observed that the elasticity is close to 0.75 and introduced

$$dX_t = (a - bX_t)dt + \sigma X_t^\gamma dW_t, \ \gamma \in [1/2, \infty).$$

The case $\frac{1}{2} < \gamma < 1$:

$$E(X_t) = \frac{a}{b} + (x - \frac{a}{b})e^{-bt} \to \frac{a}{b}, \ t \to \infty, \ \ b > 0$$

$$E(X_t) = \frac{a}{b} + (x - \frac{a}{b})e^{-bt} \to \infty, \ t \to \infty, \ \ b < 0$$

$$E(X_t) = x + at \to \infty, \ t \to \infty, \ \ b = 0.$$

The lack of first moment indicates that for certain parameter values the model can capture very large fluctuations in the data, which will reflect also in the maxima.

Stationary density of this process is

$$h(x) = \frac{2}{A\sigma^2} x^{-2\gamma} \exp\left(-\frac{2}{\sigma^2}\left(\frac{a}{2\gamma-1}x^{-(2\gamma-1)} + \frac{b}{2-2\gamma}x^{2-2\gamma}\right)\right)$$

for some constant $A > 0$.

Condition of Theorem 11.2 holds.

$\overline{F}(x) \sim bxh(x) \sim Bx^{2(1-\gamma)}\overline{H}(x),\ x \to \infty$

$F \in MDA(\Lambda)$ with norming constants

$$a_t = \frac{\sigma^2}{2b}\left(\frac{\sigma^2(1-\gamma)}{b}\log t\right)^{\frac{2\gamma-1}{2-2\gamma}},$$

$$b_t = \left(\frac{\sigma^2(1-2\gamma)}{b}\log t\right)^{\frac{1}{2-2\gamma}}\left(1 - \frac{2\gamma-1}{(2-2\gamma)^2}\frac{\log\left(\frac{\sigma^2(1-\gamma)}{b}\log t\right)}{\log t}\right)$$
$$+a_t \log\left(\frac{2b}{A\sigma^2}\right).$$

The case $\gamma = 1$:

The model has an explicit solution

$$X_t = e^{-(b+\frac{\sigma^2}{2})t+\sigma W_t}\left(x + a\int_0^t e^{(b+\frac{\sigma^2}{2})s-\sigma W_s}ds\right).$$

The stationary density is inverse gamma:

$$h(x) = \left(\frac{\sigma^2}{2a}\right)^{-\frac{2b}{\sigma^2}-1}\left(\Gamma\left(\frac{2b}{\sigma^2}+1\right)\right)^{-1} x^{-2b/\sigma^2-2}\exp\left(-\frac{2a}{\sigma^2}x^{-1}\right),$$

$x > 0,\ \ h \in \mathcal{R}(-2b/\sigma^2 - 2)$ and the tail $\overline{H}$ of the stationary distribution is regularly varying.

$$\overline{F}(x) \sim Bx^{-2b/\sigma^2-1},\quad x \to \infty.$$

$F \in MDA(\Phi_{1+2b/\sigma^2})$ with norming constants

$$a_t \sim Ct^{1/(1+2b/\sigma^2)},\quad b_t = 0.$$

The case $\gamma > 1$:

h has the same form as in the case $\frac{1}{2} < \gamma < 1$ and $\overline{H} \in \mathcal{R}(-2\gamma+1)$

$$\overline{F}(x) \sim (Ax)^{-1},\ x \to \infty.$$

CKLS pointed out, most plausible value of $\gamma = 1.5$. Condition of Theorem 11.2 holds.

$F \in MDA(\Phi_1)$ with norming constants

$$a_t \sim t/A, \quad b_t = 0.$$

1.11.4 Stochastic Volatility Extremes

Levy–Ornstein–Uhlenbeck Volatility

Empirical volatility changes in time and exhibits tails which are heavier than normal. Empirical volatility has upward jumps and clusters on high levels. Levy driven Ornstein–Uhlenbeck models can capture heavy tails and volatility jumps and have volatility clusters if the driving Levy process has regularly varying tails.

Black-Scholes Model The basic model for stock price with constant volatility σ is given by

$$dS_t = rS_t dt + \sigma S_t dW_t.$$

Heston Model The model for stock price with stochastic volatility V is given by

$$dS_t = rS_t dt + \sqrt{V_t} S_t dW_t$$

$$dV_t = \lambda(a - V_t)dt + \sigma\sqrt{V_t} dB_t$$

$\lambda, a, \sigma > 0, \quad \lambda a \geq \frac{\sigma^2}{2}$ a is the long run mean, λ is the rate of mean reversion. Volatility is a CIR Process, W and B are two independent Brownian motions for simplicity, they could be correlated to include *leverage*.

GARCH Model The continuous time GARCH model of Nelson is given by

$$dS_t = \sqrt{V_t} dW_t$$

$$dV_t = \lambda(a - V_t)dt + \sigma V_t dB_t.$$

Volatility is a CKLS model with elasticity $\gamma = 1$.

Barndorff-Nielsen-Shephard Model Both the stock price and the volatility have jumps in the BNS model

$$dS_t = (\mu + rS_t)dt + \sqrt{V_t} dW_t + \rho dL_{\lambda t}$$

$$dV_t = -\lambda V_t dt + \sigma dL_{\lambda t}$$

$$V_t = e^{-\lambda t} V_0 + \int_0^t e^{-\lambda(t-s)} dL_{\lambda s}$$

is a *cadlag* process.

If V_0 is independent of L and $V_0 =^d \int_0^\infty e^{-s} dL_s$, then the process is stationary. The stationary solution is

$$V_t = e^{-\lambda t} \int_{-\infty}^{t} e^{\lambda s} dL_{\lambda s}.$$

We are concerned with processes L which are heavy or semi-heavy tailed, i.e., whose tails decrease no faster than exponentially.

Define

$$M_h := \sup_{0 \le t \le h} V_t.$$

Theorem 11.4 (Fasen et al. 2006)
(a) If $L_1 \in \mathcal{S} \cap MDA(\Phi_\alpha)$, then

$$P(M_h > x) \sim (\lambda h + \alpha^{-1}) P(L_1 > x), \quad x \to \infty.$$

(b) If $L_1 \in \mathcal{S} \cap MDA(\Lambda)$, then

$$P(M_h > x) \sim \lambda h P(L_1 > x), \quad x \to \infty.$$

Running Maxima Theorem

(a) If $L_1 \in \mathcal{S} \cap MDA(\Phi_\alpha)$, then

$$P(a_{\lambda T}^{-1} M_T \le x) = e^{-x^{-\alpha}}, \quad x > 0,$$

where a_T is such that

$$\lim_{T \to \infty} T P(L_1 > a_T x) = x^{-\alpha}, \quad x > 0.$$

(b) If $L_1 \in \mathcal{S} \cap MDA(\Lambda)$, then

$$P(a_{\lambda T}^{-1}(M_T - b_{\lambda T}) \le x) = e^{-e^{-x}}, \quad x \in \mathbb{R}.$$

Example: Positive Shot Noise Process

Let L be a positive compound Poisson process

$$L_t = \sum_{j=1}^{N_t} \xi_j,$$

where $(N_t)_{t \ge 0}$ is a Poisson process on $\mathbb{R}_+$ with intensity $\mu > 0$ and jump times $(\Gamma_k)_{k \in \mathbb{N}}$. The process N is independent of the i.i.d. sequence of positive r.v.s $(\xi_k)_{k \in \mathbb{N}}$ with d.f. F. The resulting volatility process is then the positive shot noise process

$$\begin{aligned} V_t &= e^{-\lambda t} V_0 + \int_0^t e^{-\lambda(t-s)} dL_{\lambda s} \\ &= e^{-\lambda t} V_0 + \sum_{i=1}^{N_{\lambda t}} e^{-\lambda t + \Gamma_j} \xi_j. \end{aligned}$$

Running Maxima Theorem

(a) Let V be a stationary version of the OU process where L is a positive, compound Poisson process. Assume $L_1 \in \mathcal{S}(\gamma), \gamma > 0$. Then

$$\lim_{T \to \infty} P(a_{\lambda T}^{-1}(M_T - b_{\lambda T}) \leq x) = e^{-e^{-x}}, \quad x \in \mathbb{R}.$$

(b) Assume that V is a $\Gamma(\mu, \gamma)$-OU process. Then

$$\lim_{T \to \infty} P(a_{\lambda T}^{-1}(M_T - b_{\lambda T}) \leq x) = e^{-e^{-x}}, \quad x \in \mathbb{R}.$$

1.11.5 Long Memory Extremes

A stationary process with correlation function ρ exhibits long-range dependence , if there exists a $H \in (0, 1/2)$ and l is a slowly varying function such that

$$\rho(h) \sim l(h)h^{-2H}, \quad h \to \infty.$$

Long-range dependence implies that

$$\int_0^\infty \rho(h)dh = \infty.$$

Superposition of Ornstein–Uhlenbeck Processes

Barndorff-Nielsen and Shephard (2001) proposed *supOU processes* as volatility models. Empirical volatility has long memory in the sense that the empirical autocorrelation function decreases slower than exponential. The class of supOU processes can capture *extremal clusters and long-range dependence.*

The supOU process is given by

$$V_t = \int_{\mathbb{R}_+ \times \mathbb{R}} e^{-r(t-s)} I_{[0,\infty)}(t-s) d\Lambda(r, \lambda s), \quad t \geq 0,$$

where $\lambda > 0$ and Λ is an *infinitely divisible independently scattered random measure (i.d.i.s.r.m.)* which are extensions of OU type processes of the form

$$V_t = \int_{-\infty}^t e^{-\lambda(t-s)} dL_{\lambda s},$$

where $\lambda > 0$ and L is a Levy process. The time change by λ yields marginal distributions independent of λ. To guarantee that the volatility process V is positive, the Levy process L is chosen as a *subordinator*. The resulting price process has martingale term $dS_t = \sqrt{V_t}dB_t$, where B is a Brownian motion independent of L.

The *generating quadruple* (m, σ^2, ν, π) determines completely the distribution of Λ. The underlying driving Levy process

$$L_t = \Lambda(\mathbb{R}_+ \times [0, t])$$

has *generating triplet* $(m,\ \sigma^2,\ \nu)$. The underlying driving Levy process

$$L_t = \Lambda(\mathbb{R}_+ \times [0, t]),\ t \geq 0.$$

Define the probability measure $\overline{\pi}(dr) := \lambda / r\pi(dr)$ and the i.d.i.s.r.m. $\overline{\Lambda}$ with generating quadruple $(m/\lambda, \sigma^2/\lambda, \nu\lambda, \overline{\pi})$.

Thus $\overline{\pi}$ is a probability measure on $\mathbb{R}_+$ with $\lambda := \int_{\mathbb{R}+} r\overline{\pi}(dr)$. The distribution π governs the long-range dependence of the model. Essentially the measure π needs sufficient mass near 0. We write $\pi(r) := \pi((0, r])$.

Then

$$X_t = \int_{-\infty}^{\infty} e^{-rt} \int_{-\infty}^{rt} e^s d\overline{\Lambda}(r, s).$$

Then $X = V$ *a.s.*

$$dX_t = \int_{\mathbb{R}_+} \{-rX(t, dr)dt + d\overline{\Lambda}(t, r)\},$$

where

$$X(t, B) = \int_B e^{-rt} \int_{-\infty}^{rt} e^s d\overline{\Lambda}(r, \lambda s).$$

Example Let π be gamma distribution with density

$$\pi(dr) = \Gamma(2H + 1)^{-1} r^{2H} e^{-r} dr$$

for $r > 0$ and $H > 0$. Then $\lambda = 2H$ and

$$\rho(h) = \Gamma(2H)^{-1} \int_0^{\infty} r^{2H-1} e^{-r(h+1)} dr = (h + 1)^{-2H},\ h \geq 0.$$

The following theorem shows how long-range dependence can be introduced in the supOU models.

Theorem 11.5

$$\overline{\pi}(r) \sim (2H)^{-1} l(r^{-1}) r^{2H},\ \ r \to 0$$

if and only if

$$\rho(h) \sim \Gamma(2H) l(h) h^{-2H},\ \ h \to \infty.$$

Define

$$M_T := \sup_{0 \leq t \leq T} V_t.$$

Theorem 11.6 (Fasen and Kluppelberg 2007)
(a) Let $L_1 \in \mathcal{R}_{-\alpha}$ with norming constants $a_T > 0$ such that

$$\lim_{T\to\infty} TP(L_1 > a_T x) = x^{-\alpha}, \quad x > 0.$$

Then

$$\lim_{T\to\infty} P(a_{\lambda T}^{-1} M_T \leq x) = e^{-x^{-\alpha}}, \quad x > 0.$$

(b) Let $L_1 \in \mathcal{S}(\gamma) \cap MDA(\Lambda)$ with norming constants $a_T > 0$ and $b_T \in \mathbb{R}$ such that

$$\lim_{T\to\infty} TP(L_1 > a_T x + b_T) = e^{-x}, \quad x \in \mathbb{R}.$$

Then

$$\lim_{T\to\infty} P(a_{\lambda T}^{-1}(M_T - b_{\lambda T}) \leq x) = e^{Ee^{\gamma L_1 - 1} Ee^{\gamma V_0} e^{-x}}, \quad x \in \mathbb{R}.$$

Typical examples of d.f.s in $\mathcal{S}(\gamma) \cap MDA(\Lambda)$ are Generalized Inverse Gaussian (GIG), Normal Inverse Gaussian (NIG), Generalized Hyperbolic (GH), Carr-Geman-Madan-Yor(CGMY) model. All these distributions are self-decomposable, which means that they are possible stationary distributions of OU-type processes and hence also supOU processes.

1.12 Epsilon-Markov Processes

The ϵ-Markov process included many types of stochastic processes such as a strong solution of stochastic differential equation, a higher order nonlinear time series model, and a cluster process. Consider the process (Volatility, Observation)= (X, Y) given by

$$dX_t = f(X_{(t-s)-} : s \in [0, \epsilon], \theta)\ dL_t, t \geq 0, \quad X_t = X_t^0, t \in [-\epsilon, 0],$$

$$dY_t = g(X_{t-}, \theta)\ dL_t, t \geq 0,$$

where the function $f : \mathbb{D}([0, \epsilon], \mathbb{R}) \times \Theta \to \mathbb{R} \times \mathbb{R}$ (here $\mathbb{D}(I, \mathbb{R})$ is the set of cadlag functions on $I \subset \mathbb{R}$) is assumed to satisfy the regularity conditions ensuring the existence of a unique strong solution for X. For the function $g : \mathbb{R} \times \Theta \to \mathbb{R} \times \mathbb{R}$, the process $g(X_{t-}, \theta), t \geq 0$ is predictable and almost surely bounded on compacts so that the stochastic integral $g(X_{t-}, \theta) \cdot L$ makes sense. When $\epsilon = 0$, the process X is a diffusion with jumps, hence is a Markov process.

When $\epsilon > 0$, the process is a non-Markov process. Consider the SDE

$$dX_t = \left(\int_{-\epsilon}^{0} X_{t+s} dA_s \right) dt + \sigma dW_t, X_t = X_t^0, t \in [-\epsilon, 0],$$

where $A_s, s \in [-\epsilon, 0]$ is a weighting function (signed measure on $[-\epsilon, 0]$) and W is a standard Brownian motion.

Usually the latent process is assumed to have the mixing property which carries on into observation process and classical limit theorems are applicable.

Chapter 2

Sequential Monte Carlo Methods

2.1 Introduction

Stochastic volatility and long-memory models have received a lot of attention in the last two decades. In this chapter, we study the sequential Monte Carlo (SMC) methods, also known as *particle filters*, for estimation and pricing in stochastic volatility models with general noises.

Sequential learning and inference methods are important in many applications where data arrival is inherently sequential. Furthermore, one might wish to adopt a sequential processing strategy to deal with non-stationarity in hidden process, so that information from the recent past is given greater weighting than information from the distant past. Computational simplicity in the form of not having to store all the data might also constitute an additional motivating factor for sequential methods. Many real-world modeling problems involve elements of non-Gaussianity, non-linearity, and non-stationarity. Consequently, it is not usually possible to derive exact closed form estimators based upon the standard criteria of maximum likelihood method. Classical sub-optimal methods, such as extended Kalman filters and Gaussian sum approximations, are easy to implement. However, they do not take into account all the salient statistical features of the processes under consideration, thereby often leading to poor results.

The advent of cheap and massive computational power, in conjunction with some recent developments in applied statistics, have stimulated many advancements in the field of Sequential Monte Carlo (SMC) or particle filter simulation. Monte Carlo methods are very flexible in that they do not require any assumptions about the probability distributions of the data. Moreover, experimental evidence suggests that these methods lead to improved results. From a Bayesian perspective, Sequential Monte Carlo methods allow one to compute the posterior probability distributions of interest on-line. Yet, the methods can also be applied within a maximum likelihood context. As a result, they are being applied to a large number of interesting real problems such as computer vision, econometrics, medical prognosis, tracking, communications, blind deconvolution, statistical model diagnostic tools, automatic control, navigation, and neural network training among others.

J. P. N. Bishwal, *Parameter Estimation in Stochastic Volatility Models*,
https://doi.org/10.1007/978-3-031-03861-7_2

Imputation can be useful for a number of applications including estimation and variance reduction in Mote Carlo simulation. For estimation, imputation can be used to augment a data set with unequally spaced times to one with equally spaced times, thereby simplifying the form of the likelihood function and the resulting estimators. For pricing path dependent derivative securities by Monte Carlo simulation, stratifying the sample path is a well known and commonly used variance reduction technique. This requires bringing the initial value of the process to a terminal value at one or more intermediate times, a problem well suited for imputation.

Stochastic volatility model is a state space model. The aim of filtering is to estimate current value of the state given the observations to date. Particle filters approximate the posterior distribution at a point by a set of weighted particles. The particles are possible realizations of the state and the approximation of the posterior by a discrete distribution whose support is the set of particles, with the probability mass assigned to each particle being proportional to that particle's weight. For the simplest filter, these particles are propagated through time according to the dynamics of the state and reweighted at each time point to take account of the information of the observation at that time.

Particle filters have good theoretical properties and under mild regularity conditions, the rate of convergence of the particle approximation of the true posterior is $O(N^{-1/2})$, where N is the number of particles. Furthermore, the computational cost of the filter is linear in the number of particles. Using Quasi-Monte Carlo integration, which uses deterministic sequence of points, chosen to have low discrepancy, the rate of convergence to a d-dimensional integral can be improved to $O(N^{-1}\log N^{d-1})$. This faster rate of convergence is due to the dependence between the particles which means that they are more regularly spaced than independent particles. Using random quasi-Monte Carlo, the rate of convergence can be improved to $O(N^{-3/2}\log N^{(d-1)/2})$. Using regularized particle filter the rate of convergence can be improved to $O(N^{-1})$. The computational complexity of the regularized particle filter is quadratic in the number of particles. One can apply regularized particle filter to stochastic volatility models. One should also consider other particle weights in order to improve the rate of convergence.

Also it is needed to study the fundamental issue of the behavior of Bayesian posteriors for stochastic volatility models, which is a hidden Markov model for non-fractional Levy type noises. This concerns investigation of the Bernstein-von Mises phenomenon for hidden models of increasing complexity, of the behavior of model selection schemes and Bayes factors, of the behavior of posteriors of none of the proposed models is correct, and of the behavior of the posterior if the proposed models are confounded and partially unidentifiable. Bernstein-von Mises phenomenon for stochastic volatility models appears to have received no attention so far. One needs to fill the gap. The Bernstein-von Mises phenomenon gives a precise estimate for the uncertainty in the posterior and hence a good indication for the amount of data needed or simulation error to focus on. These investigations will not only have independent value but also help will help us to determine the dependence of the results of our

SMC computations, that is posterior distribution of stochastic volatility on the choice of prior distribution and model specification, to determine which priors give best balance of statistical and computational efficiency, and how to best sample the data. We shall apply the methods to real data.

It is a well known phenomenon that stock return volatility changes randomly over time. Since Robert Engle won Nobel prize in 2003, modeling and measuring stochastic volatility has been the central issue in financial econometrics which in turn emerged as a new discipline. Robert Engle introduced the autoregressive conditional heteroscedastic (ARCH) model. Its continuous counterpart, non-Gaussian processes of Ornstein–Uhlenbeck (OU) type have considerable potential as building-blocks for stochastic models of observational series from a wide range of applications. They offer the possibility of capturing important distributional deviations from Gaussianity and for flexible modeling of dependence structures.

One should study Monte Carlo estimation results for stochastic volatility models which will help pricing, hedging, calibration of derivative products (on equities, interest rate, credit risk, operational risk), and portfolio optimization. Special attention should be paid to models with jumps and long memory. Schonkwiler (2013) studied Monte Carlo methods for finance.

The development of increasingly complex financial products requires the use of advanced computational methods. This work will develop this potential, drawing on and extending powerful results from probability theory and Monte Carlo methods for applications in statistical analysis. The power of the applications of Ornstein–Uhlenbeck processes within the context of finance and econometrics will be illustrated. On the basis of empirical stylized facts , one should study continuous time stochastic volatility models for financial assets where the volatility processes are non-Gaussian Ornstein–Uhlenbeck processes driven by Levy processes. Barndorff-Nielsen et al. (2001, 2002a,b,c,d, 2003, 2004a,b, 2005a,b,c, 2006, 2006a,b,c,d) introduced and studied non-Gaussian Ornstein–Uhlenbeck stochastic volatility models. Bishwal (2008a) studied higher order theory for parameter estimation problems on Gaussian Ornstein–Uhlenbeck models and diffusion models. One should extend his ideas to non-Gaussian OU based stochastic volatility models which are challenging problems.

2.2 Stochastic Volatility Models and SMC Methods

A general form of Ornstein–Uhlenbeck process Y_t is a solution of the stochastic differential equation

$$dY_t = -\lambda Y_t dt + dZ_t,$$

where Z, with $Z_0 = 0$, is a homogeneous Levy process, i.e., a process with independent and stationary increments. Familiar special cases of Levy processes are Brownian motion and compound Poisson process. All Levy processes except for Brownian motion have jumps.

One can consider cases where Z has no Gaussian component and increments of Z are positive, implying positivity of the process Y. Volatility is an unobserved hidden process in the asset price process. Inference about the unknown volatility parameter λ is a difficult problem. One should focus on estimation of the unknown parameter λ.

The main advantage of these OU processes is that they offer plenty of analytical tractability which is not available for more standard geometric Gaussian OU processes and constant elasticity of volatility processes. For geometric OU processes, $\log Y_t$ is assumed to follow a Gaussian OU process. For constant elasticity of volatility processes

$$dY_t = -\lambda(Y_t - \theta)dt + \sigma Y_t^k dB_t,$$

where B_t is a standard Brownian motion, $k \geq 1/2$.

Stochastic volatility models in general allow aggregate returns to be heavy-tailed, skewed and exhibit volatility clustering. Returns are scale mixtures of normals, where the scaling is typically time dependent, introducing dependence in the returns.

The Black-Scholes model which models the logarithm of the asset price S_t ($\log S_t = X_t$) by the solution to the stochastic differential equation

$$dX_t = (\mu + \beta\sigma^2)dt + \sigma dW_t,$$

where W_t is a standard Brownian motion. This means that aggregate returns over intervals of length $\Delta > 0$ are

$$I_n = \int_{(n-1)\Delta}^{n\Delta} dX_t = X_{n\Delta} - X_{(n-1)\Delta}$$

implying returns are normal and independently distributed with mean of $\mu\Delta + \beta\sigma^2\Delta$ and a variance of $\Delta\sigma^2$. Unfortunately, for moderate to small values of Δ (corresponding to returns measured over 5-minute to 1-day intervals) returns are typically heavy-tailed, exhibit volatility clustering (in particular, I_n are correlated) and are skew. This means that all assumptions of Black-Scholes model are rejected by the type of data that are usually used in practice. The most successful empirical rejection of option pricing models based on this model was first pioneered by Engle (1982) who proposed autoregressive conditional heteroscedastic (ARCH) models.

A general model will be of the type

$$dX_t = (\mu + \beta Y_t)dt + Y_t dW_t,$$

where Y_t is the volatility process. Even though Y_t exhibits jumps, X_t is a continuous process for all parameter values. In the special case where $\mu =$

$\beta = 0$, a stochastic volatility model can be thought of as a subordinated Brownian motion. One should give special attention to the cases where Y_t has inverse-Gaussian distribution, Y_t has Gamma distribution, and Y_t has tempered stable distribution. One should develop discretization methods for these models and implement Monte Carlo simulation of these cases. Nonlinear filtering theory and Kitagawa algorithm, which uses a linear spline technique to approximate the filter, can also be used to estimate volatility.

Integrated volatility is an important object in option pricing in stochastic volatility model. Realized volatility is known to be a standard estimator of integrated volatility. One should provide several alternative estimators of integrated volatility. One should also study bootstrap for stochastic volatility models. Finally we will consider stochastic volatility models with fractional Brownian motion noise, which keeps the long memory.

One should study the econometrics of high-frequency financial data. The development of increasingly complex financial products requires the use of advanced statistical methods. One should study sequential Monte Carlo methods for pricing, hedging, calibration of derivative products (on equities, interest rate, credit risk), and portfolio optimization. Special attention should be paid to models in high dimension, models with jumps, and models with long memory.

Since Black and Scholes (1973) established the theory of option pricing, volatility has played an important role not only in derivative pricing but also in portfolio selection and risk management. Despite the assumption of constant volatility in Black and Scholes (1973), it is widely recognized that volatility changes over time. The stylized facts of volatility are (1) volatility changes in time, (2) volatility is random, (3) volatility has heavy tails, (4) volatility clusters on high level. Starting with Engle (1982)'s autoregressive conditional heteroskedasticity (ARCH) model, various stochastic volatility models have been proposed. On the other hand, volatility is often modeled as a parametrized diffusion coefficient of a continuous time diffusion process and then the parameters are estimated by various methods like maximum likelihood or method of moments. One should study nonparametric estimation of volatility process. In principle, the more data we can use, the more accurate the estimate will be. However, one would have technological restriction on the amount of data. Recently this kind of restriction has been removed by development of computer power and data recording system. Those kind of data are called high-frequency data. Such high-frequency data lend the validity of the method based on quadratic variation formula, that is called the realized volatility. Barndorff-Nielsen and Shephard have suggested modeling the volatility as a Levy driven Ornstein–Uhlenbeck process.

One should emphasize on model calibration and inference. The implied volatility smile phenomenon has led to the appearance of a large variety of extensions of the Black-Scholes model: the local volatility models, diffusions with stochastic volatility, jump-diffusions, and long-memory models. The model calibration is the reconstruction of model parameters from the prices of traded options. It is an inverse problem to that of option pricing and as such, typically ill-posed.

One should also emphasize on interest rate models. The calibration problem is yet more complex in the interest rate markets since in this case the empirical data that can be used includes a wider variety of financial products from standard obligations to swaptions. Efficient computational algorithms are thus most needed.

A stochastic volatility model models both the underlier's value and its volatility as stochastic processes. As with a jump-diffusion model, this has the effect of giving the underlier's value a leptokurtic distribution. Heston's (1993) model is a popular stochastic volatility model.

A jump-diffusion model adds random jumps to the geometric Brownian motion that Black and Scholes (1973) assumes for the underlier. Among other things, this has the effect of giving the underlier's value a leptokurtic distribution.

It is well accepted that estimating financial volatility is a central issue in financial theory and practice. To estimate volatility, researchers have given attention to high-frequency data. Unfortunately, high-frequency data is often contaminated my movements dictated by institutional factors of the market. Removing these institutionally driven factors, termed micro structure effects, is vital to accurately determine the price volatility. Thus high-frequency data is contaminated by market microstructure noise. The available market microstructure noise models have used i.i.d. noise with constant variance to describe market microstructure effects. In continuous time i.i.d. process is a process with extreme volatility and because of this, the observed price process is dominated by the noise process and carries no information on the volatility of the efficient price process. To overcome this problem, one should use short-memory fractional Brownian motion (with Hurst parameter less than 0.5) driving noise for market microstructure noise.

The main difficulty in modeling and forecasting financial volatility is the fact that the volatility is latent and not directly observable, a hidden Markov model. Usually researchers have used realized volatility as an observable proxy for latent volatility. Realized volatility is defined as the summation of squared intraday returns. It is a good estimate to estimate the object of interest, the integrated volatility of an asset, so is optimal to use all the available data. However, it is a common practice in measuring realized volatility to avoid the data at highest frequency available, the tick-by-tick data. Instead, researchers tend to sample intra daily returns at somewhat moderate frequencies such as 5, 10, or 30 minutes. This is due to the suspicion that the market microstructure effects become more severe as sampling frequency increases.

In a series of papers, given in the reference, by Barndorff-Nielsen and Shephard (henceforth BN-S) have developed an asymptotic theory for realized volatility-like measures. In particular, for a general stochastic volatility model, BN-S establish a central limit theorem for realized volatility over a fixed interval of time, for instance, a day, as the number of intraday returns increases to infinity. They have also showed that a CLT applies to empirical measures based on powers of intra day returns (realized power variation) and products of powers of absolute returns (e.g., bipower variation). BN-S also provide a joint asymptotic distribution theory for the realized volatility and realized bipower variation and show how to use this distribution to test for the presence of jumps in asset prices.

One should also study SMC method for realized range-based variance, a statistics that replaces every squared return of realized variance with a normalized squared range. Realized range-based variance is asymptotically mixed normal and five times more efficient than the realized variance. This is not surprising. In contrast to the estimator based on daily returns, the range utilizes the information from the entire sample path and one would expect it to be superior.

One should also study SMC method of continuous time GARCH model which is called COGARCH model. In this context one should consider heavy-tailed distributions (in the sense that all moments are not finite) of the volatility.

In the classical stochastic volatility model, the log-price $y^* = \log S$ with S being the asset price follows

$$dy^*(t) = \alpha(t)dt + \sigma(t)dW(t)$$

$$d\sigma^2(t) = \tilde{\alpha}(t)dt + \tilde{\sigma}(t)d\tilde{W}(t),$$

where σ and α are assumed to be independent of the standard Brownian motion W. The process σ is called the *instantaneous volatility* or *spot volatility* and α is called the *mean process* and the Brownian motions W and $\tilde{W}$ are allowed to be correlated. A simple example of this is

$$\alpha(t) = \mu t + \beta\sigma^{2*}(t) \quad \text{where} \quad \sigma^{2*}(t) = \int_0^t \sigma^2(u)du$$

in which case β is called the *risk premium* and σ^{2*} is called the *integrated variance*. Over an interval of time length $h > 0$, returns are defined as

$$y_i := y^*(hi) - y^*((i-1)h), \quad i = 1, 2, \cdots, T$$

which implies that

$$y_i|\alpha_i, \sigma_i^2 \sim N(\alpha_i, \sigma_i^2),$$

where $\alpha_i := \alpha(ih) - \alpha((i-1))h)$
and $\sigma_i^2 := \sigma^{2*}(ih) - \sigma^{2*}((i-1))h) = \int_{(i-1)h}^{ih} \sigma^2(u)du$.
Here σ_i^2 is called the *actual variance* and α_i is called the *actual mean*. Suppose one is interested in estimating the actual volatility σ_i using M intra-h observations. A natural candidate is the *realized volatility* given by
$\sqrt{[y_M^*]} := \sqrt{\sum_{i=1}^M y_{j,i}^2}$,

where $y_{j,i} := y^*\left((i-1)h + \frac{jh}{M}\right) - y^*\left((i-1)h + \frac{(j-1)h}{M}\right)$, $j = 1, 2, \cdots, M$. When $h \to 0$, realized volatility converges in L_2 to the integrated volatility. One should study large deviations and moderate deviations results in this case.

We will first consider *fractional Brownian motion* innovation in stochastic volatility.

A normalized fractional Brownian motion $\{W_t^H, t \geq 0\}$ with Hurst parameter $H \in (0, 1)$ is a centered Gaussian process with continuous sample paths whose covariance kernel is given by

$$E(W_t^H W_s^H) = \frac{1}{2}(s^{2H} + t^{2H} - |t-s|^{2H}), \quad s, t \geq 0.$$

The process is self-similar (scale invariant) and it can be represented as a stochastic integral with respect to standard Brownian motion. For $H = \frac{1}{2}$, the process is a standard Brownian motion. For $H \neq \frac{1}{2}$, the fBm is not a semimartingale and not a Markov process, but a Dirichlet process. The increments of the fBm are negatively correlated for $H < \frac{1}{2}$ (anti-persistent) and positively correlated for $H > \frac{1}{2}$ and in this case they display long-range dependence (persistence). The parameter H, which is also called the self-similarity parameter, measures the intensity of the long-range dependence.

Next we consider *Hermite process* as noise for stochastic volatility.

As a generalization of fractional Brownian motion we get the Hermite process of order k with Hurst parameter $H \in (\frac{1}{2}, 1)$ which is defined as a multiple Wiener–Itô integral of order k with respect to standard Brownian motion $(B(t))_{t \in \mathbb{R}}$

$$M_t^{H,k} := c(H,k) \int_{\mathbb{R}}^{k} \int_0^t \Pi_{j=1}^k (s - y_i)_+^{-(\frac{1}{2} + \frac{H-1}{2})} ds \, dB(y_1) dB(y_2) \cdots dB(y_k),$$

where $x_+ = \max(x, 0)$.

For $k = 1$ the process is fractional Brownian motion W_t^H with Hurst parameter $H \in (0, 1)$. For $k = 2$ the process is Rosenblatt process. For $k \geq 2$ the process is non-Gaussian.

The covariance kernel $R(t, s)$ is given by

$$R(t,s) := E[M_t^{H,k} M_s^{H,k}]$$

$$= c(H,k)^2 \int_0^t \int_0^s \left[(u-y)_+^{-(\frac{1}{2} + \frac{H-1}{2})} ds (v-y)_+^{-(\frac{1}{2} + \frac{H-1}{2})} dy \right]^k dudv.$$

Let

$$\beta(p,q) = \int_0^1 z^{p-1}(1-z)^{q-1} dz, \; p, q > 0$$

be the beta function.

Using the identity

$$\int_{\mathbb{R}} [(u-y)_+^{a-1} ds (v-y)_+^{a-1} dy] = \beta(a, 2a-1)|u-v|^{2a-1}$$

we have

$$\begin{aligned} R(t,s) &= c(H,k)^2 \beta\left(\frac{1}{2} - \frac{1-H}{k}, \frac{2H-2}{k}\right)^k \int_0^t \int_0^s \left(|u-v|^{\frac{2H-2}{k}}\right)^k dv du \\ &= c(H,k)^2 \frac{\beta(\frac{1}{2} - \frac{1-H}{k}, \frac{2H-2}{k})^k}{H(2H-1)} \frac{1}{2}(t^{2H} + s^{2H} - |t-s|^{2H}). \end{aligned}$$

In order to obtain $E(M_t^{(H,k)})^2 = 1$,
choose

$$c(H,k)^2 = \left(\frac{\beta(\frac{1}{2} - \frac{1-H}{k}, \frac{2H-2}{k})^k}{H(2H-1)} \right)^{-1}$$

and we have

$$R(t,s) = \frac{1}{2}(t^{2H} + s^{2H} - |t-s|^{2H}).$$

Thus the covariance structure of the Hermite process and fractional Brownian motion are the same.

The process $M_t^{(H,k)}$ is H-self-similar with stationary increments and all moments are finite.

For any $p \geq 1$,

$$E|M_t^{(H,k)} - M_s^{(H,k)}|^p \leq c(p,H,k)|t-s|^{pH}.$$

Thus the Hermite process has Holder continuous paths of order $\delta < H$.

Hermite–Ornstein–Uhlenbeck process is defined as

$$dY_t^{(H,k)} = \theta Y_t^{(H,k)} dt + \sigma dM_t^{(H,k)}$$

has unique solution

$$Y_t^{(H,k)} = e^{\theta t}\left(\xi + \sigma \int_0^t e^{-\theta u} dM_u^{(H,k)}\right).$$

In particular, if

$$X_0 = \sigma \int_{-\infty}^0 e^{-\theta u} dM_u^{(H,k)},$$

then

$$Y_t^{(H,k)} = e^{\theta t} \sigma \int_{-\infty}^t e^{-\theta u} dM_u^{(H,k)}.$$

The process $Y_t^{(H,k)}$ is called the Hermite Ornstein–Uhlenbeck process of order k.

Because the covariance structure of the Hermite process is same as the fractional Brownian motion's one, the above integral is a Wiener integral.

$$E[Y_t^{(H,k)}Y_s^{(H,k)}] = \sigma^2 \int_{-\infty}^{t}\int_{-\infty}^{s} e^{\theta(t-u)}e^{\theta(t-v)}|u-v|^{2H-2}dvdu < \infty.$$

Next we will consider *fractional Poisson process* as noise for stochastic volatility.

A fractional Poisson process $\{P_H(t), t > 0\}$ with Hurst parameter $H \in (1/2, 1)$ is defined as

$$P_H(t) = \frac{1}{\Gamma(H-\frac{1}{2})}\int_0^t u^{\frac{1}{2}-H}\left(\tau^{H-\frac{1}{2}}(\tau-u)^{H-\frac{3}{2}}d\tau\right)dq(u),$$

where $q(u) = \frac{N(u)}{\sqrt{\lambda}} - \sqrt{\lambda}u$ where $N(u)$ is a Poisson process.

The process is self-similar in the wide sense, has wide sense stationary increments, has fat-tailed non-Gaussian distribution, and exhibits long range dependence. The process converges to fractional Brownian motion in distribution. The process is self-similar in the asymptotic sense.

Strict sense, wide sense and asymptotic sense self-similarity are equivalent for fractional Brownian motion. Stock returns are far from being self-similar in strict sense.

The stochastic volatility model we consider is

$$dV_t = \mu V_t dt + \sigma V_t dP_H(t).$$

Next we will consider *Modified Tempered Stable Process* as noise for volatility. An infinitely divisible distribution is said to be α-modified tampered stable distribution (α-MTS) distribution if its Levy triplet is given by

$$\sigma^2 = 0$$

$$\nu(dx) = C\left(\frac{\lambda_+^{\alpha+\frac{1}{2}}K_{\alpha+\frac{1}{2}}\lambda_+ x}{x^{\alpha+\frac{1}{2}}}I_{x>0} + \frac{\lambda_+^{\alpha+\frac{1}{2}}K_{\alpha+\frac{1}{2}}\lambda_- x}{x^{\alpha+\frac{1}{2}}}I_{x<0}\right)dx$$

$$\gamma = \mu + C\left(\frac{\Gamma(\frac{1}{2}-\alpha)}{2^{\alpha+\frac{1}{2}}}(\lambda_+^{2\alpha-1} - \lambda_-^{2\alpha-1}) - \lambda_+^{\alpha-\frac{1}{2}}K_{\alpha-\frac{1}{2}}(\lambda_+) + \lambda_-^{\alpha-\frac{1}{2}}K_{\alpha-\frac{1}{2}}(\lambda_-)\right),$$

where $C > 0, \lambda_+, \lambda_- > 0, \mu \in \mathbb{R}, \alpha \in (-\infty, 1)\backslash\{\frac{1}{2}\}$ and $K_p(x)$ is the modified Bessel function of second kind. We denote the MTS random variable by $X \sim MTS(\alpha, C, \lambda_+, \lambda_-, \mu)$. The Levy measure $\nu(dx)$ is called the MTS Levy measure with parameter $(\alpha, C, \lambda_+, \lambda_-)$. The MTS distribution is obtained by taking a symmetric α-stable distribution by $\alpha \in (0,1)$ and multiplying by a Levy measure with $\sqrt{|x|}\lambda^{\alpha+\frac{1}{2}}K_{\alpha+\frac{1}{2}}(\lambda|x|)$ on each half of the real axis. The measure can be extended to the case $\alpha \leq 0$. If $\alpha = \frac{1}{2}$, then γ may not be defined, so it is removed. The MTS distribution was introduced by Kim et al. (2008, 2009).

The tails of the α-MTS distribution are thinner than those of the 2α-stable and fatter than those of the 2α-TS distribution. At the zero neighborhood, all three have the same asymptotic behavior. The characteristic function and Laplace transform of MTS distribution are known. The cummulants of all orders are known.

If $\lambda_+ > \lambda_-$, then the distribution is skewed to the left. If $\lambda_+ < \lambda_-$, then the distribution is skewed to the right. If $\lambda_+ = \lambda_-$, then the distribution is symmetric.

As α decreases, the distribution has fatter tails and increased peakedness. The Levy process corresponding to the MTS distribution has finite activity if $\alpha < 0$ and infinite activity if $\alpha > 0$. It has finite variation if $\alpha < \frac{1}{2}$ and infinite variation if $\alpha > \frac{1}{2}$.

With proper choice of C and μ, MTS distribution has zero mean and unit variance, and the distribution is called standard MTS distribution and denoted $X \sim stdMTS(\alpha, \lambda_+, \lambda_-)$.

The variance Gamma or CGMY process is a tampered stable process. In order to obtain a closed form solution of the European option price, CGMY used the generalized Fourier transform of the distribution of the stock price under the assumption of Markov property.

The stochastic volatility model is

$$dY_t = (\mu + \beta\sigma_t^2)dt + \sigma_t dW_t$$

$$d\sigma_t^2 = -\lambda\sigma_t^2 dt + dS_t$$

μ is the drift parameter, β is the risk premium, and S_t is a MTS process.

2.3 Spline Method in Volatility Estimation

We will study DANOVA models. DANOVA stands for ANOVA (Analysis of Variance) for Diffusions. In order to improve the realized volatility estimator of integrated volatility, we use splines. This will be the first step towards the use of splines for stochastic volatility models.

Let us consider splines in classical stochastic volatility model, the log-price $y^* = \log S$ with S being the asset price follows

$$dy^*(t) = \alpha(t)dt + \sigma(t)dW(t)$$

$$d\sigma^2(t) = \tilde{\alpha}(t)dt + \tilde{\sigma}(t)d\tilde{W}(t),$$

where σ and α are assumed to be independent of the standard Brownian motion W. The process σ is called the *instantaneous volatility* or *spot volatility* and α is called the *mean process* and W and $\tilde{W}$ are allowed to be correlated. A simple example of this is

$$\alpha(t) = \mu t + \beta\sigma^{2*}(t) \quad \text{where} \quad \sigma^{2*}(t) = \int_0^t \sigma^2(u)du$$

in which case β is called the *risk premium* and σ^{2*} is called the *integrated variance.*

Over an interval of time length $h > 0$, returns are defined as

$$y_i := y^*(hi) - y^*((i-1)h), \quad i = 1, 2, \cdots, T$$

which implies that

$$y_i|\alpha_i, \sigma_i^2 \sim N(\alpha_i, \sigma_i^2),$$

where $\alpha_i := \alpha(ih) - \alpha((i-1))h)$
and $\sigma_i^2 := \sigma^{2*}(ih) - \sigma^{2*}((i-1))h) = \int_{(i-1)h}^{ih} \sigma^2(u)du$.
Here σ_i^2 is called the *actual variance* and α_i is called the *actual mean.*
Suppose one is interested in estimating the actual volatility σ_i using M intra-h observations. A natural candidate is the *realized volatility* given by

$$\sqrt{[y_M^*]} := \sqrt{\sum_{j=1}^{M} y_{j,i}^2}$$

where $y_{j,i} := y^*\left((i-1)h + \frac{jh}{M}\right) - y^*\left((i-1)h + \frac{(j-1)h}{M}\right), \quad j = 1, 2, \cdots, M.$

Denote

$$Y_j := y^*((i-1)h + \frac{jh}{M})$$

and

$$Y_{j-1} := y^*((i-1)h + \frac{(j-1)h}{M}).$$

Thus the realized volatility is given by

$$\hat{\sigma}_i := \sqrt{\sum_{j=1}^{M} (Y_j - Y_{j-1})^2}.$$

When $h \to 0$, realized volatility converges in L_2 to the integrated volatility. We consider the fixed h case.

The realized variance is a quadratic form.

Note that the realized volatility is based on first order difference.

We introduce some new estimators:

$$\breve{\sigma}_i := \frac{1}{2}\sqrt{\sum_{j=1}^{M} [(Y_{j+1} - Y_j) - (Y_j - Y_{j-1})]^2}.$$

The above estimator is based on second order difference.

$$\bar{\sigma}_i := \sqrt{\sum_{j=n_1}^{M-n_2} \left(\sum_{k=-n_1}^{n_2} d_k Y_{k+j} \right)^2},$$

where n_1 and n_2 are non-negative integers, $n = n_1 + n_2$ is called the order, and the difference sequence $\{d_k\}_{k=-n_1,\cdots,n_2}$ satisfies $\sum_{k=-n_1}^{n_2} d_k = 0, \sum_{k=-n_1}^{n_2} d_k^2 = 1$ and $d_{-n_1}d_{-n_2} \neq 0$.

Note that for difference based estimators

$$|\bar{\sigma}_i^2 - \sigma_i^2| = O_P(M^{-1/2}).$$

To improve this error bound, we introduce the lag-k estimator

$$\tilde{\sigma}_i = \sqrt{\sum_{j=k}^{M}(Y_j - Y_{j-k})^2}, \quad k = 1, 2, \cdots, M-1.$$

We will compare the mean square error (MSE) of the estimators.

In practice, the choice of the order n and an appropriate difference sequence which minimizes the finite sample MSE is difficult.

We fit the following regression model:

$$\tilde{\sigma}_i^2 = \alpha + \beta d_k + \epsilon_k, k = 1, 2, \cdots, n.$$

Next we will consider *Kernel Estimation* of stochastic volatility using different weights to kernels.

Consider the stochastic volatility model

$$dY_t = \mu\ Y_t\ dt + \sigma_t\ \ dW_t$$

$$d\sigma_t^2 = \theta\ \sigma_t^2\ dt +\ dZ_t, \tag{1.1}$$

where $\{W_t\}$ is a standard Brownian motion, $\{Z_t\}$ is a homogeneous Inverse-Gaussian Levy process and $\theta < 0$. The integrated volatility is

$$V_T := \int_0^T \sigma_t^2 dt. \tag{1.2}$$

In IGOU model, calculation of conditional cummulants of the integrated volatility conditioned on the initial value is enough to be able to compute European style options very rapidly.

The cumulant functions of IG-OU process are given by

$$k(\theta) = \log E[e^{-\theta Z(1)}] = -\theta\delta\gamma^{-1}(1 + 2\theta\gamma^{-2})^{-1/2}.$$

$$k'(\theta) = \log E(e^{-\theta V_t}) = \delta\gamma - \delta\gamma(1 + 2\theta\gamma^{-2})^{1/2}.$$

Consider a continuously differentiable kernel K with shrinking bandwidth $h \to 0$.

$K_h(x) = \frac{1}{h}K(x/h)$, $K : \mathbb{R} \to \mathbb{R}$ is a kernel which normalize to $\int_{\mathbb{R}} K(x)\ dx = 1$.

We will consider Kernel weighted average of the quadratic variation.

In order to define the kernel volatility (KV), we use various discrete approximations of the integral in the definition (1.2) of integrated volatility.

Define a weighted sum of squares

$$M_{n,T} := \left\{ \sum_{i=1}^{n} w_{t_i} K_{t_{i-1}} \Delta Y_{t_{i-1}}^2 + \sum_{i=2}^{n+1} w_{t_i} K_{t_{i-1}} \Delta Y_{t_{i-1}}^2 \right\}, \tag{1.10}$$

where $w_{t_i} \geq 0$ is a weight function.

Denote the discrete increasing functions

$$I_{n,T} := \sum_{i=1}^{n} K_{t_{i-1}}^2 \Delta Y_{t_{i-1}}^2, \tag{1.11}$$

$$J_{n,T} := \sum_{i=2}^{n+1} K_{t_{i-1}}^2 \Delta Y_{t_{i-1}}^2 = \sum_{i=1}^{n} K_{t_i}^2 \Delta Y_{t_{i-1}}^2. \tag{1.12}$$

General weighted KV is defined as

$$\tilde{V}_{n,T} := \{M_{n,T}\}. \tag{1.13}$$

With $w_{t_i} = 1$, we obtain the forward KV as

$$\tilde{V}_{n,T,F} := \{I_{n,T}\}. \tag{1.14}$$

With $w_{t_i} = 0$, we obtain the backward KV as

$$\tilde{V}_{n,T,B} := \{J_{n,T}\}. \tag{1.15}$$

With $w_{t_i} = 0.5$, the simple symmetric AIV is defined as

$$\tilde{V}_{n,T,z} := \left\{ \frac{1}{2}[I_{n,T} + J_{n,T}] \right\} = \left\{ \sum_{i=2}^{n} K_{t_{i-1}}^2 \Delta Y_{t_{i-1}}^2 + 0.5(K_{t_0} + K_{t_n}) \Delta Y_{t_{i-1}}^2 \right\}. \tag{1.16}$$

With the weight function

$$w_{t_i} = \begin{cases} 0 & : \quad i = 1 \\ \frac{i-1}{n} & : \quad i = 2, 3, \cdots, n \\ 1 & : \quad i = n+1 \end{cases}$$

the weighted symmetric KV is defined as

$$\tilde{V}_{n,T,w} := \left\{ \sum_{i=2}^{n} K_{t_{i-1}} \Delta Y_{t_{i-1}}^2 + \sum_{i=1}^{n} K_{t_{i-1}} \Delta Y_{t_{i-1}}^2 \right\}. \tag{1.17}$$

Note that estimator (1.13) is analogous to the trapezoidal rule in numerical analysis. One can instead use the midpoint rule to define another estimator

$$\tilde{V}_{n,T,A} := \left\{ \sum_{i=1}^{n} \left(\frac{K_{t_{i-1}} + K_{t_i}}{2} \right) \Delta Y_{t_{i-1}}^2 \right\}. \tag{1.18}$$

One can use the Simpson's rule to define another estimator where the denominator is a convex combination of the denominators in (1.11) and (1.12)

$$\tilde{V}_{n,T,S} := \left\{ \frac{1}{6} \sum_{i=1}^{n} \left\{ K_{t_{i-1}} + 4\left(\frac{K_{t_{i-1}} + K_{t_i}}{2} \right) + K_{t_i} \right\} \Delta Y^2_{t_{i-1}} \right\}. \tag{1.19}$$

In general, one can generalize Simpson's rule as

$$\tilde{V}_{n,T,GS} := \left\{ \sum_{i=1}^{n} \left\{ \alpha \frac{K_{t_{i-1}} + K_{t_i}}{2} + (1-\alpha)\left(\frac{K_{t_{i-1}} + K_{t_i}}{2} \right) \right\} \Delta Y^2_{t_{i-1}} \right\} \tag{1.20}$$

for any $0 \leq \alpha \leq 1$.

The case $\alpha = 0$ produces the estimator (1.18). The case $\alpha = 1$ produces the estimator (1.17). The case $\alpha = \frac{1}{3}$ produces the estimator (1.19).

We propose a very general form of the quadrature based KV as

$$\tilde{V}_{n,T,w} := \left\{ \sum_{i=1}^{n} \sum_{j=1}^{m} \left[(1-s_j) K_{t_{i-1}} + s_j K_{t_i} \right] p_j \Delta Y^2_{t_{i-1}} \right\}, \tag{1.21}$$

where $p_j, \quad j \in \{1, 2, \cdots, m\}$ is a probability mass function of a discrete random variable S on $0 \leq s_1 < s_2 < \cdots < s_m \leq 1$ with $P(S = s_j) := p_j, \ j \in \{1, 2, \cdots, m\}$.

Denote the k-th moment of the random variable S as $\mu_k := \sum_{j=1}^{m} s_j^k p_j, \ k = 1, 2, \cdots$.

If one chooses the probability distribution as uniform distribution for which the moments are a harmonic sequence $(\mu_1, \mu_2, \mu_3, \mu_4, \mu_5, \mu_6, \cdots) = (\frac{1}{2}, \frac{1}{3}, \frac{1}{4}, \frac{1}{5}, \frac{1}{7}, \cdots)$, then there is no change in rate of convergence than second order. If one can construct a probability distribution for which the harmonic sequence is truncated at a point, then there is a rate of convergence improvement at the point of truncation.

Given a positive integer m, construct a probability mass function $p_j, \quad j \in \{1, 2, \cdots, m\}$ on $0 \leq s_1 < s_2 < \cdots < s_m \leq 1$ such that

$$\sum_{j=1}^{m} s_j^r p_j = \frac{1}{r+1}, \quad r \in \{0, \cdots, m-2\} \tag{1.22}$$

$$\sum_{j=1}^{m} s_j^{m-1} p_j \neq \frac{1}{m}. \tag{1.23}$$

Neither the probabilities p_j nor the atoms, s_j, of the distribution are specified in advance.

This problem is related to the truncated Hausdorff moment problem. I obtain examples of such probability distributions and use them to get higher order accurate (up to sixth order) KVs.

The order of approximation error (rate of convergence) of a KV is $n^{-\nu}$ where

$$\nu := \inf\left\{k : \mu_k \neq \frac{1}{1+k},\ \mu_j = \frac{1}{1+j}, j = 1, 2, \cdots, k-1\right\}. \tag{1.24}$$

I construct probability distributions satisfying these moment conditions and obtain KVs of the rate of convergence up to order 6.

Probability $p_1 = 1$ at the point $s_1 = 0$ gives the KV (1.11) for which $\mu_1 = 0$. Note that $\mu_1 \neq \frac{1}{2}$. Thus $\nu = 1$

Probability $p_1 = 1$ at the point $s_1 = 1$ gives the KV (1.12) for which $\mu_1 = 1$. Note that $\mu_1 \neq \frac{1}{2}$. Thus $\nu = 1$.

Probabilities $(p_1, p_2) = (\frac{1}{2}, \frac{1}{2})$ at the respective points $(s_1, s_2) = (0, 1)$ produce the KV $\tilde{V}_{n,T,Z}$ for which $(\mu_1, \mu_2) = (\frac{1}{2}, \frac{1}{4})$. Thus $\nu = 2$.

Probability $p_j = 1$ at the point $s_j = \frac{1}{2}$ produces the KV $\tilde{V}_{n,T,A}$ for which $(\mu_1, \mu_2) = (\frac{1}{2}, \frac{1}{2})$. Thus $\nu = 2$.

Probabilities $(p_1, p_2) = (\frac{1}{4}, \frac{3}{4})$ at the respective points $(s_1, s_2) = (0, \frac{2}{3})$ produce the asymmetric KV

$$\tilde{V}_{n,T,3} := \left\{\tfrac{1}{4}\textstyle\sum_{i=1}^{n}\left[(K_{t_{i-1}}) + 3(\frac{K_{t_{i-1}}+2K_{t_i}}{3})\right]\Delta Y_{t_{i-1}}^2\right\} \tag{1.25}$$

for which $(\mu_1, \mu_2, \mu_3) = (\frac{1}{2}, \frac{1}{3}, \frac{2}{9})$. Thus $\nu = 3$.

Probabilities $(p_1, p_2) = \frac{3}{4}, \frac{1}{4}$ at the respective points $(s_1, s_2) = (\frac{1}{3}, 1)$ produce asymmetric KV

$$\tilde{V}_{n,T,4} := \left\{\tfrac{1}{4}\textstyle\sum_{i=1}^{n}\left[3(\frac{2K_{t_{i-1}}+K_{t_i}}{3}) + (K_{t_i})\right]\Delta Y_{t_{i-1}}^2\right\} \tag{1.26}$$

for which $(\mu_1, \mu_2, \mu_3) = (\frac{1}{2}, \frac{1}{3}, \frac{10}{36})$. Thus $\nu = 3$.

Probabilities $(p_1, p_2, p_3) = (\frac{1}{6}, \frac{2}{3}, \frac{1}{6})$ at the respective points $(s_1, s_2, s_3) = (0, \frac{1}{2}, 1)$ produce the KV $\tilde{V}_{n,T,5}$ for which $(\mu_1, \mu_2, \mu_3, \mu_4) = (\frac{1}{2}, \frac{1}{3}, \frac{1}{4}, \frac{5}{25})$. Thus $\nu = 4$.

Probabilities $(p_1, p_2, p_3, p_4) = (\frac{1}{8}, \frac{3}{8}, \frac{3}{8}, \frac{1}{8})$ at the respective points $(s_1, s_2, s_3, s_4) = (0, \frac{1}{3}, \frac{2}{3}, 1)$ produce the symmetric KV

$$\tilde{V}_{n,T,5} := \left\{\tfrac{1}{8}\textstyle\sum_{i=1}^{n}\left[(K_{t_{i-1}}) + 3(\frac{2K_{t_{i-1}}+K_{t_i}}{3}) + 3(\frac{K_{t_{i-1}}+2K_{t_i}}{3}) + (K_{t_i})\right]\Delta Y_{t_{i-1}}^2\right\} \tag{1.27}$$

for which $(\mu_1, \mu_2, \mu_3, \mu_4) = (\frac{1}{2}, \frac{1}{3}, \frac{1}{4}, \frac{11}{54})$. Thus $\nu = 4$.

Probabilities $(p_1, p_2, p_3, p_4, p_5) = (\frac{1471}{24192}, \frac{6925}{24192}, \frac{1475}{12096}, \frac{2725}{12096}, \frac{5675}{24192}, \frac{1721}{24192})$ at the respective points $(s_1, s_2, s_3, s_4, s_5) = (0, \frac{1}{5}, \frac{2}{5}, \frac{3}{5}, \frac{4}{5}, 1)$ produce the asymmetric KV

$$\tilde{V}_{n,T,7}$$
$$:= \Big\{\tfrac{1}{24192}\textstyle\sum_{i=1}^{n}\Big[1471(K_{t_{i-1}}) + 6925(\frac{K_{t_{i-1}}+K_{t_i}}{5}) + 2950(\frac{2K_{t_{i-1}}+2K_{t_i}}{5})^2$$
$$+5450(\frac{3K_{t_{i-1}}+3K_{t_i}}{5}) + 5675(\frac{4K_{t_{i-1}}+4K_{t_i}}{5}) + 1721(K_{t_i})\Big]\,\Delta Y^2_{t_{i-1}}\Big\} \quad (1.28)$$

for which $(\mu_1, \mu_2, \mu_3, \mu_4, \mu_5, \mu_6) = (\frac{1}{2}, \frac{1}{3}, \frac{1}{4}, \frac{1}{5}, \frac{841}{5040})$. Thus $\nu = 5$.

Probabilities $(p_1, p_2, p_3, p_4, p_5) = (\frac{7}{90}, \frac{16}{45}, \frac{2}{15}, \frac{16}{45}, \frac{7}{90})$ at the respective points $(s_1, s_2, s_3, s_4, s_5) = (0, \frac{1}{4}, \frac{1}{2}, \frac{3}{4}, 1)$ produce the symmetric KV $\tilde{V}_{n,T,8}$ given by

$$\tilde{V}_{n,T,8}$$
$$:= \Big\{\tfrac{1}{90}\textstyle\sum_{i=1}^{n}\Big[7(K_{t_{i-1}}) + 32(\frac{3K_{t_{i-1}}+K_{t_i}}{4}) + 12(\frac{K_{t_{i-1}}+K_{t_i}}{2}) + 32(\frac{K_{t_{i-1}}+3K_{t_i}}{4})^2$$
$$+7(t_i, K_{t_i})]\,\Delta Y^2_{t_{i-1}}\Big\} \quad (1.29)$$

for which $(\mu_1, \mu_2, \mu_3, \mu_4, \mu_5, \mu_6) = (\frac{1}{2}, \frac{1}{3}, \frac{1}{4}, \frac{1}{5}, \frac{1}{6}, \frac{110}{768})$. Thus $\nu = 6$.

Probabilities $(p_1, p_2, p_3, p_4, p_5) = (\frac{19}{288}, \frac{75}{288}, \frac{50}{288}, \frac{50}{288}, \frac{75}{288}, \frac{19}{288})$ at the respective points
$(s_1, s_2, s_3, s_4, s_5) = (0, \frac{1}{5}, \frac{2}{5}, \frac{3}{5}, \frac{4}{5}, 1)$ produce symmetric KV

$$\tilde{V}_{n,T,9} :=$$
$$-\Big\{\tfrac{2}{n}\tfrac{1}{288}\textstyle\sum_{i=1}^{n}\Big[19(K_{t_{i-1}}) + 75(\frac{4K_{t_{i-1}}+K_{t_i}}{5}) + 50(\frac{3K_{t_{i-1}}+2K_{t_i}}{5})$$
$$+50(\frac{2K_{t_{i-1}}+3K_{t_i}}{5}) + 75(\frac{K_{t_{i-1}}+4K_{t_i}}{5}) + 19(K_{t_i})\Big]\,\Delta Y^2_{t_{i-1}}\Big\} \quad (1.30)$$

for which $(\mu_1, \mu_2, \mu_3, \mu_4, \mu_5, \mu_6) = (\frac{1}{2}, \frac{1}{3}, \frac{1}{4}, \frac{1}{5}, \frac{1}{6}, \frac{3219}{22500})$. Thus $\nu = 6$.

The KV $\tilde{V}_{n,T,S}$ is based on the arithmetic mean of $I_{n,T}$ and $J_{n,T}$. One can use geometric mean and harmonic mean instead. The geometric mean based symmetric KV (which is based on the ideas of partial autocorrelation) is defined as

$$\tilde{V}_{n,T,G} := \frac{1}{\sqrt{I_{n,T}\,J_{n,T}}}. \quad (1.31)$$

The harmonic mean based symmetric KV is defined as

$$\tilde{V}_{n,T,H} := \frac{1}{\frac{1}{I_{n,T}} + \frac{1}{J_{n,T}}}. \quad (1.32)$$

Note that

$$\tilde{V}_{n,T,S} \;\le\; \tilde{V}_{n,T,G} \;\le\; \tilde{V}_{n,T,H}. \quad (1.33)$$

Remarks

The following methods are important for applications:

1. *Sequential Importance Sampling with Resampling (SISR) method*
2. *Genetic Algorithm for Stochastic Volatility*
3. *Prediction based estimating function for stochastic Volatility*
4. *MCMC method for stochastic volatility model*
5. *Exact Simulation of Diffusions*

One can consider the exact algorithm proposed in Beskos and Roberts (2005) which simulates exact sample paths of diffusions. One can also consider random weight particle filter.

2.4 Multilevel Monte Carlo Method for Nonlinear SPDE

Consider two SDEs

$$dX_t = f(\theta, X_t)dt + g(X_t)dW_t,$$

$$dY_t = a(X_t)dt + b(X_t)dB_t,$$

where B_t and W_t are two independent Brownian motions. Observation: X. It is known that

$$\|X_t^n(Y_t^n(y)) - X_t(Y_t(y))\|_{L^p} = O(n^{-1/2}),$$

where $X_t^n(x), Y_t^n(y)$ are Euler schemes of $X_t(x), Y_t(y)$, respectively, of step size T/n.

We improve the bound in the following theorem.

Theorem 1.1

$$\|\hat{X}_t^n(\hat{Y}_t^n(y)) - X_t(Y_t(y))\|_{L^p} = O(n^{-1}),$$

where $\hat{X}_t^n(x), \hat{Y}_t^n(y)$ are Milstein schemes of $X_t(x), Y_t(y)$, respectively, of step size T/n, respectively.

Estimation: Observe $\hat{X}_{|\hat{Y}_t|}$ based on compound Milstein scheme. Estimate $X_{|Y_t|}$ and then obtain $\hat{\theta}_T$.

Solutions of SPDEs can be approximated by compound SDEs. Denote

$$U(t, y) := X_t(Y_t(y)).$$

Applying the Itô-Ventzel formula $U(t, y)$ satisfies a second order nonlinear SPDE:

$$dU(t,y) = \Big[\partial_y U(t,y)\frac{a(t,Y_t(y))}{\partial_y Y_t(y)} + \frac{1}{2}\left(\partial_y^2 U(t,y) - \partial_y U(t,y)\frac{\partial_y^2 Y_t(y)}{\partial_y Y_t(y)}\right)\frac{b^2(t,Y_t(y))}{(\partial_y Y_t(y))^2}$$

$$+ f(\theta,t,U(t,y)) + \partial_x g(t,U(t,y))b(t,Y_t(y))\Big]\,dt$$

$$+ \left(\partial_y U(t,y)\frac{b(t,Y_t(y))}{\partial_y Y_t(y)} + g(t,U(t,y))\right)dW_t.$$

$U(t, x)$ is the forward utility and $\partial_x U$ is the marginal utility, x is called wealth.

This helps in reducing the computational cost of multilevel Monte Carlo. Thus estimation in the above nonlinear SPDE is equivalent to observe $\hat{X}_{|\hat{Y}_t|}$ based on compound Milstein scheme. Estimate $X_{|Y_t|}$ and then obtain $\hat{\theta}_T$.

2.5 Third Order Composition Scheme for Diffusions

2.5.1 Hörmander Condition

Extending Alfonsi (2005, 2010), Rey (2017) introduced the third order scheme using cubature methods and abstract commutativity condition on the coefficients of the underlying diffusion process.

Consider the Markov diffusion process

$$dX_t = V_0(\theta, X_t)dt + V_1(X_t)\ o\ dW_t,$$

where $V_i \in \mathcal{C}_b^\infty(\mathbb{R}, \mathbb{R})$ for $i = 0, 1$, W_t is a standard Brownian motion, $o\ dW_t$ is the Stratonovich integral with respect to W_t. We consider homogeneous time grid $t_k = kT/n$ for clarity for fixed $T > 0$ and $n \in \mathbb{N}$. Let Π be the partition of the time interval $[0, T]$ with the homogeneous grid. For a test function $\mathcal{C}_b^q(\mathbb{R})$, local order of $\beta + 1$ implies global order of β, i.e.,

If

$$|E(f(X_{t_{k+1}}) - f(X_{t_k}^n)| \le \frac{C\|f\|_{q,\infty}}{n^{\beta+1}},$$

then

$$\sup_{t\in\Pi}|E(f(X_t) - f(X_t^n)| \le \frac{C\|f\|_{q,\infty}}{n^{\beta}}.$$

It means that $X_{t_k}^n$ is a weak approximation scheme of order β for the Markov process X_t.

The first order scheme is due to Bally and Talay (1996). Ninomiya and Victoir (2008) obtained the second order scheme under Hörmander type condition. Al Gerbi et al. (2018) studied asymptotics of the asymptotics of the normalized error of the Ninomiya-Victoir scheme.

For independent Bernoulli random variables ρ_k

$$\psi_k(x, w^1, w^0) = \exp(w^0 V_0) \text{ o } \exp(w^1 V_1) \text{ o o } \ldots \text{o } \exp(w^0 V_N) \text{ o } \exp(w^0 V_0)(x),$$

with $\rho_k = 1$

$$\psi_k(x, w^1, w^0) = \exp(w^0 V_0) \text{ o } \exp(w^1 V_1) \text{ o o } \ldots \text{o } \exp(w^0 V_N) \text{ o } \exp(w^0 V_0)(x),$$

with $\rho_k = -1$.

Ninomiya and Victoir (2008) used independent random variables Z_k satisfying the following moment conditions: $E(Z_k) = 0,$, $E(Z_k^3) = 0$, $E(Z_k^5) = 0$, $E(Z_k^2) = 1$, $E(Z_k^4) = 6$.

One step of the scheme is defined as

$$X_{t_{k+1}} = \psi(X_{t_k}, w_{k+1}^1, w_{k+1}^0).$$

Third Order Weak Scheme

Composition scheme: (Splitting Method)

Rey (2017) obtained third order scheme using the ellipticity condition. Let us introduce the commutativity property:

$$V_1^2 V_0 - V_0 V_1^2 = 2\widetilde{V}^2.$$

The Lie bracket between V_1^2 and V_0 is equal to $2\tilde{V}^2$ with $\tilde{V}$ a first order differential operator, i.e.,

$$[V_1^2, V_0]f = V_1^2 V_0 f - V_0 V_1^2 f = 2\widetilde{V}^2 f$$

with

$$\widetilde{V}(x) = \sqrt{|V_1(x)(V_1(x)\partial_x V_0(x) - \partial_x V_1(x) V_0(x)|}, x \in \mathbb{R}.$$

Let $\epsilon_k, , k \in \mathbb{N}$ be independent uniform random variables with values in $\{-1, 1\}$.

Let $Z_k, k \in \mathbb{N}$ be independent random variables which are bounded below by the Lebesgue measure satisfying the following moment conditions

$$E(Z_k) = E(Z_k^3) = E(Z_k^5) = E(Z_k^7) = 0$$

$$E(Z_k^2) = 1, E(Z_k^4) = 3, E(Z_k^6) = 15$$

$$\forall p \geq 1, E(|Z_k|^p) < \infty.$$

Let

$$w_k^0 = T/n, \quad w_k^1 = \sqrt{T} Z_k / \sqrt{n}.$$

Let $\rho_k, k \in \mathbb{N}$ be independent uniform random variables with values in $\{1, 2, 3\}$. Define the function

$$\psi : \{-1, 1\} \times \{1, 2, 3\} \times \mathbb{R}^3 \to \mathbb{R}.$$

For $\rho_k = 1$,

$\psi(\epsilon_k, \rho_k, x, w^1_{k+1}, w^0_{k+1}) = \exp(\epsilon_k w^0_{k+1}\widetilde{V})) \text{ o } \exp(w^0_{k+1}V_0)) \text{ o } \exp(w^1_{k+1}V_1))(x)$

For $\rho_k = 2$,

$\psi(\epsilon_k, \rho_k, x, w^1_{k+1}, w^0_{k+1}) = \exp(w^0_{k+1}V_0)) \text{ o } \exp(\epsilon_k w^0_{k+1}\widetilde{V})) \text{ o } \exp(w^1_{k+1}V_1))(x)$

For $\rho_k = 3$,

$\psi(\epsilon_k, \rho_k, x, w^1_{k+1}, w^0_{k+1}) = \exp(w^0_{k+1}V_0)) \text{ o } \exp(w^1_{k+1}V_1)) \text{ o } \exp(\epsilon_k w^0_{k+1}\widetilde{V}))(x)$

$$\psi(\epsilon_k, \rho_k, x, 0, 0) = x.$$

One step of the scheme is

$$X_{t_{k+1}} = \psi(\epsilon_k, \rho_k, X_{t_k}, w^1_{k+1}, w^0_{k+1}).$$

Theorem 1.1 For every $f \in C^8_b(\mathbb{R})$

$$\sup_{t\in\Pi} |E(f(X_t) - f(X^n_t))| \leq \frac{C\|f\|_{8,\infty}}{n^3}.$$

$$\psi(\epsilon_k, 1, x, w^1_{k+1}, w^0_{k+1}) = \exp(\epsilon_k w^0_{k+1}\widetilde{V})) \text{ o } \exp(w^0_{k+1}V_0)) \text{ o } \exp(w^1_{k+1}V_1))(x)$$
$$\psi(\epsilon_k, 2, x, w^1_{k+1}, w^0_{k+1}) = \exp(w^0_{k+1}V_0)) \text{ o } \exp(\epsilon_k w^0_{k+1}\widetilde{V})) \text{ o } \exp(w^1_{k+1}V_1))(x)$$
$$\psi(\epsilon_k, 3, x, w^1_{k+1}, w^0_{k+1}) = \exp(w^0_{k+1}V_0)) \text{ o } \exp(w^1_{k+1}V_1)) \text{ o } \exp(\epsilon_k w^0_{k+1}\widetilde{V}))(x)$$
$$\psi(-1, 1, x, w^1_{k+1}, w^0_{k+1}) = \exp(-w^0_{k+1}\widetilde{V})) \text{ o } \exp(w^0_{k+1}V_0)) \text{ o } \exp(w^1_{k+1}V_1))(x)$$
$$\psi(-1, 2, x, w^1_{k+1}, w^0_{k+1}) = \exp(w^0_{k+1}V_0)) \text{ o } \exp(-w^0_{k+1}\widetilde{V})) \text{ o } \exp(w^1_{k+1}V_1))(x)$$
$$\psi(-1, 3, x, w^1_{k+1}, w^0_{k+1}) = \exp(w^0_{k+1}V_0)) \text{ o } \exp(w^1_{k+1}V_1)) \text{ o } \exp(-w^0_{k+1}\widetilde{V}))(x)$$

$$\psi(1, 1, x, w^1_{k+1}, w^0_{k+1}) = \exp(w^0_{k+1}\widetilde{V})) \text{ o } \exp(w^0_{k+1}V_0)) \text{ o } \exp(w^1_{k+1}V_1))(x)$$
$$\psi(1, 2, x, w^1_{k+1}, w^0_{k+1}) = \exp(w^0_{k+1}V_0)) \text{ o } \exp(w^0_{k+1}\widetilde{V})) \text{ o } \exp(w^1_{k+1}V_1))(x)$$
$$\psi(1, 3, x, w^1_{k+1}, w^0_{k+1}) = \exp(w^0_{k+1}V_0)) \text{ o } \exp(w^1_{k+1}V_1)) \text{ o } \exp(w^0_{k+1}\widetilde{V}))(x)$$

For simplicity take $T = 1$ without loss of generality. For $\epsilon = -1, 1$, denote

$$\tau_0 f(x) = f(\exp(\frac{1}{n}V_0)(x)), \ \tau_1 f(x) = f(\exp(\frac{Z}{\sqrt{n}}V_1)(x)),$$

$$\widetilde{\tau}_\epsilon f(x) = f(\exp(\frac{\epsilon}{n}\widetilde{V})(x))$$

We have

$$E[f(\psi_k(x, w_{k+1}^1, w_{k+1}^0))] = \frac{1}{6}\sum_{\epsilon=-1,1} E[(\tilde{\tau}_\epsilon\tau_0\tau_1 + \tau_0\tilde{\tau}_\epsilon\tau_1 + \tau_0\tau_1\tilde{\tau}_\epsilon)f(x)]$$

$$= \frac{1}{6}E[(\tilde{\tau}_{-1}\tau_0\tau_1 + \tau_0\tilde{\tau}_{-1}\tau_1 + \tau_0\tau_1\tilde{\tau}_{-1})f(x)] + \frac{1}{6}E[(\tilde{\tau}_1\tau_0\tau_1 + \tau_0\tilde{\tau}_1\tau_1 + \tau_0\tau_1\tilde{\tau}_1)f(x)]$$

$$= f(x)+\frac{1}{n}(V_0+\frac{1}{2}V_1^2)f(x)+\frac{1}{2n^2}(V_0+\frac{1}{2}V_1^2)^2f(x)+\frac{1}{6n^3}(V_0+\frac{1}{2}V_1^2)^3f(x)+\frac{1}{n^4}Rf(x)$$

$$= f(x) + \frac{1}{n}Af(x) + \frac{1}{2n^2}A^2f(x) + \frac{1}{6n^3}A^3f(x) + \frac{1}{n^4}Rf(x),$$

where $A = V_0 + \frac{1}{2}V_1^2$ is the infinitesimal generator of the Markov process with the notation $Vf(x) = V(x)\partial f(x)$.

Define $\exp(V)(x) := \Phi_V(x, 1)$ where Φ_V soles the differential equation

$$\Phi_V(x,t) = x + \int_0^t V(\Phi_V(x,s))ds.$$

The semigroup of the Markov process which is given by $P_t^Vf(x)=f(\Phi_V(x,t))$ infinitesimal operator $A_vf(x) = Vf(x)$. In particular, $P_t^VA_V = A_VP_t^V$ reads

$$Vf(\Phi_V(x,t)) = A_VP_t^Vf = P_t^VA_Vf = V(x)\partial_x(fo\Phi_V)(x,t).$$

Using m times Dynkin's formula

$$P_t^Vf(x) = f(x) + \int_0^t P_s^VA_Vf(x)ds$$

we obtain

$$f(\Phi_V(x,t) = f(x) + f(x) + \sum_{r=1}^m \frac{t^r}{r}V^r(f(x)) + \frac{1}{m}\int_0^t (t-s)^m V^{m+1}P_s^Vf(x)ds.$$

$$E[f(X_t(x))] = f(x) + tAf(x) + \frac{t^2}{2}A^2f(x) + \frac{t^3}{6}A^3f(x) + \frac{t^4}{24}R_t'f(x).$$

Take $t = n^{-1}$. The first four terms are same. Then the difference is of the order $O(n^{-4})$.

2.5.2 Estimate of the Density and Its Derivative

Let $p_t(x,y)$ be the transition density of X_t and let $p_t^n(x,y)$ be the transition density. Let

$$X_t^{n,\alpha}(x) := \frac{1}{n^\alpha}G + X_t^n(x),$$

where G is standard normal.

Let

$$Q_t^{n,\alpha} f(x) = E\left[f\left(\frac{1}{n^\alpha} G + X_t^n(x)\right)\right]$$

$$Q_t^{n,\alpha}(x, dy) = P(X_t^{n,\alpha}(x) \in dy) = p_t^{n,\alpha}(x, y)dy.$$

$p_t^{n,\alpha}(x, y)$ is the density of the law of $X_t^{n,\alpha}(x)$.

$$P_t(x, dy) = p_t(x, y)dy.$$

Theorem 2.2 Under the ellipticity condition:

$$V_1(x)^2 \geq \lambda_\star > 0 \; \forall \; x \in \mathbb{R},$$

we have

$$\sup_{\Pi} \sup_{x,y} |\partial_x \partial_y p_t(x, y) - \partial_x \partial_y p_t^n(x, y)| \leq \frac{c}{n^{3(1-\epsilon)}}.$$

2.5.3 Examples

Example 1 Consider the SDE

$$dX_t = \theta \; dt + \frac{\sigma}{\arctan(X_t) + \pi} \circ dW_t$$

with $\sigma > 0$ and $\theta \in \mathbb{R}$. The coefficients of the SDE belong to $C_b^\infty(\mathbb{R})$ and moreover $V_1 : x \to \sigma/(\arctan(x) + \pi)$ satisfies $\inf_x V_1(x) > 2\sigma/\pi$ and the function $V_0/V_1 = \theta/V_1$ is increasing. Thus the splitting scheme is well defined. The first order operator $\widetilde{V}$ is explicit, that is,

$$\widetilde{V}(x) = \frac{\sigma\sqrt{\theta}}{\sqrt{1 + x^2}(\arctan(x) + \pi)^{3/2}}.$$

Example 2 CIR Model
Consider the Cox–Ingersoll–Ross model

$$dX_t = (\alpha - \beta X_t)dt + \sigma\sqrt{X_t}dW_t, \quad X_0 = x > 0,$$

where α, β, σ are strictly positive.

Third order scheme is obtained easily.

2.5.4 Minimum Contrast Estimation

We study parameter estimation in a diffusion process based on a third weak order scheme.

The third order minimum contrast estimator is defines as

$$\hat{\theta}_{n,T} = \arg\max_{\theta}[X_{t_{k+1}} - \psi(\epsilon_k, \rho_k, X_{t_k}, w_{k+1}^1, w_{k+1}^0)]^2.$$

Define the continuous MLE as

$$\hat{\theta}_T = \arg\max_{\theta \in \Theta} L_T,$$

where L_T is the likelihood function based on continuous observation inside $[0, T]$.

Theorem 2.1

$$|\hat{\theta}_{n,T} - \hat{\theta}_T| = O_P\left(\left(\frac{n}{T}\right)^{-3/2}\right).$$

Theorem 2.2

$$(a) \quad \hat{\theta}_n \to^P \theta \quad as \quad T \to \infty, n \to \infty, T/n \to 0.$$

$$(b) \quad \sqrt{T}(\hat{\theta}_{n,T} - \theta) \to^D \mathcal{N}(0, I^{-1}(\theta)) \; as \; T \to \infty, n \to \infty, T/\sqrt{n} \to 0.$$

Chapter 3

Parameter Estimation in the Heston Model

3.1 Introduction

We study refined asymptotics of some new estimators of the Cox–Ingersoll–Ross model and the Heston model. In order to motivate the theory, first, we obtain the rate of weak convergence of the distribution of the normalized minimum contrast estimator of the drift parameter based on continuous observation.

Then, we obtain rate of normal approximation of the approximate minimum contrast estimators when the process is densely observed at discrete time points which are of practical interest in finance.

The Cox–Ingersoll-Ross (CIR) model, also called the square-root process, is used as a short rate mean reverting model in term structure of interest rates and a stochastic volatility process in the Heston model. In view of this, it becomes necessary to estimate the unknown parameters in the model from discrete data. See the monograph (Bishwal, 2008a,b) for recent results on approximate likelihood asymptotics and approximate Bayes asymptotics for drift estimation of discretely observed diffusions based on high-frequency data. In this work, we assume constant volatility and without loss of generality assume it to be one. To estimate the drift parameter, we adopt minimum contrast method and study the accuracy of distributional approximation of the resulting estimators from continuous data and high-frequency data.

3.2 Continuous Observation

Let $(\Omega, \mathcal{F}, \{\mathcal{F}_t\}_{t\geq 0}, P)$ be a stochastic basis on which the Cox–Ingersoll–Ross process $\{X_t\}$ is defined satisfying the Itô stochastic differential equation

$$dX_t = (\alpha + \beta X_t)\, dt + \; 2\sqrt{X_t}\, dW_t, \; t \geq 0, \;\; X_0 = 1, \qquad (2.1)$$

where $\{W_t\}$ is a standard Wiener process with the filtration $\{\mathcal{F}_t\}_{t\geq 0}$, $\alpha > 0$ and $\beta < 0$ are the unknown parameters to be estimated on the basis of observations of the process $\{X_t\}$.

J. P. N. Bishwal, *Parameter Estimation in Stochastic Volatility Models*,
https://doi.org/10.1007/978-3-031-03861-7_3

The true transition density, which is the fundamental solution to the PDE

$$u_t = 2xu_{xx} + \alpha u_x - \left(\frac{\mu}{x} + \lambda x\right) u \tag{2.2},$$

is given by

$$q(t,x,y,\alpha,\beta) := -2\beta\left(\frac{y}{x}\right)^{\alpha-\frac{1}{2}} \frac{e^{(\frac{1}{2}-\alpha)\beta t}}{1-e^{\beta t}} \exp\left[\frac{2\beta(x+y)}{e^{-\beta t}-1}\right] I_\nu\left(\frac{-2\beta\sqrt{xy}}{\sinh(-\frac{1}{2}\beta t)}\right), \tag{2.3}$$

where I_ν is the modified Bessel function of first kind with index ν which is noncentral chi-square density. The invariant density as $t \to \infty$ is gamma.

Let the continuous realization $\{X_t, 0 \le t \le T\}$ be denoted by X_0^T. Let P_β^T be the measure generated on the space (C_T, B_T) of continuous functions on $[0,T]$ with the associated Borel σ-algebra B_T generated under the supremum norm by the process X_0^T, and let P_0^T be the standard Wiener measure. It is well known that when β is the true value of the parameter, $P_{\beta,\alpha}^T$ is absolutely continuous with respect to P_0^T and the Radon–Nikodym derivative (likelihood) of P_β^T with respect to P_0^T based on X_0^T is given by

$$L_T(\beta,\alpha) := \frac{dP_{\beta,\alpha}^T}{dP_0^T}(X_0^T) = \exp\left\{\int_0^T \frac{\alpha+\beta X_t}{4X_t} dX_t - \int_0^T \frac{(\alpha+\beta X_t)^2}{8X_t} dt\right\}. \tag{2.4}$$

Consider the score function, the derivative of the log-likelihood function, which is given by

$$\gamma_T(\beta,\alpha) := \left\{\int_0^T \frac{\alpha+\beta X_t}{4X_t} dX_t - \int_0^T \frac{(\alpha+\beta X_t)^2}{8X_t} dt\right\}. \tag{2.5}$$

We estimate α and β. A solution of the estimating equation $\gamma_T(\beta,\alpha) = 0$ provides the maximum likelihood estimates (MLEs)

$$\hat\beta_T := \frac{X_0 - X_T + \alpha T}{\int_0^T X_t dt},$$

$$\hat\alpha_T := \frac{\int_0^T X_t^{-1} dX_t + \beta T}{\int_0^T X_t^{-1} dt} = \frac{\log X_T - \log X_0 + \int_0^T X_t^{-1} dt + \beta T}{\int_0^T X_t^{-1} dt}.$$

It is important to note that if $\beta > 0$ and $\alpha \ge 2$, the MLE $\hat\alpha_T$ is inconsistent. It remains open to find a consistent estimator in this case.

As far as we know, the rate of normal approximation for the minimum contrast estimator has not been studied earlier. Our aim in this work is to bridge this gap.

Consider the minimum contrast estimates (MCEs)

$$\widetilde{\beta}_T := \frac{\alpha T}{\int_0^T X_t dt} = \frac{\alpha}{\overline{\overline{X_T}}},$$

where

$$\overline{X_T} = \frac{1}{T}\int_0^T X_t dt$$

and

$$\widetilde{\alpha}_T := \frac{\beta T}{\int_0^T X_t^{-1} dt} = \frac{\beta}{\overline{X_T^{-1}}},$$

where

$$\overline{X_T^{-1}} = \frac{1}{T}\int_0^T X_t^{-1} dt.$$

Note that using the Skorohod embedding of martingale, which has been the one of the basic tools for normal approximation of martingales, will not give a rate better than $O(T^{-1/4})$ (see Borokov (1973)). To obtain the rate of normal approximation of the order $O(T^{-1/2})$, we adopt the Fourier method followed by the squeezing technique of Pfanzagl (1971).

Observe that

$$\left(\frac{T\alpha}{-4\beta}\right)^{1/2} (\widetilde{\beta}_T - \beta) = \frac{\left(\frac{-4\beta}{T\alpha}\right)^{1/2} N_T}{\left(\frac{-4\beta}{T\alpha}\right) I_T} \tag{2.6}$$

and

$$\left(\frac{T\beta}{-4(\alpha-2)}\right)^{1/2} (\widetilde{\alpha}_T - \alpha) = \frac{\left(\frac{-4(\alpha-2)}{T\beta}\right)^{1/2} M_T}{\left(\frac{-4(\alpha-2)}{T\beta}\right) J_T}, \tag{2.7}$$

where

$$N_T := \alpha T - \beta I_T, \quad M_T := \beta T - \alpha J_T, \quad I_T := \int_0^T X_t dt, \text{ and } J_T := \int_0^T X_t^{-1} dt.$$

The process I_T that is energy of the CIR process which plays an important role in clustering time or activity persistence in stochastic volatility modeling.

Based on continuous time observation $\{X_t, 0 \le t \le T\}$, the continuous conditional least squares estimators of β and α are, respectively, given by

$$\beta_T := \frac{\int_0^T X_s dX_s - (X_T - X_0)\overline{X_T}}{\int_0^T (X_t - \overline{X}_T)^2 dt}, \tag{2.8}$$

$$\alpha_T := -\overline{X_T}\beta_T + T^{-1}(X_T - X_0), \tag{2.9}$$

where

$$\overline{X_T} := \int_0^T X_t dt. \tag{2.10}$$

Note that by the Itô formula

$$X_T^2 - X_0^2 = 2\int_0^T X_s dX_s + \int_0^T X_s ds. \tag{2.11}$$

Hence

$$\beta_T = \frac{T\overline{X_T}}{\int_0^T (X_t - \overline{X_T})^2 dt} + o_P(T^{-1/2})$$

$$= \frac{T\overline{X_T}}{2(\overline{X_T^2} - (\overline{X_T})^2)} + o_P(T^{-1/2}) \tag{2.12}$$

$$\alpha_T = \frac{\overline{X_T^2}}{2(\overline{X_T^2} - (\overline{X_T})^2)} + o_P(T^{-1/2}), \tag{2.13}$$

where

$$\overline{X_T^2} := \int_0^T X_t^2 dt. \tag{2.14}$$

We define the minimum contrast estimators as

$$\breve{\beta}_T := \frac{T\overline{X_T}}{2(\overline{X_T^2} - (\overline{X_T})^2)} \tag{2.15}$$

$$\breve{\alpha}_T := \frac{\overline{X_T^2}}{2(\overline{X_T^2} - (\overline{X_T})^2)}. \tag{2.16}$$

Let $\Phi(\cdot)$ denote the standard normal distribution function. Throughout the work, C denotes a generic constant (which does not depend on T and x).

Laplace and Fourier Transforms and Cameron–Martin Theorems

The Case $\beta = 0$

Kummer's confluent hypergeometric function is given by

$${}_1F_1(r, s, z) = \frac{\Gamma(s)}{\Gamma(r)\Gamma(s-r)} z^{1-s} \int_0^z e^u u^{r-1}(z-u)^{s-r-1} du. \tag{2.17}$$

The Whittaker function of first kind is given by

$$M_{s,r}(z) = z^{r+\frac{1}{2}} e^{-z/2} \; {}_1F_1(r - s + \frac{1}{2}, 2r+1, z).$$

Let $J_T := \int_0^T X_t^{-1} dt$. Then

$$E\exp(-uJ_T) = \frac{\Gamma(k + \frac{\nu}{2} + \frac{1}{2})}{\Gamma(\nu+1)} (\frac{x}{\sigma t})^{\frac{\nu}{2}+\frac{1}{2}-k} \exp(\frac{x}{\sigma t}) \; {}_1F_1(k + \frac{\nu}{2} + \frac{1}{2}, \frac{x}{\sigma t}),$$

where $\nu = \frac{1}{\sigma}\sqrt{(\alpha-\sigma)^2+4u}$.

$$\begin{aligned}
&E\exp(-uI_T - vJ_T)\\
= \quad & \frac{\Gamma(k+\frac{\nu}{2}+\frac{1}{2})}{\Gamma(\nu+1)}\left(\frac{\sqrt{\sigma u}x\coth(\sqrt{\sigma u}t)}{\sigma}\right)^{-k}\exp\left(\frac{\sqrt{\sigma u}x}{\sigma}\coth(\sqrt{\sigma u}t)\right)\\
&\times\exp\left(\frac{\sqrt{\sigma u}x}{2\sigma\sinh(\sqrt{\sigma u}t)\cosh\sqrt{\sigma u}t}\right)M_{-k,\frac{\nu}{2}}\left(\frac{\sqrt{\sigma u}x}{\sigma\sinh(\sqrt{\sigma u}t)\cosh\sqrt{\sigma u}t}\right)
\end{aligned}$$

$$\begin{aligned}
&E\exp\left(-uX_T - vI_T\right)\\
&= \left(\frac{2\sigma u}{\rho}\sinh(\frac{\rho t}{2})+\cosh(\frac{\rho t}{2})\right)^{-a/\sigma}\frac{\frac{2v}{\rho}\sinh(\frac{\rho t}{2})+u\cosh(\frac{\rho t}{2})}{\frac{2\sigma u}{\rho}\sinh(\frac{\rho t}{2})+\cosh(\frac{\rho t}{2})},
\end{aligned}$$

where $\rho := 2\sqrt{\sigma u}$.

The Case $\beta \neq 0$

The following characteristic function of I_T is closely associated with *Levy's stochastic area formula* and is well known from Brownian motion literature and also from the work of Cox et al. (1985).

Let $\phi_T(u) := E\exp(iuI_T), u \in \mathbb{R}$. Then

$$E\exp(iuI_T) = \exp\left(\frac{2iu}{\beta+\gamma\coth\frac{\gamma T}{2}}\right)\left[\cosh\frac{\gamma T}{2}+\frac{\beta}{\gamma}\sinh\frac{\gamma T}{2}\right]^{-1},$$

where $\gamma := (\beta^2-2iu)^{1/2}$.
Further

$$\lim_{T\to\infty} E\left(e^{-uI_T}\right) = \lim_{T\to\infty}\exp\left(-\frac{ab}{2\sigma}T\left(\sqrt{1+\frac{4u\sigma}{b^2}}-1\right)\right).$$

$$\begin{aligned}
&E\exp(-uX_T - vI_T)\\
= \quad & \left(\frac{2\rho e^{(b-\rho)T/2}}{2\sigma u(1-e^{-\rho T})+(\rho-b)e^{-\rho T}+(\rho+b)}\right)^{a/\sigma}\\
&\times\exp\left(\frac{u((\rho+b)e^{-\rho T}+(\rho-b)+2v(1-e^{-\rho T})}{2\sigma u(1-e^{-\rho T})+(\rho-b)e^{-\rho T}+(\rho+b)}\right),
\end{aligned}$$

where $\rho =: \sqrt{b^2 + 4\sigma v}$.

$$\begin{aligned} & E\exp(-uJ_T) \\ = & \frac{\Gamma(k + \frac{\nu}{2} + \frac{1}{2})}{\Gamma(\nu+1)}(\frac{x}{\alpha})^{-k}\beta^{\frac{\nu}{2}+\frac{1}{2}}\exp(\frac{b}{2\sigma}[at - \frac{2x}{e^{bt}-1}]) \\ & {}_1F_1(k + \frac{\nu}{2} + \frac{1}{2}, \nu + 1, \beta), \end{aligned}$$

where

$$k =: \frac{a}{2\sigma}, \quad \alpha =: \frac{be^{bt}}{\sigma(e^{bt}-1)}, \quad \nu =: \frac{1}{\sigma}\sqrt{(\alpha-\sigma)^2 + 4u\sigma},$$

and ${}_1F_1$ is Kummer's confluent hypergeometric function.

Option Pricing

For the CIR model

$$dX_t = a(b - X_t)dt + \sigma\sqrt{X_t}dW_t,$$

the price at time t of a zero coupon bond that pays \$1 at time T is given by

$$P(t,T) = E_Q\left(e^{-\int_t^T X_t dt}\right) = A(t,T)e^{-B(t,T)X_t},$$

where

$$B(t,T) = \frac{2(e^{\gamma(T-t)-1})}{(\gamma + a)(e^{\gamma(T-t)-1}) + 2\gamma},$$

$$A(t,T) = \left(\frac{2\gamma e^{(a+\gamma)(T-t)/2}}{(\gamma + a)(e^{\gamma(T-t)-1}) + 2\gamma}\right)^{2ab/\sigma^2},$$

where $\gamma = \sqrt{a^2 + 2\sigma^2}$ and Q is the risk-neutral measure.

The European call option is given by

$$\begin{aligned} C_t = {} & P(t,s)\chi^2\left(2\frac{\log(A(t,s)/K)}{B(T,s)}[\phi + \psi + B(T,s)]; \frac{4ab}{\sigma^2}, \frac{2\phi^2 X_t e^{\gamma(T-t)}}{\phi + \psi + B(T,s)}\right) \\ & -KP(t,T)\chi^2\left(2\frac{\log(A(t,s)/K)}{B(T,s)}[\phi + \psi]; \frac{4ab}{\sigma^2}, \frac{2\phi^2 X_t e^{\gamma(T-t)}}{\phi + \psi}\right), \end{aligned}$$

where

$$\phi =: \frac{2\gamma}{\sigma^2(e^{\gamma(T-t)-1})}, \quad \psi =: \frac{a+\gamma}{\sigma^2}$$

and $\chi^2(x; d, \lambda)$ is the noncentral chi-square distribution with d degrees of freedom and noncentrality parameter λ.

For the Vasicek model for short rate

$$dr_t = a(b - r_t)dt + \sigma dW_t,$$

the price of a zero coupon bond at time t maturing at time T is given by

$$P(t,T) = A(t,T)e^{-B(t,T)r(t)},$$

where

$$B(t,T) = \frac{1 - e^{-a(T-t)}}{a}$$

and

$$A(t,T) = \exp\left(\frac{(B(t,T) - T + t)(a^2 b - \sigma^2/2)}{a^2} - \frac{\sigma^2 B(t,T)^2}{4a}\right).$$

The interest rate derivative, European call option, is given by

$$C = LP(0,s)\Phi(h) - KP(0,T)\Phi(h - \sigma_p),$$

where L is the bond principal, s is the bond maturity, T is the option maturity, K is the strike price,

$$h = \frac{1}{\sigma_P} \ln \frac{LP(0,s)}{P(0,T)K} + \frac{\sigma_P}{2}$$

$$\sigma_P =: \frac{\sigma}{a}(1 - e^{-a(s-T)})\sqrt{\frac{1 - e^{-2aT}}{2a}}.$$

When $a = 0$, $\sigma_P = \sigma(s - T)\sqrt{T}$.

Rate and Esseen's Lemma

The first lemma gives the rate in an ergodic theorem and we omit the proof.

Lemma 2.1

(a) For every $\delta > 0$,

$$P\left\{\left|\left(\frac{-4\beta}{T\alpha}\right) I_T - 1\right| \geq \delta\right\} \leq CT^{-1}\delta^{-2}.$$

(b) For every $\delta > 0$,

$$P\left\{\left|\left(\frac{-4(\alpha - 2)}{T\beta}\right) J_T - 1\right| \geq \delta\right\} \leq CT^{-1}\delta^{-2}.$$

We omit the details of the proof of the next lemma which uses the bounds on the characteristic functions given later in Lemmas 2.3 and 2.7 along with Esseen's smoothing lemma.

Lemma 2.2

$$(a)\quad \sup_{x\in\mathbb{R}}\left|P\left\{\left(\frac{-4\beta}{T\alpha}\right)^{1/2} N_T \le x\right\} - \Phi(x)\right| \le C\,T^{-1/2}.$$

$$(b)\quad \sup_{x\in\mathbb{R}}\left|P\left\{\left(\frac{-4(\alpha-2)}{T\beta}\right)^{1/2} M_T \le x\right\} - \Phi(x)\right| \le C\,T^{-1/2}.$$

Lemma 2.3 Let

$$H_{T,x} := \left(\frac{-4\beta}{T\alpha}\right)^{1/2} N_T - \left(\frac{-4\beta}{T\alpha} I_T - 1\right) x.$$

Then, for $|x| \le 2(\log T)^{1/2}$ and for $|u| \le \epsilon T^{1/2}$, where ϵ is sufficiently small,

$$\left|E\exp(iuH_{T,x}) - \exp(\frac{-u^2}{2})\right| \le C\exp(\frac{-u^2}{4})(|u| + |u|^3)T^{-1/2}.$$

Proof Observe that

$$E\exp(iuI_T) = \exp\left(\frac{2iu(e^{\frac{\gamma T}{2}} - e^{\frac{-\gamma T}{2}})}{e^{\frac{\gamma T}{2}}(\gamma+\beta) + e^{-\frac{\gamma T}{2}}(\gamma-\beta)}\right)$$
$$\times 2\left[e^{\frac{\gamma T}{2}}(\gamma+\beta) + e^{-\frac{\gamma T}{2}}(\gamma-\beta)\right]^{-1}. \tag{2.18}$$

Now consider

$$\begin{aligned}
& E\exp(iuH_{T,x}) \\
= \;& E\exp\left[-iu\left(\tfrac{-4\beta}{T\alpha}\right)^{1/2} N_T - iu\left(\tfrac{-4\beta}{T\alpha}I_T - 1\right)x\right] \\
= \;& E\exp\left[-iu\left(\tfrac{-4\beta}{T\alpha}\right)^{1/2}\{\beta I_T - \alpha T\} - iu\left(\tfrac{-4\beta}{T\alpha}I_T - 1\right)x\right] \\
= \;& E\exp(z_1 I_T + z_3) \\
= \;& \exp(z_3)\phi_T(z_1),
\end{aligned} \tag{2.19}$$

where $z_1 := -iu\beta\delta_{T,x}$ and $z_3 := \frac{iuT}{2}\delta_{T,x}$ with $\delta_{T,x} := \left(\frac{-4\beta}{T\alpha}\right)^{1/2} + \frac{2x}{T}$. Note that $\phi_T(z_1)$ satisfies the conditions of (a) by choosing ϵ sufficiently small. Let $\omega_{1,T}(u), \omega_{2,T}(u)$, $\omega_{3,T}(u)$ and $\omega_{4,T}(u)$ be functions which are $O(|u|T^{-1/2}), O(|u|^2T^{-1/2})$, $O(|u|^3T^{-3/2})$, and $O(|u|^3T^{-1/2})$, respectively. Note that for the given range of values of x and u, the conditions on z_1 for the Lemma are satisfied. Further, with

$$\varpi_T(u) := 1 + iu\frac{\delta_{T,x}}{\beta} + \frac{u^2\delta_{T,x}^2}{2\beta^2},$$

$$
\begin{aligned}
\gamma &= (\beta^2 - 2z_1)^{1/2} \\
&= \beta\left[1 - \frac{z_1}{\beta^2} - \frac{z_1^2}{2\beta^4} + \frac{z_1^3}{2\beta^8} + \cdots\right] \\
&= \beta\left[1 + iu\frac{\delta_{T,x}}{\beta} + \frac{u^2\delta_{T,x}^2}{2\beta^2} + \frac{iu^3\delta_{T,x}^3}{2\beta^3} + \cdots\right] \\
&= \beta[1 + \omega_{1,T}(u) + \omega_{2,T}(u) + \omega_{3,T}(u)] \\
&= \beta\varpi_T(u) + \omega_{3,T}(u) \\
&= \beta[1 + \omega_{1,T}(u)].
\end{aligned} \tag{2.20}
$$

Thus

$$
\gamma - \beta = \omega_{1,T}, \ \gamma + \beta = 2\beta + \omega_{1,T}. \tag{2.21}
$$

Hence the above expectation equals

$$
\begin{aligned}
&\exp\left(z_3 + \tfrac{\beta T}{2}\right) \\
&\times \left[\frac{2\beta\varpi_T(u)+\omega_{3,T}(u)}{\omega_{1,T}\exp\{-\beta T\varpi_T(u)+\omega_{4,T}(u)\}+(2\beta+\omega_{1,T}(u))\exp\{\beta T\varpi_T(u)+\omega_{4,T}(u)\}}\right]^{1/2} \\
&= \left[\frac{1+\omega_{1,T}(u)}{\omega_{1,T}\exp(\chi_T(u)) + (1+\omega_{1,T}(u))\exp(\psi_T(t))}\right]^{1/2},
\end{aligned} \tag{2.22}
$$

where

$$
\begin{aligned}
\chi_T(u) &:= -\beta T\beta_T(u) + \alpha_{4,T}(u) - 2z_3 - \beta T \\
&= -2\beta T + \omega_{1,T}(u) + t^2\omega_{1,T}(u)
\end{aligned} \tag{2.23}
$$

and

$$
\begin{aligned}
\psi_T(u) &:= \beta T\varpi_T(u) + \omega_{4,T}(u) - 2z_3 - \beta T \\
&= \beta T\left[1 + iu\frac{\delta_{T,x}}{\beta} + \frac{u^2\delta_{T,x}^2}{2\beta^2}\right] + \alpha_{4,T}(u) - iuT\delta_{T,x} - \beta T \\
&= \frac{u^2 T\alpha}{-4\beta}\left[\left(\frac{-4\beta}{T\alpha}\right)^{1/2} + \frac{2x}{T}\right]^2 \\
&= u^2 + u^2\omega_{1,T}(u).
\end{aligned} \tag{2.24}
$$

Hence, for the given range of values of u, $\chi_T(u) - \psi_T(u) \leq -\beta T$.

Hence the above expectation equals

$$
\begin{aligned}
&\exp(-\tfrac{u^2}{2})(1+\omega_{1,T})^{1/2} \times \\
&\left[\omega_{1,T}\exp\{-2\beta T + \omega_{1,T} + u^2\omega_{1,T}\} + (1+\omega_{1,T}(u))\exp\{u^2\omega_{1,T}(u)\}\right]^{-1/2} \\
&= \exp(-\tfrac{u^2}{2})\left[1+\omega_{1,T})(1+\omega_{1,T}(1+\omega_{1,T})\exp\{-\beta T + \omega_{1,T} + t^2\omega_{1,T}\}\right] \\
&\times \exp(u^2\omega_{1,T}(u)).
\end{aligned} \tag{2.25}
$$

This completes the proof of the lemma. □

To obtain the rate of normal approximation for the MCE, we need the following estimate on the tail behavior of the MCE.

Lemma 2.4

$$P\left\{(\frac{T\alpha}{-4\beta})^{1/2}|\widetilde{\beta}_T - \beta| \geq 2(\log T)^{1/2}\right\} \leq CT^{-1/2}.$$

Proof Observe that

$$\begin{aligned}
& P\left\{(\frac{T\alpha}{-4\beta})^{1/2}|\widetilde{\beta}_T - \beta| \geq 2(\log T)^{1/2}\right\} \\
= \; & P\left\{\left|\frac{(\frac{-4\beta}{T\alpha})^{1/2}N_T}{(\frac{-4\beta}{T\alpha})I_T}\right| \geq 2(\log T)^{1/2}\right\} \\
\leq \; & P\left\{\left|(\frac{-4\beta}{T\alpha})^{1/2}N_T\right| \geq (\log T)^{1/2}\right\} + P\left\{\left|\frac{-4\beta}{T\alpha}I_T\right| \leq \frac{1}{2}\right\} \\
\leq \; & \left|P\left\{(\frac{-4\beta}{T\alpha})^{1/2}|N_T| \geq (\log T)^{1/2}\right\} - 2\Phi(-(\log T)^{1/2})\right| \\
& +2\Phi(-(\log T)^{1/2}) + P\left\{\left|\frac{-4\beta}{T\alpha}I_T - 1\right| \geq \frac{1}{2}\right\} \\
\leq \; & \sup_{x\in\mathbb{R}}\left|P\left\{(\frac{-4\beta}{T\alpha})^{1/2}|N_T| \geq x\right\} - 2\Phi(-x)\right| \\
\leq \; & \sup_{x\in\mathbb{R}}\left|P\left\{(\frac{-4\beta}{T\alpha})^{1/2}|N_T| \geq x\right\} - 2\Phi(-x)\right| \\
& +2\Phi(-(\log T)^{1/2}) + P\left\{\left|(\frac{-4\beta}{T\alpha})I_T - 1\right| \geq \frac{1}{2}\right\} \\
\leq \; & CT^{-1/2} + C(T\log T)^{-1/2} + CT^{-1} \\
\leq \; & CT^{-1/2}.
\end{aligned}$$

The bounds for the first and the third terms come from Lemmas 2.2 and 2.1, respectively, and that for the middle term comes from Feller (1957, p. 166). □

Lemma 2.5

$$P\left\{\left(\frac{T\beta}{4(\alpha-2)}\right)^{1/2}|\widetilde{\alpha}_T - \alpha| \geq 2(\log T)^{1/2}\right\} \leq CT^{-1/2}.$$

Proof Observe that

$$P\left\{(\frac{T\beta}{4(\alpha-2)})^{1/2}|\widetilde{\alpha}_T - \alpha| \geq 2(\log T)^{1/2}\right\}$$

$$
\begin{aligned}
&= P\left\{\left|\frac{(\frac{2\alpha}{T})^{1/2}N_T}{(\frac{4(\alpha-2)}{T\beta})I_T}\right| \geq 2(\log T)^{1/2}\right\} \\
&\leq P\left\{\left|(\frac{4(\alpha-2)}{T\beta})^{1/2}N_T\right| \geq (\log T)^{1/2}\right\} + P\left\{\left|\frac{4(\alpha-2)}{T\beta}I_T\right| \leq \frac{1}{2}\right\} \\
&\leq \left|P\left\{(\frac{4(\alpha-2)}{T\beta})^{1/2}|N_T| \geq (\log T)^{1/2}\right\} - 2\Phi(-(\log T)^{1/2})\right| \\
&\quad +2\Phi(-(\log T)^{1/2}) + P\left\{\left|\frac{4(\alpha-2)}{T\beta}I_T - 1\right| \geq \frac{1}{2}\right\} \\
&\leq \sup_{x\in\mathbb{R}}\left|P\left\{(\frac{4(\alpha-2)}{T\beta})^{1/2}|N_T| \geq x\right\} - 2\Phi(-x)\right| \\
&\leq \sup_{x\in\mathbb{R}}\left|P\left\{(\frac{4(\alpha-2)}{T\beta})^{1/2}|N_T| \geq x\right\} - 2\Phi(-x)\right| \\
&\quad +2\Phi(-(\log T)^{1/2}) + P\left\{\left|(\frac{4(\alpha-2)}{T\beta})I_T - 1\right| \geq \frac{1}{2}\right\} \\
&\leq CT^{-1/2} + C(T\log T)^{-1/2} + CT^{-1} \\
&\leq CT^{-1/2}.
\end{aligned}
$$

The bounds for the first and the third terms come from Lemmas 2.2 and 2.1, respectively, and that for the middle term comes from Feller (1957, p. 166). □

Now we are ready to obtain the uniform rate of normal approximation of the distribution of the MCEs.

Theorem 2.6

$$\sup_{x\in\mathbb{R}}\left|P\left\{\left(\frac{T\alpha}{-4\beta}\right)^{1/2}(\widetilde{\beta}_T - \beta) \leq x\right\} - \Phi(x)\right| \leq C_\beta T^{-1/2}.$$

Proof We shall consider two possibilities (i) and (ii).
(i) $|x| > 2(\log T)^{1/2}$
We shall give a proof for the case $x > 2(\log T)^{1/2}$. The proof for the case $x < -2(\log T)^{1/2}$ runs similarly. Note that

$$\left|P\left\{(\frac{T\alpha}{-4\beta})^{1/2}(\widetilde{\beta}_T - \beta) \leq x\right\} - \Phi(x)\right| \leq P\left\{(\frac{T\alpha}{-4\beta})^{1/2}(\widetilde{\beta}_T - \beta) \geq x\right\} + \Phi(-x).$$

But $\Phi(-x) \leq \Phi(-2(\log T)^{1/2}) \leq CT^{-2}$. See Feller (1957, p. 166). Moreover, by Lemma 2.4, we have

$$P\left\{(\frac{T\alpha}{-4\beta})^{1/2}(\widetilde{\beta}_T - \beta) \geq 2(\log T)^{1/2}\right\} \leq CT^{-1/2}.$$

Hence

$$\left|P\left\{(\frac{T\alpha}{-4\beta})^{1/2}(\widetilde{\beta}_T - \beta) \leq x\right\} - \Phi(x)\right| \leq CT^{-1/2}.$$

(ii) $|x| \leq 2(\log T)^{1/2}$

Let $A_T := \left\{(\frac{T\alpha}{-4\beta})^{1/2}|\widetilde{\beta}_T - \beta| \leq 2(\log T)^{1/2}\right\}$ and $B_T := \left\{\frac{I_T}{T} > c_0\right\}$,

where $0 < c_0 < \frac{1}{-4\beta}$. By Lemma 2.4, we have

$$P(A_T^c) \leq CT^{-1/2}. \tag{2.26}$$

By Lemma 2.1, we have

$$\begin{aligned} P(B_T^c) &= P\left\{\tfrac{-4\beta}{T\alpha} I_T - 1 < 2\beta c_0 - 1\right\} \\ &< P\left\{|\tfrac{-4\beta}{T\alpha} I_T - 1| > 1 - 2\beta c_0\right\} \leq CT^{-1}. \end{aligned} \tag{2.27}$$

Let b_0 be some positive number. On the set $A_T \cap B_T$ for all $T > T_0$ with $4b_0(\log T_0)^{1/2}(\frac{-4\beta}{T_0\alpha})^{1/2} \leq c_0$, we have

$$\begin{aligned} & (\tfrac{T\alpha}{-4\beta})^{1/2}(\widetilde{\beta}_T - \beta) \leq x \\ \Rightarrow\ & I_T + b_0 T(\widetilde{\beta}_T - \beta) < I_T + (\tfrac{T\alpha}{-4\beta})^{1/2} 2b_0\beta x \\ \Rightarrow\ & (\tfrac{T\alpha}{-4\beta})^{1/2}(\widetilde{\beta}_T - \beta)[I_T + b_0 T(\widetilde{\beta}_T - \beta)] < x[I_T + (\tfrac{T\alpha}{-4\beta})^{1/2} 2b_0\beta x] \\ \Rightarrow\ & (\widetilde{\beta}_T - \beta)I_T + b_0 T(\widetilde{\beta}_T - \beta)^2 < (\tfrac{-4\beta}{T\alpha})^{1/2} I_T x + 2b_0\beta x^2 \\ \Rightarrow\ & -N_T + (\widetilde{\beta}_T - \beta)I_T + b_0 T(\widetilde{\beta}_T - \beta)^2 < -N_T + (\tfrac{-4\beta}{T\alpha})^{1/2} I_T x + 2b_0\beta x^2 \\ \Rightarrow\ & 0 < -N_T + (\tfrac{-4\beta}{T\alpha})^{1/2} I_T x + 2b_0\beta x^2 \end{aligned} \tag{2.28}$$

since

$$\begin{aligned} & I_T + b_0 T(\widetilde{\beta}_T - \beta) \\ > \ & Tc_0 + b_0 T(\widetilde{\beta}_T - \beta) \\ > \ & 4b_0(\log T)^{1/2}(\tfrac{-4\beta}{T\alpha})^{1/2} - 2b_0(\log T)^{1/2}(\tfrac{-4\beta}{T\alpha})^{1/2} \\ = \ & 2b_0(\log T)^{1/2}(\tfrac{-4\beta}{T\alpha})^{1/2} > 0. \end{aligned} \tag{2.29}$$

On the other hand, on the set $A_T \cap B_T$ for all $T > T_0$ with $4b_0(\log T_0)^{1/2}(\frac{-4\beta}{T_0\alpha})^{1/2} \leq c_0$, we have

$$\begin{aligned} & (\tfrac{T\alpha}{-4\beta})^{1/2}(\widetilde{\beta}_T - \beta) > x \\ \Rightarrow\ & I_T - b_0 T(\widetilde{\beta}_T - \beta) < I_T - (\tfrac{T\alpha}{-4\beta})^{1/2} 2b_0\beta x \\ \Rightarrow\ & (\tfrac{T\alpha}{-4\beta})^{1/2}(\widetilde{\beta}_T - \beta)[I_T - b_0 T(\widetilde{\beta}_T - \beta)] > x[I_T - (\tfrac{T\alpha}{-4\beta})^{1/2} 2b_0\beta x] \\ \Rightarrow\ & (\widetilde{\beta}_T - \beta)I_T - b_0 T(\widetilde{\beta}_T - \beta)^2 > (\tfrac{-4\beta}{T\alpha})^{1/2} I_T x - 2b_0\beta x^2 \\ \Rightarrow\ & -N_T + (\widetilde{\beta}_T - \beta)I_T - b_0 T(\beta_T - \beta)^2 > -N_T + (\tfrac{-4\beta}{T\alpha})^{1/2} I_T x - 2b_0\beta x^2 \\ \Rightarrow\ & 0 > -N_T + (\tfrac{-4\beta}{T\alpha})^{1/2} I_T x - 2b_0\beta x^2 \end{aligned} \tag{2.30}$$

since

$$\begin{aligned} & I_T - b_0 T(\widetilde{\beta}_T - \beta) \\ > \ & Tc_0 - b_0 T(\widetilde{\beta}_T - \beta) \\ > \ & 4b_0(\log T)^{1/2}(\tfrac{-4\beta}{T\alpha})^{1/2} - 2b_0(\log T)^{1/2}(\tfrac{-4\beta}{T\alpha})^{1/2} \\ = \ & 2b_0(\log T)^{1/2}(\tfrac{-4\beta}{T\alpha})^{1/2} > 0. \end{aligned} \tag{2.31}$$

Hence

$$0 < -N_T + (\frac{-4\beta}{T\alpha})^{1/2} I_T x - 2b_0\beta x^2 \Rightarrow (\frac{T\alpha}{-4\beta})^{1/2}(\beta_T - \beta) \le x.$$

Letting $D^{\pm}_{T,x} := \left\{-N_T + (\frac{-4\beta}{T})^{1/2} I_T x \pm 2b_0\beta x^2 > 0\right\}$, we obtain

$$D^{-}_{T,x} \cap A_T \cap B_T \subseteq A_T \cap B_T \cap \{(\frac{T\alpha}{-4\beta})^{1/2}(\beta_T - \beta) \le x\} \subseteq D^{+}_{T,x} \cap A_T \cap B_T. \quad (2.32)$$

This gives

$$P(D^{-}_{T,x} \cap A_T \cap B_T) \le P(A_T \cap B_T \cap \{(\frac{T\alpha}{-4\beta})^{1/2}(\beta_T - \beta) \le x\}) \le P(D^{+}_{T,x} \cap A_T \cap B_T)$$

so that

$$\left| P\left(A_T \cap B_T \cap \left\{\left(\frac{T\alpha}{-4\beta}\right)^{1/2}(\beta_T - \beta) \le x\right\}\right) - \Phi(x)\right|$$

$$\le \max\left\{|P(D^{-}_{T,x} \cap A_T \cap B_T) - \Phi(x)|,\ |P(D^{+}_{T,x} \cap A_T \cap B_T) - \Phi(x)|\right\}$$

$$\le \max\left\{|P(D^{-}_{T,x}) - \Phi(x)|,\ |P(D^{+}_{T,x}) - \Phi(x)|\right\} + P(A_T \cap B_T)^c.$$

If it is shown that

$$\left|P\left\{D^{\pm}_{T,x}\right\} - \Phi(x)\right| \le CT^{-1/2} \quad (2.33)$$

for all $T > T_0$ and $|x| \le 2(\log T)^{1/2}$, then the theorem would follow from (2.31)–(2.33). We shall prove (2.33) for $D^{+}_{T,x}$. The proof for $D^{-}_{T,x}$ is analogous. Observe that

$$\begin{aligned}
& \left|P\left\{D^{+}_{T,x}\right\} - \Phi(x)\right| \\
= & \left|P\left\{(\frac{-4\beta}{T\alpha})^{1/2} N_T - \left(\frac{-4\beta}{T} I_T - 1\right) x < x + 2(\frac{-4\beta}{T})^{1/2} b_0\beta x^2\right\} - \Phi(x)\right| \\
\le & \sup_{y\in\mathbb{R}} \left|P\left\{(\frac{-4\beta}{T\alpha})^{1/2} N_T - \left(\frac{-4\beta}{T\alpha} I_T - 1\right) x \le y\right\} - \Phi(y)\right| \\
& + \left|\Phi\left(x + (\tfrac{-4\beta}{T\alpha})^{1/2} b_0\beta x^2\right) - \Phi(x)\right| \\
=: & \Delta_1 + \Delta_2.
\end{aligned} \quad (2.34)$$

Lemma 2.3 and Esseen's smoothing lemma (see Feller (1957)) immediately yield

$$\Delta_1 \le CT^{-1/2}. \quad (2.35)$$

On the other hand, for all $T > T_0$,

$$\Delta_2 \le 2(\frac{-4\beta}{T\alpha})^{1/2} b_0\beta x^2 (2\pi)^{-1/2} \exp(-\overline{x}^2/2),$$

where

$$|\overline{x} - x| \leq 2(\frac{-4\beta}{T\alpha})^{1/2} b_0 \beta x^2.$$

Since $|x| \leq 2(\log T)^{1/2}$, it follows that $|\overline{x}| > |x|/2$ for all $T > T_0$ and consequently

$$\begin{aligned} \Delta_2 &\leq 2(\frac{-4\beta}{T\alpha})^{1/2} b_0 \beta x^2 (2\pi)^{-1/2} x^2 \exp(-x^2/8) \\ &\leq CT^{-1/2}. \end{aligned} \tag{2.36}$$

From (2.34)–(2.36), we obtain

$$\left| P\left\{ D_{T,x}^{+} \right\} - \Phi(x) \right| \leq CT^{-1/2}. \tag{2.37}$$

This completes the proof of the theorem. □

Lemma 2.7 Let

$$G_{T,x} := \left(\frac{-4(\alpha-2)}{T\beta} \right)^{1/2} M_T - \left(\frac{-4(\alpha-2)}{T\beta} J_T - 1 \right) x.$$

Then, for $|x| \leq 2(\log T)^{1/2}$ and for $|u| \leq \epsilon T^{1/2}$, where ϵ is sufficiently small,

$$\left| E \exp(iuG_{T,x}) - \exp(\frac{-u^2}{2}) \right| \leq C \exp(\frac{-u^2}{4})(|u| + |u|^3) T^{-1/2}.$$

Proof Observe that

$$E \exp(-uJ_T) = \frac{\Gamma(k + \frac{\nu}{2} + \frac{1}{2})}{\Gamma(\nu+1)} (\frac{x}{\sigma t})^{\frac{\nu}{2} + \frac{1}{2} - k} \exp(\frac{x}{\sigma t}) \;\; {}_1F_1(k + \frac{\nu}{2} + \frac{1}{2}, \frac{x}{\sigma t}), \tag{2.38}$$

where $\nu = \frac{1}{\sigma}\sqrt{(\alpha-\sigma)^2 + 4u}$.
Now consider

$$\begin{aligned} & E \exp(iuG_{T,x}) \\ = \;& E \exp\left[-iu \left(\tfrac{-4(\alpha-2)}{T\beta} \right)^{1/2} M_T - iu \left(\tfrac{-4(\alpha-2)}{T\beta} J_T - 1 \right) x \right] \\ = \;& E \exp\left[-iu \left(\tfrac{-4\beta}{T\alpha} \right)^{1/2} \{\alpha J_T - \beta T\} - iu \left(\tfrac{-4(\alpha-2)}{T\beta} J_T - 1 \right) x \right] \\ = \;& E \exp(z_1 J_T + z_3) \\ = \;& \exp(z_3) \phi_T(z_1), \end{aligned} \tag{2.39}$$

where $z_1 := -iu\beta\delta_{T,x}$ and $z_3 := \frac{iuT}{2}\delta_{T,x}$ with $\delta_{T,x} := \left(\frac{-4\beta}{T\alpha} \right)^{1/2} + \frac{2x}{T}$.
Note that $\phi_T(z_1)$ satisfies the conditions of (a) by choosing ϵ sufficiently small. Let $\omega_{1,T}(u), \omega_{2,T}(u)$, $\omega_{3,T}(u)$ and $\omega_{4,T}(u)$ be functions which are $O(|u|T^{-1/2}), O(|u|^2T^{-1/2})$, $O(|u|^3T^{-3/2})$, and $O(|u|^3T^{-1/2})$, respectively. Note that for the given range of values of x and u, the conditions on z_1

for part (a) of Lemma are satisfied. Further, with

$$\varpi_T(u) := 1 + iu\frac{\delta_{T,x}}{\beta} + \frac{u^2\delta_{T,x}^2}{2\beta^2},$$

$$\begin{aligned}\gamma &= (\beta^2 - 2z_1)^{1/2}\\ &= \beta\left[1 - \tfrac{z_1}{\beta^2} - \tfrac{z_1^2}{2\beta^4} + \tfrac{z_1^3}{2\beta^8} + \cdots\right]\\ &= \beta\left[1 + iu\tfrac{\delta_{T,x}}{\beta} + \tfrac{u^2\delta_{T,x}^2}{2\beta^2} + \tfrac{iu^3\delta_{T,x}^3}{2\beta^3} + \cdots\right]\\ &= \beta[1 + \omega_{1,T}(u) + \omega_{2,T}(u) + \omega_{3,T}(u)]\\ &= \beta\varpi_T(u) + \omega_{3,T}(u)\\ &= \beta[1 + \omega_{1,T}(u)].\end{aligned} \tag{2.40}$$

Thus

$$\gamma - \beta = \omega_{1,T},\ \gamma + \beta = 2\beta + \omega_{1,T}. \tag{2.41}$$

Hence the above expectation equals

$$\begin{aligned}&\exp\left(z_3 + \tfrac{\beta T}{2}\right)\\ &\times\left[\frac{2\beta\varpi_T(u)+\omega_{3,T}(u)}{\omega_{1,T}\exp\{-\beta T\varpi_T(u)+\omega_{4,T}(u)\}+(2\beta+\omega_{1,T}(u))\exp\{\beta T\varpi_T(u)+\omega_{4,T}(u)\}}\right]^{1/2}\\ &= \left[\frac{1+\omega_{1,T}(u)}{\omega_{1,T}\exp(\chi_T(u)) + (1+\omega_{1,T}(u))\exp(\psi_T(t))}\right]^{1/2},\end{aligned} \tag{2.42}$$

where

$$\begin{aligned}\chi_T(u) &:= -\beta T\beta_T(u) + \alpha_{4,T}(u) - 2z_3 - \beta T\\ &= -2\beta T + \omega_{1,T}(u) + t^2\omega_{1,T}(u)\end{aligned} \tag{2.43}$$

and

$$\begin{aligned}\psi_T(u) &:= \beta T\varpi_T(u) + \omega_{4,T}(u) - 2z_3 - \beta T\\ &= \beta T\left[1 + iu\tfrac{\delta_{T,x}}{\beta} + \tfrac{u^2\delta_{T,x}^2}{2\beta^2}\right] + \alpha_{4,T}(u) - iuT\delta_{T,x} - \beta T\\ &= \tfrac{u^2T\beta}{-4(\alpha-2)}\left[\left(\tfrac{-4(\alpha-2)}{T\beta}\right)^{1/2} + \tfrac{2x}{T}\right]^2\\ &= u^2 + u^2\omega_{1,T}(u).\end{aligned} \tag{2.44}$$

Hence, for the given range of values of u, $\chi_T(u) - \psi_T(u) \le -\beta T$.

Hence the above expectation equals

$$\begin{aligned}&\exp(-\tfrac{u^2}{2})(1+\omega_{1,T})^{1/2}\times\\ &\left[\omega_{1,T}\exp\{-2\beta T + \omega_{1,T} + u^2\omega_{1,T}\} + (1+\omega_{1,T}(u))\exp\{u^2\omega_{1,T}(u)\}\right]^{-1/2}\\ =\ &\exp(-\tfrac{u^2}{2})\left[1+\omega_{1,T})(1+\omega_{1,T}(1+\omega_{1,T})\exp\{-\beta T + \omega_{1,T} + t^2\omega_{1,T}\}\right]\\ &\times\exp(u^2\omega_{1,T}(u)).\end{aligned} \tag{2.45}$$

This completes the proof of the lemma. □

Theorem 2.8

$$\sup_{x\in\mathbb{R}}\left|P\left\{\left(\frac{T\beta}{-4(\alpha-2)}\right)^{1/2}(\widetilde{\alpha}_T-\alpha)\le x\right\}-\Phi(x)\right|\le C_\alpha T^{-1/2}.$$

Proof We shall consider two possibilities (i) and (ii).
(i) $|x| > 2(\log T)^{1/2}$
We shall give a proof for the case $x > 2(\log T)^{1/2}$. The proof for the case $x < -2(\log T)^{1/2}$ runs similarly. Note that

$$\begin{aligned}&\left|P\left\{(\frac{T\beta}{-4(\alpha-2)})^{1/2}(\widetilde{\alpha}_T-\alpha)\le x\right\}-\Phi(x)\right|\\ &\le P\left\{(\frac{T\beta}{-4(\alpha-2)})^{1/2}(\widetilde{\alpha}_T-\alpha)\ge x\right\}+\Phi(-x).\end{aligned}$$

But $\Phi(-x)\le\Phi(-2(\log T)^{1/2})\le CT^{-2}$. See (Feller, 1957, p. 166).
Moreover, by Lemma 2.4, we have

$$P\left\{(\frac{T\beta}{-4(\alpha-2)})^{1/2}(\widetilde{\alpha}_T-\alpha)\ge 2(\log T)^{1/2}\right\}\le CT^{-1/2}.$$

Hence

$$\left|P\left\{(\frac{T\beta}{-4(\alpha-2)})^{1/2}(\widetilde{\alpha}_T-\alpha)\le x\right\}-\Phi(x)\right|\le CT^{-1/2}.$$

(ii) $|x|\le 2(\log T)^{1/2}$

Let $A_T := \left\{(\frac{T\beta}{-4(\alpha-2)})^{1/2}|\widetilde{\alpha}_T-\alpha|\le 2(\log T)^{1/2}\right\}$ and $B_T := \left\{\frac{J_T}{T} > c_0\right\}$

where $0 < c_0 < \frac{\beta}{-4(\alpha-2)}$. By Lemma 2.5, we have

$$P(A_T^c)\le CT^{-1/2}. \tag{2.46}$$

By Lemma 2.1, we have

$$\begin{aligned}P(B_T^c)&=P\left\{\tfrac{-4(\alpha-2)}{T\beta}J_T-1<-4(\alpha-2)\beta^{-1}c_0-1\right\}\\ &<P\left\{|\tfrac{-4(\alpha-2)}{T\beta}J_T-1|>1+4(\alpha-2)\beta^{-1}c_0\right\}\le CT^{-1}.\end{aligned} \tag{2.47}$$

Let b_0 be some positive number. On the set $A_T\cap B_T$ for all $T > T_0$ with $4b_0(\log T_0)^{1/2}(\frac{-4(\alpha-2)}{T_0\beta})^{1/2}\le c_0$, we have

$$\begin{aligned}&(\tfrac{T}{-4(\alpha-2)})^{1/2}(\widetilde{\alpha}_T-\alpha)\le x\\ \Rightarrow\ &J_T+b_0T(\widetilde{\alpha}_T-\alpha)<J_T+(\tfrac{T\beta}{-4(\alpha-2)})^{1/2}2b_0\alpha x\\ \Rightarrow\ &(\tfrac{T\beta}{-4(\alpha-2)})^{1/2}(\widetilde{\alpha}_T-\alpha)[J_T+b_0T(\alpha_T-\beta)]<x[J_T+(\tfrac{T\beta}{-4(\alpha-2)})^{1/2}2b_0\alpha x]\\ \Rightarrow\ &(\widetilde{\alpha}_T-\alpha)J_T+b_0T(\alpha_T-\alpha)^2<(\tfrac{-4(\alpha-2)}{T\beta})^{1/2}J_Tx+2b_0\alpha x^2\\ \Rightarrow\ &-N_T+(\widetilde{\alpha}_T-\alpha)J_T+b_0T(\widetilde{\alpha}_T-\alpha)^2<-N_T+(\tfrac{-4(\alpha-2)}{T\beta})^{1/2}J_Tx+2b_0\alpha x^2\\ \Rightarrow\ &0<-N_T+(\tfrac{-4(\alpha-2)}{T\beta})^{1/2}J_Tx+2b_0\alpha x^2\end{aligned} \tag{2.48}$$

since

$$\begin{array}{rl} & J_T + b_0 T(\widetilde{\alpha}_T - \alpha) \\ > & Tc_0 + b_0 T(\widetilde{\alpha}_T - \alpha) \\ > & 4b_0(\log T)^{1/2}(\frac{-4(\alpha-2)}{T\beta})^{1/2} - 2b_0(\log T)^{1/2}(\frac{-4(\alpha-2)}{T\beta})^{1/2} \\ = & 2b_0(\log T)^{1/2}(\frac{-4(\alpha-2)}{T\beta})^{1/2} > 0. \end{array} \tag{2.49}$$

On the other hand, on the set $A_T \cap B_T$ for all $T > T_0$ with $4b_0(\log T_0)^{1/2}(\frac{-4(\alpha-2)}{T_0\beta})^{1/2} \leq c_0$, we have

$$\begin{array}{rl} & (\frac{T}{-4(\alpha-2)})^{1/2}(\widetilde{\alpha}_T - \alpha) > x \\ \Rightarrow & J_T - b_0 T(\widetilde{\alpha}_T - \alpha) < J_T - (\frac{T\beta}{-4(\alpha-2)})^{1/2} 2b_0\beta x \\ \Rightarrow & (\frac{T\beta}{-4(\alpha-2)})^{1/2}(\widetilde{\alpha}_T - \beta)[J_T - b_0 T(\widetilde{\alpha}_T - \beta)] > x[J_T - (\frac{T\beta}{-4(\alpha-2)})^{1/2} 2b_0\alpha x] \\ \Rightarrow & (\widetilde{\alpha}_T - \alpha)J_T - b_0 T(\widetilde{\alpha}_T - \alpha)^2 > (\frac{-4(\alpha-2)}{T\beta})^{1/2} J_T x - 2b_0\alpha x^2 \\ \Rightarrow & -N_T + (\widetilde{\alpha}_T - \alpha)J_T - b_0 T(\alpha_T - \alpha)^2 > -N_T + (\frac{-4(\alpha-2)}{T\beta})^{1/2} J_T x - 2b_0\alpha x^2 \\ \Rightarrow & 0 > -N_T + (\frac{-4(\alpha-2)}{T\beta})^{1/2} J_T x - 2b_0\alpha x^2 \end{array} \tag{2.50}$$

since

$$\begin{array}{rl} & J_T - b_0 T(\widetilde{\alpha}_T - \alpha) \\ > & Tc_0 - b_0 T(\widetilde{\alpha}_T - \alpha) \\ > & 4b_0(\log T)^{1/2}(\frac{-4(\alpha-2)}{T\beta})^{1/2} - 2b_0(\log T)^{1/2}(\frac{-4(\alpha-2)}{T\beta})^{1/2} \\ = & 2b_0(\log T)^{1/2}(\frac{-4(\alpha-2)}{T\beta})^{1/2} > 0. \end{array} \tag{2.51}$$

Hence

$$0 < -N_T + (\frac{-4(\alpha-2)}{T})^{1/2} J_T x - 2b_0\alpha x^2 \Rightarrow (\frac{T\beta}{-4(\alpha-2)})^{1/2}(\alpha_T - \alpha) \leq x.$$

Letting $D^{\pm}_{T,x} := \left\{-N_T + (\frac{-4(\alpha-2)}{T\beta})^{1/2} I_T x \pm 2b_0\alpha x^2 > 0\right\}$, we obtain

$$D^{-}_{T,x} \cap A_T \cap B_T \subseteq A_T \cap B_T \cap \{(\frac{T\beta}{-4(\alpha-2)})^{1/2}(\alpha_T - \alpha) \leq x\} \subseteq D^{+}_{T,x} \cap A_T \cap B_T. \tag{2.52}$$

If it is shown that

$$\left| P\left\{D^{\pm}_{T,x}\right\} - \Phi(x)\right| \leq CT^{-1/2} \tag{2.53}$$

for all $T > T_0$ and $|x| \leq 2(\log T)^{1/2}$, then the theorem would follow from (2.51) – (2.53). We shall prove (2.53) for $D^{+}_{T,x}$. The proof for $D^{-}_{T,x}$ is analogous.

Observe that

$$\begin{aligned}
& \left|P\left\{D^{+}_{T,x}\right\} - \Phi(x)\right| \\
= & \left|P\left\{(\tfrac{-4(\alpha-2)}{T\beta})^{1/2} N_T - \left(\tfrac{-4(\alpha-2)}{T\beta} J_T - 1\right) x \right.\right. \\
& \left.\left. < x + 2(\tfrac{-4(\alpha-2)}{T\beta})^{1/2} b_0 \alpha x^2\right\} - \Phi(x)\right| \\
\le & \sup_{y\in\mathbb{R}} \left|P\left\{(\tfrac{-4(\alpha-2)}{T\beta})^{1/2} N_T - \left(\tfrac{-4(\alpha-2)}{T\beta} J_T - 1\right) x \le y\right\} - \Phi(y)\right| \\
& + \left|\Phi\left(x + (\tfrac{-4(\alpha-2)}{T\beta})^{1/2} b_0 \alpha x^2\right) - \Phi(x)\right| \\
=: & \Delta_1 + \Delta_2.
\end{aligned} \tag{2.54}$$

Lemma 2.3 and Esseen's lemma (see Feller (1957)) immediately yield

$$\Delta_1 \le CT^{-1/2}. \tag{2.55}$$

On the other hand, for all $T > T_0$,

$$\Delta_2 \le 2(\frac{-4(\alpha-2)}{T\beta})^{1/2} b_0 \alpha x^2 (2\pi)^{-1/2} \exp(-\overline{x}^2/2),$$

where

$$|\overline{x} - x| \le 2(\frac{-4(\alpha-2)}{T\beta})^{1/2} b_0 \alpha x^2.$$

Since $|x| \le 2(\log T)^{1/2}$, it follows that $|\overline{x}| > |x|/2$ for all $T > T_0$ and consequently

$$\begin{aligned}
\Delta_2 & \le 2(\frac{-4(\alpha-2)}{T\beta})^{1/2} b_0 \alpha x^2 (2\pi)^{-1/2} x^2 \exp(-x^2/8) \\
& \le CT^{-1/2}.
\end{aligned} \tag{2.56}$$

From (2.54)–(2.56), we obtain

$$\left|P\left\{D^{+}_{T,x}\right\} - \Phi(x)\right| \le CT^{-1/2}. \tag{2.57}$$

This completes the proof of the theorem. □

3.3 Discrete Observations

In finance, we have discrete observations of the interest rate process $\{X_t\}$ at times $0 = t_0 < t_1 < \cdots < t_n = T$ with $t_i - t_{i-1} = \frac{T}{n}, i = 1, 2 \cdots, n$. We assume two types of high frequency data with long observation time: (1) $T \to \infty, n \to \infty, \frac{T}{\sqrt{n}} \to 0$ and (2) $T \to \infty, n \to \infty, \frac{T}{n^{2/3}} \to 0$.

The approximate minimum contrast estimator based on discrete observations is the following:

$$\widetilde{\beta}_{n,T} := \frac{\alpha T}{\sum_{i=1}^{n} (X_{t_i} - X_{t_{i-1}})^2}. \tag{3.1}$$

Let

$$V_{n,T} := \sum_{i=1}^{n} (X_{t_i} - X_{t_{i-1}})^2. \tag{3.2}$$

When $t_i - t_{i-1} = \Delta$ fixed, the conditional least squares estimators (CLSEs) of β and α are, respectively, given by

$$\hat{\beta}_n := \frac{1}{\Delta} \log \frac{\sum_{i=1}^{n} (X_{t_i} - \overline{X}_n)(X_{t_{i-1}} - \overline{X'_n})}{\sum_{i=1}^{n} (X_{t_i} - \overline{X}_n)^2} \tag{3.3}$$

and

$$\hat{\alpha}_n := \frac{\overline{X}_n - e^{\hat{\beta}_n \Delta} \overline{X'_n}}{e^{\hat{\beta}_n \Delta} - 1} \hat{\beta}_n, \tag{3.4}$$

where

$$\overline{X}_n := \frac{1}{n} \sum_{i=1}^{n} X_{t_i}, \quad \overline{X'_n} := \frac{1}{n} \sum_{i=1}^{n} X_{t_{i-1}}. \tag{3.5}$$

Approximate maximum likelihood estimators (AMLEs) based on approximation of the continuous Girsanov likelihood are given by

$$\widetilde{\beta}_n := \frac{\sum_{i=1}^{n} X_{t_{i-1}}^{-1} (X_{t_i} - X_{t_{i-1}}) - T^{-1} X_T \sum_{i=1}^{n} X_{t_{i-1}}^{-1} (t_i - t_{i-1})}{T - \overline{X}_n \sum_{i=1}^{n} X_{t_{i-1}}^{-1} (t_i - t_{i-1})} \tag{3.6}$$

and

$$\widetilde{\alpha}_n := \frac{X_T - \overline{X}_n \sum_{i=1}^{n} X_{t_{i-1}}^{-1} (t_i - t_{i-1})}{T - \overline{X}_n \sum_{i=1}^{n} X_{t_{i-1}}^{-1} (t_i - t_{i-1})}. \tag{3.7}$$

The approximate transition density based on Dacunha-Castelle and Florens-Zmirou (1986) is given by

$$\begin{aligned} p(t, x, y, t) \quad := \quad & -\frac{1}{2} \log(2\pi t y) - \frac{2(\sqrt{y} - \sqrt{x})^2}{t} + \beta(y - x) + (\alpha - \frac{1}{4}) \log(\frac{y}{x}) \\ & -\frac{t}{2} \left[\frac{1}{3} \left\{ \beta\sqrt{x} + (\alpha - \frac{1}{4}) \frac{1}{\sqrt{x}} \right\}^2 + \frac{1}{2} \left\{ \frac{\beta}{2} - \frac{1}{2}(\alpha - \frac{1}{4}) \frac{1}{x} \right\} \right. \\ & + \frac{1}{3} \left\{ \beta\sqrt{y} + (\alpha - \frac{1}{4}) \frac{1}{\sqrt{y}} \right\}^2 + \frac{1}{2} \left\{ \frac{\beta}{2} - \frac{1}{2}(\alpha - \frac{1}{4}) \frac{1}{y} \right\} \\ & \left. + \frac{1}{3} \left\{ \beta\sqrt{x} + (\alpha - \frac{1}{4}) \frac{1}{\sqrt{x}} \right\} \left\{ \beta\sqrt{y} + (\alpha - \frac{1}{4}) \frac{1}{\sqrt{y}} \right\} \right]. \end{aligned} \tag{3.8}$$

The AMLEs of β and α are given, respectively, by

$$\beta_{n,T} := \frac{u_{1,n}w_{2,n} - u_{2,n}w_{1,n}}{v_{1,n}u_{2,n} - v_{2,n}u_{1,n}}, \tag{3.9}$$

$$\alpha_{n,T} := \frac{v_{2,n}w_{1,n} - v_{1,n}w_{2,n}}{v_{1,n}u_{2,n} - v_{2,n}u_{1,n}}, \tag{3.10}$$

where

$$\begin{aligned}
u_{1,n} &:= \frac{2n\Delta}{3} + \frac{\Delta}{6}\sum_{i=1}^{n} \frac{X_{t_{i-1}} + X_{t_i}}{\sqrt{X_{t_{i-1}}X_{t_i}}} \\
v_{1,n} &:= \frac{\Delta}{3}\sum_{i=1}^{n}\left(X_{t_{i-1}} + \sqrt{X_{t_{i-1}}X_{t_i}} + X_{t_i}\right) \\
w_{1,n} &:= \frac{n\Delta}{12} - \sum_{i=1}^{n}(X_{t_i} - X_{t_{i-1}}) - \frac{\Delta}{24}\sum_{i=1}^{n}\frac{X_{t_i} + X_{t_{i-1}}}{\sqrt{X_{t_{i-1}}X_{t_i}}} \\
u_{2,n} &:= \frac{\Delta}{3}\sum_{i=1}^{n}\left(\frac{1}{X_{t_i}} + \frac{1}{X_{t_{i-1}}} + \frac{1}{\sqrt{X_{t_{i-1}}X_{t_i}}}\right) \\
v_{2,n} &:= \frac{\Delta}{6}\left(4n + \sum_{i=1}^{n}\frac{X_{t_{i-1}} + X_{t_i}}{\sqrt{X_{t_{i-1}}X_{t_i}}}\right) \\
w_{2,n} &:= -\sum_{i=1}^{n}(\log X_{t_i} - \log X_{t_{i-1}}) - \frac{\Delta}{8}\sum_{i=1}^{n}\left(\frac{1}{X_{t_i}} + \frac{1}{X_{t_{i-1}}}\right) \\
&\quad -\frac{\Delta}{12}\sum_{i=1}^{n}\left(\frac{1}{X_{t_i}} + \frac{1}{X_{t_{i-1}}} + \frac{1}{\sqrt{X_{t_{i-1}}X_{t_i}}}\right).
\end{aligned} \tag{3.11}$$

Based on discrete observations, we define the approximate minimum contrast estimators (AMCEs) as follows:

In order to define the approximate minimum contrast estimators (AMCEs), we use various discrete approximations of the integrals in the definition (2.15) and (2.16) of MCEs.

An Euler type discrete approximation of (1.8) and (1.9) gives

$$\breve{\beta}_n = \frac{T\overline{X'_n}}{2\sum_{i=1}^{n}(X_{t_{i-1}} - \overline{X'_n})^2}, \tag{3.12}$$

$$\breve{\alpha}_n = \frac{(\overline{X'_n})^2}{2\sum_{i=1}^{n}(X_{t_{i-1}} - \overline{X'_n})^2}. \tag{3.13}$$

We will next consider weighted AMCES. Define a weighted sum of squares

$$M_{n,T} := \frac{T}{n}\left\{\sum_{i=1}^{n} w_{t_i}X_{t_{i-1}}^2 + \sum_{i=2}^{n+1} w_{t_i}X_{t_{i-1}}^2\right\}, \tag{3.14}$$

where $w_{t_i} \geq 0$ is a weight function.

Denote the discrete increasing functions as

$$I_{n,T} := \frac{T}{n}\sum_{i=1}^{n} X_{t_{i-1}}^2, \tag{3.15}$$

$$J_{n,T} := \frac{T}{n}\sum_{i=2}^{n+1} X_{t_{i-1}}^2 = \frac{T}{n}\sum_{i=1}^{n} X_{t_i}^2. \tag{3.16}$$

General weighted AMCE is defined as

$$\widetilde{\beta}_{n,T} := -\left\{\frac{2}{n}M_{n,T}\right\}^{-1}. \tag{3.17}$$

With $w_{t_i} = 1$, we obtain the forward AMCE as

$$\widetilde{\beta}_{n,T,F} := -\left\{\frac{2}{n}I_{n,T}\right\}^{-1}. \tag{3.18}$$

With $w_{t_i} = 0$, we obtain the backward AMCE as

$$\widetilde{\beta}_{n,T,B} := -\left\{\frac{2}{n}J_{n,T}\right\}^{-1}. \tag{3.19}$$

Analogous to the estimators for the discrete AR(1) model, we define the simple symmetric and weighted symmetric estimators (see Fuller (1996)).

With $w_{t_i} = 0.5$, the simple symmetric AMCE is defined as

$$\widetilde{\beta}_{n,T,z} := -\left\{\frac{1}{n}[I_{n,T} + J_{n,T}]\right\}^{-1} = -\left\{\frac{2}{n}\sum_{i=2}^{n} X_{t_{i-1}}^2 + 0.5(X_{t_0}^2 + X_{t_n}^2)\right\}^{-1}. \tag{3.20}$$

With the weight function

$$w_{t_i} = \begin{cases} 0 & : \ i = 1 \\ \frac{i-1}{n} & : \ i = 2, 3, \cdots, n \\ 1 & : \ i = n+1, \end{cases}$$

the weighted symmetric AMCE is defined as

$$\widetilde{\beta}_{n,T,w} := -\left\{\frac{2}{n}\sum_{i=2}^{n} X_{t_{i-1}}^2 + \frac{1}{n}\sum_{i=1}^{n} X_{t_{i-1}}^2\right\}^{-1}. \tag{3.21}$$

Note that estimator (1.13) is analogous to the trapezoidal rule in numerical analysis. One can instead use the midpoint rule to define another estimator

$$\widetilde{\beta}_{n,T,A} := -\left\{\frac{2}{n}\sum_{i=1}^{n}\left(\frac{X_{t_{i-1}} + X_{t_i}}{2}\right)^2\right\}^{-1}. \tag{3.22}$$

One can use Simpson's rule to define another estimator where the denominator is a convex combination of the denominators in (1.11) and (1.12)

$$\widetilde{\beta}_{n,T,S} := -\left\{\frac{1}{3n}\sum_{i=1}^{n}\left\{X_{t_{i-1}}^2 + 4\left(\frac{X_{t_{i-1}}+X_{t_i}}{2}\right)^2 + X_{t_i}^2\right\}\right\}^{-1}. \tag{3.23}$$

In general, one can generalize Simpson's rule as

$$\widetilde{\beta}_{n,T,GS} := -\left\{\frac{2}{n}\sum_{i=1}^{n}\left\{p\frac{X_{t_{i-1}}^2 + X_{t_i}^2}{2} + (1-p)\left(\frac{X_{t_{i-1}}+X_{t_i}}{2}\right)^2\right\}\right\}^{-1} \tag{3.24}$$

for any $0 \le p \le 1$.

The case $p = 0$ produces the estimator (3.22). The case $p = 1$ produces the estimator (3.21). The case $p = \frac{1}{3}$ produces the estimator (3.23).

We propose a very general form of the quadrature based estimator as

$$\widetilde{\beta}_{n,T,w} := -\left\{\frac{2}{n}\sum_{i=1}^{n}\sum_{j=1}^{m}\left[(1-s_j)X_{t_{i-1}} + s_jX_{t_i}\right]^2 p_j\right\}^{-1}, \tag{3.25}$$

where $p_j, \quad j \in \{1,2,\cdots,m\}$ is a probability mass function of a discrete random variable S on $0 \le s_1 < s_2 < \cdots < s_m \le 1$ with $P(S = s_j) := p_j, \ j \in \{1,2,\cdots,m\}$.

Denote the k-th moment of the random variable S as $\mu_k := \sum_{j=1}^{m} s_j^k p_j, \ k = 1,2,\cdots$.

If one chooses the probability distribution as uniform distribution for which the moments are a harmonic sequence $(\mu_1,\mu_2,\mu_3,\mu_4,\mu_5,\mu_6,\cdots) = (\frac{1}{2},\frac{1}{3},\frac{1}{4},\frac{1}{5},\frac{1}{6},\frac{1}{7},\cdots)$, then there is no change in rate of convergence than second order. If one can construct a probability distribution for which the harmonic sequence is truncated at a point, then there is an improvement in the rate of convergence at the point of truncation.

Given a positive integer m, construct a probability mass function $p_j, \quad j \in \{1,2,\cdots,m\}$, on $0 \le s_1 < s_2 < \cdots < s_m \le 1$ such that

$$\sum_{j=1}^{m} s_j^r p_j = \frac{1}{r+1}, \quad r \in \{0,\cdots,m-2\} \tag{3.26}$$

$$\sum_{j=1}^{m} s_j^{m-1} p_j \neq \frac{1}{m}. \tag{3.27}$$

Neither the probabilities p_j nor the atoms, s_j, of the distribution are specified in advance.

This problem is related to the truncated Hausdorff moment problem. I obtain examples of such probability distributions and use them to get higher order accurate (up to sixth order) AMCEs.

The order of approximation error (rate of convergence) of an estimator is $n^{-\nu}$ where

$$\nu := \inf\left\{k : \mu_k \neq \frac{1}{1+k},\ \mu_j = \frac{1}{1+j}, j = 1, 2, \cdots, k-1\right\}. \tag{3.28}$$

I construct probability distributions satisfying these moment conditions and obtain estimators of the rate of convergence up to order 6.

Probability $p_1 = 1$ at the point $s_1 = 0$ gives the estimator (1.11) for which $\mu_1 = 0$. Note that $\mu_1 \neq \frac{1}{2}$. Thus $\nu = 1$.

Probability $p_1 = 1$ at the point $s_1 = 1$ gives the estimator (1.12) for which $\mu_1 = 1$. Note that $\mu_1 \neq \frac{1}{2}$. Thus $\nu = 1$.

Probabilities $(p_1, p_2) = (\frac{1}{2}, \frac{1}{2})$ at the respective points $(s_1, s_2) = (0, 1)$ produce the estimator $\widetilde{\beta}_{n,T,Z}$ for which $(\mu_1, \mu_2) = (\frac{1}{2}, \frac{1}{4})$. Thus $\nu = 2$.

Probability $p_j = 1$ at the point $s_j = \frac{1}{2}$ produces the estimator $\widetilde{\beta}_{n,T,A}$ for which $(\mu_1, \mu_2) = (\frac{1}{2}, \frac{1}{2})$. Thus $\nu = 2$.

Probabilities $(p_1, p_2) = (\frac{1}{4}, \frac{3}{4})$ at the respective points $(s_1, s_2) = (0, \frac{2}{3})$ produce the asymmetric estimator

$$\widetilde{\beta}_{n,T,3} := -\left\{\frac{2}{n}\frac{1}{4}\sum_{i=1}^{n}\left[(X_{t_{i-1}})^2 + 3(\frac{X_{t_{i-1}} + 2X_{t_i}}{3})^2\right]\right\}^{-1} \tag{3.29}$$

for which $(\mu_1, \mu_2, \mu_3) = (\frac{1}{2}, \frac{1}{3}, \frac{2}{9})$. Thus $\nu = 3$.

Probabilities $(p_1, p_2) = (\frac{3}{4}, \frac{1}{4})$ at the respective points $(s_1, s_2) = (\frac{1}{3}, 1)$ produce asymmetric estimator

$$\widetilde{\beta}_{n,T,4} := -\left\{\frac{2}{n}\frac{1}{4}\sum_{i=1}^{n}\left[3(\frac{2X_{t_{i-1}} + X_{t_i}}{3})^2 + (X_{t_i})^2\right]\right\}^{-1} \tag{3.30}$$

for which $(\mu_1, \mu_2, \mu_3) = (\frac{1}{2}, \frac{1}{3}, \frac{10}{36})$. Thus $\nu = 3$.

Probabilities $(p_1, p_2, p_3) = (\frac{1}{6}, \frac{2}{3}, \frac{1}{6})$ at the respective points $(s_1, s_2, s_3) = (0, \frac{1}{2}, 1)$ produce the estimator $\widetilde{\beta}_{n,T,5}$ for which $(\mu_1, \mu_2, \mu_3, \mu_4) = (\frac{1}{2}, \frac{1}{3}, \frac{1}{4}, \frac{5}{25})$. Thus $\nu = 4$.

Probabilities $(p_1, p_2, p_3, p_4) = (\frac{1}{8}, \frac{3}{8}, \frac{3}{8}, \frac{1}{8})$ at the respective points $(s_1, s_2, s_3, s_4) = (0, \frac{1}{3}, \frac{2}{3}, 1)$ produce the symmetric estimator

$$\widetilde{\beta}_{n,T,5} := -\left\{\frac{2}{n}\frac{1}{8}\sum_{i=1}^{n}\left[(X_{t_{i-1}})^2 + 3(\frac{2X_{t_{i-1}} + X_{t_i}}{3})^2 + 3(\frac{X_{t_{i-1}} + 2X_{t_i}}{3})^2 + (X_{t_i})^2\right]\right\}^{-1} \tag{3.31}$$

for which $(\mu_1, \mu_2, \mu_3, \mu_4) = (\frac{1}{2}, \frac{1}{3}, \frac{1}{4}, \frac{11}{54})$. Thus $\nu = 4$.

Probabilities $(p_1, p_2, p_3, p_4, p_5) = (\frac{1471}{24192}, \frac{6925}{24192}, \frac{1475}{12096}, \frac{2725}{12096}, \frac{5675}{24192}, \frac{1721}{24192})$ at the respective points $(s_1, s_2, s_3, s_4, s_5) = (0, \frac{1}{5}, \frac{2}{5}, \frac{3}{5}, \frac{4}{5}, 1)$ produce the asymmetric estimator

$$\widetilde{\beta}_{n,T,7} :=$$

$$-\Big\{\frac{2}{n}\frac{1}{24192}\sum_{i=1}^{n}\Big[1471(X_{t_{i-1}})^2 + 6925(\frac{X_{t_{i-1}} + X_{t_i}}{5})^2 + 2950(\frac{2X_{t_{i-1}} + 2X_{t_i}}{5})^2$$
$$+5450(\frac{3X_{t_{i-1}}+3X_{t_i}}{5})^2 + 5675(\frac{4X_{t_{i-1}}+4X_{t_i}}{5})^2 + 1721(X_{t_i})^2\Big]\Big\}^{-1} \tag{3.32}$$

for which $(\mu_1, \mu_2, \mu_3, \mu_4, \mu_5, \mu_6) = (\frac{1}{2}, \frac{1}{3}, \frac{1}{4}, \frac{1}{5}, \frac{841}{5040})$. Thus $\nu = 5$.

Probabilities $(p_1, p_2, p_3, p_4, p_5) = (\frac{7}{90}, \frac{16}{45}, \frac{2}{15}, \frac{16}{45}, \frac{7}{90})$ at the respective points $(s_1, s_2, s_3, s_4, s_5) = (0, \frac{1}{4}, \frac{1}{2}, \frac{3}{4}, 1)$ produce the symmetric estimator $\widetilde{\beta}_{n,T,8}$ given by

$$\widetilde{\beta}_{n,T,8} := -\Big\{\frac{2}{n}\frac{1}{90}\sum_{i=1}^{n}\Big[7(X_{t_{i-1}})^2 + 32(\frac{3X_{t_{i-1}} + X_{t_i}}{4})^2 + 12(\frac{X_{t_{i-1}} + X_{t_i}}{2})^2$$
$$+32(\frac{X_{t_{i-1}}+3X_{t_i}}{4})^2 + 7(t_i, X_{t_i})^2\Big]\Big\}^{-1} \tag{3.33}$$

for which $(\mu_1, \mu_2, \mu_3, \mu_4, \mu_5, \mu_6) = (\frac{1}{2}, \frac{1}{3}, \frac{1}{4}, \frac{1}{5}, \frac{1}{6}, \frac{110}{768})$. Thus $\nu = 6$.

Probabilities $(p_1, p_2, p_3, p_4, p_5) = (\frac{19}{288}, \frac{75}{288}, \frac{50}{288}, \frac{50}{288}, \frac{75}{288}, \frac{19}{288})$ at the respective points $(s_1, s_2, s_3, s_4, s_5) = (0, \frac{1}{5}, \frac{2}{5}, \frac{3}{5}, \frac{4}{5}, 1)$ produce symmetric estimator

$$\widetilde{\beta}_{n,T,9} :=$$

$$-\Big\{\frac{2}{n}\frac{1}{288}\sum_{i=1}^{n}\Big[19(X_{t_{i-1}})^2 + 75(\frac{4X_{t_{i-1}} + X_{t_i}}{5})^2 + 50(\frac{3X_{t_{i-1}} + 2X_{t_i}}{5})^2$$
$$+50(\frac{2X_{t_{i-1}}+3X_{t_i}}{5})^2 + 75(\frac{X_{t_{i-1}}+4X_{t_i}}{5})^2 + 19(X_{t_i})^2\Big]\Big\}^{-1} \tag{3.34}$$

for which $(\mu_1, \mu_2, \mu_3, \mu_4, \mu_5, \mu_6) = (\frac{1}{2}, \frac{1}{3}, \frac{1}{4}, \frac{1}{5}, \frac{1}{6}, \frac{3219}{22500})$. Thus $\nu = 6$.

We summarize these results in the following table where the first column gives the estimator, the second column gives the support points $(s_1, s_2, \ldots, s_i)$ of the corresponding discrete probability distribution, the third column gives the corresponding probability masses $(p_1, p_2, \ldots, p_i)$, the fourth column gives the moments $(\mu_1, \mu_2, \ldots, \mu_i)$, and the last column gives the order of convergence.

Estimator	Support	Probability Mass	Moment	Order
$\widetilde{\beta}_{n,T,-1}$	0	1	0	$\nu = 1$
$\widetilde{\beta}_{n,T,0}$	1	1	1	$\nu = 1$
$\widetilde{\beta}_{n,T,1}$	$(0, 1)$	$(\frac{1}{2}, \frac{1}{2})$	$(\frac{1}{2}, \frac{1}{4})$	$\nu = 2$
$\widetilde{\beta}_{n,T,2}$	$\frac{1}{2}$	1	$(\frac{1}{2}, \frac{1}{2})$	$\nu = 2$
$\widetilde{\beta}_{n,T,3}$	$(0, \frac{2}{3})$	$(\frac{1}{4}, \frac{3}{4})$	$(\frac{1}{2}, \frac{1}{3}, \frac{2}{9})$	$\nu = 3$
$\widetilde{\beta}_{n,T,4}$	$(\frac{1}{3}, 1)$	$\frac{3}{4}, \frac{1}{4}$	$(\frac{1}{2}, \frac{1}{3}, \frac{10}{36})$	$\nu = 3$
$\widetilde{\beta}_{n,T,5}$	$(0, \frac{1}{2}, 1)$	$(\frac{1}{6}, \frac{2}{3}, \frac{1}{6})$	$(\frac{1}{2}, \frac{1}{3}, \frac{1}{4}, \frac{5}{24})$	$\nu = 4$
$\widetilde{\beta}_{n,T,6}$	$(0, \frac{1}{3}, \frac{2}{3}, 1)$	$(\frac{1}{8}, \frac{3}{8}, \frac{3}{8}, \frac{1}{8})$	$(\frac{1}{2}, \frac{1}{3}, \frac{1}{4}, \frac{11}{54})$	$\nu = 4$
$\widetilde{\beta}_{n,T,7}$	$(0, \frac{1}{5}, \frac{2}{5}, \frac{3}{5}, \frac{4}{5}, 1)$	$(\frac{1471}{24192}, \frac{6925}{24192}, \frac{1475}{12096}, \frac{2725}{12096}, \frac{5675}{24192}, \frac{1721}{24192})$	$(\frac{1}{2}, \frac{1}{3}, \frac{1}{4}, \frac{1}{5}, \frac{841}{5040})$	$\nu = 5$
$\widetilde{\beta}_{n,T,8}$	$(0, \frac{1}{4}, \frac{1}{2}, \frac{3}{4}, 1)$	$(\frac{7}{90}, \frac{16}{45}, \frac{2}{15}, \frac{16}{45}, \frac{7}{90})$	$(\frac{1}{2}, \frac{1}{3}, \frac{1}{4}, \frac{1}{5}, \frac{1}{6}, \frac{110}{768})$	$\nu = 6$
$\widetilde{\beta}_{n,T,9}$	$(0, \frac{1}{5}, \frac{2}{5}, \frac{3}{5}, \frac{4}{5}, 1)$	$(\frac{19}{288}, \frac{75}{288}, \frac{50}{288}, \frac{50}{288}, \frac{75}{288}, \frac{19}{288})$	$(\frac{1}{2}, \frac{1}{3}, \frac{1}{4}, \frac{1}{5}, \frac{1}{6}, \frac{3219}{22500})$	$\nu = 6$

$\widetilde{\beta}_{n,T,8}$ is called the *stochastic Boole's estimator.* The estimator $\widetilde{\beta}_{n,T,S}$ is based on the arithmetic mean of $I_{n,T}$ and $J_{n,T}$. One can use geometric mean and harmonic mean instead. The geometric mean based symmetric AMCE (which is based on the ideas of partial autocorrelation) is defined as

$$\widetilde{\beta}_{n,T,G} := \frac{-T/2}{\sqrt{I_{n,T}\ J_{n,T}}}. \tag{3.35}$$

The harmonic mean based symmetric AMCE is defined as

$$\widetilde{\beta}_{n,T,H} := \frac{-T}{\frac{1}{I_{n,T}} + \frac{1}{J_{n,T}}}. \tag{3.36}$$

Note that

$$\widetilde{\beta}_{n,T,H} \le \widetilde{\beta}_{n,T,G} \le \widetilde{\beta}_{n,T,S}. \tag{3.37}$$

Note that for the *stochastic Simpson's estimator*, we have

$$\begin{aligned} &\tfrac{1}{3}\textstyle\sum_{i=1}^{n}\left[X_{t_{i-1}} + \sqrt{X_{t_i}X_{t_{i-1}}} + X_{t_i}\right] \\ = \tfrac{1}{6}\textstyle\sum_{i=1}^{n}&\left[X_{t_{i-1}} + 4\left(\frac{\sqrt{X_{t_{i-1}}}+\sqrt{X_{t_i}}}{2}\right)^2 + X_{t_i}\right]. \end{aligned} \tag{3.38}$$

AMCEs for α Define a weighted sum of squares

$$M_{n,T} := \frac{T}{n}\left\{\sum_{i=1}^{n} w_{t_i} X_{t_{i-1}}^2 + \sum_{i=2}^{n+1} w_{t_i} X_{t_{i-1}}^2\right\}, \tag{3.39}$$

where $w_{t_i} \geq 0$ is a weight function.

Denote the discrete increasing functions as

$$I_{n,T} := \frac{T}{n}\sum_{i=1}^{n} X_{t_{i-1}}^2, \quad J_{n,T} := \frac{T}{n}\sum_{i=2}^{n+1} X_{t_{i-1}}^2 = \frac{T}{n}\sum_{i=1}^{n} X_{t_i}^2. \tag{3.40}$$

General weighted AMCE is defined as

$$\widetilde{\alpha}_{n,T} := -\left\{\frac{2}{n}M_{n,T}\right\}^{-1}. \tag{3.41}$$

With $w_{t_i} = 1$, we obtain the forward AMCE as

$$\widetilde{\alpha}_{n,T,F} := -\left\{\frac{2}{n}I_{n,T}\right\}^{-1}. \tag{3.42}$$

With $w_{t_i} = 0$, we obtain the backward AMCE as

$$\widetilde{\alpha}_{n,T,B} := -\left\{\frac{2}{n}J_{n,T}\right\}^{-1}. \tag{3.43}$$

Analogous to the estimators for the discrete AR(1) model, we define the simple symmetric and weighted symmetric estimators (see Fuller (1996)).

With $w_{t_i} = 0.5$, the simple symmetric AMCE is defined as

$$\widetilde{\alpha}_{n,T,z} := -\left\{\frac{1}{n}[I_{n,T} + J_{n,T}]\right\}^{-1} = -\left\{\frac{2}{n}\sum_{i=2}^{n} X_{t_{i-1}}^2 + 0.5(X_{t_0}^2 + X_{t_n}^2)\right\}^{-1}. \tag{3.44}$$

With the weight function

$$w_{t_i} = \begin{cases} 0 & : \ i = 1 \\ \frac{i-1}{n} & : \ i = 2, 3, \cdots, n \\ 1 & : \ i = n+1, \end{cases} \tag{3.45}$$

the weighted symmetric AMCE is defined as

$$\widetilde{\alpha}_{n,T,w} := -\left\{\frac{2}{n}\sum_{i=2}^{n} X_{t_{i-1}}^2 + \frac{1}{n}\sum_{i=1}^{n} X_{t_{i-1}}^2\right\}^{-1}. \tag{3.46}$$

Note that estimator (1.13) is analogous to the trapezoidal rule in numerical analysis. One can instead use the midpoint rule to define another estimator

$$\widetilde{\alpha}_{n,T,A} := -\left\{\frac{2}{n}\sum_{i=1}^{n}\left(\frac{X_{t_{i-1}} + X_{t_i}}{2}\right)^2\right\}^{-1}. \tag{3.47}$$

One can use Simpson's rule to define another estimator where the denominator is a convex combination of the denominators in (1.11) and (1.12)

$$\widetilde{\alpha}_{n,T,S} := -\left\{\frac{1}{3n}\sum_{i=1}^{n}\left\{X_{t_{i-1}}^2 + 4\left(\frac{X_{t_{i-1}} + X_{t_i}}{2}\right)^2 + X_{t_i}^2\right\}\right\}^{-1}. \tag{3.48}$$

In general, one can generalize Simpson's rule as

$$\widetilde{\alpha}_{n,T,GS} := -\left\{\frac{2}{n}\sum_{i=1}^{n}\left\{\alpha\frac{X_{t_{i-1}}^2 + X_{t_i}^2}{2} + (1-\alpha)\left(\frac{X_{t_{i-1}} + X_{t_i}}{2}\right)^2\right\}\right\}^{-1} \tag{3.49}$$

for any $0 \le \alpha \le 1$.

The case $\alpha = 0$ produces the estimator (1.18). The case $\alpha = 1$ produces the estimator (1.17). The case $\alpha = \frac{1}{3}$ produces the estimator (1.19).

We propose a very general form of the quadrature based estimator as

$$\widetilde{\alpha}_{n,T,w} := -\left\{\frac{2}{n}\sum_{i=1}^{n}\sum_{j=1}^{m}\left[(1-s_j)X_{t_{i-1}} + s_j X_{t_i}\right]^2 p_j\right\}^{-1}, \tag{3.50}$$

where $p_j, \quad j \in \{1, 2, \cdots, m\}$, is a probability mass function of a discrete random variable S on $0 \le s_1 < s_2 < \cdots < s_m \le 1$ with $P(S = s_j) := p_j, \ j \in \{1, 2, \ldots, m\}$.

Denote the k-th moment of the random variable S as $\mu_k := \sum_{j=1}^{m} s_j^k p_j, \ k = 1, 2, \ldots$.

If one chooses the probability distribution as uniform distribution for which the moments are a harmonic sequence $(\mu_1, \mu_2, \mu_3, \mu_4, \mu_5, \mu_6, \cdots) = (\frac{1}{2}, \frac{1}{3}, \frac{1}{4}, \frac{1}{5}, \frac{1}{6}, \frac{1}{7}, \cdots)$, then there is no change in rate of convergence than second order. If one can construct a probability distribution for which the harmonic sequence is truncated at a point, then there is a rate of convergence improvement at the point of truncation.

The order of approximation error (rate of convergence) of an estimator is $n^{-\nu}$ where

$$\nu := \inf\left\{k : \mu_k \ne \frac{1}{1+k}, \ \mu_j = \frac{1}{1+j}, j = 1, 2, \cdots, k-1\right\}. \tag{3.51}$$

I construct probability distributions satisfying these moment conditions and obtain estimators of the rate of convergence up to order 6.

Probability $p_1 = 1$ at the point $s_1 = 0$ gives the estimator (1.11) for which $\mu_1 = 0$. Note that $\mu_1 \ne \frac{1}{2}$. Thus $\nu = 1$.

Probability $p_1 = 1$ at the point $s_1 = 1$ gives the estimator (1.12) for which $\mu_1 = 1$. Note that $\mu_1 \ne \frac{1}{2}$. Thus $\nu = 1$.

Probabilities $(p_1, p_2) = (\frac{1}{2}, \frac{1}{2})$ at the respective points $(s_1, s_2) = (0, 1)$ produce the estimator $\widetilde{\alpha}_{n,T,Z}$ for which $(\mu_1, \mu_2) = (\frac{1}{2}, \frac{1}{4})$. Thus $\nu = 2$.

Probability $p_j = 1$ at the point $s_j = \frac{1}{2}$ produces the estimator $\widetilde{\alpha}_{n,T,A}$ for which $(\mu_1, \mu_2) = (\frac{1}{2}, \frac{1}{2})$. Thus $\nu = 2$.

Probabilities $(p_1, p_2) = (\frac{1}{4}, \frac{3}{4})$ at the respective points $(s_1, s_2) = (0, \frac{2}{3})$ produce the asymmetric estimator

$$\widetilde{\alpha}_{n,T,3} := -\left\{\frac{2}{n}\frac{1}{4}\sum_{i=1}^{n}\left[(X_{t_{i-1}})^2 + 3(\frac{X_{t_{i-1}} + 2X_{t_i}}{3})^2\right]\right\}^{-1} \tag{3.52}$$

for which $(\mu_1, \mu_2, \mu_3) = (\frac{1}{2}, \frac{1}{3}, \frac{2}{9})$. Thus $\nu = 3$.

Probabilities $(p_1, p_2) = \frac{3}{4}, \frac{1}{4}$ at the respective points $(s_1, s_2) = (\frac{1}{3}, 1)$ produce asymmetric estimator

$$\widetilde{\alpha}_{n,T,4} := -\left\{\frac{2}{n}\frac{1}{4}\sum_{i=1}^{n}\left[3(\frac{2X_{t_{i-1}} + X_{t_i}}{3})^2 + (X_{t_i})^2\right]\right\}^{-1} \tag{3.53}$$

for which $(\mu_1, \mu_2, \mu_3) = (\frac{1}{2}, \frac{1}{3}, \frac{10}{36})$. Thus $\nu = 3$.

Probabilities $(p_1, p_2, p_3) = (\frac{1}{6}, \frac{2}{3}, \frac{1}{6})$ at the respective points $(s_1, s_2, s_3) = (0, \frac{1}{2}, 1)$ produce the estimator $\widetilde{\alpha}_{n,T,5}$ for which $(\mu_1, \mu_2, \mu_3, \mu_4) = (\frac{1}{2}, \frac{1}{3}, \frac{1}{4}, \frac{5}{25})$. Thus $\nu = 4$.

Probabilities $(p_1, p_2, p_3, p_4) = (\frac{1}{8}, \frac{3}{8}, \frac{3}{8}, \frac{1}{8})$ at the respective points $(s_1, s_2, s_3, s_4) = (0, \frac{1}{3}, \frac{2}{3}, 1)$ produce the symmetric estimator

$$\widetilde{\alpha}_{n,T,5} :=$$
$$-\left\{\frac{2}{n}\frac{1}{8}\sum_{i=1}^{n}\left[(X_{t_{i-1}})^2 + 3(\frac{2X_{t_{i-1}} + X_{t_i}}{3})^2 + 3(\frac{X_{t_{i-1}} + 2X_{t_i}}{3})^2 + (X_{t_i})^2\right]\right\}^{-1} \tag{3.54}$$

for which $(\mu_1, \mu_2, \mu_3, \mu_4) = (\frac{1}{2}, \frac{1}{3}, \frac{1}{4}, \frac{11}{54})$. Thus $\nu = 4$.

Probabilities $(p_1, p_2, p_3, p_4, p_5) = (\frac{1471}{24192}, \frac{6925}{24192}, \frac{1475}{12096}, \frac{2725}{12096}, \frac{5675}{24192}, \frac{1721}{24192})$ at the respective points $(s_1, s_2, s_3, s_4, s_5) = (0, \frac{1}{5}, \frac{2}{5}, \frac{3}{5}, \frac{4}{5}, 1)$ produce the asymmetric estimator

$$\widetilde{\alpha}_{n,T,7} :=$$
$$-\left\{\frac{2}{n}\frac{1}{24192}\sum_{i=1}^{n}\left[1471(X_{t_{i-1}})^2 + 6925(\frac{X_{t_{i-1}} + X_{t_i}}{5})^2 + 2950(\frac{2X_{t_{i-1}} + 2X_{t_i}}{5})^2\right.\right.$$
$$\left.\left.+5450(\tfrac{3X_{t_{i-1}}+3X_{t_i}}{5})^2 + 5675(\tfrac{4X_{t_{i-1}}+4X_{t_i}}{5})^2 + 1721(X_{t_i})^2\right]\right\}^{-1} \tag{3.55}$$

for which $(\mu_1, \mu_2, \mu_3, \mu_4, \mu_5, \mu_6) = (\frac{1}{2}, \frac{1}{3}, \frac{1}{4}, \frac{1}{5}, \frac{841}{5040})$. Thus $\nu = 5$.

Probabilities $(p_1, p_2, p_3, p_4, p_5) = (\frac{7}{90}, \frac{16}{45}, \frac{2}{15}, \frac{16}{45}, \frac{7}{90})$ at the respective points $(s_1, s_2, s_3, s_4, s_5) = (0, \frac{1}{4}, \frac{1}{2}, \frac{3}{4}, 1)$ produce the symmetric estimator

$\widetilde{\alpha}_{n,T,8}$ given by

$$\widetilde{\alpha}_{n,T,8} := -\left\{\frac{2}{n}\frac{1}{90}\sum_{i=1}^{n}\left[7(X_{t_{i-1}})^2 + 32(\frac{3X_{t_{i-1}}+X_{t_i}}{4})^2 + 12(\frac{X_{t_{i-1}}+X_{t_i}}{2})^2 \right.\right.$$
$$\left.\left. +32(\tfrac{X_{t_{i-1}}+3X_{t_i}}{4})^2 + 7(t_i, X_{t_i})^2\right]\right\}^{-1} \quad (3.56)$$

for which $(\mu_1,\mu_2,\mu_3,\mu_4,\mu_5,\mu_6) = (\frac{1}{2},\frac{1}{3},\frac{1}{4},\frac{1}{5},\frac{1}{6},\frac{110}{768})$. Thus $\nu = 6$.

Probabilities $(p_1,p_2,p_3,p_4,p_5) = (\frac{19}{288},\frac{75}{288},\frac{50}{288},\frac{50}{288},\frac{75}{288},\frac{19}{288})$ at the respective points

$(s_1,s_2,s_3,s_4,s_5) = (0,\frac{1}{5},\frac{2}{5},\frac{3}{5},\frac{4}{5},1)$ produce symmetric estimator

$$\widetilde{\alpha}_{n,T,9} :=$$
$$-\left\{\tfrac{2}{n}\tfrac{1}{288}\sum_{i=1}^{n}\left[19(X_{t_{i-1}})^2 + 75(\frac{4X_{t_{i-1}}+X_{t_i}}{5})^2 + 50(\frac{3X_{t_{i-1}}+2X_{t_i}}{5})^2\right.\right.$$
$$\left.\left. +50(\tfrac{2X_{t_{i-1}}+3X_{t_i}}{5})^2 + 75(\tfrac{X_{t_{i-1}}+4X_{t_i}}{5})^2 + 19(X_{t_i})^2\right]\right\}^{-1}$$
$$(3.57)$$

for which $(\mu_1,\mu_2,\mu_3,\mu_4,\mu_5,\mu_6) = (\frac{1}{2},\frac{1}{3},\frac{1}{4},\frac{1}{5},\frac{1}{6},\frac{3219}{22500})$. Thus $\nu = 6$.

The estimator $\widetilde{\alpha}_{n,T,S}$ is based on the arithmetic mean of $I_{n,T}$ and $J_{n,T}$. One can use geometric mean and harmonic mean instead. The geometric mean based symmetric AMCE (which is based on the ideas of partial autocorrelation) is defined as

$$\widetilde{\alpha}_{n,T,G} := \frac{-T/2}{\sqrt{I_{n,T}\,J_{n,T}}}. \quad (3.58)$$

The harmonic mean based symmetric AMCE is defined as

$$\widetilde{\alpha}_{n,T,H} := \frac{-T}{\frac{1}{I_{n,T}}+\frac{1}{J_{n,T}}}. \quad (3.59)$$

Note that

$$\widetilde{\alpha}_{n,T,H} \le \widetilde{\alpha}_{n,T,G} \le \widetilde{\alpha}_{n,T,S}. \quad (3.60)$$

We need the following lemma in the sequel.

Lemma 3.1 Let ξ,η, and ζ be any three random variables on a probability space $(\Omega,\mathcal{F},P)$ with $P(\zeta>0)=1$. Then, for any $\epsilon>0$, we have

(a) $\sup_{x\in\mathbb{R}}|P\{\xi+\eta\le x\} - \Phi(x)| \le \sup_{x\in\mathbb{R}}|P\{\xi\le x\}-\Phi(x)| + P(|\eta|>\epsilon)+\epsilon.$

(b) $\sup_{x\in\mathbb{R}}|P\{\frac{\xi}{\zeta}\le x\} - \Phi(x)| \le \sup_{x\in\mathbb{R}}|P\{\xi\le x\}-\Phi(x)| + P\{|\zeta-1|>\epsilon\}+\epsilon.$

(c) Let $\varrho_n, \tau_n, \varrho$, and τ be random variables on the same probability space $(\Omega, \mathcal{F}, P)$ with $P(\tau_n > 0) = 1$ and $P(\tau > 0) = 1$. Suppose $|\varrho_n - \varrho| = O_P(\delta_{1n})$ and $|\tau_n - \tau| = O_P(\delta_{2n})$, where $\delta_{1n},\ \delta_{2n} \to 0$ as $n \to \infty$. Then,

$$\left|\frac{\varrho_n}{\tau_n} - \frac{\varrho}{\tau}\right| = O_P(\max(\delta_{1n}, \delta_{2n})).$$

(d) (Wick's Lemma or Feynman Diagram Formula)
Let $(\xi_1, \xi_2, \xi_3, \xi_4)$ be a Gaussian random vector with zero mean. Then

$$E(\xi_1\xi_2\xi_3\xi_4) = E(\xi_1\xi_2)E(\xi_3\xi_4) + E(\xi_1\xi_3)E(\xi_2\xi_4) + E(\xi_1\xi_4)E(\xi_2\xi_3).$$

Lemma 3.1 (a) is from Michel and Pfanzagl (1971) and Lemma 3.1 (b) and (c) are from Bishwal and Bose (2001).

Lemma 3.2
(a)

$$E|X_t|^\eta \sim \left(\frac{2}{\beta}\right)^\eta \frac{\Gamma(\alpha/2+\eta)}{\Gamma(\alpha/2)} \quad as \ \ t \to \infty.$$

Hence $\sup_{t\geq 0} E(X_t)^\eta < \infty$.
(b) *For* $q \geq 1$,*with* $0 \leq s < t$ *such that* $0 < t - s < 1$,

$$E|X_t - X_s|^q \leq C(t-s)^{q/2}.$$

(c) *For* $\alpha > 4$,

$$E|X_t^{-1} - X_s^{-1}| \leq C(t-s)^{1/2}.$$

(d)

$$E\left|I_{n,T} - I_T\right|^2 = O\left(\frac{T^4}{n^2}\right).$$

Proof By linearization, we obtain

$$d(e^{\beta t}X_t) = \alpha e^{\beta t}dt + e^{\beta t}\sqrt{2\sigma X_t}dW_t.$$

Hence

$$X_t - X_s = (\alpha/\beta - X_s)(1 - e^{-\beta(t-s)}) + \int_s^t e^{-\beta(t-s)}\sqrt{2\sigma X_u}dW_u.$$

$$E|X_t - X_s|^q \leq 2^{q-1}(1-e^{-\beta(t-s)})^q E|\alpha/\beta - X_s|^q + 2^{q-1}E|\int_s^t e^{-\beta(t-u)}\sqrt{2\sigma X_u}dW_u|^q.$$

The first term on the right side is bounded by $C(t-s)^q$ since $1 - e^{-x} \leq x$ for $x \geq 0$ and $\sup_{t\geq 0} E(X_t^q) < \infty$. For the second term by the Burkholder–Davis–Gundy inequality,

$$E\left|\int_s^t e^{-\beta(t-s)}\sqrt{2\sigma X_u}dW_u\right|^q \le E\left|\int_s^t 4e^{-\beta(t-u)}X_u du\right|^{q/2}$$
$$\le C(t-s)^{q/2-1}\int_s^t e^{-\beta q(t-u)}X_u du \le (t-s)^{q/2}$$

by Holder's inequality since the integral is bounded using $\sup_{u\ge 0} E(X_u^{q/2}) < \infty$.

For part (c), using Holder's inequality with $1/q+2/p=1$ and $2<p<\alpha/2$,

$$E|X_t^{-1} - X_s^{-1}| \le \| X_t - X_s \|_q \| X_t^{-1} \|_p \| X_s^{-1} \|_p \le C(t-s)^{1/2}.$$

(d) Let $h_i(t) := X_{t_{i-1}} - X_t$. Observe that

$$\begin{aligned}
& E|I_{n,T} - I_t|^2 \\
= \;& E\left|\sum_{i=1}^n X_{t_{i-1}}(t_i - t_{i-1}) - \int_0^T X_t\, dt\right|^2 \\
= \;& E\left|\textstyle\sum_{i=1}^n \int_{t_{i-1}}^{t_i} [X_{t_{i-1}} - X_t]\, dt\right|^2 \\
= \;& E\left|\sum_{i=1}^n \int_{t_{i-1}}^{t_i} h_i(t)\, dt\right|^2 \\
= \;& \sum_{i=1}^n E\left|\int_{t_{i-1}}^{t_i} h_i(t)\, dt\right|^2 + 2\sum_{i,j=1,\, i<j}^n E\int_{t_{i-1}}^{t_i}\int_{t_{j-1}}^{t_j} h_i(t)h_j(s)\, dt\, ds \\
=: \;& B_1 + B_2.
\end{aligned} \tag{3.61}$$

Note that

$$\begin{aligned}
& Eh_i^2(t) \\
= \;& E[X_{t_{i-1}} - X_t]^2 \\
= \;& E[\sqrt{X_{t_{i-1}}} - \sqrt{X_t}]^2[\sqrt{X_{t_{i-1}}} + \sqrt{X_t}]^2 \\
\le \;& \{E[\sqrt{X_{t_{i-1}}} - \sqrt{X_t}]^4\}^{1/2}\{E[\sqrt{X_{t_{i-1}}} + \sqrt{X_t}]^4\}^{1/2} \le C(t - t_{i-1})
\end{aligned} \tag{3.62}$$

(by (2.1) and the boundedness of the second term).
Now

$$\begin{aligned}
B_1 &= \sum_{i=1}^n E\left|\int_{t_{i-1}}^{t_i} h_i(t)\, dt\right|^2 \\
&\le \sum_{i=1}^n (t_i - t_{i-1})\int_{t_{i-1}}^{t_i} E(h_i^2(t))\, dt \\
&\le C\tfrac{T}{n}\sum_{i=1}^n \int_{t_{i-1}}^{t_i} (t - t_{i-1})\, dt \\
&= C\tfrac{T^3}{n^2}.
\end{aligned} \tag{3.63}$$

Note that

$$\begin{aligned}
& E[h_i(t)h_j(s)] \\
= \;& E(X_{t_{i-1}} - X_t)(X_{t_{j-1}} - X_s) \\
= \;& E(\sqrt{X_{t_{i-1}}} - \sqrt{X_t})(\sqrt{X_{t_{i-1}}} + \sqrt{X_t})(\sqrt{X_{t_{j-1}}} - \sqrt{X_s})(\sqrt{X_{t_{j-1}}} + \sqrt{X_s}).
\end{aligned} \tag{3.64}$$

Now, using Wick's lemma, it is easy to see that

$$B_2 \le C\frac{T^4}{n^2}.$$

Combining B_1 and B_2, the lemma follows. □

Theorem 3.3 Let $\alpha_{n,T} = T^{-1/2}(\log T)^{1/2} \bigvee \frac{T^2}{n}(\log T)^{-1}$. We have

(a) $\sup_{x\in\mathbb{R}} \left| P\left\{ \left(\frac{T\alpha}{-4\beta}\right)^{1/2} (\beta_{n,T} - \beta) \le x \right\} - \Phi(x) \right| = O(\alpha_{n,T}).$

(b) $\sup_{x\in\mathbb{R}} \left| P\left\{ I_{n,T}^{1/2}(\beta_{n,T} - \beta) \le x \right\} - \Phi(x) \right| = O(\alpha_{n,T}).$

(c) $\sup_{x\in\mathbb{R}} \left| P\left\{ \left(\frac{T\alpha}{|4\beta_{n,T}|}\right)^{1/2} (\beta_{n,T} - \beta) \le x \right\} - \Phi(x) \right| = O(\alpha_{n,T}).$

Proof

(a) It is easy to see that

$$\beta_{n,T} - \beta = \frac{Y_{n,T}}{I_{n,T}} + \beta\frac{V_{n,T}}{I_{n,T}}. \tag{3.65}$$

Hence

$$\begin{aligned}
&\sup_{x\in\mathbb{R}} \left| P\left\{ \left(\frac{T\alpha}{-4\beta}\right)^{1/2} (\beta_{n,T} - \beta) \le x \right\} - \Phi(x) \right| \\
= &\sup_{x\in\mathbb{R}} \left| P\left\{ \left(\frac{T\alpha}{-4\beta}\right)^{1/2} \frac{Y_{n,T}}{I_{n,T}} + \left(\frac{T}{-2\beta}\right)^{1/2} \beta\frac{V_{n,T}}{I_{n,T}} \le x \right\} - \Phi(x) \right| \\
\le &\sup_{x\in\mathbb{R}} \left| P\left\{ \left(\frac{T\alpha}{-4\beta}\right)^{1/2} \frac{Y_{n,T}}{I_{n,T}} \le x \right\} - \Phi(x) \right| \\
&+P\left\{ \left| \left(\frac{T\alpha}{-4\beta}\right)^{1/2} \frac{V_{n,T}}{I_{n,T}} \right| > \epsilon \right\} + \epsilon. \\
=: &K_1 + K_2 + \epsilon.
\end{aligned} \tag{3.66}$$

Note that by Lemma 3.1 (b),

$$
\begin{aligned}
K_1 &= \sup_{x\in\mathbb{R}} \left| P\left\{ \left(\frac{T\alpha}{-4\beta}\right)^{1/2} \frac{Y_{n,T}}{I_{n,T}} \le x \right\} - \Phi(x) \right| \\
&= \sup_{x\in\mathbb{R}} \left| P\left\{ \frac{\left(-\frac{4\beta}{T\alpha}\right)^{1/2} Y_{n,T}}{\left(\frac{-4\beta}{T\alpha}\right) I_{n,T}} \le x \right\} - \Phi(x) \right| \\
&\le \sup_{x\in\mathbb{R}} \left| P\left\{ \left(\frac{-4\beta}{T\alpha}\right)^{1/2} Y_{n,T} \le x \right\} - \Phi(x) \right| \\
&\quad + P\left\{ \left(\tfrac{-4\beta}{T\alpha}\right) I_{n,T} - 1 > \epsilon \right\} + \epsilon \\
&=: J_1 + J_2 + \epsilon.
\end{aligned}
\tag{3.67}
$$

$$
\begin{aligned}
J_1 &= \sup_{x\in\mathbb{R}} \left| P\left\{ \left(\frac{-4\beta}{T\alpha}\right)^{1/2} (Y_{n,T} - Y_T + Y_T) \le x \right\} - \Phi(x) \right| \\
&\le \sup_{x\in\mathbb{R}} \left| P\left\{ \left(\frac{-4\beta}{T\alpha}\right)^{1/2} Y_T \le x \right\} - \Phi(x) \right| \\
&\quad + P\left\{ \left(\frac{-4\beta}{T\alpha}\right)^{1/2} |Y_{n,T} - Y_T| > \epsilon \right\} + \epsilon \\
&\le CT^{-1/2} + \left(\frac{-4\beta}{T\alpha}\right) \frac{E|Y_{n,T} - Y_T|^2}{\epsilon^2} + \epsilon \\
&\le CT^{-1/2} + C\frac{T/n}{\epsilon^2} + \epsilon \\
&\quad \text{(by Corollary 2.3(a) and Lemma 3.2(a)).}
\end{aligned}
\tag{3.68}
$$

$$
\begin{aligned}
J_2 &= P\left\{ \left| \left(\frac{-4\beta}{T\alpha}\right) (I_{n,T} - I_T + I_T) - 1 \right| > \epsilon \right\} \\
&\le P\left\{ \left| \left(\frac{-4\beta}{T\alpha}\right) I_T - 1 \right| > \frac{\epsilon}{2} \right\} + P\left\{ \left(-\frac{4\beta}{2}\right) |I_{n,T} - I_T| > \frac{\epsilon}{2} \right\} \\
&\le C\exp\left(\frac{T\beta}{16}\epsilon^2\right) + \frac{16\beta^2}{T^2} \frac{E|I_{n,T} - I_T|^2}{\epsilon^2} \\
&\le C\exp\left(\frac{T\beta}{16}\epsilon^2\right) + C\frac{T^2/n^2}{\epsilon^2}.
\end{aligned}
\tag{3.69}
$$

Here the bound for the first term in (3.8) comes from Lemma 2.4(a) and that for the second term from Lemma 3.2(c). From the proof of Lemma 3.2(b), we have

$$
E|V_{n,T}|^2 \le C\tfrac{T^4}{n^2} \tag{3.70}
$$

$$
\begin{aligned}
K_2 &= P\left\{\left|\left(\frac{T\alpha}{-4\beta}\right)^{1/2}\beta\frac{V_{n,T}}{I_{n,T}}\right|>\epsilon\right\}\\
&= P\left\{\left|\frac{\left(\frac{-4\beta}{T\alpha}\right)^{1/2}\beta V_{n,T}}{\left(-\frac{4\beta}{T\alpha}\right)I_{n,T}}\right|>\epsilon\right\}\\
&= P\left\{\left|\left(\frac{-4\beta}{T\alpha}\right)^{1/2}\beta V_{n,T}\right|>\delta\right\}+P\left\{\left(-\frac{4\beta}{T\alpha}\right)I_{n,T}<\frac{\delta}{\epsilon}\right\}\\
&\quad \text{(where we choose } \delta=\epsilon-C\epsilon^2)\\
&\le P\left\{\left|\left(\frac{-4\beta}{T\alpha}\right)^{1/2}\beta V_{n,T}\right|>\delta\right\}+P\left\{\left|\left(-\frac{4\beta}{T\alpha}\right)I_{n,T}-1\right|>\delta_1\right\}\\
&\quad \text{(where } \delta_1=\frac{\epsilon-\delta}{\delta}=C\epsilon)\\
&\le -\frac{4\beta}{T\alpha}\beta^2\frac{E|V_{n,T}|^2}{\delta^2}+C\exp\left(\frac{T\beta}{16}\delta_1^2\right)+C\frac{T^2/n^2}{\delta_1^2}\\
&\le C\frac{T^3/n^2}{\delta^2}+C\exp\left(\frac{T\beta}{16}\delta_1^2\right)+C\frac{T^2/n^2}{\delta_1^2} \quad \text{(by (3.9) and (3.8)).}
\end{aligned}
\tag{3.71}
$$

Now from (3.4)–(3.10), since $T/n \to 0$,

$$
\begin{aligned}
&\sup_{x\in\mathbb{R}}\left|P\left\{\left(\frac{T\alpha}{-4\beta}\right)^{1/2}(\beta_{n,T}-\beta)\le x\right\}-\Phi(x)\right|\\
\le\ & CT^{-1/2}+C\exp\left(\frac{T\beta}{16}\epsilon^2\right)+C\frac{T/n}{\epsilon^2}+C\frac{T^2/n^2}{\epsilon^2}+C\frac{T^3/n^2}{\delta^2}\\
&+C\exp\left(\frac{T\beta}{16}\delta_1^2\right)+C\left(\frac{T^2/n^2}{\delta_1^2}\right)+\epsilon.
\end{aligned}
\tag{3.72}
$$

Choosing $\epsilon = T^{-1/2}(\log T)^{1/2}$, the terms of (3.11) are of the order $O(T^{-1/2}(\log T)^{1/2} \bigvee \frac{T^2}{n}(\log T)^{-1})$. This proves (a).

(b) Using the expression (3.3),

$$
\begin{aligned}
&\sup_{x\in\mathbb{R}}\left|P\left\{I_{n,T}^{1/2}(\beta_{n,T}-\beta)\le x\right\}-\Phi(x)\right|\\
=\ &\sup_{x\in\mathbb{R}}\left|P\left\{\frac{Y_{n,T}}{I_{n,T}^{1/2}}+\beta\frac{V_{n,T}}{I_{n,T}^{1/2}}\le x\right\}-\Phi(x)\right|\\
\le\ &\sup_{x\in\mathbb{R}}\left|P\left\{\frac{Y_{n,T}}{I_{n,T}^{1/2}}\le x\right\}-\Phi(x)\right|+P\left\{\left|\beta\frac{V_{n,T}}{I_{n,T}^{1/2}}\right|>\epsilon\right\}+\epsilon.\\
=:\ & H_1+H_2+\epsilon.
\end{aligned}
\tag{3.73}
$$

Note that

$$\begin{aligned} H_1 &= \sup_{x\in\mathbb{R}}\left|P\left\{\frac{Y_{n,T}-Y_T+Y_T}{I_{n,T}^{1/2}}\le x\right\}-\Phi(x)\right| \\ &= \sup_{x\in\mathbb{R}}\left|P\left\{\frac{Y_T}{I_{n,T}^{1/2}}\le x\right\}-\Phi(x)\right|+P\left\{\frac{|Y_{n,T}-Y_T|}{I_{n,T}^{1/2}}>\epsilon\right\}+\epsilon \\ &=: F_1+F_2+\epsilon. \end{aligned} \tag{3.74}$$

Now

$$\begin{aligned} F_1 &= \sup_{x\in\mathbb{R}}\left|P\left\{\frac{Y_T}{I_{n,T}^{1/2}}\le x\right\}-\Phi(x)\right| \le \sup_{x\in\mathbb{R}}\left|P\left\{\left(\frac{-4\beta}{T\alpha}\right)^{1/2} Y_T\le x\right\}-\Phi(x)\right| \\ &\quad +P\left\{\left|\left(\frac{-4\beta}{T\alpha}\right)^{1/2} I_{n,T}^{1/2}-1\right|>\epsilon\right\}+\epsilon \text{ (by Lemma 3.1(b))} \\ &\le C_T^{-1/2}+P\left\{\left|\left(\frac{-4\beta}{T\alpha}\right) I_{n,T}-1\right|>\epsilon\right\}+\epsilon \text{ (by Corollary 2.3(a))} \\ &\le CT^{-1/2}+C\exp\left(-\tfrac{T\beta}{16}\epsilon^2\right)+C\tfrac{T^2/n^2}{\epsilon^2}+\epsilon \text{ (by (3.8)).} \end{aligned} \tag{3.75}$$

On the other hand,

$$\begin{aligned} F_2 &= P\left\{\frac{|Y_{n,T}-Y_T|}{I_{n,T}^{1/2}}>\epsilon\right\} \\ &\le P\left\{\left(\frac{-4\beta}{T\alpha}\right)^{1/2}|Y_{n,T}-Y_T|>\delta\right\}+P\left\{\left|\left(\frac{-4\beta}{T\alpha}\right)^{1/2} I_{n,T}^{1/2}-1\right|>\delta_1\right\} \\ &\quad \text{(where } \delta=\epsilon-C\epsilon^2 \text{ and } \delta_1=(\epsilon-\delta)/\epsilon>0) \\ &\le \frac{\left(\frac{-4\beta}{T\alpha}\right)E|Y_{n,T}-Y_T|^2}{\delta^2}+P\left\{\left|\left(\frac{-2\beta}{T}\right) I_{n,T}-1\right|>\delta_1\right\} \\ &\le C\frac{T/n}{\delta^2}+C\exp\left(\frac{T\beta}{16}\delta_1^2\right)+C\frac{T^2/n^2}{\delta_1^2} \text{ (from Lemma 3.2(a) and (3.8)).} \end{aligned} \tag{3.76}$$

Using (3.15) and (3.14) in (3.13), we obtain

$$\begin{aligned} H_1 &= \sup_{x\in\mathbb{R}}\left|P\left\{\frac{Y_T}{I_{n,T}^{1/2}}\le x\right\}-\Phi(x)\right| \\ &\le CT^{-1/2}+C\exp\left(\frac{T\beta}{16}\epsilon^2\right)+C\frac{T/n}{\delta^2}+C\frac{T^2/n^2}{\delta_1^2} \\ &\quad +C\exp\left(\frac{T\beta}{16}\delta_1^2\right)+C\frac{T^2/n^2}{\epsilon^2}+\epsilon. \end{aligned} \tag{3.77}$$

$$
\begin{aligned}
H_2 &= P\left\{\left|\beta\frac{V_{n,T}}{I_{n,T}^{1/2}}\right| > \epsilon\right\} = P\left\{\frac{\left|\left(\frac{-4\beta}{T\alpha}\right)^{1/2}\beta V_{n,T}\right|}{\left|\left(\frac{-4\beta}{T\alpha}\right)^{1/2} I_{n,T}^{1/2}\right|} > \epsilon\right\} \\
&\le P\left\{\left|\left(\frac{-4\beta}{T\alpha}\right)^{1/2}\beta V_{n,T}\right| > \delta\right\} + P\left\{\left|\left(\frac{-4\beta}{T\alpha}\right)^{1/2} I_{n,T}^{1/2}\right| < \delta/\epsilon\right\} \\
&\le \left(-\frac{4\beta}{T\alpha}\right)\beta^2\frac{E|V_{n,T}|^2}{\delta^2} + P\left\{\left|\left(-\frac{2\beta}{T}\right)I_{n,T} - 1\right| > \delta_1\right\} \\
&\quad \text{(where } 0 < \delta < \epsilon \text{ and } \delta_1 = (\epsilon-\delta)/\epsilon = C\epsilon > 0) \\
&\le C\frac{T^2/n^2}{\delta^2} + C\exp\left(\frac{T\beta}{16}\delta_1^2\right) + C\frac{T^2/n^2}{\delta_1^2}. \quad \text{(from (3.9) and (3.8)).}
\end{aligned}
\tag{3.78}
$$

Using (3.17) and (3.16) in (3.12) and choosing $\epsilon = T^{-1/2}(\log T)^{1/2}$, the terms of (3.12) are of the order $O(T^{-1/2}(\log T)^{1/2} \bigvee \frac{T^2}{n}(\log T)^{-1})$. This proves (b).

(c) Let $D_T = \{|\beta_{n,T} - \beta| \le d_T\}$ and $d_T = CT^{-1/2}(\log T)^{1/2}$.

On the set D_T, expanding $(2|\beta_{n,T}|)^{-1/2}$, we obtain

$$
\begin{aligned}
(-2\beta_{n,T})^{-1/2} &= (-2\beta)^{-1/2}\left[1 - \frac{\beta - \beta_{n,T}}{\beta}\right]^{-1/2} \\
&= (-2\beta)^{-1/2}\left[1 + \frac{1}{2}\left(\frac{\beta - \beta_{n,T}}{\beta}\right) + O(d_T^2)\right].
\end{aligned}
$$

Then

$$
\begin{aligned}
&\sup_{x\in\mathbb{R}}\left|P\left\{\left(\frac{T\alpha}{4|\beta_{n,T}|}\right)^{1/2}(\beta_{n,T} - \beta) \le x\right\} - \Phi(x)\right| \\
\le\ &\sup_{x\in\mathbb{R}}\left|P\left\{\left(\frac{T\alpha}{4|\beta_{n,T}|}\right)^{1/2}(\beta_{n,T} - \beta) \le x, D_T\right\} - \Phi(x)\right| + P(D_T^c).
\end{aligned}
\tag{3.79}
$$

$$
\begin{aligned}
&P(D_T^c) \\
=\ &P\left\{|\beta_{n,T} - \beta| > CT^{-1/2}(\log T)^{1/2}\right\} \\
=\ &P\left\{\left(\frac{T\alpha}{-4\beta}\right)^{1/2}|\beta_{n,T} - \beta| > C(\log T)^{1/2}(-4\beta)^{-1/2}\right\} \\
\le\ &C(T^{-1/2}(\log T)^{1/2} \bigvee \frac{T^2}{n}(\log T)^{-1}) + 2(1 - \Phi\left((\log T)^{1/2}(-2\beta)^{-1/2}\right) \\
&\quad \text{(by Theorem 3.3(a))} \\
\le\ &C(T^{-1/2}(\log T)^{1/2} \bigvee \frac{T^2}{n}(\log T)^{-1}).
\end{aligned}
$$

On the set D_T,

$$
\left|\left(\frac{\beta_{n,T}}{\beta}\right)^{1/2} - 1\right| \le CT^{-1/2}(\log T)^{1/2}.
$$

Hence, upon choosing $\epsilon = CT^{-1/2}(\log T)^{1/2}, C$ large, we obtain

$$\begin{aligned}
&\left| P\left\{ \left(\frac{T\alpha}{-4\beta_{n,T}} \right)^{1/2} (\beta_{n,T} - \beta) \le x, D_T \right\} - \Phi(x) \right| \\
\le\ & \left| P\left\{ \left(\frac{T\alpha}{-4\beta} \right)^{1/2} (\beta_{n,T} - \beta) \le x, D_T \right\} - \Phi(x) \right| \\
& + P\left\{ \left| \left(\frac{\beta_{n,T}}{\beta} \right)^{1/2} - 1 \right| > \epsilon, D_T \right\} + \epsilon \\
& \quad \text{(by Lemma 3.1(b))} \\
\le\ & C(T^{-1/2}(\log T)^{1/2} \bigvee \frac{T^2}{n}(\log T)^{-1}) \\
& \text{(by Theorem 3.3(a)).} \qquad (3.80)
\end{aligned}$$

(c) follows from (3.18)–(3.20). □

Theorem 3.4

$$\sup_{x \in \mathbb{R}} \left| P\left\{ I_{n,T} \left(-\frac{4\beta}{T\alpha} \right)^{1/2} (\beta_{n,T} - \beta) \le x \right\} - \Phi(x) \right| = O\left(T^{-1/2} \bigvee \left(\frac{T}{n} \right)^{1/3} \right).$$

Proof Let $a_{n,T} = Z_{n,T} - Z_T$ and $b_{n,T} = I_{n,T} - I_T$.

By Lemma 3.2, $E|a_{n,T}|^2 = O\left(\frac{T^2}{n}\right)$ and $E|b_{n,T}|^2 = O\left(\frac{T^4}{n^2}\right)$. (3.81)

From (3.5),

$$\begin{aligned}
I_{n,T}\beta_{n,T} &= \sum_{i=1}^{n} X_{t_{i-1}} \left[X_{t_i} - X_{t_{i-1}} \right] \\
&= \int_0^T X_t dX_t + a_{n,T} \\
&= \int_0^T X_t dW_t + \beta \int_0^T X_t^2 dt + a_{n,T}.
\end{aligned}$$

Hence $I_{n,T}(\beta_{n,T} - \beta) = -\beta b_{n,T} + a_{n,T}$.

Thus

$$\begin{aligned}
&\sup_{x\in\mathbb{R}}\left|P\left\{I_{n,T}\left(-\frac{4\beta}{T\alpha}\right)^{1/2}(\beta_{n,T}-\beta)\le x\right\}-\Phi(x)\right| \\
=\ &\sup_{x\in\mathbb{R}}\left|P\left\{\left(-\frac{4\beta}{T\alpha}\right)^{1/2}[Y_T-\beta b_{n,T}+a_{n,T}]\le x\right\}-\Phi(x)\right| \\
\le\ &\sup_{x\in\mathbb{R}}\left|P\left\{\left(-\frac{4\beta}{T\alpha}\right)^{1/2}Y_T\le x\right\}-\Phi(x)\right| \\
&+P\left\{\left|\left(-\frac{4\beta}{T\alpha}\right)^{1/2}[-\beta b_{n,T}+a_{n,T}]\right|>\epsilon\right\}+\epsilon \\
\le\ & CT^{-1/2}+\left(-\frac{4\beta}{T\alpha}\right)\frac{E|-\beta b_{n,T}+a_{n,T}|^2}{\epsilon^2}+\epsilon \\
\le\ & CT^{-1/2}+C\frac{T/n}{\epsilon^2}+\epsilon \text{ (by Corollary 2.3(a) and (3.21)).}
\end{aligned}$$

Choosing $\epsilon=\left(\frac{T}{n}\right)^{1/3}$, the rate is $O\left(T^{-1/2}\bigvee\left(\frac{T}{n}\right)^{1/3}\right)$. □

Theorem 3.5

$$|\beta_{n,T}-\beta_T|=O_P\left(\frac{T^2}{n}\right)^{1/2}.$$

Proof Note that $\beta_{n,T}-\beta_T=\frac{Z_{n,T}}{I_{n,T}}-\frac{Z_T}{I_T}$.

From Lemma 3.2, it follows that $|Z_{n,T}-Z_T|=O_P\left(\frac{T^2}{n}\right)^{1/2}$ and $|I_{n,T}-I_T|=O_P\left(\frac{T^4}{n^2}\right)^{1/2}$. Now the theorem follows easily from Lemma 2.1. □

Bessel Process

If $\beta=0$, the model is two-dimensional Bessel process

$$dX_t=\alpha dt+2\sqrt{X_t}dW_t,$$

and the MLE is

$$\hat{\alpha}_T=\frac{\int_0^T X_t^{-1}dX_t}{\int_0^T X_t^{-2}dt}=\frac{\log X_T-\log X_0+2\int_0^T X_t^{-1}dt}{\int_0^T X_t^{-1}dt}.$$

In this case, we have a different rate of convergence.

Theorem 3.6 Denote $b_{n,T}:=O(\max((\log T)^{-1/2},(\frac{T^4}{n^2})(\log T)^{-2}))$. If $\beta=0$ and $\alpha>2$, then

(a) $$\sup_{x\in\mathbb{R}}\left|P\left\{\left(\frac{\log T}{4(\alpha-2)}\right)^{1/2}(\widetilde{\alpha}_{n,T,F}-\alpha)\le x\right\}-\Phi(x)\right|=O(b_{n,T}).$$

(b) $\sup_{x\in\mathbb{R}}\left|P\left\{I_{n,T}^{1/2}(\widetilde{\alpha}_{n,T,F}-\alpha)\le x\right\}-\Phi(x)\right| = O(b_{n,T})$.

(c) $\sup_{x\in\mathbb{R}}\left|P\left\{\left(\frac{\log T}{4|\widetilde{\alpha}_{n,T,F}-2|}\right)^{1/2}(\widetilde{\alpha}_{n,T,F}-\alpha)\le x\right\}-\Phi(x)\right| = O(b_{n,T})$.

Theorem 3.7 Denote $b_{n,T} := O(\max(T^{-1/2}(\log T)^{1/2}, (\frac{T^4}{n^2})(\log T)^{-1}))$. If $\beta < 0$ and $\alpha > 2$, then

(a) $\sup_{x\in\mathbb{R}}\left|P\left\{\left(-\frac{T\beta}{4(\alpha-2)}\right)^{1/2}(\widetilde{\alpha}_{n,T,F}-\alpha)\le x\right\}-\Phi(x)\right| = O(b_{n,T})$.

(b) $\sup_{x\in\mathbb{R}}\left|P\left\{I_{n,T}^{1/2}(\widetilde{\alpha}_{n,T,F}-\alpha)\le x\right\}-\Phi(x)\right| = O(b_{n,T})$.

(c) $\sup_{x\in\mathbb{R}}\left|P\left\{\left(\frac{T\beta}{4|\widetilde{\alpha}_{n,T,F}-2|}\right)^{1/2}(\widetilde{\alpha}_{n,T,F}-\alpha)\le x\right\}-\Phi(x)\right| = O(b_{n,T})$.

Proof

(a) Observe that

$$\left(-\frac{T\beta}{4(\alpha-2)}\right)^{1/2}(\widetilde{\alpha}_T-\alpha) = \frac{\left(-\frac{4(\alpha-2)}{T\beta}\right)^{1/2}Y_T}{\left(-\frac{4(\alpha-2)}{T\beta}\right)I_T}, \tag{3.82}$$

where

$$Y_T := -(\alpha-2)I_T - \frac{T\beta}{4} \quad \text{and} \quad I_T := \int_0^T X_t^2 dt.$$

Thus, we have

$$I_{n,T}\widetilde{\alpha}_{n,T,F} = Y_T + \alpha I_T.$$

Hence,

$$\begin{aligned}
&\left(-\frac{T\beta}{4(\alpha-2)}\right)^{1/2}(\widetilde{\alpha}_{n,T,F}-\alpha)\\
&= \frac{\left(-\frac{T\beta}{2\alpha}\right)^{1/2}Y_T + \alpha\left(-\frac{T\beta}{2\alpha}\right)^{1/2}(I_T - I_{n,T})}{I_{n,T}}\\
&= \frac{\left(-\frac{4(\alpha-2)}{T\beta}\right)^{1/2}Y_T + \left(-\frac{4(\alpha-2)}{T\beta}\right)^{1/2}(I_T - I_{n,T})}{\left(-\frac{4(\alpha-2)}{T\beta}\right)I_{n,T}}.
\end{aligned} \tag{3.83}$$

Further,

$$P\left\{\left|\left(\frac{-4(\alpha-2)}{T\beta}\right)(I_{n,T}-1)\right| > \epsilon\right\}$$

$$
\begin{aligned}
&= P\left\{\left|\left(\frac{-4(\alpha-2)}{T\beta}\right)(I_{n,T}-I_T+I_T)-1\right|>\epsilon\right\}\\
&\le P\left\{\left|\left(\frac{-4(\alpha-2)}{T\beta}\right)I_T-1\right|>\frac{\epsilon}{2}\right\}+P\left\{\left(-\frac{4(\alpha-2)}{T\beta}\right)|I_{n,T}-I_T|>\frac{\epsilon}{2}\right\}\\
&\le C\exp\left(\frac{T\alpha}{16}\epsilon^2\right)+\frac{16\alpha^2}{T^2}\frac{E|I_{n,T}-I_T|^2}{\epsilon^2}\\
&\le C\exp\left(\frac{T\alpha}{16}\epsilon^2\right)+C\frac{T^2/n^2}{\epsilon^2}.
\end{aligned}
$$

(3.84)

Next, observe that

$$
\begin{aligned}
&\sup_{x\in\mathbb{R}}\left|P\left\{\left(-\frac{T}{2\alpha}\right)^{1/2}(\widetilde{\alpha}_{n,T,F}-\alpha)\le x\right\}-\Phi(x)\right|\\
&= \sup_{x\in\mathbb{R}}\left|P\left\{\frac{\left(-\frac{4(\alpha-2)}{T\beta}\right)^{1/2}Y_T+\left(-\frac{4(\alpha-2)}{T}\right)^{1/2}(I_T-I_{n,T})}{\left(-\frac{2\alpha}{T}\right)I_{n,T}}\le x\right\}-\Phi(x)\right|\\
&\le \sup_{x\in\mathbb{R}}\left|P\left\{\left(-\frac{4(\alpha-2)}{T\beta}\right)^{1/2}Y_T\le x\right\}-\Phi(x)\right|\\
&\quad +P\left\{\left|\alpha\left(-\frac{4(\alpha-2)}{T\beta}\right)^{1/2}(I_{n,T}-I_T)\right|>\epsilon\right\}\\
&\quad +P\left\{\left|\left(-\frac{4(\alpha-2)}{T\beta}\right)I_{n,T}-1\right|>\epsilon\right\}+2\epsilon\\
&\le CT^{-1/2}+\alpha^2\frac{\left(-\frac{4(\alpha-2)}{T\beta}\right)E|I_{n,T}-I_T|^2}{\epsilon^2}\\
&\quad +C\exp\left(\frac{T\alpha}{4}\epsilon^2\right)+C\frac{T^2}{n^2\epsilon^2}+2\epsilon(3.85)
\end{aligned}
$$

(the bound for the third term in the right hand side of (2.4) is obtained from (2.3))

$$
\le CT^{-1/2}+C\frac{T^2}{n^2\epsilon^2}+C\exp\left(\frac{T\alpha}{4}\epsilon^2\right)+C\frac{T}{n^2\epsilon^2}+\epsilon
$$

(by Lemma 2.3(a)).

(3.86)

Choosing $\epsilon=CT^{-1/2}(\log T)^{1/2}$, the terms in the right hand side of (2.5) are of the order
$O(\max(T^{-1/2}(\log T)^{1/2},(\frac{T^4}{n^2})(\log T)^{-1}))$. □

(b) From (2.1), we have

$$
I_{n,T}^{1/2}(\widetilde{\alpha}_{n,T,F}-\alpha)=\frac{Y_T+\alpha(I_T-I_{n,T})}{I_{n,T}^{1/2}}.
$$

Then,

$$
\begin{aligned}
&\sup_{x\in\mathbb{R}}\left|P\left\{I_{n,T}^{1/2}(\widetilde{\alpha}_{n,T,F}-\alpha)\le x\right\}-\Phi(x)\right| \\
= &\sup_{x\in\mathbb{R}}\left|P\left\{\frac{Y_T}{I_{n,T}^{1/2}}+\alpha\frac{I_T-I_{n,T}}{I_{n,T}^{1/2}}\le x\right\}-\Phi(x)\right| \\
\le &\sup_{x\in\mathbb{R}}\left|P\left\{\frac{Y_T}{I_{n,T}^{1/2}}\le x\right\}-\Phi(x)\right|+P\left\{\left|\frac{\alpha(I_T-I_{n,T})}{I_{n,T}^{1/2}}\right|>\epsilon\right\}+\epsilon \\
=: &\ U_1+U_2+\epsilon. \qquad (3.87)
\end{aligned}
$$

We have from (2.3),

$$
U_1\le CT^{-1/2}+C\exp\left(\frac{T\alpha}{16}\epsilon^2\right)+C\frac{T^2}{n^2\epsilon^2}+\epsilon. \qquad (3.88)
$$

Now,

$$
\begin{aligned}
U_2 &= P\left\{|\alpha|\left|\frac{I_{n,T}-I_T}{I_{n,T}^{1/2}}\right|>\epsilon\right\} \\
&= P\left\{|\alpha|\frac{\left|\left(-\frac{4(\alpha-2)}{T}\right)^{1/2}(I_{n,T}-I_T)\right|}{\left|\left(-\frac{2\alpha}{T}\right)^{1/2}I_{n,T}^{1/2}\right|}>\epsilon\right\} \\
&\le P\left\{\left|\left(-\frac{4(\alpha-2)}{T}\right)^{1/2}\right||I_{n,T}-I_T|>\delta\right\} \\
&\quad +P\left\{\left|\left(-\frac{4(\alpha-2)}{T}\right)^{1/2}I_{n,T}^{1/2}-1\right|>\delta_1\right\} \qquad (3.89)\\
&\quad (\text{where } \delta=\epsilon-C\epsilon^2 \text{ and } \delta_1=(\epsilon-\delta)/\epsilon>0) \\
&\le \left(-\frac{4(\alpha-2)}{T}\right)\frac{E|I_{n,T}-I_T|^2}{\delta^2}+P\left\{\left|\left(-\frac{4(\alpha-2)}{T}\right)I_{n,T}-1\right|>\delta_1\right\} \\
&\le C\frac{T^3}{n^2\delta^2}+C\exp\left(\frac{T\alpha}{16}\delta_1^2\right)+C\frac{T^2}{n^2\delta_1^2}. \qquad (3.90)
\end{aligned}
$$

Here, the bound for the first term in the right hand side of (3.89) comes from Lemma 2.2(c) and that for the second term is obtained from (2.3).

Now, using the bounds (3.88) and (3.90) in (3.87) with $\epsilon = CT^{-1/2}$ $(\log T)^{1/2}$, we obtain that the terms in (3.87) are of the order $O(\max(T^{-1/2}(\log T)^{1/2}, (\frac{T^4}{n^2})(\log T)^{-1}))$. □

(c) Let

$$
G_T := \{|\widetilde{\alpha}_{n,T,F}-\alpha|\le d_T\}, \qquad \text{and} \qquad d_T := CT^{-1/2}(\log T)^{1/2}.
$$

On the set G_T, expanding $(2|\widetilde{\alpha}_{n,T,F}|)^{1/2}$, we obtain

$$(-2\widetilde{\alpha}_{n,T})^{-1/2} = (-4(\alpha-2))^{1/2}\left[1-\frac{\alpha-\widetilde{\alpha}_{n,T,F}}{\alpha}\right]^{-1/2}$$

$$= (-2\alpha)^{1/2}\left[1+\frac{1}{2}\left(\frac{\alpha-\widetilde{\alpha}_{n,T,F}}{\alpha}\right)\right]+O(d_T^2).$$

Then,

$$\begin{aligned}
&\sup_{x\in\mathbb{R}}\left|P\left\{\left(\frac{T\beta}{4|\widetilde{\alpha}_{n,T,F-2}|}\right)^{1/2}(\widetilde{\alpha}_{n,T,F}-\alpha)\le x\right\}-\Phi(x)\right| \\
\le\ &\sup_{x\in\mathbb{R}}\left\{P\left(\frac{T\beta}{2|\widetilde{\alpha}_{n,T,F-2}|}\right)^{1/2}(\widetilde{\alpha}_{n,T,F}-\alpha)\le x, G_T\right\}+P(G_T^c).
\end{aligned}$$

Now,

$$\begin{aligned}
P(G_T^c) &= P\left\{|\widetilde{\alpha}_{n,T,F}-\alpha|>CT^{-1/2}(\log T)^{1/2}\right\} \\
&= P\left\{\left(-\frac{T\beta}{4(\alpha-2)}\right)^{1/2}|\widetilde{\alpha}_{n,T,F}-\alpha|>C(\log T)^{1/2}(-4(\alpha-2))^{-1/2}\right\} \\
&\le C\max\left(T^{-1/2}(\log T)^{1/2}, \frac{T^3}{n^2}(\log T)^{-1}\right) \\
&\quad +2(1-\Phi\log T^{1/2}(-4(\alpha-2))^{-1/2}) \\
&\quad \text{(by Theorem 2.1(a))} \\
&\le C\max\left(T^{-1/2}(\log T)^{1/2}, \frac{T^3}{n^2}(\log T)^{-1}\right).
\end{aligned}$$

On the set G_T,

$$\left|\left(\frac{\widetilde{\alpha}_{n,T,F}}{\alpha}\right)^{1/2}-1\right|\le CT^{-1/2}(\log T)^{1/2}.$$

Hence, upon choosing $\epsilon = CT^{-1/2}(\log T)^{1/2}, C$ large, we obtain

$$\begin{aligned}
&\left|P\left\{\left(\frac{T\beta}{-4(\widetilde{\alpha}_{n,T,F}-2)}\right)^{1/2}(\widetilde{\alpha}_{n,T,F}-\alpha)\le x, G_T\right\}-\Phi(x)\right| \\
\le\ &\left|P\left\{\left(\frac{T\beta}{-4(\alpha-2)}\right)^{1/2}(\widetilde{\alpha}_{n,T,F}-\alpha)\le x, G_T\right\}\right| \\
&+P\left\{\left|\left(\frac{\widetilde{\alpha}_{n,T,F}}{\alpha}\right)^{1/2}-1\right|>\epsilon, G_T\right\}+\epsilon \\
&\text{(by Lemma 1.1(b))}
\end{aligned}$$

$$\leq \; C \max\left(T^{-1/2}(\log T)^{1/2}, \frac{T^4}{n^2}(\log T)^{-1}\right)$$
(by Theorem 2.1(a)). □

In the following theorem, we improve the bound on the error of normal approximation using a mixture of random and nonrandom normings. Thus asymptotic normality of the AMCEs needs $T \to \infty$ and $\frac{T}{n^{2/3}} \to 0$ which are sharper than the bound in Theorem 2.1.

Theorem 3.8

$$\begin{aligned}&\sup_{x\in\mathbb{R}}\left|P\left\{I_{n,T}\left(-\tfrac{4(\alpha-2)}{T\beta}\right)^{1/2}(\widetilde{\alpha}_{n,T,F}-\alpha)\leq x\right\}-\Phi(x)\right|\\ =\;& O\left(\max\left(T^{-1/2},\left(\tfrac{T^3}{n^2}\right)^{1/3}\right)\right).\end{aligned}$$

Proof From (2.2), we have

$$\begin{aligned}&I_{n,T}\left(-\tfrac{4(\alpha-2)}{T}\right)^{1/2}(\widetilde{\alpha}_{n,T,F}-\alpha)\\ =\;&\left(-\tfrac{4(\alpha-2)}{T\beta}\right)^{1/2}Y_T+\alpha\left(-\tfrac{4(\alpha-2)}{T\beta}\right)^{1/2}(I_T-I_{n,T}).\end{aligned}$$

Hence, by Lemmas 2.1–2.3,

$$\begin{aligned}&\sup_{x\in\mathbb{R}}\left|P\left\{I_{n,T}\left(-\tfrac{4(\alpha-2)}{T\beta}\right)^{1/2}(\widetilde{\alpha}_{n,T,F}-\alpha)\leq x\right\}-\Phi(x)\right|\\ =\;&\sup_{x\in\mathbb{R}}\left|P\left\{\left(-\tfrac{4(\alpha-2)}{T\beta}\right)^{1/2}Y_T+\alpha\left(-\tfrac{4(\alpha-2)}{T\beta}\right)^{1/2}(I_T-I_{n,T})\leq x\right\}-\Phi(x)\right|\\ \leq\;&\sup_{x\in R}\left|P\left\{\left(-\tfrac{4(\alpha-2)}{T\beta}\right)^{1/2}Y_T\leq x\right\}-\Phi(x)\right|\\ &+P\left\{\left|\alpha\left(-\tfrac{4(\alpha-2)}{T\beta}\right)^{1/2}(I_T-I_{n,T})\right|>\epsilon\right\}+\epsilon\\ \leq\;& CT^{-1/2}+C\tfrac{E|I_T-I_{n,T}|^2}{T\epsilon^2}+\epsilon\leq CT^{-1/2}+C\tfrac{T^3}{n^2\epsilon^2}+\epsilon.\end{aligned}$$

Choosing $\epsilon = (\frac{T^3}{n^2})^{1/3}$, the theorem follows. □

The following theorem gives stochastic bound on the error of approximation of the continuous MCE by AMCEs.

Theorem 3.9

$$\text{(a)}\;\; |\widetilde{\alpha}_{n,T}-\widetilde{\alpha}_T| = O_P\left(\frac{T}{n}\right)^{1/2},$$

$$\text{(b)}\;\; |\widetilde{\alpha}_{n,T,z}-\widetilde{\alpha}_T| = O_P\left(\frac{T^2}{n^2}\right)^{1/2}.$$

Proof From (1.9) and (1.14), we have $\widetilde{\alpha}_T = -\frac{T}{2I_T}$ and $\widetilde{\alpha}_{n,T} = -\frac{T}{2I_{n,T}}$. Hence, applying Lemma 1.2 with the aid of Lemma 2.3 and noting that $|\frac{I_{n,T}}{T}| = O_P(1)$ and $|\frac{I_T}{T}| = O_P(1)$, the part (a) of theorem follows. From (1.9) and (1.16), we have $\widetilde{\alpha}_T = -\frac{T}{2I_T}$ and $\widetilde{\alpha}_{n,T,z} = -\frac{T}{I_{n,t}+J_{n,T}}$.

Applying Lemma 1.2 with the aid of Lemma 2.3 and noting that $|\frac{J_{n,T}}{T}| = O_P(1)$ and $|\frac{I_T}{T}| = O_P(1)$, the part (b) of theorem follows. □

3.4 Estimators in the Supercritical CIR Process

We show that the distribution of the normalized ratio estimator of the drift parameter in the Cox–Ingersoll–Ross process observed over $[0, T]$ converges to the normal distribution with a uniform error rate of the order $O(T^{-1/2})$. Then we obtain the rate of normal approximation of the approximate minimum contrast estimator when the process is densely observed at discrete time points.

The Cox–Ingersoll–Ross model, also called the square-root process, is used as a short rate mean reverting model in term structure of interest rates. In view of this, it becomes necessary to estimate the unknown parameters in the model from discrete data. See Bishwal (2008a) for recent results on approximate likelihood asymptotics and approximate Bayes asymptotics for drift estimation of discretely observed diffusions based on high-frequency data. In this work, we assume constant volatility and without loss of generality assume it to be one. To estimate the drift parameter, we use approximate maximum likelihood estimators and study the accuracy of distributional approximation of the resulting estimators from high-frequency data.

Let $(\Omega, \mathcal{F}, \{\mathcal{F}_t\}_{t\geq 0}, P)$ be a stochastic basis on which is defined the Cox–Ingersoll–Ross process $\{X_t\}$ satisfying the Itô stochastic differential equation

$$dX_t = -(\alpha + \theta X_t)dt + \sigma\sqrt{X_t^+}\, dW_t,\ t \geq 0,\ \ X_0 = 1, \tag{4.1}$$

where $\{W_t\}$ is a standard Wiener process with the filtration $\{\mathcal{F}_t\}_{t\geq 0}$ and $\theta < 0$ is the unknown parameter to be estimated on the basis of discrete observations of the process $\{X_t\}$ at times $0 = t_0 < t_1 < \cdots < t_n = T$ with $t_i - t_{i-1} = \frac{T}{n}, i = 1, 2, \ldots, n$. We assume two types of high-frequency data with long observation time: (1) $T \to \infty, n \to \infty, \frac{T}{\sqrt{n}} \to 0$ and (2) $T \to \infty, n \to \infty, \frac{T}{n^{2/3}} \to 0$. For simplicity, we assume $\alpha = 0$ and $\sigma = 1$.

Let the continuous realization $\{X_t, 0 \leq t \leq T\}$ be denoted by X_0^T. Let P_θ^T be the measure generated on the space (C_T, B_T) of continuous functions on $[0, T]$ with the associated Borel σ-algebra B_T generated under the supremum norm by the process X_0^T, and let P_0^T be the standard Wiener measure. It is well known that when θ is the true value of the parameter, P_θ^T is absolutely continuous with respect to P_0^T, and the Radon–Nikodym derivative (likelihood) of P_θ^T with respect to P_0^T based on X_0^T is given by

$$L_T(\theta) := \frac{dP_\theta^T}{dP_0^T}(X_0^T) = \exp\left\{-\theta \int_0^T dX_t - \frac{\theta^2}{2}\int_0^T X_t dt\right\}. \tag{4.2}$$

Consider the score function, the derivative of the log-likelihood function, which is given by

$$\gamma_T(\theta) := -\int_0^T dX_t - \theta \int_0^T X_t dt. \tag{4.3}$$

A solution of the estimating equation $\gamma_T(\theta) = 0$ provides the maximum likelihood estimate (MLE)

$$\hat{\theta}_T := \frac{X_0 - X_T}{\int_0^T X_t dt}. \tag{4.4}$$

As far as we know, the rate of normal approximation for the minimum contrast estimator has not been studied earlier. Our aim in this work is to bridge this gap.

The minimum contrast estimate (MCE) is given by

$$\tilde{\theta}_T := \frac{1}{\int_0^T X_t dt}. \tag{4.5}$$

The discrete time representation of the CIR model is an AR(1) model

$$X_{i\Delta} = \gamma_0 + \gamma_1 X_{(i-1)\Delta} + \epsilon_{i\Delta},$$

where $\epsilon_{i\Delta}$ is a martingale increment sequence. We study the several estimators of the drift parameter in the supercritical case ($\theta > 0$).

The ratio estimator of θ is defined as

$$\hat{\theta}_n := -\log\left[\min_{1\le i\le n} \frac{X_{i\Delta}}{X_{(i-1)\Delta}}\right].$$

We also study the following estimator of the drift. The estimator of θ is defined as

$$\tilde{\theta}_n := -\log\left[\frac{\sum_{i=1}^n X_{i\Delta}}{\sum_{i=1}^n X_{(i-1)\Delta}}\right].$$

Another estimator of θ is defined as

$$\breve{\theta}_n := -\log\left[\frac{X_{n\Delta}}{X_{(n-1)\Delta}}\right].$$

Yet another estimator of θ is defined as

$$\bar{\theta}_n := -\frac{1}{n}\log X_{n\Delta}.$$

Another estimator following Wei and Winnicki is

$$\theta_n = -\log \frac{nX_{n\Delta} - n\Delta}{\sum_{i=1}^{n}(X_{(i-1)\Delta} + 1)}.$$

One can compare with Arato estimator and Davis estimator.

$$X_{n-1}(\breve{\theta}_n - \theta) = \sum_{i=1}^{n}(X_i - \theta X_{i-1}) = \sum_{i=1}^{n} \epsilon_i,$$

$$X_i = \gamma_0 + \gamma_1 X_{(i-1)} + \epsilon_i,$$

$$m(x,\theta) = E(X_i|X_{i-1} = x) = \gamma_0 + \gamma_1 x,$$

where

$$\gamma_0 = \frac{\alpha}{\beta}(e^{\beta\Delta} - 1) = \frac{\alpha}{\beta}(\gamma_1 - 1), \quad \gamma_1 = e^{\beta\Delta}.$$

Weighted CLSE:

$$G_n(\theta) = \sum_{i=1}^{n} w_m(X_{i-1},\theta)(X_i - m(X_{i-1},\theta))$$

$$+\sum_{i=1}^{n} w_v(X_{i-1},\theta)[X_i - m(X_{i-1},\theta))^2 - v(X_{i-1},\theta)]$$

with

$$v(x,\theta) = E[X_i - E(X_i|X_{i-1} = x)^2|X_{i-1} = x] = \sigma^2(\eta_0 + \eta_1 x).$$

$$\eta_0 = \frac{\alpha}{2\beta^2}(e^{\beta\Delta} - 1)^2, \quad \eta_1 = \frac{1}{\beta}e^{\beta\Delta}(e^{\beta\Delta} - 1).$$

Note that $G_n(\theta)$ is a martingale. Rewrite the model as

$$\frac{X_i}{(X_{i-1}+1)^{1/2}} = \gamma_1(X_{i-1}+1)^{1/2} + (\gamma_0 - \gamma_1)(X_{i-1}+1)^{-1/2} + \delta_i,$$

where

$$\delta_i := \frac{\epsilon_i}{(X_{i-1}+1)^{1/2}}.$$

Weighted CLSE

$$\sum_{i=1}^{n}\delta_i^2 = \sum_{i=1}^{n}\frac{\epsilon_i^2}{X_{i-1}+1} = \sum_{i=1}^{n}\frac{[X_i - \gamma_1(X_{i-1}+1) - (\gamma_0 - \gamma_1)]^2}{X_{i-1}+1}.$$

The reason for considering this quantity is that this quantity does not fluctuate too much even when the values of the samples X_i, $i = 0, 1, \ldots, n$, are large.

Let

$$Y_i = \frac{1}{X_{i-1}+1}, \quad \tilde{X} = \frac{1}{n}\sum_{i=1}^{n} X_{i-1}, \quad Z_i := X_{i-1}+1 = Y_i^{-1}.$$

Weighted CLSEs are given by

$$\hat{\gamma}_1 = \frac{\sum_{i=1}^{n} X_i(Y_i - \bar{Y})}{n - \bar{Y}\sum_{i=1}^{n} Y_i^{-1}}, \quad \hat{\gamma}_0 = \frac{\bar{X}\sum_{i=1}^{n} X_{i-1}Y_i - \tilde{X}\sum_{i=1}^{n} X_i Y_i}{n - \bar{Y}\sum Y_i^{-1}}.$$

Theorem 4.1
(a) In the subcritical case ($\gamma_1 < 1$), both $\hat{\gamma}_1$ and $\hat{\gamma}_0$ are strongly consistent and asymptotically normal.

(b) In the supercritical case ($\gamma_1 > 1$), $\hat{\gamma}_1$ is consistent, but $\hat{\gamma}_0$ is not consistent. $\hat{\gamma}_1$ is asymptotically normal.

(c) In the critical case ($\gamma_1 = 1$), $\hat{\gamma}_1$ is consistent and $\hat{\gamma}_0$ is weakly consistent. The estimator $\hat{\gamma}_0$ is asymptotically normal, but $\hat{\gamma}_1$ is asymptotically nonnormal.

(d) Weighted CLSEs perform better than CLSEs.

Sketch of Proof Observe that

$$I_n := \sum_{i=1}^{n} Z_i, \quad I_n^{1/2}(\hat{\gamma}_1 - \gamma_1) = \frac{A_n - B_n}{1 - C_n},$$

where

$$A_n = \frac{\sum_{i=1}^{n} \epsilon_i}{I_n^{1/2}}, \quad B_n = \frac{n\sum_{i=1}^{n} Y_i \epsilon_i}{I_n^{1/2}\sum_{i=1}^{n} Y_i}, \quad C_n = \frac{n^2}{\sum_{i=1}^{n} Y_i \sum_{i=1}^{n} Z_i}.$$

Equivalence of the Estimators

The ratio estimator (also known as the nonparametric MLE) and the weighted CLSE are equivalent. Ratio estimator is defined as

$$\bar{\gamma}_1 = \frac{\bar{X}}{\tilde{X}} = \frac{\sum_{i=1}^{n} X_i}{\sum_{i=1}^{n} X_{i-1}},$$

where

$$\bar{X} = \frac{1}{n}\sum_{i=1}^{n} X_i, \quad \tilde{X} = \frac{1}{n}\sum_{i=1}^{n} X_{i-1}$$

$$\tilde{\gamma}_1 = \frac{n\sum_{i=1}^{n} X_i X_{i-1}^{-1} - \sum_{i=1}^{n} X_i \sum_{i=1}^{n} X_{i-1}^{-1}}{n^2 - \sum_{i=1}^{n} X_i \sum_{i=1}^{n} X_{i-1}^{-1}}.$$

Theorem 4.2

$$(a)\quad \hat{\gamma}_1 = \bar{\gamma}_1 + O\left(\frac{n^2}{\sum_{i=1}^n Z_i}\right) \quad a.s.$$

$$(b)\quad \tilde{\gamma}_1 = \bar{\gamma}_1 + O\left(\frac{n^2}{\sum_{i=1}^n Z_i}\right) \quad a.s.$$

Sketch of Proof Observe that

$$C_n \to 0 \;\; a.s., \;\; \hat{\gamma}_1 - \gamma_1 =, \;\; \check{\gamma}_1 = \frac{X_n}{X_{n-1}}, \;\; I_{n-1}^{1/2}(\check{\gamma}_1 - \gamma_1) \to^D N(0, \sigma^2),$$

$$I_{n-1}(\check{\gamma}_1 - \gamma_1) = \sum_{i=1}^{I_{n-1}} \eta_i,$$

where η_i are i.i.d. with the distribution of $X_1 - \gamma_1$ and $\frac{X_n}{\gamma_1^n} \to V$, where V is nondegenerate and almost surely positive.

$$\sigma^{-1}(\gamma_1 - 1)^{-1/2}\gamma_1^{n/2}(\hat{\gamma}_1 - \gamma_1) \to V^{-1/2}\mathcal{N}(0,1),$$
$$\sigma^{-1}(\gamma_1 - 1)^{-1/2}\gamma_1^{n/2}(\check{\gamma}_1 - \gamma_1) \to V^{-1/2}\mathcal{N}(0,1),$$

where $V^{-1/2}$ is independent of $\mathcal{N}(0,1)$.

Theorem 4.3
(a) $\gamma_1^* = X_n^{1/n}$ *is an inferior estimator and* $n(\gamma_1^* - \gamma_1) \to \gamma_1 \log W$.

Random Norming
(b)

$$I_n^{1/2}(\hat{\gamma}_1 - \gamma_1) \to^D \mathcal{N}(0, \sigma^2).$$

(c)

$$I_n^{1/2}(\tilde{\gamma}_1 - \gamma_1) \to^D \mathcal{N}(0, \sigma^2).$$

Critical Case

(d) *When* $\gamma_1 = 1$, $\hat{\gamma}_1$ *and* $\hat{\gamma}_0$ *are consistent.*

$$I_n^{1/2}(\hat{\gamma}_1 - \gamma_1) \to^D \frac{N_1 - 1}{(\int_0^1 N_t dt)^{1/2}},$$

where

$$N_t = \lim_{n\to\infty} \frac{X_{[nt]}}{n}, \;\; J_n := \sum_{i=1}^n Y_i, \;\; \lim_{n\to\infty} J_n = \infty \;\; a.s., \;\; \frac{J_n}{\log n} \to^P \frac{1}{\gamma_0 - \sigma^2}.$$

If $2\gamma_0 > \sigma^2$, $J_n^{1/2}(\hat{\gamma}_0 - \gamma_0) \to^D \mathcal{N}(0, \sigma^2)$ since $(\hat{\gamma}_0 - \gamma_0) = \frac{D_n - E_n}{1 - C_n}$, where

$$D_n = \frac{I_n - n}{I_n}\frac{\sum_{i=1}^n \epsilon_i Y_i}{J_n}, \;\; E_n = \sum_{i=1}^n X_{i-1} Y_i \frac{\sum_{i=1}^n \epsilon_i}{I_n J_n}, \;\; C_n = \frac{n^2}{\sum_{i=1}^n Y_i \sum_{i=1}^n Z_i}$$

$$\left(\frac{X_n}{n}, \frac{I_n}{n^2}\right) \to^D \left(N_1, \int_0^1 N_t dt\right), \quad \frac{1}{n}\sum_{i=1}^{n} \epsilon_i \to^D N_1 - \gamma_0.$$

$$A_n \to^D \frac{N_1 - \gamma_0}{\int_0^1 N_t dt}, \quad J_n^{1/2} E_n \to^P 0, \quad J_n^{1/2} D_n \to^D \mathcal{N}(0, \sigma^2), \quad \frac{I_n - n}{I_n} \to^P 1.$$

$$(\log n)^{-1/2} \sum_{i=1}^{n} Y_i \epsilon_i \to^D \mathcal{N}\left(0, \frac{\sigma^2}{\gamma_0 - \frac{\sigma^2}{2}}\right)$$

.

Remarks

(1) If $2\gamma_0 > \sigma^2$, the process is transient.

(2) If $2\gamma_0 \leq \sigma^2$, the process is null recurrent. In this case, the limiting distribution of the WCLSE $\hat{\gamma}_0$ is unknown.

Rates of Convergence

$$\frac{\log X_n}{\log n} \to 1 \quad a.s.$$

$$(\log n)^{-1} \sum_{i=1}^{n} Y_i \to (\gamma_0 - \sigma^2/2)^{-1} \quad a.s.$$

In the *critical case*, the rates of convergence are as follows:

Theorem 4.4

$$(a) \quad \hat{\gamma}_1 - \gamma_1 = O\left(\frac{\log \log n}{n}\right) \quad a.s.$$

$$(b) \quad \hat{\gamma}_0 - \gamma_0 = O\left(\left(\frac{\log \log \log n}{\log n}\right)^{1/2}\right) \quad a.s.$$

$$(c) \quad \hat{\gamma}_0 \quad \text{is inconsistent.}$$

As opposed to the classical situation of i.i.d. observations, the nonexistence of consistent estimators is not due to identifiability. The essential problem is that for certain range of parameters, the probability measures of the process are not mutually singular for distinct parameter values.

3.5 Sequential Estimation in CIR Process

Let $(\Omega, \mathcal{F}, \{\mathcal{F}_t\}_{t\geq 0}, P)$ be a stochastic basis on which is defined the Cox–Ingersoll–Ross process $\{X_t\}$ satisfying the Itô stochastic differential equation

$$dX_t = (\alpha + \beta\, X_t)\, dt + \; 2\sqrt{X_t}\, dW_t,\; t \geq 0, \;\; X_0 = \nu, \tag{5.1}$$

where $\{W_t\}$ is a standard Wiener process with the filtration $\{\mathcal{F}_t\}_{t\geq 0}$ and α $\beta > 0$ are the unknown parameters to be estimated on the basis of continuous observation and also discrete observations of the process.

Let the continuous realization $\{X_t, 0 \leq t \leq T\}$ be denoted by X_0^T. Let P_β^T be the measure generated on the space (C_T, B_T) of continuous functions on $[0, T]$ with the associated Borel σ-algebra B_T generated under the supremum norm by the process X_0^T, and let P_0^T be the standard Wiener measure. It is well known that when β is the true value of the parameter, P_β^T is absolutely continuous with respect to P_0^T and the Radon–Nikodym derivative (likelihood) of P_β^T with respect to P_0^T based on X_0^T is given by

$$L_T(\beta) := \frac{dP_\beta^T}{dP_0^T}(X_0^T) = \exp\left\{\beta \int_0^T dX_t - \frac{\beta^2}{2}\int_0^T X_t dt\right\}. \tag{5.2}$$

Consider the score function, the derivative of the log-likelihood function, which is given by

$$\gamma_T(\beta) := \int_0^T dX_t - \beta \int_0^T X_t dt. \tag{5.3}$$

A solution of the estimating equation $\gamma_T(\beta) = 0$ provides the maximum likelihood estimate (MLE)

$$\hat{\beta}_T := \frac{X_T - X_0}{\int_0^T X_t dt}. \tag{5.4}$$

Let us consider the distribution of the continuous energy process I_T. The asymptotic distribution of the I_T process is closely related to the inverse Gaussian distribution with parameters $(T/2,\; T/2)$.

The maximum likelihood estimator of β in the diffusion branching process (5.1) is known to have different limit distributions for the three cases $\beta < 0$ (subcritical), $\beta = 0$ (critical), and $\beta > 0$ (supercritical). This section is concerned with the two cases $\beta = 0$ and $\beta > 0$. It will be useful to develop a unified approach to estimation which does not require the prior information as to whether $\beta = 0$ or $\beta > 0$. We propose to use sequential approach to achieve this. The estimators are given by

$$\tilde{\beta}_\tau = \frac{1}{h}\left[\int_0^\tau X_s dX_s - (X_\tau - X_0)\bar{X}_\tau\right], \;\; \tilde{\alpha}_\tau = -\bar{X}_\tau \tilde{\beta}_\tau + \frac{1}{\tau}(X_\tau - X_0).$$

Let $\{X_t\}$ be observed at times $0 = t_0 < t_1 < \cdots < t_n = T$ with $t_i - t_{i-1} = \frac{T}{n}, i = 1, 2, \ldots, n$. Define the one barrier stopping time

$$\tau := \inf\{i \geq 1 : X_{t_i} \geq b\},$$

where $b > 0$. Define the sequential estimator

$$\hat{\beta}_\tau = \frac{b-1}{\sum_{i=1}^{\gamma} X_{t_{i-1}}(t_i - t_{i-1})} - \frac{1}{b-1}.$$

Note that $\sum_{i=1}^{\tau} X_{t_{i-1}}(t_i - t_{i-1})$ is a complete sufficient statistic. Thus $\hat{\beta}_\tau$ is MVUE.

Let $Y_t := \sqrt{X_t}$. By the Itô formula,

$$dY_t = \left(-\frac{1}{8Y_t} - \frac{1}{2}Y_t\right) dt + \frac{1}{2} dW_t.$$

Let $Z_n = \{Z_{n,k} : k = 0, 1, 2, \ldots\}$ be a sequence of simple branching processes whose offspring distributions have means m_n and variances σ_n and are uniformly square integrable in n.

$$Z_{n,[nt]} = \sum_{k=1}^{Z_{n-1,[(n-1)t]}} \xi_{n-1,k} + Y_n,$$

where $\xi_{n-1,k}$ is the number of offsprings of the k-th individual belonging to the $(n-1)$-th generation and Y_n denotes the number of immigrants in the n-th generation. Suppose that $\{\xi_{n-1,k}\}, k = 0, 1, 2, \ldots, n = 1, 2, \ldots$, and $\{Y_n\}, n = 1, 2, \ldots$ are two independent sequences of independent and identically distributed (i.i.d.) random variables. The initial state Z_0 is a random variable which is independent of $\{\xi_{n-1,k}\}$ and $\{Y_n\}$ has an arbitrary distribution. The offspring and the immigration distributions are assumed to be unspecified with means m_n and λ_n and variances σ_n, respectively.

Define

$$X_{n,t} := \frac{1}{n} Z_{n,[nt]}.$$

Then $X_n \to X$ weakly in the sense of convergence of the corresponding probability measures on $D[0, \infty)$ if

$$\frac{1}{n} Z_{n,0} \to \nu, \ n(m_n - 1) \to \beta, \ \sigma_n \to 2 \text{ as } n \to \infty.$$

By the Levy result for quadratic variation of Brownian motion,

$$\lim_{n\to\infty} \sum_{i=1}^{n} [X_{t_i} - X_{t_{i-1}}]^2 = 2 \int_0^t X_s ds \ \ a.s.$$

$$\hat{\beta}_{n,T} = \frac{X_T - X_0}{\sum_{i=1}^{n} [X_{t_i} - X_{t_{i-1}}]^2} \to \beta_T = \frac{X_T - X_0}{\int_0^T X_t dt} \quad a.s.$$

$$\hat{\beta}_{n,\tau} = \frac{X_\tau - X_0}{\sum_{i=1}^{n\wedge[n\tau]} [X_{t_i} - X_{t_{i-1}}]^2} \to \beta_\tau = \frac{X_\tau - X_0}{\int_0^\tau X_t dt} \quad a.s.$$

Define the stopping rule

$$N := \inf\{n \geq 1 : \sum_{i=1}^{n} (X_i - \bar{X}_n)^2 \geq H\}.$$

Define the sequential estimators

$$\hat{\beta}_H := \frac{1}{\Delta} \log \frac{\sum_{i=1}^{N}(X_i - \bar{X}_N)(X_{i-1} - \bar{X}'_N)}{\sum_{i=1}^{N}(X_i - \bar{X}_N)^2}, \quad \hat{\alpha}_H := \frac{\bar{X}_N - e^{\hat{\beta}_N \Delta}\bar{X}'_N}{e^{\hat{\beta}_N \Delta} - 1}\hat{\beta}_N.$$

The discrete sequential estimators based on stopping time N are given by

$$\hat{b}_H = -\frac{1}{\Delta} \log \left(\frac{N \sum_{i=1}^{N} X_i X_{i-1}^{-1} - \sum_{i=1}^{N} X_i \sum_{i=1}^{N} X_{i-1}^{-1}}{N^2 - \sum_{i=1}^{N} X_i \sum_{i=1}^{N} X_{i-1}^{-1}} \right),$$

$$\hat{a}_H = \frac{1}{N} \sum_{i=1}^{N} X_i + \frac{e^{-\hat{b}_H \Delta}(X_N - x)}{N \hat{b}_H (1 - e^{-\hat{b}_H \Delta})}.$$

Theorem 5.1

$$\begin{aligned}
&\hat{b}_H \to b \text{ as } H \to \infty, \\
&\sqrt{H}(\hat{b}_H - b) \to^D \mathcal{N}(0,\ I_b^{-1}), \\
&\hat{a}_H \to a \text{ as } H \to \infty, \\
&\sqrt{H}(\hat{a}_H - a) \to^D \mathcal{N}(0,\ I_a^{-1}).
\end{aligned}$$

3.6 Berry–Esseen Bound for Heston Model

We develop new explicit estimators of the mean reversion parameter in the Heston model by using the minimum contrast method. We obtain the bound on the Kolmogorov distance between the distribution of the normalized estimator and the standard normal distribution.

These days high-frequency intraday data of asset returns are available. Integrated volatility plays an important role in option pricing in stochastic volatility model. Hence realized volatility that is a measure of the integrated volatility has received considerable interest in recent days empirical finance. The realized volatility is defined as the sum of squared increments of returns. Here we estimate the integrated volatility by realized volatility. Then we obtain an explicit estimator of the parameter in the volatility process based on asset price data.

Let $(\Omega, \mathcal{F}, \{\mathcal{F}_t\}_{t \geq 0}, P)$ be a stochastic basis on which is defined the Cox–Ingersoll–Ross process $\{\sigma_t^2\}$ satisfying the Itô stochastic differential equation

$$d\sigma_t^2 = (1 + 2\theta\ \sigma_t^2)\ dt + \ 2\sigma_t\ dW_t,\ t \geq 0, \tag{6.1}$$

where $\{W_t\}$ is a standard Wiener process with the filtration $\{\mathcal{F}_t\}_{t\geq 0}$ and consider the classical direct estimation problem where $\theta < 0$ is the unknown parameter to be estimated on the basis of discrete observations of the process $\{\sigma_t^2\}$ at times $0 = t_0 < t_1 < \cdots < t_n = T$ with $t_i - t_{i-1} = \frac{T}{n}, i = 1, 2, \ldots, n$. For our asymptotic framework, we assume two types of high-frequency data with long observation time: (1) $T \to \infty, n \to \infty, \frac{T}{\sqrt{n}} \to 0$ and (2) $T \to \infty, n \to \infty, \frac{T}{n^{2/3}} \to 0$.

For the moment, assume that a continuous realization $\{\sigma_t^2, 0 \leq t \leq T\}$ is denoted by ${\sigma^2}_0^T$. Let P_θ^T be the measure generated on the space (C_T, B_T) of continuous functions on $[0, T]$ with the associated Borel σ-algebra B_T generated under the supremum norm by the process V_0^T, and let P_0^T be the standard Wiener measure. It is well known that when θ is the true value of the parameter, P_θ^T is absolutely continuous with respect to P_0^T and the Radon–Nikodym derivative (likelihood) of P_θ^T with respect to P_0^T based on ${\sigma^2}_0^T$ is given by

$$L_T(\theta) := \frac{dP_\theta^T}{dP_0^T}({\sigma^2}_0^T) = \exp\left\{\theta \int_0^T d\sigma_t^2 - \frac{\theta^2}{2}\int_0^T \sigma_t^2 dt\right\}. \tag{6.2}$$

Consider the score function, the derivative of the log-likelihood function, which is given by

$$\gamma_T(\theta) := \int_0^T d\sigma_t^2 - \theta \int_0^T \sigma_t^2 dt. \tag{6.3}$$

A solution of the estimating equation $\gamma_T(\theta) = 0$ provides the maximum likelihood estimate (MLE)

$$\hat{\theta}_T := \frac{\sigma_T^2 - \sigma_0^2 - T/2}{\int_0^T \sigma_t^2 dt}. \tag{6.4}$$

Consider the minimum contrast estimate (MCE)

$$\tilde{\theta}_T := \frac{-T/2}{\int_0^T \sigma_t^2 dt}. \tag{6.5}$$

Note that the volatility which is given by the CIR process is not observed. In the following section, we obtain nonparametric estimators of the minimum contrast estimator of the mean reversion parameter in the Heston model using approximations to $\tilde{\theta}_T$ defined in (6.5).

Consider the Heston stochastic volatility model

$$dS_t = \mu\, S_t\, dt + \sqrt{X_t}\, S_t\ dW_t \tag{6.6}$$

$$dX_t = (1 + 2\theta\, X_t)\, dt + 2\sqrt{X_t}\ dZ_t, \tag{6.7}$$

where $\{W_t\}$ is a standard Brownian motion independent of another standard Brownian motion $\{Z_t\}$ and $\theta < 0$. The integrated volatility is given by

$$I_T := \int_0^T X_t dt. \tag{6.8}$$

The estimator of $\tilde{\tilde{\theta}}$ based on discrete observations of the process $\{S_t\}$ at times $0 = t_0 < t_1 < \cdots < t_n = T$ is approximate minimum contrast estimator

$$\tilde{\tilde{\theta}}_T = \frac{-T}{2\widehat{I_T}}, \tag{6.9}$$

where $\widehat{I_T}$ is nonparametric estimator of I_T based on discrete observations of the process $\{Y_t\}$ at times $0 = t_0 < t_1 < \cdots < t_n = T$.

Integrated volatility has to be estimated on the basis of discrete observations of the price process $\{S_t\}$ at times $0 = t_0 < t_1 < \cdots < t_n = T$ with $t_i - t_{i-1} = \frac{T}{n}, i = 1, 2, \ldots, n$. Let $s_t = \log S_t$ be the log-price process. Then

$$ds_t = (\mu - \frac{1}{2}X_t)dt + \sqrt{X_t}\ dW_t.$$

Thus the drift term depends on the volatility.

$$dS_t = \mu S_t dt + \sqrt{V_t} S_t dW_t^1$$
$$dV_t = \kappa(\theta - V_t)dt + \sigma\sqrt{V_t} dW_t^2$$
$$dW_t^1 dW_t^2 = \rho dt$$

$$dS_t = rS_t dt + \sqrt{S_t} d\tilde{W}_t^1$$
$$dV_t = \kappa^*(\theta^* - V_t)dt + \sigma\sqrt{V_t} d\tilde{W}_t^2$$
$$d\tilde{W}_t^1 d\tilde{W}_t^2 = \rho dt,$$

where

$$\kappa^* = \kappa + \lambda, \quad \theta^* = \frac{\kappa\theta}{\kappa + \lambda}$$

$$d\tilde{W}_t^1 = dW_t^1 + \theta_t dt, \quad d\tilde{W}_t^2 = dW_t^2 + k\sqrt{V_t} dt$$

$$\frac{dQ}{dP} = \exp\left\{-\frac{1}{2}\int_0^t (\theta_s^2 + k^2 V_s)ds - \int_0^t \theta_s dW_s^1 - \int_0^t k^2 V_s dW_s^2\right\}$$

with

$$\theta_t = \frac{\mu - r}{\sqrt{V_t}}.$$

The characteristic function of s_T under Q is given by

$$\psi_T(u) = E_Q(e^{is_T u}) = A(u,T)e^{B(u,T)},$$

where

$$\begin{aligned} A(u,T) &= e^{iu(s_0+rT)} \\ B(u,T) &= \frac{-(u^2+iu)(1-e^{\gamma T})V_0}{2\gamma-(\gamma-(\kappa-\rho\sigma ui)(1-e^{\gamma T})} \\ &- \frac{\kappa\theta}{\sigma^2}\left[2\log\left(\frac{2\gamma-(\gamma-\kappa-\rho\sigma ui)(1-e^{\gamma T})}{2\gamma}\right)\right. \\ &\left.+(\gamma-\kappa-\rho\sigma ui)T\right] \end{aligned}$$

with

$$\gamma =: \sqrt{(\kappa-\rho\sigma ui)^2+(u^2+iu)}.$$

Option Price

European option price is defined as $C = E_t^T[e^{r(T-t)}H(T)]$, where $H(T)$ is the payoff at time T.

Lemma 6.1 European call option is given by

$$C(S_t, V_t, t, T) = S_t P_1(S_t, V_t, T, K) - Ke^{-r(T-t)}P_2(S_t, V_t, T, K),$$

where

$$P_j(x, V_t, T, K) = \frac{1}{2} + \frac{1}{\pi}\int_0^\infty Re\left(\frac{e^{-iu\ln K}f_j(x, V_t, T, u)}{iu}\right)du$$

$$x = \ln S_t, \quad f_j(x, V_t, T, u) = \exp(C(T-t,u) + D(T-t,u)V_t + iux)$$

$$C(T-t,u) = ruir + \frac{a}{\sigma^2}\left[(b_j - \rho\sigma ui + d)\tau - 2\ln\left(\frac{1-ge^{dr}}{1-g}\right)\right]$$

$$D(T-t,u) = \frac{b_j-\rho\sigma ui+d}{\sigma^2}\left(\frac{1-e^{dr}}{1-ge^{dr}}\right)$$

$$g = \frac{b_j-\rho\sigma ui+d}{b_j-\rho\sigma ui-d}, \quad d = \sqrt{(\rho\sigma ui - b_j)^2 - \sigma^2(2\nu_j ui - u^2)}$$

for $j = 1, 2$, where

$$\nu_1 = \frac{1}{2}, \nu_2 = -\frac{1}{2}, a = \kappa\theta, b_1 = \kappa+\lambda-\rho\sigma, b_2 = \kappa+\lambda.$$

Denote

$$\Delta s_{t_{i-1}}^2 := (s_{t_i} - s_{t_{i-1}})^2. \tag{6.10}$$

The realized volatility is defined as

$$R_{n,T} := \sum_{i=1}^{n} \Delta s_{t_{i-1}}^2. \tag{6.11}$$

It is well known that

$$\operatorname*{P-lim}_{n\to\infty} R_{n,T} = I_T.$$

The approximate minimum contrast estimator (AMCE) is defined as

$$\tilde{\theta}_{n,T} = \frac{-T}{2R_{n,T}}. \tag{6.12}$$

For our asymptotic framework, we assume two types of high-frequency data with long observation time: (1) $T \to \infty, n \to \infty, \frac{T}{\sqrt{n}} \to 0$ and (2) $T \to \infty, n \to \infty, \frac{T}{n^{2/3}} \to 0$.
Observe that

$$\left(\frac{T}{2\theta}\right)^{1/2} (\tilde{\theta}_T - \theta) = \frac{\left(\frac{2\theta}{T}\right)^{1/2} N_T}{\left(\frac{2\theta}{T}\right) I_T}, \tag{6.13}$$

where

$$N_T := 1 - \theta I_T \quad \text{and} \quad I_T := \int_0^T X_t dt. \tag{6.14}$$

The process I_T that is energy of the CIR process which plays an important role in clustering time or activity persistence in stochastic volatility modeling.

Define

$$I_{n,T} := \frac{T}{n}\sum_{i=1}^{n} X_{t_{i-1}}, \quad J_{n,T} := \frac{T}{n}\sum_{i=2}^{n+1} X_{t_{i-1}} = \frac{T}{n}\sum_{i=1}^{n} X_{t_i}. \tag{6.15}$$

$$Y_{n,T} := \sum_{i=1}^{n} X_{t_{i-1}} [W_{t_i} - W_{t_{i-1}}] \tag{6.16}$$

$$Y_T := \int_0^T X_t dW_t \tag{6.17}$$

$$V_{n,T} := \sum_{i=1}^{n} \int_{t_{i-1}}^{t_i} X(t_{i-1})[X_t - X(t_{i-1})] dt. \tag{6.18}$$

Lemma 6.2

$$(a) \quad E\left|I_{n,T} - I_t\right|^2 = O\left(\frac{T^4}{n^2}\right).$$

$$(b) \quad E\left|R_{n,T} - I_t\right|^2 = O\left(\frac{T^3}{n^2}\right).$$

$$(c) E\left|Y_{n,T} - Y_T\right|^2 = O\left(\frac{T^2}{n}\right).$$

Proof Let $h_i(t) := X(t_{i-1}) - X_t$. Observe that

$$\begin{aligned} & E|I_{n,T} - I_t|^2 \\ = \;& E\left|\sum_{i=1}^n X(t_{i-1})(t_i - t_{i-1}) - \int_0^T X_t\,dt\right|^2 \\ = \;& E\left|\sum_{i=1}^n \int_{t_{i-1}}^{t_i} [X(t_{i-1}) - X_t]\,dt\right|^2 \\ = \;& E\left|\sum_{i=1}^n \int_{t_{i-1}}^{t_i} h_i(t)\,dt\right|^2 \\ = \;& \sum_{i=1}^n E\left|\int_{t_{i-1}}^{t_i} h_i(t)\,dt\right|^2 + 2\sum_{i,j=1,\,i<j} E\int_{t_{i-1}}^{t_i}\int_{t_{j-1}}^{t_j} h_i(t)h_j(s)\,dt\,ds \\ =: \;& B_1 + B_2. \end{aligned} \tag{6.19}$$

Note that

$$\begin{aligned} & Eh_i^2(t) \\ = \;& E[X(t_{i-1}) - X_t]^2 \\ = \;& E[\sqrt{X(t_{i-1})} - \sqrt{X_t}]^2[\sqrt{X(t_{i-1})} + \sqrt{X_t}]^2 \\ \le \;& \{E[\sqrt{X(t_{i-1})} - \sqrt{X_t}]^4\}^{1/2}\{E[\sqrt{X(t_{i-1})} + \sqrt{X_t}]^4\}^{1/2} \le C(t - t_{i-1}) \end{aligned} \tag{6.20}$$

(by (6.7) and the boundedness of the second term)

$$\begin{aligned} B_1 &= \sum_{i=1}^n E\left|\int_{t_{i-1}}^{t_i} h_i(t)\,dt\right|^2 \\ &\le \sum_{i=1}^n (t_i - t_{i-1}) \int_{t_{i-1}}^{t_i} E(h_i^2(t))\,dt \\ &\le C\tfrac{T}{n}\sum_{i=1}^n \int_{t_{i-1}}^{t_i} (t - t_{i-1})\,dt \\ &= C\tfrac{T^3}{n^2}. \end{aligned} \tag{6.21}$$

We have by Itô's formula, $\sqrt{X_t}$ is an Ornstein–Uhlenbeck process satisfying

$$d(\sqrt{X_t}) = \theta\sqrt{X_t}dt + dW_t.$$

Hence

$$\begin{aligned} & E[h_i(t)h_j(s)] \\ = \;& E(X_{t_{i-1}} - X_t)(X_{t_{j-1}} - X_s) \\ = \;& E(\sqrt{X_{t_{i-1}}} - \sqrt{X_t})(\sqrt{X_{t_{i-1}}} + \sqrt{X_t})(\sqrt{X_{t_{j-1}}} - \sqrt{X_s})(\sqrt{X_{t_{j-1}}} + \sqrt{X_s}). \end{aligned} \tag{6.22}$$

Now, using Wick's lemma, it is easy to see that

$$B_2 \le C\frac{T^4}{n^2}.$$

Combining B_1 and B_2, part (a) of the lemma follows. The proof of part (b) is analogous to part (a). We omit the details. □

In order to prove (c), let $g_i(t) = X_{t_{i-1}} - X_t$ for $t_{i-1} \le t < t_i, \quad i = 1, 2, \ldots, n$. Since
$E|X_{t_{i-1}} - X_t|^{2k} \le C(t_{i-1} - t)^k, k = 1, 2, \ldots,$
hence

$$\begin{aligned}
& E|Y_{n,T} - Y_T|^2 \\
= \; & E|\sum_{i=1}^{n} X_{t_{i-1}}[W_{t_i} - W_{t_{i-1}}] - \int_0^T X_t dW_t|^2 \\
= \; & E|\int_0^T g_i(t) dW_t|^2 = \int_0^T E(g_i^2(t))dt \\
\le \; & C \sum_{i=1}^{n} \int_{t_{i-1}}^{t_i} |t_{i-1} - t| dt \\
= \; & Cn \frac{(t_i - t_{i-1})^2}{2} = C\frac{T^2}{n}.
\end{aligned}$$

Lemma 6.3 For $\epsilon = \epsilon(T) = o(1)$,

$$P\left\{\left|\frac{2\theta}{T} I_T - 1\right| > \epsilon\right\} \le C \exp\left(-\frac{T\theta}{4}\epsilon^2\right).$$

The following theorem gives the bound on the error of normal approximation of the AMCEs. Note that part (a) uses parameter dependent nonrandom norming. While this is useful for testing hypotheses about θ, it may not necessarily give a confidence interval. The normings in parts (b) and (c) are sample dependent which can be used for obtaining a confidence interval. Following theorem shows that asymptotic normality of the AMCEs needs $T \to \infty$ and $\frac{T}{\sqrt{n}} \to 0$.

Theorem 6.3 Let $\alpha_{n,T} = T^{-1/2}(\log T)^{1/2} \bigvee \frac{T^2}{n}(\log T)^{-1}$. We have

(a) $\sup_{x \in \mathbb{R}} \left| P\left\{ \left(\frac{T}{-2\theta}\right)^{1/2} (\theta_{n,T} - \theta) \le x \right\} - \Phi(x) \right| = O(\alpha_{n,T}).$

(b) $\sup_{x \in \mathbb{R}} \left| P\left\{ I_{n,T}^{1/2} (\theta_{n,T} - \theta) \le x \right\} - \Phi(x) \right| = O(\alpha_{n,T}).$

(c) $\sup_{x \in \mathbb{R}} \left| P\left\{ \left(\frac{T}{|2\theta_{n,T}|}\right)^{1/2} (\theta_{n,T} - \theta) \le x \right\} - \Phi(x) \right| = O(\alpha_{n,T}).$

Proof

(a) It is easy to see that

$$\theta_{n,T} - \theta = \frac{Y_{n,T}}{I_{n,T}} + \theta \frac{V_{n,T}}{I_{n,T}}. \tag{6.23}$$

Hence

$$
\begin{aligned}
& \sup_{x\in\mathbb{R}}\left|P\left\{\left(\frac{T}{-2\theta}\right)^{1/2}(\theta_{n,T}-\theta)\le x\right\}-\Phi(x)\right| \\
= & \sup_{x\in\mathbb{R}}\left|P\left\{\left(\frac{T}{-2\theta}\right)^{1/2}\frac{Y_{n,T}}{I_{n,T}}+\left(\frac{T}{-2\theta}\right)^{1/2}\theta\frac{V_{n,T}}{I_{n,T}}\le x\right\}-\Phi(x)\right| \\
\le & \sup_{x\in\mathbb{R}}\left|P\left\{\left(\frac{T}{-2\theta}\right)^{1/2}\frac{Y_{n,T}}{I_{n,T}}\le x\right\}-\Phi(x)\right| \\
& +P\left\{\left|\left(\frac{T}{-2\theta}\right)^{1/2}\frac{V_{n,T}}{I_{n,T}}\right|>\epsilon\right\}+\epsilon \\
=: & K_1+K_2+\epsilon.
\end{aligned}
\tag{6.24}
$$

Note that by Lemma 3.1 (b),

$$
\begin{aligned}
K_1 &= \sup_{x\in\mathbb{R}}\left|P\left\{\left(\frac{T}{-2\theta}\right)^{1/2}\frac{Y_{n,T}}{I_{n,T}}\le x\right\}-\Phi(x)\right| \\
&= \sup_{x\in\mathbb{R}}\left|P\left\{\frac{\left(-\frac{2\theta}{T}\right)^{1/2}Y_{n,T}}{\left(\frac{-2\theta}{T}\right)I_{n,T}}\le x\right\}-\Phi(x)\right| \\
&\le \sup_{x\in\mathbb{R}}\left|P\left\{\left(\frac{-2\theta}{T}\right)^{1/2}Y_{n,T}\le x\right\}-\Phi(x)\right| \\
&\quad +P\left\{\left(\frac{-2\theta}{T}\right)I_{n,T}-1>\epsilon\right\}+\epsilon \\
&=: J_1+J_2+\epsilon.
\end{aligned}
\tag{6.25}
$$

$$
\begin{aligned}
J_1 &= \sup_{x\in\mathbb{R}}\left|P\left\{\left(\frac{-2\theta}{T}\right)^{1/2}(Y_{n,T}-Y_T+Y_T)\le x\right\}-\Phi(x)\right| \\
&\le \sup_{x\in\mathbb{R}}\left|P\left\{\left(\frac{-2\theta}{T}\right)^{1/2}Y_T\le x\right\}-\Phi(x)\right| \\
&\quad +P\left\{\left(\frac{-2\theta}{T}\right)^{1/2}|Y_{n,T}-Y_T|>\epsilon\right\}+\epsilon \\
&\le CT^{-1/2}+\left(\frac{-2\theta}{T}\right)\frac{E|Y_{n,T}-Y_T|^2}{\epsilon^2}+\epsilon \\
&\le CT^{-1/2}+C\frac{T/n}{\epsilon^2}+\epsilon \\
&\quad \text{(by Lemma 3.1(a) and Lemma 3.2(c)).}
\end{aligned}
\tag{6.26}
$$

$$
\begin{aligned}
J_2 &= P\left\{\left|\left(\frac{-2\theta}{T}\right)(I_{n,T}-I_T+I_T)-1\right|>\epsilon\right\}\\
&\le P\left\{\left|\left(\frac{-2\theta}{T}\right)I_T-1\right|>\frac{\epsilon}{2}\right\}+P\left\{\left(-\frac{2\theta}{2}\right)|I_{n,T}-I_T|>\frac{\epsilon}{2}\right\}\\
&\le C\exp\left(\frac{T\theta}{16}\epsilon^2\right)+\frac{16\theta^2}{T^2}\frac{E|I_{n,T}-I_T|^2}{\epsilon^2}\\
&\le C\exp\left(\frac{T\theta}{16}\epsilon^2\right)+C\frac{T^2/n^2}{\epsilon^2}.
\end{aligned}
\tag{6.27}
$$

Here the bound for the first term in (6.27) comes from Lemma 3.4(a) and that for the second term from Lemma 6.2(a). From the proof of Lemma 5.3.2(b), we have

$$
E|V_{n,T}|^2 \le C\frac{T^4}{n^2} \tag{6.28}
$$

$$
\begin{aligned}
K_2 &= P\left\{\left|\left(\frac{T}{-2\theta}\right)^{1/2}\theta\frac{V_{n,T}}{I_{n,T}}\right|>\epsilon\right\}=P\left\{\left|\frac{\left(\frac{-2\theta}{T}\right)^{1/2}\theta V_{n,T}}{\left(-\frac{2\theta}{T}\right)I_{n,T}}\right|>\epsilon\right\}\\
&= P\left\{\left|\left(\frac{-2\theta}{T}\right)^{1/2}\theta V_{n,T}\right|>\delta\right\}+P\left\{\left(-\frac{2\theta}{T}\right)I_{n,T}<\frac{\delta}{\epsilon}\right\}\\
&\quad\text{(where we choose } \delta=\epsilon-C\epsilon^2)\\
&\le P\left\{\left|\left(\frac{-2\theta}{T}\right)^{1/2}\theta V_{n,T}\right|>\delta\right\}+P\left\{\left|\left(-\frac{2\theta}{T}\right)I_{n,T}-1\right|>\delta_1\right\}\\
&\quad\text{(where } \delta_1=\frac{\epsilon-\delta}{\delta}=C\epsilon)\\
&\le -\frac{2\theta}{T}\theta^2\frac{E|V_{n,T}|^2}{\delta^2}+C\exp\left(\frac{T\theta}{16}\delta_1^2\right)+C\frac{T^2/n^2}{\delta_1^2}\\
&\le C\frac{T^3/n^2}{\delta^2}+C\exp\left(\frac{T\theta}{16}\delta_1^2\right)+C\frac{T^2/n^2}{\delta_1^2}\\
&\quad\text{(by (6.28) and (6.27)).}
\end{aligned}
\tag{6.29}
$$

Now from (6.23)–(6.29), since $T/n \to 0$,

$$
\begin{aligned}
&\sup_{x\in\mathbb{R}}\left|P\left\{\left(\frac{T}{-2\theta}\right)^{1/2}(\theta_{n,T}-\theta)\le x\right\}-\Phi(x)\right|\\
&\le CT^{-1/2}+C\exp\left(\frac{T\theta}{16}\epsilon^2\right)+C\frac{T/n}{\epsilon^2}+C\frac{T^2/n^2}{\epsilon^2}+C\frac{T^3/n^2}{\delta^2} \qquad (6.30)\\
&\quad +C\exp(\frac{T\theta}{16}\delta_1^2)+C(\frac{T^2/n^2}{\delta_1^2})+\epsilon.
\end{aligned}
$$

Choosing $\epsilon = T^{-1/2}(\log T)^{1/2}$, the terms of (6.30) are of the order $O(T^{-1/2}(\log T)^{1/2} \bigvee \frac{T^2}{n}(\log T)^{-1})$. This proves (a).

(b) Using the expression (6.7),

$$\begin{aligned}
&\sup_{x\in\mathbb{R}}\left|P\left\{I_{n,T}^{1/2}(\theta_{n,T}-\theta)\le x\right\}-\Phi(x)\right| \\
=\;& \sup_{x\in\mathbb{R}}\left|P\left\{\frac{Y_{n,T}}{I_{n,T}^{1/2}}+\theta\frac{V_{n,T}}{I_{n,T}^{1/2}}\le x\right\}-\Phi(x)\right| \\
\le\;& \sup_{x\in\mathbb{R}}\left|P\left\{\frac{Y_{n,T}}{I_{n,T}^{1/2}}\le x\right\}-\Phi(x)\right|+P\left\{\left|\theta\frac{V_{n,T}}{I_{n,T}^{1/2}}\right|>\epsilon\right\}+\epsilon. \\
=:\;& H_1+H_2+\epsilon.
\end{aligned}\tag{6.31}$$

Note that

$$\begin{aligned}
H_1 =\;& \sup_{x\in\mathbb{R}}\left|P\left\{\frac{Y_{n,T}-Y_T+Y_T}{I_{n,T}^{1/2}}\le x\right\}-\Phi(x)\right| \\
=\;& \sup_{x\in\mathbb{R}}\left|P\left\{\frac{Y_T}{I_{n,T}^{1/2}}\le x\right\}-\Phi(x)\right|+P\left\{\frac{|Y_{n,T}-Y_T|}{I_{n,T}^{1/2}}>\epsilon\right\}+\epsilon \\
=:\;& F_1+F_2+\epsilon.
\end{aligned}\tag{6.32}$$

$$\begin{aligned}
F_1 =\;& \sup_{x\in\mathbb{R}}\left|P\left\{\frac{Y_T}{I_{n,T}^{1/2}}\le x\right\}-\Phi(x)\right| \\
\le\;& \sup_{x\in\mathbb{R}}\left|P\left\{\left(\frac{-2\theta}{T}\right)^{1/2}Y_T\le x\right\}-\Phi(x)\right| \\
&+P\left\{\left|\left(\frac{-2\theta}{T}\right)^{1/2}I_{n,T}^{1/2}-1\right|>\epsilon\right\}+\epsilon \text{ (by Lemma 3.1(b))} \\
\le\;& CT^{-1/2}+P\left\{\left|\left(\frac{-2\theta}{T}\right)I_{n,T}-1\right|>\epsilon\right\}+\epsilon \text{ (by Lemma 3.1(a))} \\
\le\;& CT^{-1/2}+C\exp\left(-\frac{T\theta}{16}\epsilon^2\right)+C\frac{T^2/n^2}{\epsilon^2}+\epsilon \text{ (by (6.27)).}
\end{aligned}\tag{6.33}$$

On the other hand,

$$\begin{aligned}
F_2 =\;& P\left\{\frac{|Y_{n,T}-Y_T|}{I_{n,T}^{1/2}}>\epsilon\right\} \\
\le\;& P\left\{\left(\frac{-2\theta}{T}\right)^{1/2}|Y_{n,T}-Y_T|>\delta\right\}+P\left\{\left|\left(\frac{-2\theta}{T}\right)^{1/2}I_{n,T}^{1/2}-1\right|>\delta_1\right\} \\
&\text{(where } \delta=\epsilon-C\epsilon^2 \textit{ and } \delta_1=(\epsilon-\delta)/\epsilon>0) \\
\le\;& \frac{\left(\frac{-2\theta}{T}\right)E|Y_{n,T}-Y_T|^2}{\delta^2}+P\left\{\left|\left(\frac{-2\theta}{T}\right)I_{n,T}-1\right|>\delta_1\right\} \\
\le\;& C\frac{T/n}{\delta^2}+C\exp\left(\frac{T\theta}{16}\delta_1^2\right)+C\frac{T^2/n^2}{\delta_1^2} \text{ (from Lemma 3.2(c) and (6.27)).}
\end{aligned}\tag{6.34}$$

Using (6.34) and (6.33) in (6.32), we obtain

$$\begin{aligned} H_1 &= \sup_{x\in\mathbb{R}} \left| P\left\{ \frac{Y_T}{I_{n,T}^{1/2}} \le x \right\} - \Phi(x) \right| \\ &\le CT^{-1/2} + C\exp\left(\frac{T\theta}{16}\epsilon^2\right) + C\frac{T/n}{\delta^2} + C\frac{T^2/n^2}{\delta_1^2} \\ &\quad + C\exp(\frac{T\theta}{16}\delta_1^2) + C\frac{T^2/n^2}{\epsilon^2} + \epsilon. \end{aligned} \tag{6.35}$$

$$\begin{aligned} H_2 &= P\left\{ \left| \theta\frac{V_{n,T}}{I_{n,T}^{1/2}} \right| > \epsilon \right\} = P\left\{ \frac{\left| \left(\frac{-2\theta}{T}\right)^{1/2} \theta V_{n,T} \right|}{\left| \left(\frac{-2\theta}{T}\right)^{1/2} I_{n,T}^{1/2} \right|} > \epsilon \right\} \\ &\le P\left\{ \left| \left(\frac{-2\theta}{T}\right)^{1/2} \theta V_{n,T} \right| > \delta \right\} + P\left\{ \left| \left(\frac{-2\theta}{T}\right)^{1/2} I_{n,T}^{1/2} \right| < \delta/\epsilon \right\} \\ &\le \left(-\frac{2\theta}{T}\right)\theta^2 \frac{E|V_{n,T}|^2}{\delta^2} + P\left\{ \left| \left(-\frac{2\theta}{T}\right) I_{n,T} - 1 \right| > \delta_1 \right\} \\ &\quad \text{(where } 0 < \delta < \epsilon \text{ and } \delta_1 = (\epsilon-\delta)/\epsilon = C\epsilon > 0) \\ &\le C\frac{T^2/n^2}{\delta^2} + C\exp\left(\frac{T\theta}{16}\delta_1^2\right) + C\frac{T^2/n^2}{\delta_1^2} \quad \text{(from (6.28) and (6.27)).} \end{aligned} \tag{6.36}$$

Using (6.36) and (6.35) in (6.31) and choosing $\epsilon = T^{-1/2}(\log T)^{1/2}$, the terms of (6.31) are of the order $O(T^{-1/2}(\log T)^{1/2} \bigvee \frac{T^2}{n}(\log T)^{-1})$. This proves (b).

(c) Let $D_T = \{|\theta_{n,T} - \theta| \le d_T\}$ and $d_T = CT^{-1/2}(\log T)^{1/2}$.

On the set D_T, expanding $(2|\theta_{n,T}|)^{-1/2}$, we obtain

$$\begin{aligned} (-2\theta_{n,T})^{-1/2} &= (-2\theta)^{-1/2}\left[1 - \frac{\theta-\theta_{n,T}}{\theta}\right]^{-1/2} \\ &= (-2\theta)^{-1/2}\left[1 + \frac{1}{2}\left(\frac{\theta-\theta_{n,T}}{\theta}\right) + O(d_T^2)\right]. \end{aligned}$$

Then

$$\begin{aligned} &\sup_{x\in\mathbb{R}} \left| P\left\{ \left(\frac{T}{2|\theta_{n,T}|}\right)^{1/2} (\theta_{n,T}-\theta) \le x \right\} - \Phi(x) \right| \\ &\le \sup_{x\in\mathbb{R}} \left| P\left\{ \left(\frac{T}{2|\theta_{n,T}|}\right)^{1/2} (\theta_{n,T}-\theta) \le x, D_T \right\} - \Phi(x) \right| + P(D_T^c). \end{aligned} \tag{6.37}$$

$$
\begin{aligned}
& P(D_T^c) \\
= \;& P\left\{|\theta_{n,T}-\theta| > CT^{-1/2}(\log T)^{1/2}\right\} \\
= \;& P\left\{\left(\frac{T}{-2\theta}\right)^{1/2}|\theta_{n,T}-\theta| > C(\log T)^{1/2}(-2\theta)^{-1/2}\right\} \\
\leq \;& C(T^{-1/2}(\log T)^{1/2} \bigvee \frac{T^2}{n}(\log T)^{-1}) + 2(1-\Phi\left((\log T)^{1/2}(-2\theta)^{-1/2}\right) \\
& \text{(by Theorem 6.3(a))} \\
\leq \;& C(T^{-1/2}(\log T)^{1/2} \bigvee \frac{T^2}{n}(\log T)^{-1}).
\end{aligned}
$$

On the set D_T,

$$
\left|\left(\frac{\theta_{n,T}}{\theta}\right)^{1/2} - 1\right| \leq CT^{-1/2}(\log T)^{1/2}.
$$

Hence, upon choosing $\epsilon = CT^{-1/2}(\log T)^{1/2}, C$ large, we obtain

$$
\begin{aligned}
& \left|P\left\{\left(\frac{T}{-2\theta_{n,T}}\right)^{1/2}(\theta_{n,T}-\theta) \leq x, D_T\right\} - \Phi(x)\right| \\
\leq \;& \left|P\left\{\left(\frac{T}{-2\theta}\right)^{1/2}(\theta_{n,T}-\theta) \leq x, D_T\right\} - \Phi(x)\right| \\
& + P\left\{\left|\left(\frac{\theta_{n,T}}{\theta}\right)^{1/2} - 1\right| > \epsilon, D_T\right\} + \epsilon \\
& \text{(by Lemma 3.1(b))} \\
\leq \;& C(T^{-1/2}(\log T)^{1/2} \bigvee \frac{T^2}{n}(\log T)^{-1}) \\
& \text{(by Theorem 5.3.3(a))}.
\end{aligned} \tag{6.38}
$$

(c) follows from (6.36)–(6.38). □

Theorem 6.4

$$
\sup_{x\in\mathbb{R}}\left|P\left\{I_{n,T}\left(-\frac{2\theta}{T}\right)^{1/2}(\theta_{n,T}-\theta) \leq x\right\} - \Phi(x)\right| = O\left(T^{-1/2}\bigvee\left(\frac{T}{n}\right)^{1/3}\right).
$$

Proof Let $a_{n,T} = Z_{n,T} - Z_T$ and $b_{n,T} = I_{n,T} - I_T$.

$$
\text{By Lemma 3.2,} \quad E|a_{n,T}|^2 = O\left(\frac{T^2}{n}\right) \text{ and } E|b_{n,T}|^2 = O\left(\frac{T^4}{n^2}\right). \tag{6.39}
$$

From (6.9),

$$
\begin{aligned}
I_{n,T}\theta_{n,T} &= \sum_{i=1}^{n} X_{t_{i-1}}\left[X_{t_i} - X_{t_{i-1}}\right] = \int_0^T X_t dX_t + a_{n,T} \\
&= \int_0^T X_t dW_t + \theta\int_0^T X_t^2 dt + a_{n,T}.
\end{aligned}
$$

Hence $I_{n,T}(\theta_{n,T} - \theta) = -\theta b_{n,T} + a_{n,T}$.
Thus

$$
\begin{aligned}
&\sup_{x\in\mathbb{R}} \left| P\left\{ I_{n,T}\left(-\frac{2\theta}{T}\right)^{1/2} (\theta_{n,T} - \theta) \le x \right\} - \Phi(x) \right| \\
= &\sup_{x\in\mathbb{R}} \left| P\left\{ \left(-\frac{2\theta}{T}\right)^{1/2} [Y_T - \theta b_{n,T} + a_{n,T}] \le x \right\} - \Phi(x) \right| \\
\le &\sup_{x\in\mathbb{R}} \left| P\left\{ \left(-\frac{2\theta}{T}\right)^{1/2} Y_T \le x \right\} - \Phi(x) \right| \\
&+ P\left\{ \left| \left(-\frac{2\theta}{T}\right)^{1/2} [-\theta b_{n,T} + a_{n,T}] \right| > \epsilon \right\} + \epsilon \\
\le &\ CT^{-1/2} + \left(-\frac{2\theta}{T}\right) \frac{E| - \theta b_{n,T} + a_{n,T}|^2}{\epsilon^2} + \epsilon \\
\le &\ CT^{-1/2} + C\frac{T/n}{\epsilon^2} + \epsilon \text{ (by Lemma 3.1(a) and (6.39))}.
\end{aligned}
$$

Choosing $\epsilon = \left(\frac{T}{n}\right)^{1/3}$, the rate is $O\left(T^{-1/2} \bigvee \left(\frac{T}{n}\right)^{1/3}\right)$. □

Theorem 6.5

$$|\theta_{n,T} - \theta_T| = O_P(\frac{T^2}{n})^{1/2}.$$

Proof Note that $\theta_{n,T} - \theta_T = \frac{Z_{n,T}}{I_{n,T}} - \frac{Z_T}{I_T}$. From Lemma 3.2, it follows that $|Z_{n,T} - Z_T| = O_P(\frac{T^2}{n})^{1/2}$ and $|I_{n,T} - I_T| = O_P(\frac{T^4}{n^2})^{1/2}$. Now the theorem follows easily from Lemma 3.1(c) □

Remarks

(1) The rate in Theorem 6.3 is uniform over compact subsets of the parameter space.

(2) The rates of normal approximation of the conditional least squares estimator and an approximate maximum likelihood estimator when the O-U process is observed at discrete time points in $[0, T]$ have been studied in Bishwal and Bose (2001). The rate of normal approximation for some approximate minimum contrast estimators based on discrete observations of the O-U process has been studied in Bishwal (2006a). See Bishwal (2008a) for results on approximate likelihood asymptotics and approximate Bayes asymptotics for drift estimation of discretely observed diffusions based on high-frequency data.

(3) It remains to investigate the nonuniform rates of convergence to normality which are more useful.

3.7 Method of Moments Estimation in Heston Model

We study the method of moments estimator for the parameters of the Heston stochastic volatility model based on equally spaced observations. Using the fact that the increments of a Heston process are strongly mixing with an exponential rate, we show that the resulting estimators are consistent and asymptotically normal. The model is given by

$$dY_t = \sqrt{X_t} dW_t + \rho dZ_{\theta t}$$

$$dX_t = (1 - 2\theta X_t)dt + 2\sqrt{X_t} dW_t,$$

where Z_t is a Levy process independent of X_0.

Invariant distribution of X_t is Gamma (α, β) with density function

$$f(x) = \frac{\beta^\alpha}{\Gamma(\alpha)} x^{\alpha-1} e^{-\alpha x}.$$

Let $y_i := Y_{i\Delta} - Y_{i-1\Delta}, i = 1, 2, \ldots, n$.

The cummulants satisfy

$$k_{y_1}^{(1)} = \theta\rho\Delta k_G^{(1)}. \tag{7.1}$$

$$k_{y_1}^{(2)} = \Delta k_G^{(1)} + 2\theta\rho^2\Delta k_G^{(2)}. \tag{7.2}$$

We know that

$$k_G^{(1)} = \frac{\alpha}{\beta}, \quad k_G^{(2)} = \frac{\alpha(\alpha+1)}{\beta^2}.$$

$$\mathcal{L}(X_0) = \mathcal{L}(X_t) = \mathcal{GIG}(\lambda, \delta, \sqrt{\alpha^2 - \beta^2}).$$

Inverting (7.1) and (7.2) and replacing the cummulants by their sample cummulants, we obtain the explicit method of moments estimators

$$\hat{\theta}_n = \frac{2\alpha^3(\alpha+1)}{\beta^4\Delta} \frac{(\frac{1}{n}\sum_{i=1}^n (Y_{i\Delta} - Y_{i-1\Delta}))^2}{\frac{1}{n}\sum_{i=1}^n (Y_{i\Delta} - Y_{i-1\Delta})^2 - \Delta\frac{1}{n}\sum_{i=1}^n (Y_{i\Delta} - Y_{i-1\Delta})}$$

$$\hat{\rho}_n = \frac{\beta^3\Delta}{2\alpha^2(\alpha+1)} \frac{\frac{1}{n}\sum_{i=1}^n (Y_{i\Delta} - Y_{i-1\Delta})^2 - \Delta\frac{1}{n}\sum_{i=1}^n (Y_{i\Delta} - Y_{i-1\Delta})}{\frac{1}{n}\sum_{i=1}^n (Y_{i\Delta} - Y_{i-1\Delta})}.$$

Theorem 7.1

$$\begin{aligned}
&\hat{\theta}_n \to \theta_0 \quad a.s. \quad as \;\; n \to \infty, \\
&\hat{\rho}_n \to \rho_0 \quad a.s. \quad as \;\; n \to \infty \\
&\sqrt{n}(\hat{\theta}_n - \theta_0) \to^{\mathcal{D}} N(0, I^{-1}(\theta_0)) \;\; as \;\; n \to \infty, \\
&\sqrt{n}(\hat{\rho}_n - \rho_0) \to^{\mathcal{D}} N(0, I^{-1}(\rho_0)) \;\; as \;\; n \to \infty,
\end{aligned}$$

where $I(\theta_0)$ and $I(\theta_0)$ are the corresponding Fisher information.

Sketch of Proof Birkoff Ergodic Theorem and Mixing CLT (see Durrett (2019)) give the results.

In the case of -1 correlation, the moment estimator is $\theta_T = -\frac{Y_T}{T}$.

Next in order to obtain the minimum contrast estimators, observe that the mixture distribution is variance Gamma Q_θ GH.

$$\mathcal{L}(Y_t) = \mathcal{GH}(\lambda, \frac{\alpha}{\sqrt{t}}, \frac{\beta}{\sqrt{t}}, \sqrt{t}\delta, \mu t),$$

$$\mathcal{L}(Y_{t+1} - Y_t) = \mathcal{GH}(\lambda, \alpha, \beta, \delta, \mu).$$

The density function of the GH distribution is given by

$$d_{\mathcal{GH}(\lambda,\alpha,\beta,\delta,\mu)}(x) =$$

$$\frac{(\alpha^2-\beta^2)^{\frac{\lambda}{2}}}{\sqrt{2\pi}\alpha^{\lambda-\frac{1}{2}}\delta^\lambda K_\lambda(\delta\sqrt{\alpha})}(\delta^2+(x-\mu)^2)^{(\lambda-\frac{1}{2})/2}e^{\beta(x-\mu)}K_{\lambda-\frac{1}{2}}(\alpha\sqrt{\delta^2+(x-\mu)^2}).$$

Here K_ν denotes the modified Bessel function of third kind with index ν. When $\theta = -\frac{1}{2}$, GIG becomes $IG(\delta, \gamma)$. Let $q(x, \theta)$ be the density function of Y_t. The contrast function is defined as $u(x, \theta) = -\log q(x, \theta)$. The corresponding minimum contrast estimator is defined as $\tilde{\theta}_n = \arg\min u(x, \theta)$.

Chapter 4

Fractional Ornstein–Uhlenbeck Processes, Levy–Ornstein–Uhlenbeck Processes, and Fractional Levy–Ornstein–Uhlenbeck Processes

4.1 Introduction

One obtains the Ornstein–Uhlenbeck–Gamma process when the Brownian component of the Ornstein–Uhlenbeck process is contaminated by a Gamma process. This chapter introduces some new estimators of the drift parameter in the Ornstein–Uhlenbeck–Gamma process based on discretely sampled data and obtains rates of weak convergence of the distributions of the estimators to the standard normal distribution using random, nonrandom, and mixed normings.

This chapter studies fractional Levy–Ornstein–Uhlenbeck (FLOU) processes based on continuous and discrete observations. First it shows that based on continuous observation, the distribution of the normalized minimum contrast estimator of the drift parameter in the fractional Ornstein–Uhlenbeck process observed over $[0, T]$ converges to the standard normal distribution with a uniform error rate of the order $O(T^{-1+H})$ where $H > 1/2$ is the Hurst exponent of the fractional Brownian motion driving the Ornstein–Uhlenbeck process. Second, it studies hidden FLOU processes and also studies the behavior of maximum likelihood estimator. Then, based on discrete observations, it introduces several approximate minimum contrast estimators and studies their rate of weak convergence to normal distributions. Finally, it studies estimators for pure jump FLOU processes.

4.2 Continuous Sampling

Consider a probability space $(\Omega, \mathcal{F}, \mathbb{P})$ on which all random variables and processes below are defined. A normalized fractional Brownian motion $\{W_t^H, t \geq$

J. P. N. Bishwal, *Parameter Estimation in Stochastic Volatility Models*,
https://doi.org/10.1007/978-3-031-03861-7_4

0} with Hurst parameter $H \in (0,1)$ is a centered Gaussian process with continuous sample paths whose covariance kernel is given by

$$E(W_t^H W_s^H) = \frac{1}{2}(s^{2H} + t^{2H} - |t-s|^{2H}), \quad s,t \geq 0.$$

The process is self-similar (scale invariant) and it can be represented as a stochastic integral with respect to standard Brownian motion. For $H = \frac{1}{2}$, the process is a standard Brownian motion. For $H \neq \frac{1}{2}$, the fBm is not a semimartingale and not a Markov process, but a Dirichlet process. The increments of the fBm are negatively correlated for $H < \frac{1}{2}$ and positively correlated for $H > \frac{1}{2}$, and in this case they display long-range dependence. The parameter H which is also called the self-similarity parameter measures the intensity of the long-range dependence. Note that the estimation of the parameter H based on observation of fractional Brownian motion has already been paid some attention, see, e.g., Peltier and Levy Vehel (1994) and the references therein. However, we assume H to be known.

Hence, for $H \neq \frac{1}{2}$, the classical theory of stochastic integration with respect to semimartingales is not applicable to stochastic integration with respect to fBm. Now there exist several approaches to stochastic integration with respect to fBm. See, for instance, classical Riemann sum approach (Lin 1995, Dai and Heyde 1996, Kleptsyna et al. 1998a), Malliavin calculus approach (Coutin and Decreusefond 1997; Decreusefond and Ustunel 1998, 1999, Alos et al. 2000, 2001), Wick product approach (Duncan et al. 2000), pathwise calculus (Zahle 1998, Ruzmaikina 2000), and Dirichlet calculus (Lyons and Zhang 1994).

The fractional Ornstein–Uhlenbeck process is an extension of Ornstein–Uhlenbeck process with fractional Brownian motion (fBm) driving term. In finance, it is known as fractional Vasicek model and is being extensively used these days as one-factor short-term interest rate model which takes into account the long-memory effect of the interest rate. The model parameter is usually unknown and must be estimated from data.

Define

$$\begin{aligned}
\kappa_H &:= 2H\Gamma(3/2-H)\Gamma(H+1/2),\\
k_H(t,s) &:= \kappa_H^{-1}(s(t-s))^{\frac{1}{2}-H}\\
\lambda_H &:= \frac{2H\Gamma(3-2H)\Gamma(H+\frac{1}{2})}{\Gamma(3/2-H)}\\
v_t \equiv v_t^H &:= \lambda_H^{-1}t^{2-2H}\\
M_t^H &:= \int_0^t k_H(t,s)dW_s^H.
\end{aligned}$$

From Norros et al. (1999), it is well known that M_t^H is a Gaussian martingale, called the fundamental martingale whose variance function $\langle M^H\rangle_t$ is v_t^H.

Moreover, the natural filtration of the martingale M^H coincides with the natural filtration of the fBm W^H since

$$W_t^H := \int_0^t K(t,s)dM_s^H$$

holds for $H \in (1/2, 1)$ where

$$K_H(t,s) := H(2H-1)\int_s^t r^{h-\frac{1}{2}}(r-s)^{H-\frac{3}{2}}, \quad 0 \le s \le t,$$

and for $H = 1/2$, the convention $K_{1/2} \equiv 1$ is used.

Define

$$Q_t := \frac{d}{dv_t}\int_0^t k_H(t,s)X_s ds.$$

It is easy to see that

$$Q_t = \frac{\lambda_H}{2(2-2H)}\left\{t^{2H-1}Z_t + \int_0^t r^{2H-1}dZ_s\right\}.$$

Define the process $Z = (Z_t, t \in [0,T])$ by

$$Z_t := \int_0^t k_H(t,s)dX_s.$$

The following facts are known from Kleptsyna and Le Breton (2002):

(i) Z is the fundamental semimartingale associated with the process X.
(ii) Z is a $(\mathcal{F}_t)$ -semimartingale with the decomposition

$$Z_t = \theta\int_0^t Q_s dv_s + M_t^H.$$

(iii) X admits the representation

$$X_t = \int_0^t K_H(t,s)dZ_s.$$

(iv) The natural filtration $(\mathcal{Z}_t)$ of Z and $(\mathcal{X}_t)$ of X coincide.

4.2.1 Directly Observed FOU Process

Let $(\Omega, \mathcal{F}, \{\mathcal{F}_t\}_{t\ge 0}, P)$ be a stochastic basis on which is defined the Ornstein–Uhlenbeck process X_t satisfying the Itô stochastic differential equation

$$dX_t = \theta X_t dt + dW_t^H, \quad t \ge 0, \quad X_0 = 0, \tag{1.1}$$

where $\{W_t^H\}$ is a fractional Brownian motion with $H > 1/2$ with the filtration $\{\mathcal{F}_t\}_{t\geq 0}$ and $\theta > 0$ is the unknown parameter to be estimated on the basis of continuous observation of the process $\{X_t\}$ on the time interval $[0, T]$.

Let the realization $\{X_t, 0 \leq t \leq T\}$ be denoted by X_0^T. Let P_θ^T be the measure generated on the space (C_T, B_T) of continuous functions on $[0, T]$ with the associated Borel σ-algebra B_T generated under the supremum norm by the process X_0^T and P_0^T be the standard Wiener measure. Applying Girsanov type formula for fBm, when θ is the true value of the parameter, P_θ^T is absolutely continuous with respect to P_0^T and the Radon–Nikodym derivative (likelihood) of P_θ^T with respect to P_0^T based on X_0^T is given by

$$L_T(\theta) := \frac{dP_\theta^T}{dP_0^T}(X_0^T) = \exp\left\{\theta \int_0^T Q_t dZ_t - \frac{\theta^2}{2}\int_0^T Q_t^2 dv_t\right\}. \tag{1.2}$$

Consider the score function, the derivative of the log-likelihood function, which is given by

$$R_T(\theta) := \int_0^T Q_t dZ_t - \theta \int_0^T Q_t^2 dv_t. \tag{1.3}$$

A solution of $R_T(\theta) = 0$ provides the maximum likelihood estimate (MLE)

$$\hat{\theta}_T := \frac{\int_0^T Q_t dZ_t}{\int_0^T Q_t^2 dv_t}. \tag{1.4}$$

Strictly speaking, $\hat{\theta}_T$ is not the maximum likelihood estimate of θ since $\hat{\theta}_T$ may take negative values, whereas the parameter θ is assumed to be strictly positive. For an exact definition of MLE, see Kutoyants (1994a,b). Nevertheless, we use this terminology.

Kleptsyna and Le Breton (2002) showed that $\hat{\theta}_T$ is strongly consistent estimator of θ. Using the Fourier method, we prove a Berry–Esseen type theorem for the estimator $\tilde{\theta}_T$ which gives the asymptotic normality.

Beyond the first order asymptotics, for the classical OU process, Bishwal (2000a) obtained the rate of normal approximation of the order $O(T^{-1/2})$ for the MLE, thereby improving $O(T^{-1/5})$ rate of Mishra and Prakasa Rao (1985a) and $O(T^{-1/2}(\log T)^{1/2})$ rate of Bishwal and Bose (1995). As far as the order of T is concerned, $O(T^{-1/2})$ rate is the sharpest rate one could expect since the norming is $T^{1/2}$. Bishwal (2000b) obtained the rate of normal approximation of the order $O(T^{-1/2})$ for posterior distributions and Bayes estimators, with smooth priors and loss functions. Bishwal (2000a,b) used parameter dependent nonrandom norming. Bishwal (2001) obtained normal approximation of the order $O(T^{-1/2})$ for the MLE and Bayes estimators using two different random normings which are useful for computation of a confidence interval.

Using the fractional Itô formula, the score function $R_T(\theta)$ can be written as

$$R_T(\theta) = \frac{1}{2}\left[\frac{\lambda_H}{(2-2H)} Z_T \int_0^T t^{2H-1} dZ_t - T\right] - \theta \int_0^T Q_t^2 dv_t. \tag{1.5}$$

Consider the contrast function

$$c_T(\theta) = -\frac{T}{2} - \theta \int_0^T Q_t^2 dv_t \tag{1.6}$$

and the minimum contrast estimate (MCE)

$$\tilde{\theta}_T := \frac{-T}{2\int_0^T Q_t^2 dv_t.} \tag{1.7}$$

M-estimator is reduced to the minimum contrast estimator. Note that the MCE does not involve stochastic integral unlike the MLE.

Minimum contrast estimation in diffusion processes was studied by Lanksa (1979) and M-estimation in diffusion processes was studied by Yoshida (1988).

Note that using the Skorohod embedding of martingale, which has, since long, been the basic tool for normal approximation of martingales, will not give a rate better than $O(T^{-1/4})$ (see Borokov 1973). To obtain the rate of normal approximation of the order $O(T^{-1/2})$, we adopt the Fourier method followed by the squeezing technique of Pfanzagl (1971) and Michel and Pfanzagl (1971) developed for the minimum contrast estimator in the i.i.d. case.

Observe that

$$\left(\frac{T}{-2\theta}\right)^{1-H} (\tilde{\theta}_T - \theta) = \frac{\left(\frac{-2\theta}{T}\right)^{1-H} J_T}{\left(\frac{2\theta}{T}\right)^{2-2H} I_T} \tag{1.8}$$

where

$$J_T := \theta I_T - \frac{T}{2} \quad \text{and} \quad I_T := \int_0^T Q_t^2 dv_t.$$

We study the large deviations of the estimator $\tilde{\theta}_T$.

For the standard Ornstein–Uhlenbeck process, large deviations of $\hat{\theta}_T$ and $\tilde{\theta}_T$ were obtained in Florens-Landais and Pham (1999). In particular, it was shown that the large deviation probabilities of $\hat{\theta}_T$ are identical to those of $\tilde{\theta}_T$ for $\theta \le \theta_0/3$ but weaker for $\theta > \theta_0/3$, where θ_0 is the true value of the parameter. We generalize this result to the fractional Ornstein–Uhlenbeck case.

Theorem 2.1 $\tilde{\theta}_T$ satisfies the large deviations principle with good rate function $I(\theta)$ given by

$$I(\theta) = \begin{cases} -\frac{(\theta-\theta_0)^2}{4\theta} & : \ \theta \le \frac{\theta_0}{3} \\ 2\theta - \theta_0 & : \ \theta > \frac{\theta_0}{3}. \end{cases}$$

Let $\Phi(\cdot)$ denote the standard normal distribution function. Throughout the chapter, C denotes a generic constant (which does not depend on T and x). We have not tried to estimate the constant in the bound on normal approximation.

We start with some preliminary lemmas.

Lemma 2.1 For every $\delta > 0$,

$$P\left\{\left|\left(\frac{2\theta}{T}\right)^{2-2H} I_T - 1\right| \geq \delta\right\} \leq CT^{-1}\delta^{-2}.$$

Lemma 2.2

$$\sup_{x\in\mathbb{R}}\left|P\left\{\left(\frac{2\theta}{T}\right)^{1-H} J_T \leq x\right\} - \Phi(x)\right| \leq C\,T^{-1/2}.$$

The following lemma gives the error rate on the difference of the characteristic function of the denominator of the MCE and the normal characteristic function.

Lemma 2.3

(a) Let $\phi_T(z_1) := E\exp(z_1 I_T),\ z_1 \in \mathbb{C}$. Then $\phi_T(z_1)$ exists for $|z_1| \leq \delta$, for some $\delta > 0$, and is given by

$$\phi_T(z_1) = \left\{\frac{4\sin\pi H\sqrt{\theta^2 - 2z_1}e^{-\theta T}}{\pi T D_T^H(\theta; \sqrt{\theta^2 - 2z_1}}\right\}^{1/2}$$

and we choose the principal branch of the square root, where

$$\begin{aligned} D_T^H(\theta;\alpha) &:= [\alpha\cosh(\frac{\alpha}{2}T) - \theta\sinh(\frac{\alpha}{2}T)]^2 J_{-H}(\frac{\alpha}{2}T)J_{H-1}(\frac{\alpha}{2}T) \\ &- [\alpha\sinh(\frac{\alpha}{2}T) - \theta\cosh(\frac{\alpha}{2}T)]^2 J_{1-H}(\frac{\alpha}{2}T)J_H(\frac{\alpha}{2}T) \end{aligned}$$

for $\alpha > 0$ and J_ν is the modified Bessel function of first kind of order ν.

(b) Let

$$\psi_{T,x} := \left(\frac{2\theta}{T}\right)^{1-H} J_T - \left((\frac{2\theta}{T})^{2-2H} I_T - 1\right)x.$$

Then for $|x| \leq 2(\log T)^{1-H}$ and for $|u| \leq \epsilon T^{1/2}$, where ϵ is sufficiently small,

$$\left|E\exp(iu\psi_{T,x}) - \exp(\frac{-u^2}{2})\right| \leq C\exp(\frac{-u^2}{4})(|u| + |u|^3)T^{-1+H}.$$

Proof(a) Kleptsyna and Le Breton (2002) proved the following Cameron–Martin type formula for $a > 0$

$$E\exp(-aI_T) = \left\{\frac{4\sin\pi H\sqrt{\theta^2+2a}e^{-\theta T}}{\pi T D_T^H(\theta;\sqrt{\theta^2+2a}}\right\}^{1/2}.$$

For $H = 1/2$, this formula reduces to Novikov's formula given in Liptser and Shiryayev (1978). By analytic continuation, this formula extends to the complex plane $z_1 \in \mathbb{C}$.

(b) Observe that

$$\begin{aligned}
&E\exp(iu\psi_{T,x})\\
=\ & E\exp\left[-iu\left(\frac{2\theta}{T}\right)^{1-H} J_T - iu\left((\frac{2\theta}{T})^{2-2H} I_T - 1\right)x\right]\\
=\ & E\exp\left[-iu\left(\frac{2\theta}{T}\right)^{1-H}\left\{\theta I_T - \frac{T}{2}\right\} - iu\left((\frac{2\theta}{T})^{2-2H} I_T - 1\right)x\right]\\
=\ & E\exp(z_1 I_T + z_3)\\
=\ & \exp(z_3)\phi_T(z_1),
\end{aligned}$$

where $z_1 := -iu\theta\delta_{T,x}$, and $z_3 := \frac{iuT}{2}\delta_{T,x}$ with $\delta_{T,x} := \left(\frac{2\theta}{T}\right)^{1-H} + \frac{2x}{T}$. Note that $\phi_T(z_1)$ satisfies the conditions of (a) by choosing ϵ sufficiently small. Let $\alpha_{1,T}(u), \alpha_{2,T}(u), \alpha_{3,T}(u)$, and $\alpha_{4,T}(u)$ be functions which are $O(|u|T^{-1+H}), O(|u|^2T^{-1+H}), O(|u|^3T^{-3/2})$, and $O(|u|^3T^{-1/2})$, respectively. Note that for the given range of values of x and u, the conditions on z_1 for part (a) of Lemma are satisfied. Further, with

$$\beta_T(u) := 1 + iu\frac{\delta_{T,x}}{\theta} + \frac{u^2\delta_{T,x}^2}{2\theta^2},$$

we have

$$\begin{aligned}
&(\theta^2 - 2z_1)^{1/2}\\
=\ & \theta\left[1 - \frac{z_1}{\theta^2} - \frac{z_1^2}{2\theta^4} + \frac{z_1^3}{2\theta^8} + \cdots\right]\\
=\ & \theta\left[1 + iu\frac{\delta_{T,x}}{\theta} + \frac{u^2\delta_{T,x}^2}{2\theta^2} + \frac{iu^3\delta_{T,x}^3}{2\theta^3} + \cdots\right]\\
=\ & \theta[1 + \alpha_{1,T}(u) + \alpha_{2,T}(u) + \alpha_{3,T}(u)]\\
=\ & \theta\beta_T(u) + \alpha_{3,T}(u)\\
=\ & \theta[1 + \alpha_{1,T}(u)].
\end{aligned}$$

Thus

$$\gamma - \theta = \alpha_{1,T},\ \gamma + \theta = 2\theta + \alpha_{1,T}.$$

Hence the above expectation equals

$$\exp\left(z_3 + \tfrac{\theta T}{2}\right) \times$$
$$\left[\frac{2\theta\beta_T(u)+\alpha_{3,T}(u)}{\alpha_{1,T}\exp\{-\theta T\beta_T(u)+\alpha_{4,T}(u)\}+(2\theta+\alpha_{1,T}(u))\exp\{\theta T\beta_T(u)+\alpha_{4,T}(u)\}}\right]^{1/2}$$
$$= \left[\frac{1+\alpha_{1,T}(u)}{\alpha_{1,T}\exp(\chi_T(u)) + (1+\alpha_{1,T}(u))\exp(\psi_T(t))}\right]^{1-H},$$

where

$$\begin{aligned}\chi_T(u) &= -\theta T\beta_T(u) + \alpha_{4,T}(u) - 2z_3 - \theta T\\ &= -2\theta T + \alpha_{1,T}(u) + t^2\alpha_{1,T}(u)\end{aligned}$$

and

$$\begin{aligned}\psi_T(u) &= \theta T\beta_T(u) + \alpha_{4,T}(u) - 2z_3 - \theta T\\ &= \theta T\left[1 + iu\frac{\delta_{T,x}}{\theta} + \frac{u^2\delta_{T,x}^2}{2\theta^2}\right] + \alpha_{4,T}(u) - iuT\delta_{T,x} - \theta T\\ &= \frac{u^2T}{2\theta}\left[\left(\frac{2\theta}{T}\right)^{1/2} + \frac{2x}{T}\right]^2\\ &= u^2 + u^2\alpha_{1,T}(u).\end{aligned}$$

Hence, for the given range of values of u, $\chi_T(t) - \psi_T(u) \leq -\theta T$.

Hence the above expectation equals

$$\exp(-\tfrac{u^2}{2})(1+\alpha_{1,T})^{1-H}$$
$$\left[\alpha_{1,T}\exp\{-2\theta T + \alpha_{1,T} + t^2\alpha_{1,T}\} + (1+\alpha_{1,T}(t))\exp\{t^2\alpha_{1,T}(t)\}\right]^{-1/2}$$
$$= \exp(-\tfrac{u^2}{2})\left[1+\alpha_{1,T})(1+\alpha_{1,T}(1+\alpha_{1,T})\exp\{-\theta T + \alpha_{1,T} + t^2\alpha_{1,T}\}\right] \times$$
$$\exp(u^2\alpha_{1,T}(u)).$$

This completes the proof of the lemma. □

Kleptsyna and Le Breton (2002) obtained the following Cameron–Martin type formula:

Denote $V_T := \int_0^T Q_s^2 dv_s$. Let $\phi_T^H(u) := E\exp(-uV_T),\ u > 0$. Then $\phi_T^H(u)$ is given by

$$\phi_T^H(u) = \left\{\frac{4\sin\pi H\sqrt{\theta^2+2u}e^{-\theta T}}{\pi T D_T^H(\theta;\sqrt{\theta^2+2u})}\right\}^{1/2}, \tag{1.9}$$

where

$$\begin{aligned}D_T^H(\theta;\alpha) := &\ [\alpha\cosh(\tfrac{\alpha}{2}T) - \theta\sinh(\tfrac{\alpha}{2}T)]^2 J_{-H}(\tfrac{\alpha}{2}T)J_{H-1}(\tfrac{\alpha}{2}T)\\ &-[\alpha\sinh(\tfrac{\alpha}{2}T) - \theta\cosh(\tfrac{\alpha}{2}T)]^2 J_{1-H}(\tfrac{\alpha}{2}T)J_H(\tfrac{\alpha}{2}T)\end{aligned}$$

for $\alpha > 0$ and J_ν is the modified Bessel function of first kind of order ν.

We generalize the above Cameron–Martin type formula to obtain joint generating function of U_T and V_T.

Lemma 2.3 (a) Let $\Psi_T^H(z_1, z_2) := E\exp(z_1 V_T + z_2 U_T), z_1, z_2 \in \mathbb{C}$. Then $\Psi_T^H(z_1, z_2)$ exists for $|z_i| \le \delta$, $1 = 1{,}2$ for some $\delta > 0$ and is given by

$$\Psi_T^H(z_1, z_2) = \exp\left(\frac{-\theta T}{2}\right)\left[\frac{4(\sin \pi H)\gamma}{\pi T \mathcal{D}_T^H(\gamma; z_2)}\right]^{1/2}, \tag{1.10}$$

where $\gamma := (\theta^2 + 2z_1)^{1/2}$, and we choose the principal branch of the square root and

$$\begin{aligned}\mathcal{D}_T^H(\gamma; z_2) \quad := \quad & [\gamma \cosh(\frac{\gamma}{2}T) - (\theta - 2z_2)\sinh(\frac{\gamma}{2}T)]^2 J_{-H}(\frac{\gamma}{2}T) J_{H-1}(\frac{\gamma}{2}T) \\ & -[\gamma \sinh(\frac{\gamma}{2}T) - (\theta - 2z_2)\cosh(\frac{\gamma}{2}T)]^2 J_{1-H}(\frac{\gamma}{2}T) J_H(\frac{\gamma}{2}T),\end{aligned}$$

and J_ν is the modified Bessel function of first kind of order ν.

Proof We use analytic continuation, fractional Itô formula (Lemma D.2.3) along with change of measure technique as in Liptser and Shiryayev (1978) and Kleptsyna and Le Breton (2002).

First assume $z_i = a_i \in \mathbb{R}, i = 1, 2$, and a_i are sufficiently small. Define $\gamma := (\theta^2 + 2z_1)^{1/2}$ and $dX_t^\gamma = \gamma X_t^\gamma + dW_t^H, X_0^\gamma = 0$. Also recall that $dX_t^\theta = \theta X_t^\theta + dW_t^H$.

Then

$$\frac{dP_\theta^T}{dP_\gamma^T}(Z_0^T) = \left\{(\theta - \gamma)\int_0^T Q_t^\gamma dZ_t^\gamma - \frac{(\theta^2 - \gamma^2)}{2}\int_0^T (Q_t^\gamma)^2 dv_t\right\}. \tag{1.11}$$

Note that

$$\Psi_T^H(a_1, a_2) := E\exp(a_1 V_T^\theta + a_2 U_T^\theta).$$

If we change the measure to that generated by Z_t^γ, then by (1.11),

$$\begin{aligned}\Psi_T^H(a_1, a_2) \quad = \quad & E\Bigg[\exp(a_1\int_0^T (Q_t^\gamma)^2 dv_t + a_2 (Z_T^\gamma)^2) \\ & -(\gamma - \theta)\int_0^T Q_t^\gamma dZ_t^\gamma - a_1\int_0^T (Q_t^\gamma)^2 dv_t\Bigg] \\ = \quad & E\exp\left[a_2 U_T^\gamma - (\gamma - \theta)\int_0^T Q_t^\gamma dZ_t^\gamma\right].\end{aligned} \tag{1.12}$$

By fractional Itô formula (Lemma D.2.3),

$$\int_0^T Q_s^\gamma dZ_s^\gamma = \frac{1}{2}\left[\frac{\lambda_H}{2-2H} Z_T^\gamma \int_0^T t^{2H-1} dZ_s^\gamma - T\right]. \tag{1.13}$$

Using this in (1.12), we have

$$\Psi_T^H(a_1,a_2) = E\exp\left[\left(a_2 - \frac{\gamma-\theta}{2}\right)\frac{\lambda_H}{2-2H} Z_T^\gamma \int_0^T t^{2H-1} dZ_s^\gamma - \left(\frac{\gamma-\theta}{2}\right)T\right]. \tag{1.14}$$

Let $\lambda_H^* := \frac{\lambda_H}{2-2H}$.

Note that $(Z_t^\gamma, \int_0^T t^{2H-1} dZ_s^\gamma)'$ is a Gaussian pair with covariance matrix denoted by $R^\gamma(T)$.

Thus

$$\Psi_T^H(a_1,a_2) = \exp\left(\frac{(\gamma-\theta)T}{2}\right)\left\{\det[I + \frac{(\gamma-\theta)\lambda_H^*}{2} JR^\gamma(T)]\right\}^{-1/2}, \tag{1.15}$$

where the 2×2 matrices I and J are defined by

$$I = \begin{pmatrix} 1 & 0 \\ 0 & 1 \end{pmatrix}$$

and

$$J = \begin{pmatrix} 0 & 1 \\ 1 & 0 \end{pmatrix}.$$

$R^\gamma(T)$ satisfies the Lyapunov differential equation

$$\dot{R}^\gamma(T) = \frac{\gamma}{2}\{A_H(T)R^\gamma(T) + R^\gamma(T)A'_H(T)\} + \frac{1}{\lambda_H^*} B_H(T);\ T \geq 0;\ R^\gamma(0) = 0, \tag{1.16}$$

where

$$A_H(s) = \begin{pmatrix} 1 & s^{1-2H} \\ s^{2H-1} & 1 \end{pmatrix}$$

and

$$B_H(s) = \begin{pmatrix} s^{1-2H} & 1 \\ 1 & s^{2H-1} \end{pmatrix}.$$

Now solving for $R^\gamma(T)$ as in Kleptsyna and Le Breton (2002),

$$\begin{aligned}
&\det[I + \frac{(\gamma-\theta)\lambda_H^*}{2} JR^\gamma(T)] \\
&= \frac{\gamma\pi T}{4\pi \sin \pi H}\left\{\left[1 + \frac{\gamma-\theta}{\gamma} e^{\frac{\gamma T}{2}} \sinh \frac{\gamma T}{2}\right]^2 J_{-H}(\tfrac{\gamma}{2}T) J_{H-1}(\tfrac{\gamma}{2}T)\right. \\
&\left. - \left[1 - \frac{\gamma-\theta}{\gamma} e^{\frac{\gamma T}{2}} \cosh \frac{\gamma T}{2}\right]^2 J_{1-H}(\tfrac{\gamma}{2}T) J_H(\tfrac{\gamma}{2}T)\right\},
\end{aligned}$$

where J_ν is the modified Bessel function of first kind of order ν.

Substituting this into (1.15) yields (1.10) for z_i's real, around a neighborhood of zero. Thus there is no problem of existence of $\Psi_T^H(z_1, z_2)$ around zero in $\mathbb{C}$, and since we have shown that the moment generating function exists, $\Psi_T^H(z_1, z_2)$. On the other hand, (1.10) defines an analytic function in the relevant domain and agrees with $\Psi_T^H(z_1, z_2)$ for z_1, z_2 real. □

To obtain the rate of normal approximation for the MCE, we need the following estimate on the tail behavior of the MCE.

Lemma 2.4

$$P\left\{(\frac{T}{2\theta})^{1-H}|\tilde{\theta}_T - \theta| \geq 2(\log T)^{1-H}\right\} \leq CT^{-1+H}.$$

Proof Observe that

$$\begin{aligned}
& P\left\{(\frac{T}{2\theta})^{1-H}|\tilde{\theta}_T - \theta| \geq 2(\log T)^{1-H}\right\} \\
= \; & P\left\{\left|\frac{(\frac{2\theta}{T})^{1-H}J_T}{(\frac{2\theta}{T})^{2-2H}I_T}\right| \geq 2(\log T)^{1-H}\right\} \\
\leq \; & P\left\{\left|(\frac{2\theta}{T})^{1-H}J_T\right| \geq (\log T)^{1-H}\right\} + P\left\{\left|\frac{2\theta}{T}I_T\right| \leq \frac{1}{2}\right\} \\
\leq \; & \left|P\left\{(\frac{2\theta}{T})^{1-H}|J_T| \geq (\log T)^{1-H}\right\} - 2\Phi(-(\log T)^{1-H})\right| \\
& +2\Phi(-(\log T)^{1-H}) + P\left\{\left|\frac{2\theta}{T}I_T - 1\right| \geq \frac{1}{2}\right\} \\
\leq \; & \sup_{x\in\mathbb{R}}\left|P\left\{(\frac{2\theta}{T})^{1-H}|J_T| \geq x\right\} - 2\Phi(-x)\right| \\
\leq \; & \sup_{x\in\mathbb{R}}\left|P\left\{(\frac{2\theta}{T})^{1-H}|J_T| \geq x\right\} - 2\Phi(-x)\right| \\
& +2\Phi(-(\log T)^{1-H}) + P\left\{\left|(\frac{2\theta}{T})^{2-2H}I_T - 1\right| \geq \frac{1}{2}\right\} \\
\leq \; & CT^{-1+H} + C(T\log T)^{-1+H} + CT^{-2+2H} \\
\leq \; & CT^{-1+H}.
\end{aligned}$$

The bounds for the first and the third terms come from Lemma 2.2 and Lemma 2.1, respectively, and that for the middle term comes from Feller (1957, p. 166). □

Now we are ready to obtain the uniform rate of normal approximation of the distribution of the MCE.

Theorem 2.5

$$\sup_{x\in\mathbb{R}}\left|P\left\{(\frac{T}{2\theta})^{1-H}(\tilde{\theta}_T-\theta)\le x\right\}-\Phi(x)\right|\le C_\theta T^{-1+H}.$$

Proof We shall consider two possibilities (i) and (ii).
(i) $|x|>2(\log T)^{1-H}$.
We shall give a proof for the case $x>2(\log T)^{1-H}$. The proof for the case $x<-2(\log T)^{1-H}$ runs similarly. Note that

$$\left|P\left\{(\frac{T}{2\theta})^{1-H}(\tilde{\theta}_T-\theta)\le x\right\}-\Phi(x)\right|\le P\left\{(\frac{T}{2\theta})^{1-H}(\tilde{\theta}_T-\theta)\ge x\right\}+\Phi(-x).$$

But $\Phi(-x)\le\Phi(-2(\log T)^{1-H})\le CT^{-2(1-H)}$. See Feller (1957, p. 166).
Moreover, by Lemma 2.4, we have

$$P\left\{(\frac{T}{2\theta})^{1-H}(\tilde{\theta}_T-\theta)\ge 2(\log T)^{1/2}\right\}\le CT^{-1+H}.$$

Hence

$$\left|P\left\{(\frac{T}{2\theta})^{1-H}(\tilde{\theta}_T-\theta)\le x\right\}-\Phi(x)\right|\le CT^{-1+H}.$$

(ii) $|x|\le 2(\log T)^{1-H}$

Let $A_T:=\left\{(\frac{T}{2\theta})^{1-H}|\tilde{\theta}_T-\theta|\le 2(\log T)^{1-H}\right\}$ and $B_T:=\left\{\frac{I_T}{T}>c_0\right\}$,

where $0<c_0<\frac{1}{2\theta}$. By Lemma 2.4, we have

$$P(A_T^c)\le CT^{-1+H}. \tag{2.1}$$

By Lemma 2.1, we have

$$\begin{aligned}P(B_T^c)&=P\left\{(\frac{2\theta}{T})^{2-2H}I_T-1<2\theta c_0-1\right\}\\&<P\left\{|\frac{2\theta}{T}I_T-1|>1-2\theta c_0\right\}\le CT^{-1}.\end{aligned} \tag{2.2}$$

Let b_0 be some positive number. On the set $A_T\cap B_T$ for all $T>T_0$ with $4b_0(\log T_0)^{1-H}(\frac{2\theta}{T_0})^{1-H}\le c_0$, we have

$$\begin{aligned}&(\frac{T}{2\theta})^{1-H}(\tilde{\theta}_T-\theta)\le x\\
\Rightarrow\ &I_T+b_0T(\tilde{\theta}_T-\theta)<I_T+(\frac{T}{2\theta})^{1-H}2b_0\theta x\\
\Rightarrow\ &(\frac{T}{2\theta})^{1-H}(\tilde{\theta}_T-\theta)[I_T+b_0T(\theta_T-\theta)]<x[I_T+(\frac{T}{2\theta})^{1-H}2b_0\theta x]\\
\Rightarrow\ &(\tilde{\theta}_T-\theta)I_T+b_0T(\theta_T-\theta)^2<(\frac{2\theta}{T})^{1-H}I_Tx+2b_0\theta x^2\\
\Rightarrow\ &-J_T+(\tilde{\theta}_T-\theta)I_T+b_0T(\tilde{\theta}_T-\theta)^2<-J_T+(\frac{2\theta}{T})^{1-H}I_Tx+2b_0\theta x^2\\
\Rightarrow\ &0<-J_T+(\frac{2\theta}{T})^{1-H}I_Tx+2b_0\theta x^2\end{aligned}$$

since

$$\begin{aligned} & I_T + b_0 T(\theta_T - \theta) \\ > \ & Tc_0 + b_0 T(\theta_T - \theta) \\ > \ & 4b_0(\log T)^{1-H}(\frac{2\theta}{T})^{1-H} - 2b_0(\log T)^{1-H}(\frac{2\theta}{T})^{1-H} \\ = \ & 2b_0(\log T)^{1-H}(\frac{2\theta}{T})^{1-H} > 0. \end{aligned}$$

On the other hand, on the set $A_T \cap B_T$ for all $T > T_0$ with $4b_0(\log T_0)^{1-H}(\frac{2\theta}{T_0})^{1-H} \le c_0$, we have

$$\begin{aligned} & (\frac{T}{2\theta})^{1-H}(\theta_T - \theta) > x \\ \Rightarrow \ & I_T - b_0 T(\tilde{\theta}_T - \theta) < I_T - (\frac{T}{2\theta})^{1-H} 2b_0\theta x \\ \Rightarrow \ & (\frac{T}{2\theta})^{1-H}(\tilde{\theta}_T - \theta)[I_T - b_0 T(\theta_T - \theta)] > x[I_T - (\frac{T}{2\theta})^{1-H} 2b_0\theta x] \\ \Rightarrow \ & (\tilde{\theta}_T - \theta)I_T - b_0 T(\tilde{\theta}_T - \theta)^2 > (\frac{2\theta}{T})^{1-H} I_T x - 2b_0\theta x^2 \\ \Rightarrow \ & -J_T + (\tilde{\theta}_T - \theta)I_T - b_0 T(\theta_T - \theta)^2 > -J_T + (\frac{2\theta}{T})^{1-H} I_T x - 2b_0\theta x^2 \\ \Rightarrow \ & 0 > -J_T + (\frac{2\theta}{T})^{1-H} I_T x - 2b_0\theta x^2 \end{aligned}$$

since

$$\begin{aligned} & I_T - b_0 T(\tilde{\theta}_T - \theta) \\ > \ & Tc_0 - b_0 T(\tilde{\theta}_T - \theta) \\ > \ & 4b_0(\log T)^{1-H}(\frac{2\theta}{T})^{1-H} - 2b_0(\log T)^{1-H}(\frac{2\theta}{T})^{1-H} \\ = \ & 2b_0(\log T)^{1-H}(\frac{2\theta}{T})^{1-H} > 0. \end{aligned}$$

Hence

$$0 < -J_T + (\frac{2\theta}{T})^{1-H} I_T x - 2b_0\theta x^2 \Rightarrow (\frac{T}{2\theta})^{1-H}(\theta_T - \theta) \le x.$$

Letting $D^{\pm}_{T,x} := \left\{-J_T + (\frac{2\theta}{T})^{1-H} I_T x \pm 2b_0\theta x^2 > 0\right\}$, we obtain

$$D^{-}_{T,x} \cap A_T \cap B_T \subseteq A_T \cap B_T \cap \left\{(\frac{T}{2\theta})^{1-H}(\theta_T - \theta) \le x\right\} \subseteq D^{+}_{T,x} \cap A_T \cap B_T. \quad (2.3)$$

If it is shown that

$$\left|P\left\{D^{\pm}_{T,x}\right\} - \Phi(x)\right| \le CT^{-1+H} \quad (2.4)$$

for all $T > T_0$ and $|x| \le 2(\log T)^{1-H}$, then the theorem would follow from (2.1)–(2.3).

We shall prove (2.4) for $D^+_{T,x}$. The proof for $D^-_{T,x}$ is analogous. Observe that

$$\begin{aligned}
& \left|P\left\{D^+_{T,x}\right\} - \Phi(x)\right| \\
= & \left|P\left\{(\frac{2\theta}{T})^{1-H} J_T - \left((\frac{2\theta}{T})^{2-2H} I_T - 1\right) x \right.\right. \\
& \left. < x + 2(\tfrac{2\theta}{T})^{1-H} b_0 \theta x^2\right\} - \Phi(x)\Big| \\
\leq & \sup_{y\in\mathbb{R}} \left|P\left\{(\frac{2\theta}{T})^{1-H} J_T - \left(\frac{2\theta}{T} I_T - 1\right) x \leq y\right\} - \Phi(y)\right| \\
& + \left|\Phi\left(x + (\tfrac{2\theta}{T})^{1-H} b_0 \theta x^2\right) - \Phi(x)\right| \\
=: & \Delta_1 + \Delta_2.
\end{aligned} \tag{2.5}$$

Lemma 2.3(b) and Esseen's lemma immediately yield

$$\Delta_1 \leq CT^{-1+H}. \tag{2.6}$$

On the other hand, for all $T > T_0$,

$$\Delta_2 \leq 2(\frac{2\theta}{T})^{1-H} b_0 \theta x^2 (2\pi)^{-1/2} \exp(-\overline{x}^2/2),$$

where

$$|\overline{x} - x| \leq 2(\frac{2\theta}{T})^{1-H} b_0 \theta x^2.$$

Since $|x| \leq 2(\log T)^{1-H}$, it follows that $|\overline{x}| > |x|/2$ for all $T > T_0$ and consequently

$$\begin{aligned}
\Delta_2 & \leq 2(\frac{2\theta}{T})^{1-H} b_0 \theta x^2 (2\pi)^{-1/2} x^2 \exp(-x^2/8) \\
& \leq CT^{-1+H}.
\end{aligned} \tag{2.7}$$

From (2.5)–(2.7), we obtain

$$\left|P\left\{D^+_{T,x}\right\} - \Phi(x)\right| \leq CT^{-1+H}.$$

This completes the proof of the theorem. □

4.2.2 Precise Berry–Esseen Bound in the Ornstein –Uhlenbeck Process

Consider the O-U model

$$dX_t = \theta X_t dt + dW_t$$

with $\theta < 0$. The MCE is given by

$$\tilde{\theta} = \frac{T}{2\int_0^T X_t^2 dt}.$$

Note that

$$\phi_T(z_1) = \exp\left(\frac{\theta T}{2}\right)\left[\frac{2\gamma}{(\gamma-\theta)e^{-\gamma T}+(\gamma+\theta)e^{\gamma T}}\right]^{1/2},$$

where $\gamma = (\theta^2 - 2z_1)^{1/2}$, and we choose the principal branch of the square root.

It is well known that $\tilde{\theta}_T$ is strongly consistent and asymptotically $\mathcal{N}(0, 2\theta)$ distributed as $T \to \infty$ (see Lanksa 1979). The large deviations of $\hat{\theta}_T$ and $\tilde{\theta}_T$ were obtained in Florens-Landais and Pham (1999). In particular, it was shown that the large deviation probabilities of $\hat{\theta}_T$ are identical to those of $\tilde{\theta}_T$ for $\theta \geq \theta_0/3$ but weaker for $\theta < \theta_0/3$, where θ_0 is the true value of the parameter. However, as far as the rate of normal approximation is concerned, we show that $\hat{\theta}_T$ and $\tilde{\theta}_T$ have the same normal approximation rate of the order $O(T^{-1/2})$. Note that $T^{1/2}(\hat{\theta}_T - \theta)$ asymptotically $\mathcal{N}(0, 2\theta)$ distributed as $T \to \infty$.

The previous theorem does not give the precise estimate of the constant in the inequality. The following theorem gives a precise estimate of the constant.

Theorem 2.6 We have for all $T \geq T_1$,

$$\begin{aligned}&\sup_{x\in\mathbb{R}}\left|P\left\{(\tfrac{T}{2\theta})^{1/2}(\tilde{\theta}_T-\theta)\leq x\right\}-\Phi(x)\right|\\ &\leq \left[0.6751 + 0.4(\log T)^{-1/2} + a(\theta)T^{-1/2}\right]T^{-1/2},\end{aligned}$$

where $T_1 \in \mathbb{R}^+$ such that for all $T \geq T_1$, $\quad T^{-1}\log T \leq (b(\theta))^2$, where

$$\begin{aligned} b(\theta) &:= \frac{a_2 a_3^{-1}(a_3-1)}{2\theta}\epsilon(\theta)\\ a(\theta) &:= (1-a_2)^{-2}a_3^2(1-a_3)^{-2}\epsilon(\theta)2\theta - (1-a_2)^{-2},\end{aligned}$$

$a_1,\ a_2,\ a_3$ are constants (not depending on θ) satisfying $a_1 > 1,\ 0 < a_2 < 1, a_3 > 1$.

Proof Denote

$$\begin{aligned} F_{T,\theta} &:= \left\{\left(\frac{2\theta}{T}\right)I_T > a_2\right\},\\ G_{T,\theta,\epsilon} &:= \left\{\frac{N_T}{T} \leq \epsilon(-\epsilon a_1 + \frac{a_2}{2\theta})\right\},\\ Q_{T,\theta} &:= \left\{\left(\frac{2\theta}{T}\right)^{1/2} N_T > (\log T)^{1/2}\right\},\\ d(\theta) &:= \min\left(\epsilon(\theta),\ \frac{a_2}{2\theta a_1 a_3}\right),\\ R_{T,\theta} &:= F_{T,\theta}\cap G_{T,\theta,d(\theta)},\\ \gamma &:= a_2^{-1}a_3(a_3-1)^{-1},\end{aligned}$$

where $a_1 > 0,\quad 0 < a_2 < 1$, and $a_3 > 1$.

We obtain for all $\epsilon \in (0, \epsilon(\theta)]$,

$$F_{T,\theta} \cap G_{T,\theta,\epsilon} \subset \{|\tilde{\theta}_T - \theta| < \epsilon\}.$$

We shall consider two possibilities (i) and (ii).
(i) $|x| > \gamma(\log T)^{1/2}$
We shall give a proof for the case $x < -\gamma(\log T)^{1/2}$. The proof for the case $x > \gamma(\log T)^{1/2}$ runs similarly.
Observe that

$$\begin{aligned}
&\left\{(\frac{T}{2\theta})^{1/2}(\tilde{\theta}_T - \theta) \le x\right\} \\
\subset\; &\left\{(\frac{T}{2\theta})^{1/2}(\tilde{\theta}_T - \theta) \le -\gamma(\log T)^{1/2}\right\} \cap F_{T,\theta} \cap G_{T,\theta,d(\theta)} \\
\cup\; & F^c_{T,\theta} \cup G^c_{T,\theta,d(\theta)} \\
\subset\; & Q_{T,\theta} \cup F^c_{T,\theta} \cup G^c_{T,\theta,d(\theta)}.
\end{aligned}$$

Hence

$$\begin{aligned}
&P\left\{(\frac{T}{2\theta})^{1/2}(\tilde{\theta}_T - \theta) \le x\right\} \\
\le\; & P(Q_{T,\theta}) + P(F^c_{T,\theta}) + P(G^c_{T,\theta,d(\theta)}).
\end{aligned}$$

By Lemma 2.2 above and lemma 2 in Feller (1957, p.166), we have

$$\begin{aligned}
P(Q_{T,\theta}) &= |P(Q_{T,\theta}) - \Phi(-(\log T)^{1/2})| + \Phi(-(\log T)^{1/2}) \\
&\le [0.6751 \;+\; 0.4(\log T)^{-1/2}]T^{-1/2}.
\end{aligned}$$

Further, for $\gamma > 1$,

$$\begin{aligned}
\Phi(x) &\le \Phi(-\gamma(\log T)^{1/2}) \\
&\le \Phi(-(\log T)^{1/2}) \\
&\le (2\pi)^{-1/2}(\log T)^{-1/2}T^{-1/2}.
\end{aligned}$$

Thus

$$\begin{aligned}
&\left|P\left\{(\frac{T}{2\theta})^{1/2}(\tilde{\theta}_T - \theta) \le x\right\} - \Phi(x)\right| \\
\le\; & [0.6751 \;+\; 0.4(\log T)^{-1/2}]T^{-1/2} + P(F^c_{T,\theta}) + P(G^c_{T,\theta,d(\theta)}).
\end{aligned}$$

(ii) $|x| \le \gamma(\log T)^{1/2}$
For all $T \in \mathbb{R}^+$ with $T^{-1/2}(\log T)^{1/2} < \gamma^{-1}(2\theta)^{-1/2}d(\theta)$, we have

$$\begin{aligned}
&\left\{(\frac{T}{2\theta})^{1/2}(\tilde{\theta}_T - \theta) \le x\right\} \cap F_{T,\theta} \cap G_{T,\theta,d(\theta)} \\
=&\left\{N_T + (\frac{2\theta}{T})^{1/2} I_T x > 0\right\} \cap F_{T,\theta} \cap G_{T,\theta,d(\theta)}
\end{aligned}$$

by arguments similar to that in the proof of Theorem 2.5. From Lemma 2.2, we have

$$|P\{N_T > 0\} - \Phi(0)| \leq 0.6751\, T^{-1/2}.$$

Hence

$$\begin{aligned} &\left|P\left\{(\frac{T}{2\theta})^{1/2}(\tilde{\theta}_T - \theta) \leq x\right\} - \Phi(x)\right| \\ \leq\ & 0.6751\, T^{-1/2} + P(F^c_{T,\theta} \cup G^c_{T,\theta,d(\theta)}) + |\Phi(0) - \Phi(x)|. \end{aligned}$$

This completes the proof of the theorem. □

Note that

$$\left|P\left\{(\frac{T}{2\theta})^{1/2}(\tilde{\theta}_T - \theta) \leq x\right\} - \Phi(x)\right| \leq P\left\{(\frac{T}{2\theta})^{1/2}(\tilde{\theta}_T - \theta) \geq x\right\} + \Phi(-x).$$

But $\Phi(-x) \leq \Phi(-2(\log T)^{1/2}) \leq CT^{-2}$. See Feller (1957, p. 166). Moreover, by Lemma 2.4, we have

$$P\left\{(\frac{T}{2\theta})^{1/2}(\tilde{\theta}_T - \theta) \geq 2(\log T)^{1/2}\right\} \leq CT^{-1/2}.$$

Hence

$$\left|P\left\{(\frac{T}{2\theta})^{1/2}(\tilde{\theta}_T - \theta) \leq x\right\} - \Phi(x)\right| \leq CT^{-1/2}.$$

(ii) $|x| \leq \gamma(\log T)^{1/2}$

Let $A_{T,1} := \left\{(\frac{T}{2\theta})^{1/2}|\tilde{\theta}_T - \theta| \leq a_1(\log T)^{1/2}\right\}$ and $B_{T,1} := \left\{\frac{I_T}{T} > a_2 c_0\right\}$,

where $0 < c_0 < \frac{1}{2\theta}$. By Lemma 2.4, we have

$$P(A^c_T) \leq CT^{-1/2}. \tag{2.1}$$

By Lemma 2.1, we have

$$P(B^c_T) = P\left\{\frac{2\theta}{T}I_T - 1 < 2\theta c_0 - 1\right\} < P\left\{|\frac{2\theta}{T}I_T - 1| > 1 - 2\theta c_0\right\} \leq CT^{-1}. \tag{2.2}$$

Let b_0 be some positive number. On the set $A_T \cap B_T$ for all $T > T_0$ with $4b_0(\log T_0)^{1/2}(\frac{2\theta}{T_0})^{1/2} \le c_0$, we have

$$
\begin{aligned}
& (\frac{T}{2\theta})^{1/2}(\tilde{\theta}_T - \theta) \le x \\
\Rightarrow\ & I_T + b_0 T(\tilde{\theta}_T - \theta) < I_T + (\frac{T}{2\theta})^{1/2} 2b_0\theta x \\
\Rightarrow\ & (\frac{T}{2\theta})^{1/2}(\tilde{\theta}_T - \theta)[I_T + b_0 T(\theta_T - \theta)] < x[I_T + (\frac{T}{2\theta})^{1/2} 2b_0\theta x] \\
\Rightarrow\ & (\tilde{\theta}_T - \theta) I_T + b_0 T(\theta_T - \theta)^2 < (\frac{2\theta}{T})^{1/2} I_T x + 2b_0\theta x^2 \\
\Rightarrow\ & -N_T + (\tilde{\theta}_T - \theta) I_T + b_0 T(\tilde{\theta}_T - \theta)^2 < -N_T + (\frac{2\theta}{T})^{1/2} I_T x + 2b_0\theta x^2 \\
\Rightarrow\ & 0 < -N_T + (\frac{2\theta}{T})^{1/2} I_T x + 2b_0\theta x^2
\end{aligned}
$$

since

$$
\begin{aligned}
& I_T + b_0 T(\theta_T - \theta) \\
> \ & T c_0 + b_0 T(\theta_T - \theta) \\
> \ & 4b_0(\log T)^{1/2}(\frac{2\theta}{T})^{1/2} - 2b_0(\log T)^{1/2}(\frac{2\theta}{T})^{1/2} \\
= \ & 2b_0(\log T)^{1/2}(\frac{2\theta}{T})^{1/2} > 0.
\end{aligned}
$$

On the other hand, on the set $A_T \cap B_T$ for all $T > T_0$ with $4b_0(\log T_0)^{1/2}(\frac{2\theta}{T_0})^{1/2} \le c_0$, we have

$$
\begin{aligned}
& (\frac{T}{2\theta})^{1/2}(\theta_T - \theta) > x \\
\Rightarrow\ & I_T - b_0 T(\tilde{\theta}_T - \theta) < I_T - (\frac{T}{2\theta})^{1/2} 2b_0\theta x \\
\Rightarrow\ & (\frac{T}{2\theta})^{1/2}(\tilde{\theta}_T - \theta)[I_T - b_0 T(\theta_T - \theta)] > x[I_T - (\frac{T}{2\theta})^{1/2} 2b_0\theta x] \\
\Rightarrow\ & (\tilde{\theta}_T - \theta) I_T - b_0 T(\tilde{\theta}_T - \theta)^2 > (\frac{2\theta}{T})^{1/2} I_T x - 2b_0\theta x^2 \\
\Rightarrow\ & -N_T + (\tilde{\theta}_T - \theta) I_T - b_0 T(\theta_T - \theta)^2 > -N_T + (\frac{2\theta}{T})^{1/2} I_T x - 2b_0\theta x^2 \\
\Rightarrow\ & 0 > -N_T + (\frac{2\theta}{T})^{1/2} I_T x - 2b_0\theta x^2
\end{aligned}
$$

since

$$
\begin{aligned}
& I_T - b_0 T(\tilde{\theta}_T - \theta) \\
> \ & T c_0 - b_0 T(\tilde{\theta}_T - \theta)
\end{aligned}
$$

$$\begin{aligned} &> 4b_0(\log T)^{1/2}(\frac{2\theta}{T})^{1/2} - 2b_0(\log T)^{1/2}(\frac{2\theta}{T})^{1/2} \\ &= 2b_0(\log T)^{1/2}(\frac{2\theta}{T})^{1/2} > 0. \end{aligned}$$

Hence

$$0 < -N_T + (\frac{2\theta}{T})^{1/2} I_T x - 2b_0\theta x^2 \Rightarrow (\frac{T}{2\theta})^{1/2}(\theta_T - \theta) \le x.$$

Letting $D^{\pm}_{T,x} := \left\{-N_T + (\frac{2\theta}{T})^{1/2} I_T x \pm 2b_0\theta x^2 > 0\right\}$, we obtain

$$D^-_{T,x} \cap A_T \cap B_T \subseteq A_T \cap B_T \cap \left\{(\frac{T}{2\theta})^{1/2}(\theta_T - \theta) \le x\right\} \subseteq D^+_{T,x} \cap A_T \cap B_T. \quad (2.3)$$

If it is shown that

$$\left|P\left\{D^{\pm}_{T,x}\right\} - \Phi(x)\right| \le CT^{-1/2} \quad (2.4)$$

for all $T > T_0$ and $|x| \le 2(\log T)^{1/2}$, then the theorem would follow from (2.1)–(2.3).

We shall prove (2.4) for $D^+_{T,x}$. The proof for $D^-_{T,x}$ is analogous. Observe that

$$\begin{aligned} &\left|P\left\{D^+_{T,x}\right\} - \Phi(x)\right| \\ = \; &\left|P\left\{(\frac{2\theta}{T})^{1/2} N_T - \left(\frac{2\theta}{T} I_T - 1\right) x < x + 2(\frac{2\theta}{T})^{1/2} b_0\theta x^2\right\} - \Phi(x)\right| \\ \le \; &\sup_{y\in\mathbb{R}} \left|P\left\{(\frac{2\theta}{T})^{1/2} N_T - \left(\frac{2\theta}{T} I_T - 1\right) x \le y\right\} - \Phi(y)\right| \\ &+ \left|\Phi\left(x + (\tfrac{2\theta}{T})^{1/2} b_0\theta x^2\right) - \Phi(x)\right| \\ =: \; &\Delta_1 + \Delta_2. \end{aligned} \quad (2.5)$$

Lemma 2.3(b) and Esseen's lemma immediately yield

$$\Delta_1 \le CT^{-1/2}. \quad (2.6)$$

On the other hand, for all $T > T_0$,

$$\Delta_2 \le 2(\frac{2\theta}{T})^{1/2} b_0\theta x^2 (2\pi)^{-1/2} \exp(-\overline{x}^2/2),$$

where

$$|\overline{x} - x| \le 2(\frac{2\theta}{T})^{1/2} b_0\theta x^2.$$

Since $|x| \le 2(\log T)^{1/2}$, it follows that $|\bar{x}| > |x|/2$ for all $T > T_0$ and consequently

$$\begin{aligned} \Delta_2 &\le 2(\frac{2\theta}{T})^{1/2} b_0\theta x^2 (2\pi)^{-1/2} x^2 \exp(-x^2/8) \\ &\le CT^{-1/2}. \end{aligned} \quad (2.7)$$

From (2.5)–(2.7), we obtain

$$\left|P\left\{D^{+}_{T,x}\right\} - \Phi(x)\right| \leq CT^{-1/2}.$$

Remarks

(1) The rate in Theorem 2.5 is uniform over compact subsets of the parameter space.

(2) The rates of normal approximation of the conditional least squares estimator and an approximate maximum likelihood estimator when the O-U process is observed at discrete time points in $[0,T]$ have been studied in Bishwal and Bose (2001). The rate of normal approximation for some approximate minimum contrast estimators based on discrete observations of the O-U process has been studied in Bishwal (2003). See Bishwal (2008a) for recent results on approximate likelihood asymptotics and approximate Bayes asymptotics for drift estimation of discretely observed diffusions based on high-frequency data.

(3) It remains to investigate the nonuniform rates of convergence to normality which are more useful.

4.2.3 Partially Observed Fractional Processes

The problem of optimal filtering of a signal when the noise is a fractional Brownian motion was studied by (Kleptsyna et al., 1998a, 1998b) (Le Breton, 1998a, 1998b), Coutin and Decreusefond (1999), and Kleptsyna et al. (1998, 2000).

As far as estimation of unknown parameter is concerned, maximum likelihood estimator (MLE) of the constant drift parameter of a fractional Brownian motion was obtained by Decreusefond and Ustunel (1998) who developed stochastic analysis of the fBm in a Malliavin calculus framework. Norros et al. (1999) studied the properties of the MLE of the constant drift parameter of fBm using martingale tools. They showed that the MLE is unbiased and normally distributed. They also showed that the MLE is strongly consistent and proved a law of the iterated logarithm as $T \to \infty$. The problem was generalized by Le Breton (1998a,b) to a stochastic differential equation driven by fBm with drift and the diffusion coefficient being nonrandom functions and the unknown parameter in the drift coefficient. He obtained the best linear unbiased estimator (BLUE) of the drift parameter which coincides with the MLE. He also obtained the least squares estimator (LSE) and compared the relative efficiency of the LSE and the BLUE.

Our aim now is to study the properties of the MLE of a parameter appearing nonlinearly in drift coefficient of a nonlinear stochastic differential equation driven by fBm.

The section is organized as follows: first, we prepare model, assumptions, and preliminaries. Second, we give the main results of the section on strong consistency and asymptotic normality of the MLE. Then we verify the main results for a partially observed fractional linear system.

On the probability space $(\Omega, \mathcal{F}, \mathbb{P})$, consider the stochastic differential equations

$$\begin{aligned} dY_t &= \theta f(t, X_t)\, dt + g(t, X_t) dW_t^H, \\ dX_t &= a(t, X_t)\, dt + b(t, X_t)\, dV_t^H, \ t \in [0, T], \\ & X_0 = \eta, \ Y_0 = \xi, \end{aligned}$$

where $W^H, H \in (\frac{1}{2}, 1)$ and $V^H, H \in (\frac{1}{2}, 1)$ are independent fractional Brownian motions.

Here $\theta \in \Theta$ open in $\mathbb{R}$ is the unknown parameter to be estimated on the basis of observation $\{X_t, 0 \leq t \leq T\}$.

We assume that the functions $g : [0, T] \times \mathbb{R} \to \mathbb{R}^+$ and $f : [0, T] \times \mathbb{R} \to \mathbb{R}$ are known and it satisfies

(A1) $|f(t, x)| + |g(t, x)| \leq K(1 + |x|)$ for all $t \in [0, T]$,

(A2)$|f(t, x) - f(t, y)| + |g(t, x) - g(t, y)| \leq K(|x - y|)$ for all $t \in [0, T]$,

for some constant $K > 0$.

We assume that the functions $a : [0, T] \times \mathbb{R} \to \mathbb{R}$ and $b : [0, T] \times \mathbb{R} \to \mathbb{R}$ are known and satisfy

(A3) (i) $|a(t, x)| \leq K_1(1 + |x|)$ for all $t \in [0, T]$.

(ii)$|b(t, x)| \leq K_1(1 + |x|)$ for all $t \in [0, T]$

for some constant $K_1 > 0$.

(A4) (i)$|a(t, x) - a(t, y)| \leq K_1(|x - y|)$ for all $t \in [0, T]$.

(ii) $|b(t, x) - b(t, y)| \leq K_1(|x - y|)$ for all $t \in [0, T]$

for some constant $K_1 > 0$.

Under the conditions (A1)–(A4), it is known that there exists a unique solution of the SDEs (see Lin 1995 and Kallianpur 1980).

In order to obtain the MLE, we proceed as follows.

Consider the canonical space of the process (X, Y). Let $\Omega = \mathcal{C}([0, T]; \mathbb{R}^2)$ be the space of continuous functions from $[0, T]$ into $\mathbb{R}^2$. Consider also the canonical process $(X, W^*) = ((X_t, W_t^*), t \in [0, T])$ on Ω, where $(X_t, W_t^*)(x, y)$ $= (x_t, y_t)$ for any $(x, y) \in \Omega$.

The probability $\tilde{\mathbb{P}}$ denotes the unique probability measure on Ω such that defining the variable ξ by $\xi = W_0^*$ and $\tilde{W} = (\tilde{W}_t), t \in [0, T]$ by $\tilde{W}_t = W_t^* - W_0^*, t \in [0, T]$, the pair (X, ξ) is independent of $\tilde{W}$ and the process $\tilde{W}$ is a fBm with Hurst parameter H.

The canonical filtration on Ω is $(\mathcal{F}_t, t \in [0, T])$, where $\mathcal{F}_t = \sigma(\{(X_s, W_s^*),$ $0 \leq s \leq t\} \bigvee \mathcal{N}, \mathcal{N}$ denoting the set of null sets of $(\Omega, \tilde{\mathbb{P}})$.

Define for all continuous functions $x = (x_t, t \in [0, T])$ the function $h(\theta, x)$ on $[0, T]$ by

$$h(\theta, x)(t) := \frac{f(\theta, t, x)}{g(t, x)}, t \in [0, T].$$

Consider, for any $t \in [0, T]$, the function $k_{h(\theta,x)}^t = (k_{h(\theta,x)}^t(s), 0 < s < t)$ defined from Theorem 2.1 with $h(\theta, x)$ in place of h.

Define the processes $N = (N_t, t \in [0,T])$ and $\langle N\rangle = (\langle N\rangle_t, t \in [0,T])$ from Theorem 2.1 by plugging in the process $h(\theta, x)$ in place of h, i.e.,

$$N_t := N_t^{h(\theta,X)}, \quad \langle N\rangle_t := \langle N^{h(\theta,X)}\rangle_t.$$

Notice that N_t and $\langle N\rangle_t$ depend only on the values of $X^{(t)} = (X_s, 0 \leq s \leq t)$.

Define the $(\mathcal{F}_t)$-adapted processes $\langle N, N^*\rangle = (\langle N, N^*\rangle_t, t \in [0,T])$ and $q(\theta, X) = (q_t(\theta, X), t \in [0,T])$ by

$$\langle N, N^*\rangle_t := \langle N^{h(\theta,X)}, N^*\rangle_t = \int_0^t k_*^t(s) h(\theta, X)(s) ds, t \in [0,T])$$

and

$$q_t(\theta, X) := q_t^{h(\theta,X)} = \frac{d\langle N, N^*\rangle_t}{d\langle N^*\rangle_t}, t \in [0,T].$$

Let $\tilde{q}_t(X) := \frac{q_t(\theta,X)}{\theta}$. Define for $0 \leq t \leq T$, the processes

$$\tilde{N}_t(\theta, x) := \int_0^t k_{h(\theta,x)}^t(s) d\tilde{W}_s^H, \quad \langle \tilde{N}\rangle_t(\theta, x) := \int_0^t h(\theta, x)(s) k_h^t(s) ds.$$

$\tilde{N}_t(\theta, x)$ is a Gaussian martingale under $\tilde{\mathbb{P}}$. Define

$$\Lambda_t(\theta, x) := \exp\{\tilde{N}_t(\theta, x) - \frac{1}{2}\langle \tilde{N}\rangle_t(\theta, x)\}, t \in [0,T].$$

Define for any $t \in [0,T]$, $\Lambda_t(\theta) := \Lambda_t(\theta, X)$. Let $\mathbb{P} := \Lambda_T(\theta)\tilde{\mathbb{P}}$. Let $\mathcal{Y}_t := \sigma(\{Y_s, 0 \leq s \leq t\}), t \in [0,T]$. Define the optimal filter $\pi_t(\phi) := \mathbb{E}[\phi(X_t)|\mathcal{Y}_t], t \in [0,T]$ and the unnormalized filter $\sigma_t(\phi) := \tilde{\mathbb{E}}[\phi(X_t)\Lambda_t|\mathcal{Y}_t], t \in [0,T]$. Then the *Kallianpur–Striebel formula* holds: for all $t \in [0,T]$, $\tilde{\mathbb{P}}$ and $\mathbb{P}$ almost surely

$$\pi_t(\phi) = \frac{\sigma_t(\phi)}{\sigma_t(1)}.$$

Introduce the processes Z and Z^* defined by

$$Z_t := \int_0^t k_{h(\theta,X)}^t(s) g^{-1}(s) dY_s, t \in [0,T]$$

and

$$Z_t^* := \int_0^t k_*^t(s) g^{-1}(s) dY_s, t \in [0,T].$$

The processes Z and Z^* are semimartingales with the following decomposition:

$$Z_t^* = \langle N, N^*\rangle_t + N_t^*, t \in [0,T],$$
$$Z_t = \langle N\rangle_t + N_t, t \in [0,T].$$

From the above, we can write

$$Z_t^* = \int_0^t q_s(\theta, X) d\langle N^* \rangle_s + N_t^*, t \in [0, T]),$$

and

$$Z_t = \int_0^t q_s^2(\theta, X) d\langle N^* \rangle_s + \int_0^t q_s(\theta, X) dN_s^*, t \in [0, T]).$$

Hence we get the integral representation of Z in terms of Z^*:

$$Z_t = \int_0^t q_s(\theta, X) dZ_t^*, t \in [0, T].$$

Introduce the process $\nu = (\nu_t, t \in [0, T])$ defined by

$$\nu_t = Z_t^* - \int_0^t \pi_s(q) d\langle N^* \rangle_s, t \in [0, T],$$

which plays the role of *innovation process* in the usual situation where the noises are Brownian motions.

Recall the notation $\pi_s(q) = \mathbb{E}[q_s(\theta, X) | \mathcal{Y}_s], s \in [0, t]$.

The particular case of unnormalized filter is

$$\tilde{\Lambda}_t(\theta) := \sigma_t(1) = \tilde{\mathbb{E}}[\Lambda_t | \mathcal{Y}_t], t \in [0, T].$$

By Proposition 3 in Kleptsyna et al. (2000), we have

$$\tilde{\Lambda}_T(\theta, \mathcal{Y}_t) = \exp\left\{ \theta \int_0^T \pi_s(\tilde{q}) dZ_s^* - \frac{\theta^2}{2} \int_0^T \pi_s^2(\tilde{q}) d\langle N^* \rangle_s \right\}.$$

Thus the maximum likelihood estimator (MLE) of θ is given by

$$\begin{aligned} \hat{\theta}_T &:= \arg\max_{\theta \in \Theta} \tilde{\Lambda}_T(\theta) \\ &= \frac{\int_0^T \pi_s(\tilde{q}) dZ_s^*}{\int_0^T \pi_s^2(\tilde{q}) d\langle N^* \rangle_s}. \end{aligned}$$

We shall prove the strong consistency and asymptotic normality of the MLE with a random normalization.

Theorem 3.1 *Under the conditions (A1)–(A4), $\hat{\theta}_T \longrightarrow \theta$ a.s. as $T \to \infty$, i.e., $\hat{\theta}_T$ is a strongly consistent estimator of θ.*

Moreover,

$$\limsup_T \frac{A_T^{1/2} |\hat{\theta}_T - \theta|}{(2 \log\log A_T)^{1/2}} = 1 \text{ a.s.},$$

where $A_T = \int_0^T \pi_s^2(\tilde{q}) d\langle N^* \rangle_s$.

Proof Note that

$$dZ_t^* = \pi_t(q)d\langle N^*\rangle_t + d\nu_t.$$

From Lemma 3 in Kleptsyna et al. (1998), we have that ν is a continuous Gaussian $(\mathcal{Y}_t, \mathbb{P})$ martingale such that $\langle \nu \rangle = \langle N^* \rangle$.

Hence

$$\begin{aligned}\hat{\theta}_T &= \frac{\int_0^T \pi_s(\tilde{q})dZ_s^*}{\int_0^T \pi_s^2(\tilde{q})d\langle N^*\rangle_s} \\ &= \theta + \frac{\int_0^T \pi_s(\tilde{q})d\nu_s}{\int_0^T \pi_s^2(\tilde{q})d\langle N^*\rangle_s}.\end{aligned}$$

Now by the strong law of large numbers for continuous martingales (see Revuz and Yor 1991 or Theorem 2.6.10) in Liptser and Shiryayev 1989), the second term in r.h.s. converges to zero a.s. as $T \to \infty$. Hence strong consistency follows. Since the deviation $\hat{\theta}_T - \theta$ is obtained from a stochastic integral with respect to a continuous martingale, the LIL follows from Corollary 1.1.12 of Revuz and Yor (1991). □

Theorem 3.2 *Under the conditions (A1)–(A4),*

$$\int_0^T (\pi_s^2(\tilde{q})d\langle N^*\rangle_s)^{1/2}(\hat{\theta}_T - \theta) \overset{\mathcal{D}}{\to} \mathcal{N}(0,1) \quad \text{as } T \to \infty.$$

Proof Note that

$$\left(\int_0^T \pi_s^2(\tilde{q})d\langle N^*\rangle_s\right)^{1/2}(\hat{\theta}_T - \theta) = \frac{\int_0^T \pi_s(\tilde{q})d\nu_s}{\sqrt{\int_0^T \pi_s^2(\tilde{q})d\langle N^*\rangle_s}}.$$

By the central limit theorem for stochastic integrals with respect to Gaussian martingales (see Liptser and Shiryayev 1989), the r.h.s. above converges in distribution to $\mathcal{N}(0,1)$ as $T \to \infty$. □

Consider the following affine stochastic system with fBm observation noise:

$$\begin{aligned} dY_t &= \theta X_t dt + dW_t^H, \\ dX_t &= -X_t dt + dV_t^H, \; t \in [0,T], \\ & Y_0 = 0, \quad X_0 = 0, \end{aligned}$$

where $\theta < 0$. Note that here the signal process is also a fractional O-U process.

Let $\hat{X}_t := E(X_t|\mathcal{Y}_t)$ and $\gamma_t := E([X_t - \hat{X}_t]^2|\mathcal{Y}_t)$. When $H = \frac{1}{2}$, the noises are the standard Brownian motions and the system of Filtering equations reduces to the Kalman–Bucy system, see Liptser and Shiryayev (1978). The filter $\hat{X}_t$ satisfies the filtering SDE

$$d\hat{X}_t = -\hat{X}_t dt + \theta\gamma_t d\nu_t,\ t \in [0,T], \hat{X}_0 = 0,$$

and γ_t satisfies the deterministic differential equation, known as the Riccati equation

$$d\gamma_t = dt - 2\gamma_t dt - \theta^2\gamma_t^2 d\langle N^*\rangle_t, \gamma_0 = 0.$$

As $t \to \infty, \gamma_t \to \gamma_{\frac{1}{2}}$, where $\gamma_{\frac{1}{2}} := \frac{-1+\sqrt{1+\theta^2}}{\theta^2}$. Now we make a simplifying assumption commonly made in the literature (see Kutoyants 1994a, p. 103). We assume that the system has reached the steady state, i.e., we assume that $\hat{X}^t$ satisfies the SDE

$$d\hat{X}_t = -\sqrt{1+\theta^2}\hat{X}_t dt + \theta\gamma_{\frac{1}{2}} d\nu_t,\ t \in [0,T], \hat{X}_0 = 0,$$

which generates the filter

$$\hat{X}_t = \theta\gamma_{\frac{1}{2}} \int_0^t e^{-\sqrt{1+\theta^2}(t-s)} dY_s.$$

Thus the MLE can be now explicitly calculated and it satisfies the asymptotic properties mentioned in Theorems 3.1 and 3.2.

When $H \in (1/2, 1)$, $\gamma_t \to \gamma_H$, where

$$\gamma_H := \int_{\mathbb{R}} g_H(u+h) g_H(u) du \sim C_1 h^{2(H-1)}.$$

$$V_t^H := \int_0^t K(t,s) dM_s^H$$

holds for $H \in (1/2, 1)$.

Define the process $Z = (Z_t, t \in [0,T])$ by

$$Z_t := \int_0^t k_H(t,s) dX_s.$$

Define the process $Z^0 = (Z_t, t \in [0,T])$ by

$$Z_t := \int_0^t k_H(t,s) dY_s.$$

$$Z_t = -\int_0^t Q_s dv_s + M_t^H,\quad Z_t^0 = \theta \int_0^t Q_s dv_s + N_t^H.$$

N is the fundamental martingale associated with W^H.

The following facts are known from Kleptsyna et al. (2000):

(i) Z is the *signal fundamental semimartingale* associated with the process X.

(ii) Z is a $(\mathcal{F}_t)$ -semimartingale with the decomposition

$$Z_t = \theta \int_0^t Q_s dv_s + M_t^H.$$

(iii) X admits the representation

$$X_t = \int_0^t K_H(t,s) dZ_s.$$

In particular, we have

$$X_t = \int_0^t K_H(t,s) Q_s dv_s + \int_0^t K_h(t,s) dM_s^H.$$

(iv) Y admits the representation

$$X_t = \int_0^t K_H(t,s) dZ_s^o.$$

(v) The natural filtration $(\mathcal{Z}_t)$ of Z and $(\mathcal{X}_t)$ of X coincide.
(vi) The natural filtration $(\mathcal{Z}^{\prime}{}_t)$ of Z^o and $(\mathcal{Y}_t)$ of Y coincide.
Define

$$Q_t := \frac{d}{dv_t} \int_0^t k_H(t,s) X_s ds.$$

$$Q_t = \frac{\lambda_H}{2(2-2H)} \left\{ t^{2H-1} Z_t + \int_0^t r^{2H-1} dZ_s \right\}.$$

Define the auxiliary process

$$\hat{Q}_t := E[Q_t | \mathcal{Y}_t].$$

$$\nu_t = Z_t^0 - \theta \int_0^t \hat{Q}_s dv_s.$$

The likelihood is given by

$$\tilde{\Lambda}_T(\theta, \mathcal{Y}_t) = \exp\left\{ \theta \int_0^T \hat{Q}_s dZ_s^* - \frac{\theta^2}{2} \int_0^T \hat{Q}_s^2 dv_s \right\}.$$

Thus the maximum likelihood estimator (MLE) of θ is given by

$$\begin{aligned} \hat{\theta}_T &:= \arg\max_{\theta \in \Theta} \tilde{\Lambda}_T(\theta) \\ &= \frac{\int_0^T \hat{Q}_s dZ_s^0}{\int_0^T \hat{Q}_s^2 dv_s}. \end{aligned}$$

Hence

$$\hat{\theta}_T = \frac{\int_0^T \hat{Q}_s dZ_s^0}{\int_0^T \hat{Q}_s^2 dv_s} = \theta + \frac{\int_0^T \hat{Q}_s d\nu_s}{\int_0^T \hat{Q}_s^2 dv_s}.$$

4.3 Discrete Sampling

4.3.1 Maximum Likelihood Estimation in Fractional Diffusions Driven by Perturbed Random Walks

Maximum likelihood estimation of the drift parameter in fractional diffusion is studied by using random walk approximation of the fractional Brownian motion extending the idea of Bertin et al. (2011). Asymptotic normality of the MLE is obtained.

4.3.2. Drifted Fractional Brownian Motion

Consider the drifted fractional Brownian motion

$$dY_t = \theta dt + dB_t^H, \quad t \in [0,T],$$

where (B_t^H) is a fractional Brownian motion with Hurst parameter $H \in (1/2, 1)$ where the parameter $\theta \in \mathbb{R}$ is unknown which needs to be estimated.

Replacing the fractional Brownian motion by its associated disturbed random walk, Bertin et al. (2011) considered Euler type discretization of the above model given by

$$Y_{t_{j+1}} = Y_{t_j} + \theta(t_{j+1} - t_j) + B_{t_{j+1}}^{H,N} - B_{t_j}^{H,N},$$

where $t_j = \frac{j}{N}, j = 0, 1, 2, \ldots, N^\alpha$ and $Y_{t_0} = Y_0 = 0$ and

$$B_t^{H,N} = \sqrt{N} \sum_{i=1}^{[Nt]} \left(\int_{\frac{i-1}{N}}^{\frac{i}{N}} K^H \left(\frac{[Nt]}{N}, s \right) ds \right) \xi_i,$$

where the ξ_i are i.i.d. standard normal random variables and

$$K^H(t,s) = c_H s^{1/2-H} \int_s^t (u-s)^{H-3/2} u^{H-1/2} du$$

and $c_H = \left(\frac{H(2H-1)}{\beta(2-2H, H-1/2)} \right)^{1/2}$, where $\beta(\cdot,\cdot)$ is the beta function. Assuming that H is known, Bertin et al. (2011) obtained maximum likelihood estimator (MLE) $\hat{\theta}_N$ of θ based on discrete observations $\{Y_{t_j}, t_j = \frac{j}{N}, j = 0, 1, 2, \ldots, N^\alpha\}$. Let $\Delta = t_{j+1} - t_j$. For the drifted Brownian motion, the MLE is given by

$$\hat{\theta}_N = \frac{1}{N\Delta} \sum_{j=1}^{N} (Y_j - Y_{j-1}) = \frac{Y_N}{N\Delta}.$$

For the drifted fractional Brownian motion,

$$\hat{\theta}_N = N\frac{\sum_{i,j=1}^{N^\alpha} j\Gamma_{i,j=1}^{-1} Y_i}{\sum_{i,j}^{N^\alpha} ij\Gamma_{i,j}^{-1}},$$

where $\Gamma_{i,j} = Cov\left(B^H_{\frac{i}{N}}, B^H_{\frac{i}{N}}\right)$. They showed that the MLE is unbiased and L^2-consistent when $\alpha > 1$.

The MLE has the mean square error bound which is given by

$$E|\hat{\theta}_N - \theta|^2 \leq CN^{(2-2H)(1-\alpha)},$$

where C is a positive constant.

Here we show the asymptotic normality.

Theorem 2.1

$$N^{(H-1)(1-\alpha)}(\hat{\theta}_N - \theta) \to^D N(0, I^{-1}(\theta)).$$

When $H = 3/4$ and $\alpha = 2$, the rate is of the order $N^{9/8}$.

4.3.3 Fractional Ornstein–Uhlenbeck Process

Then we consider the fractional Ornstein–Uhlenbeck process.

$$dY_t = \theta Y_t dt + dB_t^H, \quad t \in [0, T].$$

Replacing the fractional Brownian motion by its associated disturbed random walk, the Euler discretization of the above model is given by

$$Y_{t_{j+1}} = Y_{t_j} + \theta Y_{t_j}(t_{j+1} - t_j) + B_{t_{j+1}}^{H,N} - B_{t_j}^{H,N},$$

where the MLE is given by

$$\hat{\theta}_N = \frac{1}{\Delta}\log N\frac{\sum_{i,j}^{N^\alpha} j\Gamma_{i,j}^{-1} Y_i Y_j}{\sum_{i,j}^{N^\alpha} ij\Gamma_{i,j}^{-1} Y_i^2}.$$

Theorem 3.1

$$N^{(H-1)(1-\alpha)}(\hat{\theta}_N - \theta) \to^D N(0, I^{-1}(\theta)).$$

4.3.4. Fractional Cox–Ingersoll–Ross Process

Then we consider fractional Cox–Ingersoll–Ross process.

$$dY_t = \theta Y_t dt + \sqrt{Y_t} dB_t^H, \quad t \in [0, T]$$

Replacing the fractional Brownian motion by its associated disturbed random walk, the Euler discretization of the above model is given by

$$Y_{t_{j+1}} = Y_{t_j} + \theta Y_{t_j}(t_{j+1} - t_j) + \sqrt{Y_{t_j}}(B^{H,N}_{t_{j+1}} - B^{H,N}_{t_j})$$

where The MLE is given by

$$\hat{\theta}_N = \frac{1}{\Delta} \log N \frac{\sum_{i,j}^{N^\alpha} j\Gamma_{i,j}^{-1} Y_i \sum_{i,j}^{N^\alpha} j\Gamma_{i,j}^{-1} Y_i^{-1}}{\sum_{i,j}^{N^\alpha} ij\Gamma_{i,j}^{-1} Y_i^2}.$$

Theorem 4.1

$$N^{(H-1)(1-\alpha)}(\hat{\theta}_N - \theta) \to^D N(0, I^{-1}(\theta)).$$

4.3.5 Approximate Minimum Contrast Estimators

We introduce some new approximate minimum contrast estimators of the drift parameter in the Ornstein–Uhlenbeck process based on discretely sampled data and obtain rates of weak convergence of the distributions of the estimators to the standard normal distribution using random, nonrandom, and mixed normings.

The fractional Ornstein–Uhlenbeck process, also known as the fractional Vasicek model in finance, can be used in modeling short-term interest rates, see Hog and Frederiksen (2006). In view of this, it becomes necessary to estimate the unknown parameters in the model from discrete data. See Bishwal (2008a) for recent results on approximate likelihood asymptotics and approximate Bayes asymptotics for drift estimation of discretely observed diffusions based on high-frequency data. Here we assume constant volatility and without loss of generality assume it to be one. To estimate the drift parameter, we use an alternative to the maximum likelihood estimator and study the accuracy of distributional approximation of the resulting estimators from high-frequency data.

Let $(\Omega, \mathcal{F}, \{\mathcal{F}_t\}_{t\geq 0}, P)$ be a stochastic basis on which is defined the Ornstein–Uhlenbeck process $\{X_t\}$ satisfying the Itô stochastic differential equation

$$dX_t = \theta\ X_t\ dt + \ dW_t,\ t \geq 0, \quad X_0 = 0, \tag{1.1}$$

where $\{W_t\}$ is a standard Wiener process with the filtration $\{\mathcal{F}_t\}_{t\geq 0}$ and $\theta < 0$ is the unknown parameter to be estimated on the basis of discrete observations of the process $\{X_t\}$ at times $0 = t_0 < t_1 < \cdots < t_n = T$ with $t_i - t_{i-1} = \frac{T}{n}, i = 1, 2 \cdots, n$. We assume two types of high-frequency data with long observation time:
(1) $T \to \infty, n \to \infty, \frac{T}{\sqrt{n}} \to 0$ and (2) $T \to \infty, n \to \infty, \frac{T}{n^{2/3}} \to 0$.

Let the continuous realization $\{X_t, 0 \le t \le T\}$ be denoted by X_0^T. Let P_θ^T be the measure generated on the space (C_T, B_T) of continuous functions on $[0,T]$ with the associated Borel σ-algebra B_T generated under the supremum norm by the process X_0^T, and let P_0^T be the standard Wiener measure. It is well known that when θ is the true value of the parameter, P_θ^T is absolutely continuous with respect to P_0^T, and the Radon–Nikodym derivative (likelihood) of P_θ^T with respect to P_0^T based on X_0^T is given by

$$L_T(\theta) := \frac{dP_\theta^T}{dP_0^T}(X_0^T) = \exp\left\{\theta \int_0^T X_t dX_t - \frac{\theta^2}{2}\int_0^T X_t^2 dt\right\}. \tag{1.2}$$

Consider the score function, the derivative of the log-likelihood function, which is given by

$$\gamma_T(\theta) := \int_0^T X_t dX_t - \theta \int_0^T X_t^2 dt. \tag{1.3}$$

A solution of the estimating equation $\gamma_T(\theta) = 0$ provides the conditional maximum likelihood estimate (MLE)

$$\hat{\theta}_T := \frac{\int_0^T X_t dX_t}{\int_0^T X_t^2 dt}. \tag{1.4}$$

Strictly speaking, $\hat{\theta}_T$ is not the maximum likelihood estimate of θ since $\hat{\theta}_T$ may take positive values, whereas the parameter θ is assumed to be strictly negative. For an exact definition of MLE, see Kutoyants (1994a,b). Nevertheless, this definition of MLE is widely used. It is well known that $\hat{\theta}_T$ is strongly consistent and $T^{1/2}(\hat{\theta}_T - \theta)$ asymptotically $\mathcal{N}(0, -2\theta)$ distributed as $T \to \infty$ (see Kutoyants 1984a,b). Bishwal (2000a) obtained weak convergence bound of the order $O(T^{-1/2})$ for the MLE. Bishwal (2000b) obtained weak convergence bound of the order $O(T^{-1/2})$ for the Bayes estimators, with smooth priors and loss functions. These two papers used parameter dependent nonrandom norming. Bishwal (1999a,b) obtained weak convergence bound of the order $O(T^{-1/2})$ for the MLE and Bayes estimators using two different random normings which are useful for computation of a confidence interval. Bishwal and Bose (2001) obtained weak convergence bound for two approximate maximum likelihood estimators (AMLEs).

Note that in the stationary case where X_0 has $\mathcal{N}(0, -1/2\theta)$ distribution, the exact log-likelihood function is given by

$$l_T(\theta) := \frac{1}{2}\log\left(-\frac{\theta}{\pi}\right) - \frac{\theta^2}{2}\int_0^T X_t^2 dt + \left(\frac{\theta}{2}\right)[X_T^2 + X_0^2] + \frac{\theta T}{2}.$$

In this case, the likelihood equation has two solutions:

$$\check{\theta}_{T,1} = \frac{[X_T^2 + X_0^2 - T] + \{[X_T^2 + X_0^2]^2 - \pi \int_0^T X_t^2 dt\}^{1/2}}{4\int_0^T X_t^2 dt}, \tag{1.5}$$

$$\breve{\theta}_{T,2} = \frac{[X_T^2 + X_0^2 - T] - \{[X_T^2 + X_0^2]^2 - \pi \int_0^T X_t^2 dt\}^{1/2}}{4 \int_0^T X_t^2 dt}. \tag{1.6}$$

As an alternative to maximum likelihood method and to obtain estimators with higher-order accuracy, our aim is to use contrast functions. Using the Itô formula, the score function $\gamma_T(\theta)$ can be written as

$$\gamma_T(\theta) = \frac{1}{2} X_T^2 - \int_0^T (\theta X_t^2 + \frac{1}{2}) dt. \tag{1.7}$$

Consider the contrast function

$$M_T(\theta) = -\int_0^T (\theta X_t^2 + \frac{1}{2}) dt \tag{1.8}$$

and the minimum contrast estimate (MCE)

$$\tilde{\theta}_T := -\frac{T}{2} \left\{ \int_0^T Q_t^2 dt \right\}^{-1}. \tag{1.9}$$

M-estimator is reduced to the minimum contrast estimator. It is well known that $\tilde{\theta}_T$ is strongly consistent and asymptotically $\mathcal{N}(0, \; -2\theta)$ distributed as $T \to \infty$ (see Lanksa 1979). The large deviations of $\hat{\theta}_T$ and $\tilde{\theta}_T$ were obtained in Florens-Landais and Pham (1999). In particular, it was shown that the large deviation probabilities of $\hat{\theta}_T$ are identical to those of $\tilde{\theta}_T$ for $\theta \leq \theta_0/3$ but weaker for $\theta > \theta_0/3$, where θ_0 is the true value of the parameter. However, as far as the rate of convergence to normality is concerned, Bishwal (2002) showed that $\hat{\theta}_T$ and $\tilde{\theta}_T$ have the same Berry–Esseen bound of the order $O(T^{-1/2})$.

In this chapter, we obtain weak convergence bounds for several approximate minimum contrast estimators (AMCEs) which are simpler and robust. In order to define the approximate minimum contrast estimators (AMCEs), we use various discrete approximations of the integral in the definition (1.9) of MCE.

Note that for equally spaced data,

$$v_{t_i} - v_{t_{i-1}} = \lambda_H^{-1} \left(\frac{T}{n} \right)^{2-2H} [i^{2-2H} - (i-1)^{2-2H}].$$

We have

$$\begin{aligned}
Q_t &= \frac{d}{dv_t}\int_0^t k_H(t,s)X_s ds \\
&= \kappa_H^{-1}\frac{d}{dv_t}\int_0^t s^{1/2-H}(t-s)^{1/2-H}X_s ds \\
&= \kappa_H^{-1}\lambda_H t^{2H-1}\frac{d}{dt}\int_0^t s^{1/2-H}(t-s)^{1/2-H}X_s ds \\
&= \kappa_H^{-1}\lambda_H t^{2H-1}\int_0^t \frac{d}{dt}s^{1/2-H}(t-s)^{1/2-H}X_s ds \\
&= \kappa_H^{-1}\lambda_H t^{2H-1}\int_0^t s^{1/2-H}(t-s)^{-1/2-H}X_s ds. \qquad (1.18)
\end{aligned}$$

The process Q depends continuously on X, and therefore, the discrete observations of X do not allow one to obtain the discrete observations of Q. The process Q can be approximated by

$$\tilde{Q}_n = \kappa_H^{-1}\lambda_H n^{2H-1}\sum_{j=0}^{n-1} j^{1/2-H}(n-j)^{-1/2-H}X_j. \qquad (1.19)$$

It is easy to show that $\tilde{Q}_n \to Q_t$ almost surely as $n \to \infty$, see Tudor and Viens (2007).

Define a new partition $0 \le u_1 < u_2 < u_3 < \cdots < u_{m_i} = t_i,\ i = 1, 2, \cdots, n$. Define

$$\tilde{Q}_{t_i} = \kappa_H^{-1}\lambda_H t_i^{2H-1}\sum_{j=1}^{m_i} u_j^{1/2-H}(u_{m_i}-u_j)^{-1/2-H}X_{u_j}(u_j-u_{j-1}), \qquad (1.20)$$

$i = 1, 2, \cdots, n$.
It is easy to show that $\tilde{Q}_{t_i} \to Q_t$ almost surely as $m_i \to \infty$ for each $i = 1, 2, \cdots, n$.

We use this approximate observation in place of $Q_{t_i},\ i = 1, 2, \cdots, n$ in the calculation of our estimators. Now we define the weighted approximate minimum contrast estimators (AMCEs).

Define a weighted sum of squares

$$G_{n,T} := \left\{\sum_{i=1}^{n} w_{t_i}Q_{t_{i-1}}^2(v_{t_i}-v_{t_{i-1}}) + \sum_{i=2}^{n+1} w_{t_i}Q_{t_{i-1}}^2(v_{t_i}-v_{t_{i-1}})\right\}, \qquad (1.10)$$

where $w_{t_i} \ge 0$ is a weight function.

Denote the discrete increasing functions

$$I_{n,T} := \sum_{i=1}^{n} Q_{t_{i-1}}^2(v_{t_i}-v_{t_{i-1}}), \qquad (1.11)$$

$$J_{n,T} := \sum_{i=2}^{n+1} Q_{t_{i-1}}^2 (v_{t_i} - v_{t_{i-1}}) = \sum_{i=1}^{n} Q_{t_i}^2 (v_{t_i} - v_{t_{i-1}}). \tag{1.12}$$

General weighted AMCE is defined as

$$\tilde{\theta}_{n,T} := -\left\{\frac{2}{T} G_{n,T}\right\}^{-1}. \tag{1.13}$$

With $w_{t_i} = 1$, we obtain the forward AMCE as

$$\tilde{\theta}_{n,T,F} := -\left\{\frac{2}{T} I_{n,T}\right\}^{-1}. \tag{1.14}$$

With $w_{t_i} = 0$, we obtain the backward AMCE as

$$\tilde{\theta}_{n,T,B} := -\left\{\frac{2}{T} J_{n,T}\right\}^{-1}. \tag{1.15}$$

Analogous to the estimators for the discrete AR (1) model, we define the simple symmetric and weighted symmetric estimators (see Fuller 1996).

With $w_{t_i} = 0.5$, the simple symmetric AMCE is defined as

$$\tilde{\theta}_{n,T,z} := -\left\{\frac{1}{T}[I_{n,T} + J_{n,T}]\right\}^{-1}. \tag{1.16}$$

Note that

$$\tilde{\theta}_{n,T,z} = -\left\{\frac{2}{T}\sum_{i=2}^{n} Q_{t_{i-1}}^2 (v_{t_i} - v_{t_{i-1}}) + 0.5(Q_{t_0}^2 + Q_{t_n}^2)(v_{t_i} - v_{t_{i-1}})\right\}^{-1}.$$

With the weight function

$$w_{t_i} = \begin{cases} 0 & : \; i = 1 \\ \frac{i-1}{n} & : \; i = 2, 3, \cdots, n \\ 1 & : \; i = n+1, \end{cases}$$

the weighted symmetric AMCE is defined as

$$\tilde{\theta}_{n,T,w} := -\left\{\frac{2}{T}\sum_{i=2}^{n} Q_{t_{i-1}}^2 (v_{t_i} - v_{t_{i-1}}) + \frac{1}{n}\sum_{i=1}^{n} Q_{t_{i-1}}^2 (v_{t_i} - v_{t_{i-1}})\right\}^{-1}. \tag{1.17}$$

Note that estimator (1.13) is analogous to the trapezoidal rule in numerical analysis. One can instead use the midpoint rule to define another estimator

$$\tilde{\theta}_{n,T,A} := -\left\{\frac{2}{T}\sum_{i=1}^{n}\left(\frac{Q_{t_{i-1}} + Q_{t_i}}{2}\right)^2 (v_{t_i} - v_{t_{i-1}})\right\}^{-1}. \tag{1.18}$$

One can use Simpson's rule to define another estimator where the denominator is a convex combination of the denominators in (1.11) and (1.12)

$$\tilde{\theta}_{n,T,S} := -\left\{\frac{1}{3T}\sum_{i=1}^{n}\left\{Q_{t_{i-1}}^2 + 4\left(\frac{Q_{t_{i-1}}+Q_{t_i}}{2}\right)^2 + Q_{t_i}^2\right\}(v_{t_i} - v_{t_{i-1}})\right\}^{-1}. \tag{1.19}$$

In general, one can generalize Simpson's rule as

$$\tilde{\theta}_{n,T,GS} :=$$

$$-\left\{\frac{2}{T}\sum_{i=1}^{n}\left\{\alpha\frac{Q_{t_{i-1}}^2+Q_{t_i}^2}{2} + (1-\alpha)\left(\frac{Q_{t_{i-1}}+Q_{t_i}}{2}\right)^2\right\}(v_{t_i} - v_{t_{i-1}})\right\}^{-1} \tag{1.20}$$

for any $0 \le \alpha \le 1$.

The case $\alpha = 0$ produces the estimator (1.18). The case $\alpha = 1$ produces the estimator (1.17). The case $\alpha = \frac{1}{3}$ produces the estimator (1.19).

I propose a very general form of the quadrature based estimator as

$$\tilde{\theta}_{n,T,w} := -\left\{\frac{2}{T}\sum_{i=1}^{n}\sum_{j=1}^{m}\left[(1-s_j)Q_{t_{i-1}} + s_jQ_{t_i}\right]^2 p_j(v_{t_i} - v_{t_{i-1}})\right\}^{-1}, \tag{1.21}$$

where $p_j, \quad j \in \{1, 2, \cdots, m\}$, is a probability mass function of a discrete random variable S on $0 \le s_1 < s_2 < \cdots < s_m \le 1$ with $P(S = s_j) := p_j, \ j \in \{1, 2, \cdots, m\}$.

Denote the k-th moment of the random variable S as $\mu_k := \sum_{j=1}^{m} s_j^k p_j, \ k = 1, 2, \cdots$.

If one chooses the probability distribution as uniform distribution for which the moments are a harmonic sequence $(\mu_1, \mu_2, \mu_3, \mu_4, \mu_5, \mu_6, \cdots) = (\frac{1}{2}, \frac{1}{3}, \frac{1}{4}, \frac{1}{5}, \frac{1}{7}, \cdots)$, then there is no change in rate of convergence than second order. If one can construct a probability distribution for which the harmonic sequence is truncated at a point, then there is a rate of convergence improvement at the point of truncation.

Given a positive integer m, construct a probability mass function $p_j, \ \ j \in \{1, 2, \cdots, m\}$, on $0 \le s_1 < s_2 < \cdots < s_m \le 1$ such that

$$\sum_{j=1}^{m} s_j^r p_j = \frac{1}{r+1}, \ r \in \{0, \cdots, m-2\} \tag{1.22}$$

$$\sum_{j=1}^{m} s_j^{N-1} p_j \neq \frac{1}{m}. \tag{1.23}$$

Neither the probabilities p_j nor the atoms, s_j, of the distribution are specified in advance.

This problem is related to the truncated Hausdorff moment problem. I obtain examples of such probability distributions and use them to get higher-order accurate (up to sixth order) AMCEs.

The order of approximation error (rate of convergence) of an estimator is $n^{-\nu}$ where

$$\nu := \inf\left\{k : \mu_k \neq \frac{1}{1+k},\ \mu_j = \frac{1}{1+j}, j = 1, 2, \cdots, k-1\right\}. \tag{1.24}$$

I construct probability distributions satisfying these moment conditions and obtain estimators of the rate of convergence up to order 6.

Probability $p_1 = 1$ at the point $s_1 = 0$ gives the estimator (1.11) for which $\mu_1 = 0$. Note that $\mu_1 \neq \frac{1}{2}$. Thus $\nu = 1$.

Probability $p_1 = 1$ at the point $s_1 = 1$ gives the estimator (1.12) for which $\mu_1 = 1$. Note that $\mu_1 \neq \frac{1}{2}$. Thus $\nu = 1$.

Probabilities $(p_1, p_2) = (\frac{1}{2}, \frac{1}{2})$ at the respective points $(s_1, s_2) = (0, 1)$ produce the estimator $\tilde{\theta}_{n,T,Z}$ for which $(\mu_1, \mu_2) = (\frac{1}{2}, \frac{1}{4})$. Thus $\nu = 2$.

Probability $p_j = 1$ at the point $s_j = \frac{1}{2}$ produces the estimator $\tilde{\theta}_{n,T,A}$ for which $(\mu_1, \mu_2) = (\frac{1}{2}, \frac{1}{2})$. Thus $\nu = 2$.

Probabilities $(p_1, p_2) = (\frac{1}{4}, \frac{3}{4})$ at the respective points $(s_1, s_2) = (0, \frac{2}{3})$ produce the asymmetric estimator

$$\tilde{\theta}_{n,T,3} := -\left\{\tfrac{2}{T}\tfrac{1}{4}\textstyle\sum_{i=1}^{n}\left[(Q_{t_{i-1}})^2 + 3(\tfrac{Q_{t_{i-1}}+2Q_{t_i}}{3})^2\right]\right\}^{-1} \tag{1.25}$$

for which $(\mu_1, \mu_2, \mu_3) = (\frac{1}{2}, \frac{1}{3}, \frac{2}{9})$. Thus $\nu = 3$.

Probabilities $(p_1, p_2) = \frac{3}{4}, \frac{1}{4}$ at the respective points $(s_1, s_2) = (\frac{1}{3}, 1)$ produce asymmetric estimator

$$\tilde{\theta}_{n,T,4} := -\left\{\tfrac{2}{T}\tfrac{1}{4}\textstyle\sum_{i=1}^{n}\left[3(\tfrac{2Q_{t_{i-1}}+Q_{t_i}}{3})^2 + (Q_{t_i})^2\right](v_{t_i} - v_{t_{i-1}})\right\}^{-1} \tag{1.26}$$

for which $(\mu_1, \mu_2, \mu_3) = (\frac{1}{2}, \frac{1}{3}, \frac{10}{36})$. Thus $\nu = 3$.

Probabilities $(p_1, p_2, p_3) = (\frac{1}{6}, \frac{2}{3}, \frac{1}{6})$ at the respective points $(s_1, s_2, s_3) = (0, \frac{1}{2}, 1)$ produce the estimator $\tilde{\theta}_{n,T,5}$ for which $(\mu_1, \mu_2, \mu_3, \mu_4) = (\frac{1}{2}, \frac{1}{3}, \frac{1}{4}, \frac{5}{25})$. Thus $\nu = 4$.

Probabilities $(p_1, p_2, p_3, p_4) = (\frac{1}{8}, \frac{3}{8}, \frac{3}{8}, \frac{1}{8})$ at the respective points $(s_1, s_2, s_3, s_4) = (0, \frac{1}{3}, \frac{2}{3}, 1)$ produce the symmetric estimator

$$\begin{aligned}\tilde{\theta}_{n,T,5} := -\Big\{\tfrac{2}{T}\tfrac{1}{8}\textstyle\sum_{i=1}^{n}\Big[(Q_{t_{i-1}})^2 + 3(\tfrac{2Q_{t_{i-1}}+Q_{t_i}}{3})^2 \\ +3(\tfrac{Q_{t_{i-1}}+2Q_{t_i}}{3})^2 + (Q_{t_i})^2\Big](v_{t_i} - v_{t_{i-1}})\Big\}^{-1}\end{aligned} \tag{1.27}$$

for which $(\mu_1, \mu_2, \mu_3, \mu_4) = (\frac{1}{2}, \frac{1}{3}, \frac{1}{4}, \frac{11}{54})$. Thus $\nu = 4$.

Probabilities $(p_1, p_2, p_3, p_4, p_5) = (\frac{1471}{24192}, \frac{6925}{24192}, \frac{1475}{12096}, \frac{2725}{12096}, \frac{5675}{24192}, \frac{1721}{24192})$ at the respective points $(s_1, s_2, s_3, s_4, s_5) = (0, \frac{1}{5}, \frac{2}{5}, \frac{3}{5}, \frac{4}{5}, 1)$ produce the asymmetric estimator

$$\begin{aligned}\tilde{\theta}_{n,T,7} := -\Big\{\tfrac{2}{T}\tfrac{1}{24192}\textstyle\sum_{i=1}^{n}\Big[1471(Q_{t_{i-1}})^2 + 6925(\tfrac{Q_{t_{i-1}}+Q_{t_i}}{5})^2 \\ +2950(\tfrac{2Q_{t_{i-1}}+2Q_{t_i}}{5})^2 + 5450(\tfrac{3Q_{t_{i-1}}+3Q_{t_i}}{5})^2 \\ +5675(\tfrac{4Q_{t_{i-1}}+4Q_{t_i}}{5})^2 + 1721(Q_{t_i})^2\Big](v_{t_i} - v_{t_{i-1}})\Big\}^{-1}\end{aligned} \tag{1.28}$$

for which $(\mu_1, \mu_2, \mu_3, \mu_4, \mu_5, \mu_6) = (\frac{1}{2}, \frac{1}{3}, \frac{1}{4}, \frac{1}{5}, \frac{841}{5040})$. Thus $\nu = 5$.

Probabilities $(p_1, p_2, p_3, p_4, p_5) = (\frac{7}{90}, \frac{16}{45}, \frac{2}{15}, \frac{16}{45}, \frac{7}{90})$ at the respective points $(s_1, s_2, s_3, s_4, s_5) = (0, \frac{1}{4}, \frac{1}{2}, \frac{3}{4}, 1)$ produce the symmetric estimator $\tilde{\theta}_{n,T,8}$ given by

$$\begin{aligned}\tilde{\theta}_{n,T,8} := -\Big\{\tfrac{2}{T}\tfrac{1}{90}\textstyle\sum_{i=1}^{n}\Big[7(Q_{t_{i-1}})^2 + 32(\tfrac{3Q_{t_{i-1}}+Q_{t_i}}{4})^2 \\ +12(\tfrac{Q_{t_{i-1}}+Q_{t_i}}{2})^2 + 32(\tfrac{Q_{t_{i-1}}+3Q_{t_i}}{4})^2 + 7(t_i, Q_{t_i})^2\Big](v_{t_i} - v_{t_{i-1}})\Big\}^{-1}\end{aligned} \tag{1.29}$$

for which $(\mu_1, \mu_2, \mu_3, \mu_4, \mu_5, \mu_6) = (\frac{1}{2}, \frac{1}{3}, \frac{1}{4}, \frac{1}{5}, \frac{1}{6}, \frac{110}{768})$. Thus $\nu = 6$.

Probabilities $(p_1, p_2, p_3, p_4, p_5) = (\frac{19}{288}, \frac{75}{288}, \frac{50}{288}, \frac{50}{288}, \frac{75}{288}, \frac{19}{288})$ at the respective points
$(s_1, s_2, s_3, s_4, s_5) = (0, \frac{1}{5}, \frac{2}{5}, \frac{3}{5}, \frac{4}{5}, 1)$ produce symmetric estimator

$$\begin{aligned}\tilde{\theta}_{n,T,9} := -\Big\{\tfrac{2}{T}\tfrac{1}{288}\textstyle\sum_{i=1}^{n}\Big[19(Q_{t_{i-1}})^2 + 75(\tfrac{4Q_{t_{i-1}}+Q_{t_i}}{5})^2 \\ +50(\tfrac{3Q_{t_{i-1}}+2Q_{t_i}}{5})^2 + 50(\tfrac{2Q_{t_{i-1}}+3Q_{t_i}}{5})^2 \\ +75(\tfrac{Q_{t_{i-1}}+4Q_{t_i}}{5})^2 + 19(Q_{t_i})^2\Big](v_{t_i} - v_{t_{i-1}})\Big\}^{-1}\end{aligned} \tag{1.30}$$

for which $(\mu_1, \mu_2, \mu_3, \mu_4, \mu_5, \mu_6) = (\frac{1}{2}, \frac{1}{3}, \frac{1}{4}, \frac{1}{5}, \frac{1}{6}, \frac{3219}{22500})$. Thus $\nu = 6$.

The estimator $\tilde{\theta}_{n,T,S}$ is based on the arithmetic mean of $I_{n,T}$ and $J_{n,T}$. One can use geometric mean and harmonic mean instead. The geometric mean based symmetric AMCE (which is based on the ideas of partial autocorrelation) is defined as

$$\tilde{\theta}_{n,T,G} := \frac{-T/2}{\sqrt{I_{n,T}\, J_{n,T}}} \tag{1.31}$$

The harmonic mean based symmetric AMCE is defined as

$$\tilde{\theta}_{n,T,H} := \frac{-T}{\frac{1}{I_{n,T}} + \frac{1}{J_{n,T}}}. \tag{1.32}$$

Note that

$$\tilde{\theta}_{n,T,H} \le \tilde{\theta}_{n,T,G} \le \tilde{\theta}_{n,T,S}. \tag{1.33}$$

It is a desirable property of an estimator to lie inside the parameter space. The estimators $\tilde{\theta}_{n,T,H}$ and $\tilde{\theta}_{n,T,G}$ may lie inside the parameter space even if the estimator $\tilde{\theta}_{n,T,F}$ or $\tilde{\theta}_{n,T,B}$ may not.

Note that the symmetric estimators use both the end points of observations. These estimators are sufficient statistics. So one can take advantage of the Rao–Blackwell theorem. If one excludes the end points of observations (as in the forward and the backward estimators), then the estimators will not be sufficient statistics.

From a time series viewpoint, a discretely observed stationary fractional Ornstein–Uhlenbeck model is a nonlinear (in the parameter) autoregressive model given by

$$Q_{t_i} = \exp(\theta(v_{t_i} - v_{t_{i-1}})) + \epsilon_i, \quad i \geq 1 \tag{1.34}$$

where

$$\epsilon_i := \int_{v_{t_{i-1}}}^{v_{t_i}} \exp\{\theta(v_{t_i} - u)\} dW_u^H \tag{1.36}$$

which are independent $\mathcal{N}\left(0, \frac{e^{2\theta(v_{t_i} - v_{t_{i-1}})} - 1}{2\theta}\right)$ random variables. The errors are heteroscedastic if the data points are unequally spaced, but as we assume equally spaced data points, the errors are homoscedastic and i.i.d. $\mathcal{N}(0, V_{n,T}(\theta))$ where

$$V_{n,T}(\theta) := \frac{e^{2\theta \frac{T}{n}} - 1}{2\theta}. \tag{1.37}$$

The log-likelihood is

$$l_{n,T}(\theta) := -\frac{n}{2} \log 2\pi - \frac{n}{2} V_{n,T}(\theta) - \frac{1}{2V_{n,T}(\theta)} \sum_{i=1}^{n} \left(Q_{t_i} - e^{\frac{T}{n}\theta} Q_{t_{i-1}}\right)^2. \tag{1.38}$$

The exact maximum likelihood estimator based on equally spaced discrete observations is explicit and is given by

$$\hat{\theta}_{n,T} := \frac{n}{T} \log \frac{\sum_{i=1}^{n} Q_{t_{i-1}}^2}{\sum_{i=1}^{n} Q_{t_i} Q_{t_{i-1}}}. \tag{1.39}$$

Note that

$$\breve{\theta}_{n,T} = 1 - \frac{n}{T} \exp\{\frac{T}{n} \hat{\theta}_{n,T}\}.$$

As far as the rate of convergence of this estimator to the continuous conditional MLE is concerned, it is of first order.

The Euler estimator that is the conditional least squares estimator is given by

$$\breve{\theta}_{n,T} := \frac{n}{T} \frac{\sum_{i=1}^{n} Q_{t_{i-1}} (Q_{t_i} - Q_{t_{i-1}})}{\sum_{i=1} Q_{t_{i-1}}^2}. \tag{1.40}$$

Another contrast function, which is the least squares quadratic, is

$$\kappa_{n,T}(\theta) := \frac{2\theta}{e^{2\theta\frac{T}{n}} - 1} \sum_{i=1}^{n} \left(Q_{t_i} - e^{\frac{T}{n}\theta} Q_{t_{i-1}} \right)^2 . \tag{1.41}$$

The solution of $\kappa'_{n,T}(\theta) = 0$ provides an AMCE but does not have an explicit form.

The simple symmetric approximate MLE was introduced and studied in Mishra and Bishwal (1995) for a more general diffusion process. Stochastic bound between the difference the AMLE and the continuous MLE was obtained.

In this chapter, we obtain the rate of weak convergence to normal distribution of some AMCEs using different normings. We also obtain stochastic bound on the difference of the AMCEs and their continuous counterpart MCE when T is fixed.

Let $\Phi(\cdot)$ denote the standard normal distribution function. Throughout the chapter, C denotes a generic constant (perhaps depending on θ, but not on anything else).

We need the following lemmas in the sequel.

Lemma 2.1 (Bishwal 2000a) For every $\delta > 0$,

$$P\left\{ \left| \frac{2\theta}{T} I_T - 1 \right| \geq \delta \right\} \leq C T^{-1} \delta^{-2}.$$

Lemma 2.2 (Bishwal 2000a)

$$\sup_{x \in \mathbb{R}} \left| P\left\{ \left(-\frac{2\theta}{T} \right)^{1/2} \left(\theta I_T - \frac{T}{2} \right) \leq x \right\} - \Phi(x) \right| \leq C T^{-1/2}.$$

Lemma 2.3 (Wick's Lemma) Let $(\xi_1, \xi_2, \xi_3, \xi_4)$ be a Gaussian random vector with zero mean. Then

$$E(\xi_1\xi_2\xi_3\xi_4) = E(\xi_1\xi_2)E(\xi_3\xi_4) + E(\xi_1\xi_3)E(\xi_2\xi_4) + E(\xi_1\xi_4)E(\xi_2\xi_3).$$

Lemma 2.4

$$(a) \quad E|I_{n,T} - I_T|^2 = O\left(\frac{T^4}{n^2}\right)^{2-2H}.$$

$$(b) \quad E\left|\frac{I_{n,T} + J_{n,T}}{2} - I_T\right|^2 = O\left(\frac{T^4}{n^2}\right)^{2-2H}.$$

Proof Let $g(t) := Q_{t_{i-1}} - Q_t$ for $t_{i-1} \leq t < t_i, \quad i = 1, 2, \ldots, n$. Since $E|Q_{t_{i-1}} - Q_t|^{2k} \leq C(t_{i-1} - t)^k, k = 1, 2, \ldots$ (3.1)
(by Gikhman and Skorohod (1972, p. 48)), hence

$$
\begin{aligned}
& E|Y_{n,T} - Y_T|^2 \\
= \;& E|\sum_{i=1}^{n} X_{t_{i-1}}[W_{t_i} - W_{t_{i-1}}] - \int_0^T X_t dW_t|^2 \\
= \;& E|\int_0^T g(t) dW_t|^2 \\
= \;& \int_0^T E(g^2(t)) dt \\
\leq \;& C \sum_{i=1}^{n} \int_{t_{i-1}}^{t_i} |t_{i-1} - t| dt \\
= \;& Cn \frac{(t_i - t_{i-1})^2}{2} = C\frac{T^2}{n}.
\end{aligned}
$$

$$
X_{t_i} - X_{t_{i-1}} = \int_{t_{i-1}}^{t_i} \theta X_t dt + [W_{t_i} - W_{t_{i-1}}].
$$

We obtain

$$
\begin{aligned}
& E|Z_{n,T} - Z_T|^2 \\
= \;& E|\sum_{i=1}^{n} X_{t_{i-1}}[X_{t_i} - X_{t_{i-1}}] - \int_0^T X_t dW_t|^2 \\
= \;& E|\sum_{i=1}^{n} \int_{t_{i-1}}^{t_i} \theta X_t X_{t_{i-1}} dt + \sum_{i=1}^{n} X_{t_{i-1}}[W_{t_i} - W_{t_{i-1}}] \\
& - \int_0^T \theta X_t^2 dt - \int_0^T X_t dW_t|^2 \\
\leq \;& 2E|\sum_{i=1}^{n} X_{t_{i-1}}[W_{t_i} - W_{t_{i-1}}] - \int_0^T X_t dW_t|^2 \\
& + 2\theta^2 E|\sum_{i=1}^{n} \int_{t_{i-1}}^{t_i} X_t[X_{t_{i-1}} - X_t] dt|^2. \\
=: \;& N_1 + N_2.
\end{aligned}
$$

N_1 is $O(\frac{T^2}{n})$ by Lemma 5.3.2(a). To estimate N_2, let $f_i(t) = X_t[X_{t_{i-1}} - X_t]$ for $t_{i-1} \leq t < t_i, \ i = 1, 2, \ldots, n$. Then

$$
\begin{aligned}
& E|\sum_{i=1}^{n} \int_{t_{i-1}}^{t_i} f(t) dt|^2 \\
= \;& \sum_{i=1}^{n} E|\int_{t_{i-1}}^{t_i} f_i(t) dt|^2 + 2 \sum_{i,j=1, i<j}^{n} E\left[\int_{t_{i-1}}^{t_i} f_i(t) dt \int_{t_{i-1}}^{t_i} f_j(s) ds\right] \\
=: \;& M_1 + M_2.
\end{aligned}
$$

By the boundedness of $E(X_t^4)$ and (2.1), we have

$$\begin{aligned}
& E(f^2(t)) \\
= \;& E\{X_t^2[X_{t_{i-1}} - X_t]^2\} \\
\leq \;& \{E(X_t^4)\}^{1/2}\{E[X_{t_{i-1}} - X_t]^4\}^{1/2} \\
\leq \;& C(t_{i-1} - t)
\end{aligned} \tag{3.2}$$

$$\begin{aligned}
M_1 &= \sum_{i=1}^n E|\int_{t_{i-1}}^{t_i} f(t)dt|^2 \\
&\leq \sum_{i=1}^n (t_i - t_{i-1}) \int_{t_{i-1}}^{t_i} E(f^2(t))dt \\
&\leq C\frac{T}{n}\sum_{i=1}^n \int_{t_{i-1}}^{t_i} (t - t_{i-1})dt \\
&\leq C\frac{T}{n}\sum_{i=1}^n (t_i - t_{i-1})^2 = C\frac{T^3}{n^2}.
\end{aligned}$$

$$\begin{aligned}
M_2 &= 2\sum_{i,j=1,i<j}^n E\int_{t_{i-1}}^{t_i}\int_{t_{j-1}}^{t_i} [f_i(t)f_j(s)]dtds \\
&= 2\sum_{i,j=1,i<j}^n \int_{t_{i-1}}^{t_i}\int_{t_{j-1}}^{t_j} E[f_i(t)f_j(s)]dtds.
\end{aligned}$$

Now by Lemma 2.3,

$$\begin{aligned}
& E[f_i(t)f_j(s)] \\
= \;& E[X_t(X_{t_{i-1}} - X_t)X_s(X(t_{j-1}) - X_s)] \\
= \;& E[X_t(X_{t_{i-1}} - X_t)]E[X_s(X(t_{j-1}) - X_s)] \\
+ \;& E[X_tX_s]E[(X_{t_{i-1}} - X_t)(X(t_{j-1}) - X_s)] \\
+ \;& E[X_t(X(t_{j-1}) - X_s)]E[X_s(X_{t_{i-1}} - X_t)] \\
=: \;& A_1 + A_2 + A_3.
\end{aligned}$$

Note that

$$X_t = \int_0^t e^{\theta(t-u)}dW_u, t \geq 0.$$

Let $a = e^\theta$. For $s \geq t$, we have

$$\begin{aligned}
& E(X_tX_s) \\
= \;& E(\int_0^t e^{\theta(t-u)}dW_u)(\int_0^s e^{\theta(s-u)}dW_u) \\
= \;& \int_0^t e^{\theta(t+s-2u)}du \\
= \;& \frac{1}{2\theta}[a^{s+t} - a^{s-t}].
\end{aligned}$$

Now

$$\begin{aligned}
& E(X_t - X_{t_{i-1}})(X_s - X(t_{j-1})) \\
= \;& E(X_tX_s) - E(X_tX(t_{j-1})) - E(X_{t_{i-1}}X_s) + E(X(t_{i-1}X(t_{j-1})) \\
= \;& \frac{1}{2\theta}(a^s - a^{t_{j-1}})[(a^t - a^{t_{i-1}}) + (a^{-t_{i-1}} - a^{-t})] \\
= \;& \frac{1}{2\theta}(s - t_{j-1})a^{t^*}[(t - t_{i-1}a^{t^{**}} + (t - t_{i-1})a^{-t^{***}}] \\
& (\text{where} t_{j-1} < t^* < s, t_{i-1} < t^{**}, t^{***} < t) \\
\leq \;& \frac{1}{2\theta}(s - t_{j-1})a^t(t - t_{i-1})a^{t_{i-1}} + (s - t_{j-1})a^t(t - t_{i-1})a^{-t}] \\
\leq \;& C(s - t_{j-1})(t - t_{i-1}).
\end{aligned}$$

Thus $A_2 \leq C(s - t_{j-1})(t - t_{i-1})$ since $|E(X_tX_s)|$ is bounded.

Next

$$\begin{aligned}
& |E[X_t(X(t_{i-1} - X_t)]| \\
= \;& \frac{1}{2|\theta|}[a^{t+t_{i-1}} - a^{t-t_{i-1}} - a^{2t} + 1] \\
= \;& \frac{1}{2|\theta|}a^t[a^{t_{i-1}} - a^{-t_{i-1}} - a^t + a^{-t}] \\
\leq \;& \frac{1}{2|\theta|}a^t(t - t_{i-1})[a^{t_{i-1}} + a^{-t}] \\
\leq \;& C(t - t_{i-1})
\end{aligned}$$

and

$$\begin{aligned}
& |E[X_s(X_s - X(t_{j-1})]| \\
= \;& \frac{1}{2|\theta|}[a^{2s} - 1 - a^{s+t_{j-1}} + a^{s+t_{j-1}}] \\
= \;& \frac{1}{2|\theta|}a^s[a^s - a^{-s} - a^{t_{j-1}} + a^{-t_{j-1}}] \\
\leq \;& \frac{1}{2|\theta|}a^s(s - t_{j-1})[a^{t_{j-1}} + a^{-s}] \\
\leq \;& C(s - t_{j-1}).
\end{aligned}$$

Thus $A_1 \leq C(s - t_{j-1})(t - t_{i-1})$.

Next

$$\begin{aligned}
& |E[X_t(X_s - X(t_{j-1})]| \\
= \;& \frac{1}{2|\theta|}[a^{s+t} - a^{s-t} - a^{t+t_{j-1}} + a^{t_{j-1}-t}] \\
= \;& \frac{1}{2|\theta|}a^t(a^s - a^{t_{j-1}})
\end{aligned}$$

$$
\begin{aligned}
&\leq \frac{1}{2|\theta|}a^t(1-a^{-2t})(s-t_{j-1})a^t \\
&\leq (a^{2t}-1)(s-t_{j-1}) \\
&\leq C(s-t_{j-1})
\end{aligned}
$$

and

$$
\begin{aligned}
& |E[X_s(X_t - X_{t_{i-1}}]| \\
= \; & \frac{1}{2|\theta|}[a^{t+s} - a^{s-t} - a^{s+t_{i-1}} + a^{s-t_{i-1}}] \\
= \; & \frac{1}{2|\theta|}a^s[a^t - a^{-t} - a^{t_{i-1}} + a^{-t_{i-1}}] \\
\leq \; & \frac{1}{2|\theta|}a^s(t-t_{i-1})[a^{t_{i-1}} + a^{-t}] \\
\leq \; & C(t-t_{i-1}).
\end{aligned}
$$

Thus $A_3 \leq C(s-t_{j-1})(t-t_{i-1})$.
Hence $E[f_i(t)f_j(s)] \leq C(s-t_{j-1})(t-t_{i-1})$.

Thus

$$
\begin{aligned}
M_2 &= 2\sum_{i,j=1,i<j}^{n}\int_{t_{i-1}}^{t_i}\int_{t_{j-1}}^{t_j} E[f_i(t)f_j(s)]dtds \\
&\leq C\sum_{i,j=1,i<j}^{n}\int_{t_{i-1}}^{t_i}\int_{t_{j-1}}^{t_j}(t-t_{i-1})(s-t_{j-1})dtds \\
&= C\sum_{i,j=1,i<j}^{n}(t_{i-1}-t_i)^2(t_{j-1}-t_i)^2 \\
&= Cn^2(\frac{T}{n})^4 = C\frac{T^4}{n^2}.
\end{aligned}
$$

Hence, N_2 is $O(\frac{T^3}{n^2})$.
Let $h_i(t) := X_{t_{i-1}}^2 - X_t^2$.

$$
\begin{aligned}
& E|I_{n,T} - I_t|^2 \\
= \; & E|\sum_{i=1}^{n} X_{t_{i-1}}^2(t_i - t_{i-1}) - \int_0^T X_t^2 dt|^2 \\
= \; & E|\sum_{i=1}^{n}\int_{t_{i-1}}^{t_i}[X_{t_{i-1}}^2 - X_t^2]dt|^2 \\
= \; & E|\sum_{i=1}^{n}\int_{t_{i-1}}^{t_i} h_i(t)dt|^2
\end{aligned}
$$

$$
\begin{aligned}
&= \sum_{i=1}^{n} E|\int_{t_{i-1}}^{t_i} h_i(t)dt|^2 + 2 \sum_{i,j=1,i<j}^{n} E \int_{t_{i-1}}^{t_i} \int_{t_{j-1}}^{t_j} h_i(t)h_j(s)dtds \\
&=: B_1 + B_2.
\end{aligned}
$$

$$
\begin{aligned}
Eh^2(t) &= E[X_{t_{i-1}}^2 - X_t^2]^2 \\
&= E[X_{t_{i-1}} - X_t]^2[X_{t_{i-1}} + X_t]^2 \\
&\leq \{E[X_{t_{i-1}} - X_t]^4\}^{1/2}\{\{E[X_{t_{i-1}} + X_t]^4\}^{1/2} \\
&\leq C(t - t_{i-1})
\end{aligned}
$$

(by (2.1) and the boundedness of the second term)

$$
\begin{aligned}
B_1 &= \sum_{i=1}^{n} E|\int_{t_{i-1}}^{t_i} h_i(t)dt|^2 \\
&\leq \sum_{i=1}^{n}(t_i - t_{i-1}) \int_{t_{i-1}}^{t_i} E(h_i^2(t))dt \\
&\leq C\frac{T}{n}\sum_{i=1}^{n} \int_{t_{i-1}}^{t_i} (t - t_{i-1})dt \\
&= C\frac{T^3}{n^2}.
\end{aligned}
$$

Note that

$$
\begin{aligned}
&E[h_i(t)h_j(s)] \\
&= E(X_{t_{i-1}}^2 - X_t^2)(X_{t_{j-1}}^2 - X_s^2) \\
&= E(X_{t_{i-1}} - X_t)(X_{t_{i-1}} + X_t)(X_{t_{j-1}} - X_s)(X_{t_{j-1}} + X_s).
\end{aligned}
$$

Now using Lemma 5.3.1 and proceeding similar to the estimation of M_2, it is easy to see that

$$
B_2 \leq C\frac{T^4}{n^2}.
$$

Combining B_1 and B_2, (c) follows. □

The following theorem gives the bound on the error of normal approximation of the AMCEs. Note that part (a) uses parameter dependent nonrandom norming. While this is useful for testing hypotheses about θ, it may not necessarily give a confidence interval. The normings in parts (b) and (c) are sample dependent which can be used for obtaining a confidence interval. The following theorem shows that asymptotic normality of the AMCEs needs $T \to \infty$ and $\frac{T}{\sqrt{n}} \to 0$.

Theorem 2.1 Denote $b_{n,T} := O(\max(T^{-1/2}(\log T)^{1/2}, (\frac{T^4}{n^2})(\log T)^{-1}))$.

$$
\text{(a)} \quad \sup_{x\in\mathbb{R}} \left| P\left\{ \left(-\frac{T}{2\theta}\right)^{1/2} (\tilde{\theta}_{n,T,F} - \theta) \leq x \right\} - \Phi(x) \right| = O(b_{n,T}).
$$

$$\text{(b)}\quad \sup_{x\in\mathbb{R}}\left|P\left\{I_{n,T}^{1/2}(\tilde{\theta}_{n,T,F}-\theta)\leq x\right\}-\Phi(x)\right|=O(b_{n,T}).$$

$$\text{(c)}\quad \sup_{x\in\mathbb{R}}\left|P\left\{\left(\frac{T}{2|\tilde{\theta}_{n,T,F}|}\right)^{1/2}(\tilde{\theta}_{n,T,F}-\theta)\leq x\right\}-\Phi(x)\right|=O(b_{n,T}).$$

Proof

(a) Observe that

$$\left(-\frac{T}{2\theta}\right)^{1/2}(\tilde{\theta}_T-\theta)=\frac{\left(-\frac{2\theta}{T}\right)^{1/2}Y_T}{\left(-\frac{2\theta}{T}\right)I_T}, \tag{2.1}$$

where

$$Y_T:=-\theta I_T-\frac{T}{2}\quad\text{and}\quad I_T:=\int_0^T Q_t^2dt.$$

Thus, we have

$$I_{n,T}\tilde{\theta}_{n,T,F}=-\frac{T}{2}=Y_T+\theta I_T.$$

Hence,

$$\begin{aligned}
&\left(-\frac{T}{2\theta}\right)^{1/2}(\tilde{\theta}_{n,T,F}-\theta)\\
=\;&\frac{\left(-\frac{T}{2\theta}\right)^{1/2}Y_T+\theta\left(-\frac{T}{2\theta}\right)^{1/2}(I_T-I_{n,T})}{I_{n,T}}\\
=\;&\frac{\left(-\frac{2\theta}{T}\right)^{1/2}Y_T+\left(-\frac{2\theta}{T}\right)^{1/2}(I_T-I_{n,T})}{\left(-\frac{2\theta}{T}\right)I_{n,T}}.
\end{aligned}\tag{2.2}$$

Further,

$$\begin{aligned}
&P\left\{\left|\left(\frac{-2\theta}{T}\right)(I_{n,T}-1)\right|>\epsilon\right\}\\
=\;&\left\{\left|\left(\frac{-2\theta}{T}\right)(I_{n,T}-I_T+I_T)-1\right|>\epsilon\right\}\\
\leq\;&P\left\{\left|\left(\frac{-2\theta}{T}\right)I_T-1\right|>\frac{\epsilon}{2}\right\}+P\left\{\left(-\frac{2\theta}{T}\right)|I_{n,T}-I_T|>\frac{\epsilon}{2}\right\}\\
\leq\;&C\exp\left(\frac{T\theta}{16}\epsilon^2\right)+\frac{16\theta^2}{T^2}\frac{E|I_{n,T}-I_T|^2}{\epsilon^2}\\
\leq\;&C\exp\left(\frac{T\theta}{16}\epsilon^2\right)+C\frac{T^2/n^2}{\epsilon^2}.
\end{aligned}\tag{2.3}$$

Next, observe that

$$
\begin{aligned}
&\sup_{x\in\mathbb{R}}\left|P\left\{\left(-\frac{T}{2\theta}\right)^{1/2}(\tilde{\theta}_{n,T,F}-\theta)\le x\right\}-\Phi(x)\right| \\
= &\sup_{x\in\mathbb{R}}\left|P\left\{\frac{\left(-\frac{2\theta}{T}\right)^{1/2}Y_T+\left(-\frac{2\theta}{T}\right)^{1/2}(I_T-I_{n,T})}{\left(-\frac{2\theta}{T}\right)I_{n,T}}\le x\right\}-\Phi(x)\right| \\
\le &\sup_{x\in\mathbb{R}}\left|P\left\{\left(-\frac{2\theta}{T}\right)^{1/2}Y_T\le x\right\}-\Phi(x)\right| \\
&+P\left\{\left|\theta\left(-\frac{2\theta}{T}\right)^{1/2}(I_{n,T}-I_T)\right|>\epsilon\right\} \\
&+P\left\{\left|\left(-\frac{2\theta}{T}\right)I_{n,T}-1\right|>\epsilon\right\}+2\epsilon \\
\le &\; CT^{-1/2}+\theta^2\frac{\left(-\frac{2\theta}{T}\right)E|I_{n,T}-I_T|^2}{\epsilon^2} \\
&+C\exp\left(\frac{T\theta}{4}\epsilon^2\right)+C\frac{T^2}{n^2\epsilon^2}+2\epsilon \qquad (2.4)
\end{aligned}
$$

(the bound for the third term in the right hand side of (2.4) is obtained from (2.3))

$$
\le CT^{-1/2}+C\frac{T^2}{n^2\epsilon^2}+C\exp\left(\frac{T\theta}{4}\epsilon^2\right)+C\frac{T}{n^2\epsilon^2}+\epsilon \qquad (2.5)
$$

(by Lemma 2.3(a)).

Choosing $\epsilon = CT^{-1/2}(\log T)^{1/2}$, the terms in the right hand side of (2.5) are of the order
$O(\max(T^{-1/2}(\log T)^{1/2}, (\frac{T^4}{n^2})(\log T)^{-1}))$. □

(b) From (2.1), we have

$$
I_{n,T}^{1/2}(\tilde{\theta}_{n,T,F}-\theta)=\frac{Y_T+\theta(I_T-I_{n,T})}{I_{n,T}^{1/2}}.
$$

Then,

$$
\begin{aligned}
&\sup_{x\in\mathbb{R}}\left|P\left\{I_{n,T}^{1/2}(\tilde{\theta}_{n,T,F}-\theta)\le x\right\}-\Phi(x)\right| \\
= &\sup_{x\in\mathbb{R}}\left|P\left\{\frac{Y_T}{I_{n,T}^{1/2}}+\theta\frac{I_T-I_{n,T}}{I_{n,T}^{1/2}}\le x\right\}-\Phi(x)\right| \\
\le &\sup_{x\in\mathbb{R}}\left|P\left\{\frac{Y_T}{I_{n,T}^{1/2}}\le x\right\}-\Phi(x)\right|+P\left\{\left|\frac{\theta(I_T-I_{n,T})}{I_{n,T}^{1/2}}\right|>\epsilon\right\}+\epsilon
\end{aligned}
$$

$$=: \quad U_1 + U_2 + \epsilon. \tag{2.6}$$

We have from (2.3),

$$U_1 \leq CT^{-1/2} + C\exp\left(\frac{T\theta}{16}\epsilon^2\right) + C\frac{T^2}{n^2\epsilon^2} + \epsilon. \tag{2.7}$$

Now,

$$\begin{aligned} U_2 &= P\left\{|\theta|\left|\frac{I_{n,T} - I_T}{I_{n,T}^{1/2}}\right| > \epsilon\right\} = P\left\{|\theta|\frac{\left|\left(-\frac{2\theta}{T}\right)^{1/2}(I_{n,T} - I_T)\right|}{\left|\left(-\frac{2\theta}{T}\right)^{1/2} I_{n,T}^{1/2}\right|} > \epsilon\right\} \\ &\leq P\left\{\left|\left(-\frac{2\theta}{T}\right)^{1/2}\right| |I_{n,T} - I_T| > \delta\right\} + P\left\{\left|\left(-\frac{2\theta}{T}\right)^{1/2} I_{n,T}^{1/2} - 1\right| > \delta_1\right\} \end{aligned} \tag{2.8}$$

(where $\delta = \epsilon - C\epsilon^2$ and $\delta_1 = (\epsilon - \delta)/\epsilon > 0$)

$$\begin{aligned} &\leq \left(-\frac{2\theta}{T}\right)\frac{E|I_{n,T} - I_T|^2}{\delta^2} + P\left\{\left|\left(-\frac{2\theta}{T}\right) I_{n,T} - 1\right| > \delta_1\right\} \\ &\leq C\frac{T^3}{n^2\delta^2} + C\exp\left(\frac{T\theta}{16}\delta_1^2\right) + C\frac{T^2}{n^2\delta_1^2}. \end{aligned} \tag{2.9}$$

Here, the bound for the first term in the right hand side of (2.7) comes from Lemma 2.2(c) and that for the second term is obtained from (2.3).

Now, using the bounds (2.7) and (2.9) in (2.6) with $\epsilon = CT^{-1/2}(\log T)^{1/2}$, we obtain that the terms in (2.6) are of the order $O(\max(T^{-1/2}(\log T)^{1/2}, (\frac{T^4}{n^2})(\log T)^{-1}))$. □

(c) Let

$$G_T := \left\{|\tilde{\theta}_{n,T,F} - \theta| \leq d_T\right\}, \qquad \text{and} \qquad d_T := CT^{-1/2}(\log T)^{1/2}.$$

On the set G_T, expanding $(2|\tilde{\theta}_{n,T}|)^{1/2}$, we obtain

$$\begin{aligned} (-2\tilde{\theta}_{n,T})^{-1/2} &= (-2\theta)^{1/2}\left[1 - \frac{\theta - \tilde{\theta}_{n,T,F}}{\theta}\right]^{-1/2} \\ &= (-2\theta)^{1/2}\left[1 + \frac{1}{2}\left(\frac{\theta - \tilde{\theta}_{n,T,F}}{\theta}\right)\right] + O(d_T^2). \end{aligned}$$

Then,

$$\sup_{x\in\mathbb{R}}\left|P\left\{\left(\frac{T}{2|\tilde{\theta}_{n,T,F}|}\right)^{1/2}(\tilde{\theta}_{n,T,F} - \theta) \leq x\right\} - \Phi(x)\right|$$

$$\leq \sup_{x\in\mathbb{R}}\left\{P\left(\frac{T}{2|\tilde{\theta}_{n,T}|}\right)^{1/2}(\tilde{\theta}_{n,T,F}-\theta)\leq x, G_T\right\}+P(G_T^c).$$

Now,

$$\begin{aligned}
P(G_T^c) &= P\left\{|\tilde{\theta}_{n,T,F}-\theta|>CT^{-1/2}(\log T)^{1/2}\right\}\\
&= P\left\{\left(-\frac{T}{2\theta}\right)^{1/2}|\tilde{\theta}_{n,T,F}-\theta|>C(\log T)^{1/2}(-2\theta)^{-1/2}\right\}\\
&\leq C\max\left(T^{-1/2}(\log T)^{1/2},\frac{T^3}{n^2}(\log T)^{-1}\right)\\
&\quad +2(1-\Phi\log T^{1/2}(-2\theta)^{-1/2})\\
&\quad \text{(by Theorem 2.1(a))}\\
&\leq C\max\left(T^{-1/2}(\log T)^{1/2},\frac{T^3}{n^2}(\log T)^{-1}\right).
\end{aligned}$$

On the set G_T,

$$\left|\left(\frac{\tilde{\theta}_{n,T,F}}{\theta}\right)^{1/2}-1\right|\leq CT^{-1/2}(\log T)^{1/2}.$$

Hence, upon choosing $\epsilon=CT^{-1/2}(\log T)^{1/2}, C$ large, we obtain

$$\begin{aligned}
&\left|P\left\{\left(\frac{T}{-2\tilde{\theta}_{n,T,F}}\right)^{1/2}(\tilde{\theta}_{n,T,F}-\theta)\leq x, G_T\right\}-\Phi(x)\right|\\
\leq\ &\left|P\left\{\left(\frac{T}{-2\theta}\right)^{1/2}(\tilde{\theta}_{n,T,F}-\theta)\leq x, G_T\right\}\right|\\
&+P\left\{\left|\left(\frac{\tilde{\theta}_{n,T,F}}{\theta}\right)^{1/2}-1\right|>\epsilon, G_T\right\}+\epsilon\\
&\text{(by Lemma 3.3.1(b))}\\
\leq\ &C\max\left(T^{-1/2}(\log T)^{1/2},\frac{T^4}{n^2}(\log T)^{-1}\right)\\
&\text{(by Theorem 2.1(a)).}
\end{aligned}$$

□

In the following theorem, we improve the bound on the error of normal approximation using a mixture of random and nonrandom normings. Thus asymptotic normality of the AMCEs needs $T\to\infty$ and $\frac{T}{n^{2/3}}\to 0$ which are sharper than the bound in Theorem 2.1.

Theorem 2.2

$$\sup_{x\in\mathbb{R}}\left|P\left\{I_{n,T}\left(-\frac{2\theta}{T}\right)^{1/2}(\tilde{\theta}_{n,T,F}-\theta)\le x\right\}-\Phi(x)\right|$$
$$=O\left(\max\left(T^{-1/2},\left(\frac{T^3}{n^2}\right)^{1/3}\right)\right).$$

Proof From (2.2), we have

$$I_{n,T}\left(-\frac{2\theta}{T}\right)^{1/2}(\tilde{\theta}_{n,T,F}-\theta)=\left(-\frac{2\theta}{T}\right)^{1/2}Y_T+\theta\left(-\frac{2\theta}{T}\right)^{1/2}(I_T-I_{n,T}).$$

Hence, by Lemmas 2.1–2.3,

$$\begin{aligned}
&\sup_{x\in\mathbb{R}}\left|P\left\{I_{n,T}\left(-\frac{2\theta}{T}\right)^{1/2}(\tilde{\theta}_{n,T,F}-\theta)\le x\right\}-\Phi(x)\right|\\
=\;&\sup_{x\in\mathbb{R}}\left|P\left\{\left(-\frac{2\theta}{T}\right)^{1/2}Y_T+\theta\left(-\frac{2\theta}{T}\right)^{1/2}(I_T-I_{n,T})\le x\right\}-\Phi(x)\right|\\
\le\;&\sup_{x\in R}\left|P\left\{\left(-\frac{2\theta}{T}\right)^{1/2}Y_T\le x\right\}-\Phi(x)\right|\\
&+P\left\{\left|\theta\left(-\frac{2\theta}{T}\right)^{1/2}(I_T-I_{n,T})\right|>\epsilon\right\}+\epsilon\\
\le\;&CT^{-1/2}+C\frac{E|I_T-I_{n,T}|^2}{T\epsilon^2}+\epsilon\le CT^{-1/2}+C\frac{T^3}{n^2\epsilon^2}+\epsilon.
\end{aligned}$$

Choosing $\epsilon=(\frac{T^3}{n^2})^{1/3}$, the theorem follows. □

The following theorem gives stochastic bound on the error of approximation of the continuous MCE by AMCEs.

Theorem 2.3

$$(a)\quad |\tilde{\theta}_{n,T}-\tilde{\theta}_T|=O_P\left(\frac{T}{n}\right)^{1/2}.$$

$$(b)\quad |\tilde{\theta}_{n,T,z}-\tilde{\theta}_T|=O_P\left(\frac{T^2}{n^2}\right)^{1/2}.$$

Proof From (1.9) and (1.14), we have

$$\tilde{\theta}_T=-\frac{T}{2I_T},\quad \tilde{\theta}_{n,T}=-\frac{T}{2I_{n,T}}.$$

Hence, applying Lemma 3.3.1 with the aid of Lemma 2.3(a) and noting that $|\frac{I_{n,T}}{T}| = O_P(1)$ and $|\frac{I_T}{T}| = O_P(1)$, the part (a) of theorem follows.

From (1.9) and (1.16), we have

$$\tilde{\theta}_T = -\frac{T}{2I_T}, \quad \tilde{\theta}_{n,T,z} = -\frac{T}{I_{n,t} + J_{n,T}}.$$

Applying Lemma 3.3.1 with the aid of Lemma 2.3(b) and noting that $|\frac{J_{n,T}}{T}| = O_P(1)$ and $|\frac{I_T}{T}| = O_P(1)$, the part (b) of theorem follows. □

Remarks

(1) The bounds in Theorems 2.1 and 2.2 are uniform over compact subsets of the parameter space.

(2) It remains to investigate the nonuniform rates of convergence to normality which are more useful.

(3) Weak convergence rate for AMCE remains open when $T \to \infty$ and $\frac{T}{n} \to 0$.

(4) Weighted estimators converge faster than the unweighted estimators if the weights are chosen according to (1.22) and (1.23).

4.4 Ornstein–Uhlenbeck–Gamma Process

We study rates of convergence of two approximate maximum likelihood estimators (AMLEs) to the continuous MLE of the parameter appearing linearly in the drift coefficient of Itô SDE when T remains fixed, T being the length of the observation time interval $[0, T]$. As far as we know no results are known about the rate of convergence of AMLEs to the true value of the parameter as $T \to \infty$. For the Ornstein–Uhlenbeck–Gamma process, we obtain Berry–Esseen type bounds using both random and nonrandom normings of the two AMLEs $\theta_{n,T}$ and $\theta_{n,T,1}$. First we identify one AMLE $\theta_{n,T}$ as the conditional least squares estimator (CLSE). We show that the other AMLE $\theta_{n,T,1}$ has sharper Berry–Esseen bound than that for $\theta_{n,T}$. The Berry–Esseen bounds of the AMLEs $\theta_{n,T}$ and $\theta_{n,T,1}$ show that the *rapidly increasing experimental design* (RIED) condition, i.e., $T \to \infty$ and $\frac{T}{n^{1/2}} \to 0$, is not essential for its asymptotic normality of AMLEs.

The section is organized as follows: in Sect. 4.4.1, we prepare notations and preliminaries. Section 4.4.1 contains the Berry–Esseen type bounds for one AMLE. Section 4.4.2 contains the Berry–Esseen type bounds for another AMLE. Section 4.4.3 introduces some approximate minimum contrast estimators and Sect. 4.4.4 studies the Berry–Esseen type bounds for those estimators.

Let $\{X_t, t \geq 0\}$ be the Ornstein–Uhlenbeck–Gamma process satisfying the stochastic differential equation

$$dX_t = \theta X_t dt + dW_t + dZ_t,\ t \geq 0, X_0 = 0, \tag{2.1}$$

where $\{W_t,\ t \geq 0\}$ is a standard Wiener process, $\{Z_t,\ t \geq 0\}$ is a two sided gamma process with known parameters (α, β), α is the shape parameter, β is the scale parameter, that is, the Levy measure is of the form

$$f(z) = \frac{1}{2}\alpha e^{-\beta|z|}|z|^{-1}.$$

Thus the Levy process here has infinitely many jumps in any finite time interval since $\int f(z)dz = \infty$.

Let $\theta \in (-\infty, 0)$ be the unknown parameter to be estimated from the observations of the process $\{X_t, t \geq 0\}$. Levy driven O-U processes are ergodic. Recall that based on complete observation of $\{X_t\}$ on $[0, T]$, the log-likelihood function and the maximum likelihood estimator (MLE) are given, respectively, by

$$L_T = \theta \int_0^T X_t dX_t - \frac{\theta^2}{2} \int_0^T X_t^2 dt \tag{2.2}$$

and

$$\theta_T = \frac{\int_0^T X_t dX_t}{\int_0^T X_t^2 dt}. \tag{2.3}$$

We assume that the process $\{X_t\}$ is observed at the points $0 \leq t_0 < t_1 < \ldots < t_n = T$ with $\Delta t_i = t_i - t_{i-1} = \frac{T}{n}, i = 1, 2, \ldots, n$. For simplicity, only we assume equidistant time points.

Define

$$\Delta X_{t_i} := X_{t_i} - X_{t_{i-1}},\ \ \Delta X_t := X_t - X_{t-}.$$

We assume $t_i = ih,\ T = nh,\ h \to 0,\ nh \to \infty$. We need the rapid experimental design $nh^{1+\delta} \to 0$ for a constant $\delta \in (0, 1)$. The jump part of data will be split into two parts: one is a small jump part whose jumps are smaller than a positive value $h^\rho, \rho \in (0,\ 1/2)$ and the other a large jump part whose jumps are larger than h^ρ.

By this splitting method, we can judge from the increments of the data, whether large jumps occurred or only small jumps occurred in that interval. In the proposed estimators, we ignore the parts of the small jumps and we make use of other parts.

We will often use the following indicator in the sequel $I_{\{|\Delta X_{t_i}| \leq h^\rho\}}$ as we pick up the data with X_{t_i} and ΔX_{t_i} with $\{|\Delta X_{t_i}| \leq h^\rho\}$. There does not exist large jump in an interval with $\{|\Delta X_{t_i}| \leq h^\rho\}$.

The conditional least squares estimator (CLSE) based on $X_{t_0}, \ldots, X_{t_n}$ is defined as

$$\theta_{n,T} := \arg\inf_\theta \sum_{i=1}^n \frac{[(X_{t_i} - X_{t_{i-1}})I_{\{|\Delta X_{t_i}| \leq h^\rho\}} - \theta X_{t_{i-1}} I_{\{|\Delta X_{t_i}| \leq h^\rho\}} \Delta t_i]^2}{\Delta t_i} \tag{2.4}$$

and is given by

$$\theta_{n,T} = \frac{\sum_{i=1}^{n} X_{t_{i-1}}(X_{t_i} - X_{t_{i-1}}) I_{\{|\Delta X_{t_i}| \le h^\rho\}}}{\sum_{i=1}^{n} X_{t_{i-1}}^2 I_{\{|\Delta X_{t_i}| \le h^\rho\}} \Delta t_i}. \tag{2.5}$$

Note that the estimator $\theta_{n,T}$ may be viewed as an approximate maximum likelihood estimator (AMLE1) which maximizes the approximate log-likelihood given by

$$L_{n,T}(\theta) = \theta \sum_{i=1}^{n} X_{t_{i-1}}(X_{t_i} - X_{t_{i-1}}) I_{\{|\Delta X_{t_i}| \le h^\rho\}} - \frac{\theta^2}{2} \sum_{i=1}^{n} X_{t_{i-1}}^2 I_{\{|\Delta X_{t_i}| \le h^\rho\}} \Delta t_i. \tag{2.6}$$

$L_{n,T}$ is obtained by an Itô type approximation of the stochastic integral and rectangular rule approximation of the ordinary integral in L_T.

Using the Itô formula for the stochastic integral and rectangular rule approximation for the ordinary integral in (2.2), we obtain the approximate likelihood

$$L_{n,T,1}(\theta) = \frac{\theta}{2}(X_T^2 I_{\{|\Delta X_T| \le h^\rho\}} - T) - \frac{\theta^2}{2} \sum_{i=1}^{n} X_{t_{i-1}}^2 I_{\{|\Delta X_{t_i}| \le h^\rho\}} \Delta t_i. \tag{2.7}$$

Note that if we transform the Itô integral to the FS integral in L_T and use the trapezoidal approximation of the FS integral and rectangle rule approximation for the ordinary in L_T, then also we obtain the approximate likelihood $L_{n,T,1}$.

Maximizing $L_{n,T,1}$ provides another approximate maximum likelihood estimate (AMLE2) $\theta_{n,T,1}$ given by

$$\theta_{n,T,1} = \frac{\frac{1}{2}(X_T^2 I_{\{|\Delta X_T| \le h^\rho\}} - T)}{\sum_{i=1}^{n} X_{t_{i-1}}^2 I_{\{|\Delta X_{t_i}| \le h^\rho\}} \Delta t_i}. \tag{2.8}$$

In this chapter, we obtain the Berry–Esseen bounds, i.e., rates of convergence to normality of the two estimators given in (2.5) and (2.8).

Note that

$$\int z^2 f(z) dz = \frac{\alpha}{\beta^2}, \quad \int z^4 f(z) dz = \frac{6\alpha}{\beta^4}.$$

We have two different estimators of α and β:

$$\alpha_{n,T} := \frac{6T \left[\sum_{i=1}^{n} (X_{t_i} - X_{t_{i-1}} - h X_{t_{i-1}} \theta_{n,T})^2 I_{\{|\Delta X_{t_i}| > h^\rho\}}\right]^2}{\sum_{i=1}^{n} (X_{t_i} - X_{t_{i-1}} - h X_{t_{i-1}} \theta_{n,T})^4 I_{\{|\Delta X_{t_i}| > h^\rho\}}}.$$

$$\alpha_{n,T,1} := \frac{6T\left[\sum_{i=1}^{n}(X_{t_i} - X_{t_{i-1}} - hX_{t_{i-1}}\theta_{n,T,1})^2 I_{\{|\Delta X_{t_i}|>h^\rho\}}\right]^2}{\sum_{i=1}^{n}(X_{t_i} - X_{t_{i-1}} - hX_{t_{i-1}}\theta_{n,T,1})^4 I_{\{|\Delta X_{t_i}|>h^\rho\}}}.$$

$$\beta_{n,T} := \sqrt{\frac{6T\sum_{i=1}^{n}(X_{t_i} - X_{t_{i-1}} - hX_{t_{i-1}}\theta_{n,T})^2 I_{\{|\Delta X_{t_i}|\le h^\rho\}}}{\sum_{i=1}^{n}(X_{t_i} - X_{t_{i-1}} - hX_{t_{i-1}}\theta_{n,T})^4 I_{\{|\Delta X_{t_i}|>h^\rho\}}}}.$$

$$\beta_{n,T,1} := \sqrt{\frac{6T\sum_{i=1}^{n}(X_{t_i} - X_{t_{i-1}} - hX_{t_{i-1}}\theta_{n,T,1})^2 I_{\{|\Delta X_{t_i}|>h^\rho\}}}{\sum_{i=1}^{n}(X_{t_i} - X_{t_{i-1}} - hX_{t_{i-1}}\theta_{n,T,1})^4 I_{\{|\Delta X_{t_i}|>h^\rho\}}}}.$$

The organization of the chapter is as follows: in Sect. 3, we obtain the Berry–Esseen bounds the estimator $\theta_{n,T}$. Using purely nonrandom and various random normings (both sample dependent and parameter dependent), we obtain different Berry–Esseen bounds for the estimator $\theta_{n,T}$. Then we obtain probabilistic bounds on $|\theta_{n,T} - \theta_T|$.

In Sect. 4, we derive Berry–Esseen bounds for the estimators $\theta_{n,T,1}$ using purely random and various random normings (both sample dependent and parameter dependent). Then we find probabilistic bounds on $|\theta_{n,T,1} - \theta_T|$.

In Sect. 5, we introduce minimum contrast estimators, and in Section 6, we study the Berry–Esseen bounds for the minimum contrast estimators.

We use the following notations : $\Phi(\cdot)$ denotes the standard normal distribution function. C is a generic constant (which may depend on the parameter) throughout the chapter. P and E denote probability and expectation under the true value of the parameter.

Introduce the following notations :

$$Y_{n,T} := \sum_{i=1}^{n} X(t_{i-1})\left[(W_{t_i} - W(t_{i-1})) + (Z_{t_i} - Z(t_{i-1}))\right] I_{\{|\Delta X_{t_i}|\le h^\rho\}},$$

$$Y_T := \int_0^T X_t(dW_t + dZ_t),$$

$$J_{n,T} := \sum_{i=1}^{n} X(t_{i-1})\left[X_{t_i} - X(t_{i-1})\right] I_{\{|\Delta X_{t_i}|\le h^\rho\}},\ J_T := \int_0^T X_t dX_t,$$

$$I_{n,T} := \sum_{i=1}^{n} X^2(t_{i-1}) I_{\{|\Delta X_{t_i}|\le h^\rho\}}(t_i - t_{i-1}),\ I_T := \int_0^T X_t^2 dt,$$

$$V_{n,T} := \sum_{i=1}^{n}\int_{t_{i-1}}^{t_i} X_{t_{i-1}}\left[X_t - X_{t_{i-1}}\right] I_{\{|\Delta X_{t_i}|\le h^\rho\}} dt.$$

We shall use the following lemma in the sequel whose proof is elementary.

Lemma 2.1 *Let Q_n, R_n, Q and R be random variables on the same probability space $(\Omega, \mathcal{F}, P)$ with $P(R > 0) > 0$. Suppose $|Q_n - Q| = O_P(\delta_{1n})$ and $|R_n - R| = O_P(\delta_{1n})$, where $\delta_{1n}, \delta_{2n} \to 0$ as $n \to \infty$. Then*

$$\left|\frac{Q_n}{Q} - \frac{R_n}{R}\right| = O_P(\delta_{1n} \bigvee \delta_{2n}).$$

4.4.1 Berry–Esseen Type Bounds for AMLE1

Lemma 3.1 We have

$$E\left[\sup_{s,t\in[t_{i-1},t_i]} |X_t - X_s|^k\right] \le Ch^{k/2\wedge\{1+\rho(k-2)\}}.$$

Proof We can obtain the bound by explicit calculation by separating the Poisson part and the Gaussian part.

Lemma 3.2

(a) $E\,|Y_{n,T} - Y_T|^2 = O(\frac{T^2}{n})$.

(b) $E\,|J_{n,T} - J_T|^2 = O(\frac{T^2}{n})$.

(c) $E\,|I_{n,T} - I_T|^2 = O(\frac{T^4}{n^2})$.

Proof Let $g_i(t) = (X_{t_{i-1}} - X_t)I_{\{|\Delta X_{t_i}|\le h^\rho\}}$ for $t_{i-1} \le t < t_i,\quad i = 1, 2, \ldots, n$. Since

$$E|X_{t_{i-1}} - X_t|^{2k} \le C(t_{i-1} - t)^k, k = 1, 2, \ldots \tag{3.1}$$

by Wick's lemma, hence

$$\begin{aligned}
&E|Y_{n,T} - Y_T|^2\\
=\;& E|\sum_{i=1}^{n} X_{t_{i-1}}[(W_{t_i} - W_{t_{i-1}}) + (Z_{t_i} - Z_{t_{i-1}})] - \int_0^T X_t(dW_t + dZ_t)|^2\\
=\;& E|\int_0^T g_i(t)(dW_t + dZ_t)|^2 = \int_0^T E(g_i^2(t))dt\\
\le\;& C\sum_{i=1}^{n}\int_{t_{i-1}}^{t_i} |t_{i-1} - t|dt\\
=\;& Cn\frac{(t_i - t_{i-1})^2}{2} = C\frac{T^2}{n}.
\end{aligned}$$

This completes the proof of (a).

Using (2.1) and the fact that

$$X_{t_i} - X_{t_{i-1}} = \int_{t_{i-1}}^{t_i} \theta X_t dt + W_{t_i} - W_{t_{i-1}} + Z_{t_i} - Z_{t_{i-1}},$$

we obtain

$$\begin{aligned}
& E|J_{n,T} - J_T|^2 \\
= \; & E|\sum_{i=1}^{n} X_{t_{i-1}}[X_{t_i} - X_{t_{i-1}}] - \int_0^T X_t(dW_t + dZ_t)|^2 \\
= \; & E|\sum_{i=1}^{n} \int_{t_{i-1}}^{t_i} \theta X_t X_{t_{i-1}} dt + \sum_{i=1}^{n} X_{t_{i-1}}[(W_{t_i} - W_{t_{i-1}}) + (Z_{t_i} - Z_{t_{i-1}})] \\
& - \int_0^T \theta X_t^2 dt - \int_0^T X_t dW_t|^2 \\
\le \; & 2E|\sum_{i=1}^{n} X_{t_{i-1}}[(W_{t_i} - W_{t_{i-1}}) + (Z_{t_i} - Z_{t_{i-1}})] - \int_0^T X_t d(W_t + Z_t)|^2 \\
& + 2\theta^2 E|\sum_{i=1}^{n} \int_{t_{i-1}}^{t_i} X_t[X_{t_{i-1}} - X_t] dt|^2. \\
=: \; & N_1 + N_2.
\end{aligned}$$

N_1 is $O(\frac{T^2}{n})$ by Lemma 3.2(a). To estimate N_2, let

$$\psi_i(t) := X_t[X_{t_{i-1}} - X_t] I_{\{|\Delta X_{t_i}| \le h^\rho\}}$$

for $t_{i-1} \le t < t_i$, $i = 1, 2, \ldots, n$. Then

$$\begin{aligned}
& E|\sum_{i=1}^{n} \int_{t_{i-1}}^{t_i} \psi_i(t) dt|^2 \\
= \; & \sum_{i=1}^{n} E|\int_{t_{i-1}}^{t_i} \psi_i(t) dt|^2 + 2 \sum_{i,j=1, i<j}^{n} E\left[\int_{t_{i-1}}^{t_i} \psi_i(t) dt \int_{t_{i-1}}^{t_i} \psi_j(s) ds\right] \\
=: \; & M_1 + M_2.
\end{aligned}$$

By the boundedness of $E(X_t^4)$ and (2.1), we have

$$\begin{aligned}
& E(\psi_i^2(t)) \\
= \; & E\{X_t^2[(X_{t_{i-1}} - X_t) I_{\{|\Delta X_{t_i}| \le h^\rho\}}]^2\} \\
\le \; & \{E(X_t^4)\}^{1/2} \{E[(X_{t_{i-1}} - X_t) I_{\{|\Delta X_{t_i}| \le h^\rho\}}]^4\}^{1/2} \\
\le \; & C(t_{i-1} - t).
\end{aligned} \tag{3.2}$$

Note that

$$\begin{aligned}
M_1 \; = \; & \sum_{i=1}^{n} E|\int_{t_{i-1}}^{t_i} \psi_i(t) dt|^2 \le \sum_{i=1}^{n} (t_i - t_{i-1}) \int_{t_{i-1}}^{t_i} E(\psi^2(t)) dt \\
\le \; & C\frac{T}{n} \sum_{i=1}^{n} \int_{t_{i-1}}^{t_i} (t - t_{i-1}) dt \le C\frac{T}{n} \sum_{i=1}^{n} (t_i - t_{i-1})^2 = C\frac{T^3}{n^2}
\end{aligned}$$

and

$$M_2 \; = \; 2 \sum_{i,j=1, i<j}^{n} E \int_{t_{i-1}}^{t_i} \int_{t_{j-1}}^{t_i} [\psi_i(t) \psi_j(s)] dt ds$$

$$= 2 \sum_{i,j=1,i<j}^{n} \int_{t_{i-1}}^{t_i} \int_{t_{j-1}}^{t_j} E[\psi_i(t)\psi_j(s)]dtds.$$

By Wick's lemma, we have

$$\begin{aligned}
& E[\psi_i(t)\psi_j(s)] \\
= \ & E[X_t(X_{t_{i-1}} - X_t)I_{\{|\Delta X_{t_i}| \le h^\rho\}} X_s(X_{t_{j-1}} - X_s)I_{\{|\Delta X_{t_j}| \le h^\rho\}}] \\
= \ & E[X_t(X_{t_{i-1}} - X_t)I_{\{|\Delta X_{t_i}| \le h^\rho\}}]E[X_s(X_{t_{j-1}} - X_s)I_{\{|\Delta X_{t_j}| \le h^\rho\}}] \\
+ \ & E[X_tX_s]E[(X_{t_{i-1}} - X_t)I_{\{|\Delta X_{t_i}| \le h^\rho\}}(X_{t_{j-1}} - X_s)I_{\{|\Delta X_{t_i}| \le h^\rho\}}] \\
+ \ & E[X_t(X(t_{j-1}) - X_s)I_{\{|\Delta X_{t_j}| \le h^\rho\}}]E[X_s(X_{t_{i-1}} - X_t)I_{\{|\Delta X_{t_i}| \le h^\rho\}}] \\
=: \ & A_1 + A_2 + A_3.
\end{aligned}$$

Note that

$$X_t = \int_0^t e^{\theta(t-u)} d(W_u + Z_u), t \ge 0.$$

Let $a := e^\theta$. For $s \ge t$, we have

$$\begin{aligned}
& E(X_tX_s) \\
= \ & E(\int_0^t e^{\theta(t-u)} d(W_u + Z_u))(\int_0^s e^{\theta(s-u)} d(W_u + Z_u)) \\
= \ & \int_0^t e^{\theta(t+s-2u)} du = \frac{1}{2\theta}[a^{s+t} - a^{s-t}].
\end{aligned}$$

Observe that

$$\begin{aligned}
& E(X_t - X_{t_{i-1}})I_{\{|\Delta X_{t_i}| \le h^\rho\}}(X_s - X_{t_{j-1}})I_{\{|\Delta X_{t_j}| \le h^\rho\}} \\
= \ & E(X_tX_s) - E(X_tX_{t_{j-1}}) - E(X_{t_{i-1}}X_s) + E(X_{t_{i-1}}X_{t_{j-1}}) \\
= \ & \frac{1}{2\theta}(a^s - a^{t_{j-1}})[(a^t - a^{t_{i-1}}) + (a^{-t_{i-1}} - a^{-t})] \\
= \ & \frac{1}{2\theta}(s - t_{j-1})a^{t^*}[(t - t_{i-1}a^{t^{**}} + (t - t_{i-1})a^{-t^{***}}] \\
& (\text{where } t_{j-1} < t^* < s, t_{i-1} < t^{**}, t^{***} < t) \\
\le \ & \frac{1}{2\theta}(s - t_{j-1})a^t(t - t_{i-1})a^{t_{i-1}} + (s - t_{j-1})a^t(t - t_{i-1})a^{-t}] \\
\le \ & C(s - t_{j-1})(t - t_{i-1}).
\end{aligned}$$

Thus $A_2 \le C(s - t_{j-1})(t - t_{i-1})$ since $|E(X_tX_s)|$ is bounded.

Next

$$\begin{aligned}
& |E[X_t(X_{t_{i-1}} - X_t)I_{\{|\Delta X_{t_i}| \le h^\rho\}}]| \\
= \ & \frac{1}{2|\theta|}[a^{t+t_{i-1}} - a^{t-t_{i-1}} - a^{2t} + 1]
\end{aligned}$$

$$\begin{aligned}
&= \frac{1}{2|\theta|}a^t[a^{t_{i-1}} - a^{-t_{i-1}} - a^t + a^{-t}] \\
&\leq \frac{1}{2|\theta|}a^t(t - t_{i-1})[a^{t_{i-1}} + a^{-t}] \\
&\leq C(t - t_{i-1})
\end{aligned}$$

and

$$\begin{aligned}
&|E[X_s(X_s - X_{t_{j-1}})I_{\{|\Delta X_{t_j}| \leq h^\rho\}}]| \\
&= \frac{1}{2|\theta|}[a^{2s} - 1 - a^{s+t_{j-1}} + a^{s+t_{j-1}}] \\
&= \frac{1}{2|\theta|}a^s[a^s - a^{-s} - a^{t_{j-1}} + a^{-t_{j-1}}] \\
&\leq \frac{1}{2|\theta|}a^s(s - t_{j-1})[a^{t_{j-1}} + a^{-s}] \\
&\leq C(s - t_{j-1}).
\end{aligned}$$

Thus $A_1 \leq C(s - t_{j-1})(t - t_{i-1})$.

Next

$$\begin{aligned}
&|E[X_t(X_s - X_{t_{j-1}})I_{\{|\Delta X_{t_j}| \leq h^\rho\}}]| \\
&= \frac{1}{2|\theta|}[a^{s+t} - a^{s-t} - a^{t+t_{j-1}} + a^{t_{j-1}-t}] \\
&= \frac{1}{2|\theta|}a^t(a^s - a^{t_{j-1}}) \\
&\leq \frac{1}{2|\theta|}a^t(1 - a^{-2t})(s - t_{j-1})a^t \\
&\leq (a^{2t} - 1)(s - t_{j-1}) \\
&\leq C(s - t_{j-1})
\end{aligned}$$

and

$$\begin{aligned}
&|E[X_s(X_t - X_{t_{i-1}})I_{\{|\Delta X_{t_i}| \leq h^\rho\}}]| \\
&= \frac{1}{2|\theta|}[a^{t+s} - a^{s-t} - a^{s+t_{i-1}} + a^{s-t_{i-1}}] \\
&= \frac{1}{2|\theta|}a^s[a^t - a^{-t} - a^{t_{i-1}} + a^{-t_{i-1}}] \\
&\leq \frac{1}{2|\theta|}a^s(t - t_{i-1})[a^{t_{i-1}} + a^{-t}] \\
&\leq C(t - t_{i-1}).
\end{aligned}$$

Thus $A_3 \leq C(s - t_{j-1})(t - t_{i-1})$.
Hence $E[f_i(t)f_j(s)] \leq C(s - t_{j-1})(t - t_{i-1})$.

Thus

$$
\begin{aligned}
M_2 &= 2\sum_{i,j=1,i<j}^{n}\int_{t_{i-1}}^{t_i}\int_{t_{j-1}}^{t_j} E[f_i(t)f_j(s)]dtds \\
&\le C\sum_{i,j=1,i<j}^{n}\int_{t_{i-1}}^{t_i}\int_{t_{j-1}}^{t_j}(t-t_{i-1})(s-t_{j-1})dtds \\
&= C\sum_{i,j=1,i<j}^{n}(t_{i-1}-t_i)^2(t_{j-1}-t_i)^2 \\
&= Cn^2(\frac{T}{n})^4 = C\frac{T^4}{n^2}.
\end{aligned}
$$

Hence, N_2 is $O(\frac{T^3}{n^2})$. Combining N_1 and N_2 completes the proof of (b). We next prove (c).

Let $\chi_i(t) := (X_{t_{i-1}}^2 - X_t^2, t_{i-1})I_{\{|\Delta X_{t_i}|\le h^\rho\}} \le t < t_i, i = 1, 2, \ldots, n$. Then

$$
\begin{aligned}
& E|I_{n,T} - I_t|^2 \\
=\ & E|\sum_{i=1}^{n} X_{t_{i-1}}^2 I_{\{|\Delta X_{t_i}|\le h^\rho\}}(t_i - t_{i-1}) - \int_0^T X_t^2 dt|^2 \\
=\ & E|\sum_{i=1}^{n}\int_{t_{i-1}}^{t_i}[(X_{t_{i-1}}^2 - X_t^2)I_{\{|\Delta X_{t_i}|\le h^\rho\}}]dt|^2 \\
=\ & E|\sum_{i=1}^{n}\int_{t_{i-1}}^{t_i}\chi_i(t)dt|^2 \\
=\ & \sum_{i=1}^{n} E|\int_{t_{i-1}}^{t_i}\chi_i(t)dt|^2 + 2\sum_{i,j=1,i<j}^{n} E\int_{t_{i-1}}^{t_i}\int_{t_{j-1}}^{t_j}\chi_i(t)\chi_j(s)dtds \\
=:\ & B_1 + B_2.
\end{aligned}
$$

$$
\begin{aligned}
E\chi_i^2(t) &= E[(X_{t_{i-1}}^2 - X_t^2)I_{\{|\Delta X_{t_i}|\le h^\rho\}}]^2 \\
&= E[(X_{t_{i-1}} - X_t)I_{\{|\Delta X_{t_i}|\le h^\rho\}}]^2[(X_{t_{i-1}} + X_t)I_{\{|\Delta X_{t_i}|\le h^\rho\}}]^2 \\
&\le \{E[(X_{t_{i-1}} - X_t)I_{\{|\Delta X_{t_i}|\le h^\rho\}}]^4\}^{1/2} \\
&\quad \times\{\{E[(X_{t_{i-1}} + X_t)I_{\{|\Delta X_{t_i}|\le h^\rho\}}]^4\}^{1/2} \\
&\le C(t - t_{i-1})
\end{aligned}
$$

(by (3.1) and the boundedness of the second term)

$$
\begin{aligned}
B_1 &= \sum_{i=1}^{n} E|\int_{t_{i-1}}^{t_i}\chi_i(t)dt|^2 \le \sum_{i=1}^{n}(t_i - t_{i-1})\int_{t_{i-1}}^{t_i} E(\chi_i^2(t))dt \\
&\le C\frac{T}{n}\sum_{i=1}^{n}\int_{t_{i-1}}^{t_i}(t - t_{i-1})dt = C\frac{T^3}{n^2}.
\end{aligned}
$$

Note that

$$
E[\chi_i(t)\chi_j(s)]
$$

$$
\begin{aligned}
&= E((X_{t_{i-1}}^2 - X_t^2)I_{\{|\Delta X_{t_i}|\le h^\rho\}})((X_{t_{j-1}}^2 - X_s^2)I_{\{|\Delta X_{t_j}|\le h^\rho\}}) \\
&= E((X_{t_{i-1}} - X_t)I_{\{|\Delta X_{t_i}|\le h^\rho\}})((X_{t_{i-1}} + X_t)I_{\{|\Delta X_{t_i}|\le h^\rho\}}) \\
&\quad ((X_{t_{j-1}} - X_s)I_{\{|\Delta X_{t_j}|\le h^\rho\}})((X_{t_{j-1}} + X_s)I_{\{|\Delta X_{t_j}|\le h^\rho\}}).
\end{aligned}
$$

Now using Wick's lemma and proceeding similar to the estimation of M_2, it is easy to see that

$$B_2 \le C\frac{T^4}{n^2}.$$

Combining B_1 and B_2, (c) follows. □

Theorem 3.3 *Let* $\alpha_{n,T} := \max(T^{-1/2}(\log T)^{1/2}, \frac{T^2}{n}(\log T)^{-1}, \frac{T^4}{n^2}(\log T)^{-1})$. *We have*

$$(a) \sup_{x\in\mathbb{R}}\left|P\left\{\left(\frac{T}{-2\theta}\right)^{1/2}(\theta_{n,T}-\theta)\le x\right\} - \Phi(x)\right| = O(\alpha_{n,T}).$$

$$(b) \sup_{x\in\mathbb{R}}\left|P\left\{I_{n,T}^{1/2}(\theta_{n,T}-\theta)\le x\right\} - \Phi(x)\right| = O(\alpha_{n,T}).$$

$$(c) \sup_{x\in\mathbb{R}}\left|P\left\{\left(\frac{T}{|2\theta_{n,T}|}\right)^{1/2}(\theta_{n,T}-\theta)\le x\right\} - \Phi(x)\right| = O(\alpha_{n,T}).$$

Proof

(a) It is easy to see that $\theta_{n,T} - \theta = \frac{Y_{n,T}}{I_{n,T}} + \theta\frac{V_{n,T}}{I_{n,T}}$. (3.3)

$$
\begin{aligned}
&\text{Hence } \sup_{x\in\mathbb{R}}\left|P\left\{\left(\frac{T}{-2\theta}\right)^{1/2}(\theta_{n,T}-\theta)\le x\right\} - \Phi(x)\right| \\
= &\sup_{x\in\mathbb{R}}\left|P\left\{\left(\frac{T}{-2\theta}\right)^{1/2}\frac{Y_{n,T}}{I_{n,T}} + \left(\frac{T}{-2\theta}\right)^{1/2}\theta\frac{V_{n,T}}{I_{n,T}}\le x\right\} - \Phi(x)\right| \\
\le &\sup_{x\in\mathbb{R}}\left|P\left\{\left(\frac{T}{-2\theta}\right)^{1/2}\frac{Y_{n,T}}{I_{n,T}}\le x\right\} - \Phi(x)\right| \qquad (3.4)\\
&+P\left\{\left|\left(\frac{T}{-2\theta}\right)^{1/2}\frac{V_{n,T}}{I_{n,T}}\right| > \epsilon\right\} + \epsilon. \\
=: &K_1 + K_2 + \epsilon.
\end{aligned}
$$

Note that by Lemma 2.1(b),

$$\begin{aligned} K_1 &= \sup_{x\in\mathbb{R}} \left| P\left\{ \left(\frac{T}{-2\theta}\right)^{1/2} \frac{Y_{n,T}}{I_{n,T}} \le x \right\} - \Phi(x) \right| \\ &= \sup_{x\in\mathbb{R}} \left| P\left\{ \frac{\left(-\frac{2\theta}{T}\right)^{1/2} Y_{n,T}}{\left(\frac{-2\theta}{T}\right) I_{n,T}} \le x \right\} - \Phi(x) \right| \\ &\le \sup_{x\in\mathbb{R}} \left| P\left\{ \left(\frac{-2\theta}{T}\right)^{1/2} Y_{n,T} \le x \right\} - \Phi(x) \right| \\ &\quad + P\left\{ \left(\tfrac{-2\theta}{T}\right) I_{n,T} - 1 > \epsilon \right\} + \epsilon \\ &=: J_1 + J_2 + \epsilon. \end{aligned} \tag{3.5}$$

$$\begin{aligned} J_1 &= \sup_{x\in\mathbb{R}} \left| P\left\{ \left(\frac{-2\theta}{T}\right)^{1/2} (Y_{n,T} - Y_T + Y_T) \le x \right\} - \Phi(x) \right| \\ &\le \sup_{x\in\mathbb{R}} \left| P\left\{ \left(\frac{-2\theta}{T}\right)^{1/2} Y_T \le x \right\} - \Phi(x) \right| \\ &\quad + P\left\{ \left(\frac{-2\theta}{T}\right)^{1/2} |Y_{n,T} - Y_T| > \epsilon \right\} + \epsilon \\ &\le CT^{-1/2} + \left(\frac{-2\theta}{T}\right) \frac{E|Y_{n,T} - Y_T|^2}{\epsilon^2} + \epsilon \\ &\le CT^{-1/2} + C\frac{T/n}{\epsilon^2} + \epsilon \\ &\quad \text{(by Corollary 2.3(a) and Lemma 3.2(a)).} \end{aligned} \tag{3.6}$$

$$\begin{aligned} J_2 &= P\left\{ \left| \left(\frac{-2\theta}{T}\right) (I_{n,T} - I_T + I_T) - 1 \right| > \epsilon \right\} \\ &\le P\left\{ \left| \left(\frac{-2\theta}{T}\right) I_T - 1 \right| > \frac{\epsilon}{2} \right\} + P\left\{ \left(-\frac{2\theta}{2}\right) |I_{n,T} - I_T| > \frac{\epsilon}{2} \right\} \\ &\le C\exp\left(\frac{T\theta}{16}\epsilon^2\right) + \frac{16\theta^2}{T^2} \frac{E|I_{n,T} - I_T|^2}{\epsilon^2} \\ &\le C\exp\left(\frac{T\theta}{16}\epsilon^2\right) + C\frac{T^2/n^2}{\epsilon^2}. \end{aligned} \tag{3.8}$$

Here the bound for the first term in (3.8) comes from Lemma 2.4(a) and that for the second term from Lemma 3.2(c). From the proof of Lemma 3.2(b), we have

$$E|V_{n,T}|^2 \le C\frac{T^4}{n^2}. \tag{3.9}$$

Next

$$
\begin{aligned}
K_2 &= P\left\{\left|\left(\frac{T}{-2\theta}\right)^{1/2}\theta\frac{V_{n,T}}{I_{n,T}}\right|>\epsilon\right\}\\
&= P\left\{\left|\frac{\left(\frac{-2\theta}{T}\right)^{1/2}\theta V_{n,T}}{\left(-\frac{2\theta}{T}\right)I_{n,T}}\right|>\epsilon\right\}\\
&= P\left\{\left|\left(\frac{-2\theta}{T}\right)^{1/2}\theta V_{n,T}\right|>\delta\right\}+P\left\{\left(-\frac{2\theta}{T}\right)I_{n,T}<\frac{\delta}{\epsilon}\right\}\\
&\quad \text{(where we choose } \delta=\epsilon-C\epsilon^2\text{)}\\
&\le P\left\{\left|\left(\frac{-2\theta}{T}\right)^{1/2}\theta V_{n,T}\right|>\delta\right\}+P\left\{\left|\left(-\frac{2\theta}{T}\right)I_{n,T}-1\right|>\delta_1\right\}\\
&\quad \text{(where } \delta_1=\frac{\epsilon-\delta}{\delta}=C\epsilon\text{)}\\
&\le -\frac{2\theta}{T}\theta^2\frac{E|V_{n,T}|^2}{\delta^2}+C\exp\left(\frac{T\theta}{16}\delta_1^2\right)+C\frac{T^2/n^2}{\delta_1^2}\\
&\le C\frac{T^3/n^2}{\delta^2}+C\exp\left(\frac{T\theta}{16}\delta_1^2\right)+C\frac{T^2/n^2}{\delta_1^2}\quad \text{(by (3.9) and (3.8)).}
\end{aligned}
\tag{3.10}
$$

Now combining bounds from $J_1, J_2, K_1,$ and K_2, we have since $T/n \to 0$

$$
\begin{aligned}
&\sup_{x\in\mathbb{R}}\left|P\left\{\left(\frac{T}{-2\theta}\right)^{1/2}(\theta_{n,T}-\theta)\le x\right\}-\Phi(x)\right|\\
\le\ & CT^{-1/2}+C\exp\left(\frac{T\theta}{16}\epsilon^2\right)+C\frac{T/n}{\epsilon^2}+C\frac{T^2/n^2}{\epsilon^2}+C\frac{T^3/n^2}{\delta^2}\\
&+C\exp\left(\frac{T\theta}{16}\delta_1^2\right)+C\left(\frac{T^2/n^2}{\delta_1^2}\right)+\epsilon.
\end{aligned}
\tag{3.11}
$$

Choosing $\epsilon = CT^{-1/2}(\log T)^{1/2}$, C large, the terms of (3.11) are of the order $O(\max(T^{-1/2}(\log T)^{1/2}, \frac{T^2}{n}(\log T)^{-1}, \frac{T^4}{n^2}(\log T)^{-1}))$. This proves (a).

(b) Using the expression (3.3), we have

$$
\begin{aligned}
&\sup_{x\in\mathbb{R}}\left|P\left\{I_{n,T}^{1/2}(\theta_{n,T}-\theta)\le x\right\}-\Phi(x)\right|\\
=\ &\sup_{x\in\mathbb{R}}\left|P\left\{\frac{Y_{n,T}}{I_{n,T}^{1/2}}+\theta\frac{V_{n,T}}{I_{n,T}^{1/2}}\le x\right\}-\Phi(x)\right|\\
\le\ &\sup_{x\in\mathbb{R}}\left|P\left\{\frac{Y_{n,T}}{I_{n,T}^{1/2}}\le x\right\}-\Phi(x)\right|+P\left\{\left|\theta\frac{V_{n,T}}{I_{n,T}^{1/2}}\right|>\epsilon\right\}+\epsilon.\\
=:\ &H_1+H_2+\epsilon.
\end{aligned}
\tag{3.12}
$$

$$\begin{aligned}
H_1 &= \sup_{x\in\mathbb{R}}\left|P\left\{\frac{Y_{n,T}-Y_T+Y_T}{I_{n,T}^{1/2}}\le x\right\}-\Phi(x)\right| \\
&= \sup_{x\in\mathbb{R}}\left|P\left\{\frac{Y_T}{I_{n,T}^{1/2}}\le x\right\}-\Phi(x)\right|+P\left\{\frac{|Y_{n,T}-Y_T|}{I_{n,T}^{1/2}}>\epsilon\right\}+\epsilon. \\
&=: F_1+F_2+\epsilon.
\end{aligned} \tag{3.13}$$

$$\begin{aligned}
F_1 &= \sup_{x\in\mathbb{R}}\left|P\left\{\frac{Y_T}{I_{n,T}^{1/2}}\le x\right\}-\Phi(x)\right|\le \sup_{x\in\mathbb{R}}\left|P\left\{\left(\frac{-2\theta}{T}\right)^{1/2}Y_T\le x\right\}-\Phi(x)\right| \\
&\quad +P\left\{\left|\left(\frac{-2\theta}{T}\right)^{1/2}I_{n,T}^{1/2}-1\right|>\epsilon\right\}+\epsilon \text{ (by Lemma 2.1(b))} \\
&\le CT^{-1/2}+P\left\{\left|\left(\frac{-2\theta}{T}\right)I_{n,T}-1\right|>\epsilon\right\}+\epsilon \text{ (by Corollary 2.3(a))} \\
&\le CT^{-1/2}+C\exp\left(-\tfrac{T\theta}{16}\epsilon^2\right)+C\tfrac{T^2/n^2}{\epsilon^2}+\epsilon \ \text{ (by (3.8))}.
\end{aligned} \tag{3.14}$$

On the other hand,

$$\begin{aligned}
F_2 &= P\left\{\frac{|Y_{n,T}-Y_T|}{I_{n,T}^{1/2}}>\epsilon\right\} \\
&\le P\left\{\left(\frac{-2\theta}{T}\right)^{1/2}|Y_{n,T}-Y_T|>\delta\right\}+P\left\{\left|\left(\frac{-2\theta}{T}\right)^{1/2}I_{n,T}^{1/2}-1\right|>\delta_1\right\} \\
&\quad \text{(where } \delta=\epsilon-C\epsilon^2 \text{ and } \delta_1=(\epsilon-\delta)/\epsilon>0) \\
&\le \frac{\left(\frac{-2\theta}{T}\right)E|Y_{n,T}-Y_T|^2}{\delta^2}+P\left\{\left|\left(\frac{-2\theta}{T}\right)I_{n,T}-1\right|>\delta_1\right\} \\
&\le C\frac{T/n}{\delta^2}+C\exp\left(\frac{T\theta}{16}\delta_1^2\right)+C\frac{T^2/n^2}{\delta_1^2} \ \text{ (from Lemma 3.2(a) and (3.8))}.
\end{aligned} \tag{3.15}$$

Using (3.15) and (3.14) in (3.13), we obtain

$$\begin{aligned}
H_1 &= \sup_{x\in\mathbb{R}}\left|P\left\{\frac{Y_T}{I_{n,T}^{1/2}}\le x\right\}-\Phi(x)\right| \\
&\le CT^{-1/2}+C\exp\left(\frac{T\theta}{16}\epsilon^2\right)+C\frac{T/n}{\delta^2}+C\frac{T^2/n^2}{\delta_1^2} \\
&\quad +C\exp\left(\frac{T\theta}{16}\delta_1^2\right)+C\frac{T^2/n^2}{\epsilon^2}+\epsilon.
\end{aligned} \tag{3.16}$$

$$
\begin{aligned}
H_2 &= P\left\{\left|\theta\frac{V_{n,T}}{I_{n,T}^{1/2}}\right| > \epsilon\right\} = P\left\{\frac{\left|\left(\frac{-2\theta}{T}\right)^{1/2}\theta V_{n,T}\right|}{\left|\left(\frac{-2\theta}{T}\right)^{1/2} I_{n,T}^{1/2}\right|} > \epsilon\right\} \\
&\le P\left\{\left|\left(\frac{-2\theta}{T}\right)^{1/2}\theta V_{n,T}\right| > \delta\right\} + P\left\{\left|\left(\frac{-2\theta}{T}\right)^{1/2} I_{n,T}^{1/2}\right| < \delta/\epsilon\right\} \\
&\le \left(-\frac{2\theta}{T}\right)\theta^2\frac{E|V_{n,T}|^2}{\delta^2} + P\left\{\left|\left(-\frac{2\theta}{T}\right)I_{n,T} - 1\right| > \delta_1\right\} \\
&\quad \text{(where } 0<\delta<\epsilon \text{ and } \delta_1 = (\epsilon-\delta)/\epsilon = C\epsilon > 0) \\
&\le C\frac{T^3/n^2}{\delta^2} + C\exp\left(\frac{T\theta}{16}\delta_1^2\right) + C\frac{T^2/n^2}{\delta_1^2} \quad \text{(from (3.9) and (3.8)).}
\end{aligned} \tag{3.17}
$$

Using (3.17) and (3.16) in (3.12) and choosing $\epsilon = CT^{-1/2}(\log T)^{1/2}$, C large, the terms of (3.12) are of the order $O(\max(T^{-1/2}(\log T)^{1/2}, \frac{T^2}{n}(\log T)^{-1}, \frac{T^4}{n^2}(\log T)^{-1}))$. This proves (b).
(c) Let $D_T = \{|\theta_{n,T}-\theta| \le d_T\}$ and $d_T = CT^{-1/2}(\log T)^{1/2}$.
On the set D_T, expanding $(2|\theta_{n,T}|)^{-1/2}$, we obtain

$$
\begin{aligned}
(-2\theta_{n,T})^{-1/2} &= (-2\theta)^{-1/2}\left[1 - \frac{\theta-\theta_{n,T}}{\theta}\right]^{-1/2} \\
&= (-2\theta)^{-1/2}\left[1 + \frac{1}{2}\left(\frac{\theta-\theta_{n,T}}{\theta}\right) + O(d_T^2)\right].
\end{aligned}
$$

Then

$$
\begin{aligned}
&\sup_{x\in\mathbb{R}}\left|P\left\{\left(\frac{T}{2|\theta_{n,T}|}\right)^{1/2}(\theta_{n,T}-\theta) \le x\right\} - \Phi(x)\right| \\
\le\ &\sup_{x\in\mathbb{R}}\left|P\left\{\left(\frac{T}{2|\theta_{n,T}|}\right)^{1/2}(\theta_{n,T}-\theta) \le x, D_T\right\} - \Phi(x)\right| + P(D_T^c).
\end{aligned} \tag{3.18}
$$

$$
\begin{aligned}
&P(D_T^c) \\
=\ &P\left\{|\theta_{n,T}-\theta| > CT^{-1/2}(\log T)^{1/2}\right\} \\
=\ &P\left\{\left(\frac{T}{-2\theta}\right)^{1/2}|\theta_{n,T}-\theta| > C(\log T)^{1/2}(-2\theta)^{-1/2}\right\} \\
\le\ &C(\max(T^{-1/2}(\log T)^{1/2}, \frac{T^2}{n}(\log T)^{-1}, \frac{T^4}{n^2}(\log T)^{-1}) \\
&+2(1-\Phi\left((\log T)^{1/2}(-2\theta)^{-1/2}\right) \quad \text{(by Theorem 3.3(a))} \\
\le\ &C(\max(T^{-1/2}(\log T)^{1/2}, \frac{T^2}{n}(\log T)^{-1}, \frac{T^4}{n^2}(\log T)^{-1})).
\end{aligned}
$$

On the set D_T,

$$
\left|\left(\frac{\theta_{n,T}}{\theta}\right)^{1/2} - 1\right| \le CT^{-1/2}(\log T)^{1/2}.
$$

Hence upon choosing $\epsilon = CT^{-1/2}(\log T)^{1/2}, C$ large, we obtain

$$\begin{aligned}
&\left| P\left\{ \left(\frac{T}{-2\theta_{n,T}}\right)^{1/2} (\theta_{n,T} - \theta) \le x, D_T \right\} - \Phi(x) \right| \\
\le\ & \left| P\left\{ \left(\frac{T}{-2\theta}\right)^{1/2} (\theta_{n,T} - \theta) \le x, D_T \right\} - \Phi(x) \right| \\
& + P\left\{ \left| \left(\tfrac{\theta_{n,T}}{\theta}\right)^{1/2} - 1 \right| > \epsilon, D_T \right\} + \epsilon \\
& \quad \text{(by Lemma 3.3.1(b))} \\
\le\ & C(\max(T^{-1/2}(\log T)^{1/2}, \tfrac{T^2}{n}(\log T)^{-1}, \tfrac{T^4}{n^2}(\log T)^{-1})) \\
& \text{(by Theorem 3.3(a))}.
\end{aligned} \tag{3.20}$$

(c) follows from (3.18)–(3.20). □

Theorem 3.4

$$\sup_{x\in\mathbb{R}} \left| P\left\{ I_{n,T}\left(-\frac{2\theta}{T}\right)^{1/2} (\theta_{n,T} - \theta) \le x \right\} - \Phi(x) \right| = O\left(T^{-1/2} \bigvee \left(\frac{T}{n}\right)^{1/3} \right).$$

Proof Let $a_{n,T} := Z_{n,T} - Z_T$ and $b_{n,T} := I_{n,T} - I_T$.

$$\text{By Lemma 3.2,} \quad E|a_{n,T}|^2 = O\left(\frac{T^2}{n}\right) \text{ and } E|b_{n,T}|^2 = O\left(\frac{T^4}{n^2}\right). \tag{3.21}$$

From (3.5), we have

$$\begin{aligned}
I_{n,T}\theta_{n,T} &= \sum_{i=1}^{n} X_{t_{i-1}}\left[X_{t_i} - X_{t_{i-1}}\right] = \int_0^T X_t dX_t + a_{n,T} \\
&= \int_0^T X_t d(W_t + Z_t) + \theta \int_0^T X_t^2 dt + a_{n,T}.
\end{aligned}$$

Hence $I_{n,T}(\theta_{n,T} - \theta) = -\theta b_{n,T} + a_{n,T}$.
Thus

$$\begin{aligned}
& \sup_{x\in\mathbb{R}} \left| P\left\{ I_{n,T}\left(-\frac{2\theta}{T}\right)^{1/2} (\theta_{n,T} - \theta) \le x \right\} - \Phi(x) \right| \\
=\ & \sup_{x\in\mathbb{R}} \left| P\left\{ \left(-\frac{2\theta}{T}\right)^{1/2} [Y_T - \theta b_{n,T} + a_{n,T}] \le x \right\} - \Phi(x) \right| \\
\le\ & \sup_{x\in\mathbb{R}} \left| P\left\{ \left(-\frac{2\theta}{T}\right)^{1/2} Y_T \le x \right\} - \Phi(x) \right| \\
& + P\left\{ \left| \left(-\frac{2\theta}{T}\right)^{1/2} [-\theta b_{n,T} + a_{n,T}] \right| > \epsilon \right\} + \epsilon \\
\le\ & CT^{-1/2} + \left(-\frac{2\theta}{T}\right) \frac{E|-\theta b_{n,T} + a_{n,T}|^2}{\epsilon^2} + \epsilon \\
\le\ & CT^{-1/2} + C\frac{T/n}{\epsilon^2} + \epsilon \text{ (by Corollary 1.2.3(a) and (3.21))}.
\end{aligned}$$

Choosing $\epsilon = \left(\frac{T}{n}\right)^{1/3}$, the rate is $O\left(T^{-1/2} \bigvee \left(\frac{T}{n}\right)^{1/3}\right)$. □

Theorem 3.5

$$|\theta_{n,T} - \theta_T| = O_P(\frac{T^2}{n})^{1/2}.$$

Proof Note that $\theta_{n,T} - \theta_T = \frac{Z_{n,T}}{I_{n,T}} - \frac{Z_T}{I_T}$. From Lemma 3.2, it follows that $|Z_{n,T} - Z_T| = O_P(\frac{T^2}{n})^{1/2}$ and $|I_{n,T} - I_T| = O_P(\frac{T^4}{n^2})^{1/2}$. Now the theorem follows easily from Lemma 2.1. □

4.4.2 Berry–Esseen Type Bounds for AMLE2

Theorem 4.1 *Let* $\beta_{n,T} = O\left(T^{-1/2}(\log T)^{1/2} \bigvee \frac{T^4}{n^2}(\log T)^{-1}\right)$.

(a) $\sup_{x\in\mathbb{R}} \left| P\left\{ \left(-\frac{T}{2\theta}\right)^{1/2} (\theta_{n,T,1} - \theta) \le x \right\} - \Phi(x) \right| = O(\beta_{n,T})$,

(b) $\sup_{x\in\mathbb{R}} \left| P\left\{ I_{n,T}^{1/2}(\theta_{n,T,1} - \theta) \le x \right\} - \Phi(x) \right| = O(\beta_{n,T})$,

(c) $\sup_{x\in\mathbb{R}} \left| P\left\{ \left(\frac{T}{2|\theta_{n,T,1}|}\right)^{1/2} (\theta_{n,T,1} - \theta) \le x \right\} - \Phi(x) \right| = O(\beta_{n,T})$.

Proof

(a) From (1.8), we have

$$\begin{aligned} I_{n,T}\theta_{n,T,1} &= \frac{1}{2}(X_T^2 - T) = \int_0^T X_t dX_t \\ &= \int_0^T X_t d(W_t + Z_t) + \theta \int_0^T X_t^2 dt \\ &= Y_T + \theta I_T. \end{aligned}$$

Thus

$$\begin{aligned} \left(-\tfrac{T}{2\theta}\right)^{1/2} (\theta_{n,T,1} - \theta) &= \frac{(-\frac{T}{2\theta})^{1/2} Y_T + \theta(-\frac{T}{2\theta})^{1/2}(I_T - I_{n,T})}{I_{n,T}} \\ &= \frac{(-\frac{2\theta}{T})^{1/2} Y_T + (-\frac{2\theta}{T})^{1/2}(I_T - I_{n,T})}{(-\frac{2\theta}{T}) I_{n,T}}. \end{aligned} \qquad (4.1)$$

Now

$$
\begin{aligned}
& \sup_{x\in\mathbb{R}} \left| P\left\{ \left(-\frac{T}{2\theta}\right)^{1/2} (\theta_{n,T,1} - \theta) \le x \right\} - \Phi(x) \right| \\
= & \sup_{x\in\mathbb{R}} \left| P\left\{ \frac{(-\frac{2\theta}{T})^{1/2} Y_T + (-\frac{2\theta}{T})^{1/2}(I_T - I_{n,T})}{(-\frac{2\theta}{T}) I_{n,T}} \le x \right\} - \Phi(x) \right| \\
\le & \sup_{x\in\mathbb{R}} \left| P\left\{ (-\frac{2\theta}{T})^{1/2} Y_T \le x \right\} - \Phi(x) \right| + P\left\{ \left| \theta(-\frac{2\theta}{T})^{1/2}(I_{n,T} - I_T) \right| > \epsilon \right\} \\
& + P\left\{ \left| \left(-\frac{2\theta}{T}\right) I_{n,T} - 1 \right| > \epsilon \right\} + 2\epsilon \\
\le & \; CT^{-1/2} + \theta^2 \frac{(-\frac{2\theta}{T}) E\left| I_{n,T} - I_T \right|^2}{\epsilon^2} + C\exp(\frac{T\theta}{4}\epsilon^2) + C\frac{T^2}{n^2\epsilon^2} + 2\epsilon
\end{aligned} \tag{4.2}
$$

(the bound for the third term in the r.h.s. of (4.2) is from (3.8)).

$$
\le CT^{-1/2} + C\frac{T^3}{n^2\epsilon^2} + C\exp(\frac{T\theta}{4}\epsilon^2) + C\frac{T^2}{n^2\epsilon^2} + \epsilon \tag{4.3}
$$

(by Lemma 3.2(c)).

Choosing $\epsilon = CT^{-1/2}(\log T)^{1/2}$, the terms in the r.h.s. of (4.3) are of the order $O(T^{-1/2}(\log T)^{1/2} \bigvee \frac{T^4}{n^2}(\log T)^{-1})$.

(b) From (4.1), we have

$$
I_{n,T}^{1/2}(\theta_{n,T,1} - \theta) = \frac{Y_T + \theta(I_T - I_{n,T})}{I_{n,T}^{1/2}}.
$$

Now

$$
\begin{aligned}
& \sup_{x\in\mathbb{R}} \left| P\left\{ I_{n,T}^{1/2}(\theta_{n,T,1} - \theta) \le x \right\} - \Phi(x) \right| \\
= & \sup_{x\in\mathbb{R}} \left| P\left\{ \frac{Y_T}{I_{n,T}^{1/2}} + \theta\frac{I_T - I_{n,T}}{I_{n,T}^{1/2}} \le x \right\} - \Phi(x) \right| \\
\le & \sup_{x\in\mathbb{R}} \left| P\left\{ \frac{Y_T}{I_{n,T}^{1/2}} \le x \right\} - \Phi(x) \right| + P\left\{ \left| \frac{\theta(I_T - I_{n,T})}{I_{n,T}^{1/2}} \right| > \epsilon \right\} + \epsilon \\
=: & \; U_1 + U_2 + \epsilon.
\end{aligned} \tag{4.4}
$$

We have from (3.8),

$$
U_1 \le CT^{-1/2} + C\exp(\frac{T\theta}{16}\epsilon^2) + C\frac{T^2}{n^2\epsilon^2} + \epsilon. \tag{4.5}
$$

Now

$$\begin{aligned}
U_2 &= P\left\{|\theta|\left|\frac{I_{n,T}-I_T}{I_{n,T}^{1/2}}\right|>\epsilon\right\}=P\left\{|\theta|\frac{|(-\frac{2\theta}{T})^{1/2}(I_{n,T}-I_T)|}{|(-\frac{2\theta}{T})^{1/2}I_{n,T}^{1/2}|}>\epsilon\right\}\\
&\le P\left\{\left|(-\frac{2\theta}{T})^{1/2}\right||I_{n,T}-I_T|>\delta\right\}+P\left\{\left|(-\frac{2\theta}{T})^{1/2}I_{n,T}^{1/2}-1\right|>\delta_1\right\}\\
&\quad \text{(where } \delta=\epsilon-C\epsilon^2 \text{ and } \delta_1=(\epsilon-\delta)/\epsilon>0)\\
&\le (-\frac{2\theta}{T})\frac{E|I_{n,T}-I_T|^2}{\delta^2}+P\left\{\left|(-\frac{2\theta}{T})I_{n,T}-1\right|>\delta_1\right\} \qquad (4.6)\\
&\le C\frac{T^3}{n^2\delta^2}+C\exp(\frac{T\theta}{16}\delta_1^2)+C\frac{T^2/n^2}{\delta_1^2}. \qquad (4.7)
\end{aligned}$$

Here the bound for the first term in the r.h.s. of (4.6) comes from Lemma 3.2(c) and that for the second term is from J_2 in (3.8).

Now using the bounds (4.5) and (4.7) in (4.4) with $\epsilon = CT^{-1/2}(\log T)^{1/2}$, we obtain that the terms in (4.4) are of the order $O(T^{-1/2}(\log T)^{1/2}\bigvee \frac{T^4}{n^2}(\log T)^{-1})$.

(c) Let $G_T = \{|\theta_{n,T,1}-\theta|\le d_T\}$ and $d_T = CT^{-1/2}(\log T)^{1/2}$. On the set G_T, expanding $(2|\theta_{n,T,1}|)^{1/2}$, we obtain

$$\begin{aligned}
(-2\theta_{n,T,1})^{-1/2} &= (-2\theta)^{1/2}\left[1-\frac{\theta-\theta_{n,T,1}}{\theta}\right]^{-1/2}\\
&= (-2\theta)^{1/2}[1+\frac{1}{2}(\frac{\theta-\theta_{n,T,1}}{\theta})+O(d_T^2).
\end{aligned}$$

Then

$$\begin{aligned}
&\sup_{x\in\mathbb{R}}\left|P\left\{(\frac{T}{2|\theta_{n,T,1}|})^{1/2}(\theta_{n,T,1}-\theta)\le x\right\}-\Phi(x)\right|\\
\le\ &\sup_{x\in\mathbb{R}}\left\{P(\frac{T}{2|\theta_{n,T,1}|})^{1/2}(\theta_{n,T,1}-\theta)\le x,\ G_T\right\}+P(G_T^c).
\end{aligned}$$

Now

$$\begin{aligned}
&P(G_T^c)\\
=\ &P\left\{|\theta_{n,T,1}-\theta|>CT^{-1/2}(\log T)^{1/2}\right\}\\
=\ &P\left\{\left(-\frac{T}{2\theta}\right)^{1/2}|\theta_{n,T,1}-\theta|>C(\log T)^{1/2}(-2\theta)^{-1/2}\right\}\\
\le\ &C\left(T^{-1/2}(\log T)^{1/2}\bigvee\frac{T^4}{n^2}(\log T)^{-1})+2(1-\Phi\log T^{1/2}(-2\theta)^{-1/2}\right)\\
&\text{(by Theorem 4.1(a))}\\
\le\ &C(T^{-1/2}(\log T)^{1/2}\bigvee\frac{T^4}{n^2}(\log T)^{-1}).
\end{aligned}$$

On the set G_T,

$$\left|\left(\frac{\theta_{n,T,1}}{\theta}\right)^{1/2}-1\right| \leq CT^{-1/2}(\log T)^{1/2}.$$

Hence upon choosing $\epsilon = CT^{-1/2}(\log T)^{1/2}, C$ large,

$$\begin{aligned}
&\left|P\left\{\left(\frac{T}{-2\theta_{n,T,1}}\right)^{1/2}(\theta_{n,T,1}-\theta) \leq x, G_T\right\} - \Phi(x)\right| \\
\leq\ & \left|P\left\{\left(\frac{T}{-2\theta}\right)^{1/2}(\theta_{n,T,1}-\theta) \leq x, G_T\right\} - \Phi(x)\right| \\
&+P\left\{\left|\left(\frac{\theta_{n,T,1}}{\theta}\right)^{1/2}-1\right| > \epsilon, G_T\right\} + \epsilon \\
&\text{(by Lemma 3.3.1(b))} \\
\leq\ & C(T^{-1/2}(\log T)^{1/2} \bigvee \tfrac{T^4}{n^2}(\log T)^{-1}) \quad \text{(by Theorem 4.1(a)).} \qquad \square
\end{aligned}$$

Theorem 4.2

$$\sup_{x\in\mathbb{R}}|P\left\{I_{n,T}(-\frac{2\theta}{T})^{1/2}(\theta_{n,T,1}-\theta) \leq x\right\} - \Phi(x)| = O(T^{-1/2} \bigvee (\frac{T^3}{n^2})^{1/3}).$$

Proof From (4.1), we have

$$I_{n,T}\left(-\frac{2\theta}{T}\right)^{1/2}(\theta_{n,T,1}-\theta) = (-\frac{2\theta}{T})^{1/2}Y_T + \theta(-\frac{2\theta}{T})^{1/2}(I_T - I_{n,T}).$$

Hence by Corollary 1.2.3(a) and Lemma 3.2(c),

$$\begin{aligned}
&\sup_{x\in\mathbb{R}}|P\left\{I_{n,T}(-\frac{2\theta}{T})^{1/2}(\theta_{n,T,1}-\theta) \leq x\right\} - \Phi(x)| \\
=\ & \sup_{x\in\mathbb{R}}|P\left\{(-\frac{2\theta}{T})^{1/2}Y_T + \theta(-\frac{2\theta}{T})^{1/2}(I_T - I_{n,T}) \leq x\right\} - \Phi(x)| \\
\leq\ & \sup_{x\in\mathbb{R}}|P\left\{(-\frac{2\theta}{T})^{1/2}Y_T \leq x\right\} - \Phi(x)| + \\
& P\left\{|\theta(-\frac{2\theta}{T})^{1/2}(I_T - I_{n,T})| > \epsilon\right\} + \epsilon \\
\leq\ & CT^{-1/2} + C\frac{E|I_T - I_{n,T}|^2}{T\epsilon^2} + \epsilon \leq CT^{-1/2} + C\frac{T^3}{n^2\epsilon^2} + \epsilon.
\end{aligned}$$

Choosing $\epsilon = (\frac{T^3}{n^2})^{1/3}$, the theorem follows. $\square$

Theorem 4.3 $|\theta_{n,T,1} - \theta_T| = O_P(\frac{T^2}{n})$.
Proof We have from (2.3) $\theta_T = Z_T/I_T$. By the Itô formula, it is easy to see that

$$\theta_{n,T,1} = Z_T/I_{n,T}.$$

Hence applying Lemma 3.7 with the aid of Lemma 3.2(c), the theorem follows.

□

Remarks

(1) The bounds in Theorems 3.3, 3.4, 4.1, and 4.2 are uniform over compact subsets of the parameter space.

(2) Theorems 3.3 and 3.4 are useful for testing hypothesis about θ. They do not necessarily give confidence intervals. Theorem 3.3(b) and (c) are useful for computation of a confidence interval.

(3) It may appear from Theorem 3.3(a) that $\frac{T}{n^{1/2}} \to 0$ is essential for the asymptotic normality of $\theta_{n,T}$. However it is not the case. Consider the bound given in (3.11). For this to converge to 0, it is necessary that $\frac{T}{n} \to 0$. Let $\gamma_n = \frac{T}{n}$. It is easy to see from this bound that if there exists an ϵ_n such that $n\gamma_n\epsilon_n^2 \to \infty$ and $\gamma^{-1}\epsilon_n^2 \to \infty$, then normal convergence holds and a rate may also be found out. Consider, in particular, $T = n^\tau, \tau > 0$. Then, $\gamma_n = n^{\tau-1}$. Hence we require $n^\tau\epsilon_n^2 \to \infty$ and $n^{1-\tau}\epsilon_n^2 \to \infty$. This is possible for any $\tau > 1$. Thus if $T \approx n^\tau, \tau > 1$, then normal convergence in Theorem 3.3 (a) holds. Similar comment also holds for Theorem 3.3(b) and (c) and Theorem 4.1.

(4) If $\frac{T}{n^{1/2}} \to 0$ and $T \to \infty$, then to obtain bound of the order $T^{-1/2}(\log T)^{1/2}$ in Theorem 3.3, one needs $n > T^{5/2}(\log T)^{-3/2}$. To obtain bound of the order $T^{-1/2}(\log T)^{1/2}$ in Theorem 4.1, one needs $n > T^{9/4}(\log T)^{-3/4}$. To obtain bound of the order $T^{-1/2}$ in Theorem 3.4, one needs $n \geq T^{10}$. To obtain bound of the order $T^{-1/2}$ in Theorem 4.2, one needs $n > T^{9/4}$. To obtain bound of the order $\frac{1}{T}$ in Theorem 3.5, one needs $n \geq T$.

(5) The norming in Theorems 3.4 and 4.2 is random which also contains the unknown parameter. It would be interesting to find the order $(T^{-1/2} \bigvee (T/n)^{1/3})$ and $(T^{-1/2} \bigvee (\frac{T^3}{n^2})^{1/3})$, respectively, for the estimators $\theta_{n,T}$ and $\theta_{n,T,1}$ with completely random or completely nonrandom norming. The problem of obtaining rates of convergence of the CLSE in the case of non-linear drift remains open.

(6) The Berry–Esseen bounds for the estimator $\theta_{n,T,1}$ in Theorem 4.1 are sharper than the corresponding bound for $\theta_{n,T}$ in Theorem 3.3. The Berry–Esseen bounds for the estimator $\theta_{n,T,1}$ in Theorem 4.2 are sharper than the corresponding bound for $\theta_{n,T}$ in Theorem 3.4. Also $\theta_{n,T,1}$ converges to θ_T at a faster rate than the rate at which $\theta_{n,T}$ converges to θ_T for fixed T when $n \to \infty$. This shows that $\theta_{n,T,1}$ is a better estimator of θ than $\theta_{n,T}$. The extension of this problem to nonlinear drift case now remains open. The Berry–Esseen bounds in Theorems 3.4 and 4.2 show that *rapidly increasing experimental design* condition ($T \to \infty$ and $\frac{T}{\sqrt{n}} \to 0$) is not essential for the asymptotic normality of the estimators.

(7) In this chapter, we have only studied the rates of convergence of AMLEs in the stable case, i.e., when $\theta < 0$. It would be interesting to investigate the rates of convergence of AMLEs in the explosive case, i.e., when $\theta > 0$ and

the unstable (critical) case, i.e., when $\theta = 0$. Note that in the explosive case with nonrandom norming, the limit distribution is Cauchy, but with random norming the limit distribution is normal.

(8) Here we have obtained uniform rates of convergence to normality. It would be interesting to obtain nonuniform rates of convergence to normality which are more useful.

(9) In a Bayesian framework based on the two given approximate likelihoods, it would be interesting to investigate the rate of convergence of approximate posterior distributions and approximate Bayes estimators.

4.4.3 Berry–Esseen Bounds for AMCEs

As an alternative to maximum likelihood method and to obtain estimators with higher-order accuracy our aim is to use contrast functions. Using Itô formula, the score function $\gamma_T(\theta)$ can be written as

$$\gamma_T(\theta) = \frac{1}{2}X_T^2 - \int_0^T (\theta X_t^2 + \frac{1}{2})dt. \tag{1.7}$$

Consider the estimating function

$$M_T(\theta) = -\int_0^T (\theta X_t^2 + \frac{1}{2})dt \tag{1.8}$$

and the minimum contrast estimate (MCE)

$$\tilde{\theta}_T := -\frac{T}{2}\left\{\int_0^T X_t^2 dt\right\}^{-1}. \tag{1.9}$$

M-estimator is reduced to the minimum contrast estimator. It is well known that $\tilde{\theta}_T$ is strongly consistent and asymptotically $\mathcal{N}(0, \ -2\theta)$ distributed as $T \to \infty$ (see Lanksa 1979). The large deviations of $\hat{\theta}_T$ and $\tilde{\theta}_T$ were obtained in Florens-Landais and Pham (1999). In particular, it was shown that the large deviation probabilities of $\hat{\theta}_T$ are identical to those of $\tilde{\theta}_T$ for $\theta \leq \theta_0/3$ but weaker for $\theta > \theta_0/3$, where θ_0 is the true value of the parameter. However, as far as the rate of convergence to normality is concerned, Bishwal (2003) showed that $\hat{\theta}_T$ and $\tilde{\theta}_T$ have the same Berry–Esseen bound of the order $O(T^{-1/2})$.

In this chapter, we obtain weak convergence bounds for several approximate minimum contrast estimators (AMCEs) which are simpler and robust. In order to define the approximate minimum contrast estimators (AMCEs), we use various discrete approximations of the integral in the definition (1.9) of MCE.

Define a weighted sum of squares

$$M_{n,T} := \frac{T}{n}\left\{\sum_{i=1}^{n} w_{t_i} X_{t_{i-1}}^2 I_{\{|\Delta X_{t_i}|\le h^\rho\}} + \sum_{i=2}^{n+1} w_{t_i} X_{t_{i-1}}^2 I_{\{|\Delta X_{t_i}|\le h^\rho\}}\right\}, \quad (1.10)$$

where $w_{t_i} \ge 0$ is a weight function.

Denote the discrete increasing functions

$$I_{n,T} := \frac{T}{n}\sum_{i=1}^{n} X_{t_{i-1}}^2 I_{\{|\Delta X_{t_i}|\le h^\rho\}}, \quad (1.11)$$

$$J_{n,T} := \frac{T}{n}\sum_{i=2}^{n+1} X_{t_{i-1}}^2 I_{\{|\Delta X_{t_i}|\le h^\rho\}} = \frac{T}{n}\sum_{i=1}^{n} X_{t_i}^2. \quad (1.12)$$

General weighted AMCE is defined as

$$\tilde{\theta}_{n,T} := -\left\{\frac{2}{n} M_{n,T}\right\}^{-1}. \quad (1.13)$$

With $w_{t_i} = 1$, we obtain the forward AMCE as

$$\tilde{\theta}_{n,T,F} := -\left\{\frac{2}{n} I_{n,T}\right\}^{-1}. \quad (1.14)$$

With $w_{t_i} = 0$, we obtain the backward AMCE as

$$\tilde{\theta}_{n,T,B} := -\left\{\frac{2}{n} J_{n,T}\right\}^{-1}. \quad (1.15)$$

Analogous to the estimators for the discrete AR (1) model, we define the simple symmetric and weighted symmetric estimators (see Fuller 1996).

With $w_{t_i} = 0.5$, the simple symmetric AMCE is defined as

$$\tilde{\theta}_{n,T,z} := -\left\{\frac{1}{n}[I_{n,T} + J_{n,T}]\right\}^{-1}$$

$$= -\left\{\frac{2}{n}\sum_{i=2}^{n} X_{t_{i-1}}^2 I_{\{|\Delta X_{t_i}|\le h^\rho\}} + 0.5(X_{t_0}^2 + X_{t_n}^2)\right\}^{-1}. \quad (1.16)$$

With the weight function

$$w_{t_i} = \begin{cases} 0 & : \quad i = 1 \\ \frac{i-1}{n} & : \quad i = 2, 3, \cdots, n \\ 1 & : \quad i = n+1, \end{cases}$$

the weighted symmetric AMCE is defined as

$$\tilde{\theta}_{n,T,w} := -\left\{\frac{2}{n}\sum_{i=2}^{n} X_{t_{i-1}}^2 I_{\{|\Delta X_{t_i}|\le h^\rho\}} + \frac{1}{n}\sum_{i=1}^{n} X_{t_{i-1}}^2 I_{\{|\Delta X_{t_i}|\le h^\rho\}}\right\}^{-1}. \quad (1.17)$$

Note that estimator (1.13) is analogous to the trapezoidal rule in numerical analysis. One can instead use the midpoint rule to define another estimator

$$\tilde{\theta}_{n,T,A} := -\left\{\frac{2}{n}\sum_{i=1}^{n}\left(\frac{X_{t_{i-1}} I_{\{|\Delta X_{t_i}|\le h^\rho\}} + X_{t_i} I_{\{|\Delta X_{t_i}|\le h^\rho\}}}{2}\right)^2\right\}^{-1}. \quad (1.18)$$

One can use Simpson's rule to define another estimator where the denominator is a convex combination of the denominators in (1.11) and (1.12)

$$\begin{aligned}\tilde{\theta}_{n,T,S} := -\Bigg\{\frac{1}{3n}\sum_{i=1}^{n}\Bigg\{X_{t_{i-1}}^2 + 4\left(\frac{X_{t_{i-1}} I_{\{|\Delta X_{t_i}|\le h^\rho\}} + X_{t_i} I_{\{|\Delta X_{t_i}|\le h^\rho\}}}{2}\right)^2 \\ + X_{t_i}^2 I_{\{|\Delta X_{t_i}|\le h^\rho\}}\Bigg\}\Bigg\}^{-1}.\end{aligned} \quad (1.19)$$

In general, one can generalize Simpson's rule as

$$\begin{aligned}\tilde{\theta}_{n,T,GS} := -\Bigg\{\frac{2}{n}\sum_{i=1}^{n}\Bigg\{\alpha\frac{X_{t_{i-1}}^2 I_{\{|\Delta X_{t_i}|\le h^\rho\}} + X_{t_i}^2 I_{\{|\Delta X_{t_i}|\le h^\rho\}}}{2} \\ +(1-\alpha)\left(\frac{X_{t_{i-1}} I_{\{|\Delta X_{t_i}|\le h^\rho\}} + X_{t_i} I_{\{|\Delta X_{t_i}|\le h^\rho\}}}{2}\right)^2\Bigg\}\Bigg\}^{-1}\end{aligned} \quad (1.20)$$

for any $0 \le \alpha \le 1$.

The case $\alpha = 0$ produces the estimator (1.18). The case $\alpha = 1$ produces the estimator (1.17). The case $\alpha = \frac{1}{3}$ produces the estimator (1.19).

We propose a very general form of the quadrature based estimator as

$$\tilde{\theta}_{n,T,w} := -\left\{\frac{2}{n}\sum_{i=1}^{n}\sum_{j=1}^{m}\left[(1-s_j)X_{t_{i-1}} I_{\{|\Delta X_{t_i}|\le h^\rho\}} + s_j X_{t_i} I_{\{|\Delta X_{t_i}|\le h^\rho\}}\right]^2 p_j\right\}^{-1}, \quad (1.21)$$

where $p_j,\ \ j \in \{1,2,\cdots,m\}$, is a probability mass function of a discrete random variable S on $0 \le s_1 < s_2 < \cdots < s_m \le 1$ with $P(S = s_j) := p_j,\ j \in \{1,2,\cdots,m\}$.

Denote the k-th moment of the random variable S as $\mu_k := \sum_{j=1}^{m} s_j^k p_j,\ k = 1,2,\cdots$.

If one chooses the probability distribution as uniform distribution for which the moments are a harmonic sequence $(\mu_1,\mu_2,\mu_3,\mu_4,\mu_5,\mu_6,\cdots) = (\frac{1}{2},\frac{1}{3},\frac{1}{4},\frac{1}{5},\frac{1}{7},\cdots)$, then there is no change in the rate of convergence than second order. If one can construct a probability distribution for which the harmonic sequence is truncated at a point, then there is a rate of convergence improvement at the point of truncation.

Given a positive integer m, construct a probability mass function p_j, $j \in \{1, 2, \cdots, m\}$, on $0 \leq s_1 < s_2 < \cdots < s_m \leq 1$ such that

$$\sum_{j=1}^{m} s_j^r p_j = \frac{1}{r+1}, \quad r \in \{0, \cdots, m-2\} \tag{1.22}$$

$$\sum_{j=1}^{m} s_j^{m-1} p_j \neq \frac{1}{m}. \tag{1.23}$$

Neither the probabilities p_j nor the atoms, s_j, of the distribution are specified in advance.

This problem is related to the truncated Hausdorff moment problem. I obtain examples of such probability distributions and use them to get higher-order accurate (up to sixth order) AMCEs.

The order of approximation error (rate of convergence) of an estimator is $n^{-\nu}$ where

$$\nu := \inf \left\{ k : \mu_k \neq \frac{1}{1+k}, \ \mu_j = \frac{1}{1+j}, j = 1, 2, \cdots, k-1 \right\}. \tag{1.24}$$

We construct probability distributions satisfying these moment conditions and obtain estimators of the rate of convergence up to order 6.

Probability $p_1 = 1$ at the point $s_1 = 0$ gives the estimator (1.11) for which $\mu_1 = 0$. Note that $\mu_1 \neq \frac{1}{2}$. Thus $\nu = 1$.

Probability $p_1 = 1$ at the point $s_1 = 1$ gives the estimator (1.12) for which $\mu_1 = 1$. Note that $\mu_1 \neq \frac{1}{2}$. Thus $\nu = 1$.

Probabilities $(p_1, p_2) = (\frac{1}{2}, \frac{1}{2})$ at the respective points $(s_1, s_2) = (0, 1)$ produce the estimator $\tilde{\theta}_{n,T,Z}$ for which $(\mu_1, \mu_2) = (\frac{1}{2}, \frac{1}{4})$. Thus $\nu = 2$.

Probability $p_j = 1$ at the point $s_j = \frac{1}{2}$ produces the estimator $\tilde{\theta}_{n,T,A}$ for which $(\mu_1, \mu_2) = (\frac{1}{2}, \frac{1}{2})$. Thus $\nu = 2$.

Probabilities $(p_1, p_2) = (\frac{1}{4}, \frac{3}{4})$ at the respective points $(s_1, s_2) = (0, \frac{2}{3})$ produce the asymmetric estimator

$$\tilde{\theta}_{n,T,3} := -\left\{ \tfrac{2}{n}\tfrac{1}{4} \textstyle\sum_{i=1}^{n} \left[(X_{t_{i-1}})^2 + 3(\frac{X_{t_{i-1}} I_{\{|\Delta X_{t_i}| \leq h^\rho\}} + 2X_{t_i} I_{\{|\Delta X_{t_i}| \leq h^\rho\}}}{3})^2 \right] \right\}^{-1} \tag{1.25}$$

for which $(\mu_1, \mu_2, \mu_3) = (\frac{1}{2}, \frac{1}{3}, \frac{2}{9})$. Thus $\nu = 3$.

Probabilitiesc $(p_1, p_2) = \frac{3}{4}, \frac{1}{4}$ at the respective points $(s_1, s_2) = (\frac{1}{3}, 1)$ produce asymmetric estimator

$$\tilde{\theta}_{n,T,4} := -\left\{ \tfrac{2}{n}\tfrac{1}{4} \textstyle\sum_{i=1}^{n} \left[3(\frac{2X_{t_{i-1}} I_{\{|\Delta X_{t_i}| \leq h^\rho\}} + X_{t_i} I_{\{|\Delta X_{t_i}| \leq h^\rho\}}}{3})^2 + (X_{t_i})^2 \right] \right\}^{-1} \tag{1.26}$$

for which $(\mu_1, \mu_2, \mu_3) = (\frac{1}{2}, \frac{1}{3}, \frac{10}{36})$. Thus $\nu = 3$.

Probabilities $(p_1, p_2, p_3) = (\frac{1}{6}, \frac{2}{3}, \frac{1}{6})$ at the respective points $(s_1, s_2, s_3) = (0, \frac{1}{2}, 1)$ produce the estimator $\tilde{\theta}_{n,T,5}$ for which $(\mu_1, \mu_2, \mu_3, \mu_4) = (\frac{1}{2}, \frac{1}{3}, \frac{1}{4}, \frac{5}{25})$. Thus $\nu = 4$.

Probabilities $(p_1, p_2, p_3, p_4) = (\frac{1}{8}, \frac{3}{8}, \frac{3}{8}, \frac{1}{8})$ at the respective points $(s_1, s_2, s_3, s_4) = (0, \frac{1}{3}, \frac{2}{3}, 1)$ produce the symmetric estimator

$$\begin{aligned}\tilde{\theta}_{n,T,5} := -\Bigg\{\frac{2}{n}\frac{1}{8}\sum_{i=1}^{n}\Big[(X_{t_{i-1}}I_{\{|\Delta X_{t_i}|\le h^\rho\}})^2 \\ +3(\frac{2X_{t_{i-1}}I_{\{|\Delta X_{t_i}|\le h^\rho\}} + X_{t_i}I_{\{|\Delta X_{t_i}|\le h^\rho\}}}{3})^2 \\ +3(\frac{X_{t_{i-1}}I_{\{|\Delta X_{t_i}|\le h^\rho\}} + 2X_{t_i}I_{\{|\Delta X_{t_i}|\le h^\rho\}}}{3})^2 \\ +(X_{t_i}I_{\{|\Delta X_{t_i}|\le h^\rho\}})^2\Big]\Bigg\}^{-1} \qquad (1.27)\end{aligned}$$

for which $(\mu_1, \mu_2, \mu_3, \mu_4) = (\frac{1}{2}, \frac{1}{3}, \frac{1}{4}, \frac{11}{54})$. Thus $\nu = 4$.
Probabilities $(p_1, p_2, p_3, p_4, p_5) = (\frac{1471}{24192}, \frac{6925}{24192}, \frac{1475}{12096}, \frac{2725}{12096}, \frac{5675}{24192}, \frac{1721}{24192})$ at the respective points $(s_1, s_2, s_3, s_4, s_5) = (0, \frac{1}{5}, \frac{2}{5}, \frac{3}{5}, \frac{4}{5}, 1)$ produce the asymmetric estimator

$$\begin{aligned}\tilde{\theta}_{n,T,7} := -\Bigg\{\frac{2}{n}\frac{1}{24192}\sum_{i=1}^{n}\Big[1471(X_{t_{i-1}}I_{\{|\Delta X_{t_i}|\le h^\rho\}})^2 \\ +6925(\frac{X_{t_{i-1}}I_{\{|\Delta X_{t_i}|\le h^\rho\}} + X_{t_i}I_{\{|\Delta X_{t_i}|\le h^\rho\}}}{5})^2 \\ +2950(\frac{2X_{t_{i-1}}I_{\{|\Delta X_{t_i}|\le h^\rho\}} + 2X_{t_i}I_{\{|\Delta X_{t_i}|\le h^\rho\}}}{5})^2 \\ +5450(\frac{3X_{t_{i-1}}I_{\{|\Delta X_{t_i}|\le h^\rho\}} + 3X_{t_i}I_{\{|\Delta X_{t_i}|\le h^\rho\}}}{5})^2 \\ +5675(\frac{4X_{t_{i-1}}I_{\{|\Delta X_{t_i}|\le h^\rho\}} + 4X_{t_i}I_{\{|\Delta X_{t_i}|\le h^\rho\}}}{5})^2 \\ +1721(X_{t_i}I_{\{|\Delta X_{t_i}|\le h^\rho\}})^2\Big]\Bigg\}^{-1} \qquad (1.28)\end{aligned}$$

for which $(\mu_1, \mu_2, \mu_3, \mu_4, \mu_5, \mu_6) = (\frac{1}{2}, \frac{1}{3}, \frac{1}{4}, \frac{1}{5}, \frac{841}{5040})$. Thus $\nu = 5$.

Probabilities $(p_1, p_2, p_3, p_4, p_5) = (\frac{7}{90}, \frac{16}{45}, \frac{2}{15}, \frac{16}{45}, \frac{7}{90})$ at the respective points $(s_1, s_2, s_3, s_4, s_5) = (0, \frac{1}{4}, \frac{1}{2}, \frac{3}{4}, 1)$ produce the symmetric estimator $\tilde{\theta}_{n,T,8}$ given by

$$\tilde{\theta}_{n,T,8} := -\Bigg\{\frac{2}{n}\frac{1}{90}\sum_{i=1}^{n}\Big[7(X_{t_{i-1}}I_{\{|\Delta X_{t_i}|\le h^\rho\}})^2$$

$$+32\Big(\frac{3X_{t_{i-1}}I_{\{|\Delta X_{t_i}|\le h^\rho\}}+X_{t_i}I_{\{|\Delta X_{t_i}|\le h^\rho\}}}{4}\Big)^2$$
$$+12\Big(\frac{X_{t_{i-1}}I_{\{|\Delta X_{t_i}|\le h^\rho\}}+X_{t_i}I_{\{|\Delta X_{t_i}|\le h^\rho\}}}{2}\Big)^2$$
$$+32\Big(\frac{X_{t_{i-1}}I_{\{|\Delta X_{t_i}|\le h^\rho\}}+3X_{t_i}I_{\{|\Delta X_{t_i}|\le h^\rho\}}}{4}\Big)^2$$
$$+7(X_{t_i}I_{\{|\Delta X_{t_i}|\le h^\rho\}})^2\Big]\Big\}^{-1} \tag{1.29}$$

for which $(\mu_1,\mu_2,\mu_3,\mu_4,\mu_5,\mu_6)=(\frac{1}{2},\frac{1}{3},\frac{1}{4},\frac{1}{5},\frac{1}{6},\frac{110}{768})$. Thus $\nu=6$.

Probabilities $(p_1,p_2,p_3,p_4,p_5)=(\frac{19}{288},\frac{75}{288},\frac{50}{288},\frac{50}{288},\frac{75}{288},\frac{19}{288})$ at the respective points
$(s_1,s_2,s_3,s_4,s_5)=(0,\frac{1}{5},\frac{2}{5},\frac{3}{5},\frac{4}{5},1)$ produce symmetric estimator

$$\tilde\theta_{n,T,9}:=-\Big\{\frac{2}{n}\frac{1}{288}\sum_{i=1}^{n}\big[19(X_{t_{i-1}})^2$$
$$+75\Big(\frac{4X_{t_{i-1}}I_{\{|\Delta X_{t_i}|\le h^\rho\}}+X_{t_i}I_{\{|\Delta X_{t_i}|\le h^\rho\}}}{5}\Big)^2$$
$$+50\Big(\frac{3X_{t_{i-1}}I_{\{|\Delta X_{t_i}|\le h^\rho\}}+2X_{t_i}I_{\{|\Delta X_{t_i}|\le h^\rho\}}}{5}\Big)^2$$
$$+50\Big(\frac{2X_{t_{i-1}}I_{\{|\Delta X_{t_i}|\le h^\rho\}}+3X_{t_i}I_{\{|\Delta X_{t_i}|\le h^\rho\}}}{5}\Big)^2$$
$$+75\Big(\frac{X_{t_{i-1}}I_{\{|\Delta X_{t_i}|\le h^\rho\}}+4X_{t_i}I_{\{|\Delta X_{t_i}|\le h^\rho\}}}{5}\Big)^2$$
$$+19(X_{t_i}I_{\{|\Delta X_{t_i}|\le h^\rho\}})^2\Big]\Big\} \tag{1.30}$$

for which $(\mu_1,\mu_2,\mu_3,\mu_4,\mu_5,\mu_6)=(\frac{1}{2},\frac{1}{3},\frac{1}{4},\frac{1}{5},\frac{1}{6},\frac{3219}{22500})$. Thus $\nu=6$.

The estimator $\tilde\theta_{n,T,S}$ is based on the arithmetic mean of $I_{n,T}$ and $J_{n,T}$. One can use geometric mean and harmonic mean instead. The geometric mean based symmetric AMCE (which is based on the ideas of partial autocorrelation) is defined as

$$\tilde\theta_{n,T,G}:=\frac{-T/2}{\sqrt{I_{n,T}\,J_{n,T}}}. \tag{1.31}$$

The harmonic mean based symmetric AMCE is defined as

$$\tilde\theta_{n,T,H}:=\frac{-T}{\frac{1}{I_{n,T}}+\frac{1}{J_{n,T}}}. \tag{1.32}$$

Note that

$$\tilde\theta_{n,T,H}\le\tilde\theta_{n,T,G}\le\tilde\theta_{n,T,S}. \tag{1.33}$$

It is a desirable property of an estimator to lie inside the parameter space. The estimators $\tilde\theta_{n,T,H}$ and $\tilde\theta_{n,T,G}$ may lie inside the parameter space even if the estimator $\tilde\theta_{n,T,F}$ or $\tilde\theta_{n,T,B}$ may not.

Note that the symmetric estimators use both the end points of observations. These estimators are sufficient statistics. So one can take advantage of the Rao–Blackwell theorem. If one excludes the end points of observations (as in the forward and the backward estimators), then the estimators will not be sufficient statistics.

From a time series viewpoint, a discretely observed stationary Ornstein–Uhlenbeck model is a nonlinear (in the parameter) autoregressive gamma (ARG) model given by

$$X_{t_i} = \exp(\theta(t_i - t_{i-1})) + \epsilon_i, \quad i \geq 1, \tag{1.34}$$

where

$$\epsilon_i := \int_{t_{i-1}}^{t_i} \exp\{\theta(t_i - u)\} d(W_u + Z_u), \tag{1.36}$$

which are independent $\mathcal{N}\left(0, \frac{e^{2\theta(t_i - t_{i-1})} - 1}{2\theta}\right)$ random variables. The errors are heteroscedastic if the data points are unequally spaced, but as we assume equally spaced data points, the errors are homoscedastic and i.i.d. $\mathcal{N}(0, V_{n,T}(\theta))$ where

$$V_{n,T}(\theta) := \frac{e^{2\theta\frac{T}{n}} - 1}{2\theta}. \tag{1.37}$$

The log-likelihood is

$$l_{n,T}(\theta) := -\frac{n}{2}\log 2\pi - \frac{n}{2} V_{n,T}(\theta) - \frac{1}{2V_{n,T}(\theta)} \sum_{i=1}^{n} \left(X_{t_i} - e^{\frac{T}{n}\theta} X_{t_{i-1}}\right)^2. \tag{1.38}$$

The exact maximum likelihood estimator based on equally spaced discrete observations is explicit and is given by

$$\hat{\theta}_{n,T} := \frac{n}{T} \log \frac{\sum_{i=1}^{n} X_{t_{i-1}}^2 I_{\{|\Delta X_{t_i}| \leq h^\rho\}}}{\sum_{i=1}^{n} X_{t_i} X_{t_{i-1}}}. \tag{1.39}$$

Note that

$$\breve{\theta}_{n,T} = 1 - \frac{n}{T} \exp\{\frac{T}{n} \hat{\theta}_{n,T}\}.$$

As far as the rate of convergence of this estimator to the continuous conditional MLE is concerned, it is of first order.

The Euler estimator that is the conditional least squares estimator is given by

$$\breve{\theta}_{n,T} := \frac{n}{T} \frac{\sum_{i=1}^{n} X_{t_{i-1}} I_{\{|\Delta X_{t_i}| \leq h^\rho\}} (X_{t_i} - X_{t_{i-1}}) I_{\{|\Delta X_{t_i}| \leq h^\rho\}}}{\sum_{i=1} X_{t_{i-1}}^2}. \tag{1.40}$$

Another contrast function, which is the least squares quadratic, is given by

$$\kappa_{n,T}(\theta) := \frac{2\theta}{e^{2\theta\frac{T}{n}} - 1} \sum_{i=1}^{n} \left(X_{t_i} I_{\{|\Delta X_{t_i}| \leq h^\rho\}} - e^{\frac{T}{n}\theta} X_{t_{i-1}} I_{\{|\Delta X_{t_i}| \leq h^\rho\}}\right)^2. \tag{1.41}$$

The solution of $\kappa'_{n,T}(\theta) = 0$ provides an AMCE but does not have an explicit form.

The simple symmetric approximate MLE was introduced and studied in Mishra and Bishwal (1995) for a more general diffusion process. Stochastic bound between the difference the AMLE and the continuous MLE was obtained.

In this chapter, we obtain the rate of weak convergence to normal distribution of some AMCEs using different normings. We also obtain stochastic bound on the difference of the AMCEs and their continuous counterpart MCE when T is fixed.

We need the following lemmas in the sequel.

Lemma 2.1 (Bishwal 2000a) For every $\delta > 0$,

$$P\left\{\left|\frac{2\theta}{T}I_T - 1\right| \geq \delta\right\} \leq CT^{-1}\delta^{-2}.$$

Lemma 2.2 (Bishwal 2000a)

$$\sup_{x\in\mathbb{R}}\left|P\left\{\left(-\frac{2\theta}{T}\right)^{1/2}\left(\theta I_T - \frac{T}{2}\right) \leq x\right\} - \Phi(x)\right| \leq CT^{-1/2}.$$

Lemma 2.3

$$(a)\quad E|I_{n,T} - I_T|^2 = O\left(\frac{T^4}{n^2}\right).$$

$$(b)\quad E|\frac{I_{n,T} + J_{n,T}}{2} - I_T|^2 = O\left(\frac{T^4}{n^2}\right).$$

Part (a) is from Bishwal and Bose (2001). The proof of part (b) is analogous to part (a). We omit the details.

The following theorem gives the bound on the error of normal approximation of the AMCEs. Note that part (a) uses parameter dependent nonrandom norming. While this is useful for testing hypotheses about θ, it may not necessarily give a confidence interval. The normings in parts (b) and (c) are sample dependent which can be used for obtaining a confidence interval. The following theorem shows that asymptotic normality of the AMCEs needs $T \to \infty$ and $\frac{T}{\sqrt{n}} \to 0$.

Theorem 2.1 Denote $b_{n,T} := O(\max(T^{-1/2}(\log T)^{1/2}, (\frac{T^4}{n^2})(\log T)^{-1}))$.

$$\text{(a)}\quad \sup_{x\in\mathbb{R}}\left|P\left\{\left(-\frac{T}{2\theta}\right)^{1/2}(\tilde{\theta}_{n,T,F} - \theta) \leq x\right\} - \Phi(x)\right| = O(b_{n,T}).$$

$$\text{(b)}\quad \sup_{x\in\mathbb{R}}\left|P\left\{I_{n,T}^{1/2}(\tilde{\theta}_{n,T,F} - \theta) \leq x\right\} - \Phi(x)\right| = O(b_{n,T}).$$

$$\text{(c)} \quad \sup_{x\in\mathbb{R}} \left| P\left\{ \left(\frac{T}{2|\tilde{\theta}_{n,T,F}|} \right)^{1/2} (\tilde{\theta}_{n,T,F} - \theta) \le x \right\} - \Phi(x) \right| = O(b_{n,T}).$$

Proof
(a) Observe that

$$\left(-\frac{T}{2\theta}\right)^{1/2} (\tilde{\theta}_T - \theta) = \frac{\left(-\frac{2\theta}{T}\right)^{1/2} Y_T}{\left(-\frac{2\theta}{T}\right) I_T}, \tag{2.1}$$

where

$$Y_T := -\theta I_T - \frac{T}{2} \quad \text{and} \quad I_T := \int_0^T X_t^2 dt.$$

Thus, we have

$$I_{n,T}\tilde{\theta}_{n,T,F} = -\frac{T}{2} = Y_T + \theta I_T.$$

Hence,

$$\begin{aligned}
&\left(-\frac{T}{2\theta}\right)^{1/2} (\tilde{\theta}_{n,T,F} - \theta) \\
= \;& \frac{\left(-\frac{T}{2\theta}\right)^{1/2} Y_T + \theta \left(-\frac{T}{2\theta}\right)^{1/2} (I_T - I_{n,T})}{I_{n,T}} \\
= \;& \frac{\left(-\frac{2\theta}{T}\right)^{1/2} Y_T + \left(-\frac{2\theta}{T}\right)^{1/2} (I_T - I_{n,T})}{\left(-\frac{2\theta}{T}\right) I_{n,T}}.
\end{aligned} \tag{2.2}$$

Further,

$$\begin{aligned}
& P\left\{ \left| \left(\frac{-2\theta}{T} \right) (I_{n,T} - 1) \right| > \epsilon \right\} \\
= \;& \left\{ \left| \left(\frac{-2\theta}{T} \right) (I_{n,T} - I_T + I_T) - 1 \right| > \epsilon \right\} \\
\le \;& P\left\{ \left| \left(\frac{-2\theta}{T} \right) I_T - 1 \right| > \frac{\epsilon}{2} \right\} + P\left\{ \left(-\frac{2\theta}{T} \right) |I_{n,T} - I_T| > \frac{\epsilon}{2} \right\} \\
\le \;& C \exp\left(\frac{T\theta}{16} \epsilon^2 \right) + \frac{16\theta^2}{T^2} \frac{E|I_{n,T} - I_T|^2}{\epsilon^2} \\
\le \;& C \exp\left(\frac{T\theta}{16} \epsilon^2 \right) + C \frac{T^2/n^2}{\epsilon^2}.
\end{aligned} \tag{2.3}$$

Next, observe that

$$\sup_{x\in\mathbb{R}} \left| P\left\{ \left(-\frac{T}{2\theta} \right)^{1/2} (\tilde{\theta}_{n,T,F} - \theta) \le x \right\} - \Phi(x) \right|$$

$$
\begin{aligned}
&= \sup_{x\in\mathbb{R}} \left| P\left\{ \frac{\left(-\frac{2\theta}{T}\right)^{1/2} Y_T + \left(-\frac{2\theta}{T}\right)^{1/2}(I_T - I_{n,T})}{\left(-\frac{2\theta}{T}\right) I_{n,T}} \le x \right\} - \Phi(x) \right| \\
&\le \sup_{x\in\mathbb{R}} \left| P\left\{ \left(-\frac{2\theta}{T}\right)^{1/2} Y_T \le x \right\} - \Phi(x) \right| \\
&\quad + P\left\{ \left| \theta \left(-\frac{2\theta}{T}\right)^{1/2} (I_{n,T} - I_T) \right| > \epsilon \right\} \\
&\quad + P\left\{ \left| \left(-\frac{2\theta}{T}\right) I_{n,T} - 1 \right| > \epsilon \right\} + 2\epsilon \\
&\le CT^{-1/2} + \theta^2 \frac{\left(-\frac{2\theta}{T}\right) E|I_{n,T} - I_T|^2}{\epsilon^2} \\
&\quad + C\exp\left(\frac{T\theta}{4}\epsilon^2\right) + C\frac{T^2}{n^2\epsilon^2} + 2\epsilon \qquad (2.4)
\end{aligned}
$$

(the bound for the third term in the right hand side of (2.4) is obtained from (2.4))

$$
\le CT^{-1/2} + C\frac{T^2}{n^2\epsilon^2} + C\exp\left(\frac{T\theta}{4}\epsilon^2\right) + C\frac{T}{n^2\epsilon^2} + \epsilon \qquad (2.5)
$$

(by Lemma 2.3(a)).

Choosing $\epsilon = CT^{-1/2}(\log T)^{1/2}$, the terms in the right hand side of (2.5) are of the order
$O(\max(T^{-1/2}(\log T)^{1/2}, (\frac{T^4}{n^2})(\log T)^{-1}))$. □

(b) From (2.1), we have

$$
I_{n,T}^{1/2}(\tilde{\theta}_{n,T,F} - \theta) = \frac{Y_T + \theta(I_T - I_{n,T})}{I_{n,T}^{1/2}}.
$$

Then,

$$
\begin{aligned}
&\sup_{x\in\mathbb{R}} \left| P\left\{ I_{n,T}^{1/2}(\tilde{\theta}_{n,T,F} - \theta) \le x \right\} - \Phi(x) \right| \\
&= \sup_{x\in\mathbb{R}} \left| P\left\{ \frac{Y_T}{I_{n,T}^{1/2}} + \theta\frac{I_T - I_{n,T}}{I_{n,T}^{1/2}} \le x \right\} - \Phi(x) \right| \\
&\le \sup_{x\in\mathbb{R}} \left| P\left\{ \frac{Y_T}{I_{n,T}^{1/2}} \le x \right\} - \Phi(x) \right| + P\left\{ \left| \frac{\theta(I_T - I_{n,T})}{I_{n,T}^{1/2}} \right| > \epsilon \right\} + \epsilon \\
&=: U_1 + U_2 + \epsilon. \qquad (2.6)
\end{aligned}
$$

We have from (2.3),

$$
U_1 \le CT^{-1/2} + C\exp\left(\frac{T\theta}{16}\epsilon^2\right) + C\frac{T^2}{n^2\epsilon^2} + \epsilon. \qquad (2.7)
$$

Now,

$$\begin{aligned}
U_2 &= P\left\{|\theta|\left|\frac{I_{n,T}-I_T}{I_{n,T}^{1/2}}\right| > \epsilon\right\} = P\left\{|\theta|\frac{\left|\left(-\frac{2\theta}{T}\right)^{1/2}(I_{n,T}-I_T)\right|}{\left|\left(-\frac{2\theta}{T}\right)^{1/2} I_{n,T}^{1/2}\right|} > \epsilon\right\} \\
&\le P\left\{\left|\left(-\frac{2\theta}{T}\right)^{1/2}\right| |I_{n,T}-I_T| > \delta\right\} + P\left\{\left|\left(-\frac{2\theta}{T}\right)^{1/2} I_{n,T}^{1/2} - 1\right| > \delta_1\right\} \\
&\quad \text{(where } \delta = \epsilon - C\epsilon^2 \text{ and } \delta_1 = (\epsilon-\delta)/\epsilon > 0) \qquad (2.8) \\
&\le \left(-\frac{2\theta}{T}\right)\frac{E|I_{n,T}-I_T|^2}{\delta^2} + P\left\{\left|\left(-\frac{2\theta}{T}\right) I_{n,T} - 1\right| > \delta_1\right\} \\
&\le C\frac{T^3}{n^2\delta^2} + C\exp\left(\frac{T\theta}{16}\delta_1^2\right) + C\frac{T^2}{n^2\delta_1^2}. \qquad (2.9)
\end{aligned}$$

Here, the bound for the first term in the right hand side of (2.7) comes from Lemma 2.2(c) and that for the second term is obtained from (2.3).

Now, using the bounds (2.7) and (2.9) in (2.6) with $\epsilon = CT^{-1/2}(\log T)^{1/2}$, we obtain that the terms in (2.6) are of the order $O(\max(T^{-1/2}(\log T)^{1/2}, (\frac{T^4}{n^2})(\log T)^{-1}))$. □

(c) Let

$$G_T := \left\{|\tilde{\theta}_{n,T,F} - \theta| \le d_T\right\}, \qquad \text{and} \qquad d_T := CT^{-1/2}(\log T)^{1/2}.$$

On the set G_T, expanding $(2|\tilde{\theta}_{n,T}|)^{1/2}$, we obtain

$$\begin{aligned}
(-2\tilde{\theta}_{n,T})^{-1/2} &= (-2\theta)^{1/2}\left[1 - \frac{\theta - \tilde{\theta}_{n,T,F}}{\theta}\right]^{-1/2} \\
&= (-2\theta)^{1/2}\left[1 + \frac{1}{2}\left(\frac{\theta - \tilde{\theta}_{n,T,F}}{\theta}\right)\right] + O(d_T^2).
\end{aligned}$$

Then,

$$\begin{aligned}
&\sup_{x\in\mathbb{R}}\left|P\left\{\left(\frac{T}{2|\tilde{\theta}_{n,T,F}|}\right)^{1/2}(\tilde{\theta}_{n,T,F} - \theta) \le x\right\} - \Phi(x)\right| \\
\le\ &\sup_{x\in\mathbb{R}}\left\{P\left(\frac{T}{2|\tilde{\theta}_{n,T}|}\right)^{1/2}(\tilde{\theta}_{n,T,F} - \theta) \le x, G_T\right\} + P(G_T^c).
\end{aligned}$$

Now,

$$P(G_T^c) = P\left\{|\tilde{\theta}_{n,T,F} - \theta| > CT^{-1/2}(\log T)^{1/2}\right\}$$

$$
\begin{aligned}
&= P\left\{\left(-\frac{T}{2\theta}\right)^{1/2}|\tilde{\theta}_{n,T,F}-\theta| > C(\log T)^{1/2}(-2\theta)^{-1/2}\right\}\\
&\le C\max\left(T^{-1/2}(\log T)^{1/2}, \frac{T^3}{n^2}(\log T)^{-1}\right)\\
&\quad +2(1-\Phi\log T^{1/2}(-2\theta)^{-1/2})\\
&\quad \text{(by Theorem 2.1(a))}\\
&\le C\max\left(T^{-1/2}(\log T)^{1/2}, \frac{T^3}{n^2}(\log T)^{-1}\right).
\end{aligned}
$$

On the set G_T,

$$
\left|\left(\frac{\tilde{\theta}_{n,T,F}}{\theta}\right)^{1/2} - 1\right| \le CT^{-1/2}(\log T)^{1/2}.
$$

Hence, upon choosing $\epsilon = CT^{-1/2}(\log T)^{1/2}, C$ large, we obtain

$$
\begin{aligned}
&\left|P\left\{\left(\frac{T}{-2\tilde{\theta}_{n,T,F}}\right)^{1/2}(\tilde{\theta}_{n,T,F}-\theta) \le x, G_T\right\} - \Phi(x)\right|\\
\le\ &\left|P\left\{\left(\frac{T}{-2\theta}\right)^{1/2}(\tilde{\theta}_{n,T,F}-\theta) \le x, G_T\right\}\right|\\
&+P\left\{\left|\left(\frac{\tilde{\theta}_{n,T,F}}{\theta}\right)^{1/2} - 1\right| > \epsilon, G_T\right\} + \epsilon\\
&\text{(by Lemma 3.3.1(b))}\\
\le\ &C\max\left(T^{-1/2}(\log T)^{1/2}, \frac{T^4}{n^2}(\log T)^{-1}\right)\\
&\text{(by Theorem 2.1(a)).}
\end{aligned}
$$

□

In the following theorem, we improve the bound on the error of normal approximation using a mixture of random and nonrandom normings. Thus asymptotic normality of the AMCEs needs $T \to \infty$ and $\frac{T}{n^{2/3}} \to 0$ which are sharper than the bound in Theorem 2.1.

Theorem 2.2

$$
\sup_{x\in\mathbb{R}}\left|P\left\{I_{n,T}\left(-\frac{2\theta}{T}\right)^{1/2}(\tilde{\theta}_{n,T,F}-\theta) \le x\right\} - \Phi(x)\right|
$$
$$
= O\left(\max\left(T^{-1/2}, \left(\frac{T^3}{n^2}\right)^{1/3}\right)\right).
$$

Proof From (2.2), we have

$$I_{n,T}\left(-\frac{2\theta}{T}\right)^{1/2}(\tilde{\theta}_{n,T,F}-\theta)=\left(-\frac{2\theta}{T}\right)^{1/2}Y_T+\theta\left(-\frac{2\theta}{T}\right)^{1/2}(I_T-I_{n,T}).$$

Hence, by Lemmas 2.1–2.3,

$$\begin{aligned}
&\sup_{x\in\mathbb{R}}\left|P\left\{I_{n,T}\left(-\frac{2\theta}{T}\right)^{1/2}(\tilde{\theta}_{n,T,F}-\theta)\le x\right\}-\Phi(x)\right|\\
=&\sup_{x\in\mathbb{R}}\left|P\left\{\left(-\frac{2\theta}{T}\right)^{1/2}Y_T+\theta\left(-\frac{2\theta}{T}\right)^{1/2}(I_T-I_{n,T})\le x\right\}-\Phi(x)\right|\\
\le&\sup_{x\in R}\left|P\left\{\left(-\frac{2\theta}{T}\right)^{1/2}Y_T\le x\right\}-\Phi(x)\right|\\
&+P\left\{\left|\theta\left(-\frac{2\theta}{T}\right)^{1/2}(I_T-I_{n,T})\right|>\epsilon\right\}+\epsilon\\
\le&\ CT^{-1/2}+C\frac{E|I_T-I_{n,T}|^2}{T\epsilon^2}+\epsilon\le CT^{-1/2}+C\frac{T^3}{n^2\epsilon^2}+\epsilon.
\end{aligned}$$

Choosing $\epsilon=(\frac{T^3}{n^2})^{1/3}$, the theorem follows. □

The following theorem gives stochastic bound on the error of approximation of the continuous MCE by AMCEs.

Theorem 2.3

$$(a)\ \ |\tilde{\theta}_{n,T}-\tilde{\theta}_T|=O_P\left(\frac{T}{n}\right)^{1/2}.$$

$$(b)\ \ |\tilde{\theta}_{n,T,z}-\tilde{\theta}_T|=O_P\left(\frac{T^2}{n^2}\right)^{1/2}.$$

Proof From (1.9) and (1.14), we have

$$\tilde{\theta}_T=-\frac{T}{2I_T},\quad \tilde{\theta}_{n,T}=-\frac{T}{2I_{n,T}}.$$

Hence, applying Lemma 3.3.1(c) with the aid of Lemma 2.3(a) and noting that $|\frac{I_{n,T}}{T}|=O_P(1)$ and $|\frac{I_T}{T}|=O_P(1)$, the part (a) of theorem follows.

From (1.9) and (1.16), we have

$$\tilde{\theta}_T=-\frac{T}{2I_T},\quad \tilde{\theta}_{n,T,z}=-\frac{T}{I_{n,t}+J_{n,T}}.$$

Applying Lemma 3.3.1(c) with the aid of Lemma 2.3(b) and noting that $|\frac{J_{n,T}}{T}| = O_P(1)$ and $|\frac{I_T}{T}| = O_P(1)$, the part (b) of theorem follows. □

Remarks

(1) The bounds in Theorems 2.1 and 2.2 are uniform over compact subsets of the parameter space.

(2) It remains to investigate the nonuniform rates of convergence to normality which are more useful.

(3) Weak convergence rate for AMCE remains open when $T \to \infty$ and $\frac{T}{n} \to 0$.

(4) Weighted estimators converge faster than the unweighted estimators if the weights are chosen according to (1.22) and (1.23).

4.5 Fractional Levy–Ornstein–Uhlenbeck Process

We generalize fractional Ornstein–Uhlenbeck process to include non-normal innovations. First we study the asymptotic behavior of the ratio estimator of the drift parameter in fractional Levy–Ornstein–Uhlenbeck process with Hurst parameter bigger than half. This model is interesting as it preserves both jumps and long memory. The ratio estimator is shown to be consistent and asymptotically has Weibull distribution for large sample size. Second, we consider the fractional Hermite–Ornstein–Uhlenbeck process and show that the maximum likelihood estimator is strongly consistent.

Recently, long-memory processes, i.e., processes with slowly decaying autocorrelation and processes with jumps, have received attention in finance, engineering, and physics. The simplest continuous time long-memory process is the fractional Brownian motion discovered by Kolmogorov (1940) and later on studied by Levy (1948) and Mandelbrot and Van Ness (1968). Continuous time long-memory jump process is a fractional Levy process. Hence the fractional Levy process can also be called the *Kolmogorov–Levy process*.

A normalized fractional Brownian motion $\{W_t^H, t \geq 0\}$ with Hurst parameter $H \in (0, 1)$ is a centered Gaussian process with continuous sample paths whose covariance kernel is given by

$$E(W_t^H W_s^H) = \frac{1}{2}(s^{2H} + t^{2H} - |t - s|^{2H}), \quad s, t \geq 0.$$

The process is self-similar (scale invariant) and it can be represented as a stochastic integral with respect to standard Brownian motion. For $H = \frac{1}{2}$, the process is a standard Brownian motion. For $H \neq \frac{1}{2}$, the fBm is not a semimartingale and not a Markov process, but a Dirichlet process. The increments of the fBm are negatively correlated for $H < \frac{1}{2}$ and positively correlated for $H > \frac{1}{2}$, and in this case they display long-range dependence. The parameter H, which is also called the self-similarity parameter, measures the intensity of the long-range dependence. The ARIMA(p,d,q) with autoregressive part of order p, moving average part of order q, and fractional difference parameter $d \in (0,\ 0.5)$ process converge in Donsker sense to fBm.

Levy driven processes of Ornstein–Uhlenbeck type has been extensively studied over the last few years and widely used in finance, see Barndorff-Nielsen and Shephard (2001). FLOU process generalizes fOU process to include jumps.

The fractional Levy–Ornstein–Uhlenbeck (fLOU) process is an extension of fractional Ornstein–Uhlenbeck process with fractional Levy motion (fLM) driving term. In finance, it could be useful as a generalization of fractional Vasicek model, as one-factor short-term interest rate model which could take into account the long-memory effect and jump of the interest rate. This process was introduced by Marquardt (2006a,b) who also introduced fractional Levy process in Marquardt (2006a,b). Using suitable transformation of the process, one can obtain a nonlinear stationary process satisfying a fractional SDE, see Buchmann and Kluppelberg (2005). Benth (2003) used the process as temperature and obtained weather derivative arbitrage free pricing formulas for European and average type options. Cheridito et al. obtained the fOU process as a Lamperti transformation of the fBM. The model parameter is usually unknown and must be estimated from data.

Fractional Levy process is defined as

$$M_{H,t} = \frac{1}{\Gamma(H+\frac{1}{2})}\int_{\mathbb{R}}[(t-s)_+^{H-1/2} - (-s)_+^{H-1/2}]dM_s, \quad t \in \mathbb{R},$$

where $M_t, t \in \mathbb{R}$ is a Levy process on $\mathbb{R}$ with $E(M_1) = 0,\ E(M_1^2) < \infty$ and without Brownian component.

Properties

$$(1)\ \ cov(M_{H,t}, M_{H,s}) = \frac{E(L(1))^2}{2\Gamma(2H+1)\sin(\pi H)}[|t|^{2H} + |s|^{2H} - |t-s|^{2H}].$$

(2) M_H is not a martingale. For a large class of Levy processes, M_H is neither a semimartingale.

(3)M_H is Hölder continuous of any order β less than $H - \frac{1}{2}$.

(4) M_H has stationary increments.

(5) M_H is symmetric.

(6) M_H is self-similar.

(7) M_H has infinite total variation on compacts.

(8) The FIMA process is defined as

$$Y_H(t) = \int_{-\infty}^{t} g_H(t-u)L(du), \ t \in \mathbb{R},$$

where

$$g_H(t) = \frac{1}{\Gamma(H-\frac{1}{2}}\int_0^t g(t-s)s^{H-\frac{3}{2}}ds, \ t \in \mathbb{R}$$

and the kernel g is the kernel of a short-memory moving average process.

The process $Y_H(t)$ can be written as

$$Y_H(t) = \int_{-\infty}^{t} g(t-u)dM_{H,u}, \ t \in \mathbb{R}.$$

We assume the following conditions on the kernel $g : \mathbb{R} \to \mathbb{R}$:
(A1) $g(t) = 0$ for all $t < 0$ (causality).
(A2) $|g(t)| \leq Ce^{-ct}$ for some constants $C > 0$ and $c > 0$ (short memory).
The FIMA (fractionally integrated moving average) process is stationary and is infinite divisible.

Consider the kernel

$$g(t-s) = \sigma e^{\theta(t-s)} I_{(0,\infty)}(t-s);$$

then

$$g_H(t) = \frac{\sigma}{\Gamma(H-\frac{1}{2})} \int_0^{\infty} e^{\theta(t-s)} I_{(0,\infty)}(t-s) s^{H-\frac{3}{2}} ds, \ t \in \mathbb{R}.$$

Note that

$$U_t^{H,\theta,\sigma} = \int_{\mathbb{R}} g_H(t-u)L(du), \ t \in \mathbb{R}$$

is the fractional Levy–Ornstein–Uhlenbeck (FLOU) process satisfying the fractional Langevin equation

$$dU_t = \theta U_t dt + \sigma dM_{H,t}, \ t \in \mathbb{R}.$$

The process has long memory. The autocovariance function of the FLOU process is given by

$$\gamma_H(h) = \frac{\Gamma(2H+1)}{2(1+\theta^2)^H} \left[1 + \frac{\sqrt{1+\theta^2}-1}{\sqrt{1+\theta^2}+1} \sin \pi H \right].$$

Let $(\Omega, \mathcal{F}, \{\mathcal{F}_t\}_{t \geq 0}, P)$ be a stochastic basis on which is defined the Ornstein–Uhlenbeck process X_t satisfying the Itô stochastic differential equation

$$dX_t = \theta X_t dt + dM_t^H, \quad t \geq 0, \quad X_0 = 0,$$

where $\{M_t^H\}$ is a fractional Levy motion with $H > 1/2$ with the filtration $\{\mathcal{F}_t\}_{t \geq 0}$ and $\theta > 0$ is the unknown parameter to be estimated on the basis of continuous observation of the process $\{X_t\}$ on the time interval $[0, T]$. Note that the SDE has the solution

$$X_t = \int_{-\infty}^{t} e^{\theta(t-s)} dM_s.$$

This process is stationary and is a process with long memory. It can be shown that X_{t_i} is a stationary discrete time AR (1) process with autoregression coefficient $\phi \in (0,1)$ with the following representation:

$$X_{t_i} = \phi X_{t_{i-1}} + \epsilon_{t_{i-1}},$$

where

$$\phi = e^{-\theta h}, \quad \epsilon_{t_{i-1}} = \int_{t_{i-1}}^{t_i} e^{-\theta(t_i - u)} dM_u.$$

We specialize to the case when L is a gamma process.

FLP is a generalization of fBM. FLP is not a martingale and not even a semimartingale in general. FLP is a natural counterpart to fBM. It is Hölder continuous. It has stationary increments. It is symmetric. It is self-similar. Fractional stable motion is a special case of fLP. In general, it is not a semimartingale.

The fractional Levy–Ornstein–Uhlenbeck (FLOU) process X was introduced in Marquardt (2006a,b). FLOU process is a LOU process which includes long memory.

We study the ratio estimator of the drift. The ratio estimator of θ is defined as

$$\hat{\theta}_n := -\log\left[\min_{1\le i\le n} \frac{X_{i\Delta}}{X_{(i-1)\Delta}}\right].$$

This estimator is motivated by the extreme value theory for the correlation parameter of an AR(1) process whose innovation distribution is positive. See Davis and McCormick (1989). In the case of exponential AR(1) process, it coincides with the maximum likelihood estimator. See Nielsen and Shephard (2003).

The weak consistency of the ratio estimator in the LOU process was studied in Jongbloed et al. (2005). The strong consistency and asymptotic Weibullness were studied in Brockwell et al. (2007) in the case of Gamma innovations. In this note, we obtain strong consistency and asymptotic Weibullness of the ratio estimator for the fLOU process.

We will need the following result in the sequel whose proof can be found in Davis and McCormick (1989).

Proposition 5.1 Let

$$X_t = \phi X_{t-1} + Z_t, \ t = 0, \pm 1, \pm 2, \ldots.$$

Let $1 > \phi > 0$. Let the distribution F of Z_t be concentrated on $[0,\infty)$ and be of regular variation at zero. Let

$$R_n := \bigwedge_{i=1}^{n} \frac{X_i}{X_{i-1}}$$

and

$$a_n = F^-(\frac{1}{n}) := \inf\{x : F(x) \geq \frac{1}{n}\}.$$

Then

$$R_n \to \phi \text{ a.s. as } n \to \infty$$

and

$$\lim_{n\to\infty} P\left\{a_n^{-1}(R_n - \phi)C_\alpha \leq x\right\} = 1 - \exp(-x^\alpha),\ x \geq 0$$

where

$$C_\alpha := (EX_1^\alpha)^{1/\alpha}.$$

We also study the following estimator of the drift. The estimator of θ is defined as

$$\tilde{\theta}_n := -\log\left[\frac{\sum_{i=1}^n X_{i\Delta}}{\sum_{i=1}^n X_{(i-1)\Delta}}\right].$$

We obtain the strong consistency and Weibull limit distribution of the ratio estimator.

Theorem 5.1

$$\hat{\theta}_n \to \theta_0 \quad \text{a.s.} \quad \text{as } n \to \infty.$$

Proof Once we show that the distribution F of ϵ is regularly varying with exponent α, the result follows from Proposition 1.1.

Theorem 5.2

$$k_n(\tilde{\theta}_n - \theta_0) \to^{\mathcal{D}} G_\alpha,$$

where G_α is the Weibull distribution function

$$G_\alpha(x) = 1 - \exp\{-x^\alpha\}, \text{ if } x \geq 0, \quad G_\alpha(x) = 0 \text{ if } x < 0.$$

Proof We use Proposition 5.1.

Theorem 5.3

$$\tilde{\theta}_n \to \theta_0 \quad \text{a.s.} \quad \text{as } n \to \infty.$$

Proof Once we show that the distribution F of ϵ is regularly varying with exponent α, the result follows from Proposition 1.1.

Theorem 5.4

$$k_n(\tilde{\theta}_n - \theta_0) \to^{\mathcal{D}} G_\alpha,$$

where G_α is the Weibull distribution function

$$G_\alpha(x) = 1 - \exp\{-x^\alpha\}, \text{ if } x \geq 0, \quad G_\alpha(x) = 0 \text{ if } x < 0.$$

MS (modified stable) law come by exponential and power tempering (tilting) from one of the positive k-stable laws ($0 < k < 1$). Using the MS distribution as mixing distribution for normal variance mean mixtures yields the class of NMS (normal modified stable) laws. NTS laws are normal mean variance mixture with TS mixing. MS laws are infinitely divisible and self-decomposable. For $k = 1/2$, MS becomes the GIG law and NMS becomes the GH law. The NG (normal gamma) class includes the VG processes. RLPE processes are CGMY processes. NG and RLPE processes are normal modified stable law.

An infinitely divisible distribution is said to be α-modified tampered stable (α-MTS) distribution if its Levy triplet is given by

$$\sigma^2 = 0$$

$$\nu(dx) = C\left(\frac{\lambda_+^{\alpha+\frac{1}{2}} K_{\alpha+\frac{1}{2}}\lambda_+ x}{x^{\alpha+\frac{1}{2}}} I_{x>0} + \frac{\lambda_+^{\alpha+\frac{1}{2}} K_{\alpha+\frac{1}{2}}\lambda_- x}{x^{\alpha+\frac{1}{2}}} I_{x<0}\right) dx$$

$$\gamma = \mu + C\left(\frac{\Gamma(\frac{1}{2}-\alpha)}{2^{\alpha+\frac{1}{2}}}(\lambda_+^{2\alpha-1} - \lambda_-^{2\alpha-1}) - \lambda_+^{\alpha-\frac{1}{2}} K_{\alpha-\frac{1}{2}}(\lambda_+) + \lambda_-^{\alpha-\frac{1}{2}} K_{\alpha-\frac{1}{2}}(\lambda_-)\right),$$

where $C > 0, \lambda_+, \lambda_- > 0, \mu \in \mathbb{R}, \alpha \in (-\infty, 1)\backslash\{\frac{1}{2}\}$, and $K_p(x)$ is the modified Bessel function of second kind. We denote the MTS random variable by $X \sim MTS(\alpha, C, \lambda_+, \lambda_-, \mu)$. The Levy measure $\nu(dx)$ is called the MTS Levy measure with parameter $(\alpha, C, \lambda_+, \lambda_-)$.

The MTS distribution is obtained by taking a symmetric α-stable distribution by $\alpha \in (0, 1)$ and multiplying by a Levy measure with $\sqrt{|x|}\lambda^{\alpha+\frac{1}{2}}$ $K_{\alpha+\frac{1}{2}}(\lambda|x|)$ on each half of the real axis. The measure can be extended to the case $\alpha \leq 0$. If $\alpha = \frac{1}{2}$, then γ may not be defined, so it is removed. The MTS distribution was introduced by Kim et al. (2009).

The tails of the α-MTS distribution are thinner than those of the 2α-stable and fatter than those of the 2α-TS distribution. At the zero neighborhood, all three have the same asymptotic behavior.

The characteristic function and the Laplace transform of MTS distribution are known. The cumulants of all orders are known. If $\lambda_+ > \lambda_-$, then the distribution is skewed to the left. If $\lambda_+ < \lambda_-$, then the distribution is skewed to the right. If $\lambda_+ = \lambda_-$, then the distribution is symmetric. As α decreases, the distribution has fatter tails and increased peakedness. The Levy process corresponding to the MTS distribution has finite activity if $\alpha < 0$ and infinite activity if $\alpha > 0$. It has finite variation if $\alpha < \frac{1}{2}$ and infinite variation if $\alpha > \frac{1}{2}$.

With proper choice of C and μ, MTS distribution has zero mean and unit variance, and the distribution is called the standard MTS distribution and denoted $X \sim stdMTS(\alpha, \lambda_+, \lambda_-)$.

CGMY process is a tampered stable process. In order to obtain a closed form solution of the European option price, CGMY used the generalized Fourier transform of the distribution of the stock price under the assumption of Markov property.

$$dY_t = (\mu + \beta\sigma_t^2)dt + \sigma_t dW_t$$
$$d\sigma_t^2 = -\lambda\sigma_t^2 dt + dZ_t,$$

where μ is the drift parameter, β is the risk premium, and Z_t is a MTS process.

Examples: Fractional IGOU Process and Fractional Gamma-OU Process

We consider the special cases when L is an IGOU process. L has the IGOU density with parameters δ and γ. The model is given by

$$dY_t = (\mu + \beta X_t)dt + X_t dW_t$$

$$dX_t = -\theta X_t dt + dZ_t,$$

where Z_t is a Levy process independent of X_0.

Invariant distribution is GIG.

$$\mathcal{L}(X_0) = \mathcal{L}(X_t) = \mathcal{GIG}(\lambda, \delta, \sqrt{\alpha^2 - \beta^2}).$$

Mixture distribution is Q_θ which is GH.

$$\mathcal{L}(Y_t) = \mathcal{GH}(\lambda, \frac{\alpha}{\sqrt{t}}, \frac{\beta}{\sqrt{t}}, \sqrt{t}\delta, \mu t),$$

$$\mathcal{L}(Y_{t+1} - Y_t) = \mathcal{GH}(\lambda, \alpha, \beta, \delta, \mu).$$

Let $q(x, \theta)$ is a density function of Y_t. The contrast function is defined as $u(x, \theta) = -\log q(x, \theta)$. MCE is defined as $\hat{\theta}_T = \arg\inf_\theta u(x, \theta)$.

The density function of the GH distribution is given by

$$d_{\mathcal{GH}(\lambda,\alpha,\beta,\delta,\mu)}(x) =$$

$$\frac{(\alpha^2 - \beta^2)^{\frac{\lambda}{2}}}{\sqrt{2\pi}\alpha^{\lambda-\frac{1}{2}}\delta^\lambda K_\lambda(\delta\sqrt{\alpha})}(\delta^2 + (x - \mu)^2)^{(\lambda-\frac{1}{2})/2} e^{\beta(x-\mu)} K_{\lambda-\frac{1}{2}}(\alpha\sqrt{\delta^2 + (x - \mu)^2}).$$

Here K_ν denotes the modified Bessel function of third kind with index ν. When $\lambda = -\frac{1}{2}$, GIG becomes $IG(\delta, \gamma)$.

In the case of -1 correlation, the moment estimator is given by $\theta_T = -\frac{Y_T}{T}$.

Then we consider the fractional Gamma-OU process. The limit distribution is Weibull.

4.5.1 Fractional Hermite OU Processes

As another generalization of fractional Brownian motion we obtain the Hermite process of order k with Hurst parameter $H \in (\frac{1}{2}, 1)$, which is defined as a multiple Wiener–Itô integral of order k with respect to the standard Brownian motion $(B(t))_{t\in\mathbb{R}}$.

$$Z_t^{H,k} := c(H,k)\int_{\mathbb{R}}^{k}\int_0^t \Pi_{j=1}^k (s-y_i)_+^{-(\frac{1}{2}+\frac{H-1}{2})} ds\, dB(y_1)dB(y_2)\cdots dB(y_k),$$

where $x_+ = \max(x,0)$.

For $k=1$, the process is fractional Brownian motion W_t^H with Hurst parameter $H \in (0,1)$. For $k=2$, the process is Rosenblatt process. For $k \geq 2$, the process is non-Gaussian which is more interesting from the point of finance. Hence the fractional Hermite process can also be called the *Kolmogorov–Hermite process.*

The covariance kernel $R(t,s)$ is given by

$$R(t,s) := E[Z_t^{H,k} Z_s^{H,k}]$$

$$= c(H,k)^2 \int_0^t \int_0^s \left[(u-y)_+^{-(\frac{1}{2}+\frac{H-1}{2})} ds (v-y)_+^{-(\frac{1}{2}+\frac{H-1}{2})} dy\right]^k dudv.$$

Let

$$\beta(p,q) = \int_0^1 z^{p-1}(1-z)^{q-1} dz,\ p,q>0$$

be the beta function.

Using the identity

$$\int_{\mathbb{R}} [(u-y)_+^{a-1} ds (v-y)_+^{a-1} dy] = \beta(a, 2a-1)|u-v|^{2a-1},$$

we have

$$\begin{aligned} R(t,s) &= c(H,k)^2 \beta\left(\frac{1}{2} - \frac{1-H}{k}, \frac{2H-2}{k}\right)^k \int_0^t \int_0^s \left(|u-v|^{\frac{2H-2}{k}}\right)^k dvdu \\ &= c(H,k)^2 \frac{\beta(\frac{1}{2} - \frac{1-H}{k}, \frac{2H-2}{k})^k}{H(2H-1)} \frac{1}{2}(t^{2H} + s^{2H} - |t-s|^{2H}). \end{aligned}$$

In order to obtain $E(Z_t^{(H,k)})^2 = 1$, choose

$$c(H,k)^2 = \left(\frac{\beta(\frac{1}{2} - \frac{1-H}{k}, \frac{2H-2}{k})^k}{H(2H-1)}\right)^{-1},$$

and we have

$$R(t,s) = \frac{1}{2}(t^{2H} + s^{2H} - |t-s|^{2H}).$$

Thus the covariance structure of the Hermite process and the fractional Brownian motion are the similar.

The process $Z_t^{(H,k)}$ is H-self-similar with stationary increments and all moments are finite.

For any $p \geq 1$,

$$E|Z_t^{(H,k)} - Z_s^{(H,k)}|^p \leq c(p,H,k)|t-s|^{pH}.$$

Thus the Hermite process has Hölder continuous paths of order $\delta < H$.

The fractional Hermite–Ornstein–Uhlenbeck process is given by

$$dY_t^{(H,k)} = \theta Y_t^{(H,k)} dt + \sigma dZ_t^{(H,k)}$$

and has unique solution

$$Y_t^{(H,k)} = e^{\theta t}\left(\xi + \sigma \int_0^t e^{-\theta u} dZ_u^{(H,k)}\right).$$

In particular, if

$$X_0 = \sigma \int_{-\infty}^0 e^{-\theta u} Z_u^{(H,k)},$$

then

$$Y_t^{(H,k)} = e^{\theta t}\sigma \int_{-\infty}^t e^{-\theta u} dZ_u^{(H,k)}.$$

The process $Y_t^{(H,k)}$ is called the fractional Hermite–Ornstein–Uhlenbeck process of order k. Because the covariance structure of the Hermite process is the same as the fractional Brownian motion's one, the above integral is a Wiener integral.

$$E[Y_t^{(H,k)} Y_s^{(H,k)}] = \sigma^2 \int_{-\infty}^t \int_{-\infty}^s e^{\theta(t-u)} e^{\theta(t-v)} |u-v|^{2H-2} dv du < \infty.$$

We obtain the strong consistency of the MLE of the parameter θ. First we obtain Cameron–Matin type formula.

Lemma 5.1 Let $I_T := \int_0^T Q_t^2 dv_t$ and $\phi_T(a) := E\exp(aI_T)$, $a \in \mathbb{R}$. Then $\phi_T(z_1)$ exists for $|z_1| \leq \delta$, for some $\delta > 0$, and is given by

$$\phi_T(a) = \left\{ \frac{4 \sin \pi H \sqrt{\theta^2 - 2z_1} e^{-\theta T}}{\pi T D_T^H(\theta; \sqrt{\theta^2 - 2z_1}} \right\}^{1/2},$$

and we choose the principal branch of the square root, where

$$\begin{aligned} D_T^H(\theta;\alpha) &:= [\alpha \cosh(\frac{\alpha}{2}T) - \theta \sinh(\frac{\alpha}{2}T)]^2 J_{-H}(\frac{\alpha}{2}T) J_{H-1}(\frac{\alpha}{2}T) \\ &- [\alpha \sinh(\frac{\alpha}{2}T) - \theta \cosh(\frac{\alpha}{2}T)]^2 J_{1-H}(\frac{\alpha}{2}T) J_H(\frac{\alpha}{2}T) \end{aligned}$$

for $\alpha > 0$ and J_ν is the modified Bessel function of first kind of order ν.

4.5.2 Fractional Poisson–Ornstein–Uhlenbeck Process

A fractional Poisson process $\{W_H(t), t > 0\}$ with Hurst parameter $H \in (1/2, 1)$ is defined as

$$W_H(t) = \frac{1}{\Gamma(H-\frac{1}{2})} \int_0^t u^{\frac{1}{2}-H} \left(\int_0^u \tau^{H-\frac{1}{2}} (\tau - u)^{H-\frac{3}{2}} d\tau \right) dq(u)$$

where $q(u) = \frac{N(u)}{\sqrt{\lambda}} - \sqrt{\lambda} u$.

$N(u)$ is a Poisson process. The process is self-similar in the wide sense, has wide sense stationary increments, has fat-tailed non-Gaussian distribution, and exhibits long-range dependence. The process converges to fractional Brownian motion in distribution. The process is self-similar in the asymptotic sense. Strict sense, wide sense, and asymptotic sense self-similarity are equivalent for fractional Brownian motion. Stock returns are far from being self-similar in strict sense.

The stochastic volatility model we consider is

$$dS_t = \sqrt{V}_t dB(t)$$

$$dV_t = \mu V_t dt + \sigma V_t dW_H(t).$$

Poisson Shot Noise

One can use the Girsanov theorem here and study the properties of MLE.

4.6 FIECOGARCH Process

We suggest quasi-maximum likelihood estimators for the parameters of a fractionally integrated exponential GARCH, in short FIECOGARCH process based on equally spaced observations and randomly spaced observations. First we deal with a compound Poisson FIECOGARCH process and study the asymptotic behavior of the quasi-maximum likelihood estimator. Then using the fact that the increments of a general FIECOGARCH process are strongly mixing with an exponential rate, we show that the resulting estimators are consistent and asymptotically normal.

To account for the strong persistence in volatility, which sometimes is observed in empirical data, we study the FIECOGARCH (fractionally integrated exponential continuous time generalized autoregressive conditional heteroscedastic) process. The log-price process has jumps and the log-volatility process has long memory. The long-memory effect introduced in the log-volatility process propagates to the volatility process. Thus the model here captures the two stylized facts of long memory and jumps. Continuous time long-memory process in volatility has already received some attention, see Comte and Renault (1998) where they used fractional Ornstein–Uhlenbeck process as the stochastic volatility process and Comte et al. (2012) where they used fractional CIR square-root process as the stochastic volatility process.

Sub-fractional Levy Process is a generalization of sub-fractional Brownian motion and is defined as

$$M_{H,t} = \frac{1}{\Gamma(H+\frac{1}{2})} \int_{\mathbb{R}} [(t-s)_+^{H-1/2} - (-s)_+^{H-1/2}] dM_s, \quad t \in \mathbb{R},$$

where $H \in (0.5, 1)$, $u_+ = \max(u, 0)$, $u_- = \max(-u, 0)$ and $\{M_t, t \in \mathbb{R}\}$ is a Levy process on $\mathbb{R}$ with $E(M_1) = 0$, $E(M_1^2) < \infty$ and without Brownian component.

Properties of Sub-fractional Levy Process

(1) The covariance of the SFLP is given by

$$\mathrm{Cov}(M_{H,t}, M_{H,s}) = \frac{E[L(1)^2]}{\Gamma(2H+1)\sin(\pi H)} [s^{2H} + t^{2H} - \frac{1}{2}\{(s+t)^{2H} + |s-t|^{2H}\}].$$

(2) M_H is not a martingale. For a large class of Levy processes, M_H is neither a semimartingale.

If the Levy process M is of finite activity, then the corresponding fractional Levy process M_H is of finite variation. In the case when M is not of finite activity, the corresponding fractional Levy process is not a semimartingale.

(3) M_H is Hölder continuous of any order β less than $H - \frac{1}{2}$.

(4) M_H has non-stationary increments.

(5) M_H is symmetric.

(6) M is self-similar.

(7) M_H has infinite total variation on compacts.

(8) M_H is not self-similar.

(9) M_H has long memory.

(10) Fractional Levy–Ornstein–Uhlenbeck (FLOU) Process.

The FIMA (fractionally integrated moving average) process is defined as

$$Y_H(t) = \int_{-\infty}^{t} g_H(t-u) L(du), \ t \in \mathbb{R},$$

where

$$g_H(t) = \frac{1}{\Gamma(H-\frac{1}{2}} \int_0^t g(t-s) s^{H-\frac{3}{2}} ds, \ t \in \mathbb{R},$$

and the kernel g is the kernel of a short-memory moving average process.

The process $Y_H(t)$ can be written as

$$Y_H(t) = \int_{-\infty}^{t} g(t-u) dM_{H,u}, \ t \in \mathbb{R}.$$

We assume the following conditions on the kernel $g : \mathbb{R} \to \mathbb{R}$:
(A1) $g(t) = 0$ for all $t < 0$ (causality).
(A2) $|g(t)| \leq Ce^{-ct}$ for some constants $C > 0$ and $c > 0$ (short memory).
The FIMA process is stationary and is infinite divisible.
Consider the kernel

$$g(t - s) = \sigma e^{\theta(t-s)} I_{(0,\infty)}(t - s);$$

then

$$g_H(t) = \frac{\sigma}{\Gamma(H - \frac{1}{2})} \int_0^{\infty} e^{\theta(t-s)} I_{(0,\infty)}(t - s) s^{H-\frac{3}{2}} ds, \ t \in \mathbb{R}$$

is the Riemann–Liouville fractional integral of order $H - \frac{1}{2}$ of the kernel function $g(t)$. Note that the log-volatility process

$$V_{H,t}^{\theta,\sigma} = \int_{\mathbb{R}} g_H(t - u) L(du), \ t \in \mathbb{R},$$

is the fractional Levy–Ornstein–Uhlenbeck (FLOU) process satisfying the fractional Langevin equation

$$dV_{H,t} = \theta V_{H,t} dt + \sigma dM_{H,t}, \ t \in \mathbb{R}.$$

The autocovariance function of the FLOU process is given by

$$\gamma_H(h) = \int_{\mathbb{R}} g_H(u + h) g_H(u) du \sim C_1 h^{2(H-1)}.$$

The FIECOGARCH process is given by

$$dY_{H,t} = \sqrt{X_{H,t}} dL_t$$

$$X_{H,t} = e^{\mu + \beta V_{H,t-}}$$

$$dV_{H,t} = \theta V_{H,t} dt + dM_{H,t},$$

where L_t is a Levy process independent of $M_{H,t}$, which is a fractional Levy process of integration parameter $H \in (0.5, \ 1)$.

We estimate μ, β, and θ by the quasi-maximum likelihood method.

Equivalently,

$$dY_{H,t} = \mu dt + \sigma e^{X_{H,t}} dL_t$$

$$dX_{H,t} = \theta X_{H,t} dt + dM_{H,t}.$$

This model was obtained by Haug and Czado (2006a,b) who called them fractionally integrated exponential GARCH, in short FIECOGARCH process.

The log-volatility process $\log(\sigma_{H,t}^2) = X_{H,t}$ is strictly stationary and the covariance function is of the form

$$\text{Cov}(\log(\sigma_{H,t+h}^2), \log(\sigma_{H,t}^2) \sim C_1 h^{2(H-1)}$$

when $h \to \infty$.

The autocorrelation function of log-volatility and volatility process decay at a hyperbolic rate.

Moments and autocovariance function of the price process of lag r

$$S_{H,t}^{(r)} = S_{H,t} - S_{H,t-r} = \int_{(t-r,t]} \sigma_{H,s} dL_s$$

are given by

$$E(S_{H,t}^{(r)})^2 = e^{\mu + \Psi_{H,\infty}(1)} r E(L_1^2),$$

where

$$\Psi_{H,\infty}(k) = \int_0^\infty \Psi_M(g_H(s))k) ds,$$

where $\Psi_M(u) = \psi_M(-iu)$ and

$$\psi_L(u) = -\tau_L^2 \frac{u^2}{2} + \int_{\mathbb{R}} (e^{iux} - 1 - iux)\nu_L(dx)$$

is the Levy symbol and

$$\text{Cov}((S_{H,t}^{(r)})^2, (S_{H,t+h}^{(r)})^2) = E(L_1^2) \int_h^{h+r} Cov((S_{H,t}^{(r)})^2, \sigma_s^2) ds.$$

We will treat the special case of compound Poisson FIECOGARCH process first. We will estimate the parameters by quasi-maximum likelihood estimation method.

A compound Poisson FIECOGARCH process is driven by the Levy process L with Levy symbol

$$\psi_L(u) = -\frac{u^2}{2} + \int_{\mathbb{R}} (e^{iux} - 1)\Phi_{0,1/\lambda}(dx),$$

where $\Phi_{0,1/\lambda}$ is a normal distribution with mean 0 and variance $1/\lambda$. This means that L is the sum of a standard Brownian motion W and a compound Poisson process

$$J_t = \sum_{k=1}^{N_t} Z_k, \ t \geq 0,$$

where N_t is an independent Poisson process with intensity $\lambda > 0$ and jump times $(T_k)_{k \in \mathbb{Z}}$. The Poisson process is also independent from the i.i.d. sequence of jump sizes $(Z_k)_{k \in \mathbb{Z}}$ with $Z_1 \sim N(0, 1/\lambda)$. The Levy process M in this case is given by

$$M_t = \sum_{k=1}^{N_t} (\theta Z_k + \gamma |Z_k|) - Ct, \ t > 0$$

with

$$C := \gamma \int_{\mathbb{R}} |x| \lambda \Phi_{0,1/\lambda}(dx) = \sqrt{\frac{2\lambda}{\pi}} \gamma.$$

The stationary log-volatility is of the form

$$\log(\sigma^2_{H,t_+}) = \mu + \int_{-\infty}^{t} \beta e^{-a_1(t-s)} dM_{H,s}, t > 0.$$

The log-price process is given by

$$S_t = \sum_{k=1}^{N_t} \sigma_{T_k} Z_k, \ t \geq 0, \ S_0 = 0.$$

We observe S at n consecutive jump times $0 = T_0 < T_1 < \ldots < T_n < T < T_{n+1}, n \in \mathbb{Z}$ over the time interval $[0, T]$. The state process X_H has then the following autoregressive representation:

$$\begin{aligned} \beta X_{H,T_i} &= \beta e^{-\alpha \Delta T_i} X_{H,T_{i-1}} + \sum_{k=N_{T_{i-1}}+1}^{N_{T_i}} \beta e^{-\alpha(T_i - T_k)} [\theta Z_k + \gamma |Z_k|] \\ &- \int_{T_{i-1}}^{T_i} \beta e^{(T_i - s)} C ds \\ &= \beta e^{-\alpha \Delta T_i} X_{H,T_{i-1}} + \beta \theta Z_i + \beta \left(|Z_i| - \frac{C}{\alpha}(1 - e^{-\alpha \Delta T_i}) \right), \end{aligned}$$

where $\Delta T_i := T_i - T_{i-1}, \ i = 1, 2, \ldots, n$ and $N_{T_{i-1}} + 1 = N_{T_i} = i$.

Thus

$$\log(\sigma^2_{H,T_i}) = \mu + \beta e^{-\alpha \Delta T_i} X_{T_{i-1}} - \beta \frac{C}{\alpha}(1 - e^{-\alpha \Delta T_i})$$

since

$$\beta X_{H,T_i-} = \beta X_{H,T_{i-1}} - \beta \frac{C}{\alpha}(1 - e^{-\alpha \Delta T_i})$$

as X_{T_i-} does not include the jump at time T_i, and its corresponding jump size is Z_i.

The compound Poisson process has finite activity. The observation of the log-price process is given by

$$S_{T_i} = \sum_{k=1}^{N_{T_i}} \sigma_{T_k} Z_k = S_{H,T_{i-1}} + \sigma_{H,T_i} Z_i.$$

We study the parameter estimation in two steps. The rate λ of the Poisson process N can be estimated given the jump times T_i, and therefore it is done at a first step. Since we observe the total number of jumps n of the Poisson process N over the T intervals of length one, the MLE of λ is given by

$$\hat{\lambda}_n := \frac{n}{T}.$$

Proposition 6.1 We have

$$\hat{\lambda}_n \to \lambda \text{ a.s. as } \quad n \to \infty.$$

$$n^{1-H}(\hat{\lambda}_n - \lambda) \to^{\mathcal{D}} \mathcal{N}(0,\ e^{\lambda}(1-e^{-\lambda})) \text{ as } \ n \to \infty.$$

Proof Let U_i be the number of jumps in the interval $(i-1, i]$. Then $U_i,\ i = 1, 2, \ldots, n$, are i.i.d. Poisson distributed with parameter λ. Since Φ is continuous, we have $I_{\{0\}}(U_i) = I_{\{0\}}(S^{(1)}_{H,i})$, a.s. $i = 1, 2, \ldots, n$. Note that

$$\frac{1}{n}\sum_{i=1}^{n} I_{\{0\}}(S^{(1)}_{H,i}) \to^{a.s.} E(I_{\{0\}}U_1) = P(U_1 = 0) = e^{-\lambda} \text{ as } \ n \to \infty.$$

LLN and CLT and the delta method applied to the sequence $I_{\{0\}}(S^{(1)}_{H,i}),\ i = 1, 2, \ldots, n$, give the results.

□

The CLT result above allows us to construct confidence interval for the jump rate λ. A $100(1-\alpha)\%$ confidence interval for λ is given by

$$\left[\frac{n}{T} - q_{1-\frac{\alpha}{2}}\sqrt{\frac{1}{n} - \frac{1}{T}},\ \frac{n}{T} + q_{1-\frac{\alpha}{2}}\sqrt{\frac{1}{n} - \frac{1}{T}}\right],$$

where $q_{1-\frac{\alpha}{2}}$ is the $(1-\frac{\alpha}{2})$-quantile of the standard normal distribution.

To estimate the remaining parameters (α, θ, μ), we use the quasi-maximum likelihood estimation procedure in conditionally heteroscedastic time series models developed by Straumann (2005).

The conditional likelihood has the representation

$$\mathcal{L}(\vartheta|S^{\Delta}_{H}, \lambda) := -\frac{n}{2}\log(2\pi) - \frac{1}{2}\left(\sum_{i=1}^{n}\log(\sigma^2_{T_i}/\lambda) - \sum_{i=1}^{n}\frac{(S^{\Delta T_i}_{H,T_i})^2}{\sigma^2_{T_i}/\lambda}\right).$$

The quasi-log-likelihood function for $\vartheta = (\alpha, \theta, \gamma, \mu)$ given $S^{\Delta}_{H} := (S^{\Delta T_1}_{H,T_1}, S^{\Delta T_2}_{H,T_2}, \ldots, S^{\Delta T_n}_{H,T_n})$, and the MLE $\hat{\lambda}_n$ is defined as

$$L(\vartheta|G^{\Delta}, \hat{\lambda}_n) := -\frac{1}{2}\sum_{i=1}^{n}\log(\hat{\sigma}^2_{H,T_i}(\vartheta, \hat{\lambda}_n)) - \frac{1}{2}\sum_{i=1}^{n}\frac{(S^{\Delta T_i}_{H,T_i})^2}{\hat{\sigma}^2_{H,T_i}(\vartheta, \hat{\lambda}_n)/\hat{\lambda}_n},$$

where the estimates of the volatility is given by

$$\hat{\sigma}^2_{H,T_i}(\vartheta, \lambda_n) := \exp(\mu + e^{-\alpha\Delta T_i}X_{H,T_{i-1}}(\vartheta, \lambda) - \hat{C}\Delta T_i), \imath = 1, 2, \ldots, n,$$

and the estimates of the state space is given by the recursion

$$\hat{X}_{H,T_i} = e^{-\alpha \Delta T_i} \hat{X}_{T_{i-1}} + \theta \frac{S_{H,T_i}}{\hat{\sigma}_{T_i}(\vartheta, \lambda)} + \left(\frac{S_{H,T_i}}{\hat{\sigma}_{T_i}(\vartheta, \lambda)} - \hat{C} \Delta T_i \right).$$

Note that $E(|W|) = \sqrt{\frac{2}{\pi\lambda}}$, $W \sim N(0, 1/\lambda)$.

Here the approximation $(1 - e^{-z}) \approx z$ for small z is used. The recursion needs a starting value $\hat{X}_{H,0}$ which will be the mean value zero of the stationary distribution of X.

QMLE of ϑ is defined as

$$\hat{\vartheta}_n := \arg\max_{\vartheta \in \Theta} L(\vartheta | S^{\Delta}, \hat{\lambda}_n).$$

The parametric estimator of volatility at jump times $0 = T_0 < T_1 < \ldots < T_n < T < T_{n+1}, n \in \mathbb{Z}$, is given by

$$\hat{\sigma}^2_{H,T_i}(\hat{\vartheta}_n, \hat{\lambda}_n) := \exp(\hat{\mu}_n + e^{-\hat{\alpha}_n \Delta T_i} X_{H,T_{i-1}}(\hat{\vartheta}_n, \hat{\lambda}_n) - \hat{C} \Delta T_i), \; i = 1, 2, \ldots, n,$$

which uses the QMLE $\hat{\vartheta}_n$.

Theorem 6.2

$$\hat{\vartheta}_n \to \vartheta_0 \text{ a.s. as } n \to \infty.$$

$$n^{1-H}(\hat{\vartheta}_n - \vartheta) \to^{\mathcal{D}} N(0, I^{-1}(\vartheta_0)).$$

Proof The discrete time process $S^{(r)}_{nr}, n \in \mathbb{Z}$, process is strong mixing with geometric rate, and since strict stationarity of σ^2 implied strict stationarity of $S^{(r)}_{nr}, n\mathbb{Z}$ is also ergodic. Hence application of the Birkoff ergodic theorem and central limit theorem for strongly mixing process proves the theorem. □

4.6.1 Variance Gamma FIECOGARCH Model

We study the Variance Gamma FIECOGARCH model. Again the driving process has no Brownian component. The characteristic function is given by

$$E(e^{iuL_t}) = \left(1 + \frac{u^2}{2C}\right)^{-tC}.$$

The Levy measure L has the Lebesgue density

$$\nu_L(dx) = \frac{C}{|x|} \exp\left(-(2C)^{1/2}|x|\right) dx, \quad x \neq 0.$$

We assume $E(L_1^{8+2\delta}) < \infty$ and $E(\sigma_1^{8+2\delta}) < \infty$.

Fractional Levy Process

When the Levy process M is a VG process, the process M_H is called a fVG process. M has infinite activity. Thus the fVG process is not a semimartingale.

General FIECOGARCH Process

We estimate μ, σ, and θ by the method of moments.

$$\hat{\mu}_n := \sum_{i=1}^{n} (Y_{H,i} - Y_{H,i-1})^2.$$

Then we study the fractional Levy Heston model:

$$dY_{H,t} = \mu dt + \sigma X_{H,t} dL_t$$

$$dX_{H,t} = \theta X_{H,t} dt + \gamma \sqrt{X_{H,t}} dM_{H,t}.$$

We estimate μ, σ, and θ by the quasi-maximum likelihood method.

Theorem 6.3

$$\hat{\mu}_n \to \mu_0 \text{ a.s. as } n \to \infty.$$

$$\hat{\gamma}_n \to \gamma_0 \text{ a.s. as } n \to \infty.$$

$$\hat{\theta}_n \to \theta_0 \text{ a.s. as } n \to \infty.$$

Theorem 6.4

$$n^{1-H}(\hat{\mu}_n - \mu) \to^{\mathcal{D}} N(0, I^{-1}(\mu_0)).$$

$$n^{1-H}(\hat{\gamma}_n - \gamma) \to^{\mathcal{D}} N(0, I^{-1}(\gamma_0)).$$

$$n^{1-H}(\hat{\theta}_n - \theta_0) \to^{\mathcal{D}} N(0, I^{-1}(\theta_0)).$$

The observation times $0 = t_1 < t_2 < \ldots < t_n = T$ are random. For example, they have Poisson distribution with parameter λ. That means that the inter arrival times $t_i - t_{i-1}$ have exponential distribution with parameter λ.

The jump times are $T_1, T_2, \ldots, T_{n_T}$, where

$$n_T := \max\{k \in \mathbb{N} : T_k \leq T\}.$$

$\{T_k\}$ are the compound Poisson process $J_t = \sum_{k=1}^{N_t} Z_k$ and Z_k are i.i.d. The log-price process is given by

$$Y_{H,t} = \int_0^t X_{H,s} dW_s + \sum_{k=1}^{N_t} X_{H,T_k} Z_k.$$

The log-volatility process is given by

$$\log X_{H,t+}^2 = \mu + \sum_{k=\infty,k\neq 0}^{N_t} \beta e^{\alpha(t-T_k)}[Z_k + \gamma|Z_k|],$$

i.e.,

$$\log X_{H,t+}^2 = \mu + \int_{-\infty}^{t} \beta e^{\alpha(t-T_k)} dM_s,$$

where M_s is a Levy process. If X_H^2 is strictly stationary, then Y_H has strictly stationary increments.

4.7 Fractional Gamma and Fractional Inverse Gaussian Ornstein–Uhlenbeck Process

It is well known that the suitably parametrized autoregressive (AR) process with Gaussian error has the continuous limit the Vasicek model. Wolfe (1982) studied continuous analogue of the stochastic difference equation of AR type with Levy type innovations whose limit is a Levy driven OU Process. Gourieroux and Jasiak (2006) studied autoregressive gamma (ARG) process and showed that its continuous time limit is the Cox–Ingersoll–Ross (CIR) model. Thus the stationary ARG process is a discretized version of the CIR process. Gourieroux and Jasiak (2006) studied pseudo-maximum likelihood estimation in autoregressive gamma (ARG) process. This process can also be used for application in series of squared returns and intertrade durations for high-frequency data, i.e., it is a stochastic duration model. ARG model also fits a series of volumes per trade, which is an alternative proxy for liquidity. This is different from gamma autoregressive (GAR) process studied in Sim (1990) and Gaver and Lewis (1980) where just the noise of the linear autoregressive process is Gamma distributed. For intertrade durations, the most popular model is autoregressive conditional duration (ACD) model introduced by Engle and Russell (1998).

Recall that the autoregressive gamma (ARG) process can be written as

$$X_t = \sum_{j=1}^{N_t} U_j + \epsilon_t,$$

where N_t is a Poisson process, U_j are independent and identically distributed Gamma distributed random variables with shape parameter 1 and scale parameter c_t, independent of N, and ϵ_t are gamma distributed random variables with shape parameter ν and scale parameter c_t.

We propose the discretized version of the Heston model as

$$y_i = \mu + \sqrt{x_i}\, \xi_i$$

$$x_i = \sum_{j=1}^{N_t} U_{i,j} + \epsilon_i,$$

where y_i is the return, y_i is the volatility, ξ_i are normally distributed, ϵ_i are gamma distributed, (ξ_i, ϵ_i) has the correlation ρ, and U_j are independent χ_1^2 distributed.

We consider the special cases when L is a GOU process. L has the GOU density with parameters δ and γ. The stochastic volatility model has the form

$$dY_t = (\mu + \beta X_t)dt + \sqrt{X_t}dW_t + \rho dZ_t,$$

$$dX_t = -\theta X_t dt + dZ_t,$$

where Z_t is a Levy process independent of X_0. The invariant distribution is Gamma:

$$\mathcal{L}(X_0) = \mathcal{L}(X_t) = \mathcal{G}(\theta, \delta, \sqrt{\alpha^2 - \beta^2}).$$

Mixture distribution is variance Gamma denoted by Q_θ.

$$\mathcal{L}(Y_t) = \mathcal{VG}(\theta, \frac{\alpha}{\sqrt{t}}, \frac{\beta}{\sqrt{t}}, \sqrt{t}\delta, \mu t).$$

$$\mathcal{L}(Y_{t+1} - Y_t) = \mathcal{VG}(\theta, \alpha, \beta, \delta, \mu).$$

We study minimum contrast estimator and method of moments estimator for the parameters of fractional Gamma-Ornstein–Uhlenbeck (fGOU) stochastic volatility model based on equally spaced observations. Using the fact that the increments of a fractional Gamma-OU process are strongly mixing with an exponential rate, we show that the resulting estimators are consistent and asymptotically normal. The model is given by

$$dY_t = \sqrt{X_t}dW_t + \rho dZ_{\theta t}$$

$$dX_t = -\theta X_t dt + dZ_{\theta t},$$

where Z_t is a fractional Levy process independent of X_0. Invariant distribution is Gamma (α, β) with density function

$$f(x) = \frac{\beta^\alpha}{\Gamma(\alpha)} x^{\alpha-1} e^{-\alpha x}.$$

Let $y_i := Y_{i\Delta} - Y_{i-1\Delta}, i = 1, 2, \ldots, n$.
The cummulants satisfy

$$k_{y_1}^{(1)} = \theta\rho\Delta k_G^{(1)}. \tag{7.1}$$

$$k_{y_1}^{(2)} = \Delta k_G^{(1)} + 2\theta\rho^2\Delta k_G^{(2)}. \tag{7.2}$$

We know that

$$k_G^{(1)} = \frac{\alpha}{\beta}, \quad k_G^{(2)} = \frac{\alpha(\alpha+1)}{\beta^2}.$$

$$\mathcal{L}(X_0) = \mathcal{L}(X_t) = \mathcal{GIG}(\lambda, \delta, \sqrt{\alpha^2 - \beta^2}).$$

Inverting (7.1) and (7.2) and replacing the cummulants by their sample cummulants, we obtain the explicit method of moments estimators given by

$$\hat{\theta}_n = \frac{2\alpha^3(\alpha+1)}{\beta^4\Delta} \frac{(\frac{1}{n}\sum_{i=1}^n (Y_{i\Delta} - Y_{i-1\Delta}))^2}{\frac{1}{n}\sum_{i=1}^n (Y_{i\Delta} - Y_{i-1\Delta})^2 - \Delta\frac{1}{n}\sum_{i=1}^n (Y_{i\Delta} - Y_{i-1\Delta})},$$

$$\hat{\rho}_n = \frac{\beta^3\Delta}{2\alpha^2(\alpha+1)} \frac{\frac{1}{n}\sum_{i=1}^n (Y_{i\Delta} - Y_{i-1\Delta})^2 - \Delta\frac{1}{n}\sum_{i=1}^n (Y_{i\Delta} - Y_{i-1\Delta})}{\frac{1}{n}\sum_{i=1}^n (Y_{i\Delta} - Y_{i-1\Delta})}.$$

Next in order to obtain the minimum contrast estimators, observe that the mixture distribution is variance Gamma Q_θ GH.

$$\mathcal{L}(Y_t) = \mathcal{GH}(\lambda, \frac{\alpha}{\sqrt{t}}, \frac{\beta}{\sqrt{t}}, \sqrt{t}\delta, \mu t).$$

$$\mathcal{L}(Y_{t+1} - Y_t) = \mathcal{GH}(\lambda, \alpha, \beta, \delta, \mu).$$

The density function of the GH distribution is

$$d_{\mathcal{GH}(\lambda,\alpha,\beta,\delta,\mu)}(x) =$$

$$\frac{(\alpha^2-\beta^2)^{\frac{\lambda}{2}}}{\sqrt{2\pi}\alpha^{\lambda-\frac{1}{2}}\delta^\lambda K_\lambda(\delta\sqrt{\alpha})}(\delta^2 + (x-\mu)^2)^{(\lambda-\frac{1}{2})/2} e^{\beta(x-\mu)} K_{\lambda-\frac{1}{2}}(\alpha\sqrt{\delta^2 + (x-\mu)^2}).$$

Here K_ν denotes the modified Bessel function of third kind with index ν. When $\theta = -\frac{1}{2}$, GIG becomes $IG(\delta, \gamma)$.

In the case of -1 correlation, the moment estimator is $\theta_T = -\frac{Y_T}{T}$.

Let $q(x, \theta)$ be a density function of Y_t. The contrast function is defined as $u(x, \theta) = -\log q(x, \theta)$. The corresponding minimum contrast estimators are given by $\bar{\theta}_n$ and $\bar{\rho}_n$.

The moment estimators satisfy the following:

Theorem 7.1

$$\begin{aligned}
&\hat{\theta}_n \to \theta_0 \quad a.s. \quad as \;\; n \to \infty,\\
&\hat{\rho}_n \to \rho_0 \quad a.s. \quad as \;\; n \to \infty,\\
&\sqrt{n}(\hat{\theta}_n - \theta_0) \to^{\mathcal{D}} N(0, I^{-1}(\theta_0)) \;\; as \;\; n \to \infty,\\
&\sqrt{n}(\hat{\rho}_n - \rho_0) \to^{\mathcal{D}} N(0, I^{-1}(\rho_0)) \;\; as \;\; n \to \infty,
\end{aligned}$$

where $I(\theta_0)$ and $I(\theta_0)$ are the corresponding Fisher information.

Sketch of Proof Application of Birkoff Ergodic Theorem and Mixing CLT (see Durrett 2019 give the results. We omit the details.

Next we consider fractional inverse Gaussian process Z, which is a fractional Levy process with parameters γ and δ and Hurst parameter $H > 0.5$. The process Z is the sum of two independent Levy processes $Z = Z^{(1)} + Z^{(2)}$, where $\mathcal{L}(X_0) = IG(\delta/2, \gamma)$ and $Z^{(2)}$ is a compound Poisson process given by $Z^{(2)} = \gamma^{-2} \sum_{j=1}^{N_t} u_j$ with N being a Poisson process with intensity $\delta\gamma/2$ and u_j is a sequence of independent and identically χ_1^2-distributed random variables independent of N. The processes Z and X have infinitely many jumps in any finite time interval, and hence they are infinite activity processes.

Invariant distribution is Generalized Inverse Gaussian (GIG).

$$\mathcal{L}(X_0) = \mathcal{L}(X_t) = \mathcal{GIG}(\lambda, \delta, \sqrt{\alpha^2 - \beta^2}).$$

Mixture distribution is Q_θ which is Generalized Hyperbolic (GH).

$$\mathcal{L}(Y_t) = \mathcal{GH}(\theta, \frac{\alpha}{\sqrt{t}}, \frac{\beta}{\sqrt{t}}, \sqrt{t}\delta, \mu t).$$

$$\mathcal{L}(Y_{t+1} - Y_t) = \mathcal{GH}(\lambda, \alpha, \beta, \delta, \mu).$$

Inverse Gaussian pdf is given by

$$f(x) = \frac{\delta e^{\delta\gamma}}{\sqrt{2\pi x^{3/2}}} e^{-\frac{1}{2}(\delta^2 x^{-1} + \gamma^2 x))}, \delta, \gamma > 0, x > 0.$$

The cummulant functions of IG-OU process are given by

$$k_Z(u) = \log E[e^{-uZ(1)}] = -u\delta\gamma^{-1}(1 + 2u\gamma^{-2})^{-1/2}.$$

$$k_V(u) = \log E(e^{-uV_t}) = \delta\gamma - \delta\gamma(1 + 2u\gamma^{-2})^{1/2}.$$

The cummulants satisfy

$$k_{y_1}^{(1)} = \theta\rho\Delta k_{FIG}^{(1)}. \tag{7.3}$$

$$k_{y_1}^{(2)} = \Delta k_{FIG}^{(1)} + 2\theta\rho^2 \Delta k_{FIG}^{(2)}. \tag{7.4}$$

Inverting (7.3) and (7.4) and replacing the cummulants by their sample cummulants, we obtain the explicit method of moments estimators.

The moment estimators of ρ and θ are given by

$$\hat{\rho}_n := \frac{\gamma(\gamma s_y^2 - \Delta\delta)}{2\bar{y}}, \quad \hat{\theta}_n := \frac{\gamma\bar{y}}{\Delta\delta\hat{\rho}_n},$$

where

$$s_y^2 := \frac{1}{n}\sum_{j=1}^{n}(y_j - \bar{y})^2 = \frac{1}{n}\sum_{j=1}^{n} y_j^2 - (\bar{y})^2,$$

$$\bar{y} := \frac{1}{n}\sum_{j=1}^{n} y_j, \quad y_j := Y_{j\Delta} - Y_{(j-1).\Delta}$$

Let $\vartheta = (\rho, \theta)$ and $\hat{\vartheta}_n = (\hat{\rho}_n, \hat{\theta}_n)$. The moment estimators satisfy the following:

Theorem 7.2

$$\hat{\theta}_n \to \theta_0 \text{ a.s. as } n \to \infty.$$

$$\hat{\rho}_n \to \theta_0 \text{ a.s. as } n \to \infty.$$

$$n^{1-H}(\hat{\theta}_n - \theta_0) \to^{\mathcal{D}} N(0, I^{-1}(\theta_0)) \text{ as } n \to \infty.$$

$$n^{1-H}(\hat{\rho}_n - \rho_0) \to^{\mathcal{D}} N(0, I^{-1}(\rho_0)) \text{ as } n \to \infty.$$

Sketch of Proof Application of Birkoff Ergodic Theorem and Mixing CLT (see Durrett 2019) give the results. We omit the details.

In order to construct the estimating functions for the standard inverse Gaussian process ($H = 0.5$), we use the first and second cumulants which are given, respectively, by

$$\kappa_{y_1}^{(1)} = \theta\rho\Delta\kappa_{IG}^{(1)}, \quad \kappa_{y_1}^{(2)} = \Delta\kappa_{IG}^{(1)} + 2\theta\rho^2\Delta\kappa_{IG}^{(2)}. \tag{7.5}$$

Inverting these cummulants and replacing the cummulants by their sample quantities, we obtain the explicit the moment estimators of ρ and λ.

The moment estimators of ρ and λ are given by

$$\hat{\rho}_n := \frac{\gamma(\gamma s_y^2 - \Delta\delta)}{2\bar{y}}, \quad \hat{\theta}_n := \frac{\gamma\bar{y}}{\Delta\delta\hat{\rho}_n},$$

where

$$s_y^2 := \frac{1}{n}\sum_{j=1}^{n}(y_j - \bar{y})^2 = \frac{1}{n}\sum_{j=1}^{n} y_j^2 - (\bar{y})^2,$$

$$\bar{y} := \frac{1}{n}\sum_{j=1}^{n} y_j, \quad y_j := Y_{j\Delta} - Y_{(j-1)\Delta}.$$

Let $\vartheta = (\rho, \theta)$ and $\hat{\vartheta}_n = (\hat{\rho}_n, \tilde{\theta}_n)$. We have the following properties of the estimators:

Theorem 7.3 For fixed $\Delta > 0$ as $n \to \infty$,

$$(a) \quad \hat{\vartheta}_n \to \vartheta_0 \text{ a.s. as } n \to \infty.$$

$$(b) \quad \sqrt{n}(\hat{\vartheta}_n - \vartheta_0) \to^{\mathcal{D}} \mathcal{N}_2(0, (2\lambda\rho^2\Delta^2\delta^2\gamma^{-4})^{-2}D(\vartheta_0)) \text{ as } n \to \infty,$$

where $D(\vartheta_0)$ is the limiting covariance matrix.

Remark In the IG-OU stochastic volatility model, for the case of $\rho = -1, \gamma = 1, \delta = 1$, the moment estimator is given by $\hat{\theta}_{n\Delta} := -\frac{Y_{n\Delta}-Y_0}{n\Delta}$. Thus the parameter λ can be estimated by just the two terminal observations.

The moment estimators are sensitive to outliers since they are based on mean and standard deviation of the sample data. In order to incorporate outliers and model misspecifications, we consider robust estimators. The robust estimators of ρ and λ are given by

$$\tilde{\rho}_n := \frac{\gamma(\gamma a_y - \Delta\delta)}{2\tilde{y}}, \quad \tilde{\lambda}_n := \frac{\gamma\tilde{y}}{\Delta\delta\tilde{\rho}_n},$$

where

$$a_y := \frac{1}{n}\sum_{j=1}^{n}|y_j - \tilde{y}|$$

is the sample mean absolute deviation from median,

$$\tilde{y} := \text{median of } \{y_j, 1 \leq j \leq n\},$$

which is defined as

$$\tilde{y} = \begin{cases} \frac{y_k + y_{k+1}}{2} & : \quad n = 2k \\ y_{k+1} & : \quad n = 2k+1 \end{cases}$$

Let $\vartheta = (\rho, \theta)$ and $\tilde{\vartheta}_n = (\tilde{\rho}_n, \tilde{\theta}_n)$. We have the following properties of the estimators. By using the standard theory of order statistics (see Theorem 5.9 and 5.21 in Van der Vaart 2000) and mixing property of the process, along with the Glivenko–Cantelli argument and the delta method, we obtain the following:

Theorem 7.4 For fixed $\Delta > 0$ as $n \to \infty$,

$$(a) \quad \tilde{\vartheta}_n \to \vartheta_0 \text{ a.s.} \quad \text{as } n \to \infty.$$

$$(b) \quad \sqrt{n}(\tilde{\vartheta}_n - \vartheta_0) \to^{\mathcal{D}} \mathcal{N}_2(0, \frac{\pi}{2}(2\theta\rho^2\Delta^2\delta^2\gamma^{-4})^{-2}D(\vartheta_0)) \text{ as } n \to \infty,$$

where $D(\vartheta_0)$ is the limiting covariance matrix.

Chapter 5

Inference for General Semimartingales and Self-similar Processes

5.1 Introduction

In this chapter, we give conditions for local asymptotic mixed normality (LAMN) when the observed process is a semimartingale and the observation time increases to infinity with decreasing time interval on a discrete observation scheme. As a consequence we obtain asymptotic efficiency of various estimators of the model.

We also study asymptotic behavior of Type II error of Neyman-Pearson tests when the error tends to zero. An exponential second order expansion is derived provided the log-likelihood process satisfies the central limit theorem (CLT). Finally, we extend the problem to sequential experiments using stopping rules.

Luschgy (1992) studied local asymptotic mixed normality for semimartingales. Luschgy (1993) studied second order behavior of Neyman-Pearson tests for diffusions. Luschgy (1994a) studied asymptotic expansions for error probabilities for tests. Luschgy (1994b) studied asymptotic behavior of Neyman-Pearson tests for autoregressive processes. Luschgy (1994c) studied asymptotic inference for semimartingale models with singular parameter points. Luschgy (1995) studied local asymptotic quadraticity for stochastic process based on stopping times.

5.2. Continuous Sampling of Semimartingales

Let $(\Omega, \mathcal{F}, (P_\theta : \theta \in \Theta))$ with $\Theta \subset \mathbb{R}^k$ be a statistical experiment and let $X = (X_t)_{t \geq 0}$ be a $\mathbb{R}^d$-valued process defined on Ω and assume that under each P_θ, X is a semimartingale with respect to the filtration $(\mathcal{F}_t)_{t \geq 0}$ generated by X. The experiment $(\Omega, \mathcal{F}_t, (P_\theta | \mathcal{F}_t : \theta \in \Theta))$ corresponds to the observation of X up to time t.

For every $\theta \in \Theta$, X is a P_θ-semimartingale with triplet of predictable characteristics $(B(\theta), C, \nu(\theta))$ relative to some fixed Borel measurable truncation function $h : \mathbb{R}^d \to \mathbb{R}^d$. Here $B(\theta)$ is the drift characteristic, C is the quadratic characteristic defined by $C = \langle X^c(\theta) \rangle$, where $X^c(\theta)$ is the continuous martingale part of X, $\nu(\theta)$ is the compensator of the random measure μ on $\mathbb{R}_+ \times \mathbb{R}^d$ associated with the jumps of X by

J. P. N. Bishwal, *Parameter Estimation in Stochastic Volatility Models*,
https://doi.org/10.1007/978-3-031-03861-7_5

$$\mu = \sum_{t \geq 0} \varepsilon_{(t,\Delta X_t)} I_{\{\Delta X_t \neq 0\}}$$

with $\Delta X_t = X_t - X_{t-}, \quad \Delta X_0 = 0$.

We denote $H \bullet Z_t = \int_0^t H_s dZ_s$ and $U \star \eta_t = \int_0^t \int_{\mathbb{R}} U(s,x)\eta(ds,dx)$ denote the (ordinary or stochastic) integral of H and U with respect to the semimartingale Z and the random measure η.

We have the following representation of X under P_θ:

$$X_t = X_0 + B_t(\theta) + X_t^c(\theta) + \int_0^t \int_{|x| \leq 1} x d(\mu - \nu(\theta)) + \int_0^t \int_{|x|>1} x d\mu.$$

Thus

$$X(\theta)_t = X(\theta)_t^c + X(\theta)_t^d.$$

The purely discontinuous local martingale part of X can be written as

$$X_t^d = \int_0^t \int_{|x| \leq 1} x d(\mu - \nu(\theta)).$$

Here $d\mu = \mu(\omega, dt, dx)$ is the random jump measure of X, i.e.,

$$\mu(\omega, (0,t], \gamma) = \sum_{0<s\leq t} I_{(\Delta X_s \in \Gamma)}, \quad \Gamma \in \mathcal{B}(\mathbb{R}\backslash\{0\})$$

and $d\nu = \nu(\omega, dt, dx)$ is the compensator.

$$\sum_{0<s\leq t} I_{(|\Delta X_s|>1)} = \int_0^t \int_{|x|>1} x d\mu.$$

We define LAMN, asymptotic efficiency, asymptotic differentiability, quadratic mean differentiability (q.m.d.), LAQ, LABF.

Definition 1 (Le Cam and Yang (1990), Jeganathan (1995)) Let $\mathcal{E}_n = (\Omega_n, \mathcal{F}_n, (P_\theta^n, \theta \in \Theta)), n \geq 1$ be a sequence of statistical experiments, where Θ is an open subset of $\mathbb{R}$. We denote by

$$\Lambda_{\eta\theta}^n = \log\left(\frac{dP_\eta^n}{dP_\theta^n}\right)$$

the log-likelihood between η and θ at stage n.

We say that the sequence $\mathcal{E}_n$ satisfies the *local asymptotically quadratic* (LAQ) condition at a point $\theta \in \Theta$ if there are random variables Δ_n and Γ_n defined on $(\Omega_n, \mathcal{F}_n), \Gamma_n > 0$ a.s. $[P_\theta^n]$ and a positive numerical sequence $\phi_n \downarrow 0$ such that for each bounded sequence of numbers u_n,

$$\Lambda_{\theta+\phi_n u_n,\theta}^n - (u_n\Delta_n - \frac{1}{2}u_n^2\Gamma_n) \xrightarrow{P_\theta^n} 0$$

and

$$(\Delta_n, \Gamma_n) \to (\Delta, \Gamma) \text{ in } P_\theta^n - \text{distribution},$$

where Δ and Γ are random variables on a measurable space $(\Omega, \mathcal{F}, P)$ with $\Gamma > 0$ a.s. (P) and

$$E_P \exp(u\Delta - \frac{1}{2}u^2\Gamma) = 1.$$

The sequence of experiments is called *locally asymptotically Brownian functional* (LABF) if $\Delta = \int_0^1 F_s dW_s$ and $\Gamma = \int_0^1 F_s^2 ds$ with W a standard Brownian motion and F a predictable process with respect to some filtration in $\mathcal{F}$. It is called *locally asymptotically mixed normal* (LAMN) if $\Delta = \Gamma^{1/2} W_1$ with W_1 standard normal variable independent of Γ and *locally asymptotically normal* (LAN) if, in addition, Γ is nonrandom.

Definition 2 (Gushchin (1995)) Let the sequence $(\mathcal{E}_n)$ satisfy LAQ condition at a point $\theta \in \Theta$. A sequence of estimators $\hat{\theta}_n$ is called *asymptotically centering* (AC) sequence at θ if $\phi_n^{-1}(\hat{\theta}_n - \theta) - \Gamma_n^{-1}\Delta_n \stackrel{P_\theta^n}{\to} 0$.

Asymptotically centering estimators are also called *asymptotic maximum likelihood estimators.*

Let $(\Omega^n, \mathcal{F}^n, \{\mathcal{F}_t^n\}_{t \geq 0}, \{P_\theta^n, \theta \in \Theta\})$, where Θ is an open subset of $\mathbb{R}$, be a sequence of filtered statistical experiments. Let $P_{\theta t}^n$ be the restriction of P_θ^n to $\mathcal{F}_t^n$.

The log-likelihood process $\Lambda_{\eta\theta}^n$ at stage n between η and θ is defined by

$$\Lambda_{\eta\theta t}^n = \log\left(\frac{dP_{\eta t}^n}{dP_{\theta t}^n}\right).$$

Definition 3 (Greenwood and Wefelmeyer (1993)) We call the sequence of filtered experiments *functionally LAQ* at θ if there exist processes Δ_n and $\Gamma_n > 0$ a.s. and a positive numerical sequence $\phi_n \downarrow 0$ such that for each bounded sequence of numbers u_n and all $t > 0$,

$$\sup_{s \leq t} \left|\Lambda_{\theta+\phi_n u_n, \theta, s}^n - (u_n \Delta_s^n - \frac{1}{2}u_n^2 \Gamma_s^n)\right| \stackrel{P_\theta^n}{\to} 0$$

and

$$(\Delta_t^n, \Gamma_t^n) \to (\Delta_t, \Gamma_t) \text{ in } P_\theta^n - \text{distribution},$$

where Δ_t and Γ_t are processes on a filtered probability space $(\Omega, \mathcal{F}, \{\mathcal{F}_t\}_{t \geq 0}, P)$ with $\Gamma_t > 0$ a.s. and

$$E_P \exp(u\Delta_t - \frac{1}{2}u^2\Gamma_t) = 1 \text{ for all } t \geq 0.$$

We call the sequence of filtered models *functionally LABF* if $\Delta_t = \int_0^t F_s dW_s$ and $\Gamma_t = \int_0^t F_s^2 ds$ where W is a standard Brownian motion and F is a predictable process.

Proposition 4.1 (Greenwood and Wefelmeyer (1993)) Suppose the sequence of filtered models is functionally LAQ at θ, with Δ a continuous local martingale, Γ equal to the quadratic variation $\langle\Delta\rangle$ of Δ, and Γ strictly increasing to infinity a.s. Let ω be a bounded, symmetric, bowl shaped loss function on the real line, and let $\hat{\theta}^n$ be a sequence of estimator processes. For each $n \in \mathbb{N}$ and $t \geq 0$ let

$$T_t^n = \inf\{s : \Gamma_s^n > t\}.$$

Then

$$\lim_{b\to\infty} \liminf_{n\to\infty} \sup_{|u|\leq b} E_{\theta+\phi_n u}\omega(\phi_n^{-1}(\hat{\theta}^n_{T_t^n} - (\theta + \phi_n u))) \geq E\omega(W_{1/t}). \square$$

If a filtered model satisfies the functionally LAN condition, then the filtered model time changed by T^n satisfies the functionally LAQ condition, which is evident from the following result in Greenwood and Wefelmeyer (1993).

Lemma 4.1 Under the assumptions of Theorem 4.1 for each bounded sequence of numbers u_n and all $t > 0$,

$$\sup_{s\leq t}\left|\Lambda^n_{\theta+\phi_n u_n,\theta,T_s^n} - (u_n\Delta^n_{T_s^n} - \frac{1}{2}u_n^2 s)\right| \stackrel{P_\theta^n}{\to} 0$$

and

$$\Delta^n_{T^n} \to W_1 \text{ in } P_\theta^n - \text{distribution.} \square$$

If the assertions of Lemma 4.1 hold then the sequence of models is said to satisfy the *functionally LAN* condition.

Definition 4 (Greenwood and Wefelmeyer, 1993) If the sequence of filtered models is functionally LAQ at θ, we call a sequence of estimator process $\hat{\theta}^n$ *functionally AC* at θ, if for all $t > 0$,

$$\sup_{s\leq t}\left|\phi_n^{-1}(\hat{\theta}_s^n - \theta) - \Delta_s^n(\Gamma_s^n)^{-1}\right| \stackrel{P_\theta^n}{\to} 0.$$

The filtered model observed up to T_t^n is LAN, hence by the classical theory, the local asymptotic minimax theorem applies and $\hat{\theta}^n_{T_t^n}$ are *locally asymptotically minimax* if and only if they are *AC*.

Let us return to our problem.

Likelihood:

$$B_t(\tau) = B_t(\theta) + \int_0^t c_s\beta_s(\theta,\tau)^2 dF_s + \int_0^t \int_{|x|\leq 1} (Y(\theta,\tau) - 1)hd(\nu),$$

where

$$Y(\theta,\tau) = \frac{d\nu(\tau)}{d\nu(\theta)}$$

is the jump density.

Assume that $\nu(\theta, \{t\} \times \mathbb{R}) = 0$ for every $\theta \in \Theta$ and $t \geq 0$. This condition is equivalent to the quasi-left continuity of X under P_θ. There are continuous adapted increasing process F with $F_0 = 0$ and a predictable process c such that $C = c \bullet F$.

By Girsanov theorem, the log-likelihood is given by

$$\begin{aligned} & \log L(\theta,\tau) \\ = \; & \int \beta(\theta,\tau) dX_t^c(\theta) - \tfrac{1}{2}\int c\beta(\theta,\tau)^2 dF \\ & + \int_{|x|\leq 1}(Y(\theta,\tau)-1)d(\mu-\nu) + \int_{|x|>1}(\log Y(\theta,\tau) - Y(\theta,\tau)+1)d\mu \end{aligned}$$

and $L_0(\theta,\tau) = 1$. We denote by $L'(\theta)$ is a local martingale and its quadratic variation is $[L'(\theta)]$.

Neyman-Pearson Tests

An exponential second order expansion is derived provided the log-likelihood process satisfies the CLT.

For $\alpha \in [0,1]$, the *most powerful test* of level (Type I Error) α for testing the null hypothesis $H_0 : \{P_t\}$ against the alternative hypothesis $H_1 : \{Q_t\}$ is given by

$$\varphi_{\alpha,t} = I_{\{dQ_t/dP_t > d_{\alpha,t}\}} + \gamma_{\alpha,t}\; I_{\{dQ_t/dP_t = d_{\alpha,t}\}}$$

with

$$E_{P_t}[\varphi_{\alpha,t}] = \alpha, \quad d_{\alpha,t} \in (0,\infty), \quad \gamma_{\alpha,t} \in [0,1].$$

$d_{\alpha,t}$ is the critical value. The probability of Type II Error is defined as

$$\beta_t(\alpha) = E_{Q_t}(1 - \varphi_{\alpha,t})$$

satisfies $\lim_{t\to\infty} \beta_t(\alpha) = 0$.

Let K is the Kullback information for Levy measures defined by

$$K(L(P), L(Q)) = \int (Y \log Y - Y + 1) dL(Q).$$

Let

$$K := K(L(P), L(Q)) + \frac{1}{2} c\beta(\theta,\tau)^2.$$

Then

$$\log \frac{dP_t}{dQ_t} - Kt = \beta(\theta,\tau) + \int_0^t \int \log Y d(\mu - \nu) =: N_t.$$

The process N is a P-martingale with $E_P(N_t^2) < \infty$ and the bracket process

$$\langle N\rangle_t = |c^{1/2}\beta(\theta,\tau)|^2 + \int (\log Y)^2 dLP)(t).$$

We have

$$\begin{aligned}\log \beta_t(\alpha_t) &= -t[K(L(P),L(Q)) + \tfrac{1}{2}c\beta(\theta,\tau)^2)] \\ &\quad -t^{-1/2}[\textstyle\int (\log Y)^2 dL(P) + c^{1/2}\beta(\theta,\tau)^2]^{1/2}\Phi^{-1}(\alpha) + o(t^{-1/2}).\end{aligned}$$

Note that $K(P_t,Q_t) = Kt$ and $Var_P \log \frac{dP_t}{dQ_t} = E_P(N_t^2) = E_P\langle N\rangle_t = \langle N\rangle_1 t$.

If $L(P) = L(Q)$,

$$\begin{aligned}\log \beta_t(\alpha_t) &= -t|c^{1/2}\beta(\theta,\tau)|^2/2 + \tfrac{1}{2}c\beta(\theta,\tau)^2)] \\ &\quad -t^{-1/2}|c^{1/2}\beta(\theta,\tau)|\Phi^{-1}(\alpha) + o(t^{-1/2}).\end{aligned}$$

Natural Exponential Families of Levy Processes:

$$\frac{dL(X_t|P_\theta)}{dL(X_t|R)}(x) = e^{\theta x - f(\theta)t}.$$

Example 1.1 Arato-Kolmogorov-Sinai's two-dimensional Gaussian diffusion process of Geophysics. (extend to fractional: chaos expansion approach)

The instantaneous axis of earth's rotation is displaced with respect to the minor axis of the terrestrial ellipsoid. This displacement consists of a periodic part and a fluctuation part. The following process has been used for latter fluctuation.

$$dX_1 = -\theta_1 X_1 dt - \theta_2 X_2 dt + dW_1$$

$$dX_2 = \theta_2 X_1 dt - \theta_1 X_2 dt + dW_2$$

Let $Y = X_1 + iX_2, \theta = \theta_1 - i\theta_2, W = W_1 + iW_2$

$$dY = -\theta Y dt + dW.$$

Under P_θ, let $X = (X_1, X_2)^\mathrm{T}$ be the unique strong solution of the SDE

$$dX_t = A(\theta)X_t dt + dW_t(\theta), \quad X_0 = 0,$$

where

$$A(\theta) = \begin{pmatrix} -\theta_1 & -\theta_2 \\ \theta_2 & -\theta_1 \end{pmatrix}.$$

Here $W(\theta)$ is a two-dimensional Winer process and $\Theta = \mathbb{R}^2$. By Itô's formula

$$X_t = e^{A(\theta)t}\int_0^t e^{-A(\theta)t} dW_s(\theta)$$

where

$$e^{A(\theta)t} = e^{-\theta_1 t}\begin{pmatrix} \cos\theta_2 t & -\sin\theta_2 t \\ \sin\theta_2 t & \cos\theta_2 t \end{pmatrix}.$$

The information process is given by

$$[L'(\theta)]_t = \langle L'(\theta)\rangle_t = \int_0^t |X_s|^2 ds I_2, \quad \theta\in\Theta.$$

We have

$$EX_tX_t^{\mathrm{T}} = \begin{cases} \dfrac{1-e^{-2\theta_1 t}}{2\theta_1} I_2 & : \ \theta_1 \neq 0 \\ tI_2 & : \ \theta_1 = 0 \end{cases},$$

where I_2 is the 2×2 unit matrix which gives the expected Fisher information

$$E\langle L'(\theta)\rangle_t = \begin{cases} \dfrac{e^{-2\theta_1 t}+2\theta_1 t - 1}{2\theta_1^2} I_2 & : \ \theta_1 \neq 0 \\ t^2 I_2 & : \ \theta_1 = 0 \end{cases}.$$

The eigenvalues of $A(\theta)$ are $-\theta_1 + i\theta_2$ and $-\theta_1 - i\theta_2$.
In the explosive case $\theta_1 < 0$, the localization rate is

$$\delta_{1,T} = \delta_{2,T} = -\sqrt{2}\theta_1 e^{\theta_1 T}.$$

The process $N_t(\theta) := e^{-A(\theta)t}X_t$ is a zero mean square integrable Gaussian martingale,

$$E_\theta(e^{-A(\theta)t}X_t)(e^{-A(\theta)t)}X_t)^T = \frac{e^{2\theta_1 t}-1}{2\theta_1} I_2.$$

By the martingale convergence theorem, if $\theta_1 < 0$,

$$e^{-A(\theta)t)}X_t \to \left(-\frac{1}{2\theta_1}\right)^{1/2} V(\theta) \quad P_\theta\text{-a.s.}$$

with $\mathcal{L}(V(\theta|P_\theta) = N(0, I_2)$. Since $|N_t(\theta)|^2 = e^{2\theta_1 t}|X_t|^2$, this gives

$$-\theta_1 e^{2\theta_1 t}|X_t|^2 \to |V(\theta)|^2/2 \quad P_\theta\text{-a.s.}$$

and by the integral version of Toeplitz lemma

$$2\theta_1^2 e^{2\theta_1 T}\int_0^T |X_s|^2 ds \to |V(\theta)|^2/2 \quad P_\theta\text{-a.s.}$$

Thus $G(\theta) = \frac{1}{2}|V(\theta)|^2 I_2$.
LAMN is satisfied for every θ with $\theta_1 < 0$.
In the stable case, $\theta_1 > 0$, the ergodic theorem gives

$$\frac{1}{T}\int_0^T |X_s|^2 ds \to \lim_{s\to\infty} E_\theta|X_s|^2 = \frac{1}{\theta_1} \quad P_\theta\text{-a.s.}$$

The localization rate is

$$\delta_{1,T} = \delta_{2,T} = \left(\frac{\theta_1}{T}\right)^{1/2}.$$

LAN is satisfied for every θ with $\theta_1 > 0$.

On the line $\theta_1 = 0$, a singularity occurs. The estimating equation is

$$A(\theta) = \begin{pmatrix} -\theta_1 & -\theta_2 \\ \theta_2 & -\theta_1 \end{pmatrix}$$

and

$$L'_t(\theta) = \begin{pmatrix} -X_1 \bullet X_{1,t} - X_2 \bullet X_{2,t} \\ X_1 \bullet X_{2,t} - X_2 \bullet X_{1,t} \end{pmatrix} - \langle L'(\theta)\rangle_t \theta = \begin{pmatrix} 0 \\ 0 \end{pmatrix}.$$

Since $-X_1 \bullet X_{1,t} - X_2 \bullet X_{2,t} = -\frac{1}{2}|X_t|^2 + t$, the MLE is given by

$$\hat{\theta}_t = \left(\int_0^t |X_s|^2 ds\right)^{-1} \begin{pmatrix} -\frac{1}{2}|X_t|^2 + t \\ X_1 \bullet X_{2,t} - X_2 \bullet X_{1,t} \end{pmatrix}.$$

The MLE is *regular and efficient* for every θ with $\theta_1 \neq 0$.

Theorem 1.1 *(Stable Case)*
The limiting distribution of $\delta_t^{-1}(\hat{\theta}_t - \theta)$ under P_θ is $N(0, I_2)$ distribution if $\theta_1 > 0$.

Theorem 1.2 *(Explosive Case)*
The limiting distribution of $\delta_t^{-1}(\hat{\theta}_t - \theta)$ under P_θ is the bivariate t_2-distribution if $\theta_1 < 0$.

So in the explosive case, $\delta_{1,t}^{-1}(\hat{\theta}_{1,t} - \theta)$ and $\delta_{2,t}^{-1}(\hat{\theta}_{2,t} - \theta)$ are not *asymptotically independent* in contrast to the stable case.

Theorem 1.3 *(Stable and Explosive Case)*
Under the random norming, the limiting distribution of $\left(\int_0^t |X_s|^2 ds\right)^{1/2} (\hat{\theta}_t - \theta)$ is $N(0, I_2)$ for every θ with $\theta_1 \neq 0$.

5.3 Discrete Sampling of Semimartingales

The process X is observed at discrete time points $t_1 < t_2 < \ldots < t_n$.

The approximate log-likelihood is given by

$$\begin{aligned} & \log \tfrac{dQ_t^n}{dP_t^n} \\ = \ & \log L_n(\theta,\tau) \\ = \ & \sum \beta(\theta,\tau)(X_{t_i}^c(\theta) - X_{t_{i-1}}^c(\theta)) - \tfrac{1}{2}\sum c\beta(\theta,\tau)^2(F_{t_i} - F_{t_{i-1}}) \\ & + \int_{|x|\le 1}(Y(\theta,\tau)-1)d(\mu-\nu) + \int_{|x|>1}(\log Y(\theta,\tau) - Y(\theta,\tau)+1)d\mu. \end{aligned}$$

$L_n'(\theta)$ is a local martingale and its quadratic variation is $[L_n'(\theta)]$

Neyman-Pearson Tests

For $\alpha \in [0,1]$, the most powerful test of level (Type I Error) α for testing the null hypothesis $H_0 : \{P_t^n\}$ against the alternative hypothesis $H_1 : \{Q_t^n\}$ is given by

$$\varphi_{\alpha,t} = I_{\{dQ_t^n/dP_t^n > d_{\alpha,t,n}\}} + \gamma_{\alpha,t}\, I_{\{dQ_t^n/dP_t^n = d_{\alpha,t,n}\}}$$

with

$$E_{P_t^n}\varphi_{\alpha,t,n} = \alpha, \quad d_{\alpha,t,n} \in (0,\infty), \quad \gamma_{\alpha,t,n} \in [0,1].$$

The Type II Error is defined as

$$\beta_t^n(\alpha) = E_{Q_t^n}(1 - \varphi_{\alpha,t,n})$$

and satisfies $\lim_{t\to\infty}\beta_t^n(\alpha) = 0$.

Hodges-Lehmann Test

A globally optimal test called Hodges-Lehmann test is a test φ_t such that the limit $t^{-1}\log E_{(\theta,\xi_2)}(1-\varphi_t)$ exists and is equal to $-K(\theta)$ for every $\theta \in \Theta$.

We arrive at a test which is a of asymptotic level α and globally optimal. We have

$$\begin{aligned} & \log \beta_t^n(\alpha_t) \\ = \ & -t[K(L(P^n), L(Q^n)) + \tfrac{1}{2}c\beta(\theta,\tau)^2)] \\ & -t^{-1/2}[\textstyle\int(\log Y)^2 dL(P) + |c^{1/2}\beta(\theta,\tau)|^2]^{1/2}\Phi^{-1}(\alpha) + o(t^{-1/2}), \end{aligned}$$

where K is the Kullback-Leibler information for Levy measures defined by

$$K(L(P^n), L(Q^n)) = \int (Y\log Y - Y + 1)dL(Q^n).$$

Let

$$K := K(L(P^n), L(Q^n)) + \frac{1}{2}c\beta(\theta,\tau)^2.$$

Then

$$\log\frac{dP_t^n}{dQ_t^n} - Kt = \beta(\theta,\tau)X_t^c(P) + \int_0^t\int \log Y d(\mu-\nu) =: N_t^n.$$

The process N^n is a P^n-martingale with $E_P((N_t^n)^2) < \infty$ and the bracket process

$$\langle N^n\rangle_t = |c^{1/2}\beta(\theta,\tau)|^2 + \int (\log Y)^2 dL(P)(t).$$

It follows from the CLT for martingales that

$$\mathcal{L}(t^{-1/2}N_t^n|P) \Rightarrow \mathcal{N}(0,\langle N^n\rangle_t).$$

If $L(P^n) = L(Q^n)$, then

$$X_t^c(Q) = X_t^c(P) + c\beta(\theta,\tau)t$$

$$\log\frac{dP_t^n}{dQ_t^n} = \beta(\theta,\tau)X_t^c(P) + \frac{1}{2}c\beta^2(\theta,\tau)t.$$

$$\mathcal{L}(t^{-1/2}[\log\frac{dP_t^n}{dQ_t^n} - \frac{1}{2}c\beta^2(\theta,\tau)t]|P) = N(0,c\beta^2(\theta,\tau).$$

$$\begin{aligned} &\log\beta_t^n(\alpha_t)\\ = \; &-t|c^{1/2}\beta(\theta,\tau)|^2/2 + \tfrac{1}{2}c\beta(\theta,\tau)^2)]\\ &-t^{-1/2}|c^{1/2}\beta(\theta,\tau)|\Phi^{-1}(\alpha) + o(t^{-1/2}).\end{aligned}$$

Natural Exponential Families of Levy Processes:

$$\frac{dL(X_t^n|P_\theta^n)}{dL(X_t^n|R^n)}(x) = e^{\theta x - f(\theta)t}.$$

Example 2.1 Arato-Kolmogorov-Sinai's two-dimensional Gaussian diffusion process of Geophysics

The AMLE is given by

$$\hat{\theta}_{n,t} = \left(\sum |X_{t_i}|^2\Delta t_i\right)^{-1}\begin{pmatrix} -\frac{1}{2}|X_{t_i}|^2 + t \\ X_1 \star X_{2,t} - X_2 \star X_{1,t}\end{pmatrix}$$

where $\star$ represents sum corresponding to the integral $\bullet$.
The AMLE is *regular* and *efficient* for every θ with $\theta_1 \neq 0$.

Theorem 2.4 *(Stable Case) The limiting distribution of $\delta_{n,t}^{-1}(\hat{\theta}_{n,t} - \theta)$ under P_θ is $N(0, I_2)$ distribution if $\theta_1 > 0$.*

Theorem 2.5 *(Explosive Case) The limiting distribution of $\delta_{n,t}^{-1}(\hat{\theta}_{n,t} - \theta)$ under P_θ is the bivariate t_2-distribution if $\theta_1 < 0$.*

So in the explosive case, $\delta_{1,t}^{-1}(\hat{\theta}_{1,t} - \theta)$ and $\delta_{2,t}^{-1}(\hat{\theta}_{2,t} - \theta)$ are not *asymptotically independent* in contrast to the stable case.

Theorem 2.6 *(Stable and Explosive Case) Under the random norming, the limiting distribution of* $\left(\sum |X_{t_i}|^2 \Delta t_i\right)^{1/2} (\hat{\theta}_t - \theta)$ *is* $N(0, I_2)$ *for every* θ *with* $\theta_1 \neq 0$.

The information process is given by

$$[L'(\theta)]_{n,t} = \langle L'(\theta)\rangle_{n,t} = \sum_{1}^{n} |X_i|^2 \Delta t_i I_2, \quad \theta \in \Theta.$$

The expected Fisher information is

$$EX_t X_t^{\mathrm{T}} = \begin{cases} \dfrac{1 - e^{-2\theta_1 t}}{2\theta_1} I_2 & : \quad \theta_1 \neq 0 \\ tI_2 & : \quad \theta_1 = 0 \end{cases}$$

which gives the expected Fisher information

$$E\langle L'(\theta)\rangle_t = \begin{cases} \dfrac{e^{-2\theta_1 t} + 2\theta_1 t - 1}{2\theta_1^2} I_2 & : \quad \theta_1 \neq 0 \\ t^2 I_2 & : \quad \theta_1 = 0 \end{cases}.$$

The eigenvalues of $A(\theta)$ are $-\theta_1 + i\theta_2$ and $-\theta_1 - i\theta_2$.
In the explosive case $\theta_1 < 0$, the localization rate is

$$\delta_{1,T} = \delta_{2,T} = -\sqrt{2}\theta_1 e^{\theta_1 T}.$$

The process $N_t(\theta) := e^{-A(\theta)t} X_t$ is a zero mean square integrable Gaussian martingale,

$$E_\theta(e^{-A(\theta)t} X_t)(e^{-A(\theta)t)} X_t)^T = \frac{e^{2\theta_1 t} - 1}{2\theta_1} I_2.$$

By the martingale convergence theorem, if $\theta_1 < 0$,

$$e^{-A(\theta)t)} X_t \to \left(-\frac{1}{2\theta_1}\right)^{1/2} V(\theta) \quad P_\theta\text{-a.s.}$$

with $\mathcal{L}(V(\theta|P_\theta) = N(0, I_2)$. Since $|N_t(\theta)|^2 = e^{2\theta_1 t}|X_t|^2$, this gives

$$-\theta_1 e^{2\theta_1 t}|X_t|^2 \to |V(\theta)|^2/2 \quad P_\theta\text{-a.s.}$$

and by the integral version of Toeplitz lemma

$$2\theta_1^2 e^{2\theta_1 T} \sum_{i=1}^{n} |X_i|^2 \Delta t_i \to |V(\theta)|^2/2 \quad P_\theta\text{-a.s.}$$

Thus

$$G(\theta) = \frac{1}{2}|V(\theta)|^2 I_2.$$

LAMN is satisfied for every θ with $\theta_1 < 0$. In the stable case, $\theta_1 > 0$, the ergodic theorem gives

$$\frac{1}{T}\int_0^T |X_s|^2 ds \to \lim_{s\to\infty} E_\theta |X_s|^2 = \frac{1}{\theta_1} \quad P_\theta\text{-a.s.}$$

The localization rate is

$$\delta_{1,T} = \delta_{2,T} = \left(\frac{\theta_1}{T}\right)^{1/2}.$$

LAN is satisfied for every θ with $\theta_1 > 0$. On the line $\theta_1 = 0$, a singularity occurs.

The estimating equation is

$$A(\theta) = \begin{pmatrix} -\theta_1 & -\theta_2 \\ \theta_2 & -\theta_1 \end{pmatrix}$$

and

$$L'_{n,t}(\theta) = \begin{pmatrix} -X_1 \bullet X_{1,t} - X_2 \bullet X_{2,t} \\ X_1 \bullet X_{2,t} - X_2 \bullet X_{1,t} \end{pmatrix} - \langle L'(\theta)\rangle_{n,t}\theta = \begin{pmatrix} 0 \\ 0 \end{pmatrix}.$$

Since $-X_1 \bullet X_{1,t} - X_2 \bullet X_{2,t} = -\frac{1}{2}|X_t|^2 + t$, the AMLE is given by

$$\hat{\theta}_{n,t} = \left(\sum_1^n |X_i|^2 \Delta t_i\right)^{-1} \begin{pmatrix} -\frac{1}{2}|X_t|^2 + t \\ X_1 \bullet X_{2,t} - X_2 \bullet X_{1,t} \end{pmatrix}.$$

The AMLE is *regular and efficient* for every θ with $\theta_1 \neq 0$.

5.4 Asymptotics of the Log-likelihood Function

Theorem 4.3 *Let $\alpha_t \to \alpha$, $\alpha \in (0,1)$ and $\theta_1 \neq \tau_1$.*

(i) We have with $\tau_1 > 0$,

$$\lim_{t\to\infty} t^{-1} L_t(\tau,\theta) = K(\tau,\theta).$$

$$\frac{L_t(\tau,\theta) - Kt}{\sqrt{t}} \to N(0,H).$$

(ii) We have with $\tau_1 = 0$ and $m(\tau_2) = 0$,

$$\lim_{t\to\infty} t^{-2} L_t(\tau,\theta) = \frac{1}{2}\theta_1^2 \sigma^2(\tau_2) \int_0^1 W_s^2 ds.$$

(iii) If $\tau_1 = 0$ *and* $m(\tau_2) \neq 0$,

$$\lim_{t\to\infty} t^{-5/2}(L_t(\tau,\theta) - \bar{K}t^3) \to N(0, \bar{H}),$$

where

$$\bar{K}(\tau,\theta) = \frac{1}{6}\theta_1^2 m(\tau_2)^2, \quad \bar{H}(\tau,\theta) = \frac{2}{15}\theta_1^4 m(\tau_2)^2\sigma^2(\tau_2).$$

(iv) We have with $\tau_1 < 0$,

$$\lim_{t\to\infty} e^{2\tau_1 t} L_t(\tau,\theta) = -\frac{(\theta_1 - \tau_1)^2}{4\tau_1} Y(\tau_1)^2$$

where G_τ *is the d.f. of* $Y(\tau_1)^2$ *with* $Y(\tau_1) = \int_0^\infty e^{\tau_1 s} dV_s(\tau_1)$ *under* P_τ. *The distribution of* $Y(\tau_1)$ *is infinitely divisible with a finite second moment and Levy characteristics* $(-(m(\tau_2)/\tau_1),\ -(1/2\tau_1),\ \bar{L}(\tau))$ *without truncation, where*

$$\bar{L}(\tau)(A) = \int_0^\infty L(\tau_2)(\exp(-\tau_1 s)A)ds, \quad A \in \mathcal{B}(\mathbb{R}).$$

Example 4.1 *Hypothesis Testing for Levy Driven Ornstein–Uhlenbeck Process*

Let X_t be described by the SDE

$$dX_t = -\theta_1 X_t dt + dV_t(\theta_2), \ X_0 = 0,$$

where $\theta_1 \in \Theta_1 = \mathbb{R}$ and $V(\theta_2)$ is a process with stationary independent increments with $V_0(\theta_2) = 0$ and Levy characteristics $(b(\theta_2)t, ct, L(\theta_2)t)$ depending on a parameter θ_2 from an arbitrary set Θ_2. We assume that the trajectories of $V(\theta_2)$ are right continuous with left limits. This model provides the natural discrete analogue of AR(1) models with i.i.d. innovations.

For $\theta = (\theta_1, \theta_2) \in \Theta = \Theta_1 \times \Theta_2$, let $P_{\theta,t}$ be the distribution of X when observed up to time t.

The asymptotic behavior of the probability of type II error can be obtained from the asymptotics of the likelihood ratio process.

Under P_θ, the process $V(\theta_1)$ defined by $V_t(\theta_1) = X_t + \theta_1 \int_0^t X_s ds$ has independent stationary increments and Levy characteristics $(b(\theta_2)t, ct, L(\theta_2)t)$ relative to some fixed continuous bounded truncation function $h(x)$ with compact support satisfying $h(x) = x$ in a neighborhood of zero, that is,

$$E_\theta \exp(izV(\theta_1))$$

$$= \exp\left(t\left[izb(\theta_2) - \frac{1}{2}cz^2 + \int(\exp(izx) - 1 - izh(x)L(\theta_2)(dx)\right]\right),$$

$z \in \mathbb{R}$, where $b(\theta_2) \in \mathbb{R}, c > 0$, and $L(\theta_2)$ is a Levy measure on $\mathbb{R}$ satisfying $L(\theta_2)(\{0\} = 0$ and $\int(x^2 \wedge 1)L(\theta_2)(dx) < \infty$. (See Jacod and Shiryayev (1987).)

Thus X is a solution of the above SDE with respect to $V(\theta_1)$. Assume for simplicity that $c = 1$.

Now fix $\tau, \theta \in \Theta$ such that $P_\tau \neq P_\theta$.

Let $m(\tau_2) = E_\tau(V_1(\tau_1))$ and $\sigma^2(\tau_2) = \mathrm{Var}_\tau V_1(\tau_1)$.

Then

$$m(\tau_2) = b(\tau_2) + \int (x - h(x))L(\tau_2)(dx),$$

$$\sigma^2(\tau_2) = 1 + \int x^2 L(\tau_2)dx.$$

We call $(m(\tau_2)t,\ t,\ L(\tau_2)t)$ Levy characteristics of $V(\tau_1)$ without truncation.

Define the Kullback- Leibler information of Levy measures by

$$K(L(\tau_2, \theta_2) = \int (f(\theta_2, \tau_2) - 1 - \log f(\theta_2, \tau_2))dL(\tau_2).$$

The basic regularity condition is the following: We assume the conditions (A1)–(A4).

(A1) $L(\tau_2)$ and $L(\theta_2)$ are mutually absolutely continuous and $\int (f(\tau_2, \theta_2))^{1/2} - 1)^2 dL(\theta_2) < \infty$ where $f(\tau_2, \theta_2) = dL(\tau_2)/dL(\theta_2)$.

(A2) $K((L(\tau_2), L(\theta_2)) < \infty$.

(A3) $\int (\log f(\tau_2, \theta_2))^2 dL(\tau_2) < \infty$.

(A4) $\int x^2 L(\tau_2)(dx) < \infty$.

The log-likelihood ratio admits the representation

$$L_t(\tau, \theta) = -L_t(\theta, \tau) = Z_t + \int_0^t [(\theta_1 - \tau_1)X_s + a(\tau_2, \theta_2)]^2 ds + K(L(\tau_2), L(\theta_2))t,$$

where

$$Z_t = \int_0^t [(\theta_1 - \tau_1)X_{s-} + a(\tau_2, \theta_2)]dX_s^c(\tau) + \int_0^t \int \log f(\tau_2, \theta_2) d(\mu - \nu(\tau_2)).$$

Z is a P_τ-martingale with second moments and $Z_0 = 0$.

The Kullback-Leibler information $K(P_{\tau,t}, P_{\theta,t}) = E_\tau L_t(\tau, \theta)$ of $P_{\tau,t}$ with respect to $P_{\theta,t}$ is finite under the above conditions.

Proposition 2.1 *(a) Assume (A1)–(A4)and $\theta_1 \neq \tau_1$. Then as $t \to \infty$*
(i) We have with $\tau_1 > 0$,

$$K(P_{\tau,t}, P_{\theta,t}) = K(\tau, \theta)t + O(1).$$

(ii) We have with $\tau_1 = 0$ and $m(\tau_2) = 0$,

$$K(P_{\tau,t}, P_{\theta,t}) = \frac{1}{4}\theta_1^2 \sigma^2(\tau_2)t^2 + O(t).$$

(iii) We have with $\tau_1 = 0$ *and* $m(\tau_2) \neq 0$,

$$K(P_{\tau,t}, P_{\theta,t}) = \frac{1}{6}\theta_1^2 m(\tau_2)^2 t^3 + O(t^2).$$

(iv) We have with $\tau_1 < 0$,

$$K(P_{\tau,t}, P_{\theta,t}) = \frac{1}{2}(\theta_1 - \tau_1)^2 \left(\frac{\sigma^2(\tau_2)}{4\tau_1^2} - \frac{m(\tau_2)^2}{2\tau_1^3}\right) e^{-2\tau_1 t} + O(e^{-\tau_1 t}),$$

where

$$K(\tau, \theta) = \frac{1}{2}(\theta_1 - \tau_1)^2 \left(\frac{\sigma^2(\tau_2)}{2\tau_1} + \frac{m(\tau_2)^2}{\tau_1^2}\right) + (\theta_1 - \tau_1)a(\tau_2, \theta_2)\frac{m(\tau_2)}{\tau_1}$$

$$+K(L(\tau_2), L(\theta_2)) + \frac{1}{2}a(\tau_2, \theta_2)^2.$$

(b) Assume (A1)–(A3) and $\theta_1 = \tau_1$, *then*

$$K(P_{\tau,t}, P_{\theta,t}) = \left[K(L(\tau_2), L(\theta_2)) + \frac{1}{2}a(\tau_2, \theta_2)^2\right] t.$$

Define a P_τ martingale M by $M_t = V_t - mt$. Then $E_\tau M_t^2 = \langle M \rangle_t = \sigma^2$ and integration by parts under P_τ yields

$$X_t = \frac{m}{\tau_1}(1 - e^{-\tau_1 t}) + e^{-\tau_1 t} \int_0^t e^{\tau_1 s} dM_s$$

if $\tau_1 \neq 0$.

The decomposition of M into a continuous martingale and purely discontinuous martingale is equal to

$$M_t = X_t^c + \int_0^t \int x(\mu - \nu)(ds, dx).$$

The P_τ Markov process X has a unique invariant distribution π with finite second moment which is infinitely divisible with Levy characteristics $(m/\tau_1, 1/2\tau_1, L_\pi)$ without truncation, where the Levy measure L_π is given by

$$L_\pi(A) = \int_0^\infty L(e^{s\tau_1})A)ds, \quad A \in \mathcal{B}(\mathbb{R}).$$

Hence

$$\int x d\pi(x) = \frac{m}{\tau_1},$$

$$\int (x - \int x d\pi(x))^2 d\pi(x) = \frac{1}{2\tau_1} + \int x^2 dL_\pi(x) = \frac{\sigma^2}{2\tau_1}$$

which yields

$$\int x^2 d\pi(x) = \frac{m^2}{\tau_1^2} + \frac{\sigma^2}{2\tau_1}.$$

From the ergodic theorem it follows

$$\frac{1}{t}\int_0^t X_s ds \to \frac{m}{\tau_1} \quad a.s.$$

$$\frac{1}{t}\int_0^t X_s^2 ds \to \frac{m^2}{\tau_1^2} + \frac{\sigma^2}{2\tau_1} \quad a.s.$$

Thus

$$\frac{1}{t}(L_t(\tau,\theta) - Z_t) \to K \ a.s.$$

From the SLLN for martingales

$$\frac{1}{t}Z_t \to 0 \quad a.s.$$

For $\tau, \theta \in \Theta$ with $P_\tau \neq P_\theta$, consider the Neyman-Person problem of testing the hypothesis $\{P_{\tau,t}\}$ against the alternative $\{P_{\theta,t}\}$.

The Probability of Type II Error is defined as

$$\beta_t(\alpha) := \min\{E_\theta(1-\varphi_t) : E_\tau\varphi_t \leq \alpha.\}$$

If F is the cdf of a random variable on $\mathbb{R}$, then F^{-1} denotes its quantile function defined as

$$F^{-1}(p) := \inf(\{z \in \mathbb{R} : f(z) \geq p\}), p \in (0,1).$$

We have the following results about the asymptotic behavior of the Probability of Type II Error under the assumption that $\alpha_t \to \alpha$ when $t \to \infty$.

Theorem 4.2 Let $\alpha_t \to \alpha$, as $t \to \infty$, $\alpha \in (0,1)$ and $\theta_1 \neq \tau_1$.

(i) We have with $\tau_1 > 0$,

$$\lim_{t\to\infty} t^{-1}\log\beta_t(\alpha_t) = K(\tau,\theta).$$

(ii) We have with $\tau_1 = 0$ *and* $m(\tau_2) = 0$,

$$\lim_{t\to\infty} t^{-2}\log\beta_t(\alpha_t) = -\frac{1}{2}\theta_1^2\sigma^2(\tau_2)F^{-1}(\alpha),$$

where F *is the cdf of* $\int_0^1 W_s^2 ds$ *and* W *is a standard Brownian motion.*

(iii) We have with $\tau_1 = 0$ *and* $m(\tau_2) \neq 0$,

$$\lim_{t\to\infty} t^{-5/2} \log \beta_t(\alpha_t) = -\bar{K}(\tau,\theta)t^{1/2} - \bar{H}(\tau,\theta)^{1/2}\Phi^{-1}(\alpha),$$

where

$$\bar{K}(\tau,\theta) = \frac{1}{6}\theta_1^2 m(\tau_2)^2, \quad \bar{H}(\tau,\theta) = \frac{2}{15}\theta_1^4 m(\tau_2)^2\sigma^2(\tau_2).$$

(iv) We have with $\tau_1 < 0$,

$$\lim_{t\to\infty} e^{2\tau_1 t} \log \beta_t(\alpha_t) = \frac{(\theta_1 - \tau_1)^2}{4\tau_1} G_\tau^{-1}(\alpha),$$

where G_τ *is the c.d.f. of* $Y(\tau_1)^2$ *with* $Y(\tau_1) = \int_0^\infty e^{\tau_1 s} dV_s(\tau_1)$ *under* P_τ. *The distribution of* $Y(\tau_1)$ *is infinitely divisible with a finite second moment and Levy characteristics* $(-(m(\tau_2)/\tau_1),\ -(1/2\tau_1),\ \bar{L}(\tau))$ *without truncation where*

$$\bar{L}(\tau)(A) = \int_0^\infty L(\tau_2)(\exp(-\tau_1 s)A)ds, \quad A \in \mathcal{B}(\mathbb{R}).$$

5.5 Random Observation Period: Sequential Inference

Let $\nu_i(ds,dx) = ds \otimes L_i(dx),\ i = 1,2$. The density of Y depends on t and

$$K_t(\nu_0,\nu_1) = tK(L_0,L_1).$$

Let the first passage time be given by

$$\tau_t := \inf\{s \geq 0 : \frac{1}{2}\langle M\rangle_s + sK(L_0,L_1) > t\}.$$

Observation of the process X up to time τ_t corresponds to $\mathcal{F}_{\tau_t}$.

We have

$$Z_t = \tilde{N}_t + \frac{1}{2}\langle M\rangle_t + K_t(\nu_0,\nu_1),$$

$$\tilde{N}_t = M_t + \int_0^t \int \log Y d(\mu - \nu_0).$$

The log-likelihood is given by

$$\log \frac{dP_0}{dP_1}|\mathcal{F}_{\tau_t} = Z_{\tau_t} = N_t + t,$$

where $N_t = \tilde{N}_{\eta_t}$. The process N is a local square integrable martingale under the filtration $\mathbb{F}_\tau = (\mathcal{F}_{\tau_t})_{t\geq 0}$.

Proposition 5.1 *Assume* $\frac{\tau_t}{r_2^2(t)} \to \gamma$ *in* P_0*-probability,* $\frac{t}{r_2^2(t)} \to \delta$*, where* $\gamma, \delta \in [0, \infty)$ *and* $\frac{t}{r_2(t)} \to \infty$*. Then*

$$\mathcal{L}(r_2^{-1}(t)(Z_{\tau_t} - t)|P_0) \to N(0, H),$$

where

$$H := 2\delta + \gamma \left[\int (\log Y)^2 dL_0 - 2K(L_0, L_1)\right].$$

Proof We have

$$\langle \tilde{N} \rangle_t = 2\left[\frac{1}{2}\langle M \rangle_t + tK(L_0, L_1)\right] + t\left[\int (\log Y)^2 dL_0 - 2K(L_0, L_1)\right].$$

Therefore,

$$\langle N \rangle_t = \langle \tilde{N} \rangle_{\tau_t} = 2t + \tau_t \left[\int (\log Y)^2 dL_0 - 2K(L_0, L_1)\right].$$

This implies

$$\frac{\langle N \rangle_t}{r_2(t)^2} \to^{P_0} H$$

in P_0-probability. Furthermore,

$$\frac{\tau_t}{r_2^2(t)} \int (\log Y)^2 I_{\{|\log Y| > r_2(t)\epsilon\}} dL_0 \to 0$$

in P_0-probability. The assertion now follows from CLT for martingales.

The stopping time is based on the observed Fisher information process exceeding a prescribed level and the asymptotic is for level $\to \infty$.

$$\tau(st) = \tau_t(s).$$

Asymptotic Dubins-Schwarz-Dambis Theorem

σ_t is a deterministic time change such that $t \mapsto \sigma_t(s)$ increases to infinity for every s and let $t \mapsto 1/\delta_t^2$ be continuous and strictly increasing to infinity.

Case (1) If $\theta_1 > 0$, let $\delta_{1,t} = t^{-1/2}$. If $\theta_1 > 0$, choose $\delta_{2,t} = t^{-1/2}$. Then

$$t^{-1} \int_0^{st} X_r^2 dr \to a(\theta)s$$

for every $s \geq 0$ P_θ-a.s.
where

$$a(\theta) = \frac{(E_\theta V_1(\theta_1))^2}{\theta_1^2} + \frac{Var_\theta(V_1(\theta_1)}{2\theta_1}.$$

This implies

$$t^{-1}\tau_t(s) \to \frac{s}{a(\theta)}$$

for every $s \geq 0$ P_θ-a.s.

The model is functionally LAN.

Case (2) If $\theta_1 = 0$, let $\delta_{2,t} = t^{-1/6}$. From the SLLN for Levy processes

$$t^{-1}\int_0^{st^{1/3}} X_r^2 dr \to \frac{1}{3}(E_\theta V_1(\theta))^2 s$$

for every $s \geq 0$ P_θ-a.s.

Hence

$$t^{-1/3}\tau_t(s) \to (3(E_\theta V_1(\theta))^{-2}s)^{1/3}$$

for every $s \geq 0$ P_θ-a.s.

The model is functionally LAN.

Case (3) If $\theta_1 < 0$, choose $\delta_{2,t} = (\log(1+t))^{1/2}$

$$e^{2\theta_1 t}\int_0^t X_r^2 dr \to U$$

P_θ-a.s. where U is a random variable with $U > 0$ a.s., one obtains

$$t^{-1}\int_0^{\sigma_t(s)} X_r^2 dr \to Us$$

for every $s \geq 0$ P_θ-a.s. where $\sigma_t = \sigma(st)$ and σ denotes the inverse function of $t \mapsto e^{-2\theta_1} - 1$. Hence $\sigma_t^{-1}(\tau_t(s)) \to s/U$ P_θ-a.s. This implies $-2\theta_1\tau_t(s) - \log(1+t) \to \log(s/U)$. Hence

$$(\log(1+t))^{-1}\tau_t(s) \to -\frac{1}{2\theta_1} \quad P_\theta - \text{a.s.}$$

The model is functionally LAN. $E_\theta(\tau(t)) = O(t)$.

Discrete Sampling

Let $\nu_i(ds, dx) = ds \otimes L_i(dx)$, $i = 1, 2$. The density of Y depends on t and

$$K_t(\nu_0, \nu_1) = tK(L_0, L_1).$$

Let the first passage time be given by

$$\tau_{n,t} := \inf\{s \geq 0 : \frac{1}{2}\langle M^n\rangle_s + sK(L_0, L_1)\}.$$

Observation of the process X up to time $\tau_{n,t}$ corresponds to $\mathcal{F}_{\tau_{n,t}}$.

We have

$$Z_{n,t} = \tilde{N}_t^n + \frac{1}{2}\langle M^n \rangle_t + K_t^n(\nu_0, \nu_1),$$

$$\tilde{N}_t^n = M_t^n + \sum_{i=1}^n \int \log Y d(\mu - \nu_0).$$

The log-likelihood is given by

$$\log \frac{dP_0}{dP_1} | \mathcal{F}_{\tau_{n,t}} = Z_{\tau_{n,t}} = N_t^n + t,$$

where $N_t^n = \tilde{N}_{\tau_t}^n$. The process N^n is a local square integrable martingale under the filtration $\mathbb{F}_\tau^n = (\mathcal{F}_{\tau_{n,t}})_{t \geq 0}$.

Proposition 5.2 *Assume* $\frac{\tau_{n,t}}{r_2^2(n,t)} \to \gamma$ *in* P_0*-probability,* $\frac{t}{r_2^2(n,t)} \to \delta$*, where* $\gamma, \delta \in [0, \infty)$ *and* $\frac{t}{r_2(n,t)} \to \infty$*. Then*

$$\mathcal{L}(r_2^{-1}(n,t)(Z_{\tau_t}^n - t) | P_0) \to N(0, H),$$

where

$$H := 2\delta + \gamma \left[\int (\log Y)^2 dL_0 - 2K(L_0, L_1) \right].$$

Proof We have

$$\langle \tilde{N}^n \rangle_t = 2 \left[\frac{1}{2} \langle M^n \rangle_t + tK(L_0, L_1) \right] + t \left[\int (\log Y)^2 dL_0 - 2K(L_0, L_1) \right].$$

Therefore,

$$\langle N^n \rangle_t = \langle \tilde{N}^n \rangle_{\tau_t} = 2t + \tau_t \left[\int (\log Y)^2 dL_0 - 2K(L_0, L_1) \right].$$

This implies

$$\frac{\langle N^n \rangle_t}{r_2^2(t)} \to^{P_0} H.$$

Furthermore,

$$\frac{\tau_t}{r_2^2(t)} \int (\log Y)^2 I_{\{|\log Y| > r_2(t)\epsilon\}} dL_0 \to^{P_0} 0.$$

The assertion now follows from CLT for martingales.

O-U Process Based on Stopping Times

Let X_t be described by the SDE

$$dX_t = -\theta_i X_t dt + dV_t(\theta_i), \ X_0 = 0,$$

where $\theta_i \in \mathbb{R}$ and $V(\theta_i)$ is a process with stationary independent increments with $V_0(\theta_i) = 0$ and Levy characteristics $(b(\theta_i)t, ct, L(\theta_i)t)$ depending on a parameter θ_i.

Proposition 5.3 *Assume* $\theta_0 < 0$ *and* $\int x^2 dL_0(x) < \infty$. *Then*

$$\sqrt{t}(Z_{\tau_t} - t) \to^{\mathcal{D}} \mathcal{N}(0, 2).$$

Proof By martingale convergence theorem

$$e^{\theta_0 t} X_t \to \int_0^\infty e^{\theta_0 s} dV_s(\theta_0).$$

This implies

$$e^{2\theta_0 t}\left[\frac{1}{2}\langle M\rangle_t + tK(L_0, L_1)\right] \to -\frac{(\theta_0 - \theta_1)^2}{4\theta_0}\left(\int_0^\infty e^{\theta_0 s} dV_s(\theta_0)\right)^2 =: U \ \ P_0 - a.s.$$

and, therefore,

$$e^{2\theta_0 t} t \to U \ \ P_0 - a.s.$$

One obtains

$$\frac{\tau_t}{t} \to 0 \ \ P_0 - a.s.$$

The assertion follows from Proposition 5.2.

5.6 Fractional and Sub-Fractional Levy O-U Process

Let X_t be described by the SDE

$$dX_t = -\theta_i X_t dt + dV_t^H(\theta_i), \ X_0 = 0,$$

where $\theta_i \in \mathbb{R}$ and $V(\theta_i)$ is a process with stationary dependent increments with $V_0(\theta_i) = 0$ and fractional Levy characteristics $(b(\theta_i)t, ct, L(\theta_i)t)$ depending on a parameter θ_i and H is the Hurst parameter in $[0.5, 1]$. This process has jumps and long memory and is a natural extension of the fractional Brownian motion.

Proposition 6.1 *Assume* $\theta_0 < 0$ *and* $\int x^2 dL_0(x) < \infty$. *Then*

$$\sqrt{t}(Z_{\tau_t}^H - t) \to^{\mathcal{D}} \mathcal{N}(0, 2).$$

Proof By fractional martingale convergence theorem

$$e^{\theta_0 t}X_t \to \int_0^\infty e^{\theta_0 s}dV_s(\theta_0).$$

This implies

$$e^{2\theta_0 t}\left[\frac{1}{2}\langle M\rangle_t + tK(L_0, L_1)\right] \to -\frac{(\theta_0-\theta_1)^2}{4\theta_0}\left(\int_0^\infty e^{\theta_0 s}dV_s(\theta_0)\right)^2 =: U \ \ P_0 - a.s.$$

and, therefore,

$$e^{2\theta_0 t}t \to U \ \ P_0 - a.s.$$

One obtains

$$\frac{\tau_t}{t} \to 0 \ \ P_0 - a.s.$$

The assertion follows from fractional extension of Proposition 5.2.

Let X_t be described by the SDE

$$dX_t = -\theta_i X_t dt + dV_t^H(\theta_i), \ X_0 = 0,$$

where $\theta_i \in \mathbb{R}$, $i = 1, 2$ and $V(\theta_i)$ is a process with nonstationary dependent increments with $V_0(\theta_i) = 0$ and fractional Levy characteristics $(b(\theta_i)t, ct, L(\theta_i)t)$ depending on a parameter θ_i and H is the Hurst parameter in $[0.5, 1]$. This process has jumps and medium range memory and is a natural extension of the sub-fractional Brownian motion.

Proposition 7.1 *Assume* $\theta_0 < 0$ *and* $\int x^2 dL_0(x) < \infty$. *Then*

$$\sqrt{t}(Z^H_{\tau_t} - t) \to^{\mathcal{D}} \mathcal{N}(0, 2).$$

Proof By fractional martingale convergence theorem

$$e^{\theta_0 t}X_t \to \int_0^\infty e^{\theta_0 s}dV_s(\theta_0).$$

This implies

$$e^{2\theta_0 t}\left[\frac{1}{2}\langle M\rangle_t + tK(L_0, L_1)\right] \to -\frac{(\theta_0-\theta_1)^2}{4\theta_0}\left(\int_0^\infty e^{\theta_0 s}dV_s(\theta_0)\right)^2 =: U \ \ P_0 - a.s.$$

and, therefore,

$$e^{2\theta_0 t}t \to U \ \ P_0 - a.s.$$

One obtains

$$\frac{\tau_t}{t} \to 0 \ \ P_0 - a.s.$$

The assertion follows from sub-fractional extension of Proposition 5.2.

Chapter 6

Estimation in Gamma-Ornstein –Uhlenbeck Stochastic Volatility Model

6.1 Introduction

We generalize the Ornstein–Uhlenbeck process to include non-normal innovations. First we study the asymptotic behavior of the ratio estimator of the drift parameter in Gamma-Ornstein–Uhlenbeck volatility process based on observations of the price process. This model captures the stylized facts as it preserves both jumps in the volatility process. We study the behavior of the moment estimators.

Recently, processes with jumps have received attention in finance, engineering, and physics. Levy driven processes of Ornstein–Uhlenbeck type have been extensively studied over the last few years and widely used in finance, see Barndorff-Nielsen and Shephard (2001). LOU process generalizes OU process to include jumps.

6.2 Gamma-Ornstein–Uhlenbeck Model

The Levy–Ornstein–Uhlenbeck (LOU) process is an extension of Ornstein–Uhlenbeck process with Levy process driving term. In finance, it is useful as a generalization of Vasicek model, as one-factor short-term interest rate model which could take into account the jump of the interest rate.

We consider the special cases when L is a GOU process. L has the GOU density with parameters δ and γ. The stochastic volatility model has the form

$$\begin{aligned} dY_t &= (\mu + \beta X_t)dt + X_t dW_t + \rho dZ_t \\ dX_t &= -\theta X_t dt + dZ_t. \end{aligned}$$

Z_t is a Levy process independent of X_0.

Invariant distribution is Gamma.

$$\mathcal{L}(X_0) = \mathcal{L}(X_t) = \mathcal{G}(\lambda, \delta, \sqrt{\alpha^2 - \beta^2}).$$

Mixture distribution is variance Gamma denoted by Q_θ .

$$\mathcal{L}(Y_t) = \mathcal{VG}(\lambda, \frac{\alpha}{\sqrt{t}}, \frac{\beta}{\sqrt{t}}, \sqrt{t}\delta, \mu t).$$

J. P. N. Bishwal, *Parameter Estimation in Stochastic Volatility Models*, https://doi.org/10.1007/978-3-031-03861-7_6

$$\mathcal{L}(Y_{t+1} - Y_t) = \mathcal{VG}(\lambda, \alpha, \beta, \delta, \mu).$$

Let $q(x, \theta)$ be a density function of Y_t.
The contrast function is defined as

$$u(x, \theta) = -\log q(x, \theta).$$

MCE is defined as

$$\hat{\theta}_T = \arg\inf_\theta u(x, \theta).$$

The density function of the GH distribution is

$$d_{\mathcal{GH}(\lambda,\alpha,\beta,\delta,\mu)}(x) =$$

$$\frac{(\alpha^2 - \beta^2)^{\frac{\lambda}{2}}}{\sqrt{2\pi}\alpha^{\lambda-\frac{1}{2}}\delta^\lambda K_\lambda(\delta\sqrt{\alpha})}(\delta^2 + (x-\mu)^2)^{(\lambda-\frac{1}{2})/2} e^{\beta(x-\mu)} K_{\lambda-\frac{1}{2}}(\alpha\sqrt{\delta^2 + (x-\mu)^2}).$$

Here K_ν denotes the modified Bessel function of third kind with index ν.
When $\lambda = -\frac{1}{2}$, GIG becomes $IG(\delta, \gamma)$.
In the case of -1 correlation, the moment estimator is

$$\theta_T = -\frac{Y_T}{T}.$$

Then we consider the Gamma-OU process.

Let $(\Omega, \mathcal{F}, \{\mathcal{F}_t\}_{t\geq 0}, P)$ be a stochastic basis on which is defined the Ornstein–Uhlenbeck process X_t satisfying the Itô stochastic differential equation

$$dX_t = \theta X_t dt + dZ_t, \quad t \geq 0, \quad X_0 = 0, \tag{1.1}$$

where $\{Z_t\}$ is a Gamma process with the filtration $\{\mathcal{F}_t\}_{t\geq 0}$ and $\theta > 0$ is the unknown parameter to be estimated on the basis of continuous observation of the process $\{Y_t\}$ on the time interval $[0, T]$.

$$X_t = \int_{-\infty}^t e^{\theta(t-s)} dZ_s.$$

This process is stationary.

It can be shown that X_{t_i} is a stationary discrete time AR (1) process with autoregression coefficient $\phi \in (0, 1)$ with the following representation:

$$X_{t_i} = \phi X_{t_{i-1}} + \epsilon_{t_{i-1}},$$

where

$$\phi = e^{-\theta h}$$

and

$$\epsilon_{t_{i-1}} = \int_{t_{i-1}}^{t_i} e^{-\theta(t_i - u)} dZ_u.$$

6.3 Method of Moments Estimators

The moment estimators are defined as

$$\hat{\theta}_n := \frac{\frac{1}{n^2}\left[\sum_{i=1}^n (Y_{i\Delta} - Y_{(i-1)\Delta})\right]^2}{\frac{1}{n^2}\sum_{i=1}^n (Y_{i\Delta} - Y_{(i-1)\Delta})^2 - \frac{\Delta}{n}\left[\sum_{i=1}^n (Y_{i\Delta} - Y_{(i-1)\Delta})\right]} \frac{2a^3(a+1)}{b^4\Delta}.$$

$$\hat{\rho}_n := \frac{\frac{1}{n^2}\sum_{i=1}^n (Y_{i\Delta} - Y_{(i-1)\Delta})^2 - \frac{\Delta}{n}\left[\sum_{i=1}^n (Y_{i\Delta} - Y_{(i-1)\Delta})\right]}{\frac{1}{n^2}\left[\sum_{i=1}^n (Y_{i\Delta} - Y_{(i-1)\Delta})\right]} \frac{b^3\Delta}{2a^2(a+1)}.$$

We also study the estimators

$$\breve{\theta}_n := \bigwedge_{i=1}^n \frac{\hat{X}_{i\Delta}}{\hat{X}_{(i-1)\Delta}}.$$

We also study the following estimator of the drift. The estimator of θ is defined as

$$\tilde{\theta}_n := -\log\left[\frac{\sum_{i=1}^n \hat{X}_{i\Delta}}{\sum_{i=1}^n \hat{X}_{(i-1)\Delta}}\right],$$

where $\hat{X}$ is the estimator of X based on observations of Y.

We obtain the strong consistency and asymptotic normality distribution of the ratio estimator. Using the mixing property, we obtain the following:

Theorem 2.1

$$\hat{\theta}_n \to \theta_0 \quad \text{a.s.} \quad \text{as } n \to \infty.$$

Theorem 2.2

$$\sqrt{n}(\hat{\theta}_n - \theta_0) \to^{\mathcal{D}} \mathcal{N}(0,\ J^{-1}(\theta_0)).$$

Theorem 2.3

$$\hat{\rho}_n \to \rho_0 \quad \text{a.s.} \quad \text{as } n \to \infty.$$

Theorem 2.4

$$\sqrt{n}(\hat{\rho}_n - \rho_0) \to^{\mathcal{D}} \mathcal{N}(0,\ I^{-1}(\rho_0)).$$

The FIECOGARCH process is defined as

$$dY_{d,t} = \sqrt{X_{d,t}}dL_t$$

$$X_{d,t} = e^{\mu+\beta U_{d,t-}}$$

$$dU_{d,t} = \theta U_{d,t}dt + dM_{d,t},$$

where L_t is a Levy process independent of $M_{d,t}$ which is a fractional Levy process of integration parameter $d \in (0, 0.5)$.

Bishwal (2007a) estimated μ, β, and θ by the method of moments. Equivalently,

$$dY_{d,t} = \mu dt + \sigma e^{X_{d,t}} dL_t$$

$$dX_{d,t} = \theta X_{d,t}dt + dM_{d,t}.$$

This model was obtained by Haug and Czado (2006a,b) who called them fractionally integrated exponential GARCH process, in short FIECOGARCH process. Bishwal (2007a) estimated μ, σ, and θ by the method of moments.

Bishwal (2007a) also studied the fractional Levy Heston model:

$$dY_{d,t} = \mu dt + \sigma X_{d,t} dL_t$$

$$dX_{d,t} = \theta X_{d,t}dt + \gamma\sqrt{X_{d,t}}dM_{d,t}.$$

Bishwal (2007a) estimated μ, σ, and θ by the minimum contrast method. The case where M is a fractional Brownian motion was studied by Comte et al. (2012).

In the stationary case, one can construct fOU process in another way by time change. The Lamperti transform provided one-to-one correspondence between a stationary process and a H-self-similar process. Recall that an ordinary OU process is a Lamperti transform of Brownian motion. If (B_t) is a standard Brownian motion, then $\frac{e^{-\theta t}}{\sqrt{2\theta}} B_{e^{2\theta t}}$ is an OU process. Similarly, one can represent fOU process as a time-changed fBM:

$$X_t = \frac{e^{-2H\theta t}}{\sqrt{4H\theta}} W^H_{e^{2\theta t}}$$

It is a zero mean Gaussian process with covariance kernel

$$\begin{aligned} C_t = E(X_s X_{t+s}) &= \frac{1}{8H\theta}\left[e^{2H\theta t} + e^{-2H\theta t} - |e^{\theta t} - e^{-\theta t}|^{2H}\right] \\ &= \frac{1}{8H\theta}\left[\cosh 2H\theta t - 2^{2H-1}(\sinh \theta t)^{2H}\right]. \end{aligned}$$

For $H = 1/2$, this reduces to the covariance of the ordinary OU process $\frac{1}{2\theta}e^{-\theta t}$. The covariance structure shows that the fOU process is locally asymptotically stationary for small time lag and is a sum of mutually independent Gaussian Markov processes for large time lag.

Recently, long-memory processes, i.e., processes with slowly decaying autocorrelation and processes with jumps, have received attention in finance, engineering, and physics. The simplest continuous time long-memory process is the fractional Brownian motion discovered by Kolmogorov (1940) and later on studied by Levy (1948) and Mandelbrot and Van Ness (1968). Continuous time long-memory jump process is the fractional Levy process. Hence, the fractional Levy process can also be called the *Kolmogorov–Levy process.*

A normalized fractional Brownian motion $\{W_t^H, t \geq 0\}$ with Hurst parameter $H \in (0,1)$ is a centered Gaussian process with continuous sample paths whose covariance kernel is given by

$$E(W_t^H W_s^H) = \frac{1}{2}(s^{2H} + t^{2H} - |t-s|^{2H}), \quad s,t \geq 0.$$

The process is self-similar (scale invariant) and it can be represented as a stochastic integral with respect to standard Brownian motion. For $H = \frac{1}{2}$, the process is a standard Brownian motion. For $H \neq \frac{1}{2}$, the fBm is not a semimartingale and not a Markov process, but a Dirichlet process. The increments of the fBm are negatively correlated for $H < \frac{1}{2}$ and positively correlated for $H < \frac{1}{2}$, and in this case they display long-range dependence. The parameter H which is also called the self-similarity parameter measures the intensity of the long range dependence. The ARIMA(p,d,q) with autoregressive part of order p, moving average part of order q, and fractional difference parameter $d \in (0,\ 0.5)$ process converge in Donsker sense to fBm.

This process was introduced by Marquardt (2006b,b) who also introduced the fractional Levy process in Marquardt (2006b,b). Using suitable transformation of the process, one can obtain a nonlinear stationary process satisfying a fractional SDE, see Buchmann and Kluppelberg (2005). Benth (2003) used the process as temperature and obtained weather derivative arbitrage free pricing formulas for European and average type options. Cheridito *et al.* obtained the fOU process as a Lamperti transformation of the fBM. The model parameter is usually unknown and must be estimated from data.

The ratio estimator is shown to be consistent and asymptotically has Weibull distribution for large sample size. Second, we consider the fractional Hermite–Ornstein–Uhlenbeck process and show that the maximum likelihood estimator is strongly consistent.

The fractional Levy Process is defined as

$$M_{H,t} = \frac{1}{\Gamma(H+\frac{1}{2})} \int_{\mathbb{R}} [(t-s)_+^{H-1/2} - (-s)_+^{H-1/2}] dM_s, \quad t \in \mathbb{R},$$

where $M_t, t \in \mathbb{R}$ is a Levy process on $\mathbb{R}$ with $E(M_1) = 0,\ E(M_1^2) < \infty$ and without Brownian component.

Properties

(1) $cov(M_{H,t}, M_{H,s}) = \dfrac{E(L(1)^2}{2\Gamma(2H+1)\sin(\pi H)}[|t|^{2H} + |s|^{2H} - |t-s|^{2H}].$

(2) M_H is not a martingale. For a large class of Levy processes, M_d is neither a semimartingale.

(3)M_H is Hölder continuous of any order β less than $H - \frac{1}{2}$.
(4) M_H has stationary increments.
(5) M_H is symmetric.
(6) M is self-similar.
(7) M_H is not self-similar.
(8) M_H has infinite total variation on compacts.
The FIMA process is defined as

$$Y_H(t) = \int_{-\infty}^{t} g_H(t-u)L(du),\ t \in \mathbb{R},$$

where

$$g_H(t) = \frac{1}{\Gamma(H-\frac{1}{2}}\int_0^t g(t-s)s^{H-\frac{3}{2}}ds,\ t \in \mathbb{R}$$

and the kernel g is the kernel of a short-memory moving average process.
The process $Y_H(t)$ can be written as

$$Y_H(t) = \int_{-\infty}^{t} g(t-u)dM_{H,u},\ t \in \mathbb{R}.$$

We assume the following conditions on the kernel $g : \mathbb{R} \to \mathbb{R}$:
(A1) $g(t) = 0$ for all $t < 0$ (causality).
(A2) $|g(t)| \leq Ce^{-ct}$ for some constants $C > 0$ and $c > 0$ (short memory).
FIMA (fractionally integrated moving average) process is stationary and is infinite divisible.

Consider the kernel

$$g(t-s) = \sigma e^{\theta(t-s)} I_{(0,\infty)}(t-s);$$

then

$$g_H(t) = \frac{\sigma}{\Gamma(H-\frac{1}{2}}\int_0^\infty e^{\theta(t-s)} I_{(0,\infty)}(t-s)s^{H-\frac{3}{2}}ds,\ t \in \mathbb{R}.$$

Note that

$$U_t^{H,\theta,\sigma} = \int_{\mathbb{R}} g_H(t-u)L(du),\ t \in \mathbb{R},$$

is the fractional Levy–Ornstein–Uhlenbeck (FLOU) process satisfying the fractional Langevin equation

$$dU_t = \theta U_t dt + \sigma dM_{H,t},\ t \in \mathbb{R}.$$

FLP is a generalization of fBM. FLP is not a martingale and not even a semimartingale in general. FLP is a natural counterpart to fBM. It is H'older continuous. It has stationary increments. It is symmetric. It is self-similar. Fractional stable motion is a special case of fLP.

In general, it is not a semimartingale.

The fractional Levy–Ornstein–Uhlenbeck (FLOU) process X was introduced in Marquardt (2006b). FLOU process is a LOU process which includes long memory.

We study the ratio estimator of the drift. The ratio estimator of θ is defined as

$$\hat{\theta}_n := -\log \left[\min_{1 \le i \le n} \frac{X_{i\Delta}}{X_{(i-1)\Delta}} \right].$$

This estimator is motivated by the extreme value theory for the correlation parameter of an AR(1) process whose innovation distribution is positive. See Davis and McCormick (1989). In the case of exponential AR(1) process, it coincides with the maximum likelihood estimator. See Nielsen and Shephard (2003).

The weak consistency of the ratio estimator in the LOU process was studied in Jongbloed et al. (2005). The strong consistency and asymptotic Weibullness were studied in Brockwell et al. (2007) in the case of Gamma innovations. In this note, we obtain strong consistency and asymptotic Weibullness of the ratio estimator for the fLOU process.

We will need the following result in the sequel whose proof can be found in Davis and McCormick (1989).

Proposition 1.1 Let

$$X_t = \phi X_{t-1} + Z_t, \ t = 0, \pm 1, \pm 2, \ldots.$$

Let $\phi > 0$. Let the distribution F of Z_t be concentrated on $[0, \infty)$ and be of regular variation at zero. Let

$$R_n := \bigwedge_{i=1}^{n} \frac{X_i}{X_{i-1}}$$

and

$$a_n = F^{-}(\frac{1}{n}) := \inf\{x : F(x) \ge \frac{1}{n}\}.$$

Then

$$R_n \to \phi \text{ a.s. as } n \to \infty$$

and

$$\lim_{n \to \infty} P\left\{a_n^{-1}(R_n - \phi)C_\alpha \le x\right\} = 1 - \exp(-x^\alpha), \ x \ge 0,$$

where

$$C_\alpha := (EX_1^\alpha)^{1/\alpha}.$$

6.4 Fractional MS-OU and OU-MS Processes

MS (modified stable) law evolve by exponential and power tempering (tilting) from one of the positive k-stable laws ($0 < k < 1$). Using the MS distribution as mixing distribution for normal variance mean mixtures yields the class of NMS (normal modified stable) laws.

NTS laws are normal mean variance mixture with TS mixing.

MS laws are infinitely divisible and self-decomposable.

For $k = 1/2$, MS becomes the GIG law and NMS becomes the GH law.

The NG (normal gamma) class includes the VG processes.

RLPE processes are CGMY processes.

NG and RLPE processes follow normal modified stable law.

Examples: Fractional IGOU Process and Fractional Gamma-OU Process

We consider the special cases when L is an IGOU process. L has the IGOU density with parameters δ and γ.

$$dY_t = (\mu + \beta X_t)dt + X_t dW_t$$

$$dX_t = -\theta X_t dt + dZ_t.$$

Z_t is a Levy process independent of X_0.

Invariant distribution is GIG.

$$\mathcal{L}(X_0) = \mathcal{L}(X_t) = \mathcal{GIG}(\lambda, \delta, \sqrt{\alpha^2 - \beta^2}).$$

Mixture distribution is Q_θ GH.

$$\mathcal{L}(Y_t) = \mathcal{GH}(\lambda, \frac{\alpha}{\sqrt{t}}, \frac{\beta}{\sqrt{t}}, \sqrt{t}\delta, \mu t).$$

$$\mathcal{L}(Y_{t+1} - Y_t) = \mathcal{GH}(\lambda, \alpha, \beta, \delta, \mu).$$

Let $q(x, \theta)$ be a density function of Y_t.

The contrast function is defined as

$$u(x, \theta) = -\log q(x, \theta).$$

MCE is defined as

$$\hat{\theta}_T = \arg\inf_\theta u(x, \theta).$$

The density function of the GH distribution is

$$d_{\mathcal{GH}(\lambda,\alpha,\beta,\delta,\mu)}(x) =$$

$$\frac{(\alpha^2 - \beta^2)^{\frac{\lambda}{2}}}{\sqrt{2\pi}\alpha^{\lambda-\frac{1}{2}}\delta^\lambda K_\lambda(\delta\sqrt{\alpha})}(\delta^2 + (x-\mu)^2)^{(\lambda-\frac{1}{2})/2} e^{\beta(x-\mu)} K_{\lambda-\frac{1}{2}}(\alpha\sqrt{\delta^2 + (x-\mu)^2}).$$

Here K_ν denotes the modified Bessel function of third kind with index ν.

When $\lambda = -\frac{1}{2}$, GIG becomes $IG(\delta, \gamma)$.

In the case of -1 correlation, the moment estimator is

$$\theta_T = -\frac{Y_T}{T}.$$

Then we consider the fractional Gamma-OU process. It can be shown that the limit distribution is Weibull.

6.5 Fractional Hermite OU Processes

As another generalization of fractional Brownian motion we obtain the Hermite process of order k with Hurst parameter $H \in (\frac{1}{2}, 1)$ which is defined as a multiple Wiener–Itô integral of order k with respect to standard Brownian motion $(B(t))_{t\in\mathbb{R}}$

$$Z_t^{H,k} := c(H,k) \int_{\mathbb{R}}^{k} \int_0^t \Pi_{j=1}^k (s - y_i)_+^{-(\frac{1}{2}+\frac{H-1}{2})} ds\, dB(y_1) dB(y_2) \cdots dB(y_k),$$

where $x_+ = \max(x, 0)$.

For $k = 1$, the process is fractional Brownian motion W_t^H with Hurst parameter $H \in (0,1)$. For $k = 2$, the process is Rosenblatt process. For $k \geq 2$, the process is non-Gaussian which is more interesting from the point of finance. Hence fractional Hermite process can also be called the *Kolmogorov–Hermite process.*

The covariance kernel $R(t,s)$ is given by

$$R(t,s) := E[Z_t^{H,k} Z_s^{H,k}]$$

$$= c(H,k)^2 \int_0^t \int_0^s \left[(u-y)_+^{-(\frac{1}{2}+\frac{H-1}{2})} ds (v-y)_+^{-(\frac{1}{2}+\frac{H-1}{2})} dy \right]^k dudv.$$

Let

$$\beta(p,q) = \int_0^1 z^{p-1}(1-z)^{q-1} dz, \ p,q > 0$$

be the beta function.

Using the identity

$$\int_{\mathbb{R}} [(u-y)_+^{a-1} ds (v-y)_+^{a-1} dy] = \beta(a, 2a-1)|u-v|^{2a-1},$$

we have

$$\begin{aligned} R(t,s) &= c(H,k)^2 \beta\left(\frac{1}{2} - \frac{1-H}{k}, \frac{2H-2}{k}\right)^k \int_0^t \int_0^s \left(|u-v|^{\frac{2H-2}{k}}\right)^k dvdu \\ &= c(H,k)^2 \frac{\beta(\frac{1}{2} - \frac{1-H}{k}, \frac{2H-2}{k})^k}{H(2H-1)} \frac{1}{2}(t^{2H} + s^{2H} - |t-s|^{2H}). \end{aligned}$$

In order to obtain $E(Z_t^{(H,k)})^2 = 1$,
choose

$$c(H,k)^2 = \left(\frac{\beta(\frac{1}{2} - \frac{1-H}{k}, \frac{2H-2}{k})^k}{H(2H-1)} \right)^{-1},$$

and we have

$$R(t,s) = \frac{1}{2}(t^{2H} + s^{2H} - |t-s|^{2H}).$$

Thus the covariance structure of the Hermite process and the fractional Brownian motion are the similar.

The process $Z_t^{(H,k)}$ is H-self-similar with stationary increments and all moments are finite.

For any $p \geq 1$,

$$E|Z_t^{(H,k)} - Z_s^{(H,k)}|^p \leq c(p,H,k)|t-s|^{pH}.$$

Thus the Hermite process has H'older continuous paths of order $\delta < H$.

The fractional Hermite–Ornstein–Uhlenbeck process is given by

$$dY_t^{(H,k)} = \theta Y_t^{(H,k)} dt + \sigma dZ_t^{(H,k)}$$

and has the unique solution

$$Y_t^{(H,k)} = e^{\theta t} \left(\xi + \sigma \int_0^t e^{-\theta u} dZ_u^{(H,k)} \right).$$

In particular, if

$$X_0 = \sigma \int_{-\infty}^0 e^{-\theta u} Z_u^{(H,k)},$$

then

$$Y_t^{(H,k)} = e^{\theta t} \sigma \int_{-\infty}^t e^{-\theta u} dZ_u^{(H,k)}.$$

The process $Y_t^{(H,k)}$ is called the fractional Hermite Ornstein–Uhlenbeck process of order k.

Because the covariance structure of the Hermite process is the same as the fractional Brownian motion's one, the above integral is a Wiener integral.

$$E[Y_t^{(H,k)} Y_s^{(H,k)}] = \sigma^2 \int_{-\infty}^t \int_{-\infty}^s e^{\theta(t-u)} e^{\theta(t-v)} |u-v|^{2H-2} dv du < \infty.$$

We obtain the strong consistency of the MLE of the parameter θ.
First we obtain the Cameron–Martin type formula.

Lemma 3.1 Let $I_T := \int_0^T Q_t^2 dv_t$ and $\phi_T(a) := E\exp(aI_T)$, $a \in \mathbb{R}$. Then $\phi_T(z_1)$ exists for $|z_1| \le \delta$, for some $\delta > 0$, and is given by

$$\phi_T(a) = \left\{ \frac{4\sin \pi H \sqrt{\theta^2 - 2z_1} e^{-\theta T}}{\pi T D_T^H(\theta; \sqrt{\theta^2 - 2z_1}} \right\}^{1/2},$$

and we choose the principal branch of the square root, where

$$\begin{aligned} D_T^H(\theta;\alpha) \quad &:= \quad [\alpha \cosh(\frac{\alpha}{2}T) - \theta \sinh(\frac{\alpha}{2}T)]^2 J_{-H}(\frac{\alpha}{2}T) J_{H-1}(\frac{\alpha}{2}T) \\ &- \quad [\alpha \sinh(\frac{\alpha}{2}T) - \theta \cosh(\frac{\alpha}{2}T)]^2 J_{1-H}(\frac{\alpha}{2}T) J_H(\frac{\alpha}{2}T) \end{aligned}$$

for $\alpha > 0$ and J_ν is the modified Bessel function of first kind of order ν.

Chapter 7

Berry–Esseen Inequalities for the Functional Ornstein–Uhlenbeck-Inverse -Gamma Process

7.1 Introduction

State space transform (SST) of regular diffusions was first studied by Karlin and Taylor (1981, Theorem 2.1, page 173). A standard tool in the theory of diffusions is the concept of scale function (see Itô and McKean (1974) and Revuz and Yor (1991)) which turns one dimensional diffusions into local martingales.

A function f is called a state space transform (SST) if it is continuous and strictly increasing: the open interval $I = f(\mathbb{R})$ is called the state space.

Let X_t be a stationary O-U process satisfying

$$dX_t = -\gamma X_t dt + dW_t,\ t > 0,\ X_0 = \int_{-\infty}^{0} e^{\gamma s} dW_s$$

with $\gamma > 0$. Let $Y_t := f(X_t)$ where f is a monotone transform which satisfies

$$dY_t = \mu(Y_t)dt + \sigma(Y_t)dW_t.$$

Buchmann and Kluppelberg (2006) answered what functions of μ and σ allow for the solution Y to be stationary. Buchmann and Kluppelberg (2005) studied maxima of diffusions using the state space transform.

Let (I, μ, σ) be proper and the friction coefficient (FC) $\gamma > 0$. The distribution of Y_t has a Lebesgue density $p(\cdot)$ where

$$p(\cdot) = \frac{\sqrt{\gamma}}{\sqrt{\pi\sigma(\cdot)^2}} \exp\left[-\frac{1}{\gamma}\left(\frac{\mu(\cdot)}{\sigma(\cdot)}\right)^2\right].$$

Using the SST $f(x) = \sigma x - \alpha/\beta, x \in \mathbb{R}$, for $\beta < 0, \sigma > 0$ and $\alpha \in \mathbb{R}$ and setting $\gamma = -\beta$, the process $f(X_t)$ is a Vasicek process satisfying

$$dV_t = \mu(V_t)dt + \sigma dW_t,$$

where $\mu(x) = \alpha + \beta x$.

We obtain the Squared Ornstein–Uhlenbeck-Inverse-Gaussian process when the Brownian component of the Squared Ornstein–Uhlenbeck process is contaminated by an IG process. The chapter introduces some new estimators

J. P. N. Bishwal, *Parameter Estimation in Stochastic Volatility Models*,
https://doi.org/10.1007/978-3-031-03861-7_7

of the drift parameter in the Squared Ornstein–Uhlenbeck-Inverse-Gaussian (SOUIG) process based on discretely sampled data and obtains rates of weak convergence of the distributions of the estimators to the standard normal distribution using random, nonrandom, and mixed normings. These are discretely observed branching diffusions with infinite activity jumps.

Diffusions with jumps have become increasingly important in finance. Option pricing with Levy type jump processes have attracted lot of attention in the last decade. Vasicek model is very popular in modeling short-term interest rates and pricing bonds. Squared Ornstein–Uhlenbeck-Inverse-Gaussian (SOUIG) model is a diffusion with jump process which is a generalization of CIR model with a jump component. There are two types of jump activity in finance: process with finite jump activity (e.g., a Poisson process) and processes with infinite jump activity (e.g., an Inverse-Gaussian process). In this chapter, we focus on the infinite activity jump process. We estimate mean reversion rate in an interest rate model from discrete observations which is an important problem in fixed income market.

Specifically, we study rates of convergence of two approximate maximum likelihood estimators (AMLEs) to the continuous MLE of the parameter appearing linearly in the drift coefficient of Itô SDE with jump when T remains fixed, T being the length of the observation time interval $[0, T]$. As far as we know no results are known about the rate of convergence of AMLEs to the true value of the parameter as $T \to \infty$. For the Ornstein–Uhlenbeck-Gamma process, we obtain Berry–Esseen type bounds using both random and nonrandom normings of the two AMLEs $\theta_{n,T}$ and $\theta_{n,T,1}$. First we identify one AMLE $\theta_{n,T}$ as the conditional least squares estimator (CLSE). We show that the other AMLE $\theta_{n,T,1}$ has sharper Berry–Esseen bound than that for $\theta_{n,T}$. The Berry–Esseen bounds of the AMLEs $\theta_{n,T}$ and $\theta_{n,T,1}$ show that the *rapidly increasing experimental design* (RIED) condition, i.e., $T \to \infty$ and $\frac{T}{n^{1/2}} \to 0$ is not essential for its asymptotic normality of AMLEs. For recent studies on Berry–Esseen rates of estimators in diffusion processes see Bishwal (2007a, 2008a, 2009a, 2010a).

The chapter is organized as follows: In Sect. 2 we prepare notations and preliminaries. Section 3 contains Berry–Esseen type bounds for one AMLE. Section 4 contains Berry–Esseen type bounds for another AMLE. Section 5 introduces some approximate minimum contrast estimators and in Sect. 6 studies the Berry–Esseen type bounds for those estimators.

7.2 Approximate Maximum Likelihood Estimators

Let $\{S_t, t \geq 0\}$ be the Squared Ornstein–Uhlenbeck-Inverse-Gaussian process satisfying the stochastic differential equation, which is a Cox–Ingersoll–Ross model with jumps,

$$dS_t = (1 + 2\theta S_t)dt + 2\sqrt{S_t}dW_t + dZ_t,\ t \geq 0, S_0 = s_0, \tag{2.1}$$

where $\{W_t,\ t \geq 0\}$ is a standard Wiener process, $\{Z_t,\ t \geq 0\}$ is a two sided inverse-Gaussian process with known parameters (γ, δ), γ is the shape parameter, δ is the scale parameter, that is, the Levy measure is of the form

$$f(z) = \frac{\delta e^{\gamma\delta}}{\sqrt{2\pi z^3}} \exp\{-\frac{1}{2}(\delta^2 z^{-1} + \gamma^2 z)\},$$

where δ and γ are positive parameters.

Thus the Levy process here has infinitely many jumps in any finite time interval since $\int f(z)dz = \infty$. Let $\theta \in (-\infty, 0)$ be the unknown parameter to be estimated from the observations of the process $\{X_t, t \geq 0\}$. Levy driven S-O-U processes are ergodic.

Consider the process $(S_t) = (X_t^2)$. By Itô formula,

$$dX_t = \theta X_t dt + dW_t + dZ_t.$$

Recall that based on complete observation of $\{X_t\}$ on $[0, T]$ the log-likelihood function and the maximum likelihood estimator (MLE) are given, respectively, by

$$L_T = \theta \int_0^T X_t dX_t - \frac{\theta^2}{2} \int_0^T X_t^2 dt \tag{2.2}$$

and

$$\theta_T = \frac{\int_0^T X_t dX_t}{\int_0^T X_t^2 dt.} \tag{2.3}$$

We assume that the process $\{X_t\}$ is observed at the points $0 \leq t_0 < t_1 < \ldots < t_n = T$ with $\Delta t_i = t_i - t_{i-1} = \frac{T}{n}, i = 1, 2, \ldots, n$. For simplicity only we assume equidistant time points.

Define

$$\Delta X_{t_i} := X_{t_i} - X_{t_{i-1}}.$$

$$\Delta X_t := X_t - X_{t-}.$$

We assume $t_i = ih,\ T = nh,\ h \to 0,\ nh \to \infty$.

We need the rapid experimental design $nh^{1+\delta} \to 0$ for a constant $\delta \in (0, 1)$.

The jump part of data will be split into two parts: one is a small jump part whose jump are smaller than a positive value $h^\rho, \rho \in (0,\ 1/2)$ and the other a large jump part whose jumps are larger than h^ρ.

By this splitting method we can judge from the increments of the data, whether large jumps occurred or only small jumps occurred in that interval. In the proposed estimators we ignore the parts of the small jumps and we make use of other parts.

We will often use the following indicator in the sequel $I_{\{|\Delta X_{t_i}|\le h^\rho\}}$ as we pick up the data with X_{t_i} and ΔX_{t_i} with $\{|\Delta X_{t_i}| \le h^\rho\}$. There does not exist large jump in an interval with $\{|\Delta X_{t_i}| \le h^\rho\}$.

The conditional least squares estimator (CLSE) based on $X_{t_0}, \dots, X_{t_n}$ is defined as

$$\theta_{n,T} := \arg\inf_{\theta} \sum_{i=1}^{n} \frac{[(X_{t_i} - X_{t_{i-1}})I_{\{|\Delta X_{t_i}|\le h^\rho\}} - \theta X_{t_{i-1}} I_{\{|\Delta X_{t_i}|\le h^\rho\}}\Delta t_i]^2}{\Delta t_i} \tag{2.4}$$

and is given by

$$\theta_{n,T} = \frac{\sum_{i=1}^{n} X_{t_{i-1}}(X_{t_i} - X_{t_{i-1}})I_{\{|\Delta X_{t_i}|\le h^\rho\}}}{\sum_{i=1}^{n} X_{t_{i-1}}^2 I_{\{|\Delta X_{t_i}|\le h^\rho\}}\Delta t_i}. \tag{2.5}$$

Note that the estimator $\theta_{n,T}$ may be viewed as an approximate maximum likelihood estimator (AMLE1) which maximizes the approximate log-likelihood given by

$$L_{n,T}(\theta) = \theta \sum_{i=1}^{n} X_{t_{i-1}}(X_{t_i} - X_{t_{i-1}})I_{\{|\Delta X_{t_i}|\le h^\rho\}} - \frac{\theta^2}{2}\sum_{i=1}^{n} X_{t_{i-1}}^2 I_{\{|\Delta X_{t_i}|\le h^\rho\}}\Delta t_i. \tag{2.6}$$

$L_{n,T}$ is obtained by an Itô type approximation of the stochastic integral and rectangular rule approximation of the ordinary integral in L_T.

Using Itô formula for the stochastic integral and rectangular rule approximation for the ordinary integral in (2.2) we obtain the approximate likelihood

$$L_{n,T,1}(\theta) = \frac{\theta}{2}(X_T^2 I_{\{|\Delta X_T|\le h^\rho\}} - T) - \frac{\theta^2}{2}\sum_{i=1}^{n} X_{t_{i-1}}^2 I_{\{|\Delta X_{t_i}|\le h^\rho\}}\Delta t_i. \tag{2.7}$$

Note that if we transform the Itô integral to the FS integral in L_T use the trapezoidal approximation of the FS integral and rectangle rule approximation for the ordinary in L_T, then also we obtain the approximate likelihood $L_{n,T,1}$.

Maximizing $L_{n,T,1}$ provides another approximate maximum likelihood estimate (AMLE2) $\theta_{n,T,1}$ given by

$$\theta_{n,T,1} = \frac{\frac{1}{2}(X_T^2 I_{\{|\Delta X_T|\le h^\rho\}} - T)}{\sum_{i=1}^{n} X_{t_{i-1}}^2 I_{\{|\Delta X_{t_i}|\le h^\rho\}}\Delta t_i}. \tag{2.8}$$

In this chapter, we obtain the Berry–Esseen bounds, i.e., rates of convergence to normality of the two estimators given in (2.5) and (2.8).

Note that

$$\int z^2 f(z)dz = \frac{\delta}{\gamma^3}$$

and

$$\int z^4 f(z)dz = \frac{15\delta}{\gamma^7}.$$

We have the following estimators of δ and γ.

$$\gamma_{n,T} := \left\{ \frac{15\sum_{i=1}^n (X_{t_i} - X_{t_{i-1}} - hX_{t_{i-1}}\theta_{n,T})^2 I_{\{|\Delta X_{t_i}|>h^\rho\}}}{\sum_{i=1}^n (X_{t_i} - X_{t_{i-1}} - hX_{t_{i-1}}\theta_{n,T})^4 I_{\{|\Delta X_{t_i}|>h^\rho\}}} \right\}^{1/4}.$$

$$\delta_{n,T} := \left\{ \frac{15^3 \left[\sum_{i=1}^n (X_{t_i} - X_{t_{i-1}} - hX_{t_{i-1}}\theta_{n,T})^2 I_{\{|\Delta X_{t_i}|\le h^\rho\}}\right]^7}{\left[\sum_{i=1}^n (X_{t_i} - X_{t_{i-1}} - hX_{t_{i-1}}\theta_{n,T})^4 I_{\{|\Delta X_{t_i}|>h^\rho\}}\right]^3} \right\}^{1/4}.$$

The organization of the chapter is as follows : In Sect. 3 we obtain the Berry–Esseen bounds the estimator $\theta_{n,T}$. Using purely nonrandom and various random normings (both sample dependent and parameter dependent) we obtain different Berry–Esseen bounds for the estimator $\theta_{n,T}$. Then we obtain probabilistic bounds on $|\theta_{n,T} - \theta_T|$.

In Sect. 4 we derive Berry–Esseen bounds for the estimators $\theta_{n,T,1}$ using purely random and various random normings (both sample dependent and parameter dependent). Then we find probabilistic bounds on $|\theta_{n,T,1} - \theta_T|$.

In Sect. 5 we introduce minimum contrast estimators, and in Sect. 6, we study the Berry–Esseen bounds for the minimum contrast estimators.

We use the following notations : $\Phi(\cdot)$ denotes the standard normal distribution function. C is a generic constant (which may depend on the parameter) throughout the Chapter. P and E denote probability and expectation under the true value of the parameter.

Introduce the following notations :

$$Y_{n,T} := \sum_{i=1}^n X(t_{i-1}) \left[(W_{t_i} - W(t_{i-1})) + (Z_{t_i} - Z(t_{i-1}))\right] I_{\{|\Delta X_{t_i}|\le h^\rho\}},$$

$$Y_T := \textstyle\int_0^T X_t(dW_t + dZ_t),$$

$$J_{n,T} := \sum_{i=1}^n X(t_{i-1}) \left[X_{t_i} - X(t_{i-1})\right] I_{\{|\Delta X_{t_i}|\le h^\rho\}}, \; J_T := \int_0^T X_t dX_t,$$

$$I_{n,T} := \sum_{i=1}^n X^2(t_{i-1}) I_{\{|\Delta X_{t_i}|\le h^\rho\}}(t_i - t_{i-1}), \; I_T := \int_0^T X_t^2 dt,$$

$$V_{n,T} := \sum_{i=1}^n \int_{t_{i-1}}^{t_i} X_{t_{i-1}} \left[X_t - X_{t_{i-1}}\right] I_{\{|\Delta X_{t_i}|\le h^\rho\}} dt.$$

We shall use the following lemma in the sequel whose proof is elementary.

Lemma 2.1 *Let Q_n, R_n, Q and R be random variables on the same prob. space $(\Omega, \mathcal{F}, P)$ with $P(R > 0) > 0$. Suppose $|Q_n - Q| = O_P(\delta_{1n})$ and $|R_n - R| = O_P(\delta_{1n})$ where $\delta_{1n}, \delta_{2n} \to 0$ as $n \to \infty$. Then*

$$\left|\frac{Q_n}{Q} - \frac{R_n}{R}\right| = O_P(\delta_{1n} \bigvee \delta_{2n}).$$

7.3 Berry–Esseen Bounds for AMLE1

Lemma 3.1 We have

$$E\left[\sup_{s,t\in[t_{i-1},t_i]} |X_t - X_s|^k\right] \le Ch^{k/2\wedge\{1+\rho(k-2)\}},\ i = 1, 2, \ldots, n.$$

Proof We can obtain the bound by explicit calculation by separating the Poisson part and the Gaussian part.

Lemma 3.2

(a) $E\left|Y_{n,T} - Y_T\right|^2 = O(\frac{T^2}{n})$,

(b) $E\left|J_{n,T} - J_T\right|^2 = O(\frac{T^2}{n})$,

(c) $E\left|I_{n,T} - I_T\right|^2 = O(\frac{T^4}{n^2})$.

Proof Let $g_i(t) = (X_{t_{i-1}} - X_t)I_{\{|\Delta X_{t_i}| \le h^\rho\}}$ for $t_{i-1} \le t < t_i,\quad i = 1, 2, \ldots, n$. Since

$$E|X_{t_{i-1}} - X_t|^{2k} \le C(t_{i-1} - t)^k, k = 1, 2, \ldots \tag{3.1}$$

by Wick's lemma, hence

$$\begin{aligned}
& E|Y_{n,T} - Y_T|^2 \\
= \ & E|\sum_{i=1}^{n} X_{t_{i-1}}[(W_{t_i} - W_{t_{i-1}}) + (Z_{t_i} - Z_{t_{i-1}})] - \int_0^T X_t(dW_t + dZ_t)|^2 \\
= \ & E|\int_0^T g_i(t)(dW_t + dZ_t)|^2 \\
= \ & \int_0^T E(g_i^2(t))dt \\
\le \ & C\sum_{i=1}^{n}\int_{t_{i-1}}^{t_i} |t_{i-1} - t|dt \\
= \ & Cn\frac{(t_i - t_{i-1})^2}{2} = C\frac{T^2}{n}.
\end{aligned}$$

This completes the proof of (a).

Using (2.1) and the fact that

$$X_{t_i} - X_{t_{i-1}} = \int_{t_{i-1}}^{t_i} \theta X_t dt + W_{t_i} - W_{t_{i-1}} + Z_{t_i} - Z_{t_{i-1}}$$

we obtain

$$\begin{aligned}
& E|J_{n,T} - J_T|^2 \\
= \;& E|\sum_{i=1}^{n} X_{t_{i-1}}[X_{t_i} - X_{t_{i-1}}] - \int_0^T X_t(dW_t + dZ_t)|^2 \\
= \;& E|\sum_{i=1}^{n}\int_{t_{i-1}}^{t_i} \theta X_t X_{t_{i-1}} dt + \sum_{i=1}^{n} X_{t_{i-1}}[(W_{t_i} - W_{t_{i-1}}) + (Z_{t_i} - Z_{t_{i-1}})] \\
& - \int_0^T \theta X_t^2 dt - \int_0^T X_t dW_t|^2 \\
\le \;& 2E|\sum_{i=1}^{n} X_{t_{i-1}}[(W_{t_i} - W_{t_{i-1}}) + (Z_{t_i} - Z_{t_{i-1}})] - \int_0^T X_t d(W_t + Z_t)|^2 \\
& + 2\theta^2 E|\sum_{i=1}^{n}\int_{t_{i-1}}^{t_i} X_t[X_{t_{i-1}} - X_t]dt|^2. \\
=: \;& N_1 + N_2.
\end{aligned}$$

N_1 is $O(\frac{T^2}{n})$ by Lemma 3.2(a). To estimate N_2 let

$$\psi_i(t) := X_t[X_{t_{i-1}} - X_t]I_{\{|\Delta X_{t_i}| \le h^\rho\}}$$

for $t_{i-1} \le t < t_i,\ i = 1, 2, \ldots, n$. Then

$$\begin{aligned}
& E|\sum_{i=1}^{n}\int_{t_{i-1}}^{t_i} \psi_i(t)dt|^2 \\
= \;& \sum_{i=1}^{n} E|\int_{t_{i-1}}^{t_i} \psi_i(t)dt|^2 + 2\sum_{i,j=1, i<j}^{n} E\left[\int_{t_{i-1}}^{t_i} \psi_i(t)dt \int_{t_{i-1}}^{t_i} \psi_j(s)ds\right] \\
=: \;& M_1 + M_2.
\end{aligned}$$

By the boundedness of $E(X_t^4)$ and (2.1) we have

$$\begin{aligned}
& E(\psi_i^2(t)) \\
= \;& E\{X_t^2[(X_{t_{i-1}} - X_t)I_{\{|\Delta X_{t_i}| \le h^\rho\}}]^2\} \\
\le \;& \{E(X_t^4)\}^{1/2}\{E[(X_{t_{i-1}} - X_t)I_{\{|\Delta X_{t_i}| \le h^\rho\}}]^4\}^{1/2} \\
\le \;& C(t_{i-1} - t).
\end{aligned} \tag{3.2}$$

Note that

$$
\begin{aligned}
M_1 &= \sum_{i=1}^{n} E|\int_{t_{i-1}}^{t_i} \psi_i(t)dt|^2 \\
&\le \sum_{i=1}^{n}(t_i - t_{i-1})\int_{t_{i-1}}^{t_i} E(\psi^2(t))dt \\
&\le C\frac{T}{n}\sum_{i=1}^{n}\int_{t_{i-1}}^{t_i}(t - t_{i-1})dt \\
&\le C\frac{T}{n}\sum_{i=1}^{n}(t_i - t_{i-1})^2 = C\frac{T^3}{n^2}
\end{aligned}
$$

and

$$
\begin{aligned}
M_2 &= 2\sum_{i,j=1,i<j}^{n} E\int_{t_{i-1}}^{t_i}\int_{t_{j-1}}^{t_i}[\psi_i(t)\psi_j(s)]dtds \\
&= 2\sum_{i,j=1,i<j}^{n}\int_{t_{i-1}}^{t_i}\int_{t_{j-1}}^{t_j} E[\psi_i(t)\psi_j(s)]dtds.
\end{aligned}
$$

By Wick's lemma, we have

$$
\begin{aligned}
& E[\psi_i(t)\psi_j(s)] \\
= \;& E[X_t(X_{t_{i-1}} - X_t)I_{\{|\Delta X_{t_i}|\le h^\rho\}}X_s(X_{t_{j-1}} - X_s)I_{\{|\Delta X_{t_j}|\le h^\rho\}}] \\
= \;& E[X_t(X_{t_{i-1}} - X_t)I_{\{|\Delta X_{t_i}|\le h^\rho\}}]E[X_s(X_{t_{j-1}} - X_s)I_{\{|\Delta X_{t_j}|\le h^\rho\}}] \\
+ \;& E[X_tX_s]E[(X_{t_{i-1}} - X_t)I_{\{|\Delta X_{t_i}|\le h^\rho\}}(X_{t_{j-1}} - X_s)I_{\{|\Delta X_{t_i}|\le h^\rho\}}] \\
+ \;& E[X_t(X(t_{j-1}) - X_s)I_{\{|\Delta X_{t_j}|\le h^\rho\}}]E[X_s(X_{t_{i-1}} - X_t)I_{\{|\Delta X_{t_i}|\le h^\rho\}}] \\
=: \;& A_1 + A_2 + A_3.
\end{aligned}
$$

Note that

$$
X_t = \int_0^t e^{\theta(t-u)}d(W_u + Z_u), t \ge 0.
$$

Let $a := e^\theta$. For $s \ge t$, we have

$$
\begin{aligned}
& E(X_tX_s) \\
= \;& E(\int_0^t e^{\theta(t-u)}d(W_u + Z_u))(\int_0^s e^{\theta(s-u)}d(W_u + Z_u)) \\
= \;& \int_0^t e^{\theta(t+s-2u)}du = \frac{1}{2\theta}[a^{s+t} - a^{s-t}].
\end{aligned}
$$

Observe that

$$
E(X_t - X_{t_{i-1}})I_{\{|\Delta X_{t_i}|\le h^\rho\}}(X_s - X_{t_{j-1}})I_{\{|\Delta X_{t_j}|\le h^\rho\}}
$$

$$
\begin{aligned}
&= E(X_tX_s) - E(X_tX_{t_{j-1}}) - E(X_{t_{i-1}}X_s) + E(X_{t_{i-1}}X_{t_{j-1}}) \\
&= \frac{1}{2\theta}(a^s - a^{t_{j-1}})[(a^t - a^{t_{i-1}}) + (a^{-t_{i-1}} - a^{-t})] \\
&= \frac{1}{2\theta}(s - t_{j-1})a^{t^*}[(t - t_{i-1}a^{t^{**}} + (t - t_{i-1})a^{-t^{***}}] \\
&\quad (\text{where } t_{j-1} < t^* < s, t_{i-1} < t^{**}, t^{***} < t) \\
&\leq \frac{1}{2\theta}(s - t_{j-1})a^t(t - t_{i-1})a^{t_{i-1}} + (s - t_{j-1})a^t(t - t_{i-1})a^{-t}] \\
&\leq C(s - t_{j-1})(t - t_{i-1}).
\end{aligned}
$$

Thus $A_2 \leq C(s - t_{j-1})(t - t_{i-1})$ since $|E(X_tX_s)|$ is bounded.

Next

$$
\begin{aligned}
&|E[X_t(X_{t_{i-1}} - X_t)I_{\{|\Delta X_{t_i}| \leq h^\rho\}}]| \\
&= \frac{1}{2|\theta|}[a^{t+t_{i-1}} - a^{t-t_{i-1}} - a^{2t} + 1] \\
&= \frac{1}{2|\theta|}a^t[a^{t_{i-1}} - a^{-t_{i-1}} - a^t + a^{-t}] \\
&\leq \frac{1}{2|\theta|}a^t(t - t_{i-1})[a^{t_{i-1}} + a^{-t}] \\
&\leq C(t - t_{i-1})
\end{aligned}
$$

and

$$
\begin{aligned}
&|E[X_s(X_s - X_{t_{j-1}})I_{\{|\Delta X_{t_j}| \leq h^\rho\}}]| \\
&= \frac{1}{2|\theta|}[a^{2s} - 1 - a^{s+t_{j-1}} + a^{s+t_{j-1}}] \\
&= \frac{1}{2|\theta|}a^s[a^s - a^{-s} - a^{t_{j-1}} + a^{-t_{j-1}}] \\
&\leq \frac{1}{2|\theta|}a^s(s - t_{j-1})[a^{t_{j-1}} + a^{-s}] \\
&\leq C(s - t_{j-1}).
\end{aligned}
$$

Thus $A_1 \leq C(s - t_{j-1})(t - t_{i-1})$.

Next

$$
\begin{aligned}
&|E[X_t(X_s - X_{t_{j-1}})I_{\{|\Delta X_{t_j}| \leq h^\rho\}}]| \\
&= \frac{1}{2|\theta|}[a^{s+t} - a^{s-t} - a^{t+t_{j-1}} + a^{t_{j-1}-t}] \\
&= \frac{1}{2|\theta|}a^t(a^s - a^{t_{j-1}}) \\
&\leq \frac{1}{2|\theta|}a^t(1 - a^{-2t})(s - t_{j-1})a^t
\end{aligned}
$$

$$
\begin{aligned}
&\leq (a^{2t}-1)(s-t_{j-1})\\
&\leq C(s-t_{j-1})
\end{aligned}
$$

and

$$
\begin{aligned}
& |E[X_s(X_t - X_{t_{i-1}})I_{\{|\Delta X_{t_i}|\le h^\rho\}}]|\\
= & \frac{1}{2|\theta|}[a^{t+s} - a^{s-t} - a^{s+t_{i-1}} + a^{s-t_{i-1}}]\\
= & \frac{1}{2|\theta|}a^s[a^t - a^{-t} - a^{t_{i-1}} + a^{-t_{i-1}}]\\
\leq & \frac{1}{2|\theta|}a^s(t-t_{i-1})[a^{t_{i-1}} + a^{-t}]\\
\leq & C(t-t_{i-1}).
\end{aligned}
$$

Thus $A_3 \leq C(s-t_{j-1})(t-t_{i-1})$. Hence $E[f_i(t)f_j(s)] \leq C(s-t_{j-1})(t-t_{i-1})$.

Thus

$$
\begin{aligned}
M_2 &= 2\sum_{i,j=1,i<j}^{n}\int_{t_{i-1}}^{t_i}\int_{t_{j-1}}^{t_j} E[f_i(t)f_j(s)]dtds\\
&\leq C\sum_{i,j=1,i<j}^{n}\int_{t_{i-1}}^{t_i}\int_{t_{j-1}}^{t_j}(t-t_{i-1})(s-t_{j-1})dtds\\
&= C\sum_{i,j=1,i<j}^{n}(t_{i-1}-t_i)^2(t_{j-1}-t_i)^2\\
&= Cn^2(\frac{T}{n})^4 = C\frac{T^4}{n^2}.
\end{aligned}
$$

Hence, N_2 is $O(\frac{T^3}{n^2})$. Combining N_1 and N_2 completes the proof of (b). We next prove (c).

Let $\chi_i(t) := (X_{t_{i-1}}^2 - X_t^2, t_{i-1})I_{\{|\Delta X_{t_i}|\le h^\rho\}} \le t < t_i, i = 1, 2, \ldots, n$. Then

$$
\begin{aligned}
& E|I_{n,T} - I_t|^2\\
= & E|\sum_{i=1}^{n} X_{t_{i-1}}^2 I_{\{|\Delta X_{t_i}|\le h^\rho\}}(t_i - t_{i-1}) - \int_0^T X_t^2 dt|^2\\
= & E|\sum_{i=1}^{n}\int_{t_{i-1}}^{t_i}[(X_{t_{i-1}}^2 - X_t^2)I_{\{|\Delta X_{t_i}|\le h^\rho\}}]dt|^2\\
= & E|\sum_{i=1}^{n}\int_{t_{i-1}}^{t_i}\chi_i(t)dt|^2\\
= & \sum_{i=1}^{n}E|\int_{t_{i-1}}^{t_i}\chi_i(t)dt|^2 + 2\sum_{i,j=1,i<j}^{n}E\int_{t_{i-1}}^{t_i}\int_{t_{j-1}}^{t_j}\chi_i(t)\chi_j(s)dtds\\
=: & B_1 + B_2.
\end{aligned}
$$

$$
\begin{aligned}
E\chi_i^2(t) &= E[(X_{t_{i-1}}^2 - X_t^2)I_{\{|\Delta X_{t_i}|\le h^\rho\}}]^2 \\
&= E[(X_{t_{i-1}} - X_t)I_{\{|\Delta X_{t_i}|\le h^\rho\}}]^2[(X_{t_{i-1}} + X_t)I_{\{|\Delta X_{t_i}|\le h^\rho\}}]^2 \\
&\le \{E[(X_{t_{i-1}} - X_t)I_{\{|\Delta X_{t_i}|\le h^\rho\}}]^4\}^{1/2} \\
&\quad \times\{\{E[(X_{t_{i-1}} + X_t)I_{\{|\Delta X_{t_i}|\le h^\rho\}}]^4\}^{1/2} \\
&\le C(t - t_{i-1})
\end{aligned}
$$

(by (3.1) and the boundedness of the second term)

$$
\begin{aligned}
B_1 &= \sum_{i=1}^{n} E|\int_{t_{i-1}}^{t_i} \chi_i(t)dt|^2 \\
&\le \sum_{i=1}^{n}(t_i - t_{i-1})\int_{t_{i-1}}^{t_i} E(\chi_i^2(t))dt \le C\frac{T}{n}\sum_{i=1}^{n}\int_{t_{i-1}}^{t_i}(t - t_{i-1})dt \\
&= C\frac{T^3}{n^2}.
\end{aligned}
$$

Note that

$$
\begin{aligned}
& E[\chi_i(t)\chi_j(s)] \\
&= E((X_{t_{i-1}}^2 - X_t^2)I_{\{|\Delta X_{t_i}|\le h^\rho\}})((X_{t_{j-1}}^2 - X_s^2)I_{\{|\Delta X_{t_j}|\le h^\rho\}}) \\
&= E((X_{t_{i-1}} - X_t)I_{\{|\Delta X_{t_i}|\le h^\rho\}})((X_{t_{i-1}} + X_t)I_{\{|\Delta X_{t_i}|\le h^\rho\}}) \\
&\quad ((X_{t_{j-1}} - X_s)I_{\{|\Delta X_{t_j}|\le h^\rho\}})((X_{t_{j-1}} + X_s)I_{\{|\Delta X_{t_j}|\le h^\rho\}}).
\end{aligned}
$$

Now using Wick's lemma and proceeding similar to the estimation of M_2 it is easy to see that

$$B_2 \le C\frac{T^4}{n^2}.$$

Combining B_1 and B_2, (c) follows. □

Theorem 3.3 *Let* $\alpha_{n,T} := \max(T^{-1/2}(\log T)^{1/2}, \frac{T^2}{n}(\log T)^{-1}, \frac{T^4}{n^2}(\log T)^{-1})$. *We have,*

$$\text{(a)} \sup_{x\in\mathbb{R}}\left|P\left\{\left(\frac{T}{-2\theta}\right)^{1/2}(\theta_{n,T} - \theta) \le x\right\} - \Phi(x)\right| = O(\alpha_{n,T}).$$

$$\text{(b)} \sup_{x\in\mathbb{R}}\left|P\left\{I_{n,T}^{1/2}(\theta_{n,T} - \theta) \le x\right\} - \Phi(x)\right| = O(\alpha_{n,T}).$$

$$\text{(c)} \sup_{x\in\mathbb{R}}\left|P\left\{\left(\frac{T}{|2\theta_{n,T}|}\right)^{1/2}(\theta_{n,T} - \theta) \le x\right\} - \Phi(x)\right| = O(\alpha_{n,T}).$$

Proof

(a) It is easy to see that

$$\theta_{n,T} - \theta = \frac{Y_{n,T}}{I_{n,T}} + \theta\frac{V_{n,T}}{I_{n,T}}. \tag{3.3}$$

Hence

$$
\begin{aligned}
& \sup_{x\in\mathbb{R}}\left|P\left\{\left(\frac{T}{-2\theta}\right)^{1/2}(\theta_{n,T}-\theta)\le x\right\}-\Phi(x)\right| \\
= & \sup_{x\in\mathbb{R}}\left|P\left\{\left(\frac{T}{-2\theta}\right)^{1/2}\frac{Y_{n,T}}{I_{n,T}}+\left(\frac{T}{-2\theta}\right)^{1/2}\theta\frac{V_{n,T}}{I_{n,T}}\le x\right\}-\Phi(x)\right| \\
\le & \sup_{x\in\mathbb{R}}\left|P\left\{\left(\frac{T}{-2\theta}\right)^{1/2}\frac{Y_{n,T}}{I_{n,T}}\le x\right\}-\Phi(x)\right| \\
& +P\left\{\left|\left(\frac{T}{-2\theta}\right)^{1/2}\frac{V_{n,T}}{I_{n,T}}\right|>\epsilon\right\}+\epsilon. \\
=: & K_1+K_2+\epsilon.
\end{aligned} \tag{3.4}
$$

Note that by Lemma 2.1 (b)

$$
\begin{aligned}
K_1 &= \sup_{x\in\mathbb{R}}\left|P\left\{\left(\frac{T}{-2\theta}\right)^{1/2}\frac{Y_{n,T}}{I_{n,T}}\le x\right\}-\Phi(x)\right| \\
&= \sup_{x\in\mathbb{R}}\left|P\left\{\frac{\left(-\frac{2\theta}{T}\right)^{1/2}Y_{n,T}}{\left(\frac{-2\theta}{T}\right)I_{n,T}}\le x\right\}-\Phi(x)\right| \\
&\le \sup_{x\in\mathbb{R}}\left|P\left\{\left(\frac{-2\theta}{T}\right)^{1/2}Y_{n,T}\le x\right\}-\Phi(x)\right| \\
&\quad +P\left\{\left(\frac{-2\theta}{T}\right)I_{n,T}-1>\epsilon\right\}+\epsilon \\
&=: J_1+J_2+\epsilon.
\end{aligned} \tag{3.5}
$$

$$
\begin{aligned}
J_1 &= \sup_{x\in\mathbb{R}}\left|P\left\{\left(\frac{-2\theta}{T}\right)^{1/2}(Y_{n,T}-Y_T+Y_T)\le x\right\}-\Phi(x)\right| \\
&\le \sup_{x\in\mathbb{R}}\left|P\left\{\left(\frac{-2\theta}{T}\right)^{1/2}Y_T\le x\right\}-\Phi(x)\right| \\
&\quad +P\left\{\left(\frac{-2\theta}{T}\right)^{1/2}|Y_{n,T}-Y_T|>\epsilon\right\}+\epsilon \\
&\le CT^{-1/2}+\left(\frac{-2\theta}{T}\right)\frac{E|Y_{n,T}-Y_T|^2}{\epsilon^2}+\epsilon \\
&\le CT^{-1/2}+C\frac{T/n}{\epsilon^2}+\epsilon \\
&\quad \text{(by Corollary 2.3(a) and Lemma 3.2(a).)}
\end{aligned} \tag{3.6}
$$

$$
\begin{aligned}
J_2 &= P\left\{\left|\left(\frac{-2\theta}{T}\right)(I_{n,T}-I_T+I_T)-1\right|>\epsilon\right\} \\
&\le P\left\{\left|\left(\frac{-2\theta}{T}\right)I_T-1\right|>\frac{\epsilon}{2}\right\}+P\left\{\left(-\frac{2\theta}{2}\right)|I_{n,T}-I_T|>\frac{\epsilon}{2}\right\} \\
&\le C\exp\left(\frac{T\theta}{16}\epsilon^2\right)+\frac{16\theta^2}{T^2}\frac{E|I_{n,T}-I_T|^2}{\epsilon^2} \\
&\le C\exp\left(\frac{T\theta}{16}\epsilon^2\right)+C\frac{T^2/n^2}{\epsilon^2}.
\end{aligned} \tag{3.8}
$$

Here the bound for the first term in (3.8) comes from Lemma 2.4(a) and that for the second term from Lemma 3.2(c). From the proof of Lemma 3.2(b) we have

$$E|V_{n,T}|^2 \leq C\frac{T^4}{n^2}. \tag{3.9}$$

Next

$$\begin{aligned}
K_2 &= P\left\{\left|\left(\frac{T}{-2\theta}\right)^{1/2}\theta\frac{V_{n,T}}{I_{n,T}}\right| > \epsilon\right\} \\
&= P\left\{\left|\frac{\left(\frac{-2\theta}{T}\right)^{1/2}\theta V_{n,T}}{\left(-\frac{2\theta}{T}\right)I_{n,T}}\right| > \epsilon\right\} \\
&= P\left\{\left|\left(\frac{-2\theta}{T}\right)^{1/2}\theta V_{n,T}\right| > \delta\right\} + P\left\{\left(-\frac{2\theta}{T}\right)I_{n,T} < \frac{\delta}{\epsilon}\right\} \\
&\quad \text{(where we choose } \delta = \epsilon - C\epsilon^2) \\
&\leq P\left\{\left|\left(\frac{-2\theta}{T}\right)^{1/2}\theta V_{n,T}\right| > \delta\right\} + P\left\{\left|\left(-\frac{2\theta}{T}\right)I_{n,T} - 1\right| > \delta_1\right\} \\
&\quad \text{(where } \delta_1 = \frac{\epsilon-\delta}{\delta} = C\epsilon) \\
&\leq -\frac{2\theta}{T}\theta^2\frac{E|V_{n,T}|^2}{\delta^2} + C\exp\left(\frac{T\theta}{16}\delta_1^2\right) + C\frac{T^2/n^2}{\delta_1^2} \\
&\leq C\frac{T^3/n^2}{\delta^2} + C\exp\left(\frac{T\theta}{16}\delta_1^2\right) + C\frac{T^2/n^2}{\delta_1^2} \quad \text{(by (3.9) and (3.8)).}
\end{aligned} \tag{3.10}$$

Now combining bounds from $J_1, J_2, K_1,$ and K_2, we have since $T/n \to 0$

$$\begin{aligned}
&\sup_{x\in\mathbb{R}}\left|P\left\{\left(\frac{T}{-2\theta}\right)^{1/2}(\theta_{n,T}-\theta) \leq x\right\} - \Phi(x)\right| \\
\leq\ & CT^{-1/2} + C\exp\left(\frac{T\theta}{16}\epsilon^2\right) + C\frac{T/n}{\epsilon^2} + C\frac{T^2/n^2}{\epsilon^2} + C\frac{T^3/n^2}{\delta^2} \\
&+C\exp\left(\frac{T\theta}{16}\delta_1^2\right) + C\left(\frac{T^2/n^2}{\delta_1^2}\right) + \epsilon.
\end{aligned} \tag{3.11}$$

Choosing $\epsilon = CT^{-1/2}(\log T)^{1/2}$, C large, the terms of (3.11) are of the order $O(\max(T^{-1/2}(\log T)^{1/2}, \frac{T^2}{n}(\log T)^{-1}, \frac{T^4}{n^2}(\log T)^{-1}))$. This proves (a).

(b) Using the expression (3.3), we have

$$\begin{aligned}
&\sup_{x\in\mathbb{R}}\left|P\left\{I_{n,T}^{1/2}(\theta_{n,T}-\theta) \leq x\right\} - \Phi(x)\right| \\
=\ & \sup_{x\in\mathbb{R}}\left|P\left\{\frac{Y_{n,T}}{I_{n,T}^{1/2}} + \theta\frac{V_{n,T}}{I_{n,T}^{1/2}} \leq x\right\} - \Phi(x)\right| \\
\leq\ & \sup_{x\in\mathbb{R}}\left|P\left\{\frac{Y_{n,T}}{I_{n,T}^{1/2}} \leq x\right\} - \Phi(x)\right| + P\left\{\left|\theta\frac{V_{n,T}}{I_{n,T}^{1/2}}\right| > \epsilon\right\} + \epsilon. \\
=:\ & H_1 + H_2 + \epsilon.
\end{aligned} \tag{3.12}$$

Note that

$$
\begin{aligned}
H_1 &= \sup_{x\in\mathbb{R}} \left| P\left\{ \frac{Y_{n,T} - Y_T + Y_T}{I_{n,T}^{1/2}} \le x \right\} - \Phi(x) \right| \\
&= \sup_{x\in\mathbb{R}} \left| P\left\{ \frac{Y_T}{I_{n,T}^{1/2}} \le x \right\} - \Phi(x) \right| + P\left\{ \frac{|Y_{n,T} - Y_T|}{I_{n,T}^{1/2}} > \epsilon \right\} + \epsilon. \\
&=: F_1 + F_2 + \epsilon.
\end{aligned}
\tag{3.13}
$$

Now

$$
\begin{aligned}
F_1 &= \sup_{x\in\mathbb{R}} \left| P\left\{ \frac{Y_T}{I_{n,T}^{1/2}} \le x \right\} - \Phi(x) \right| \\
&\le \sup_{x\in\mathbb{R}} \left| P\left\{ \left(\frac{-2\theta}{T}\right)^{1/2} Y_T \le x \right\} - \Phi(x) \right| \\
&\quad + P\left\{ \left| \left(\frac{-2\theta}{T}\right)^{1/2} I_{n,T}^{1/2} - 1 \right| > \epsilon \right\} + \epsilon \text{ (by Lemma 2.1(b))} \\
&\le C_T^{-1/2} + P\left\{ \left| \left(\frac{-2\theta}{T}\right) I_{n,T} - 1 \right| > \epsilon \right\} + \epsilon \text{ (by Corollary 2.3(a))} \\
&\le CT^{-1/2} + C\exp\left(-\tfrac{T\theta}{16}\epsilon^2\right) + C\tfrac{T^2/n^2}{\epsilon^2} + \epsilon. \text{ (by (3.8))}
\end{aligned}
\tag{3.14}
$$

On the other hand,

$$
\begin{aligned}
F_2 &= P\left\{ \frac{|Y_{n,T} - Y_T|}{I_{n,T}^{1/2}} > \epsilon \right\} \\
&\le P\left\{ \left(\frac{-2\theta}{T}\right)^{1/2} |Y_{n,T} - Y_T| > \delta \right\} + P\left\{ \left| \left(\frac{-2\theta}{T}\right)^{1/2} I_{n,T}^{1/2} - 1 \right| > \delta_1 \right\} \\
&\quad \text{(where } \delta = \epsilon - C\epsilon^2 \text{ and } \delta_1 = (\epsilon - \delta)/\epsilon > 0) \\
&\le \frac{\left(\frac{-2\theta}{T}\right) E|Y_{n,T} - Y_T|^2}{\delta^2} + P\left\{ \left| \left(\frac{-2\theta}{T}\right) I_{n,T} - 1 \right| > \delta_1 \right\} \\
&\le C\frac{T/n}{\delta^2} + C\exp\left(\frac{T\theta}{16}\delta_1^2\right) + C\frac{T^2/n^2}{\delta_1^2} \text{ (from Lemma 3.2(a) and (3.8)).}
\end{aligned}
\tag{3.15}
$$

Using (3.15) and (3.14) in (3.13), we obtain

$$
\begin{aligned}
H_1 &= \sup_{x\in\mathbb{R}} \left| P\left\{ \frac{Y_T}{I_{n,T}^{1/2}} \le x \right\} - \Phi(x) \right| \\
&\le CT^{-1/2} + C\exp\left(\frac{T\theta}{16}\epsilon^2\right) + C\frac{T/n}{\delta^2} + C\frac{T^2/n^2}{\delta_1^2} \\
&\quad + C\exp\left(\frac{T\theta}{16}\delta_1^2\right) + C\frac{T^2/n^2}{\epsilon^2} + \epsilon.
\end{aligned}
\tag{3.16}
$$

$$
\begin{aligned}
H_2 &= P\left\{\left|\theta\frac{V_{n,T}}{I_{n,T}^{1/2}}\right| > \epsilon\right\} = P\left\{\frac{\left|\left(\frac{-2\theta}{T}\right)^{1/2}\theta V_{n,T}\right|}{\left|\left(\frac{-2\theta}{T}\right)^{1/2} I_{n,T}^{1/2}\right|} > \epsilon\right\} \\
&\le P\left\{\left|\left(\frac{-2\theta}{T}\right)^{1/2}\theta V_{n,T}\right| > \delta\right\} + P\left\{\left|\left(\frac{-2\theta}{T}\right)^{1/2} I_{n,T}^{1/2}\right| < \delta/\epsilon\right\} \\
&\le \left(-\frac{2\theta}{T}\right)\theta^2\frac{E|V_{n,T}|^2}{\delta^2} + P\left\{\left|\left(-\frac{2\theta}{T}\right)I_{n,T} - 1\right| > \delta_1\right\} \\
&\quad \text{(where } 0 < \delta < \epsilon \text{ and } \delta_1 = (\epsilon-\delta)/\epsilon = C\epsilon > 0) \\
&\le C\frac{T^3/n^2}{\delta^2} + C\exp\left(\frac{T\theta}{16}\delta_1^2\right) + C\frac{T^2/n^2}{\delta_1^2} \quad \text{(from (3.9) and (3.8)).}
\end{aligned} \tag{3.17}
$$

Using (3.17) and (3.16) in (3.12) and choosing $\epsilon = CT^{-1/2}(\log T)^{1/2}, C$ large, the terms of (3.12) are of the order $O(\max(T^{-1/2}(\log T)^{1/2}, \frac{T^2}{n}(\log T)^{-1}, \frac{T^4}{n^2}(\log T)^{-1}))$. This proves (b).

(c) Let $D_T = \{|\theta_{n,T} - \theta| \le d_T\}$ and $d_T = CT^{-1/2}(\log T)^{1/2}$.

On the set D_T, expanding $(2|\theta_{n,T}|)^{-1/2}$, we obtain

$$
\begin{aligned}
(-2\theta_{n,T})^{-1/2} &= (-2\theta)^{-1/2}\left[1 - \frac{\theta - \theta_{n,T}}{\theta}\right]^{-1/2} \\
&= (-2\theta)^{-1/2}\left[1 + \frac{1}{2}\left(\frac{\theta - \theta_{n,T}}{\theta}\right) + O(d_T^2)\right].
\end{aligned}
$$

Then

$$
\begin{aligned}
&\sup_{x\in\mathbb{R}}\left|P\left\{\left(\frac{T}{2|\theta_{n,T}|}\right)^{1/2}(\theta_{n,T} - \theta) \le x\right\} - \Phi(x)\right| \\
\le\ &\sup_{x\in\mathbb{R}}\left|P\left\{\left(\frac{T}{2|\theta_{n,T}|}\right)^{1/2}(\theta_{n,T} - \theta) \le x, D_T\right\} - \Phi(x)\right| + P(D_T^c).
\end{aligned} \tag{3.18}
$$

$$
\begin{aligned}
&P(D_T^c) \\
=\ &P\left\{|\theta_{n,T} - \theta| > CT^{-1/2}(\log T)^{1/2}\right\} \\
=\ &P\left\{\left(\frac{T}{-2\theta}\right)^{1/2}|\theta_{n,T} - \theta| > C(\log T)^{1/2}(-2\theta)^{-1/2}\right\} \\
\le\ &C(\max(T^{-1/2}(\log T)^{1/2}, \frac{T^2}{n}(\log T)^{-1}, \frac{T^4}{n^2}(\log T)^{-1}) \\
&+2(1 - \Phi\left((\log T)^{1/2}(-2\theta)^{-1/2}\right) \quad \text{(by Theorem 3.3(a))} \\
\le\ &C(\max(T^{-1/2}(\log T)^{1/2}, \frac{T^2}{n}(\log T)^{-1}, \frac{T^4}{n^2}(\log T)^{-1})).
\end{aligned}
$$

On the set D_T,

$$
\left|\left(\frac{\theta_{n,T}}{\theta}\right)^{1/2} - 1\right| \le CT^{-1/2}(\log T)^{1/2}.
$$

Hence upon choosing $\epsilon = CT^{-1/2}(\log T)^{1/2}, C$ large we obtain

$$\begin{aligned}
&\left| P\left\{ \left(\frac{T}{-2\theta_{n,T}}\right)^{1/2} (\theta_{n,T}-\theta) \le x, D_T \right\} - \Phi(x) \right| \\
\le\ & \left| P\left\{ \left(\frac{T}{-2\theta}\right)^{1/2} (\theta_{n,T}-\theta) \le x, D_T \right\} - \Phi(x) \right| \\
&+P\left\{ \left| \left(\frac{\theta_{n,T}}{\theta}\right)^{1/2} - 1 \right| > \epsilon, D_T \right\} + \epsilon \\
&\quad \text{(by Lemma 3.3.1(b))} \\
\le\ & C(\max(T^{-1/2}(\log T)^{1/2}, \tfrac{T^2}{n}(\log T)^{-1}, \tfrac{T^4}{n^2}(\log T)^{-1})) \\
&\quad \text{(by Theorem 3.3(a))}.
\end{aligned} \tag{3.20}$$

(c) follows from (3.18)–(3.20). □

Theorem 3.4

$$\sup_{x\in\mathbb{R}} \left| P\left\{ I_{n,T}\left(-\frac{2\theta}{T}\right)^{1/2} (\theta_{n,T}-\theta) \le x \right\} - \Phi(x) \right| = O\left(T^{-1/2} \bigvee \left(\frac{T}{n}\right)^{1/3} \right).$$

Proof Let $a_{n,T} := Z_{n,T} - Z_T, \quad b_{n,T} := I_{n,T} - I_T.$

By Lemma 3.2 $E|a_{n,T}|^2 = O\left(\frac{T^2}{n}\right)$ and $E|b_{n,T}|^2 = O\left(\frac{T^4}{n^2}\right)$. (3.21)

From (3.5), we have

$$\begin{aligned}
I_{n,T}\theta_{n,T} &= \sum_{i=1}^{n} X_{t_{i-1}} \left[X_{t_i} - X_{t_{i-1}}\right] = \int_0^T X_t dX_t + a_{n,T} \\
&= \int_0^T X_t d(W_t + Z_t) + \theta \int_0^T X_t^2 dt + a_{n,T}.
\end{aligned}$$

Hence $I_{n,T}(\theta_{n,T}-\theta) = -\theta b_{n,T} + a_{n,T}$.

Thus

$$
\begin{aligned}
& \sup_{x\in\mathbb{R}} \left| P\left\{ I_{n,T} \left(-\frac{2\theta}{T}\right)^{1/2} (\theta_{n,T}-\theta) \le x \right\} - \Phi(x) \right| \\
= & \sup_{x\in\mathbb{R}} \left| P\left\{ \left(-\frac{2\theta}{T}\right)^{1/2} [Y_T - \theta b_{n,T} + a_{n,T}] \le x \right\} - \Phi(x) \right| \\
\le & \sup_{x\in\mathbb{R}} \left| P\left\{ \left(-\frac{2\theta}{T}\right)^{1/2} Y_T \le x \right\} - \Phi(x) \right| \\
& + P\left\{ \left| \left(-\frac{2\theta}{T}\right)^{1/2} [-\theta b_{n,T} + a_{n,T}] \right| > \epsilon \right\} + \epsilon \\
\le & \; CT^{-1/2} + \left(-\frac{2\theta}{T}\right) \frac{E|-\theta b_{n,T} + a_{n,T}|^2}{\epsilon^2} + \epsilon \\
\le & \; CT^{-1/2} + C\frac{T/n}{\epsilon^2} + \epsilon \text{ (by Corollary 1.2.3(a) and (3.21))}.
\end{aligned}
$$

Choosing $\epsilon = \left(\frac{T}{n}\right)^{1/3}$, the rate is $O\left(T^{-1/2} \bigvee \left(\frac{T}{n}\right)^{1/3}\right)$. □

Theorem 3.5

$$|\theta_{n,T} - \theta_T| = O_P(\frac{T^2}{n})^{1/2}.$$

Proof Note that $\theta_{n,T} - \theta_T = \frac{Z_{n,T}}{I_{n,T}} - \frac{Z_T}{I_T}$. From Lemma 3.2 it follows that $|Z_{n,T} - Z_T| = O_P(\frac{T^2}{n})^{1/2}$ and $|I_{n,T} - I_T| = O_P(\frac{T^4}{n^2})^{1/2}$. Now the theorem follows easily from the from the Lemma 2.1. □

7.4 Berry–Esseen Bounds for AMLE2

Theorem 4.1 *Let* $\beta_{n,T} = O\left(T^{-1/2}(\log T)^{1/2} \bigvee \frac{T^4}{n^2}(\log T)^{-1}\right)$.

(a) $\sup_{x\in\mathbb{R}} \left| P\left\{ \left(-\frac{T}{2\theta}\right)^{1/2} (\theta_{n,T,1}-\theta) \le x \right\} - \Phi(x) \right| = O(\beta_{n,T})$,

(b) $\sup_{x\in\mathbb{R}} \left| P\left\{ I_{n,T}^{1/2} (\theta_{n,T,1}-\theta) \le x \right\} - \Phi(x) \right| = O(\beta_{n,T})$,

(c) $\sup_{x\in\mathbb{R}} \left| P\left\{ \left(\frac{T}{2|\theta_{n,T,1}|}\right)^{1/2} (\theta_{n,T,1}-\theta) \le x \right\} - \Phi(x) \right| = O(\beta_{n,T})$.

Proof

(a) From (1.8), we have

$$\begin{aligned} I_{n,T}\theta_{n,T,1} &= \frac{1}{2}(X_T^2 - T) = \int_0^T X_t dX_t \\ &= \int_0^T X_t d(W_t + Z_t) + \theta \int_0^T X_t^2 dt \\ &= Y_T + \theta I_T. \end{aligned}$$

Thus

$$\begin{aligned} \left(-\tfrac{T}{2\theta}\right)^{1/2}(\theta_{n,T,1} - \theta) &= \frac{(-\frac{T}{2\theta})^{1/2} Y_T + \theta(-\frac{T}{2\theta})^{1/2}(I_T - I_{n,T})}{I_{n,T}} \\ &= \frac{(-\frac{2\theta}{T})^{1/2} Y_T + (-\frac{2\theta}{T})^{1/2}(I_T - I_{n,T})}{(-\frac{2\theta}{T}) I_{n,T}}. \end{aligned} \tag{4.1}$$

Now

$$\begin{aligned} & \sup_{x\in\mathbb{R}} \left| P\left\{ \left(-\frac{T}{2\theta}\right)^{1/2} (\theta_{n,T,1} - \theta) \le x \right\} - \Phi(x) \right| \\ = & \sup_{x\in\mathbb{R}} \left| P\left\{ \frac{(-\frac{2\theta}{T})^{1/2} Y_T + (-\frac{2\theta}{T})^{1/2}(I_T - I_{n,T})}{(-\frac{2\theta}{T}) I_{n,T}} \le x \right\} - \Phi(x) \right| \\ \le & \sup_{x\in\mathbb{R}} \left| P\left\{ (-\frac{2\theta}{T})^{1/2} Y_T \le x \right\} - \Phi(x) \right| + P\left\{ \left| \theta(-\frac{2\theta}{T})^{1/2}(I_{n,T} - I_T) \right| > \epsilon \right\} \\ & + P\left\{ \left| \left(-\frac{2\theta}{T}\right) I_{n,T} - 1 \right| > \epsilon \right\} + 2\epsilon \\ \le & \; CT^{-1/2} + \theta^2 \frac{(-\frac{2\theta}{T}) E\left|I_{n,T} - I_T\right|^2}{\epsilon^2} + C\exp(\frac{T\theta}{4}\epsilon^2) + C\frac{T^2}{n^2\epsilon^2} + 2\epsilon \end{aligned} \tag{4.2}$$

(The bound for the 3rd term in the r.h.s. of (4.2) is from (3.8).)

$$\le CT^{-1/2} + C\frac{T^3}{n^2\epsilon^2} + C\exp(\frac{T\theta}{4}\epsilon^2) + C\frac{T^2}{n^2\epsilon^2} + \epsilon. \tag{4.3}$$

(by Lemma 3.2 (c))

Choosing $\epsilon = CT^{-1/2}(\log T)^{1/2}$, the terms in the r.h.s. of (4.3) are of the order $O(T^{-1/2}(\log T)^{1/2} \bigvee \frac{T^4}{n^2}(\log T)^{-1})$.

(b) From (4.1), we have

$$I_{n,T}^{1/2}(\theta_{n,T,1} - \theta) = \frac{Y_T + \theta(I_T - I_{n,T})}{I_{n,T}^{1/2}}.$$

Now

$$
\begin{aligned}
& \sup_{x\in\mathbb{R}} \left| P\left\{ I_{n,T}^{1/2}(\theta_{n,T,1}-\theta) \le x \right\} - \Phi(x) \right| \\
= \ & \sup_{x\in\mathbb{R}} \left| P\left\{ \frac{Y_T}{I_{n,T}^{1/2}} + \theta \frac{I_T - I_{n,T}}{I_{n,T}^{1/2}} \le x \right\} - \Phi(x) \right| \\
\le \ & \sup_{x\in\mathbb{R}} \left| P\left\{ \frac{Y_T}{I_{n,T}^{1/2}} \le x \right\} - \Phi(x) \right| + P\left\{ \left| \frac{\theta(I_T - I_{n,T})}{I_{n,T}^{1/2}} \right| > \epsilon \right\} + \epsilon \\
=: \ & U_1 + U_2 + \epsilon.
\end{aligned}
\tag{4.4}
$$

We have from (3.8),

$$
U_1 \le CT^{-1/2} + C\exp(\frac{T\theta}{16}\epsilon^2) + C\frac{T^2}{n^2\epsilon^2} + \epsilon. \tag{4.5}
$$

Now

$$
\begin{aligned}
U_2 \ &= \ P\left\{ |\theta| \left| \frac{I_{n,T} - I_T}{I_{n,T}^{1/2}} \right| > \epsilon \right\} \\
&= \ P\left\{ |\theta| \frac{|(-\frac{2\theta}{T})^{1/2}(I_{n,T}-I_T)|}{|(-\frac{2\theta}{T})^{1/2} I_{n,T}^{1/2}|} > \epsilon \right\} \\
&\le \ P\left\{ \left|(-\frac{2\theta}{T})^{1/2}\right| |I_{n,T} - I_T| > \delta \right\} + P\left\{ \left| (-\frac{2\theta}{T})^{1/2} I_{n,T}^{1/2} - 1 \right| > \delta_1 \right\} \\
& \quad (\text{where } \delta = \epsilon - C\epsilon^2 \ \text{ and } \delta_1 = (\epsilon - \delta)/\epsilon > 0) \\
&\le \ (-\frac{2\theta}{T}) \frac{E|I_{n,T} - I_T|^2}{\delta^2} + P\left\{ \left| (-\frac{2\theta}{T}) I_{n,T} - 1 \right| > \delta_1 \right\}
\end{aligned}
\tag{4.6}
$$

$$
\le C\frac{T^3}{n^2\delta^2} + C\exp(\frac{T\theta}{16}\delta_1^2) + C\frac{T^2/n^2}{\delta_1^2}. \tag{4.7}
$$

Here the bound for the first term in the r.h.s. of (4.6) comes from Lemma 3.2(c) and that for the second term is from J_2 in (3.8).

Now using the bounds (4.5) and (4.7) in (4.4) with $\epsilon = CT^{-1/2}(\log T)^{1/2}$, we obtain that the terms in (4.4) is of the order $O(T^{-1/2}(\log T)^{1/2} \bigvee \frac{T^4}{n^2} (\log T)^{-1})$.

(c) Let $G_T = \{|\theta_{n,T,1} - \theta| \le d_T\}$ and $d_T = CT^{-1/2}(\log T)^{1/2}$. On the set G_T, expanding $(2|\theta_{n,T,1}|)^{1/2}$, we obtain,

$$
\begin{aligned}
(-2\theta_{n,T,1})^{-1/2} \ &= \ (-2\theta)^{1/2}\left[1 - \frac{\theta - \theta_{n,T,1}}{\theta}\right]^{-1/2} \\
&= \ (-2\theta)^{1/2}[1 + \frac{1}{2}(\frac{\theta - \theta_{n,T,1}}{\theta}) + O(d_T^2).
\end{aligned}
$$

Then

$$
\begin{aligned}
& \sup_{x\in\mathbb{R}} \left| P\left\{ (\frac{T}{2|\theta_{n,T,1}|})^{1/2}(\theta_{n,T,1} - \theta) \le x \right\} - \Phi(x) \right| \\
\le \ & \sup_{x\in\mathbb{R}} \left\{ P(\frac{T}{2|\theta_{n,T,1}|})^{1/2}(\theta_{n,T,1} - \theta) \le x, \ G_T \right\} + P(G_T^c).
\end{aligned}
$$

Now

$$\begin{aligned}
& P(G_T^c) \\
= \; & P\left\{|\theta_{n,T,1}-\theta| > CT^{-1/2}(\log T)^{1/2}\right\} \\
= \; & P\left\{\left(-\frac{T}{2\theta}\right)^{1/2}|\theta_{n,T,1}-\theta| > C(\log T)^{1/2}(-2\theta)^{-1/2}\right\} \\
\le \; & C\left(T^{-1/2}(\log T)^{1/2}\bigvee\frac{T^4}{n^2}(\log T)^{-1}) + 2(1-\Phi\log T^{1/2}(-2\theta)^{-1/2}\right) \\
& \text{(by Theorem 4.1(a))} \\
\le \; & C(T^{-1/2}(\log T)^{1/2}\bigvee\frac{T^4}{n^2}(\log T)^{-1}).
\end{aligned}$$

On the set G_T,

$$\left|\left(\frac{\theta_{n,T,1}}{\theta}\right)^{1/2}-1\right| \le CT^{-1/2}(\log T)^{1/2}$$

Hence upon choosing $\epsilon = CT^{-1/2}(\log T)^{1/2}, C$ large

$$\begin{aligned}
& \left|P\left\{\left(\frac{T}{-2\theta_{n,T,1}}\right)^{1/2}(\theta_{n,T,1}-\theta)\le x, G_T\right\}-\Phi(x)\right| \\
\le \; & \Bigg|P\left\{\left(\frac{T}{-2\theta}\right)^{1/2}(\theta_{n,T,1}-\theta)\le x, G_T\right\} \\
& +P\left\{\left|\left(\frac{\theta_{n,T}}{\theta}\right)^{1/2}-1\right|>\epsilon, G_T\right\}+\epsilon \\
& \text{(by Lemma 3.3.1 (b))} \\
\le \; & C(T^{-1/2}(\log T)^{1/2}\bigvee\tfrac{T^4}{n^2}(\log T)^{-1}) \;\; \text{(by Theorem 4.1(a))}. \qquad \square
\end{aligned}$$

Theorem 4.2

$$\sup_{x\in\mathbb{R}}|P\left\{I_{n,T}(-\frac{2\theta}{T})^{1/2}(\theta_{n,T,1}-\theta)\le x\right\}-\Phi(x)| = O(T^{-1/2}\bigvee(\frac{T^3}{n^2})^{1/3}).$$

Proof From (4.1) we have

$$I_{n,T}\left(-\frac{2\theta}{T}\right)^{1/2}(\theta_{n,T,1}-\theta) = (-\frac{2\theta}{T})^{1/2}Y_T+\theta(-\frac{2\theta}{T})^{1/2}(I_T-I_{n,T}).$$

Hence by Corollary 1.2.3(a) and Lemma 3.2(c)

$$\begin{aligned}
& \sup_{x\in\mathbb{R}} |P\left\{ I_{n,T}(-\frac{2\theta}{T})^{1/2}(\theta_{n,T,1}-\theta) \le x \right\} - \Phi(x)| \\
= & \sup_{x\in\mathbb{R}} |P\left\{ (-\frac{2\theta}{T})^{1/2} Y_T + \theta(-\frac{2\theta}{T})^{1/2}(I_T - I_{n,T}) \le x \right\} - \Phi(x)| \\
\le & \sup_{x\in\mathbb{R}} |P\left\{ (-\frac{2\theta}{T})^{1/2} Y_T \le x \right\} - \Phi(x)| \\
& + P\left\{ |\theta(-\frac{2\theta}{T})^{1/2}(I_T - I_{n,T})| > \epsilon \right\} + \epsilon \\
\le & \; CT^{-1/2} + C\frac{E|I_T - I_{n,T}|^2}{T\epsilon^2} + \epsilon \\
\le & \; CT^{-1/2} + C\frac{T^3}{n^2\epsilon^2} + \epsilon.
\end{aligned}$$

Choosing $\epsilon = (\frac{T^3}{n^2})^{1/3}$, the theorem follows. □

Theorem 4.3 $|\theta_{n,T,1} - \theta_T| = O_P(\frac{T^2}{n})$.
Proof We have from (2.3) $\theta_T = Z_T/I_T$. By Itô formula it is easy to see that

$$\theta_{n,T,1} = Z_T/I_{n,T}.$$

Hence applying Lemma 3.7 with the aid of Lemma 3.2(c) the theorem follows.
□

Remarks

(1) The bounds in Theorem 3.3, 3.4, 4.1, 4.2 are uniform over compact subsets of the parameter space.

(2) Theorems 3.3 and 3.4 are useful for testing hypothesis about θ. They do not necessarily give confidence intervals. Theorems 3.3(b) and 3.3(c) are useful for computation of a confidence interval.

(3) It may appear from Theorem 3.3 (a) that $\frac{T}{n^{1/2}} \to 0$ is essential for the asymptotic normality of $\theta_{n,T}$. However, it is not the case. Consider the bound given in (3.11). For this to converge to 0, it is necessary that $\frac{T}{n} \to 0$. Let $\gamma_n = \frac{T}{n}$. It is easy to see from this bound that if there exists an ϵ_n such that $n\gamma_n\epsilon_n^2 \to \infty$ and $\gamma^{-1}\epsilon_n^2 \to \infty$, then normal convergence holds and a rate may also be found out. Consider in particular, $T = n^\tau, \tau > 0$. Then, $\gamma_n = n^{\tau-1}$. Hence we require $n^\tau\epsilon_n^2 \to \infty$ and $n^{1-\tau}\epsilon_n^2 \to \infty$. This is possible for any $\tau > 1$. Thus if $T \approx n^\tau, \tau > 1$, then normal convergence in Theorem 3.3 (a) holds. Similar comment also holds for Theorem 3.3 (b), (c) and Theorem 4.1.

(4) If $\frac{T}{n^{1/2}} \to 0$ and $T \to \infty$, then to obtain bound of the order $T^{-1/2}$ $(\log T)^{1/2}$ in Theorem 3.3 one needs $n > T^{5/2}(\log T)^{-3/2}$. To obtain bound of the order $T^{-1/2}(\log T)^{1/2}$ in Theorem 4.1 one needs $n > T^{9/4}(\log T)^{-3/4}$. To obtain bound of the order $T^{-1/2}$ in Theorem 3.4 one needs $n \ge T^{10}$. To

obtain bound of the order $T^{-1/2}$ in Theorem 4.2 one needs $n > T^{9/4}$. To obtain bound of the order $\frac{1}{T}$ in Theorem 3.5 one needs $n \geq T$.

(5) The norming in Theorem 3.4 and 4.2 are random which also contains the unknown parameter. It would be interesting to find the order $(T^{-1/2} \bigvee (T/n)^{1/3})$ and $(T^{-1/2} \bigvee (\frac{T^3}{n^2})^{1/3})$, respectively, for the estimators $\theta_{n,T}$ and $\theta_{n,T,1}$ with completely random or completely nonrandom norming. The problem of obtaining rates of convergence of the CLSE in the case of nonlinear drift remains open.

(6) The Berry–Esseen bounds for the estimator $\theta_{n,T,1}$ in Theorem 4.1 are sharper than the corresponding bound for $\theta_{n,T}$ in Theorem 3.3. The Berry–Esseen bounds for the estimator $\theta_{n,T,1}$ in Theorem 4.2 are sharper than the corresponding bound for $\theta_{n,T}$ in Theorem 3.4. Also $\theta_{n,T,1}$ converges to θ_T at a faster rate than the rate at which $\theta_{n,T}$ converges to θ_T for fixed T when $n \to \infty$. This shows that $\theta_{n,T,1}$ is a better estimator of θ than $\theta_{n,T}$. Extension of this problem to nonlinear drift case now remains open. Berry–Esseen bounds in Theorems 3.4 and 4.2 show that *rapidly increasing experimental design* condition ($T \to \infty$ and $\frac{T}{\sqrt{n}} \to 0$) is not essential for the asymptotic normality of the estimators.

(7) In this Chapter we have only studied the rates of convergence of AMLEs in the stable case, i.e., when $\theta < 0$. It would be interesting to investigate the rates of convergence of AMLEs in the explosive case, i.e., when $\theta > 0$ and the unstable (critical) case, i.e., when $\theta = 0$. Note that in the explosive case with nonrandom norming the limit distribution is Cauchy but with random norming the limit distribution is normal.

(8) Here we have obtained uniform rates of convergence to normality. It would be interesting to obtain nonuniform rates of convergence to normality which are more useful.

(9) In a Bayesian framework based on the two given approximate likelihoods it would be interesting to investigate the rate of convergence of approximate posterior distributions and approximate Bayes estimators.

7.5 Berry–Esseen Bounds for AMCEs

As an alternative to maximum likelihood method and to obtain estimators with higher order accuracy our aim is to use contrast functions. Using Itô formula, the score function $\gamma_T(\theta)$ can be written as

$$\gamma_T(\theta) = \frac{1}{2}X_T^2 - \int_0^T (\theta X_t^2 + \frac{1}{2})dt. \tag{1.7}$$

Consider the estimating function

$$M_T(\theta) = -\int_0^T (\theta X_t^2 + \frac{1}{2})dt \tag{1.8}$$

and the minimum contrast estimate (MCE)

$$\tilde{\theta}_T := -\frac{T}{2}\left\{\int_0^T X_t^2 dt\right\}^{-1}. \tag{1.9}$$

M-estimator is reduced to the minimum contrast estimator. It is well known that $\tilde{\theta}_T$ is strongly consistent and asymptotically $\mathcal{N}(0, \ -2\theta)$ distributed as $T \to \infty$ (see Lanksa (1979)). The large deviations of $\hat{\theta}_T$ and $\tilde{\theta}_T$ were obtained in Florens-Landais and Pham (1999). In particular, it was shown the large deviation probabilities of $\hat{\theta}_T$ are identical to those of $\tilde{\theta}_T$ for $\theta \leq \theta_0/3$ but weaker for $\theta > \theta_0/3$ where θ_0 is the true value of the parameter. However, as far as the rate of convergence to normality is concerned, Bishwal (2002) showed that $\hat{\theta}_T$ and $\tilde{\theta}_T$ have the same Berry–Esseen bound of the order $O(T^{-1/2})$.

In this chapter we obtain weak convergence bounds for several approximate minimum contrast estimators (AMCEs) which are simpler and robust. In order to define the approximate minimum contrast estimators (AMCEs), we use various discrete approximations of the integral in the definition (1.9) of MCE.

Define a weighted sum of squares

$$M_{n,T} := \frac{T}{n}\left\{\sum_{i=1}^{n} w_{t_i} X_{t_{i-1}}^2 I_{\{|\Delta X_{t_i}| \leq h^\rho\}} + \sum_{i=2}^{n+1} w_{t_i} X_{t_{i-1}}^2 I_{\{|\Delta X_{t_i}| \leq h^\rho\}}\right\}, \tag{1.10}$$

where $w_{t_i} \geq 0$ is a weight function.

Denote the discrete increasing functions

$$I_{n,T} := \frac{T}{n}\sum_{i=1}^{n} X_{t_{i-1}}^2 I_{\{|\Delta X_{t_i}| \leq h^\rho\}}, \tag{1.11}$$

$$J_{n,T} := \frac{T}{n}\sum_{i=2}^{n+1} X_{t_{i-1}}^2 I_{\{|\Delta X_{t_i}| \leq h^\rho\}} = \frac{T}{n}\sum_{i=1}^{n} X_{t_i}^2. \tag{1.12}$$

General weighted AMCE is defined as

$$\tilde{\theta}_{n,T} := -\left\{\frac{2}{n} M_{n,T}\right\}^{-1}. \tag{1.13}$$

With $w_{t_i} = 1$, we obtain the forward AMCE as

$$\tilde{\theta}_{n,T,F} := -\left\{\frac{2}{n} I_{n,T}\right\}^{-1}. \tag{1.14}$$

With $w_{t_i} = 0$, we obtain the backward AMCE as

$$\tilde{\theta}_{n,T,B} := -\left\{\frac{2}{n} J_{n,T}\right\}^{-1}. \tag{1.15}$$

Analogous to the estimators for the discrete AR (1) model, we define the simple symmetric and weighted symmetric estimators (see Fuller (1996)): With $w_{t_i} = 0.5$, the simple symmetric AMCE is defined as

$$\tilde{\theta}_{n,T,z} := -\left\{\frac{1}{n}[I_{n,T} + J_{n,T}]\right\}^{-1}$$

$$= -\left\{\frac{2}{n}\sum_{i=2}^{n} X_{t_{i-1}}^2 I_{\{|\Delta X_{t_i}|\le h^{\rho}\}} + 0.5(X_{t_0}^2 + X_{t_n}^2)\right\}^{-1}. \tag{1.16}$$

With the weight function

$$w_{t_i} = \begin{cases} 0 & : \ i = 1 \\ \frac{i-1}{n} & : \ i = 2, 3, \cdots, n \\ 1 & : \ i = n+1 \end{cases}$$

the weighted symmetric AMCE is defined as

$$\tilde{\theta}_{n,T,w} := -\left\{\frac{2}{n}\sum_{i=2}^{n} X_{t_{i-1}}^2 I_{\{|\Delta X_{t_i}|\le h^{\rho}\}} + \frac{1}{n}\sum_{i=1}^{n} X_{t_{i-1}}^2 I_{\{|\Delta X_{t_i}|\le h^{\rho}\}}\right\}^{-1}. \tag{1.17}$$

Note that estimator (1.13) is analogous to the trapezoidal rule in numerical analysis. One can instead use the midpoint rule to define another estimator

$$\tilde{\theta}_{n,T,A} := -\left\{\frac{2}{n}\sum_{i=1}^{n}\left(\frac{X_{t_{i-1}} I_{\{|\Delta X_{t_i}|\le h^{\rho}\}} + X_{t_i} I_{\{|\Delta X_{t_i}|\le h^{\rho}\}}}{2}\right)^2\right\}^{-1}. \tag{1.18}$$

One can use the Simpson's rule to define another estimator where the denominator is a convex combination of the denominators in (1.11) and (1.12)

$$\tilde{\theta}_{n,T,S} := -\left\{\frac{1}{3n}\sum_{i=1}^{n}\left\{X_{t_{i-1}}^2 + 4\left(\frac{X_{t_{i-1}} I_{\{|\Delta X_{t_i}|\le h^{\rho}\}} + X_{t_i} I_{\{|\Delta X_{t_i}|\le h^{\rho}\}}}{2}\right)^2 \right.\right.$$
$$\left.\left. + X_{t_i}^2 I_{\{|\Delta X_{t_i}|\le h^{\rho}\}}\right\}\right\}^{-1}. \tag{1.19}$$

In general, one can generalize Simpson's rule as

$$\tilde{\theta}_{n,T,GS} := -\left\{\frac{2}{n}\sum_{i=1}^{n}\left\{\alpha\frac{X_{t_{i-1}}^2 I_{\{|\Delta X_{t_i}|\le h^{\rho}\}} + X_{t_i}^2 I_{\{|\Delta X_{t_i}|\le h^{\rho}\}}}{2}\right.\right.$$
$$\left.\left. +(1-\alpha)\left(\frac{X_{t_{i-1}} I_{\{|\Delta X_{t_i}|\le h^{\rho}\}} + X_{t_i} I_{\{|\Delta X_{t_i}|\le h^{\rho}\}}}{2}\right)^2\right\}\right\}^{-1} \tag{1.20}$$

for any $0 \le \alpha \le 1$.

The case $\alpha = 0$ produces the estimator (1.18). The case $\alpha = 1$ produces the estimator (1.17). The case $\alpha = \frac{1}{3}$ produces the estimator (1.19).

I propose a very general form of the quadrature based estimator as

$$\tilde{\theta}_{n,T,w} := -\left\{\frac{2}{n}\sum_{i=1}^{n}\sum_{j=1}^{m}\left[(1-s_j)X_{t_{i-1}}I_{\{|\Delta X_{t_i}|\leq h^{\rho}\}} + s_j X_{t_i} I_{\{|\Delta X_{t_i}|\leq h^{\rho}\}}\right]^2 p_j\right\}^{-1}, \tag{1.21}$$

where $p_j, \quad j \in \{1, 2, \cdots, m\}$ is a probability mass function of a discrete random variable S on $0 \leq s_1 < s_2 < \cdots < s_m \leq 1$ with $P(S = s_j) := p_j, \; j \in \{1, 2, \cdots, m\}$.
Denote the k-th moment of the random variable S as $\mu_k := \sum_{j=1}^{m} s_j^k p_j, \; k = 1, 2, \cdots$.

If one chooses the probability distribution as uniform distribution for which the moments are a harmonic sequence $(\mu_1, \mu_2, \mu_3, \mu_4, \mu_5, \mu_6, \cdots) = (\frac{1}{2}, \frac{1}{3}, \frac{1}{4}, \frac{1}{5}, \frac{1}{7}, \cdots)$, then there is no change in rate of convergence than second order. If one can construct a probability distribution for which the harmonic sequence is truncated at a point, then there is a rate of convergence improvement at the point of truncation.

Given a positive integer m, construct a probability mass function $p_j, \quad j \in \{1, 2, \cdots, m\}$ on $0 \leq s_1 < s_2 < \cdots < s_m \leq 1$ such that

$$\sum_{j=1}^{m} s_j^r p_j = \frac{1}{r+1}, \quad r \in \{0, \cdots, m-2\} \tag{1.22}$$

$$\sum_{j=1}^{m} s_j^{m-1} p_j \neq \frac{1}{m}. \tag{1.23}$$

Neither the probabilities p_j nor the atoms, s_j, of the distribution are specified in advance.

This problem is related to the truncated Hausdorff moment problem. I obtain examples of such probability distributions and use them to get higher order accurate (up to sixth order) AMCEs.

The order of approximation error (rate of convergence) of an estimator is $n^{-\nu}$ where

$$\nu := \inf\left\{k : \mu_k \neq \frac{1}{1+k}, \; \mu_j = \frac{1}{1+j}, j = 1, 2, \cdots, k-1\right\}. \tag{1.24}$$

I construct probability distributions satisfying these moment conditions and obtain estimators of the rate of convergence up to order 6.

Probability $p_1 = 1$ at the point $s_1 = 0$ gives the estimator (1.11) for which $\mu_1 = 0$. Note that $\mu_1 \neq \frac{1}{2}$. Thus $\nu = 1$

Probability $p_1 = 1$ at the point $s_1 = 1$ gives the estimator (1.12) for which $\mu_1 = 1$. Note that $\mu_1 \neq \frac{1}{2}$. Thus $\nu = 1$.

Probabilities $(p_1, p_2) = (\frac{1}{2}, \frac{1}{2})$ at the respective points $(s_1, s_2) = (0, 1)$ produces the estimator $\tilde{\theta}_{n,T,Z}$ for which $(\mu_1, \mu_2) = (\frac{1}{2}, \frac{1}{4})$. Thus $\nu = 2$.

Probability $p_j = 1$ at the point $s_j = \frac{1}{2}$ produces the estimator $\tilde{\theta}_{n,T,A}$ for which $(\mu_1, \mu_2) = (\frac{1}{2}, \frac{1}{2})$. Thus $\nu = 2$.

Probabilities $(p_1, p_2) = (\frac{1}{4}, \frac{3}{4})$ at the respective points $(s_1, s_2) = (0, \frac{2}{3})$ produce the asymmetric estimator

$$\tilde{\theta}_{n,T,3} := -\left\{\tfrac{2}{n}\tfrac{1}{4}\textstyle\sum_{i=1}^{n}\left[(X_{t_{i-1}})^2 + 3\left(\frac{X_{t_{i-1}}I_{\{|\Delta X_{t_i}|\le h^\rho\}}+2X_{t_i}I_{\{|\Delta X_{t_i}|\le h^\rho\}}}{3}\right)^2\right]\right\}^{-1} \tag{1.25}$$

for which $(\mu_1, \mu_2, \mu_3) = (\frac{1}{2}, \frac{1}{3}, \frac{2}{9})$. Thus $\nu = 3$.

Probabilities $(p_1, p_2) = \frac{3}{4}, \frac{1}{4}$ at the respective points $(s_1, s_2) = (\frac{1}{3}, 1)$ produce asymmetric estimator

$$\tilde{\theta}_{n,T,4} := -\left\{\tfrac{2}{n}\tfrac{1}{4}\textstyle\sum_{i=1}^{n}\left[3\left(\frac{2X_{t_{i-1}}I_{\{|\Delta X_{t_i}|\le h^\rho\}}+X_{t_i}I_{\{|\Delta X_{t_i}|\le h^\rho\}}}{3}\right)^2 + (X_{t_i})^2\right]\right\}^{-1} \tag{1.26}$$

for which $(\mu_1, \mu_2, \mu_3) = (\frac{1}{2}, \frac{1}{3}, \frac{10}{36})$. Thus $\nu = 3$.

Probabilities $(p_1, p_2, p_3) = (\frac{1}{6}, \frac{2}{3}, \frac{1}{6})$ at the respective points $(s_1, s_2, s_3) = (0, \frac{1}{2}, 1)$ produce the estimator $\tilde{\theta}_{n,T,5}$ for which $(\mu_1, \mu_2, \mu_3, \mu_4) = (\frac{1}{2}, \frac{1}{3}, \frac{1}{4}, \frac{5}{25})$. Thus $\nu = 4$.

Probabilities $(p_1, p_2, p_3, p_4) = (\frac{1}{8}, \frac{3}{8}, \frac{3}{8}, \frac{1}{8})$ at the respective points $(s_1, s_2, s_3, s_4) = (0, \frac{1}{3}, \frac{2}{3}, 1)$ produce the symmetric estimator

$$\begin{aligned}
\tilde{\theta}_{n,T,5} := -\Bigg\{\frac{2}{n}\frac{1}{8}\sum_{i=1}^{n}\Big[&(X_{t_{i-1}}I_{\{|\Delta X_{t_i}|\le h^\rho\}})^2 \\
&+3\Big(\frac{2X_{t_{i-1}}I_{\{|\Delta X_{t_i}|\le h^\rho\}} + X_{t_i}I_{\{|\Delta X_{t_i}|\le h^\rho\}}}{3}\Big)^2 \\
&+3\Big(\frac{X_{t_{i-1}}I_{\{|\Delta X_{t_i}|\le h^\rho\}} + 2X_{t_i}I_{\{|\Delta X_{t_i}|\le h^\rho\}}}{3}\Big)^2 \\
&+(X_{t_i}I_{\{|\Delta X_{t_i}|\le h^\rho\}})^2\Big]\Bigg\}^{-1}
\end{aligned} \tag{1.27}$$

for which $(\mu_1, \mu_2, \mu_3, \mu_4) = (\frac{1}{2}, \frac{1}{3}, \frac{1}{4}, \frac{11}{54})$. Thus $\nu = 4$.

Probabilities $(p_1, p_2, p_3, p_4, p_5) = (\frac{1471}{24192}, \frac{6925}{24192}, \frac{1475}{12096}, \frac{2725}{12096}, \frac{5675}{24192}, \frac{1721}{24192})$ at the respective points $(s_1, s_2, s_3, s_4, s_5) = (0, \frac{1}{5}, \frac{2}{5}, \frac{3}{5}, \frac{4}{5}, 1)$ produce the asymmetric estimator

$$\begin{aligned}
\tilde{\theta}_{n,T,7} := -\Bigg\{\frac{2}{n}\frac{1}{24192}\sum_{i=1}^{n}\Big[&1471(X_{t_{i-1}}I_{\{|\Delta X_{t_i}|\le h^\rho\}})^2 \\
&+6925\Big(\frac{X_{t_{i-1}}I_{\{|\Delta X_{t_i}|\le h^\rho\}} + X_{t_i}I_{\{|\Delta X_{t_i}|\le h^\rho\}}}{5}\Big)^2
\end{aligned}$$

$$+2950(\frac{2X_{t_{i-1}}I_{\{|\Delta X_{t_i}|\leq h^\rho\}}+2X_{t_i}I_{\{|\Delta X_{t_i}|\leq h^\rho\}}}{5})^2$$
$$+5450(\frac{3X_{t_{i-1}}I_{\{|\Delta X_{t_i}|\leq h^\rho\}}+3X_{t_i}I_{\{|\Delta X_{t_i}|\leq h^\rho\}}}{5})^2$$
$$+5675(\frac{4X_{t_{i-1}}I_{\{|\Delta X_{t_i}|\leq h^\rho\}}+4X_{t_i}I_{\{|\Delta X_{t_i}|\leq h^\rho\}}}{5})^2$$
$$+1721(X_{t_i}I_{\{|\Delta X_{t_i}|\leq h^\rho\}})^2\Big]\Big\}^{-1} \tag{1.28}$$

for which $(\mu_1,\mu_2,\mu_3,\mu_4,\mu_5,\mu_6)=(\frac{1}{2},\frac{1}{3},\frac{1}{4},\frac{1}{5},\frac{841}{5040})$. Thus $\nu=5$.

Probabilities $(p_1,p_2,p_3,p_4,p_5)=(\frac{7}{90},\frac{16}{45},\frac{2}{15},\frac{16}{45},\frac{7}{90})$ at the respective points $(s_1,s_2,s_3,s_4,s_5)=(0,\frac{1}{4},\frac{1}{2},\frac{3}{4},1)$ produce the symmetric estimator $\tilde{\theta}_{n,T,8}$ given by

$$\tilde{\theta}_{n,T,8}:=-\Big\{\frac{2}{n}\frac{1}{90}\sum_{i=1}^{n}\Big[7(X_{t_{i-1}}I_{\{|\Delta X_{t_i}|\leq h^\rho\}})^2$$
$$+32(\frac{3X_{t_{i-1}}I_{\{|\Delta X_{t_i}|\leq h^\rho\}}+X_{t_i}I_{\{|\Delta X_{t_i}|\leq h^\rho\}}}{4})^2$$
$$+12(\frac{X_{t_{i-1}}I_{\{|\Delta X_{t_i}|\leq h^\rho\}}+X_{t_i}I_{\{|\Delta X_{t_i}|\leq h^\rho\}}}{2})^2$$
$$+32(\frac{X_{t_{i-1}}I_{\{|\Delta X_{t_i}|\leq h^\rho\}}+3X_{t_i}I_{\{|\Delta X_{t_i}|\leq h^\rho\}}}{4})^2$$
$$+7(X_{t_i}I_{\{|\Delta X_{t_i}|\leq h^\rho\}})^2\Big]\Big\}^{-1} \tag{1.29}$$

for which $(\mu_1,\mu_2,\mu_3,\mu_4,\mu_5,\mu_6)=(\frac{1}{2},\frac{1}{3},\frac{1}{4},\frac{1}{5},\frac{1}{6},\frac{110}{768})$. Thus $\nu=6$.

Probabilities $(p_1,p_2,p_3,p_4,p_5)=(\frac{19}{288},\frac{75}{288},\frac{50}{288},\frac{50}{288},\frac{75}{288},\frac{19}{288})$ at the respective points
$(s_1,s_2,s_3,s_4,s_5)=(0,\frac{1}{5},\frac{2}{5},\frac{3}{5},\frac{4}{5},1)$ produce symmetric estimator

$$\tilde{\theta}_{n,T,9}:=-\Big\{\frac{2}{n}\frac{1}{288}\sum_{i=1}^{n}\big[19(X_{t_{i-1}})^2$$
$$+75(\frac{4X_{t_{i-1}}I_{\{|\Delta X_{t_i}|\leq h^\rho\}}+X_{t_i}I_{\{|\Delta X_{t_i}|\leq h^\rho\}}}{5})^2$$
$$+50(\frac{3X_{t_{i-1}}I_{\{|\Delta X_{t_i}|\leq h^\rho\}}+2X_{t_i}I_{\{|\Delta X_{t_i}|\leq h^\rho\}}}{5})^2$$
$$+50(\frac{2X_{t_{i-1}}I_{\{|\Delta X_{t_i}|\leq h^\rho\}}+3X_{t_i}I_{\{|\Delta X_{t_i}|\leq h^\rho\}}}{5})^2$$
$$+75(\frac{X_{t_{i-1}}I_{\{|\Delta X_{t_i}|\leq h^\rho\}}+4X_{t_i}I_{\{|\Delta X_{t_i}|\leq h^\rho\}}}{5})^2$$
$$+19(X_{t_i}I_{\{|\Delta X_{t_i}|\leq h^\rho\}})^2\Big]\Big\} \tag{1.30}$$

for which $(\mu_1,\mu_2,\mu_3,\mu_4,\mu_5,\mu_6)=(\frac{1}{2},\frac{1}{3},\frac{1}{4},\frac{1}{5},\frac{1}{6},\frac{3219}{22500})$. Thus $\nu=6$.

The estimator $\tilde{\theta}_{n,T,S}$ is based on the arithmetic mean of $I_{n,T}$ and $J_{n,T}$. One can use geometric mean and harmonic mean instead. The geometric mean based symmetric AMCE (which is based on the ideas of partial autocorrelation) is defined as

$$\tilde{\theta}_{n,T,G} := \frac{-T/2}{\sqrt{I_{n,T}\, J_{n,T}}}. \tag{1.31}$$

The harmonic mean based symmetric AMCE is defined as

$$\tilde{\theta}_{n,T,H} := \frac{-T}{\frac{1}{I_{n,T}} + \frac{1}{J_{n,T}}}. \tag{1.32}$$

Note that

$$\tilde{\theta}_{n,T,H} \le \tilde{\theta}_{n,T,G} \le \tilde{\theta}_{n,T,S}. \tag{1.33}$$

It is a desirable property of an estimator to lie inside the parameter space. The estimators $\tilde{\theta}_{n,T,H}$ and $\tilde{\theta}_{n,T,G}$ may lie inside the parameter space even if the estimator $\tilde{\theta}_{n,T,F}$ or $\tilde{\theta}_{n,T,B}$ may not.

Note that the symmetric estimators use both the end points of observations. These estimators are sufficient statistics. So one can take advantage of Rao-Blackwell theorem. If one excludes the end points of observations (as in the forward and the backward estimators), then the estimators will not be sufficient statistics.

From a time series viewpoint, a discretely observed stationary Ornstein–Uhlenbeck model is a nonlinear (in the parameter) autoregressive-gamma (ARG) model given by

$$X_{t_i} = \exp(\theta(t_i - t_{i-1})) + \epsilon_i, \quad i \ge 1, \tag{1.34}$$

where

$$\epsilon_i := \int_{t_{i-1}}^{t_i} \exp\{\theta(t_i - u)\} d(W_u + Z_u) \tag{1.36}$$

which are independent $\mathcal{N}\left(0, \frac{e^{2\theta(t_i - t_{i-1})} - 1}{2\theta}\right)$ random variables. The errors are heteroscedastic if the data points are unequally spaced, but as we assume equally spaced data points, the errors are homoscedastic and i.i.d. $\mathcal{N}(0, V_{n,T}(\theta))$ where

$$V_{n,T}(\theta) := \frac{e^{2\theta\frac{T}{n}} - 1}{2\theta}. \tag{1.37}$$

The log-likelihood is

$$l_{n,T}(\theta) := -\frac{n}{2}\log 2\pi - \frac{n}{2} V_{n,T}(\theta) - \frac{1}{2V_{n,T}(\theta)} \sum_{i=1}^{n} \left(X_{t_i} - e^{\frac{T}{n}\theta} X_{t_{i-1}}\right)^2. \tag{1.38}$$

The exact maximum likelihood estimator based on equally spaced discrete observations is explicit and is given by

$$\hat{\theta}_{n,T} := \frac{n}{T} \log \frac{\sum_{i=1}^{n} X_{t_{i-1}}^2 I_{\{|\Delta X_{t_i}| \le h^\rho\}}}{\sum_{i=1}^{n} X_{t_i} X_{t_{i-1}}}. \tag{1.39}$$

Note that

$$\breve{\theta}_{n,T} = 1 - \frac{n}{T} \exp\{\frac{T}{n}\hat{\theta}_{n,T}\}.$$

As far as the rate of convergence of this estimator to the continuous conditional MLE is concerned, it is of first order.

The Euler estimator which is the conditional least squares estimator is given by

$$\breve{\theta}_{n,T} := \frac{n}{T} \frac{\sum_{i=1}^{n} X_{t_{i-1}} I_{\{|\Delta X_{t_i}| \le h^\rho\}} (X_{t_i} - X_{t_{i-1}}) I_{\{|\Delta X_{t_i}| \le h^\rho\}}}{\sum_{i=1} X_{t_{i-1}}^2}. \tag{1.40}$$

Another contrast function, which is the least squares quadratic, is

$$\kappa_{n,T}(\theta) := \frac{2\theta}{e^{2\theta \frac{T}{n}} - 1} \sum_{i=1}^{n} \left(X_{t_i} I_{\{|\Delta X_{t_i}| \le h^\rho\}} - e^{\frac{T}{n}\theta} X_{t_{i-1}} I_{\{|\Delta X_{t_i}| \le h^\rho\}} \right)^2. \tag{1.41}$$

The solution of $\kappa'_{n,T}(\theta) = 0$ provides an AMCE but does not have an explicit form.

The simple symmetric approximate MLE was introduced and studied in Mishra and Bishwal (1995) for a more general diffusion process. Stochastic bound between the difference the AMLE and the continuous MLE was obtained.

In this chapter we obtain rate of weak convergence to normal distribution of some AMCEs using different normings. We also obtain stochastic bound on the difference of the AMCEs and their continuous counterpart MCE when T is fixed.
Let $\Phi(\cdot)$ denote the standard normal distribution function. Throughout the chapter C denotes a generic constant (perhaps depending on θ, but not on anything else).

We need the following lemmas in the sequel.

Lemma 2.1 Bishwal (2000a) For every $\delta > 0$,

$$P\left\{ \left| \frac{2\theta}{T} I_T - 1 \right| \ge \delta \right\} \le C T^{-1} \delta^{-2}.$$

Lemma 2.2 Bishwal (2000a)

$$\sup_{x \in \mathbb{R}} \left| P\left\{ \left(-\frac{2\theta}{T} \right)^{1/2} \left(\theta I_T - \frac{T}{2} \right) \le x \right\} - \Phi(x) \right| \le C T^{-1/2}.$$

Lemma 2.3

$$(a)\ \ E|I_{n,T} - I_T|^2 = O\left(\frac{T^4}{n^2}\right).$$

$$(b)\ \ E|\frac{I_{n,T}+J_{n,T}}{2} - I_T|^2 = O\left(\frac{T^4}{n^2}\right).$$

Part (a) is from Bishwal and Bose (2001). The proof of part (b) is analogous to part (a). We omit the details.

The following theorem gives the bound on the error of normal approximation of the AMCEs. Note that part (a) uses parameter dependent nonrandom norming. While this is useful for testing hypotheses about θ, it may not necessarily give a confidence interval. The normings in parts (b) and (c) are sample dependent which can be used for obtaining a confidence interval. Following theorem shows that asymptotic normality of the AMCEs need $T \to \infty$ and $\frac{T}{\sqrt{n}} \to 0$.

Theorem 2.1 Denote $b_{n,T} := O(\max(T^{-1/2}(\log T)^{1/2}, (\frac{T^4}{n^2})(\log T)^{-1}))$.

$$\text{(a)}\quad \sup_{x\in\mathbb{R}}\left|P\left\{\left(-\frac{T}{2\theta}\right)^{1/2}(\tilde{\theta}_{n,T,F}-\theta)\le x\right\} - \Phi(x)\right| = O(b_{n,T}),$$

$$\text{(b)}\quad \sup_{x\in\mathbb{R}}\left|P\left\{I_{n,T}^{1/2}(\tilde{\theta}_{n,T,F}-\theta)\le x\right\} - \Phi(x)\right| = O(b_{n,T}),$$

$$\text{(c)}\quad \sup_{x\in\mathbb{R}}\left|P\left\{\left(\frac{T}{2|\tilde{\theta}_{n,T,F}|}\right)^{1/2}(\tilde{\theta}_{n,T,F}-\theta)\le x\right\} - \Phi(x)\right| = O(b_{n,T}).$$

Proof

(a) Observe that

$$\left(-\frac{T}{2\theta}\right)^{1/2}(\tilde{\theta}_T-\theta) = \frac{\left(-\frac{2\theta}{T}\right)^{1/2} Y_T}{\left(-\frac{2\theta}{T}\right) I_T}, \tag{2.1}$$

where

$$Y_T := -\theta I_T - \frac{T}{2} \quad \text{and} \quad I_T := \int_0^T X_t^2\,dt.$$

Thus, we have

$$I_{n,T}\tilde{\theta}_{n,T,F} = -\frac{T}{2} = Y_T + \theta I_T.$$

Hence,

$$\left(-\frac{T}{2\theta}\right)^{1/2}(\tilde{\theta}_{n,T,F}-\theta)$$

$$= \frac{\left(-\frac{T}{2\theta}\right)^{1/2} Y_T + \theta \left(-\frac{T}{2\theta}\right)^{1/2} (I_T - I_{n,T})}{I_{n,T}}$$

$$= \frac{\left(-\frac{2\theta}{T}\right)^{1/2} Y_T + \left(-\frac{2\theta}{T}\right)^{1/2} (I_T - I_{n,T})}{\left(-\frac{2\theta}{T}\right) I_{n,T}} \tag{2.2}$$

Further,

$$\begin{aligned}
& P\left\{\left|\left(\frac{-2\theta}{T}\right)(I_{n,T} - 1)\right| > \epsilon\right\} \\
= & \left\{\left|\left(\frac{-2\theta}{T}\right)(I_{n,T} - I_T + I_T) - 1\right| > \epsilon\right\} \\
\le & P\left\{\left|\left(\frac{-2\theta}{T}\right) I_T - 1\right| > \frac{\epsilon}{2}\right\} + P\left\{\left(-\frac{2\theta}{T}\right) |I_{n,T} - I_T| > \frac{\epsilon}{2}\right\} \\
\le & C \exp\left(\frac{T\theta}{16}\epsilon^2\right) + \frac{16\theta^2}{T^2} \frac{E|I_{n,T} - I_T|^2}{\epsilon^2} \\
\le & C \exp\left(\frac{T\theta}{16}\epsilon^2\right) + C\frac{T^2/n^2}{\epsilon^2}. \qquad (2.3)
\end{aligned}$$

Next, observe that

$$\begin{aligned}
& \sup_{x\in\mathbb{R}} \left| P\left\{\left(-\frac{T}{2\theta}\right)^{1/2} (\tilde{\theta}_{n,T,F} - \theta) \le x\right\} - \Phi(x)\right| \\
= & \sup_{x\in\mathbb{R}} \left| P\left\{\frac{\left(-\frac{2\theta}{T}\right)^{1/2} Y_T + \left(-\frac{2\theta}{T}\right)^{1/2} (I_T - I_{n,T})}{\left(-\frac{2\theta}{T}\right) I_{n,T}} \le x\right\} - \Phi(x)\right| \\
\le & \sup_{x\in\mathbb{R}} \left| P\left\{\left(-\frac{2\theta}{T}\right)^{1/2} Y_T \le x\right\} - \Phi(x)\right| \\
& + P\left\{\left|\theta\left(-\frac{2\theta}{T}\right)^{1/2} (I_{n,T} - I_T)\right| > \epsilon\right\} \\
& + P\left\{\left|\left(-\frac{2\theta}{T}\right) I_{n,T} - 1\right| > \epsilon\right\} + 2\epsilon \\
\le & CT^{-1/2} + \theta^2 \frac{\left(-\frac{2\theta}{T}\right) E|I_{n,T} - I_T|^2}{\epsilon^2} \\
& + C \exp\left(\frac{T\theta}{4}\epsilon^2\right) + C\frac{T^2}{n^2\epsilon^2} + 2\epsilon, \qquad (2.4)
\end{aligned}$$

(the bound for the 3rd term in the right hand side of (2.4) is obtained from (2.3))

$$\le CT^{-1/2} + C\frac{T^2}{n^2\epsilon^2} + C \exp\left(\frac{T\theta}{4}\epsilon^2\right) + C\frac{T}{n^2\epsilon^2} + \epsilon \tag{2.5}$$

(by Lemma 2.3(a)).

Choosing $\epsilon = CT^{-1/2}(\log T)^{1/2}$, the terms in the right hand side of (2.5) are of the order $O(\max(T^{-1/2}(\log T)^{1/2}, (\frac{T^4}{n^2})(\log T)^{-1}))$. □

(b) From (2.1), we have

$$I_{n,T}^{1/2}(\tilde{\theta}_{n,T,F} - \theta) = \frac{Y_T + \theta(I_T - I_{n,T})}{I_{n,T}^{1/2}}.$$

Then,

$$\begin{aligned}
& \sup_{x\in\mathbb{R}} \left| P\left\{ I_{n,T}^{1/2}(\tilde{\theta}_{n,T,F} - \theta) \le x \right\} - \Phi(x) \right| \\
= \; & \sup_{x\in\mathbb{R}} \left| P\left\{ \frac{Y_T}{I_{n,T}^{1/2}} + \theta \frac{I_T - I_{n,T}}{I_{n,T}^{1/2}} \le x \right\} - \Phi(x) \right| \\
\le \; & \sup_{x\in\mathbb{R}} \left| P\left\{ \frac{Y_T}{I_{n,T}^{1/2}} \le x \right\} - \Phi(x) \right| + P\left\{ \left| \frac{\theta(I_T - I_{n,T})}{I_{n,T}^{1/2}} \right| > \epsilon \right\} + \epsilon \\
=: \; & U_1 + U_2 + \epsilon. \qquad (2.6)
\end{aligned}$$

We have from (2.3),

$$U_1 \le CT^{-1/2} + C\exp\left(\frac{T\theta}{16}\epsilon^2\right) + C\frac{T^2}{n^2\epsilon^2} + \epsilon. \qquad (2.7)$$

Now,

$$\begin{aligned}
U_2 \; = \; & P\left\{ |\theta| \left| \frac{I_{n,T} - I_T}{I_{n,T}^{1/2}} \right| > \epsilon \right\} = P\left\{ |\theta| \frac{\left| \left(-\frac{2\theta}{T}\right)^{1/2} (I_{n,T} - I_T) \right|}{\left| \left(-\frac{2\theta}{T}\right)^{1/2} I_{n,T}^{1/2} \right|} > \epsilon \right\} \\
\le \; & P\left\{ \left| \left(-\frac{2\theta}{T}\right)^{1/2} \right| |I_{n,T} - I_T| > \delta \right\} \\
& + P\left\{ \left| \left(-\frac{2\theta}{T}\right)^{1/2} I_{n,T}^{1/2} - 1 \right| > \delta_1 \right\} \qquad (2.8) \\
& \text{(where } \delta = \epsilon - C\epsilon^2 \text{ and } \delta_1 = (\epsilon - \delta)/\epsilon > 0) \\
\le \; & \left(-\frac{2\theta}{T}\right) \frac{E|I_{n,T} - I_T|^2}{\delta^2} + P\left\{ \left| \left(-\frac{2\theta}{T}\right) I_{n,T} - 1 \right| > \delta_1 \right\} \\
\le \; & C\frac{T^3}{n^2\delta^2} + C\exp\left(\frac{T\theta}{16}\delta_1^2\right) + C\frac{T^2}{n^2\delta_1^2}. \qquad (2.9)
\end{aligned}$$

Here, the bound for the first term in the right hand side of (2.7) comes from Lemma 2.2(c) and that for the second term is obtained from (2.3).

Now, using the bounds (2.7) and (2.9) in (2.6) with $\epsilon = CT^{-1/2}(\log T)^{1/2}$, we obtain that the terms in (2.6) are of the order $O(\max(T^{-1/2}(\log T)^{1/2}, (\frac{T^4}{n^2})(\log T)^{-1}))$. □

(c) Let

$$G_T := \left\{|\tilde{\theta}_{n,T,F} - \theta| \le d_T\right\}, \qquad \text{and} \qquad d_T := CT^{-1/2}(\log T)^{1/2}.$$

On the set G_T, expanding $(2|\tilde{\theta}_{n,T}|)^{1/2}$, we obtain

$$\begin{aligned}(-2\tilde{\theta}_{n,T})^{-1/2} &= (-2\theta)^{1/2}\left[1 - \frac{\theta - \tilde{\theta}_{n,T,F}}{\theta}\right]^{-1/2} \\ &= (-2\theta)^{1/2}\left[1 + \frac{1}{2}\left(\frac{\theta - \tilde{\theta}_{n,T,F}}{\theta}\right)\right] + O(d_T^2).\end{aligned}$$

Then,

$$\begin{aligned}&\sup_{x\in\mathbb{R}}\left|P\left\{\left(\frac{T}{2|\tilde{\theta}_{n,T,F}|}\right)^{1/2}(\tilde{\theta}_{n,T,F} - \theta) \le x\right\} - \Phi(x)\right| \\ \le\ &\sup_{x\in\mathbb{R}}\left\{P\left(\frac{T}{2|\tilde{\theta}_{n,T}|}\right)^{1/2}(\tilde{\theta}_{n,T,F} - \theta) \le x, G_T\right\} + P(G_T^c).\end{aligned}$$

Now,

$$\begin{aligned}P(G_T^c) &= P\left\{|\tilde{\theta}_{n,T,F} - \theta| > CT^{-1/2}(\log T)^{1/2}\right\} \\ &= P\left\{\left(-\frac{T}{2\theta}\right)^{1/2}|\tilde{\theta}_{n,T,F} - \theta| > C(\log T)^{1/2}(-2\theta)^{-1/2}\right\} \\ &\le C\max\left(T^{-1/2}(\log T)^{1/2}, \frac{T^3}{n^2}(\log T)^{-1}\right) \\ &\quad +2(1 - \Phi\log T^{1/2}(-2\theta)^{-1/2}) \\ &\quad \text{(by Theorem 2.1(a))} \\ &\le C\max\left(T^{-1/2}(\log T)^{1/2}, \frac{T^3}{n^2}(\log T)^{-1}\right).\end{aligned}$$

On the set G_T,

$$\left|\left(\frac{\tilde{\theta}_{n,T,F}}{\theta}\right)^{1/2} - 1\right| \le CT^{-1/2}(\log T)^{1/2}.$$

Hence, upon choosing $\epsilon = CT^{-1/2}(\log T)^{1/2}, C$ large, we obtain

$$\begin{aligned}
&\left|P\left\{\left(\frac{T}{-2\tilde{\theta}_{n,T,F}}\right)^{1/2}(\tilde{\theta}_{n,T,F}-\theta) \le x, G_T\right\} - \Phi(x)\right| \\
\le\ & \left|P\left\{\left(\frac{T}{-2\theta}\right)^{1/2}(\tilde{\theta}_{n,T,F}-\theta) \le x, G_T\right\}\right| \\
&+P\left\{\left|\left(\frac{\tilde{\theta}_{n,T,F}}{\theta}\right)^{1/2} - 1\right| > \epsilon, G_T\right\} + \epsilon \\
&\text{(by Lemma 3.3.1(b))} \\
\le\ & C\max\left(T^{-1/2}(\log T)^{1/2}, \frac{T^4}{n^2}(\log T)^{-1}\right) \\
&\text{(by Theorem 2.1(a))}.
\end{aligned}$$

□

In the following theorem, we improve the bound on the error of normal approximation using a mixture of random and nonrandom normings. Thus asymptotic normality of the AMCEs needs $T \to \infty$ and $\frac{T}{n^{2/3}} \to 0$ which are sharper than the bound in Theorem 2.1.

Theorem 2.2

$$\begin{aligned}
&\sup_{x\in\mathbb{R}}\left|P\left\{I_{n,T}\left(-\frac{2\theta}{T}\right)^{1/2}(\tilde{\theta}_{n,T,F}-\theta) \le x\right\} - \Phi(x)\right| \\
&\qquad = O\left(\max\left(T^{-1/2}, \left(\frac{T^3}{n^2}\right)^{1/3}\right)\right).
\end{aligned}$$

Proof From (2.2), we have

$$I_{n,T}\left(-\frac{2\theta}{T}\right)^{1/2}(\tilde{\theta}_{n,T,F}-\theta) = \left(-\frac{2\theta}{T}\right)^{1/2} Y_T + \theta\left(-\frac{2\theta}{T}\right)^{1/2}(I_T - I_{n,T}).$$

Hence, by Lemmas 2.1–2.3

$$\begin{aligned}
&\sup_{x\in\mathbb{R}}\left|P\left\{I_{n,T}\left(-\frac{2\theta}{T}\right)^{1/2}(\tilde{\theta}_{n,T,F}-\theta) \le x\right\} - \Phi(x)\right| \\
=\ & \sup_{x\in\mathbb{R}}\left|P\left\{\left(-\frac{2\theta}{T}\right)^{1/2} Y_T + \theta\left(-\frac{2\theta}{T}\right)^{1/2}(I_T - I_{n,T}) \le x\right\} - \Phi(x)\right|
\end{aligned}$$

$$
\begin{aligned}
&\leq \sup_{x\in R}\left|P\left\{\left(-\frac{2\theta}{T}\right)^{1/2} Y_T \leq x\right\} - \Phi(x)\right| \\
&\quad +P\left\{\left|\theta\left(-\frac{2\theta}{T}\right)^{1/2}(I_T - I_{n,T})\right| > \epsilon\right\} + \epsilon \\
&\leq CT^{-1/2} + C\frac{E|I_T - I_{n,T}|^2}{T\epsilon^2} + \epsilon \leq CT^{-1/2} + C\frac{T^3}{n^2\epsilon^2} + \epsilon.
\end{aligned}
$$

Choosing $\epsilon = (\frac{T^3}{n^2})^{1/3}$, the theorem follows. □

The following theorem gives stochastic bound on the error of approximation of the continuous MCE by AMCEs.

Theorem 2.3

$$(a)\ |\tilde{\theta}_{n,T} - \tilde{\theta}_T| = O_P\left(\frac{T}{n}\right)^{1/2}.$$

$$(b)\ |\tilde{\theta}_{n,T,z} - \tilde{\theta}_T| = O_P\left(\frac{T^2}{n^2}\right)^{1/2}.$$

Proof From (1.9) and (1.14), we have

$$\tilde{\theta}_T = -\frac{T}{2I_T}, \quad \tilde{\theta}_{n,T} = -\frac{T}{2I_{n,T}}.$$

Hence, applying Lemma 1.2 with the aid of Lemma 2.3(a) and noting that $|\frac{I_{n,T}}{T}| = O_P(1)$ and $|\frac{I_T}{T}| = O_P(1)$ the part (a) of theorem follows.

From (1.9) and (1.16), we have

$$\tilde{\theta}_T = -\frac{T}{2I_T}, \quad \tilde{\theta}_{n,T,z} = -\frac{T}{I_{n,t} + J_{n,T}}.$$

Applying Lemma 1.2 with the aid of Lemma 2.3(b) and noting that $|\frac{J_{n,T}}{T}| = O_P(1)$ and $|\frac{I_T}{T}| = O_P(1)$ the part (b) of theorem follows. □

Remarks

(1) The bounds in Theorems 2.1 and 2.2 are uniform over compact subsets of the parameter space.

(2) It remains to investigate the nonuniform rates of convergence to normality which are more useful.

(3) Weak convergence rate for AMCE remains open when $T \to \infty$ and $\frac{T}{n} \to 0$.

(4) Weighted estimators converge faster than the unweighted estimators if the weights are chosen according to (1.22) and (1.23).

7.6 Geometric Mean Reversion Process: Black-Karasinski Model

An alternative to squared O-U model is the following geometric O-U model with explicit solution and transition density.

$$dS_t = \alpha(\beta - \log S_t)S_t dt + \sigma S_t dW_t,$$

where $S_0 = s_0$, α, β, and σ are positive constants. This is an alternative to the CIR model which preserves positivity and stationarity.

We study distributional properties of approximate minimum contrast estimators of the unknown parameters in the Black-Karasinski model from discrete data with precise rates of convergence which are robust and efficient.

Let $(\Omega, \mathcal{F}, \{\mathcal{F}_t\}_{t\geq 0}, P)$ be a stochastic basis on which is defined the Gompertz diffusion process $\{X_t\}_{t\geq 0}$ satisfying the Itô stochastic differential equation

$$dX_t = (\alpha X_t - \beta X_t \ln X_t)\, dt + \sigma X_t dW_t,\ t \geq 0,\quad X_0 = x_0, \qquad (1.1)$$

where $\{W_t\}_{t\geq 0}$ is a standard Brownian motion with the filtration $\{\mathcal{F}_t\}_{t\geq 0}$ and $\alpha > 0, \beta > 0, \sigma > 0$ are the unknown parameters to be estimated on the basis of discrete observations of the process $\{X_t\}$ at times $0 = t_0 < t_1 < \cdots < t_n = T$ with $t_i - t_{i-1} = \frac{T}{n}, i = 1, 2 \cdots, n$. We assume equispaced sampling for simplicity. We assume two types of high-frequency data: (1) $T \to \infty, n \to \infty, \frac{T}{\sqrt{n}} \to 0$, (2) $T \to \infty, n \to \infty, \frac{T}{n^{2/3}} \to 0$.

Here X_t is the interest rate, which is measured at discrete time, α is the intrinsic growth rate of the interest rate, β is the interest rate growth acceleration factor, and σ is the diffusion coefficient. Other parameters are the plateau of the model $X_\infty = \exp(\frac{\alpha}{\beta})$, interest rate growth decay, and the first time the growth curve of the model reaches X_∞. We assume that the growth deceleration factor β does not change, while the variability of environmental conditions induces fluctuations in the intrinsic growth rate α. In finance literature, this model is known as Black-Karasinski model which is a geometric mean reverting Vasicek model used for modeling term structure of interest rates which preserves positivity of the interest rates.

Let the continuous realization $\{X_t, 0 \leq t \leq T\}$ be denoted by X_0^T. Denote $\theta := (\alpha, \beta, \sigma)$. Let P_θ^T be the measure generated on the space (C_T, B_T) of continuous functions on $[0, T]$ with the associated Borel σ-algebra B_T generated under the supremum norm by the process X_0^T and let P_0^T be the standard Wiener measure. It is well known that when θ is the true value of the parameter P_θ^T is absolutely continuous with respect to P_0^T and the Radon–Nikodym derivative (likelihood) of P_θ^T with respect to P_0^T based on the data X_0^T is given by

$$L_T(\theta) := \frac{dP_\theta^T}{dP_0^T}(X_0^T)$$

$$= \exp\left\{\frac{1}{\sigma^2}\int_0^T (\alpha - \beta \ln X_t)X_t^{-1}dX_t - \frac{1}{2\sigma^2}\int_0^T (\alpha - \beta \ln X_t)^2 dt\right\}. \qquad (1.2)$$

Consider the log-likelihood function, which is given by

$$l_T(\theta) := \frac{1}{\sigma^2}\left\{\int_0^T (\alpha - \beta \ln X_t) X_t^{-1} dX_t - \frac{1}{2}\int_0^T (\alpha - \beta \ln X_t)^2 dt\right\}. \tag{1.3}$$

A solution of the estimating equation $l'_T(\theta) = 0$ provides the conditional maximum likelihood estimators (MLEs)

$$\widehat{\alpha}_T := \frac{1}{T}\left\{\int_0^T X_t^{-1} dX_t + \hat{\beta}_T \int_0^T \ln X_t dt\right\}, \tag{1.4}$$

$$\widehat{\beta}_T := \frac{(\int_0^T X_t^{-1} dX_t)(\int_0^T \ln X_t dt) - T(\int_0^T X_t^{-1} \ln X_t dX_t)}{\int_0^T \ln^2 X_t dX_t - (\int_0^T \ln X_t dX_t)^2}, \tag{1.5}$$

$$\widehat{\sigma}_T^2 := \frac{\sum_{i=1}^n (X_{t_i} - X_{t_{i-1}})^2}{\sum_{i=1}^n X_{t_{i-1}}^2 (t_i - t_{i-1})}. \tag{1.6}$$

As an alternative to maximum likelihood method and to obtain robust estimators with higher efficiency we use contrast functions. Suppose α and σ are known, for simplicity let $\alpha = 1/2,\ \sigma = 1,\ x_0 = 1$ and our aim is to estimate the acceleration parameter β. Using Itô formula, the score function $l'_T(\theta)$ can be written as

$$\gamma_T(\beta) := -\frac{1}{2}\ln^2 X_T - \int_0^T (\beta \ln^2 X_t - \frac{1}{2}) dt. \tag{1.7}$$

Using a contrast function which is related to the negative derivative of the log-likelihood function, we consider the estimating function,

$$M_T(\beta) := \int_0^T (\beta \ln^2 X_t - \frac{1}{2}) dt. \tag{1.8}$$

Then the minimum contrast estimate (MCE) of β which is the solution of $M_T(\beta) = 0$ is given by

$$\widetilde{\beta}_T := \frac{T}{2}\left\{\int_0^T \ln^2 X_t dt\right\}^{-1} =: \frac{T}{2I_T}, \tag{1.9}$$

where $I_T := \int_0^T \ln^2 X_t dt$. Hence $M_T(\beta) = \beta I_T - \frac{T}{2}$. Sometimes we will denote this by just M_T. We find several discrete approximations of the MCE. Define a weighted approximation of I_T:

$$K_{n,T} := \frac{T}{n}\{\sum_{i=1}^n w_{t_i} \ln^2 X_{t_{i-1}} + \sum_{i=1}^n (1 - w_{t_i}) \ln^2 X_{t_i}\}, \tag{1.10}$$

where $w_{t_i} \geq 0$ is a weight function. Denote the forward and backward approximations of I_T:

$$I_{n,T} := \frac{T}{n}\sum_{i=1}^{n} \ln^2 X_{t_{i-1}}, \tag{1.11}$$

$$J_{n,T} := \frac{T}{n}\sum_{i=1}^{n} \ln^2 X_{t_i}. \tag{1.12}$$

General weighted AMCE is defined as

$$\widetilde{\beta}_{n,T} := \{\frac{2}{T}K_{n,T}\}^{-1}. \tag{1.13}$$

With $w_{t_i} = 1$ in (1.10), we obtain the forward AMCE as

$$\widetilde{\beta}_{n,T,F} := \{\frac{2}{T}I_{n,T}\}^{-1}. \tag{1.14}$$

With $w_{t_i} = 0$ in (1.10), we obtain the backward AMCE as

$$\widetilde{\beta}_{n,T,B} := \{\frac{2}{T}J_{n,T}\}^{-1}. \tag{1.15}$$

With $w_{t_i} = 0.5$ in (1.10), the simple symmetric AMCE is defined as

$$\widetilde{\beta}_{n,T,S} := \{\frac{1}{T}[I_{n,T} + J_{n,T}]\}^{-1} = \{\frac{2}{T}\sum_{i=2}^{n} \ln^2 X_{t_{i-1}} + 0.5(\ln^2 X_{t_0} + \ln^2 X_{t_n})\}^{-1}. \tag{1.16}$$

Define the weighted symmetric estimators: With the weight function

$$w_{t_i} = \begin{cases} 0 & : \quad i = 1 \\ \frac{i-1}{n} & : \quad i = 2, 3, \cdots, n \\ 1 & : \quad i = n+1 \end{cases}$$

the weighted symmetric AMCE is defined as

$$\widetilde{\beta}_{n,T,w} := \{\frac{2}{T}\sum_{i=2}^{n} \ln^2 X_{t_{i-1}} + \frac{1}{n}\sum_{i=1}^{n} \ln^2 X_{t_{i-1}}\}^{-1}.$$

Note that estimator (1.16) is analogous to the trapezoidal rule in numerical analysis. One can instead use the midpoint rule to define another estimator $\widetilde{\beta}_{n,T,A} := \{\frac{2}{T}\sum_{i=1}^{n} \ln^2(\frac{X_{t_{i-1}}+X_{t_i}}{2})\}^{-1}$. One can further use the Simpson's rule to define another estimator where the denominator is a convex combination of the trapezoidal and midpoint estimators,

$$\widetilde{\beta}_{n,T,N} := \{\frac{1}{3T}\sum_{i=1}^{n}\{\ln^2 X_{t_{i-1}} + 4\ln^2(\frac{X_{t_{i-1}} + X_{t_i}}{2}) + \ln^2 X_{t_i}\}\}^{-1}.$$

The AMCE has several good properties. The AMCE is simpler to calculate, in the sense that it does not involve simulation of a stochastic integral unlike AMLE. Hence AMCE is a more practical estimator. They are robust since M-estimator is reduced to the AMCE. The AMCE is efficient, see Tanaka (2013). Tanaka (2013) calculated the asymptotic relative efficiency of the minimum contrast estimator with respect to least squares estimator (LSE) and showed that MCE is asymptotically efficient while LSE is inefficient. We study the distributional properties of the AMCE. We obtain the rate of weak convergence to normal distribution of the AMCE using different normings. We also obtain stochastic bound on the difference of the AMCE and its continuous counterpart MCE when T is fixed.

7.6.2 Distributional Properties of Estimators

Let $\Phi(\cdot)$ denote the standard normal distribution function. Throughout the Chapter C denotes a generic constant (perhaps depending on β, but not on anything else). Since $\ln X_t$ is an Ornstein–Uhlenbeck process, we can use the following lemmas from Bishwal (2008a) in the sequel.

Lemma 2.1 For every $\delta > 0$, $P\{|\frac{2\beta}{T}I_T - 1| \geq \delta\} \leq CT^{-1}\delta^{-2}$.

Lemma 2.2 $\sup_{x\in\mathbb{R}} |P\{(\frac{2\beta}{T})^{1/2}M_T \leq x\} - \Phi(x)| \leq CT^{-1/2}$.

Lemma 2.3 (a) $E|I_{n,T} - I_T|^2 = O(\frac{T^4}{n^2})$, (b) $E|\frac{I_{n,T}+J_{n,T}}{2} - I_T|^2 = O(\frac{T^4}{n^2})$.

The following theorem gives the bound on the error of approximation of the distributions of the AMCE to normal distribution. Note that part (a) uses parameter dependent nonrandom norming. While this is useful for testing hypotheses about β, it may not necessarily give a confidence interval. The normings in parts (b) and (c) are sample dependent which can be used for obtaining a confidence interval. Following theorem shows that asymptotic normality of the AMCE needs the design condition $T \to \infty$ and $\frac{T}{\sqrt{n}} \to 0$.

Theorem 2.1 Denote $b_{n,T} := O(\max(T^{-1/2}(\log T)^{1/2}, (\frac{T^4}{n^2})(\log T)^{-1}))$.

$$\text{(a)} \quad \sup_{x\in\mathbb{R}} |P\{(\frac{T}{2\beta})^{1/2}(\widetilde{\beta}_{n,T,F} - \beta) \leq x\} - \Phi(x)| = O(b_{n,T}),$$

$$\text{(b)} \quad \sup_{x\in\mathbb{R}} |P\{I_{n,T}^{1/2}(\widetilde{\beta}_{n,T,F} - \beta) \leq x\} - \Phi(x)| = O(b_{n,T}),$$

$$\text{(c)} \quad \sup_{x\in\mathbb{R}} |P\{(\frac{T}{2\widetilde{\beta}_{n,T,F}})^{1/2}(\widetilde{\beta}_{n,T,F} - \beta) \leq x\} - \Phi(x)| = O(b_{n,T}).$$

Proof

(a) Observe that

$$(\frac{T}{2\beta})^{1/2}(\widetilde{\beta}_T - \beta) = -\frac{(\frac{2\beta}{T})^{1/2}M_T}{(\frac{2\beta}{T})I_T}. \tag{2.1}$$

Thus, we have $I_{n,T}\widetilde{\beta}_{n,T,F} = \frac{T}{2} = -M_T + \beta I_T$. Hence,

$$(\frac{T}{2\beta})^{1/2}(\widetilde{\beta}_{n,T,F} - \beta) = \frac{(\frac{T}{2\beta})^{1/2}M_T + \beta(\frac{T}{2\beta})^{1/2}(I_T - I_{n,T})}{I_{n,T}}$$

$$= \frac{(\frac{2\beta}{T})^{1/2}M_T + (\frac{2\beta}{T})^{1/2}(I_T - I_{n,T})}{(\frac{2\beta}{T})I_{n,T}}. \tag{2.2}$$

Further,

$$\begin{aligned}
P\{|(\frac{2\beta}{T})(I_{n,T} - 1)| > \epsilon\} &= \{|(\frac{2\beta}{T})(I_{n,T} - I_T + I_T) - 1| > \epsilon\} \\
\leq\ & P\{|(\frac{2\beta}{T})I_T - 1| > \frac{\epsilon}{2}\} + P\{(\frac{2\beta}{T})|I_{n,T} - I_T| > \frac{\epsilon}{2}\} \\
\leq\ & C\exp(-\frac{T\beta}{16}\epsilon^2) + \frac{16\beta^2}{T^2}\frac{E|I_{n,T} - I_T|^2}{\epsilon^2} \\
&\leq C\exp(-\frac{T\beta}{16}\epsilon^2) + C\frac{T^2/n^2}{\epsilon^2}. \qquad (2.3)
\end{aligned}$$

Next, observe that

$$\begin{aligned}
& \sup_{x\in\mathbb{R}}|P\{(\frac{T}{2\beta})^{1/2}(\widetilde{\beta}_{n,T,F} - \beta) \leq x\} - \Phi(x)| \\
=\ & \sup_{x\in\mathbb{R}}|P\{\frac{(\frac{2\beta}{T})^{1/2}M_T + (\frac{2\beta}{T})^{1/2}(I_T - I_{n,T})}{(\frac{2\beta}{T})I_{n,T}} \leq x\} - \Phi(x)| \\
\leq\ & \sup_{x\in\mathbb{R}}|P\{(\frac{2\beta}{T})^{1/2}M_T \leq x\} - \Phi(x)| + P\{|\beta(\frac{2\beta}{T})^{1/2}(I_{n,T} - I_T)| > \epsilon\} \\
& + P\{|(\frac{2\beta}{T})I_{n,T} - 1| > \epsilon\} + 2\epsilon \\
\leq\ & CT^{-1/2} + \beta^2\frac{(\frac{2\beta}{T})E|I_{n,T} - I_T|^2}{\epsilon^2} \\
& + C\exp(-\frac{T\beta}{4}\epsilon^2) + C\frac{T^2}{n^2\epsilon^2} + 2\epsilon, \qquad (2.4)
\end{aligned}$$

(the bound for the 3rd term in the right hand side of (2.4) is obtained from (2.3))

$$\leq\ CT^{-1/2} + C\frac{T^2}{n^2\epsilon^2} + C\exp(-\frac{T\beta}{4}\epsilon^2) + C\frac{T}{n^2\epsilon^2} + \epsilon \tag{2.5}$$

(by Lemma 2.3(a)).

Choosing $\epsilon = CT^{-1/2}(\log T)^{1/2}$, the terms in the right hand side of (2.5) are of the order
$O(\max(T^{-1/2}(\log T)^{1/2}, (\frac{T^4}{n^2})(\log T)^{-1}))$.
(b) From (2.1), we have

$$I_{n,T}^{1/2}(\widetilde{\beta}_{n,T,F} - \beta) = \frac{M_T + \beta(I_T - I_{n,T})}{I_{n,T}^{1/2}}.$$

Then,

$$\begin{aligned}
& \sup_{x\in\mathbb{R}} |P\{I_{n,T}^{1/2}(\widetilde{\beta}_{n,T,F} - \beta) \le x\} - \Phi(x)| \\
= \; & \sup_{x\in\mathbb{R}} |P\{\frac{M_T}{I_{n,T}^{1/2}} + \beta\frac{I_T - I_{n,T}}{I_{n,T}^{1/2}} \le x\} - \Phi(x)| \\
\le \; & \sup_{x\in\mathbb{R}} |P\{\frac{M_T}{I_{n,T}^{1/2}} \le x\} - \Phi(x)| + P\{|\frac{\beta(I_T - I_{n,T})}{I_{n,T}^{1/2}}| > \epsilon\} + \epsilon \\
=: \; & U_1 + U_2 + \epsilon. \qquad (2.6)
\end{aligned}$$

We have from (2.3),

$$U_1 \le CT^{-1/2} + C\exp(-\frac{T\beta}{16}\epsilon^2) + C\frac{T^2}{n^2\epsilon^2} + \epsilon. \tag{2.7}$$

Further,

$$\begin{aligned}
U_2 \; = \; & P\{\beta|\frac{I_{n,T} - I_T}{I_{n,T}^{1/2}}| > \epsilon\} = P\{\beta\frac{|(\frac{2\beta}{T})^{1/2}(I_{n,T} - I_T)|}{|(\frac{2\beta}{T})^{1/2}I_{n,T}^{1/2}|} > \epsilon\} \\
\le \; & P\{|(\frac{2\beta}{T})^{1/2}||I_{n,T} - I_T| > \delta\} + P\{|(\frac{2\beta}{T})^{1/2}I_{n,T}^{1/2} - 1| > \delta_1\} \\
& (\text{where } \delta = \epsilon - C\epsilon^2 \text{ and } \delta_1 = (\epsilon - \delta)/\epsilon > 0) \\
\le \; & (\frac{2\beta}{T})\frac{E|I_{n,T} - I_T|^2}{\delta^2} + P\{|(\frac{2\beta}{T})I_{n,T} - 1| > \delta_1\} \\
\le \; & C\frac{T^3}{n^2\delta^2} + C\exp(-\frac{T\beta}{16}\delta_1^2) + C\frac{T^2}{n^2\delta_1^2}. \qquad (2.8)
\end{aligned}$$

Here, the bound for the first term in the right hand side of (2.7) comes from Lemma 2.2(c) and that for the second term is obtained from (2.3). Now, using the bounds (2.7) and (2.8) in (2.6) with $\epsilon = CT^{-1/2}(\log T)^{1/2}$, we obtain that the terms in (2.6) are of the order
$O(\max(T^{-1/2}(\log T)^{1/2}, (\frac{T^4}{n^2})(\log T)^{-1}))$.

(c) Let $G_T := \{|\widetilde{\beta}_{n,T,F} - \beta| \le d_T\}$, $d_T := CT^{-1/2}(\log T)^{1/2}$. On the set G_T, expanding $(2|\widetilde{\beta}_{n,T,F}|)^{1/2}$, we obtain

$$(2\widetilde{\beta}_{n,T,F})^{-1/2} = (2\beta)^{1/2}[1 - \frac{\beta - \widetilde{\beta}_{n,T,F}}{\beta}]^{-1/2}$$

$$= (2\beta)^{1/2}[1 + \frac{1}{2}(\frac{\beta - \widetilde{\beta}_{n,T,F}}{\beta})] + O(d_T^2).$$

Then

$$\sup_{x\in\mathbb{R}} |P\{(\frac{T}{2\widetilde{\beta}_{n,T,F}})^{1/2}(\widetilde{\beta}_{n,T,F} - \beta) \le x\} - \Phi(x)|$$

$$\le \sup_{x\in\mathbb{R}}\{P(\frac{T}{2\widetilde{\beta}_{n,T,F}})^{1/2}(\widetilde{\beta}_{n,T,F} - \beta) \le x, G_T\} + P(G_T^c).$$

Further,

$$\begin{aligned}
P(G_T^c) &= P\{|\widetilde{\beta}_{n,T,F} - \beta| > CT^{-1/2}(\log T)^{1/2}\} \\
&= P\{(\frac{T}{2\beta})^{1/2}|\widetilde{\beta}_{n,T,F} - \beta| > C(\log T)^{1/2}(2\beta)^{-1/2}\} \\
&\le C\max(T^{-1/2}(\log T)^{1/2}, \frac{T^3}{n^2}(\log T)^{-1}) \\
&\quad +2(1 - \Phi((\log T)^{1/2})(2\beta)^{-1/2}) \\
&\quad \text{(by Theorem 2.1(a))} \\
&\le C\max(T^{-1/2}(\log T)^{1/2}, \frac{T^3}{n^2}(\log T)^{-1}).
\end{aligned}$$

On the set G_T, $|(\frac{\widetilde{\beta}_{n,T,F}}{\beta})^{1/2} - 1| \le CT^{-1/2}(\log T)^{1/2}$. Hence, upon choosing $\epsilon = CT^{-1/2}(\log T)^{1/2}, C$ large, using Lemma 3.3.1(b) and Theorem 2.1(a), we obtain

$$\begin{aligned}
&|P\{(\frac{T}{2\widetilde{\beta}_{n,T,F}})^{1/2}(\widetilde{\beta}_{n,T,F} - \beta) \le x, G_T\} - \Phi(x)| \\
\le\ & |P\{(\frac{T}{2\beta})^{1/2}(\widetilde{\beta}_{n,T,F} - \beta) \le x, G_T\}| + P\{|(\frac{\widetilde{\beta}_{n,T,F}}{\beta})^{1/2} - 1| > \epsilon, G_T\} + \epsilon \\
\le\ & C\max(T^{-1/2}(\log T)^{1/2}, \frac{T^4}{n^2}(\log T)^{-1}).
\end{aligned}$$

□

In the following theorem, we improve the bound on the error of normal approximation using a mixture of random and nonrandom normings. Thus asymptotic normality of the AMCEs needs $T \to \infty$ and $\frac{T}{n^{2/3}} \to 0$ which are sharper than the bound in Theorem 2.1.

Theorem 2.2

$$\sup_{x\in\mathbb{R}} |P\{I_{n,T}(\frac{2\beta}{T})^{1/2}(\widetilde{\beta}_{n,T,F} - \beta) \le x\} - \Phi(x)| = O(\max(T^{-1/2}, (\frac{T^3}{n^2})^{1/3})).$$

Proof From (2.2), we have

$$I_{n,T}(\frac{2\beta}{T})^{1/2}(\widetilde{\beta}_{n,T,F} - \beta) = (\frac{2\beta}{T})^{1/2}M_T + \beta(\frac{2\beta}{T})^{1/2}(I_T - I_{n,T}).$$

Hence, by Lemma 2.1–2.3

$$
\begin{aligned}
& \sup_{x\in\mathbb{R}} |P\{I_{n,T}(\frac{2\beta}{T})^{1/2}(\widetilde{\beta}_{n,T,F}-\beta) \le x\} - \Phi(x)| \\
= & \sup_{x\in\mathbb{R}} |P\{(\frac{2\beta}{T})^{1/2}M_T + \beta(\frac{2\beta}{T})^{1/2}(I_T - I_{n,T}) \le x\} - \Phi(x)| \\
\le & \sup_{x\in R} |P\{(\frac{2\beta}{T})^{1/2}M_T \le x\} - \Phi(x)| + P\{|\beta(\frac{2\beta}{T})^{1/2}(I_T - I_{n,T})| > \epsilon\} + \epsilon \\
\le & CT^{-1/2} + C\frac{E|I_T - I_{n,T}|^2}{T\epsilon^2} + \epsilon \le CT^{-1/2} + C\frac{T^3}{n^2\epsilon^2} + \epsilon.
\end{aligned}
$$

Choosing $\epsilon = (\frac{T^3}{n^2})^{1/3}$, the theorem follows. □

The following theorem gives stochastic bound on the error of approximation of the continuous MCE by AMCEs.

Theorem 2.3 *(a)* $|\widetilde{\beta}_{n,T,F} - \widetilde{\beta}_T| = O_P(\frac{T}{n})^{1/2}$, *(b)* $|\widetilde{\beta}_{n,T,S} - \widetilde{\beta}_T| = O_P(\frac{T^2}{n^2})^{1/2}$.

Proof From (1.9) and (1.14), we have $\widetilde{\beta}_T = \frac{T}{2I_T}$, $\widetilde{\beta}_{n,T,F} = \frac{T}{2I_{n,T}}$. Hence, applying Lemma 1.2 with the aid of Lemma 2.3(a) and noting that $|\frac{I_{n,T}}{T}| = O_P(1)$ and $|\frac{I_T}{T}| = O_P(1)$ the part (a) of theorem follows. From (1.9) and (1.16), we have $\widetilde{\beta}_T = \frac{T}{2I_T}$, $\widetilde{\beta}_{n,T,S} = \frac{T}{I_{n,t}+J_{n,T}}$. Applying Lemma 1.2 with the aid of Lemma 2.3(b) and noting that $|\frac{J_{n,T}}{T}| = O_P(1)$ and $|\frac{I_T}{T}| = O_P(1)$ the part (b) of theorem follows. □

Chapter 8

Maximum Quasi-Likelihood Estimation in Fractional Levy Stochastic Volatility Model

8.1 Introduction

Usually asset price process has jumps and volatility process has long memory. We study maximum quasi-likelihood estimators for the parameters of a fractionally integrated exponential GARCH, in short FIECOGARCH process based on discrete observations. We deal with a compound Poisson FIECOGARCH process and study the asymptotic behavior of the maximum quasi-likelihood estimator. We show that the resulting estimators are consistent and asymptotically normal.

Parameter estimation in GARCH models has been studied extensively in view of their wide applications. In order to incorporate long memory in the model we look at the continuous time counterpart. Most of the financial transactions are recorded these days with ultra-high frequency which are also called tick-by-tick data. The most important feature of these data is irregular spacing of observation time points. If one aggregates the data up to fixed time intervals, there is loss of information which should be avoided. Hence it is natural to model price processes in continuous time. Also volatility clusters on high level. This long-run volatility persistence is not taken into account by GARCH models. Levy processes contribute non-normality and jumps in the observed part of the model and fractional processes contribute long memory in unobserved part in the model. Based on discrete observations of the price process we estimate all the unknown parameters of the price and the volatility process by the maximum quasi-likelihood estimation. We use Gaussian quasi-likelihood as a contrast function. Maximum quasi-likelihood estimator is an application of Kalman filter in the context of maximum likelihood estimation. Even if the data are non-normal, we use the Gaussian log-likelihood.

Non-negative Ornstein–Uhlenbeck processes were first studied by Wolfe (1982) who obtained it as a limit of an autoregressive process with positive innovations. These days non-negative Ornstein–Uhlenbeck processes have received lot of attention in view of their applications as interest rate and

J. P. N. Bishwal, *Parameter Estimation in Stochastic Volatility Models*,
https://doi.org/10.1007/978-3-031-03861-7_8

stochastic volatility models, see Barndorff-Nielsen and Shephard (2001) and Barndorff-Nielsen and Shephard (2002a). Gaver and Lewis (1980) studied estimation in first order autoregressive gamma process. Davis and McCormick (1989) studied estimation for first order autoregressive process with positive or bounded innovations. Nielsen and Shephard (2003) studied likelihood analysis of a first order autoregressive model with exponential innovations. Brockwell et al. (2007) studied parameter estimation of non-negative Levy driven Ornstein–Uhlenbeck processes. Jongbloed et al. (2005) studied nonparametric estimation for Levy driven Ornstein–Uhlenbeck processes.

Parameter estimation in Gaussian Ornstein–Uhlenbeck process was extensively studied in Bishwal (2008a) and Bishwal (2010a). To account for the strong persistence in volatility, which sometimes is observed in empirical data, we study the FIECOGARCH (fractionally integrated exponential continuous time generalized autoregressive conditional heteroscedastic) process. The log-price process has jumps and the log-volatility process has long memory. The long-memory effect introduced in the log-volatility process, propagates to the volatility process. Thus the model here captures the two stylized facts of long memory and jumps. Continuous time long-memory process in volatility received some attention, see Comte and Renault (1996) where they used fractional Ornstein–Uhlenbeck process as the stochastic volatility process and Comte et al. (2012) where they used fractional CIR square-root process as the stochastic volatility process. Marquardt (2006a) studied fractional Levy processes and applied it to long memory moving average processes. Marquardt (2006b) studied multivariate fractionally integrated CARMA processes. Bender and Marquardt (2008) studied stochastic calculus for convoluted Levy processes. Haug and Czado (2006a) studied structural properties of an exponential continuous time GARCH process. Haug and Czado (2006b) studied the structural properties of FIECOGARCH model. Czado and Haug (2010) used an exponential autoregressive conditional duration model to describe the dependence structure in durations of ultra-high-frequency financial data. Haug et al. (2007) studied method of moment estimation in the COGARCH(1,1) model. COGARCH model is suitable for irregularly spaced observation times. Straumann and Mikosch (2006) studied quasi-maximum likelihood estimation in conditionally heteroscedastic time series model. Haug and Czado (2006a) studied quasi-maximum likelihood estimation and prediction in the compound Poisson ECOGARCH(1,1) model.

8.2 Fractional Levy Process

Fractional Levy Process is a generalization of fractional Brownian motion and is defined as

$$M_{H,t} = \frac{1}{\Gamma(H+\frac{1}{2})} \int_{\mathbb{R}} [(t-s)_+^{H-1/2} - (-s)_+^{H-1/2}] dM_s, \quad t \in \mathbb{R},$$

where $H \in (0.5, 1)$, $u_+ = \max(u, 0)$, $u_- = \max(-u, 0)$ and $\{M_t, t \in \mathbb{R}\}$ is a Levy process on $\mathbb{R}$ with $E(M_1) = 0$, $E(M_1^2) < \infty$ and without Brownian component.

Following are the properties of fractional Levy process:

(1) The covariance of the FLP is given by

$$\mathrm{Cov}(M_{H,t}, M_{H,s}) = \frac{E[L(1)^2]}{2\Gamma(2H+1)\sin(\pi H)}[|t|^{2H} + |s|^{2H} - |t-s|^{2H}].$$

(2) M_H is not a martingale. For a large class of Levy processes, M_d is neither a semimartingale.

If the Levy process M is of finite activity, then the corresponding fractional Levy process M_H is of finite variation. In the case when M is not of finite activity, the corresponding fractional Levy process is not a semimartingale.

(3)M_H is Hölder continuous of any order β less than $H - \frac{1}{2}$.

(4) M_H has stationary increments.

(5) M_H is symmetric.

(6) M is self-similar.

(7) M_H has infinite total variation on compacts.

(8) M_H is not self-similar.

(9) M_H has long memory.

We now introduce the Fractional Levy Ornstein–Uhlenbeck (FLOU) Process.

The FIMA (fractionally integrated moving average) process is defined as

$$Y_H(t) = \int_{-\infty}^{t} g_H(t-u)L(du), \ t \in \mathbb{R},$$

where

$$g_H(t) = \frac{1}{\Gamma(H - \frac{1}{2}} \int_0^t g(t-s)s^{H-\frac{3}{2}}ds, \ t \in \mathbb{R}$$

and the kernel g is the kernel of a short memory moving average process.

The process $Y_H(t)$ can be written as

$$Y_H(t) = \int_{-\infty}^{t} g(t-u)dM_{H,u}, \ t \in \mathbb{R}.$$

We assume the following conditions on the kernel $g : \mathbb{R} \to \mathbb{R}$
(A1) $g(t) = 0$ for all $t < 0$ (causality).
(A2) $|g(t)| \leq Ce^{-ct}$ for some constants $C > 0$ and $c > 0$ (short memory).
FIMA process is stationary and is infinite divisible.

Consider the kernel

$$g(t-s) = \sigma e^{\theta(t-s)} I_{(0,\infty)}(t-s)$$

then

$$g_H(t) = \frac{\sigma}{\Gamma(H-\frac{1}{2}} \int_0^\infty e^{\theta(t-s)} I_{(0,\infty)}(t-s) s^{H-\frac{3}{2}} ds, \ t \in \mathbb{R}.$$

Note that

$$V_{H,t}^{\theta,\sigma} = \int_{\mathbb{R}} g_H(t-u) L(du), \ t \in \mathbb{R}$$

is the fractional Levy Ornstein–Uhlenbeck (FLOU) process satisfying the fractional Vasicek model

$$dV_{H,t} = \theta V_{H,t} dt + \sigma dM_{H,t}, \ t \in \mathbb{R}.$$

8.3 Quasi-Maximum Likelihood Estimator

The FIECOGARCH process is given by

$$dY_{H,t} = \sqrt{X_{H,t}} dL_t$$

$$X_{H,t} = e^{\mu + \beta V_{H,t_-}}$$

$$dV_{H,t} = \theta V_{H,t} dt + dM_{H,t},$$

where L_t is a Levy process independent of $M_{H,t}$ which is a fractional Levy process of integration (Hurst) parameter $H \in (0.5, \ 0.75)$.

We estimate μ, β, and θ by the maximum quasi-likelihood method and study the asymptotic properties of the estimators.

Equivalently,

$$dY_{H,t} = \mu dt + \sigma e^{X_{H,t}} dL_t$$

$$dX_{H,t} = \theta X_{H,t} dt + dM_{H,t}.$$

The log-volatility process is strictly stationary and the covariance function is of the form $\mathrm{Cov}(\log(\sigma_{H,t+h}^2), \log(\sigma_{H,t}^2) \sim C_1 h^{2(H-1)}$ when $h \to \infty$.

The autocorrelation function of log-volatility and volatility process decay at a hyperbolic rate.

The autocovariance function of the price returns over a time interval of length $r > 0$

$$S_{H,t}^{(r)} = S_{H,t} - S_{H,t-r} = \int_{(t-r,t]} \sigma_{H,s} dL_s, \ t \geq r > 0$$

is given by

$$\mathrm{Cov}((S_{H,t}^{(r)})^2, (S_{H,t+h}^{(r)})^2) = E(L_1^2) \int_h^{h+r} Cov((S_{H,t}^{(r)})^2, \sigma_s^2) ds.$$

We will treat the special case of Compound Poisson FIECOGARCH process first. We will estimate the parameters by maximum quasi-likelihood estimation method.

A compound Poisson FIECOGARCH process is driven by the Levy process L with Levy symbol

$$\psi_L(u) = -\frac{u^2}{2} + \int_{\mathbb{R}} (e^{iux} - 1)\Phi_{0,1/\lambda}(dx),$$

where $\Phi_{0,1/\lambda}$ is a normal distribution with mean 0 and variance $1/\lambda$. This means that L is the sum of a standard Brownian motion W and a compound Poisson process

$$J_t = \sum_{k=1}^{N_t} Z_k, \ t \geq 0,$$

where N_t is an independent Poisson process with intensity $\lambda > 0$ and jump times $(T_k)_{k\in\mathbb{Z}}$. The Poisson process is also independent from the i.i.d. sequence of jump sizes $(Z_k)_{k\in\mathbb{Z}}$ with $Z_1 \sim N(0, 1/\lambda)$. The Levy process M in this case is given by

$$M_t = \sum_{k=1}^{N_t} (\theta Z_k + \gamma|Z_k|) - Ct, \ t > 0$$

with

$$C := \gamma \int_{\mathbb{R}} |x|\lambda\Phi_{0,1/\lambda}(dx) = \sqrt{\frac{2\lambda}{\pi}}\gamma.$$

The stationary log-volatility is of the form

$$\log(\sigma_{H,t_+}^2) = \mu + \int_{-\infty}^{t} \beta e^{-a_1(t-s)} dM_{H,s}, t > 0.$$

The log-price process is given by

$$S_t = \sum_{k=1}^{N_t} \sigma_{T_k} Z_k, \ t \geq 0, \ S_0 = 0.$$

We observe S at n consecutive jump times $0 = T_0 < T_1 < \ldots < T_n < T < T_{n+1}$, $n \in \mathbb{Z}$ over the time interval $[0, T]$. The state process X_H has then the following autoregressive representation

$$\begin{aligned}
&\beta X_{H,T_i} \\
&= \beta e^{-\alpha\Delta T_i} X_{H,T_{i-1}} + \sum_{k=N_{T_{i-1}}+1}^{N_{T_i}} \beta e^{-\alpha(T_i-T_k)}[\theta Z_k + \gamma|Z_k|] \\
&\quad - \int_{T_{i-1}}^{T_i} \beta e^{(T_i-s)} C ds \\
&= \beta e^{-\alpha\Delta T_i} X_{H,T_{i-1}} + \beta\theta Z_i + \beta\left(|Z_i| - \frac{C}{\alpha}(1-e^{-\alpha\Delta T_i})\right),
\end{aligned}$$

where $\Delta T_i := T_i - T_{i-1},\ i = 1, 2, \ldots, n$ and $N_{T_{i-1}} + 1 = N_{T_i} = i$.

Thus

$$\log(\sigma^2_{H,T_i}) = \mu + \beta e^{-\alpha\Delta T_i} X_{T_{i-1}} - \beta\frac{C}{\alpha}(1-e^{-\alpha\Delta T_i})$$

since

$$\beta X_{H,T_{i-}} = \beta X_{H,T_{i-1}} - \beta\frac{C}{\alpha}(1-e^{-\alpha\Delta T_i})$$

as $X_{T_{i-}}$ does not include the jump at time T_i and its corresponding jump size is Z_i.

The leverage effect depends on the sign of $\beta\theta$. We will also identify the leverage parameter θ.

The compound Poisson process has finite activity. The observation of the log-price process is given by

$$S_{T_i} = \sum_{k=1}^{N_{T_i}} \sigma_{T_k} Z_k = S_{H,T_{i-1}} + \sigma_{H,T_i} Z_i.$$

We study the parameter estimation in two steps. The rate λ of the Poisson process N can be estimated given the jump times T_i; therefore, it is done at a first step. Since we observe total number of jumps n of the Poisson process N over the T intervals of length one, the MLE of λ is given by

$$\hat{\lambda}_n := \frac{n}{T}.$$

Theorem 3.1 We have

$$(a)\quad \hat{\lambda}_n \to \lambda \text{ a.s. as } \quad n \to \infty.$$

$$(b)\quad n^{1-H}(\hat{\lambda}_n - \lambda) \to^{\mathcal{D}} \mathcal{N}(0,\ e^{\lambda}(1-e^{-\lambda})) \text{ as } \ n \to \infty.$$

Proof Let U_i be the number of jumps in the interval $(i-1, i]$. Then $U_i,\ i = 1, 2, \ldots, n$ are i.i.d. Poisson distributed with parameter λ. Since Φ is continuous, we have $I_{\{0\}}(U_i) = I_{\{0\}}(S^{(1)}_{H,i}),$ a.s. $i = 1, 2, \ldots, n$. Note that

$$\frac{1}{n}\sum_{i=1}^{n} I_{\{0\}}(S^{(1)}_{H,i}) \to^{a.s.} E(I_{\{0\}}U_1) = P(U_1 = 0) = e^{-\lambda} \text{ as } \ n \to \infty.$$

LLN and CLT and delta method applied to the sequence $I_{\{0\}}(S_{H,i}^{(1)})$, $i = 1, 2, \ldots, n$ give the results. □

The CLT result above allows us to construct confidence interval for the jump rate λ. A $100(1-\alpha)\%$ confidence interval for λ is given by

$$\left[\frac{n}{T} - q_{1-\frac{\alpha}{2}}\sqrt{\frac{1}{n} - \frac{1}{T}},\ \frac{n}{T} + q_{1-\frac{\alpha}{2}}\sqrt{\frac{1}{n} - \frac{1}{T}}\right],$$

where $q_{1-\frac{\alpha}{2}}$ is the $(1-\frac{\alpha}{2})$-quantile of the standard normal distribution.

To estimate the remaining parameters (α, θ, μ), we use the quasi-maximum likelihood (QML) estimation procedure as developed by Straumann (2005) in conditionally heteroscedastic time series models. Since the conditional distributions of the returns is unknown, we will use the QML approach by choosing the Gaussian quasi-likelihood as a contrast function.

The conditional log-likelihood has the representation

$$\mathcal{L}(\vartheta | S_H^{\Delta}, \lambda) := -\frac{n}{2}\log(2\pi) - \frac{1}{2}\left(\sum_{i=1}^{n}\log(\sigma_{T_i}^2/\lambda) - \sum_{i=1}^{n}\frac{(S_{H,T_i}^{\Delta T_i})^2}{\sigma_{T_i}^2/\lambda}\right).$$

The quasi log-likelihood function for $\vartheta = (\alpha, \theta, \gamma, \mu)$ given $S_H^{\Delta} := (S_{H,T_1}^{\Delta T_1}, S_{H,T_2}^{\Delta T_2}, \ldots, S_{H,T_n}^{\Delta T_n})$ and the MLE $\hat{\lambda}_n$ is defined as

$$L(\vartheta | S_H^{\Delta}, \hat{\lambda}_n) := -\frac{1}{2}\sum_{i=1}^{n}\log(\hat{\sigma}_{H,T_i}^2(\vartheta, \hat{\lambda}_n)) - \frac{1}{2}\sum_{i=1}^{n}\frac{(S_{H,T_i}^{\Delta T_i})^2}{\hat{\sigma}_{H,T_i}^2(\vartheta, \hat{\lambda}_n)/\hat{\lambda}_n},$$

where the estimates of the volatility $\sigma_{H,T_i}^2, i = 1, 2, \ldots, n$ are given by

$$\hat{\sigma}_{H,T_i}^2(\vartheta, \lambda_n) := \exp(\mu + e^{-\alpha\Delta T_i}X_{H,T_{i-1}}(\vartheta, \lambda) - \hat{C}\Delta T_i), \imath = 1, 2, \ldots, n$$

and given the parameters ϑ and λ the estimates of the state process X are given by the recursion

$$\hat{X}_{H,T_i} = e^{-\alpha\Delta T_i}\hat{X}_{T_{i-1}} + \theta\frac{S_{H,T_i}}{\hat{\sigma}_{T_i}(\vartheta, \lambda)} + \left(\frac{S_{H,T_i}}{\hat{\sigma}_{T_i}(\vartheta, \lambda)} - \hat{C}\Delta T_i\right).$$

Note that $E(|W|) = \sqrt{\frac{2}{\pi\lambda}}$, $W \sim N(0, 1/\lambda)$.

Here the approximation $(1 - e^{-z}) \approx z$ for small z is used. The recursion needs a starting value $\hat{X}_{H,0}$ which will be the mean value zero of the stationary distribution of X.

MQLE of ϑ is defined as

$$\hat{\vartheta}_n := \arg\max_{\vartheta \in \Theta} L(\vartheta | S^{\Delta}, \hat{\lambda}_n).$$

The parametric estimator of volatility at jump times $0 = T_0 < T_1 < \ldots < T_n < T < T_{n+1}, n \in \mathbb{Z}$ is given by

$$\hat{\sigma}^2_{H,T_i}(\hat{\vartheta}_n, \hat{\lambda}_n) := \exp(\hat{\mu}_n + e^{-\hat{\alpha}_n \Delta T_i} X_{H,T_{i-1}}(\hat{\vartheta}_n, \hat{\lambda}_n) - \hat{C}\Delta T_i),\ i = 1, 2, \ldots, n$$

which uses the MQLE $\hat{\vartheta}_n$.

Let $F_H(\vartheta_0)$ be the Fisher information of the log-price process where ϑ_0 is the true parameter. The following theorem gives the strong consistency and the asymptotic normality of the MQLE.

Theorem 3.2

$$\text{(a)} \quad \hat{\vartheta}_n \to \vartheta_0 \text{ a.s. as } n \to \infty.$$

$$\text{(b)} \quad n^{1/2}(\hat{\vartheta}_n - \vartheta_0) \to^{\mathcal{D}} \mathcal{N}(0, F_H^{-1}(\vartheta_0)) \text{ as } n \to \infty.$$

Proof The discrete time process S^Δ process is strong mixing with geometric rate and since strict stationarity of σ^2 implied strict stationarity of S^Δ is also ergodic. Hence application of the Birkoff ergodic theorem and central limit theorem for strongly mixing process proves the theorem. □

8.4 Conclusion

We considered the case $H \in [0.5, 0.75)$. For the case $H = 0.75$, the asymptotic distribution of the QMLE will be normal with the rate $n(\log n)^{1/2}$ and the asymptotic distribution of the MQLE will be non-Gaussian Rosenblatt distribution with rate n^{2H-1}. This shows the different behavior of the MQLE in different parts of the Hurst space even in the stationary case. In the non-stationary case, the limit distribution will be mixture normal. Thus not only the original parameter space but also the Hurst space plays an important role in obtaining the limit distribution of the QMLE in long-memory models.

We studied parameter estimation in a FIECOGARCH model from discrete data. We obtained the consistency and asymptotic normality of the maximum quasi-likelihood estimator. Here we assumed H to be known. An important problem is estimation of H, for instance, based on second order quadratic variation and remains to be investigated.

Chapter 9

Estimation in Barndorff Nielsen-Shephard Ornstein–Uhlenbeck Stochastic Volatility Models

9.1. Introduction

Models driven by Levy processes were studied by Barndorff-Nielsen and Shephard (2004a). Barndorff-Nielsen and Schmiegel (2009) studied Brownian semistationary processes with its connection to volatility and intermittency. Barndorff-Nielsen and Shephard (2002) studied normal modified stable processes. Barndorff-Nielsen and Shephard (2001) studied non-Gaussian Ornstein –Uhlenbeck models of volatility. Barndorff-Nielsen and Shephard (2002a,b,c,d) used realized volatility in estimating stochastic volatility model. Barndorff-Nielsen and Shephard (2003) studied realized power variation in estimating stochastic volatility model. Barndorff-Nielsen and Shephard (2004a,b) studied realized covariation and bipower variation. Barndorff-Nielsen and Shephard (2005b) studied power variation and time change. Barndorff-Nielsen and Shephard (2005c, 2006a) studied multipower variation. Barndorff-Nielsen and Shephard (2006a,b,c,d,e) studied testing for jumps using bipower variation. Barndorff-Nielsen and Shephard (2006d) studied time-deformed Levy process. Barndorff-Nielsen and Shephard (2007) studied variations, jumps, market frictions, and high-frequency data in financial econometrics. Barndorff-Nielsen et al. (2005) obtained a central limit theorem for realized power and bipower variations of continuous semimartingales. Barndorff-Nielsen and Shephard (2004a,b) gave a review of power variations and stochastic volatility. Barndorff-Nielsen and Shephard (2006a,b,c,d,e) gave limit theorems for bipower variation.

We study estimation and pricing in MTS-OU SV model. IG-OU model is a special case. We study estimation by method of moments (MM), least absolute deviation (LAD) method, empirical characteristic function (ECF) method, and quasi-maximum likelihood (QML) method.

J. P. N. Bishwal, *Parameter Estimation in Stochastic Volatility Models*, https://doi.org/10.1007/978-3-031-03861-7_9

9.2. Modified Tempered Stable Models

An infinitely divisible distribution is said to be α-modified tempered stable distribution (α-MTS) distribution if its Levy triplet is given by

$$\sigma^2 = 0,$$

$$\nu(dx) = C\left(\frac{\lambda_+^{\alpha+\frac{1}{2}}K_{\alpha+\frac{1}{2}}\lambda_+ x}{x^{\alpha+\frac{1}{2}}}I_{x>0} + \frac{\lambda_+^{\alpha+\frac{1}{2}}K_{\alpha+\frac{1}{2}}\lambda_- x}{x^{\alpha+\frac{1}{2}}}I_{x<0}\right)dx,$$

$$\gamma =$$

$$\mu + C\left(\frac{\Gamma(\frac{1}{2}-\alpha)}{2^{\alpha+\frac{1}{2}}}(\lambda_+^{2\alpha-1} - \lambda_-^{2\alpha-1}) - \lambda_+^{\alpha-\frac{1}{2}}K_{\alpha-\frac{1}{2}}(\lambda_+) + \lambda_-^{\alpha-\frac{1}{2}}K_{\alpha-\frac{1}{2}}(\lambda_-)\right),$$

where $C > 0, \lambda_+, \lambda_- > 0$, $\mu \in \mathbb{R}$, $\alpha \in (-\infty, 1)\backslash\{\frac{1}{2}\}$ and $K_p(x)$ is the modified Bessel function of second kind. We denote the MTS random variable by $X \sim MTS(\alpha, C, \lambda_+, \lambda_-, \mu)$. The Levy measure $\nu(dx)$ is called the MTS Levy measure with parameter $(\alpha, C, \lambda_+, \lambda_-)$.

The MTS distribution is obtained by taking a symmetric α-stable distribution with $\alpha \in (0,1)$ and multiplying by a Levy measure with $\sqrt{|x|}\lambda^{\alpha+\frac{1}{2}}K_{\alpha+\frac{1}{2}}(\lambda|x|)$ on each half of the real axis. The measure can be extended to the case $\alpha \leq 0$. If $\alpha = \frac{1}{2}$, then γ may not be defined, so it is removed. The MTS distribution was introduced by Kim et al. (2009).

The tails of the α-MTS distribution are thinner than those of the 2α-stable and fatter than those of the 2α-TS distribution. At the zero neighborhood, all three have the same asymptotic behavior.

If $\lambda_+ > \lambda_-$, then the distribution is *skewed to the left.*

If $\lambda_+ < \lambda_-$, then the distribution is *skewed to the right.*

If $\lambda_+ = \lambda_-$, then the distribution is *symmetric.*

C controls the kurtosis of the distribution. If C increases, the peakedness of the distribution increases.

As α decreases, the distribution has fatter tails and increased peakedness.

The Levy process corresponding to the MTS distribution has *finite activity* if $\alpha < 0$ and *infinite activity* if $\alpha > 0$.

It has *finite variation* if $\alpha < \frac{1}{2}$ and *infinite variation* if $\alpha > \frac{1}{2}$.

With proper choice of C and μ, MTS distribution has zero mean and unit variance, and the distribution is called standard MTS distribution and denoted $X \sim stdMTS(\alpha, \lambda_+, \lambda_-)$.

CGMY process proposed in Carr et al. (2003) is a tempered stable process. In order to obtain a closed form solution of the European option price, CGMY used the generalized Fourier transform of the distribution of the stock price under the assumption of Markov property.

The BN-S stochastic volatility model is given by

$$\begin{aligned} dY_t &= (\mu + \beta X_t)\,dt + \sqrt{X_t}\,dW_t + \rho\,dZ_t \\ dX_t &= -\theta\,X_t\,dt + dZ_t, \end{aligned}$$

where μ is the drift parameter, β is the risk premium, $\theta > 0$ is the drift of the volatility, and Z_t is a MTS process.

We estimate θ from the observations of $\{Y_t\}$ at the time points $t_k = k\Delta,\ k = 0, 1, 2, \ldots, n,\ \Delta > 0$.

$$c_m(Z) = \frac{d^m}{du^m} \log \phi_{TS}(u)|_{u=0}$$

For the tempered stable distribution $\mathrm{TS}(b, \delta, \gamma)$ where $0 < b < 1, \delta > 0, \gamma \geq 0$, the m-th cumulant is given by

$$c_m(Z) = -\delta(-2)^m \gamma^{(b-m)/b} b(b-1)\ldots(b-(m-1))$$

for $\gamma > 0$.

When $\gamma = 0$, it is positive b-stable distribution for which the moments of only order $k < b$ exist.

For $b = 1/2$, TS distribution reduces to IG distribution.

The infinite divisibility of this distribution allows one to construct the corresponding Levy process.

A Levy process $Z = (Z_t)_{t \geq 0}$ is said to be a tempered stable process if Z_1 follows a tempered stable distribution.

The tempered stable process is of *finite activity* if $\alpha < 0$ and *infinite activity* if $0 < \alpha < 2$.

The tempered stable process is of *finite variation* if $0 < \alpha < 1$ and *infinite variation* if $1 < \alpha < 2$.

The MTS-GARCH model is given by

$$\log \frac{S_t}{S_{t-1}} = r_t - d_t + \lambda_t \sigma_t - g(\sigma_t; \alpha, \lambda_+, \lambda_-) + \sigma_t \epsilon_t$$

$$\sigma_t^2 = (\alpha_0 + \alpha_1 \sigma_{t-1}^2 \epsilon_{t-1}^2 + \beta_1 \sigma_{t-1}^2) \wedge \rho,\ \epsilon_0 = 0$$

$\alpha_0, \alpha_1, \beta_1 \geq 0,\ \alpha_1 + \beta_1 < 1,\ 0 < \rho < \lambda_+^2,\ \epsilon_t \sim stdMTS(\alpha, \lambda_+, \lambda_-)$, r_t is the risk-free rate, d_t is the dividend rate, λ_t is the market price of risk, g is the characteristic exponent of the Laplace transform for the distribution $stdMTS(\alpha, \lambda_+, \lambda_-)$, i.e., $g(x; \alpha, \lambda_+, \lambda_-) = \log(E(\exp(x\epsilon_t)))$.

The characteristic function of Z is given by

$$\phi_Z(u) = \exp(iu\mu + G_R(u; \alpha, C, \lambda_+, \lambda_-) + G_I(u; \alpha, C, \lambda_+, \lambda_-)),$$

where for $u \in \mathbb{R}$

$$\begin{aligned} & G_R(u; \alpha, C, \lambda_+, \lambda_-) \\ = \;& 2^{-\frac{\alpha+3}{2}} \sqrt{\pi} C \Gamma\left(1 - \frac{\alpha}{2}\right) \left[(\lambda_+^2 + u^2)^{\frac{\alpha}{2}} - \lambda_+^{\alpha} + (\lambda_-^2 + u^2)^{\frac{\alpha}{2}} - \lambda_-^{\alpha}\right], \end{aligned}$$

$$G_I(u; \alpha, C, \lambda_+, \lambda_-)$$

$$= iuC2^{-\frac{\alpha+1}{2}}\Gamma\left(\frac{1-\alpha}{2}\right)\times$$
$$\left[\lambda_+^{\alpha-1}F\left(1,\frac{1-\alpha}{2};\frac{3}{2},;-\frac{u^2}{\lambda_+^2}\right)-\lambda_-^{\alpha-1}F\left(1,\frac{1-\alpha}{2};\frac{3}{2},;-\frac{u^2}{\lambda_-^2}\right)\right],$$

where F is the hyper-geometric function.
The value of G_I for symmetric MTS distribution is always zero.
The m-th cumulant is

$$c_m(Z)=\mu \qquad if \;\; m=1,$$

$$=2^{m-\frac{\alpha+3}{2}}\left(\frac{m-1}{2}\right)!C\Gamma\left(\frac{m-\alpha}{2}\right)(\lambda_+^{\alpha-m}-\lambda_-^{\alpha-m}) \;\; if \;\; m=3,5,7,\ldots$$

$$=2^{-\frac{\alpha+3}{2}}\sqrt{\pi}\left(\frac{m!}{\frac{m}{2}!}\right)C\Gamma\left(\frac{m-\alpha}{2}\right)(\lambda_+^{\alpha-m}+\lambda_-^{\alpha-m}) \;\; if \;\; m=2,4,6,\ldots$$

The mean, variance, skewness, and excess kurtosis are given by

$$E(Z)=c_1(Z)=\mu+2^{-\frac{\alpha+1}{2}}C\Gamma\left(\frac{1-\alpha}{2}\right)(\lambda_+^{\alpha-1}-\lambda_-^{\alpha-1})$$

$$V(Z)=c_2(Z)=2^{-\frac{\alpha+1}{2}}\sqrt{\pi}C\Gamma\left(1-\frac{\alpha}{2}\right)(\lambda_+^{\alpha-2}+\lambda_-^{\alpha-2})$$

$$s(Z)=\frac{c_3(Z)}{c_2(Z)^{3/2}}=\frac{2^{\frac{\alpha+9}{4}}\Gamma\left(\frac{3-\alpha}{2}\right)(\lambda_+^{\alpha-3}-\lambda_-^{\alpha-3})}{\pi^{3/4}C^{1/2}(\Gamma(\frac{1-\alpha}{2})(\lambda_+^{\alpha-2}+\lambda_-^{\alpha-2}))^{3/2}}$$

$$\kappa(Z)=\frac{c_4(Z)}{c_2(Z)^2}=\frac{3\cdot 2^{\frac{\alpha+3}{2}}C\Gamma\left(2-\frac{\alpha}{2}\right)(\lambda_+^{\alpha-4}+\lambda_-^{\alpha-4})}{\sqrt{\pi}C(\Gamma(\frac{1-\alpha}{2})(\lambda_+^{\alpha-2}+\lambda_-^{\alpha-2}))^2}.$$

If $\alpha\in(0,2)\backslash\{1\}$, the Levy measure of α-stable, α-TS, and α-MTS have the same asymptotic behavior at the zero neighborhood. However, the tails of the Levy measures for the α-MTS distribution are *thinner* than those of α-stable and *heavier* than those of α-TS distribution.

Multi-Step Estimation in Stable OU Model

Masuda and Uehara (2017) studied two-step estimation in ergodic Levy driven SDE

$$dX_t=a(\theta,X_t)dt+b(\beta,X_{t-})dZ_t, X_0=x_0.$$

Masuda (2019) studied multi-step estimation in stable OU Model.

$$dX_t=-\theta X_t dt+\sigma dZ_t, X_0=x_0.$$

For the LSE of θ, Hu and Long (2009) obtained

$$\left(\frac{T}{\log n}\right)^{1/\beta}(\tilde{\theta}_n - \theta_0) \to^D \frac{S'_\beta}{S''^{+}_{\beta/2}}.$$

While in Gaussian OU case, for different parts $\theta > 0$, $\theta < 0$, and $\theta = 0$, LAN, LAMN, and LABF hold, respectively, in stable case entirely different phenomena occur.

The solution of the SDE is given by

$$X_t = e^{-\theta(t-s)}X_s + \sigma \int_s^t e^{-\theta(t-s)} dZ_u.$$

Due to the stable integral property,

$$\mathcal{L}\left(\int_s^t e^{-\theta(t-s)} dZ_u\right) = S_\beta(\kappa_h(\theta)),$$

where

$$\kappa_h(\theta) = \left\{\frac{1 - e^{-\theta h}}{\theta\beta}\right\}^{1/\beta} \sim h^{1/\beta}.$$

For each $j \le n$, the transition probability is given by

$$\mathcal{L}(X_{t_j}|X_{t_{j-1}} = x) = \delta_{x\exp(-\theta h)} \star S_\beta(\kappa_h(\theta)).$$

LAMN holds for $\theta \in \mathbb{R}$ when T is fixed.

$$n^{1/\beta - 1/2}(\hat{\theta}_n - \theta) \to^D MN(0, \Gamma_0(T)^{-1}).$$

9.3 Method of Moments Estimators

The moment estimators when Z is a Gamma process are given by

$$\hat{\theta}_n := \frac{\frac{1}{n^2}\left[\sum_{i=1}^n (Y_{i\Delta} - Y_{(i-1)\Delta})\right]^2}{\frac{1}{n^2}\sum_{i=1}^n (Y_{i\Delta} - Y_{(i-1)\Delta})^2 - \frac{\Delta}{n}\left[\sum_{i=1}^n (Y_{i\Delta} - Y_{(i-1)\Delta})\right]} \frac{2a^3(a+1)}{b^4\Delta}.$$

$$\hat{\rho}_n := \frac{\frac{1}{n^2}\sum_{i=1}^n (Y_{i\Delta} - Y_{(i-1)\Delta})^2 - \frac{\Delta}{n}\left[\sum_{i=1}^n (Y_{i\Delta} - Y_{(i-1)\Delta})\right]}{\frac{1}{n^2}\left[\sum_{i=1}^n (Y_{i\Delta} - Y_{(i-1)\Delta})\right]} \frac{b^3\Delta}{2a^2(a+1)}.$$

For the MTS-OU model, the estimating functions are

$$c_1(y_1) = \lambda\rho\Delta c_1(Z)$$
$$c_2(y_1) = \Delta c_1(Z) + 2\lambda\rho^2\Delta c_2(Z)$$
$$c_3(y_1) = \Delta c_1(Z) + 2\lambda\rho^2\Delta c_2(Z)$$

$$c_4(y_1) = \Delta c_1(Z) + 2\lambda\rho^2\Delta c_2(Z)$$

$$E(y_1) = c_1(y_1) = \mu + 2^{-\frac{\alpha+1}{2}} C\Gamma\left(\frac{1-\alpha}{2}\right)(\lambda_+^{\alpha-1} - \lambda_-^{\alpha-1})$$

$$V(y_1) = c_2(y_1) = 2^{-\frac{\alpha+1}{2}}\sqrt{\pi}C\Gamma\left(1-\frac{\alpha}{2}\right)(\lambda_+^{\alpha-2} + \lambda_-^{\alpha-2})$$

$$\hat{\theta}_n := \frac{\frac{1}{n^2}\left[\sum_{i=1}^n (Y_{i\Delta} - Y_{(i-1)\Delta})\right]^2}{\frac{1}{n^2}\sum_{i=1}^n (Y_{i\Delta} - Y_{(i-1)\Delta})^2 - \frac{\Delta}{n}\left[\sum_{i=1}^n (Y_{i\Delta} - Y_{(i-1)\Delta})\right]}$$

$$[2^{-\frac{\alpha+1}{2}} C\Gamma\left(\frac{1-\alpha}{2}\right)(\lambda_+^{\alpha-1} - \lambda_-^{\alpha-1})]^2[2^{-\frac{\alpha+1}{2}}\sqrt{\pi}C\Gamma\left(1-\frac{\alpha}{2}\right)(\lambda_+^{\alpha-2} + \lambda_-^{\alpha-2})]2\Delta^{-1}.$$

$$\hat{\rho}_n := \frac{\frac{1}{n^2}\sum_{i=1}^n (Y_{i\Delta} - Y_{(i-1)\Delta})^2 - \frac{\Delta}{n}\left[\sum_{i=1}^n (Y_{i\Delta} - Y_{(i-1)\Delta})\right]}{\frac{1}{n^2}\left[\sum_{i=1}^n (Y_{i\Delta} - Y_{(i-1)\Delta})\right]}$$

$$[2^{-\frac{\alpha+1}{2}} C\Gamma\left(\frac{1-\alpha}{2}\right)(\lambda_+^{\alpha-1} - \lambda_-^{\alpha-1})2^{-\frac{\alpha+1}{2}}\sqrt{\pi}C\Gamma\left(1-\frac{\alpha}{2}\right)(\lambda_+^{\alpha-2} + \lambda_-^{\alpha-2})]^{-1}2^{-1}\Delta.$$

We obtain the consistency and asymptotic normality of the estimators. Using mixing property we obtain

Theorem 2.1

$$\hat{\theta}_n \to \theta_0 \quad \text{a.s.} \quad \text{as } n \to \infty.$$

Theorem 2.2

$$\sqrt{n}(\hat{\theta}_n - \theta_0) \to^{\mathcal{D}} \mathcal{N}(0,\ J^{-1}(\theta_0))$$

Theorem 2.3

$$\hat{\rho}_n \to \rho_0 \quad \text{a.s.} \quad \text{as } n \to \infty.$$

Theorem 2.4

$$\sqrt{n}(\hat{\rho}_n - \rho_0) \to^{\mathcal{D}} \mathcal{N}(0,\ I^{-1}(\rho_0))$$

We propose another estimator

$$\breve{\theta}_n := \bigwedge_{i=1}^n \frac{\hat{X}_{i\Delta}}{\hat{X}_{(i-1)\Delta}}.$$

We also propose the following estimator of the drift. The estimator of θ is defined as

$$\tilde{\theta}_n := -\log\left[\frac{\sum_{i=1}^n \hat{X}_{i\Delta}}{\sum_{i=1}^n \hat{X}_{(i-1)\Delta}}\right],$$

where $\hat{X}$ is the estimator of X based on observations of Y.

9.4 Robust Estimation in Inverse Gaussian Ornstein-Uhlenbeck Stochastic Volatility Model

We study the LAD estimators in the stable OU case. The Least Absolute Deviation (LAD) estimator is robust to "outlying data". The LAD estimation has a long history and is one of the popular estimation procedure robust to outliers. The LAD estimation is based on the Laplacian L^1- loss while the LSE is on the Gaussian L^2-loss. We refer to Portnoy and Koenker (1997), Knight (1998) and Koenker (2005) as well as the references therein for a detailed account and historical background on LAD estimation. For time series literature see David and Dunsmuir (1997) and Davis et al. (1992).

The least absolute deviation (LAD) estimator is defined as the minimizer $\hat{\theta}_n$ of the contrast function $\theta \to \sum_{i=1}^n |X_{t_i} - e^{-\theta h} X_{t_{i-1}}|$. For fixed T, asymptotic normality of $\hat{\theta}_n$ is achieved at the rate $n^{1/\beta - 1/2}$ where β stands for the activity index of the driving Levy process, also known as the Blumenthal and Getoor (1961) activity index defined as

$$\beta = \inf\left\{r > 0 : \int_{|z| \le 1} |z|^r \nu(dz) < \infty\right\}$$

which is the degree of small-jump fluctuations.

Now consider another least absolute deviation (LAD) estimator which is defined as the minimizer $\tilde{\theta}_n$ of the contrast function $\theta \to \sum_{i=1}^n |X_{t_i} + \theta X_{t_{i-1}} h|$.

Under infill and large time sampling design, that is when $h \to 0$ and $nh \to \infty$, asymptotic normality of $\tilde{\theta}_n$ is achieved at the rate $\sqrt{n} h^{1-1/\beta}$ where β stands for the activity index of the driving Levy process. Note that $\sqrt{n} h^{1-1/\beta} = T^{1-1/\beta} n^{(2-\beta)/(2\beta)}$. This implies that the rate of convergence is determined by the most active part of the driving Levy process, the presence of a driving Wiener part leads to $\sqrt{nh}$, which is familiar in the context of asymptotically efficient estimation of diffusions with compound Poisson jumps, while a pure jump driving Levy process leads to a faster one. As a result, when Z is a pure jump Levy process, we have a faster rate of convergence that the familiar rate $\sqrt{nh}$. It is interesting to note that rate of convergence is faster by only changing the type of loss from L^2 to L^1.

Using self-weighted LAD (SLAD) contrast function $\theta \to \sum_{i=1}^n w(X_{t_{i-1}})$ $|X_{t_i} + \theta X_{t_{i-1}}|$, for a bounded continuous weight function w, the rate of convergence can be improved to the conventional rate $\sqrt{n}$, see Masuda (2010) which extended autoregressive process with infinite variance studied in Ling (2005). An example of weight function is $w(x) = \exp(-|x|)$. The unweighted weight function corresponds to the case $w \equiv 1$.

In the Wiener case, the unweighted SLAD estimator leans to the asymptotic variance $\pi\theta_0$ whereas the asymptotic variance of the exact MLE is $2\theta_0$. Hence the asymptotic efficiency of the SALD estimator relative to the MLE is $2/\pi$. This is same as asymptotic relative efficiency for the same the sample median over sample mean in estimating the mean of i.i.d. normal samples. For the asymptotic normality of SLAD estimator one does not need the rapidly increasing experimental design $nh^2 \to 0$ which is quite inevitable while adopting contrast function based on Euler type approximation. For SLAD estimator, the weaker condition $nh^3 \to 0$ is sufficient.

The SLAD estimator converges more rapidly than the LSE as soon as $h = o(n^{-1/2}(\log n)^{1/(2-\beta)})$.

The sampling design condition for h is $h = n^{-\tau}$ where $\tau \in (0,1)$.

An interval estimator of τ is given by

$$\frac{\beta}{2(2\beta-1)} < \tau < \frac{\beta}{2\beta-1}, \text{ if } \beta > 1,$$

$$\frac{\beta}{2(2\beta-1)} < \tau < 1, \text{ if } \beta \in (2/3, 1).$$

An example of infill interval is $h = n^{-3/5}$.

Consider the SDEs

$$dY_t = (\mu + \beta X_t)dt + \sqrt{X_t}dW_t + \rho dZ_{\lambda t}$$

$$dX_t = -\lambda X_t dt + dZ_{\lambda t},$$

where Z_t is a Levy process independent of X_0 and $\mathcal{L}(X_0) = IG(\delta,\gamma)$. We suppose that δ and γ are known and we are interested in estimating $\theta_0 = (\lambda, \rho)$, where $\lambda > 0$ and $\rho < 1$.

Note that if $X_{j\Delta}, 1 \le j \le n$ were observed, then

$$\lambda_n := -\ln\left(\min_{1\le j\le n} \frac{X_{j\Delta}}{X_{(j-1)\Delta}}\right)$$

is a weakly consistent estimator of λ as $n \to \infty$ as showed by Jongbloed et al. (2005). In our case, $X_{j\Delta}, 1 \le j \le n$ are unobserved.

We can estimate the state process X by $\hat{X}$ given by the recursion

$$\hat{X}_{t_i} = e^{-\alpha\Delta t_i}\hat{X}_{t_{i-1}} + Z_{t_i} - Z_{t_{i-1}}.$$

Then we estimate λ by

$$\check{\lambda}_n := -\ln\left(\min_{1\le j\le n} \frac{\hat{X}_{j\Delta}}{\hat{X}_{(j-1)\Delta}}\right).$$

The moment estimators of ρ and λ are given by

$$\hat{\rho}_n := \frac{\gamma(\gamma s_y^2 - \Delta\delta)}{2\bar{y}},$$

where

$$s_y^2 := \frac{1}{n}\sum_{j=1}^{n}(y_j - \bar{y})^2 = \frac{1}{n}\sum_{j=1}^{n} y_j^2 - (\bar{y})^2,$$

$$\bar{y} := \frac{1}{n}\sum_{j=1}^{n} y_j, \quad y_j := Y_{j\Delta} - Y_{(j-1)\Delta}$$

and

$$\hat{\lambda}_n := \frac{\gamma\bar{y}}{\Delta\delta\hat{\rho}_n}.$$

The robust estimators of ρ and λ are given by

$$\tilde{\rho}_n := \frac{\gamma(\gamma a_y - \Delta\delta)}{2\tilde{y}}, \quad \tilde{\lambda}_n := \frac{\gamma\tilde{y}}{\Delta\delta\tilde{\rho}_n},$$

where

$$a_y := \frac{1}{n}\sum_{j=1}^{n} |y_j - \tilde{y}|$$

is the sample mean absolute deviation from median,

$$\begin{aligned}\tilde{y} &:= \text{median of } \{y_j, 1 \le j \le n\}\\ &= \frac{y_k + y_{k+1}}{2} \quad if \quad n = 2k\\ &= y_{k+1} \quad if \quad n = 2k+1.\end{aligned}$$

Let $\theta = (\rho, \lambda)$. and $\tilde{\theta}_n = (\tilde{\rho}_n, \tilde{\lambda}_n)$. We have the following properties of the estimators. Using the standard theory of order statistics and mixing property of the process, we obtain:

Theorem 2.1 For fixed $\Delta > 0$ as $n \to \infty$,

$$(a) \quad \tilde{\theta}_n \to \theta_0 \text{ a.s.}$$

$$(b) \quad \sqrt{n}(\tilde{\theta}_n - \theta_0) \to^{\mathcal{D}} N_2(0, (2\lambda\rho^2\Delta^2\delta^2\gamma^{-4})^{-2}\Sigma^{-1}(\theta_0)).$$

9.5 Quasi-Maximum Likelihood Estimation in Stable-OU Process

Quasi-likelihood estimation can be used for the nonlinear SDE

$$dX_t = \left(\theta_1 X_t + \frac{\theta_2}{1 + X_t^2}\right) dt + \exp\{\sigma_1 \cos(X_t) + \sigma_2 \sin(X_t)\}\, dZ_t,$$

where Z_t could be normal inverse-Gaussian Levy process (and symmetric generalized hyperbolic Levy process) or the 1.5-stable Levy process. Asymptotic mixed normality can be obtained. See Masuda (2019).

Locally Stable Levy Processes

Symmetric Generalized Hyperbolic Levy Process

The symmetric generalized hyperbolic distribution, denoted by $GH(\theta, \eta, \zeta)$, is infinitely divisible with characteristic function

$$u \to \left(\frac{\eta^2}{\eta^2+u^2}\right)\frac{K_\theta(\zeta\sqrt{\eta^2+u^2})}{K_\theta(\eta\zeta)},$$

where K_θ denoted the modified Bessel function of third kind with index $\theta \in \mathbb{R}$. If $\mathcal{L}(Z_1) = GH(\theta, \eta, \zeta)$, then for each $u \in \mathbb{R}$,

$$E\left(e^{-ih^{-1}Z_h}\right) = \left(\frac{(\eta h)^2}{(\eta h)^2+u^2}\right)^{\theta h/2}\frac{K_\theta((\zeta/h)\sqrt{(\eta h)^2+u^2})}{K_\theta(\eta\zeta)} \to \exp(-\zeta|u|),\ h \to 0$$

showing that Z is locally Cauchy (for $(\zeta = 1)$.

The Levy measure of $GH(\theta, \eta, 1)$ admits the density such that

$$z \to \frac{1}{\pi|z|^2}\left\{1+\frac{\pi}{2}\left(\theta+\frac{1}{2}\right)|z|+o(|z|)\right\},\ |z| \to 0.$$

The case $\theta = -1/2$ corresponds to a symmetric normal inverse-Gaussian process.

Symmetric Tempered Stable Levy Process

The Levy density is given by

$$x \to g_{0,\beta}(x)\exp(-\theta|x|),$$

where

$$g_{0,\beta}(x) := \frac{c_\beta}{|x|^{1+\beta}}, x \neq 0$$

and

$$c_\beta := \frac{1}{2}\left\{\frac{1}{\beta}\Gamma(1-\beta)\cos\left(\frac{\beta\pi}{2}\right)\right\}^{-1}$$

with

$$c_1 = \lim_{\beta\to 1} c_\beta = \pi^{-1}.$$

When $\mathcal{L}(Z_1) = TS_\beta(\theta)$, then $\mathcal{L}(h^{-1/\beta}Z_h) = TS_\beta(\theta h) \Rightarrow S_\beta$ as $h \to 0$ where S_β stands of the standard symmetric β-stable distribution corresponding to the characteristic function $\varphi_0(u) = \exp(-|u|^\beta)$, $\beta \in [1,2)$. The function ψ_h is explicitly given by

$$\psi_h(u) = \frac{1}{\pi}\left\{\theta h\log\left(1+\frac{u^2}{\theta^2h^2}\right) - 2u\arctan\left(\frac{u}{\theta h}\right)\right\},\ \beta = 1$$

$$\psi_h(u) = 2c_\beta\Gamma(-\beta)\left[(\theta^2 h^{2/\beta}+u^2)^{\beta/2}\cos\left\{\beta\arctan\left(\frac{u}{\theta h^{1/\beta}}\right)\right\}-\theta^\beta h\right],\ \beta\in(1,2)$$

For $\beta = 1$,

$$\partial_u\psi_h(u) = -\frac{2}{\pi}\arctan\left(\frac{u}{\theta h}\right).$$

$$\sqrt{n}\int |f_h(y)-\phi_\beta(y)|dy \to 0$$

if $nh^{2-\epsilon_1}\to 0$ for some $\epsilon_1 > 0$.

For $\beta \in (1,2)$,

$$\sqrt{n}\int |f_h(y)-\phi_\beta(y)|dy \to 0$$

if $nh^{2/\beta}\to 0$.

Consider the Markovian SDE

$$dX_t = a(X_t,\theta)dt + b(X_{t-},\sigma)dZ_t.$$

The Euler approximation is given by

$$X_{t_j} = X_{t_{j-1}} + a_{j-1}(\theta)h + b_{j-1}(\sigma)\Delta_j Z.$$

The quasi-likelihood function is based on the small-time stable approximation of the unknown transition density.

Now assume that the process X admits a time homogeneous Lebesgue transition density under $P_{\theta,\sigma}$, say $p_h(x,y,\theta,\sigma)dy = P_{\theta,\sigma}(X_h\in dy|X_0=x)$.

The transition Lebesgue density is given by the following two fold approximation of $\mathcal{L}(X_{t_j}|X_{t_{j-1}})$:

$$
\begin{aligned}
& p_h(X_{t_j},X_{t_{j-1}};\theta,\sigma)\\
=\ & \frac{1}{2\pi}\int \exp(-iuX_{t_j})\int e^{iuy}p_h(X_{t_{j-1}},y;\theta,\sigma))dy\\
\approx\ & \frac{1}{2\pi}\int \exp(-iuX_{t_j})E_{j-1}\left[\exp\left\{iu(X_{t_{j-1}}+a_{j-1}(\theta)h+b_{j-1}(\sigma)\Delta_j Z)\right\}\right]\\
& \text{(Euler Approximation)}\\
=\ & \frac{1}{2\pi}\int \exp(-iu(X_{t_j}-X_{t_{j-1}}-a_{j-1}(\theta)h)\varphi_h(b_{j-1}(\sigma)h^{1/\beta}u)du\\
=\ & \frac{1}{b_{j-1}(\sigma)h^{1/\beta}}\frac{1}{2\pi}\int \exp\{-iv\epsilon_j(\theta,\sigma)\}\varphi_h(v)dv\\
=\ & \frac{1}{b_{j-1}(\sigma)h^{1/\beta}}f_h(\epsilon_j(\theta,\sigma))\\
\approx\ & \frac{1}{b_{j-1}(\sigma)h^{1/\beta}}\phi_\beta(\epsilon_j(\theta,\sigma)) \qquad \text{(Locally stable approximation)}
\end{aligned}
$$

where

$$\epsilon_j(\theta,\sigma) := \frac{X_{t_j}-X_{t_{j-1}}-ha_{j-1}(\theta)}{h^{1/\beta}b_{j-1}(\sigma)} \approx h^{-1/\beta}\Delta_j Z.$$

The contrast function is given by

$$H_n(\theta,\sigma) := \sum_{j=1}^{n} \log \left(\frac{1}{b_{j-1}(\sigma)h^{1/\beta}} \right) \phi_\beta(\epsilon_j(\theta,\sigma)))$$

which is called the stable quasi-likelihood. The stable quasi-maximum likelihood estimator (SQMLE) is given by

$$(\hat{\theta}_n, \hat{\sigma}_n) \in \arg\max H_n(\theta,\sigma) = \sum_{j=1}^{n} (-\log b_{j-1}(\sigma) + \log \phi_\beta(\epsilon_j(\theta,\sigma))) .$$

The SQMLE is the non-Gaussian-stable counterpart to the Gaussian quasi-likelihood previously studied by Masuda (2013a,b,c,d).

For $\beta \in [1,2)$, we have

$$(\sqrt{n}h^{1-1/\beta}(\hat{\theta}_n - \theta_0), \sqrt{n}(\hat{\sigma}_n - \sigma_0)) \to^D MN(0, \Gamma_T(\theta_0,\sigma_0;\beta)^{-1}).$$

For $\beta \in (1,2)$, we have

$$(n^{1/\beta-1/2}(\hat{\theta}_n - \theta_0), \sqrt{n}(\hat{\sigma}_n - \sigma_0)) \to^D MN(0, \Gamma_T(\theta_0,\sigma_0;\beta)^{-1}).$$

Of special interest is the locally Cauchy case ($\beta = 1$), where the contrast function H_n is fully explicit:

$$H_n(\theta,\sigma) = -\sum_{j=1}^{n} \left\{ \log \pi h + \log b_{j-1}(\sigma) + \log(1+\epsilon_j^2(\theta,\sigma)) \right\} .$$

In this case,

$$(\sqrt{n}(\hat{\theta}_n - \theta_0), \sqrt{n}(\hat{\sigma}_n - \sigma_0)) \to^D MN(0, \Gamma_T(\theta_0,\sigma_0;\beta)^{-1}).$$

This formally extends the i.i.d. model from the location-scale Cauchy population, where we have $\sqrt{n}$-asymptotic normality for the maximum likelihood estimator. The Cauchy quasi-likelihood has been also investigated in the robust-regression literature, see Mizera and Muller (1999, 2002) for breakdown-point result in some relevant models.

This estimator is much more efficient compared to the Gaussian maximum quasi-likelihood estimator studied in Masuda (2013a,b,c,d), where the transition distribution is assumed to be approximately Gaussian.

The asymptotic efficiency in the sense of Hajek-Lecam-Jeaganathan is of primary theoretical importance. Compared with the diffusion case, asymptotic efficiency phenomena for the Levy driven SDE is less well-known. Clement et al. (2019) studied asymptotic efficiency for the stable Levy SDE.

The Gaussian QMLE is defined to be the maximizer of the function

$$\theta \to \sum_{j=1}^{n} \log \left\{ \frac{1}{\sqrt{2\pi v_{j-1}}} \exp\left(-\frac{(X_{t_j} - m_{j-1}(\theta))^2}{2v_{j-1}(\theta)} \right) \right\},$$

where $m_{j-1}(\theta)$ and $v_{j-1}(\theta)$ are the conditional mean and conditional variance, respectively, of X_{t_j} given $X_{t_0}, \ldots, X_{t_{j-1}}$.

GQL-based estimation has computational simplicity and robustness of misspecification of the noise distribution. This is not asymptotically efficient in general. One can consult Heyde (1997) for an extensive and systematic account of GQL.

9.6 Empirical Characteristic Function Estimator

Chambers and Heathcote (1981) studied ECF estimator in regression model under outlying errors which are shown to be robust. The least squares estimator (LSE) and the estimator which minimizes the empirical characteristic function (ECF) are same in the stable case. Jiang and Knight (2002) studied ECF estimator. We study estimation in Stable-OU process via empirical characteristic function method.

Next we study robust estimators for the parameters of Inverse-Gaussian Ornstein–Uhlenbeck (IGOU) stochastic volatility model based on equally spaced observations. Using the fact that the increments of a IGOU process are strongly mixing with an exponential rate, we show that the resulting estimators are consistent and asymptotically normal.

Consider the SDEs

$$dY_t = (\mu + \beta X_t)dt + \sqrt{X_t} dW_t + \rho dZ_{\lambda t}$$
$$dX_t = -\lambda X_t dt + dZ_{\lambda t},$$

where Z_t is a Levy process independent of X_0 and $\mathcal{L}(X_0) = IG(\delta, \gamma)$. We suppose that δ and γ are known and we are interested in estimating $\theta_0 = (\lambda, \rho)$, where $\lambda > 0$ and $\rho < 1$.

This model has mean reversion and leverage.

Z is the sum of two independent Levy processes $Z = Z^{(1)} + Z^{(2)}$ where $\mathcal{L}(X_0) = IG(\delta/2, \gamma)$ and $Z^{(2)}$ is a compound Poisson process given by $Z^{(2)} = \gamma^{-2} \sum_{j=1}^{N_t} u_j$ with N being a Poisson process with intensity $\delta\gamma/2$ and u_j is a sequence of independent and identically χ_1^2-distributed random variables independent of N.

Z and X have infinitely many jumps in any finite time interval.

Invariant distribution is generalized inverse Gaussian (GIG).

$$\mathcal{L}(X_0) = \mathcal{L}(X_t) = \mathcal{GIG}(\lambda, \delta, \sqrt{\alpha^2 - \beta^2}).$$

Mixture distribution is Q_θ generalized hyperbolic (GH).

$$\mathcal{L}(Y_t) = \mathcal{GH}(\lambda, \frac{\alpha}{\sqrt{t}}, \frac{\beta}{\sqrt{t}}, \sqrt{t}\delta, \mu t).$$

$$\mathcal{L}(Y_{t+1} - Y_t) = \mathcal{GH}(\lambda, \alpha, \beta, \delta, \mu).$$

Let $q(x, \theta)$ is a density function of Y_t.
The contrast function is defined as

$$u(x, \theta) = -\log q(x, \theta).$$

The behavior of the corresponding MCE is studied in Bishwal (2007a,b).
The density function of the GH distribution is

$$d_{\mathcal{GH}(\lambda,\alpha,\beta,\delta,\mu)}(x) =$$

$$\frac{(\alpha^2 - \beta^2)^{\frac{\lambda}{2}}}{\sqrt{2\pi}\alpha^{\lambda-\frac{1}{2}}\delta^{\lambda}K_{\lambda}(\delta\sqrt{\alpha})}(\delta^2 + (x-\mu)^2)^{(\lambda-\frac{1}{2})/2}e^{\beta(x-\mu)}K_{\lambda-\frac{1}{2}}(\alpha\sqrt{\delta^2 + (x-\mu)^2}).$$

Here K_ν denotes the modified Bessel function of third kind with index ν.
When $\lambda = -\frac{1}{2}$, GIG becomes $IG(\delta, \gamma)$.
In the case of $\rho = -1$ correlation, the moment estimator is $\lambda_T = -\frac{Y_T}{T}$.
X is strictly stationary and mixing. Y is strictly stationary.

Chapter 10

Parameter Estimation in Student Ornstein–Uhlenbeck Process

10.1 Introduction

We study estimation of the drift coefficient in the Ornstein–Uhlenbeck process whose marginal distribution is Student t-distribution. We also study estimation in nonergodic (Cauchy type) OU process. Leonenko et al. (2020) studied approximation of heavy-tailed fractional Pearson diffusion. Leonenko and Suvak (2010a) studied statistical inference for a student diffusion process. Leonenko and Suvak (2010b) studied statistical inference for a reciprocal gamma process.

10.2 Student O–U Process

Consider the Ornstein–Uhlenbeck process satisfying the SDE

$$dX_t = -\alpha X_t dt + dL_t,$$

where $(L_t)_{t\geq 0}$ is a Levy process with cumulant function

$$\ln E\left(e^{iuL_1}\right) = iu\mu - \lambda \frac{K_{\nu/2-1}(\lambda|u|)}{K_{\nu/2}(\lambda|u|)}, \qquad u \in \mathbb{R}\backslash\{0\},$$

where K is the modified Bessel function of the third kind defined as

$$\begin{aligned} K_\nu(x) &= \frac{1}{2}\int_0^\infty y^{\nu-1}e^{-\frac{x}{2}(y+y^{-1})}dy \\ &\equiv \frac{1}{2}\int_{-\infty}^\infty e^{\nu u - x\cosh x}du, \end{aligned}$$

where $\nu > 0$ is arbitrary. Let $\mu \in \mathbb{R}$ and $\lambda > 0$ be arbitrary. The weakly stationary solution satisfies

$$X_t = e^{-\alpha t}X_0 + e^{-\alpha t}\int_0^t e^{\alpha s}dL_s,$$

and X_t has marginal t-distribution $T(\nu, \lambda, \mu)$ with pdf

$$f(x) = \frac{1}{\lambda B(\nu/2, 1/2)[1 + ((x-\mu)/\lambda)^2]^{(\nu+1)/2}},$$

J. P. N. Bishwal, *Parameter Estimation in Stochastic Volatility Models*,
https://doi.org/10.1007/978-3-031-03861-7_10

$B(\cdot,\cdot)$ being the beta function. If $\nu > 1$, then $E(X_t) = \mu$, and if $\nu > 2$, then $corr(X_{t+\tau}, X_t) = e^{-\alpha|\tau|}$.

The SOU process (X_t) has power function decay of the distribution tails.

For a strictly increasing continuous mapping $h : \mathbb{R} \to \mathbb{R}$, consider the process $Y_t = h(X_t)$. It is obvious that $\{Y_t\}_{t\geq 0}$ is a stationary Markov process. Further, if h is twice continuously differentiable, then $\{Y_t\}_{t\geq 0}$ is a diffusion process.

Symmetric Student Diffusion

The student diffusion is given by

$$dX_t = -\alpha X_t dt + \sqrt{\frac{2\alpha}{\nu - 1}(\nu + X_t^2)}\; dW_t.$$

The invariant density is given by

$$f(x) = \frac{\Gamma((\nu+1)/2)}{\sqrt{\nu\pi}\Gamma(\nu/2)}\left(1 + \frac{1}{\nu}x^2\right)^{-(\nu+1)/2}, \quad x \in \mathbb{R},\ \nu > 0.$$

$$= \frac{1}{\sqrt{\nu}B(\nu/2, 1/2))[1 + (x/\nu)^2]^{(\nu+1)/2]}},$$

where $B(\cdot,\cdot)$ is the beta function.

Skewed Student Diffusion

$$dX_t = -\alpha X_t dt + \sqrt{\frac{2\alpha}{\nu - 1}[X_t^2 + 2\gamma\nu^{1/2}X_t + (1+\gamma^2)\nu]}\; dW_t.$$

Here γ is the skewness parameter. $\gamma = 0$ gives the symmetric Student diffusion.

Spearman's Estimator

The Spearman's rank correlation coefficient ρ_S for a pair of random variables U and V is

$$\rho_S(U, V) = \rho(F_U(U), F_V(V),$$

where F_U and F_V are the marginal distributions of U and V and ρ is the linear correlation. Spearman's rho is the linear correlation of the probability transformed random variables.

For any bivariate random vector (X_1, X_2) following meta-Gaussian distribution,

$$\rho_S(X_1, X_2) = \frac{6}{\pi}\arcsin\frac{\rho}{2}.$$

Since Spearman's rho is invariant under strictly increasing transformation of data and (X_j, X_k) is Gaussian, we have

$$\rho_S(Y_0, Y_k) = \rho_S(X_0, X_k) = \frac{6}{\pi} \arcsin \frac{\rho(X_0, X_k)}{2} = \frac{6}{\pi} \arcsin \frac{e^{-\alpha k}}{2}.$$

Hence, inverting and plugging in give the *plug-in* estimator of α based on Spearman's rank correlation coefficient as

$$\hat{\alpha}_k := -\frac{1}{k} \ln \left(2 \sin \left(\frac{\pi}{6} \hat{\rho}_S(Y_0, Y_k) \right) \right), \quad k = 1, 2, \ldots$$

where

$$\hat{\rho}_S(Y_0, Y_k) := \frac{12}{n(n^2-1)} \sum_{k=1}^{n} \left(rank(Y_0) - \frac{n+1}{2} \right) \left(rank(Y_k) - \frac{n+1}{2} \right).$$

Quantile Estimator

Let

$$g(x) := -\ln[\sin(2\pi(x - 1/4)], \quad x \in (1/4, 1/2)$$

and

$$\hat{p}_k^0 := \frac{1}{n-k} \sum_{j=0}^{n-k} I(X_j > 0, \ X_{j+k} > 0)$$

be an empirical estimate of $p_k := P(X_0 > 0, \ X_k > 0)$. Then define

$$\begin{aligned} \hat{\alpha}_k &:= \frac{1}{k} g(\hat{p}_k^0), \quad k = 1, 2, \ldots. \\ &= -\frac{1}{k} \ln[\sin(2\pi(\hat{p}_k^0 - 1/4)]. \end{aligned}$$

Borovkov and Decrouez (2013) showed that $\hat{\alpha}_k$ is strongly consistent and asymptotically normal as $n \to \infty$.

Since X_t is unobservable, $\hat{p}_k^0$ needs to be replaced with an estimator of p_k from Y_t.

Let $\hat{m}_n(Y)$ be the sample median of the sample $Y_0, Y_1, \ldots, Y_n$ and m be the median of F. $\hat{m}_n(Y) \to m$ a.s. as $n \to \infty$.

Since $\{X_j > 0\} = \{Y_j > m\}$, we have

$$\hat{p}_k^0 := \frac{1}{n-k} \sum_{j=0}^{n-k} I(Y_j > m, \ Y_{j+k} > m).$$

Since we do not know m, we use the sample median and obtain

$$\hat{p}_k := \frac{1}{n-k} \sum_{j=0}^{n-k} I(Y_j > \hat{m}_n(Y), \ Y_{j+k} > \hat{m}_n(Y)).$$

Therefore,

$$\breve{\alpha}_k := \frac{1}{k} g(\hat{p}_k), \quad k = 1, 2, \ldots,$$

will also be strongly consistent estimator of α.

Third Estimator

The process (X_t) is unobservable, but the transformed process $Y_t = h(X_t)$ is observable. Even if, Y_t is partially observed, that is, if we know the number of times the process' trajectory crossed the band between two given levels A and B, during a given time interval, then one can estimate α, provided we also know the values of the distribution function $F(y)$ of the stationary distribution of Y_t at the points A and B.

Let

$$\tau_{a,b} := \inf\{t > 0 : X_t = a | X_0 = b\}, \quad a, b \in \mathbb{R}.$$

$\tau_{a,b}$ is the time taken by the OU process (X_t) to transit from b to a.

The *excursion time* $\tau_{(a),b}$ from a to b is the time it takes the OU process starting from b to reach the point a and return back to b. Clearly,

$$\tau_{(a),b} =^D \tau_{a,b} + \tau_{b,a},$$

where $\tau_{a,b}$ and $\tau_{b,a}$ are assumed to be independent.

Set $a := h^{-1}(A), \ \ b := h^{-1}(B)$. Since $X_t \sim N(0, 1/(2\alpha))$ in the stationary regime, one has $\Phi(\sqrt{2\alpha}a) = P(X_t \leq a) = P(h(X_t) \leq A) = F(A)$, and likewise, $\Phi(\sqrt{2\alpha}b) = P(X_t \leq b) = P(h(X_t) \leq B) = F(B)$. Therefore,

$$a = \frac{\Phi^{-1}(F(A))}{\sqrt{2\alpha}} =: \frac{a^*}{\sqrt{\alpha}}, \quad b = \frac{\Phi^{-1}(F(B))}{\sqrt{2\alpha}} =: \frac{b^*}{\sqrt{\alpha}}.$$

The standard scaling argument shows that the process $X_t^* := \alpha^{1/2} X_{t/\alpha}$ satisfies the OU equation with $\alpha = 1$ and another Brownian motion, namely $W_t^* = \alpha^{1/2} W_{t/\alpha}$. Therefore,

$$\tau_{(a),b} =^D \tau^*_{(a^*),b^*},$$

where the asterisk indicates that the excursion time on the right side is for the process X_t^*, the distribution of $\tau^*_{(a^*),b^*}$ being independent of α and having a mean m_{a^*,b^*}.

Now strong Markov property implies that subsequent excursion from b to a in $\{X_t\}$ (or, equivalently, from B to A in $\{Y_t\}$) forms an i.i.d. sequence. Therefore, if $N_{A,B}(T)$ denotes the number of times $\{Y_t\}$ crosses the band between the levels A and B in a given direction during the time interval $[0, T]$, the integral renewal theorem asserts that, as $T \to \infty$,

$$\frac{N_{A,B}(T)}{T} \to^{a.s.} \frac{1}{E(\tau_{(a),b})} = \frac{\alpha}{m_{a^*,b^*}},$$

which gives another consistent estimator of α given by

$$\widetilde{\alpha} := \frac{m_{a^*,b^*} N_{A,B}}{T}.$$

10.3 Speed of Convergence for the Cauchy Approximation

The section shows that the distribution of the normalized maximum likelihood estimator of the positive drift parameter θ in the explosive Ornstein–Uhlenbeck process observed over $[0, T]$ converges to the standard Cauchy distribution with an error rate $O(e^{-\theta T})$ as $T \to \infty$. This shows that the rate of convergence is faster in the explosive case than the stationary case.

We study Itô stochastic differential equation, explosive Ornstein–Uhlenbeck process, maximum likelihood estimator, and accuracy of Cauchy approximation.

Estimating the rate in the Kolmogorov distance between two distributions has a long history in probability and statistics. The estimate could be useful in finding confidence interval and in hypothesis testing. In the i.i.d. case, the Berry–Esseen bound for minimum contrast estimators was obtained in Pfanzagl (1971) improving that from Michel and Pfanzagl (1971). Borokov (1973) obtained the rate of convergence for the invariance principle in the i.i.d. case. Hall and Heyde (1980) obtained the rate of convergence in the central limit theorem for martingales using Skorokhod embedding. Uniform rate of weak convergence for the minimum contrast estimator in the Ornstein–Uhlenbeck (O-U) process was studied in Bishwal (2010a,b,c,d). The rates of convergence of the conditional least squares estimator and an approximate maximum likelihood estimator when the O-U process is observed at discrete time points in $[0, T]$ has been studied in Bishwal and Bose (2001) in the ergodic case. In a Bayesian framework, the rates of convergence of the posterior distributions and the Bayes estimators has been studied in Bishwal (2000b) and Bishwal (2021a,b,c) for the continuous observation and discrete observations, respectively, in the ergodic case. In finance, asset price may behave in nonergodic manner, i.e., efficient market hypotheses may not hold, possibly be due to social interaction among consumers among other reasons, see Horst and Wenzelburger (2008). Parameter estimation in diffusion process is extensively studied in Bishwal (2008a) for long time asymptotics and Kutoyants (1994a,b) for vanishing noise asymptotics. We study the nonergodic Ornstein–Uhlenbeck process in this chapter and focus on the rate of convergence of the Kolmogorov distance in the long time framework.

Let $(\Omega, \mathcal{F}, \{\mathcal{F}_t\}_{t\geq 0}, P)$ be a stochastic basis on which is defined the Ornstein–Uhlenbeck process $\{X_t\}$ satisfying the Itô stochastic differential equation

$$dX_t = \theta X_t dt + dW_t, t \geq 0, \quad X_0 = 0, \tag{3.1}$$

where $\{W_t\}$ is a standard Wiener process with respect to the filtration $\{\mathcal{F}_t\}_{t\geq 0}$ and $\theta > 0$ is the unknown parameter to be estimated on the basis of continuous observation of the process $\{X_t\}$ on the time interval $[0, T]$.

Let us denote the realization $\{X_t, 0 \leq t \leq T\}$ by X_0^T. Let P_θ^T be the measure generated on the space (C_T, B_T) of continuous functions on $[0, T]$ with the associated Borel σ-algebra B_T generated under the supremum norm by the process X_0^T and P_0^T be the standard Wiener measure. It is well known that when θ is the true value of the parameter, P_θ^T is absolutely continuous with respect to P_0^T and the Radon–Nikodym derivative (likelihood) of P_θ^T with respect to P_0^T based on X_0^T is given by

$$L_T(\theta) := \frac{dP_\theta^T}{dP_0^T}(X_0^T) = \exp\left\{\theta \int_0^T X_t dX_t - \frac{\theta^2}{2}\int_0^T X_t^2 dt\right\}. \tag{3.2}$$

Maximizing the log-likelihood with respect to θ provides the maximum likelihood estimate (MLE)

$$\theta_T := \frac{\int_0^T X_t dX_t}{\int_0^T X_t^2 dt}. \tag{3.3}$$

In this transient case, we show that this estimator converges to the Cauchy distribution with an error rate $O(e^{-\theta T})$. Note that in the transient case, with random norming, specifically if one normalizes the MLE by the square root of the observed Fisher information, then the MLE converges to the normal distribution, see Feigin (1976).

$$\theta_T - \theta := \frac{\int_0^T X_t dW_t}{\int_0^T X_t^2 dt} = \frac{Z_T}{I_T}, \tag{3.4}$$

where

$$Z_T := \int_0^T X_t dW_t \quad \text{and} \quad I_T := \int_0^T X_t^2 dt. \tag{3.5}$$

Hence,

$$\frac{e^{\theta T}}{2\theta}(\theta_T - \theta) = \frac{e^{-\theta T} 2\theta Z_T}{e^{-2\theta T} 4\theta^2 I_T} = \frac{\left(e^{-2\theta T} 4\theta^2\right)^{1/2} Z_T}{e^{-2\theta T} 4\theta^2 I_T}. \tag{3.6}$$

In (3.6), the numerator of the normalized MLE is a normalized martingale that converges to the standard normal variable, and the denominator is its corresponding increasing process that converges to a chi-square random variable as $T \to \infty$ that is independent of the numerator. Hence, the ratio converges to a Cauchy distribution with parameters $(0, 1)$.

Let us introduce two Wiener integrals:

$$\xi_t := \int_0^t e^{-\theta s} dW_s \text{ and } \eta_t := \int_0^t e^{\theta s} dW_s, \ t \geq 0. \tag{3.7}$$

With this notation

$$Z_T := \int_0^T X_t dW_t = \int_0^T e^{\theta t}\xi_t dW_t \quad \text{and} \quad I_T := \int_0^T X_t^2 dt = \int_0^T e^{2\theta t}\xi_t^2 dt,$$

$$\theta_T - \theta = \frac{\int_0^T e^{\theta t}\xi_t dW_t}{\int_0^T e^{2\theta t}\xi_t^2 dt}, \tag{3.8}$$

$$\begin{aligned}
& \frac{e^{\theta T}}{2\theta}(\theta_T - \theta) \\
= \; & \frac{(e^{-2\theta T}4\theta^2)^{1/2}\int_0^T e^{\theta t}\xi_t dW_t}{e^{-2\theta T}4\theta^2\int_0^T e^{2\theta t}\xi_t^2 dt} = \frac{(e^{-2\theta T}4\theta^2)^{1/2}\int_0^T e^{\theta t}(\int_0^t e^{-\theta s}dW_s)dW_t}{e^{-2\theta T}4\theta^2\int_0^T e^{2\theta t}(\int_0^t e^{-\theta s}dW_s)^2 dt} \\
= \; & \frac{\xi_T\xi}{2\theta e^{-2\theta T}\int_0^T e^{2\theta t}\xi_t^2 dt} \times \frac{e^{-\theta T}\int_0^T e^{\theta t}dW_t}{\xi} = \frac{\xi_T\xi}{2\theta e^{-2\theta T}I_T} \times \frac{e^{-\theta T}\eta_T}{\xi} \\
=: \; & A_T^\theta \times B_T^\theta.
\end{aligned} \tag{3.9}$$

We have

$$A_T^\theta \stackrel{a.s.}{\rightarrow} 1, \tag{3.10}$$

$$B_T^\theta \stackrel{\mathcal{D}}{\rightarrow} \frac{N}{\sqrt{2\theta}\xi}, \tag{3.11}$$

where $\sqrt{2\theta}\xi = N_1$, and N_1 and N are independent standard normal random variables. Since

$$\frac{N}{\sqrt{2\theta}\xi} =^{\mathcal{D}} \mathcal{C}(1), \tag{3.12}$$

where $\mathcal{C}(1)$ is the standard Cauchy distribution, by Slutsky's theorem, we have

$$A_T^\theta \times B_T^\theta \rightarrow \mathcal{C}(1). \tag{3.13}$$

Note that

$$\xi_T \stackrel{\mathcal{D}}{\rightarrow} \xi \text{ as } T \rightarrow \infty, \tag{3.14}$$

where $\xi = \int_0^\infty e^{-\theta s}d\widetilde{W}_s$ has $\mathcal{N}(0, \frac{1}{2\theta})$ distribution with some Wiener process $\{\widetilde{W}_s, s \geq 0\}$.

Using Borel–Cantelli lemma and stochastic Fubini theorem, it can be shown that $\xi_t \rightarrow \xi$ almost surely and in $L_2(\Omega)$ as $t \rightarrow \infty$.

By integration by parts, we have

$$\begin{aligned}
& e^{-2\theta T}I_T = e^{-2\theta T}\int_0^T X_s^2 ds = e^{-2\theta T}\int_0^T e^{2\theta s}\xi_s^2 ds \\
= \; & \frac{\xi^2}{2\theta} - \frac{e^{-2\theta T}}{\theta}\int_0^T e^{2\theta s}\xi_s d\xi_s - \frac{Te^{-2\theta T}}{\theta} \\
= \; & \frac{\xi^2}{2\theta} - \frac{e^{-2\theta T}}{\theta}\int_0^T e^{\theta s}\xi_s dW_s - \frac{Te^{-2\theta T}}{\theta}.
\end{aligned} \tag{3.15}$$

This equality together with

$$E\left(\int_0^T e^{\theta s}\xi_s dW_s\right)^2 = \int_0^T e^{2\theta s}E(\xi_s^2)ds$$

$$= \frac{1}{2\theta}\int_0^T e^{2\theta s}(1-e^{-2\theta s})ds = \frac{e^{2\theta T}-1-2\theta T}{4\theta^2} \tag{3.16}$$

by the CLT for stochastic integrals provides

$$e^{-2\theta T}\int_0^T X_s^2 ds \xrightarrow{\mathcal{D}} \frac{\xi^2}{2\theta}. \tag{3.17}$$

$$e^{-2\theta T} I_T \xrightarrow{\mathcal{D}} \frac{\xi^2}{2\theta}.$$

It can be shown that

$$e^{-2\theta T} I_T \xrightarrow{a.s.} \frac{\xi^2}{2\theta}. \tag{3.18}$$

By Itô formula, we have

$$Z_T = \int_0^T X_s dW_s = \int_0^T e^{\theta s}\xi_s dW_s = \int_0^T \xi_s d\eta_s = \xi_T\eta_T - T - \int_0^T \eta_s d\xi_s$$

$$= \xi_T\eta_T - T - \int_0^T \eta_s e^{-\theta s} dW_s. \tag{3.19}$$

Hence,

$$e^{-\theta T}Z_T = e^{-\theta T}\xi_T\eta_T - e^{-\theta T}T - e^{-\theta T}\int_0^T \eta_s e^{-\theta s}dW_s. \tag{3.20}$$

Direct calculation gives

$$e^{-\theta T}\eta_T \xrightarrow{\mathcal{D}} \eta \sim \mathcal{N}\left(0, \frac{1}{2\theta}\right) \tag{3.21}$$

and

$$E(\xi\eta) = \lim_{T\to\infty} E(\xi_T\eta_T e^{-\theta T}) = \lim_{T\to\infty} Te^{-\theta T} = 0. \tag{3.22}$$

Hence,

$$e^{-\theta T}Z_T \xrightarrow{\mathcal{D}} \mathcal{N}\left(0, \frac{1}{2\theta}\right). \tag{3.23}$$

Hence, the limit distribution of the pair $(\xi_T, e^{-\theta T}\eta_T)$ is a Gaussian distribution of two independent variables. Thus,

$$e^{\theta T}\frac{\xi_T}{\eta_T} \xrightarrow{\mathcal{D}} \zeta, \tag{3.24}$$

where ζ is the standard Cauchy variable with probability density function

$$f(x) = \frac{1}{\pi(1+x^2)}, \quad x \in \mathbb{R} \tag{3.25}$$

and cdf

$$\mathcal{C}(x) = \frac{1}{2} + \frac{1}{\pi}\arctan x, \ x \in \mathbb{R} \tag{3.26}$$

and characteristic function

$$\int_{-\infty}^{\infty} e^{i\lambda x}d\mathcal{C}(x) = e^{-|\lambda|}. \tag{3.27}$$

Hence,

$$\frac{e^{\theta T}}{2\theta}(\theta_T - \theta) \xrightarrow{\mathcal{D}} \zeta. \tag{3.28}$$

We estimate the rate of convergence in this phenomenon. We need the following lemma in the sequel.

Lemma 3.1 (Esseen's Smoothing Lemma)

Let F be a non-decreasing function and H be a differentiable function of bounded variation on the real line with $F(\pm\infty) = G(\pm\infty)$. Denote the corresponding Fourier–Stieltjes transforms by $\widehat{F}$ and $\widehat{G}$, respectively. Then for all $\Lambda > 0$,

$$\sup_{x\in\mathbb{R}}|F(x) - G(x)| \le \frac{1}{\pi}\int_{-\Lambda}^{\Lambda}\frac{\left|\widehat{F}(\lambda) - \widehat{G}(\lambda)\right|}{|\lambda|}d\lambda + \frac{24}{\pi\Lambda}\sup_{x\in\mathbb{R}}|G'(x)|. \tag{3.29}$$

Proof See Petrov (1995) or Feller (1971). □

10.4 Berry–Esseen Type Bounds

Let $\Phi(\cdot)$ denote the standard normal distribution function and $\mathcal{C}(\cdot)$ denote the standard Cauchy distribution function. Throughout this chapter, C denotes a generic constant (perhaps depending on θ, but not on anything else).

We start with the Dambis–Dubins–Schwarz (DDS) theorem, see, Dambis (1965) and Dubins and Schwaz (1965). Since Z_T is a continuous time martingale, due to time change (Skorokhod embedding), $Z_T = B_{I_T}$, where B is a Brownian motion independent of W, we have

$$E(\exp(iuZ_T)) = E\exp(-\frac{u^2}{2}I_T). \tag{4.1}$$

$$E(\exp(iue^{-\theta T}\sqrt{2\theta}Z_T)) = E\exp(-\frac{u^2}{2}e^{-2\theta T}2\theta I_T). \tag{4.2}$$

$$\theta_T - \theta = \frac{B_{I_T}}{I_T}, \tag{4.3}$$

$$\frac{e^{\theta T}}{2\theta}(\theta_T - \theta) = \frac{e^{-\theta T}B_{I_T}}{e^{-2\theta T}2\theta I_T} = \frac{\left(e^{-2\theta T}4\theta^2\right)^{1/2}B_{I_T}}{e^{-2\theta T}4\theta^2 I_T} =: Y_T, \tag{4.4}$$

where

$$Y_T = \frac{e^{-\theta T}B_{I_T}}{e^{-2\theta T}2\theta I_T}. \tag{4.5}$$

Our main claim in this chapter is to show that

$$|E(e^{iuY_T}) - e^{-|u|}| \leq C|u|e^{-|u|/2}e^{-\theta T}. \tag{4.6}$$

This is done through several lemmas. Once it is shown, let

$$F(x) = P(Y_T \leq x), \tag{4.7}$$

$$\mathcal{C}(x) = \frac{1}{2} + \frac{1}{\pi}\arctan x, \ x \in \mathbb{R}. \tag{4.8}$$

Take $\Lambda = e^{\theta T}$. Then

$$\sup_{x\in\mathbb{R}}|F(x) - \mathcal{C}(x)| \leq \frac{1}{\pi}J + \frac{24}{\pi e^{\theta T}}\sup \mathcal{C}'(x) \qquad (4.9),$$

where

$$J = \frac{1}{\pi}\int_{|\lambda|\leq e^{\theta T}} \frac{\left|\widehat{F}(x) - \widehat{\mathcal{C}}(x)\right|}{|\lambda|}d\lambda. \tag{4.10}$$

Clearly

$$\sup \mathcal{C}'(x) < \infty$$

and

$$J \leq \frac{C}{e^{\theta T}}\int_{|\lambda|\leq e^{\theta T}} e^{-|\lambda|/2}d\lambda \leq \frac{C}{e^{\theta T}}\int_{-\infty}^{\infty} e^{-|\lambda|/2}d\lambda = O(e^{-\theta T}), \tag{4.11}$$

which would ultimately give

$$\sup_{x\in\mathbb{R}}|F(x) - \mathcal{C}(x)| = O(e^{-\theta T}). \tag{4.12}$$

First we start with Kolmogorov distance for Wiener chaos and its relative:

Lemma 4.1 We have the following rate of convergence for the double stochastic integral or the second Wiener chaos $\int_0^T e^{\theta t}\left(\int_0^t e^{-\theta s} dW_s\right) dW_t$):

(a) $$\sup_{x\in\mathbb{R}}\left|P\left\{e^{\theta T}\left(2\theta e^{-2\theta T}\int_0^T e^{\theta t}\left(\int_0^t e^{-\theta s}dW_s\right)dW_t\right)\le x\right\}-\Phi(x)\right|\le Ce^{-\theta T}.$$

(b) $$\sup_{x\in\mathbb{R}}\left|P\left\{e^{\theta T}\left(2\theta e^{-2\theta T}\int_0^T e^{2\theta t}\xi_t^2 dt-\xi^2\right)\le x\right\}-\Phi(x)\right|\le Ce^{-\theta T}.$$

Proof Observe that

$$Z_T=\int_0^T X_t dW_t=\int_0^T e^{\theta t}\left(\int_0^t e^{-\theta s}dW_s\right)dW_t,$$

which is a second Wiener chaos

$$\int_0^T e^{\theta t}\left(\int_0^t e^{-\theta s}dW_s\right)dW_t.$$

One can use the Stein–Malliavin method and estimate the Kolmogorov distance for Z_T. However, part (a) follows as a consequence of Lemma 4.4(c) below along with Lemma 3.1 above. Part (b) follows as a consequence of Lemma 4.2 below along with Lemma 3.1 above. □

Note that $\xi^2 \sim \chi_1^2$. The next theorem gives an exponential estimate on the rate of convergence to the chi-square distribution for energy I_T of the O–U process.

Theorem 2.1

$$\sup_{x\in\mathbb{R}}\left|P\left\{e^{-2\theta T}2\theta I_T\le x\right\}-P\left\{\xi^2\le x\right\}\right|=O(e^{-\theta T}).$$

The above theorem is a consequence of the following lemma and the *Esseen's smoothing Lemma 3.1.*

Lemma 4.2 For $|u|\le e^{\theta T}\varepsilon$, ε sufficiently small, we have

$$\left|E\exp\left(iue^{-2\theta T}2\theta I_T\right)-\frac{1}{(1-2iu)^{\frac{1}{2}}}\right|\le C(|u|+|u|^3)e^{-\theta T}.$$

Proof From Liptser and Shiryayev (1977), we have

$$E\exp\left(iue^{-2\theta T}2\theta I_T\right)=\exp\left(\frac{\theta T}{2}\right)\left[\frac{2\gamma}{(\gamma-\theta)e^{-\gamma T}+(\gamma+\theta)e^{\gamma T}}\right]^{1/2}, \quad (4.13)$$

where

$$\gamma:=\left(\theta^2-2iue^{-2\theta T}2\theta\right)^{1/2}. \quad (4.14)$$

The lemma is an easy consequence of this result. □

Lemma 4.3 For every $\delta>0$,

$$P\left\{\left|e^{-2\theta T}2\theta I_T-\xi^2\right|\geq\delta\right\}\leq Ce^{-2\theta T}\delta^{-2}.$$

Proof It is clear that

$$X_T=\int_0^T e^{-\theta(T-s)}dW_s. \quad (4.15)$$

Further,

$$\int_0^T e^{\theta(T-s)}dW_s=W_T-\theta\int_0^T e^{\theta(T-s)}W_s ds,$$

$$\xi_t=\int_0^t e^{-\theta s}dW_s=e^{-\theta t}W_t-\theta\int_0^t e^{-\theta s}dW_s,$$

$$\eta_t=\int_0^t e^{\theta s}dW_s=e^{\theta t}W_t+\theta\int_0^t e^{\theta s}dW_s.$$

Note that

$$\begin{aligned} E(X_T^2) &= \frac{1-e^{-2\theta T}}{2\theta}, \quad E(X_T^4)=\frac{3(1-e^{-2\theta T})^2}{4\theta}, \\ \text{and } E(I_T) &= \frac{2\theta T-1+e^{-2\theta T}}{4\theta^2}. \end{aligned} \quad (4.16)$$

By Itô formula (see Friedman (1975)), we have

$$I_T=\frac{X_T^2}{2\theta}-\frac{T}{2\theta}-\frac{Z_T}{\theta}. \quad (4.17)$$

By Chebyshev inequality, we have

$$\begin{aligned} & P\left\{\left|e^{-2\theta T}2\theta I_T-\xi^2\right|\geq\delta\right\} \\ \leq\ & \frac{1}{\delta^2}E\left|e^{-2\theta T}2\theta I_T-\xi^2\right|^2 \\ =\ & \frac{1}{\delta^2}E\left|e^{-2\theta T}2\theta\int_0^T e^{2\theta t}\xi_t^2dt-\xi^2\right|^2 \end{aligned}$$

$$
\begin{aligned}
&= \frac{1}{\delta^2} E \left| e^{-2\theta T} 2\theta \int_0^T e^{2\theta t} \xi_t^2 dt - \xi_T^2 + \xi_T^2 - \xi^2 \right|^2 \\
&\leq \frac{2}{\delta^2} \left[E|e^{-2\theta T} 2\theta \int_0^T e^{2\theta t} \xi_t^2 dt - \xi_T^2|^2 + E|\xi_T^2 - \xi^2|^2 \right] \\
&\leq \frac{2}{\delta^2} \left[E|e^{-2\theta T} 2\theta \int_0^T e^{2\theta t} \xi_t^2 dt - \xi_T^2|^2 + E|\xi_T - \xi|^2 E|\xi_T + \xi|^2 \right] \\
&\leq \frac{2}{\delta^2} \left[E|e^{-2\theta T} 2\theta \int_0^T e^{2\theta t} \xi_t^2 dt - \xi_T^2|^2 + \frac{e^{-2\theta T}}{\sqrt{2\theta}} \right] \\
&\leq C e^{-2\theta T} \delta^{-2} \qquad (4.18)
\end{aligned}
$$

since $E|\xi_T + \xi|^2 \leq 2E|\xi_T|^2 + 2E|\xi|^2 < \infty$.

Since

$$
E(\xi_T - \xi)^2 = \int_T^\infty \int_T^\infty e^{-\theta r} e^{-\theta s} |r-s|^{-1} dr ds = \frac{e^{-2\theta T}}{\sqrt{2\theta}}, \qquad (4.19)
$$

$$
E(\xi_T - \xi)^2 = \frac{e^{-2\theta T}}{\sqrt{2\theta}} \qquad (4.20)
$$

gives the L_2 convergence rate. Recall that

$$
I_T = \int_0^T e^{2\theta t} \xi_t^2 dt, \qquad (4.21)
$$

$$
E(\xi_T - \xi)^2 \to 0, \qquad (4.22)
$$

$$
E(\xi_t - \xi_s)^2 \leq (t-s). \qquad (4.23)
$$

We have

$$
\begin{aligned}
E\left(2\theta e^{-2\theta T} I_T - \xi^2\right)^2 &= E\left(2\theta e^{-2\theta T} \int_0^T e^{2\theta t} \xi_t^2 dt - \xi^2\right)^2 \\
&= E\left(\frac{2\theta}{e^{2\theta T}} \int_0^T e^{2\theta t} \xi_t^2 dt - \xi^2\right)^2. \qquad (4.24)
\end{aligned}
$$

Further,

$$
\lim_{T\to\infty} 2\theta e^{-2\theta T} \int_0^T e^{2\theta t} \xi_t^2 dt = \lim_{T\to\infty} \xi_T^2 = \xi^2 \quad \text{almost surely.} \qquad (4.25)
$$

$E(\xi^2) < \infty$, which implies that $P(\xi = 0) = 0$.

$$
\lim_{T\to\infty} \left[2\theta e^{-2\theta T} \int_0^T e^{2\theta t} \xi_t^2 dt - \xi_T^2 \right] = 0 \quad \text{almost surely.} \qquad (4.26)
$$

Because of the continuity of ξ_t, for every $t \geq 0$,

$$\int_0^T e^{2\theta t}\xi_t^2 dt \geq \int_{\frac{T}{2}}^T e^{2\theta t}\xi_t^2 dt \geq \frac{T}{2}e^{\theta T}\left(\inf_{\frac{T}{2}<t<T} \xi_t^2\right) \text{ almost surely.} \tag{4.27}$$

Furthermore, the continuity of ξ_t gives

$$\lim_{T\to\infty}\left(\inf_{\frac{T}{2}<t<T} \xi_t^2\right) = \xi^2 \text{ almost surely} \tag{4.28}$$

$$\lim_{T\to\infty}\int_0^T e^{2\theta t}\xi_t^2 dt = \infty \text{ almost surely.} \tag{4.29}$$

By L'Hopital rule,

$$\lim_{T\to\infty}\frac{\int_0^T e^{2\theta t}\xi_t^2 dt}{e^{2\theta T}} = \lim_{T\to\infty}\frac{\xi_T^2}{2\theta} = \frac{\xi^2}{2\theta} \text{ almost surely.} \tag{4.30}$$

$$\theta_T - \theta = \frac{\int_0^T e^{\theta t}\xi_t dW_t}{\int_0^T e^{2\theta t}\xi_t^2 dt} = \frac{\xi_T^2}{2e^{-2\theta T}\int_0^T e^{2\theta t}\xi_t^2 dt} - \theta. \tag{4.31}$$

$$\theta_T - \theta \to 0 \text{ almost surely.} \tag{4.32}$$

$$\hat{\theta}_T - \theta = \frac{\xi_T^2 - 2\theta e^{-2\theta T}\int_0^T e^{2\theta t}\xi_t^2 dt}{2e^{-2\theta T}\int_0^T e^{2\theta t}\xi_t^2 dt}. \tag{4.33}$$

It is easy to verify that

$$E\left[\xi_T^2 - 2\theta e^{-2\theta T}\int_0^T e^{2\theta t}\xi_t^2 dt\right]^2 \leq Ce^{-2\theta T}. \tag{4.34}$$

This completes the proof of the lemma. □

The following lemma (*Cameron–Martin Type Theorem*) gives the bound on the joint characteristic functions of the sufficient statistics defining the MLE:

Lemma 4.4
(a) Let $\phi_T(z_1, z_2) := E\exp(z_1 I_T + z_2 X_T^2)$, $z_1, z_2 \in \mathbb{C}$. Then $\phi_T(z_1, z_2)$ exists for $|z_i| \leq \delta$, 1 = 1,2 for some $\delta > 0$ and is given by

$$\phi_T(z_1, z_2) = \exp\left(\frac{\theta T}{2}\right)\left[\frac{2\gamma}{(\gamma - \theta + 2z_2)e^{-\gamma T} + (\gamma + \theta - 2z_2)e^{\gamma T}}\right]^{1/2},$$

where $\gamma = (\theta^2 - 2z_1)^{1/2}$, and we choose the principal branch of the square root.

(b) Let

$$H_{T,x} := \left(e^{-2\theta T}4\theta^2\right)^{1/2} Z_T - \left(e^{-2\theta T}4\theta^2 I_T - \xi^2\right)x.$$

Then for $|x| \leq 2(\log e^{2\theta T})^{1/2}$ and for $|u| \leq \epsilon e^{\theta T}$, where ϵ is sufficiently small,

$$\left| E \exp(iuH_{T,x}) - \exp(-\frac{u^2}{2}) \right| \leq C \exp(\frac{-|u|}{2})(|u| + |u|^3)e^{-\theta T}.$$

(c) For $|u| \leq \epsilon_1 e^{\theta T}$, where ϵ_1 is sufficiently small, we have as $T \to \infty$,

$$\left| E \exp\left\{iu\left(e^{-\theta T} 2\theta\right) Z_T\right\} - \exp(-\frac{u^2}{2}) \right| \leq C \exp(-\frac{|u|}{2})(|u| + |u|^3)e^{-\theta T}.$$

Part (a) is from Bishwal (2000a).

We shall prove part (b) in detail. Proof of part (c) is very similar to part (b) and will be omitted.

Proof By Itô formula,

$$Z_T = \theta I_T + \frac{X_T^2}{2} - \frac{T}{2}.$$

Note that

$$\begin{aligned}
& E \exp(itH_{T,x}) \\
= \;& E \exp\left[-it\left(e^{-2\theta T} 4\theta^2\right)^{1/2} Z_T - it\left(\left(e^{-2\theta T} 4\theta^2\right) I_T - \xi^2\right) x\right] \\
= \;& E \exp\left[-it\left(e^{-2\theta T} 4\theta^2\right)^{1/2} \left\{\theta I_T + \frac{X_T^2}{2} - \frac{T}{2}\right\} - it\left(\left(e^{-2\theta T} 4\theta^2\right) I_T - \xi^2\right) x\right] \\
= \;& E \exp(z_1 I_T + z_2 X_T^2 + z_3) \\
= \;& \exp(z_3)\phi_T(z_1, z_2),
\end{aligned}$$

where $z_1 = -it\theta\delta_{T,x}$, $z_2 = -\frac{it}{2}\left(e^{-2\theta T} 4\theta^2\right)^{1/2}$, $z_3 = \frac{itT}{2}\delta_{T,x}$, $\delta_{T,x} = \left(e^{-2\theta T} 4\theta^2\right)^{1/2} + \frac{2x}{T}$. Note that (z_1, z_2) satisfies the conditions of (a) by choosing ϵ sufficiently small. Let $\alpha_{1,T}(t), \alpha_{2,T}(t), \alpha_{3,T}(t)$, and $\alpha_{4,T}(t)$ be functions that are of the orders $O(|t|e^{-\theta T/2}), O(|t|^2 e^{-\theta T/2})$, $O(|t|^3 e^{-3\theta T/2})$, and $O(|t|^3 e^{-\theta T/2})$, respectively. Note that for the given range of values of x and t, the conditions on z_i for part (a) of Lemma are satisfied. Note also that $z_2 = \alpha_{1,T}(t)$. Further, with

$$\beta_T(t) = 1 + it\frac{\delta_{T,x}}{\theta} + \frac{t^2\delta_{T,x}^2}{2\theta^2},$$

$$\begin{aligned}
\gamma &= (\theta^2 - 2z_1)^{1/2} \\
&= \theta\left[1 - \frac{z_1}{\theta^2} - \frac{z_1^2}{2\theta^4} + \frac{z_1^3}{2\theta^8} + \cdots\right]
\end{aligned}$$

$$\begin{aligned}
&= \theta\left[1 + it\frac{\delta_{T,x}}{\theta} + \frac{t^2\delta_{T,x}^2}{2\theta^2} + \frac{it^3\delta_{T,x}^3}{2\theta^3} + \cdots\right] \\
&= \theta[1 + \alpha_{1,T}(t) + \alpha_{2,T}(t) + \alpha_{3,T}(t)] \\
&= \theta\beta_T(t) + \alpha_{3,T}(t) \\
&= \theta[1 + \alpha_{1,T}(t)].
\end{aligned}$$

Thus

$$\gamma - \theta = \alpha_{1,T}, \ \gamma + \theta = 2\theta + \alpha_{1,T}.$$

Hence, the above expectation equals

$$\begin{aligned}
&\exp\left(z_3 + \frac{\theta T}{2}\right) \\
&\times \left[\frac{2\theta\beta_T(t) + \alpha_{3,T}(t)}{\alpha_{1,T}\exp\{-\theta T\beta_T(t) + \alpha_{4,T}(t)\} + (2\theta + \alpha_{1,T}(t))\exp\{\theta T\beta_T(t) + \alpha_{4,T}(t)\}}\right]^{1/2} \\
= &\left[\frac{1 + \alpha_{1,T}(t)}{\alpha_{1,T}\exp(\chi_T(t)) + (1 + \alpha_{1,T}(t))\exp(\psi_T(t))}\right]^{1/2},
\end{aligned}$$

where

$$\begin{aligned}
\chi_T(t) &:= -\theta T\beta_T(t) + \alpha_{4,T}(t) - 2z_3 - \theta T \\
&= -2\theta T + \alpha_{1,T}(t) + t^2\alpha_{1,T}(t)
\end{aligned}$$

and

$$\begin{aligned}
\psi_T(t) &:= \theta T\beta_T(t) + \alpha_{4,T}(t) - 2z_3 - \theta e^{\theta T} \\
&= \theta T\left[1 + it\frac{\delta_{T,x}}{\theta} + \frac{t^2\delta_{T,x}^2}{2\theta^2}\right] + \alpha_{4,T}(t) - ite^{\theta T}\delta_{T,x} - \theta e^{\theta T} \\
&= \frac{t^2 e^{\theta T}}{2\theta}\left[\left(\frac{4\theta^2}{e^{2\theta T}}\right)^{1/2} + \frac{2x}{e^{\theta T}}\right]^2 \\
&= t^2 + t^2\alpha_{1,T}(t).
\end{aligned}$$

Hence, for the given range of values of t,

$$\chi_T(t) - \psi_T(t) \leq -\theta e^{\theta T}.$$

Hence, the above expectation equals

$$\begin{aligned}
&\exp(-\frac{t^2}{2})(1 + \alpha_{1,T})^{1/2} \\
&\times \left[\alpha_{1,T}\exp\{-2\theta e^{\theta T} + \alpha_{1,T} + t^2\alpha_{1,T}\} + (1 + \alpha_{1,T}(t))\exp\{t^2\alpha_{1,T}(t)\}\right]^{-1/2} \\
= &\exp(-\frac{t^2}{2})\left[1 + \alpha_{1,T})(1 + \alpha_{1,T}(1 + \alpha_{1,T})\exp\{-\theta e^{\theta T} + \alpha_{1,T} + t^2\alpha_{1,T}\}\right] \\
&\times \exp(t^2\alpha_{1,T}(t).
\end{aligned}$$

Lemmas 4.4 (c) and 4.2, respectively, give the Berry–Esseen rate for Z_T and I_T immediately by using the Esseen's lemma 3.1.

Corollary 4.1

$$\text{(a)} \quad \sup_{x\in\mathbb{R}} \left| P\left\{ \left(\frac{4\theta^2}{e^{2\theta T}}\right)^{1/2} Z_T \le x \right\} - \Phi(x) \right| \le Ce^{-\theta T}.$$

$$\text{(b)} \quad \sup_{x\in\mathbb{R}} \left| P\left\{ \left(\frac{4\theta^2}{e^{2\theta T}}\right)^{1/2} \left(\theta I_T - \xi^2 \frac{e^{\theta T}}{2}\right) \le x \right\} - \Phi(x) \right| \le Ce^{-\theta T}.$$

Remark Though this was basically shown in Lemma 2.1, here we obtain Kolmogorov distance for a martingale and Kolmogorov distance for its quadratic variation through Cameron–Matin type results that are generalization of Levy area formula. In Lemma 2.1, one could go directly to the Stein–Malliavin way through Wiener chaos expansion that does not depend on any martingale characteristics.

Before we prove the results on the Berry–Esseen bound on the Kolmogorov distance for the MLE with random norming, we need the following large deviations of the MLE.

Lemma 4.5

$$P\left\{ \left(\frac{e^{2\theta T}}{4\theta^2}\right)^{1/2} |\theta_T - \theta| \ge 2(2\theta T)^{1/2} \right\} \le Ce^{-\theta T}.$$

Proof Observe that

$$\begin{aligned}
& P\left\{ \left(\frac{e^{2\theta T}}{4\theta^2}\right)^{1/2} |\theta_T - \theta| \ge 2(2\theta T)^{1/2} \right\} \\
= \; & P\left\{ \left| \frac{\left(\frac{4\theta^2}{e^{2\theta T}}\right)^{1/2} Z_T}{\left(\frac{2\theta}{e^{2\theta T}}\right) I_T} \right| \ge 2(2\theta T)^{1/2} \right\} \\
\le \; & P\left\{ \left| \left(\frac{4\theta^2}{e^{2\theta T}}\right)^{1/2} Z_T \right| \ge (\theta T)^{1/2} \right\} + P\left\{ \left| \frac{2\theta}{e^{2\theta T}} I_T \right| \le \frac{1}{2} \right\} \\
\le \; & \left| P\left\{ \left(\frac{4\theta^2}{e^{2\theta T}}\right)^{1/2} |Z_T| \ge (\theta T)^{1/2} \right\} - 2\Phi(-(2\theta T)^{1/2}) \right| \\
& + 2\Phi(-(2\theta T)^{1/2}) + P\left\{ \left| \frac{2\theta}{e^{2\theta T}} I_T - \xi^2 \right| \ge \frac{1}{2} \right\}
\end{aligned}$$

$$
\begin{aligned}
&\leq \sup_{x\in\mathbb{R}}\left|P\left\{\left(\frac{4\theta^2}{e^{2\theta T}}\right)^{1/2}|Z_T|\geq x\right\}-2\Phi(-x)\right| \\
&\leq \sup_{x\in\mathbb{R}}\left|P\left\{\left(\frac{4\theta^2}{e^{2\theta T}}\right)^{1/2}|Z_T|\geq x\right\}-2\Phi(-x)\right| \\
&\quad +2\Phi(-(2\theta T)^{1/2})+P\left\{\left|\left(\frac{4\theta^2}{e^{2\theta T}}\right)I_T-\xi^2\right|\geq\frac{1}{2}\right\} \\
&\leq Ce^{-\theta T}+C(e^{2\theta T}2\theta T)^{-1/2}+C(e^{2\theta T})^{-1} \\
&\leq Ce^{-\theta T}.
\end{aligned}
$$

The bounds for the first and third terms come from Corollary 4.1(a) and Lemma 4.3, respectively, and that for the middle term comes from (Feller, 1957, p. 166).

□

We are now in a position to obtain the Berry–Esseen bound of the order $O(e^{-\theta T})$ on the Kolmogorov distance for the MLE.

Theorem 4.2

$$\sup_{x\in\mathbb{R}}\left|P\left\{\left(\frac{e^{\theta T}}{2\theta}\right)(\theta_T-\theta)\leq x\right\}-\mathcal{C}(x)\right|=O(e^{-\theta T}).$$

Proof We shall consider two possibilities: (i) $|x|>2(\theta T)^{1/2}$ and (ii) $|x|\leq 2(\theta T)^{1/2}$.

(i) We shall give a proof for the case $x>2(\theta T)^{1/2}$. The proof for the case $x<-2(\theta T)^{1/2}$ runs similarly. Note that

$$\left|P\left\{\left(\frac{e^{\theta T}}{2\theta}\right)(\theta_T-\theta)\leq x\right\}-\mathcal{C}(x)\right|\leq P\left\{\left(\frac{e^{\theta T}}{2\theta}\right)(\theta_T-\theta)\geq x\right\}+\mathcal{C}(-x).$$

But $\mathcal{C}(-x)\leq\mathcal{C}(-2(\theta T)^{1/2})\leq Ce^{-2\theta T}$. Moreover by Lemma 4.5, we have

$$P\left\{\left(\frac{e^{\theta T}}{2\theta}\right)(\theta_T-\theta)\geq 2(\theta T)^{1/2}\right\}\leq Ce^{-\theta T/2}.$$

Hence,

$$\left|P\left\{\left(\frac{e^{\theta T}}{2\theta}\right)^{1/2}(\theta_T-\theta)\leq x\right\}-\mathcal{C}(x)\right|\leq Ce^{-\theta T/2}.$$

(ii)

Let $A_T:=\left\{\left(\frac{e^{\theta T}}{2\theta}\right)|\theta_T-\theta|\leq 2(\theta T)^{1/2}\right\}$ and $B_T:=\left\{\frac{I_T}{e^{\theta T}}>c_0\right\}$,

where $0<c_0<\frac{1}{2\theta}$. By Lemma 4.5, we have

$$P(A_T^c)\leq Ce^{-\theta T}. \tag{4.35}$$

By Lemma 2.3, we have

$$P(B_T^c) = P\left\{\frac{2\theta}{e^{\theta T}} I_T - \xi^2 < 2\theta c_0 - \xi^2\right\}$$

$$< P\left\{\left|\frac{2\theta}{e^{\theta T}} I_T - \xi^2\right| > \xi^2 - 2\theta c_0\right\} \leq Ce^{-\theta T}. \tag{4.36}$$

Let b_0 be some positive number. For $\omega \in A_T \cap B_T$ and for all $T > T_0$ with $4b_0(2\theta T_0)^{1/2}(\frac{2\theta}{e^{\theta T_0}})^{1/2} \leq c_0$, we have

$$\begin{aligned}
& \left(\frac{e^{\theta T}}{2\theta}\right)(\theta_T - \theta) \leq x \\
\Rightarrow \quad & I_T + b_0 e^{\theta T}(\theta_T - \theta) < I_T + \left(\frac{e^{\theta T}}{2\theta}\right) 2b_0\theta x \\
\Rightarrow \quad & \left(\frac{e^{\theta T}}{2\theta}\right)(\theta_T - \theta)[I_T + b_0 e^{\theta T}(\theta_T - \theta)] < x[I_T + \left(\frac{e^{\theta T}}{2\theta}\right) 2b_0\theta x] \\
\Rightarrow \quad & (\theta_T - \theta) I_T + b_0 T(\theta_T - \theta)^2 < \left(\frac{2\theta}{e^{\theta T}}\right) I_T x + 2b_0\theta x^2 \\
\Rightarrow \quad & Z_T + (\theta_T - \theta) I_T + b_0 e^{\theta T}(\theta_T - \theta)^2 < Z_T + \left(\frac{2\theta}{e^{\theta T}}\right) I_T x + 2b_0\theta x^2 \\
\Rightarrow \quad & 0 < Z_T + \left(\frac{2\theta}{e^{\theta T}}\right) I_T x + 2b_0\theta x^2
\end{aligned}$$

since

$$\begin{aligned}
& I_T + b_0 e^{\theta T}(\theta_T - \theta) \\
> \ & e^{\theta T} c_0 + b_0 e^{\theta T}(\theta_T - \theta) \\
> \ & 4b_0(\theta T)^{1/2}\left(\frac{2\theta}{e^{\theta T}}\right) - 2b_0(\theta T)^{1/2}(\frac{2\theta}{e^{\theta T}}) \\
= \ & 2b_0(\theta T)^{1/2}\left(\frac{2\theta}{e^{\theta T}}\right) > 0.
\end{aligned}$$

Hence, for $\omega \in A_T \cap B_T$,

$$\left(\frac{e^{\theta T}}{2\theta}\right)(\theta_T - \theta) \leq x \Rightarrow Z_T + \left(\frac{2\theta}{e^{\theta T}}\right) I_T x + 2b_0\theta x^2 > 0.$$

On the other hand, for $\omega \in A_T \cap B_T$ and for all $T > T_0$ with $4b_0(2\theta T_0)^{1/2}$ $(\frac{2\theta}{e^{\theta T_0}}) \leq c_0$, we have

$$\left(\frac{e^{\theta T}}{2\theta}\right)(\theta_T - \theta) > x$$

$$\Rightarrow \quad I_T - b_0 e^{\theta T}(\theta_T - \theta) < I_T - \left(\frac{e^{\theta T}}{2\theta}\right) 2b_0\theta x$$

$$\Rightarrow \quad \left(\frac{e^{\theta T}}{2\theta}\right)(\theta_T - \theta)[I_T - b_0 e^{\theta T}(\theta_T - \theta)] > x[I_T - \left(\frac{e^{\theta T}}{2\theta}\right) 2b_0\theta x]$$

$$\Rightarrow \quad (\theta_T - \theta)I_T - b_0 e^{\theta T}(\theta_T - \theta)^2 > \left(\frac{2\theta}{e^{\theta T}}\right) I_T x - 2b_0\theta x^2$$

$$\Rightarrow \quad Z_T + (\theta_T - \theta)I_T - b_0 e^{\theta T}(\theta_T - \theta)^2 > Z_T + \left(\frac{2\theta}{e^{\theta T}}\right) I_T x - 2b_0\theta x^2$$

$$\Rightarrow \quad 0 > Z_T + \left(\frac{2\theta}{e^{\theta T}}\right) I_T x - 2b_0\theta x^2$$

since

$$\begin{aligned} & I_T - b_0 e^{\theta T}(\theta_T - \theta) \\ > \; & e^{\theta T} c_0 - b_0 e^{\theta T}(\theta_T - \theta) \\ > \; & 4b_0(\theta T)^{1/2}\left(\frac{2\theta}{e^{\theta T}}\right) - 2b_0(\theta T)^{1/2}\left(\frac{2\theta}{e^{\theta T}}\right) \\ = \; & 2b_0(\theta T)^{1/2}\left(\frac{2\theta}{e^{\theta T}}\right) > 0. \end{aligned}$$

Hence, for $\omega \in A_T \cap B_T$,

$$0 < Z_T + \left(\frac{2\theta}{e^{\theta T}}\right) I_T x - 2b_0\theta x^2 \Rightarrow \left(\frac{e^{\theta T}}{2\theta}\right)(\theta_T - \theta) \le x.$$

We use the squeezing method developed in Pfanzagl (1971) for the i.i.d. case instead of the splitting method of Michel and Pfanzagl (1971).

Let us introduce the piecewise quadratic random functions involving the martingale and quadratic variation part of $\theta_T - \theta$:

$$g^{\pm}(x) := Z_T + (\frac{2\theta}{e^{\theta T}})I_T x \pm 2b_0\theta x^2.$$

Let us introduce the events

$$D^{\pm}_{T,x} := \left\{ Z_T + (\frac{2\theta}{e^{\theta T}})I_T x \pm 2b_0\theta x^2 > 0 \right\}.$$

Thus we have

$$D^{-}_{T,x} \cap A_T \cap B_T \subseteq A_T \cap B_T \cap \left\{ \left(\frac{e^{\theta T}}{2\theta}\right)(\theta_T - \theta) \le x \right\} \subseteq D^{+}_{T,x} \cap A_T \cap B_T. \quad (4.37)$$

This gives

$$P(D^{-}_{T,x} \cap A_T \cap B_T) \le P(A_T \cap B_T \cap \{\left(\frac{e^{\theta T}}{2\theta}\right)(\theta_T - \theta) \le x\}) \le P(D^{+}_{T,x} \cap A_T \cap B_T)$$

so that

$$\left|P\left(A_T \cap B_T \cap \left\{\left(\frac{e^{\theta T}}{2\theta}\right)(\theta_T - \theta) \le x\right\}\right) - \mathcal{C}(x)\right|$$
$$\le \max\left\{|P(D^-_{T,x} \cap A_T \cap B_T) - \mathcal{C}(x)|,\ |P(D^+_{T,x} \cap A_T \cap B_T) - \mathcal{C}(x)|\right\}$$
$$\le \max\left\{|P(D^-_{T,x}) - \mathcal{C}(x)|,\ |P(D^+_{T,x}) - \mathcal{C}(x)|\right\} + P(A_T \cap B_T)^c.$$

From (4.35) and (4.36),

$$P(A_T \cap B_T)^c \le Ce^{-\theta T}$$

for all $T > T_0$ and $|x| \le 2(\theta T)^{1/2}$. If it is shown that

$$\left|P\left\{D^\pm_{T,x}\right\} - \mathcal{C}(x)\right| \le Ce^{-\theta T} \tag{4.38}$$

for all $T > T_0$ and $|x| \le 2(\theta T)^{1/2}$, then the theorem would follow from (4.35)–(4.38).

We shall prove (4.38) for $D^+_{T,x}$. The proof for $D^-_{T,x}$ is analogous. Note that

$$\begin{aligned}
&\left|P\left\{D^+_{T,x}\right\} - \mathcal{C}(x)\right| \\
= \quad &\left|P\left\{-(\frac{2\theta}{e^{\theta T}})Z_T - \left(\frac{2\theta}{e^{\theta T}}I_T - \xi^2\right)x < x + 2\left(\frac{2\theta}{e^{\theta T}}\right)b_0\theta x^2\right\} - \mathcal{C}(x)\right| \\
\le \quad &\sup_{y\in\mathbb{R}}\left|P\left\{-\left(\frac{2\theta}{e^{\theta T}}\right)Z_T - \left(\frac{2\theta}{e^{\theta T}}I_T - \xi^2\right)x \le y\right\} - \mathcal{C}(y)\right| \\
&+ \left|\mathcal{C}\left(x + \left(\frac{2\theta}{e^{\theta T}}\right)b_0\theta x^2\right) - \mathcal{C}(x)\right| \\
=: \quad &\Delta_1 + \Delta_2.
\end{aligned} \tag{4.39}$$

Lemma 2.4(b) and Esseen's Smoothing Lemma 1.1 immediately yield

$$\Delta_1 \le Ce^{-\theta T}. \tag{4.40}$$

On the other hand, for all $T > T_0$,

$$\Delta_2 \le 2(\frac{2\theta}{e^{\theta T}})b_0\theta x^2(2\pi)^{-1/2}\exp(-\overline{x}^2/2),$$

where

$$|\overline{x} - x| \le 2\left(\frac{2\theta}{e^{\theta T}}\right)b_0\theta x^2.$$

Since $|x| \le 2(\theta T)^{1/2}$, it follows that $|\overline{x}| > |x|/2$ for all $T > T_0$, and consequently,

$$\begin{aligned}
\Delta_2 \quad &\le \quad 2\left(\frac{2\theta}{e^{\theta T}}\right)b_0\theta x^2(2\pi)^{-1/2}x^2\exp(-x^2/8) \\
&\le \quad Ce^{-\theta T}.
\end{aligned} \tag{4.41}$$

From (4.39)–(4.41), we obtain

$$\left|P\left\{D_{T,x}^{+}\right\} - \mathcal{C}(x)\right| \leq Ce^{-\theta T}.$$

This completes the proof of the theorem. □

Concluding Remarks

(1) The bound in Theorem 4.2 is uniform over compact subsets of the parameter space Θ.

(2) The bound in Theorem 4.2 is optimal and cannot be improved further.

(3) Note that in the critical case, i.e., when $\theta = 0$, the MlE has a distribution concentrated on a half line, precisely the distribution of the ratio of a noncentral chi-square to the to the sum of chi-squares. Note that the behavior of the O–U process depends on both the initial condition $X_0 = X^0$ and the parameter space. Classically, it has been assumed that X^0 either has a normal distribution or has a nonzero constant and $\theta < 0$, which makes the process stationary with Gaussian invariant distribution. If $X^0 = 0$ is with $\theta < 0$, then the process is asymptotically stationary and ergodic. In above two cases, the model satisfies the LAN (local asymptotic normality) property. With X^0 a nonzero constant and $\theta > 0$, the process is transient and satisfies the LAMN (local asymptotic mixed normality) property. With $\theta = 0$, the process is nonstationary and satisfies the LABF (local asymptotic Brownian functional) property. For all $\theta \in \mathbb{R}$, the model satisfies the LABF property, see Bishwal (2018b) for the definitions of these LAN, LAMN, and LABF properties. Bishwal (2018b) has shown that MLE from sequential sampling based on a stopping rule unifies the three properties and makes them LAN.

(4) It remains to study the Kolmogorov distance for Bayes estimator from both continuous and discrete observations and approximate maximum likelihood estimator from discrete observations in the nonergodic case.

(5) Extension to multidimensional process and to multiparameter case remains to be investigated.

(6) It remains to investigate the nonuniform rates of convergence to Cauchy distribution that are more useful.

Chapter 11

Berry–Esseen Asymptotics for Pearson Diffusions

11.1 Introduction

Forman and Sørensen (2008) studied a statistically tractable class of diffusions called Pearson diffusions whose finite dimensional distributions are Pearson class of distributions.

11.2 Pearson Diffusions

Using Malliavin calculus along with Stein's equation, the chapter shows that the distribution of the maximum likelihood estimator of the drift parameter in the Pearson diffusion process observed over $[0, T]$ converges to the standard normal distribution with an uniform error rate of the order $O(T^{-1/2})$. Then based on discrete observations, it obtains martingale estimation function estimators and studies their rate of weak convergence to normal distribution.

Pearson diffusion is a stationary solution to the stochastic differential equation of the form

$$dX_t = -\theta(X_t - \mu)dt + \sqrt{2\theta(aX_t^2 + bX_t + c)}dW_t,$$

where $\theta > 0$ and a, b, and c are such that the square root is well defined when X_t is in the state space.

This is a special case of the following equation

$$dX_t = -\theta(X_t - \mu)dt + \sigma(X_t)dW_t,$$

where $\sigma^2(X_t) = 2\theta(aX_t^2 + bX_t + c)$.

This is a diffusion process with given invariant measure ν that admits a density p with respect to Lebesgue measure. Assume that the density p is continuous, bounded, has finite variance and p is strictly positive in the interval $(l, u)(-\infty \leq l \leq u \leq \infty)$ is zero outside (l, u). Let μ be the expectation of the measure ν and

$$\sigma^2(x) = \frac{2\int_l^x (\mu - y)p(y)dy}{p(x)} = \frac{2\mu F(x) - 2\int_l^x yp(y)dy}{p(x)}, \ x \in (l, u),$$

J. P. N. Bishwal, *Parameter Estimation in Stochastic Volatility Models*,
https://doi.org/10.1007/978-3-031-03861-7_11

where $F(x) = \int_{-\infty}^{x} p(y)dy$ is the distribution function associated with the density p.

The SDE with the above diffusion coefficient has a Markovian weak solution with $E(\sigma^2(X)) < \infty$ when $X \sim \nu$. The solution is ergodic with invariant density p.

Student Diffusion

The student diffusion is given by

$$dX_t = -\theta X_t dt + \sqrt{\frac{2\theta}{\nu - 1}(\nu + X_t^2)}dW_t.$$

The invariant density is

$$f(x) = \frac{\Gamma((\nu+1)/2)}{\sqrt{\nu\pi}\Gamma(\nu/2)}\left(1 + \frac{1}{\nu}x^2\right)^{-(\nu+1)/2}, \ x \in \mathbb{R}, \ \nu > 0.$$

$$= \frac{1}{\sqrt{\nu}B(\nu/2, 1/2))[1 + (x/\nu)^2]^{(\nu+1)/2]}}.$$

Skew Student Diffusion

$$dX_t = -\theta X_t dt + \sqrt{\frac{2\theta}{\nu - 1}[X_t^2 + 2\gamma\nu^{1/2}X_t + (1+\gamma^2)\nu]}dW_t$$

γ is the skewness parameter. $\gamma = 0$ gives the symmetric Student diffusion.

Gamma Diffusion

$$dX_t = -\theta(X_t - \mu)dt + \sqrt{2\theta X_t}dW_t$$

Snedecor Diffusion

$$dX_t = -\theta(X_t - \mu)dt + \sqrt{2\theta a X_t(X_t + 1)}dW_t$$

Jacobi Diffusion

$$dX_t = -\theta(X_t - \mu)dt + \sqrt{2\theta a X_t(X_t - 1)}dW_t$$

Uniform Diffusion

$$dX_t = -\theta(X_t - \frac{1}{2})dt + \sqrt{\theta X_t(1 - X_t)}dW_t$$

Logistic Diffusion

$$dX_t = -\theta[1 - 2\mu + (1-\mu)e^{X_t} - \mu e^{-1} - 8a\cosh^4(X_t/2)]dt + 2\sqrt{-a\theta}\cosh(X_t/2)dW_t$$

11.3 Fractional Student Ornstein–Uhlenbeck Process

The fractional student diffusion is given by

$$dX_t = -\theta X_t dt + \sqrt{\frac{2\theta}{\nu - 1}(\nu + X_t^2)}dW_t^H,$$

where W_t^H is a fractional Brownian motion. We study estimation of the drift coefficient in the FOU process whose marginal distribution is Student t-distribution.

Consider the fractional Ornstein–Uhlenbeck process satisfying the SDE

$$dX_t = -\alpha X_t dt + dL_t^H,$$

where L_t^H is a fractional Levy process whose Levy driver (in the integral representation) $(L_t)_{t\geq 0}$ is a Levy process with cumulant function

$$\ln E\left(e^{iuL_1}\right) = iu\mu - \lambda \frac{K_{\nu/2-1}(\lambda|u|)}{K_{\nu/2}(\lambda|u|)}, \qquad u \in \mathbb{R}\backslash\{0\},$$

where K is the modified Bessel function of the third kind defined as

$$K_\nu(x) = \frac{1}{2}\int_0^\infty y^{\nu-1}e^{-\frac{x}{2}(y+y^{-1})}dy \equiv \frac{1}{2}\int_{-\infty}^\infty e^{\nu u - x\cosh x}du,$$

where $\nu > 0$ is arbitrary. Let $\mu \in \mathbb{R}$ and $\lambda > 0$ be arbitrary. The weakly stationary solution satisfies

$$X_t = e^{-\alpha t}X_0 + e^{-\alpha t}\int_0^t e^{\alpha s}dL_s$$

and X_t has marginal t-distribution $T(\nu, \lambda, \mu)$ with pdf

$$f(x) = \frac{1}{\lambda B(\nu/2, 1/2)[1 + ((x-\mu)/\lambda)^2]^{(\nu+1)/2}},$$

$B(\cdot,\cdot)$ being the beta function. If $\nu > 1$, then $E(X_t) = \mu$, and if $\nu > 2$, then $corr(X_{t+\tau}, X_t) = e^{-\alpha|\tau|}$. The FSOU process (X_t) has power function decay of the distribution tails. For a strictly increasing continuous mapping $h : \mathbb{R} \to \mathbb{R}$ consider the process $Y_t = h(X_t)$. It is obvious that $\{Y_t\}_{t\geq 0}$ is a stationary Markov process. Further, if h is twice continuously differentiable, then $\{Y_t\}_{t\geq 0}$ is a diffusion process.

Fractional Symmetric Student Fractional Diffusion

The student diffusion is given by

$$dX_t = -\alpha X_t dt + \sqrt{\frac{2\alpha}{\nu - 1}(\nu + X_t^2)}\ dW_t^H.$$

The invariant density is given by

$$f(x) = \frac{\Gamma((\nu+1)/2)}{\sqrt{\nu\pi}\Gamma(\nu/2)}\left(1 + \frac{1}{\nu}x^2\right)^{-(\nu+1)/2}, \ x \in \mathbb{R},\ \nu > 0.$$

$$= \frac{1}{\sqrt{\nu}B(\nu/2, 1/2))[1 + (x/\nu)^2]^{(\nu+1)/2]}},$$

where $B(\cdot,\cdot)$ is the beta function.

Fractional Skewed Student Fractional Diffusion

$$dX_t = -\alpha X_t dt + \sqrt{\frac{2\alpha}{\nu - 1}[X_t^2 + 2\gamma\nu^{1/2}X_t + (1+\gamma^2)\nu]}\ dW_t^H$$

Here γ is the skewness parameter. $\gamma = 0$ gives the symmetric Student diffusion.

11.4 Estimators

Spearman's Estimator

The Spearman's rank correlation coefficient ρ_S for a pair of random variables U and V is

$$\rho_S(U, V) = \rho(F_U(U), F_V(V),$$

where F_U and F_V are marginal distributions of U and V and ρ is the linear correlation. Spearman's rho is the linear correlation of the probability transformed random variables.
For any bivariate random vector (X_1, X_2) following meta-Gaussian distribution

$$\rho_S(X_1, X_2) = \frac{6}{\pi}\arcsin\frac{\rho}{2}.$$

Since Spearman's rho is invariant under strictly increasing transformation of data and (X_j, X_k) is Gaussian, we have

$$\rho_S(Y_0, Y_k) = \rho_S(X_0, X_k) = \frac{6}{\pi}\arcsin\frac{\rho(X_0, X_k)}{2} = \frac{6}{\pi}\arcsin\frac{e^{-\alpha k}}{2}.$$

Hence by inverting and plugging in gives the *plug-in* estimator of α based on Spearman's rank correlation coefficient as

$$\hat{\alpha}_k := -\frac{1}{k}\ln\left(2\sin\left(\frac{\pi}{6}\hat{\rho}_S(Y_0, Y_k)\right)\right), \quad k = 1, 2, \dots$$

where

$$\hat{\rho}_S(Y_0, Y_k) := \frac{12}{n(n^2-1)}\sum_{k=1}^{n}\left(rank(Y_0) - \frac{n+1}{2}\right)\left(rank(Y_k) - \frac{n+1}{2}\right)$$

Quantile Estimator

Let

$$g(x) := -\ln[\sin(2\pi(x - 1/4)], \quad x \in (1/4, 1/2)$$

and

$$\hat{p}_k^0 := \frac{1}{n-k}\sum_{j=0}^{n-k} I(X_j > 0,\ X_{j+k} > 0)$$

be an empirical estimate of $p_k := P(X_0 > 0,\ X_k > 0)$. Then define

$$\hat{\alpha}_k := \frac{1}{k}g(\hat{p}_k^0) = -\frac{1}{k}\ln[\sin(2\pi(\hat{p}_k^0 - 1/4)], \quad k = 1, 2, \dots.$$

Borokov and Decrouez (2013) showed that $\hat{\alpha}_k$ is strongly consistent and asymptotically normal as $n \to \infty$. Since X_t is unobservable, $\hat{p}_k^0$ needs to be replaced with an estimator of p_k from Y_t.

Let $\hat{m}_n(Y)$ is the sample median of the sample $Y_0, Y_1, \dots, Y_n$ and m is the median of F. $\hat{m}_n(Y) \to m$ a.s. as $n \to \infty$.

Since $\{X_j > 0\} = \{Y_j > m\}$, we have

$$\hat{p}_k^0 := \frac{1}{n-k}\sum_{j=0}^{n-k} I(Y_j > m,\ Y_{j+k} > m).$$

Since we do not know m, we use the sample median and obtain

$$\hat{p}_k := \frac{1}{n-k}\sum_{j=0}^{n-k} I(Y_j > \hat{m}_n(Y),\ Y_{j+k} > \hat{m}_n(Y)).$$

Therefore,

$$\breve{\alpha}_k := \frac{1}{k}g(\hat{p}_k), \quad k = 1, 2, \dots$$

will also be strongly consistent estimator of α.

Third Estimator

The process (X_t) is unobservable, but the transformed process $Y_t = h(X_t)$ is observable. Even if, Y_t is partially observed, that is, if we know the number of times the process' trajectory crossed the band between two given levels A and B, during a given time interval, then one can estimate α, provided we also know the values of the distribution function $F(y)$ of the stationary distribution of Y_t at the points A and B.

Let

$$\tau_{a,b} := \inf\{t > 0 : X_t = a | X_0 = b\}, \quad a, b \in \mathbb{R}.$$

$\tau_{a,b}$ is the times it takes the OU process (X_t) to transit from b to a.

The *excursion time* $\tau_{(a),b}$ from a to b is the time it takes the OU process starting from b to reach the point a and return back to b. Clearly,

$$\tau_{(a),b} =^D \tau_{a,b} + \tau_{b,a},$$

where $\tau_{a,b}$ and $\tau_{b,a}$ are assumed to be independent.

Set $a := h^{-1}(A), \;\; b := h^{-1}(B)$. Since $X_t \sim N(0, 1/(2\alpha))$ in the stationary regime, one has $\Phi(\sqrt{2\alpha}a) = P(X_t \le a) = P(h(X_t) \le A) = F(A)$ and likewise $\Phi(\sqrt{2\alpha}b) = P(X_t \le b) = P(h(X_t) \le B) = F(B)$. Therefore,

$$a = \frac{\Phi^{-1}(F(A))}{\sqrt{2\alpha}} =: \frac{a^*}{\sqrt{\alpha}}, \quad b = \frac{\Phi^{-1}(F(B))}{\sqrt{2\alpha}} =: \frac{b^*}{\sqrt{\alpha}}.$$

The standard scaling argument shows that the process $X_t^* := \alpha^{1/2} X_{t/\alpha}$ satisfies the OU equation with $\alpha = 1$ and another Brownian motion, namely $W_t^* = \alpha^{1/2} W_{t/\alpha}$. Therefore,

$$\tau_{(a),b} =^D \tau^*_{(a^*),b^*},$$

where the asterisk indicates that the excursion time on the right side is for the process X_t^*, the distribution of $\tau^*_{(a^*),b^*}$ being independent of α and having a mean m_{a^*,b^*}.

Now strong Markov property implies that subsequent excursion from b to a in $\{X_t\}$ (or, equivalently, from B to A in $\{Y_t\}$) form an i.i.d. sequence. Therefore, if $N_{A,B}(T)$ denotes the number of times $\{Y_t\}$ crosses the band between the levels A and B in a given direction during the time interval $[0, T]$, the integral renewal theorem asserts that, as $T \to \infty$,

$$\frac{N_{A,B}(T)}{T} \to^{a.s.} \frac{1}{E(\tau_{(a),b})} = \frac{\alpha}{m_{a^*,b^*}}$$

which gives another consistent estimator of α given by

$$\widetilde{\alpha} := \frac{m_{a^*,b^*} N_{A,B}}{T}.$$

Chapter 12

Bayesian Maximum Likelihood Estimation in Fractional Stochastic Volatility Model

12.1 Introduction

In mathematical finance, it is well accepted that volatility of a stock price is a stochastic process, not a constant. It is also known that volatility has long memory and clusters on high level. One way of modeling long memory is superposition of Ornstein–Uhlenbeck (supOU) processes as volatility models. The class of supOU processes can capture extremal clusters and long-range dependence. We consider volatility as a continuous model satisfying a stochastic differential equation driven by a persistent fractional Brownian motion. Long memory in volatility is a stylized fact in finance due to volatility clustering and persistence. Empirical study shows that volatility has long memory in the sense that the empirical autocorrelation function decreases slower than exponential. Hence parameter estimation in stochastic volatility model with long memory is an important problem in mathematical finance. But the difficulty arises from the fact that volatility is latent as it is not observed in the market. Hence the parameters must be estimated from the corresponding stock price observations. Parameter estimation in directly observed stochastic differential equations is extensively studied in Bishwal (2008a). Models with fractional Brownian motion and fractional Levy processes as diving terms have attracted recent attention, see Bishwal (2011b,c,i, 2017b,c).

Asymptotic behavior of estimators in partially observed stochastic differential system was first studied by Kutoyants (1994a,b) when the state and the observation equations are driven by ordinary Brownian motions when the observation time becomes large. The problem was generalized by Kallianpur and Selukar (1991) to multiparameter case. The problem was studied by Campillo and Le Gland (1989), Konecky (1990), James and Le Gland (1995), Kutoyants and Pohlman (1994), and Kutoyants (1984a,b, 1994a,b) when the intensity of noise becomes small.

On the other hand, for the case of directly observed system, the study of asymptotic properties of several estimators of a parameter in the drift coefficient of the Itô stochastic differential equation driven by a standard

J. P. N. Bishwal, *Parameter Estimation in Stochastic Volatility Models*,
https://doi.org/10.1007/978-3-031-03861-7_12

Brownian motion based on observations of the corresponding diffusion process continuously inside a time interval $[0, T]$ or at discrete time points in $[0, T]$ is now well developed; see Kutoyants (1984a,b, 1994a,b).

Recently, long memory processes or stochastic models having long-range dependence phenomena have been paid a lot of attention in view of their applications in hydrology, computer networks, and finance (see Beran (1994)). While parameter estimation in discrete time models having long-range dependence like the autoregressive fractionally integrated moving average (ARFIMA) models has already been paid some attention, this problem for continuous time models is not well settled.

The fractional Brownian motion (fBm, in short), which provides a suitable generalization of the Brownian motion, is one of the simplest stochastic processes exhibiting long-range dependence. It was introduced by Kolmogorov (1940) and later on studied by Levy (1948) and Mandelbrot and Van Ness (1968).

Consider a probability space $(\Omega, \mathcal{F}, \mathbb{P})$ on which all random variables and processes below are defined. A normalized fractional Brownian motion $\{W_t^H, t \geq 0\}$ with Hurst parameter $H \in (0, 1)$ is a centered Gaussian process with continuous sample paths whose covariance kernel is given by

$$E(W_t^H W_s^H) = \frac{1}{2}(s^{2H} + t^{2H} - |t-s|^{2H}), \quad s, t \geq 0.$$

The process is self-similar (scale invariant) and it can be represented as a stochastic integral with respect to standard Brownian motion. For $H = \frac{1}{2}$, the process is a standard Brownian motion. For $H \neq \frac{1}{2}$, the fBm is not a semimartingale and not a Markov process, but a Dirichlet process. The increments of the fBm are negatively correlated for $H < \frac{1}{2}$ and positively correlated for $H > \frac{1}{2}$, and in this case they display long-range dependence. The parameter H, which is also called the self-similarity parameter, measures the intensity of the long-range dependence. Note that the estimation of the parameter H based on observation of fractional Brownian motion has already been paid some attention, see, e.g., Peltier and Levy Vehel (1994) and the references there in. However, we assume H to be known.

Hence, for $H \neq \frac{1}{2}$, the classical theory of stochastic integration with respect to semimartingales is not applicable to stochastic integration with respect to fBm. Now there exist several approaches to stochastic integration with respect to fBm. See, for instance, classical Riemann sum approach Dai and Heyde (1996); Kleptsyna et al. (1998); Lin (1995), Malliavin calculus approach (Decreusefond and Ustunel, 1998, 1999), Coutin and Decreusefond (1997), Alos et al. (2000, 2001), Wick product approach (Duncan et al., 2000), pathwise calculus (Ruzmaikina, 2000; Zahle, 1998), and Dirichlet calculus (Lyons and Zhang, 1994).

The problem of optimal filtering of a signal when the noise is a fractional Brownian motion was studied by Kleptsyna et al. (1998a,b), Le Breton (1998a,b), Coutin and Decreusefond (1999), and Kleptsyna et al. (1998, 2000).

As far as the estimation of unknown parameter is concerned, the maximum likelihood estimator (MLE) of the constant drift parameter of a fractional Brownian motion was obtained by Decreusefond and Ustunel (1998) who developed stochastic analysis of the fBm in a Malliavin calculus framework. Norros et al. (1999) studied the properties of the MLE of the constant drift parameter of fBm using martingale tools. They showed that the MLE is unbiased and normally distributed. They also showed that the MLE is strongly consistent and proved a law of the iterated logarithm as $T \to \infty$. The problem was generalized by Le Breton (1998a) to a stochastic differential equation driven by fBm with drift and the diffusion coefficient being nonrandom functions and the unknown parameter in the drift coefficient. He obtained the best linear unbiased estimator (BLUE) of the drift parameter which coincides with the MLE. He also obtained the least squares estimator (LSE) and compared the relative efficiency of the LSE and the BLUE.

Our aim in this chapter is to study the properties of the MLE of a parameter appearing linearly in drift coefficient of a nonlinear stochastic differential equation driven by fBm when the signal process is a nonlinear diffusion process.

We study the strong consistency and asymptotic normality of the maximum likelihood estimator of a parameter in a stochastic volatility model when the observed process is driven by fractional Brownian motion and volatility is driven by standard Brownian motion. We illustrate the results for the fractional Heston model.

The chapter is organized as follows: in Sect. 2, we prepare model, assumptions, and preliminaries. In Sect. 3, we give the main results of the chapter on strong consistency and asymptotic normality of the MLE. In Sect. 4, we verify the main results for a partially observed fractional linear system.

12.2 Fractional Stochastic Volatility Model

On the probability space $(\Omega, \mathcal{F}, \mathbb{P})$, consider the stochastic differential equations

$$\begin{aligned} dY_t &= \theta f(t, X_t)\, dt + g(X_t) dW_t^H, \\ dX_t &= a(t, X_t)\, dt + b(t, X_t)\, dV_t, \;\; t \in [0, T], \\ & \quad Y_0 = \xi, X_0 = \eta, \end{aligned}$$

where $W^H, H \in (\frac{1}{2}, 1)$ is the fBm defined above and V is a standard Brownian motion independent of W^H such that the pair (η, ξ) is independent of (V, W^H). We assume that the conditional distribution of η given ξ is some fixed π_0.

Here, $\theta \in \Theta$ open in $\mathbb{R}$ is the unknown parameter to be estimated on the basis of observation $\{Y_t, 0 \le t \le T\}$.

We assume that the functions $g : [0,T] \to \mathbb{R}^+$ and $f : [0,T] \times \mathbb{R} \to \mathbb{R}$ are known and satisfy

(A1) $|f(t,x)| \leq K(1+|x|)$ for all $t \in [0,T]$.

(A2)$|f(t,x) - f(t,y)| \leq K(|x-y|)$ for all $t \in [0,T]$,

for some constant $K > 0$.

(A2)$|g(x) - g(y)| \leq K_1(|x-y|)$ for all $t \in [0,T]$,

for some constant $K > 0, K_1 > 0$.

We assume that the functions $a : [0,T] \times \mathbb{R} \to \mathbb{R}$ and $b : [0,T] \times \mathbb{R} \to \mathbb{R}$ are known and satisfy

(A3) (i) $|a(t,x)| \leq K_2(1+|x|)$ for all $t \in [0,T]$,

(ii)$|b(t,x)| \leq K_2(1+|x|)$ for all $t \in [0,T]$

for some constant $K_2 > 0$.

(A4) (i)$|a(t,x) - a(t,y)| \leq K_1(|x-y|)$ for all $t \in [0,T]$,

(ii) $|b(t,x) - b(t,y)| \leq K_1(|x-y|)$ for all $t \in [0,T]$

for some constant $K_1 > 0$.

Under the conditions (A1)–(A4), it is known that there exists a unique solution of the SDEs (see Lin, 1995 and Kallianpur, 1980).

We shall use the following result of Kleptsyna et al. (1998) in the sequel.

Theorem 2.1 *Let h be a continuous function from $[0,T]$ to $\mathbb{R}$. Define for $0 < t \leq T$, the function $k_h^t = (k_h^t(s), 0 < s < t)$ by*

$$k_h^t(s) := -\rho_H^{-1} s^{\frac{1}{2}-H} \frac{d}{ds} \int_s^t d\omega \omega^{2H-1} (\omega - s)^{\frac{1}{2}-H} \frac{d}{d\omega} \int_0^\omega dz z^{\frac{1}{2}-H} (\omega - z)^{\frac{1}{2}-H} h(z),$$

where

$$\rho_H = \Gamma^2(\frac{3}{2} - H)\Gamma(2H+1)\sin \pi H.$$

Define for $0 \leq t \leq T$,

$$N_t^h := \int_0^t k_h^t(s) dW_s^H, \quad \langle N^h \rangle_t := \int_0^t h(s) k_h^t(s) ds.$$

Then the process $\{N_t^h, 0 \leq t \leq T\}$ is a Gaussian martingale with variance function $\{\langle N^h \rangle_t, 0 \leq t \leq T\}$. For $h = 1$, the function k_h^t is $k_*^t(s) := \tau_H^{-1}(s(t-s))^{\frac{1}{2}-H}$, where $\tau_H := 2H\Gamma(\frac{3}{2}-H)\Gamma(H+\frac{1}{2})$. Then the corresponding Gaussian martingale is given by

$$N_t^* = \int_0^t k_*^t(s) dW_s^H, \quad \text{with } \langle N^* \rangle_t = \int_0^t k_*^t(s) ds = \lambda_H^{-1} t^{2-2H},$$

where

$$\lambda_H = \frac{2H\Gamma(3-2H)\Gamma(H+\frac{1}{2})}{\Gamma(\frac{3}{2}-H)}.$$

In order to obtain the MLE, we proceed as follows: consider the canonical space of the process (X, Y). Let $\Omega = \mathcal{C}([0,T];\mathbb{R}^2)$ be the space of continuous functions from $[0,T]$ into $\mathbb{R}^2$. Consider also the canonical process $(X, W^*) = ((X_t, W_t^*), t \in [0,T])$ on Ω where $(X_t, W_t^*)(x,y) = (x_t, y_t)$ for any $(x,y) \in \Omega$.

The probability $\tilde{\mathbb{P}}$ denotes the unique probability measure on Ω such that defining the variable ξ by $\xi = W_0^*$ and $\tilde{W} = (\tilde{W}_t), t \in [0,T]$, by $\tilde{W}_t = W_t^* - W_0^*, t \in [0,T]$, the pair (X, ξ) is independent of $\tilde{W}$ and the process $\tilde{W}$ is a fBm with Hurst parameter H.

The canonical filtration on Ω is $(\mathcal{F}_t, t \in [0,T])$, where $\mathcal{F}_t = \sigma(\{(X_s, W_s^*), 0 \leq s \leq t\} \bigvee \mathcal{N}$, where $\mathcal{N}$ denotes the set of null sets of $(\Omega, \tilde{\mathbb{P}})$. Define for all continuous functions $x = (x_t, t \in [0,T])$ the function $h(\theta, x)$ on $[0,T]$ by

$$h(\theta, x)(t) := \frac{\theta f(t, X_t)}{g(X_t)}, \quad t \in [0,T].$$

Consider, for any $t \in [0,T]$, the function $k_{h(\theta,x)}^t = (k_{h(\theta,x)}^t(s), 0 < s < t)$ defined from Theorem 2.1 with $h(\theta, x)$ in place of h. Define the processes $N = (N_t, t \in [0,T])$ and $\langle N \rangle = (\langle N \rangle_t, t \in [0,T])$ from Theorem 2.1 by plugging in the process $h(\theta, x)$ in place of h, i.e.,

$$N_t := N_t^{h(\theta,X)}, \quad \langle N \rangle_t := \langle N^{h(\theta,X)} \rangle_t.$$

Notice that N_t and $\langle N \rangle_t$ depend only on the values of $X^{(t)} = (X_s, 0 \leq s \leq t)$. Define the $(\mathcal{F}_t)$-adapted processes $\langle N, N^* \rangle = (\langle N, N^* \rangle_t, t \in [0,T])$ and $q(\theta, X) = (q_t(\theta, X), t \in [0,T])$ by

$$\langle N, N^* \rangle_t := \langle N^{h(\theta,X)}, N^* \rangle_t = \int_0^t k_*^t(s) h(\theta, X)(s) ds, t \in [0,T])$$

and

$$q_t(\theta, X) := q_t^{h(\theta,X)} = \frac{d\langle N, N^* \rangle_t}{d\langle N^* \rangle_t}, t \in [0,T].$$

Let $\tilde{q}_t(X) := \frac{q_t(\theta,X)}{\theta}$. Define for $0 \leq t \leq T$, the processes

$$\tilde{N}_t(\theta, x) := \int_0^t k_{h(\theta,x)}^t(s) d\tilde{W}_s^H, \quad \langle \tilde{N} \rangle_t(\theta, x) := \int_0^t h(\theta, x)(s) k_h^t(s) ds.$$

$\tilde{N}_t(\theta, x)$ is a Gaussian martingale under $\tilde{\mathbb{P}}$. Define

$$\Lambda_t(\theta, x) := \exp\{\tilde{N}_t(\theta, x) - \frac{1}{2} \langle \tilde{N} \rangle_t(\theta, x)\}, t \in [0,T].$$

Define for any $t \in [0,T]$,

$$\Lambda_t(\theta) := \Lambda_t(\theta, X).$$

Let $\mathbb{P} := \Lambda_T(\theta)\tilde{\mathbb{P}}$. Let $\mathcal{Y}_t := \sigma(\{Y_s, 0 \leq s \leq t\}), t \in [0,T]$. Define the optimal filter $\pi_t(\phi) := \mathbb{E}[\phi(X_t)|\mathcal{Y}_t], t \in [0,T]$, and the unnormalized filter $\sigma_t(\phi) :=$

$\tilde{\mathbb{E}}[\phi(X_t)\Lambda_t|\mathcal{Y}_t], t \in [0,T]$. Then the *Kallianpur–Striebel formula* holds: for all $t \in [0,T], \tilde{\mathbb{P}}$ and $\mathbb{P}$ almost surely

$$\pi_t(\phi) = \frac{\sigma_t(\phi)}{\sigma_t(1)}.$$

Introduce the processes Z and Z^* defined by

$$Z_t := \int_0^t k^t_{h(\theta,X)}(s) g^{-1}(X_s) dY_s, \quad Z^*_t := \int_0^t k^t_*(s) g^{-1}(X_s) dY_s, t \in [0,T].$$

The processes Z and Z^* are semimartingales with the following decomposition:

$$Z^*_t = \langle N, N^* \rangle_t + N^*_t, t \in [0,T],$$

$$Z_t = \langle N \rangle_t + N_t, t \in [0,T].$$

From the above, we can write

$$Z^*_t = \int_0^t q_s(\theta, X) d\langle N^* \rangle_s + N^*_t, t \in [0,T]),$$

and

$$Z_t = \int_0^t q^2_s(\theta, X) d\langle N^* \rangle_s + \int_0^t q_s(\theta, X) dN^*_s, t \in [0,T]).$$

Hence we get the integral representation of Z in terms of Z^*:

$$Z_t = \int_0^t q_s(\theta, X) dZ^*_t, t \in [0,T].$$

Introduce the process $\nu = (\nu_t, t \in [0,T])$ defined by

$$\nu_t = Z^*_t - \int_0^t \pi_s(q) d\langle N^* \rangle_s, t \in [0,T],$$

which plays the role of *innovation process* in the usual situation where the noises are Brownian motions.

Recall the notation $\pi_s(q) = \mathbb{E}[q_s(\theta, X)|\mathcal{Y}_s], s \in [0,t]$. The particular case of unnormalized filter is

$$\tilde{\Lambda}_t(\theta) := \sigma_t(1) = \tilde{\mathbb{E}}[\Lambda_t|\mathcal{Y}_t], t \in [0,T].$$

By Proposition 3 in Kleptsyna et al. (2000), we have

$$\tilde{\Lambda}_T(\theta, \mathcal{Y}_t) = \exp\left\{\theta \int_0^T \pi_s(\tilde{q}) dZ^*_s - \frac{\theta^2}{2} \int_0^T \pi^2_s(\tilde{q}) d\langle N^* \rangle_s\right\}.$$

Thus the maximum likelihood estimator (MLE) of θ is given by

$$\hat{\theta}_T \quad := \quad \arg\max_{\theta\in\Theta} \tilde{\Lambda}_T(\theta) = \frac{\int_0^T \pi_s(\tilde{q})dZ_s^*}{\int_0^T \pi_s^2(\tilde{q})d\langle N^*\rangle_s}.$$

In a linear state-space system, it is well known that Kalman filter is an algorithm for the exact computation of the conditional p.d.f. of the signal given the observation (and the initial conditions). Except for some very specific cases, such as the fractional Cox–Ingersoll–Ross (fCIR) model which has noncetral chi-square transition density, exact computation of the likelihood function is not possible in nonlinear models. Since we have a nonlinear state-space type model, we seek numerical or simulated methods. One such computational algorithm is Kitagawa's algorithm to compute the conditional p.d.f. recursively.

Kitagawa (1987) suggested a *linear spline technique* for approximating the nonlinear filter. The basic idea is to use the relevant probability density functions for each period by piecewise linear functions. The accuracy of density approximation depends on the number of nodes used for the piecewise linear approximation and the number of nodes is limited by the computational demands. We use the Kitagawa algorithm to compute the MLE. We approximate the filter

$$\pi_s(\widetilde{q}) = \mathbb{E}[\widetilde{q}_s(\theta, \sigma_s^2)|\mathcal{Y}_s]$$

using the Kitagawa algorithm given below.

The *prediction-update algorithm* proceeds as follows:

Prediction Step

Recall that for $s \in [0, T]$, $\mathcal{Y}_s := \sigma(\{Y_u, 0 \leq u \leq s\})$. The prediction step determines the conditional probability density function.

Let $p(\sigma_s^2|\sigma_u^2)$ be the conditional density of σ_s^2 given σ_u^2 where $u < s$. Then

$$p[\widetilde{q}_s(\theta, \sigma_s^2)|\mathcal{Y}_u] = \int p(\widetilde{q}_s(\theta, \sigma_s^2)|\sigma_u^2)p[\widetilde{q}_s(\theta, \sigma_u^2)|\mathcal{Y}_s]d\sigma_u^2.$$

Update Step

If observation becomes available at the forecast time, the update step combines this additional information with the estimate from the prediction step.

Suppose that a new observation Y_s becomes available. This additional information can be used to produce an updated estimate of the predicted state. This new estimate is summarized by the conditional density function $\widetilde{p}[\widetilde{q}_s(\theta, \sigma_s^2)|\mathcal{Y}_s]$.

Let $r(Y_s|\sigma_s^2, \mathcal{Y}_u)$ be the conditional density of Y_s given σ_s^2 and $\mathcal{Y}_u$, $\quad u < s$.

By Bayes theorem,

$$\widetilde{p}[\widetilde{q}_s(\theta, \sigma_s^2)|\mathcal{Y}_s] = \frac{r(Y_s|\sigma_s^2, \mathcal{Y}_u)p[\widetilde{q}_s(\theta, \sigma_s^2)|\mathcal{Y}_u]}{\int r(Y_s|\sigma_s^2, \mathcal{Y}_u)p[\widetilde{q}_s(\theta, \sigma_s^2)|\mathcal{Y}_u]d\sigma_s^2}$$

and

$$\pi_s(\widetilde{q}) = \mathbb{E}[\widetilde{q}_s(\theta, \sigma_s^2)|\mathcal{Y}_s] = \int \widetilde{q}_s(\theta, \sigma_s^2)\widetilde{p}[\widetilde{q}_s(\theta, \sigma_s^2)|\mathcal{Y}_s]d\sigma_s^2.$$

The update and the prediction step yield the general filtering solution.

12.3 Bayesian Maximum Likelihood Estimation

We shall prove the strong consistency and asymptotic normality of the Bayesian MLE with a random normalization.

Theorem 3.1 *Under the conditions (A1)–(A4), $\hat{\theta}_T \longrightarrow \theta$ a.s. as $T \to \infty$, i.e., $\hat{\theta}_T$ is a strongly consistent estimator of θ.*

Moreover,

$$\limsup_{T} \frac{A_T^{1/2}|\hat{\theta}_T - \theta|}{(2\log\log A_T)^{1/2}} = 1 \text{ a.s.},$$

where $A_T = \int_0^T \pi_s^2(\widetilde{q})d\langle N^*\rangle_s$.

Proof Note that

$$dZ_t^* = \pi_t(q)d\langle N^*\rangle_t + d\nu_t.$$

From Lemma 3 in Kleptsyna et al. (1998), we have ν is a continuous Gaussian $(\mathcal{Y}_t, \mathbb{P})$ martingale such that $\langle \nu \rangle = \langle N^* \rangle$.

Hence

$$\hat{\theta}_T = \frac{\int_0^T \pi_s(\widetilde{q})dZ_s^*}{\int_0^T \pi_s^2(\widetilde{q})d\langle N^*\rangle_s} = \theta + \frac{\int_0^T \pi_s(\widetilde{q})d\nu_s}{\int_0^T \pi_s^2(\widetilde{q})d\langle N^*\rangle_s}.$$

Now by the strong law of large numbers for continuous martingales (see Revuz and Yor (1991) or Theorem 2.6.10) in Liptser and Shiryayev (1989)), the second term in r.h.s. converges to zero a.s. as $T \to \infty$. Hence strong consistency follows. Since the deviation $\hat{\theta}_T - \theta$ is obtained from a stochastic integral with respect to a continuous martingale, the LIL follows from Corollary 1.1.12 of Revuz and Yor (1991). □

Theorem 3.2 *Under the conditions (A1)–(A4),*

$$\left(\int_0^T \pi_s^2(\widetilde{q})d\langle N^*\rangle_s\right)^{1/2}(\hat{\theta}_T - \theta) \xrightarrow{\mathcal{D}} \mathcal{N}(0,1) \quad \text{as } T \to \infty.$$

Proof Note that

$$\left(\int_0^T \pi_s^2(\tilde{q})d\langle N^*\rangle_s\right)^{1/2}(\hat{\theta}_T - \theta) = \frac{\int_0^T \pi_s(\tilde{q})d\nu_s}{\sqrt{\int_0^T \pi_s^2(\tilde{q})d\langle N^*\rangle_s}}.$$

By the central limit theorem for stochastic integrals with respect to Gaussian martingales (see Liptser and Shiryayev, 1989), the r.h.s. above converges in distribution to $\mathcal{N}(0,1)$ as $T \to \infty$. □

12.4 Fractional Heston Model

Consider the fractional version of the Heston model with fBm observation noise

$$\begin{aligned} dY_t &= \theta X_t dt + \sqrt{X_t} dW_t^H, \\ dX_t &= -X_t dt + \sqrt{X_t} dV_t, \ t \in [0,T], \\ & Y_0 = 0, X_0 = 0, \end{aligned}$$

where $\theta < 0$. Note that here the volatility is the familiar square-root process.

Let $\hat{X}_t := E(X_t|\mathcal{Y}_t)$ and $\gamma_t := E([X_t - \hat{X}_t]^2|\mathcal{Y}_t)$.

Then $\hat{X}_t$ satisfies the SDE

$$d\hat{X}_t = -\hat{X}_t dt + \theta\gamma_t d\nu_t, \ t \in [0,T], \hat{X}_0 = 0,$$

and γ_t satisfies the deterministic differential equation, known as the Riccati equation

$$d\gamma_t = dt - 2\gamma_t dt - \theta^2\gamma_t^2 d\langle N^*\rangle_t, \gamma_0 = 0.$$

As $t \to \infty, \gamma_t \to \gamma_\theta$, where $\gamma_\theta := \frac{-1+\sqrt{1+\theta^2}}{\theta^2}$. Now we make a simplifying assumption commonly made in the literature (see Kutoyants, 1994a,b, p. 103). We assume that the system has reached the steady state, i.e., we assume that $\hat{X}^t$ satisfies the SDE

$$d\hat{X}_t = -\hat{X}_t dt + \theta\gamma_\theta d\nu_t, \ t \in [0,T], \hat{X}_0 = 0.$$

Now it is easy to verify that

$$\hat{X}_t = \theta\gamma_\theta \int_0^t \exp\{\sqrt{1+\theta^2}(t-s)\}dY_s.$$

Thus the MLE can be now explicitly calculated and it satisfies the asymptotic properties mentioned in Theorems 3.1 and 3.2.

Chapter 13

Berry–Esseen–Stein–Malliavin Theory for Fractional Ornstein–Uhlenbeck Process

13.1 Introduction

Using Malliavin calculus along with Stein's equation, the chapter shows that the distribution of the normalized minimum contrast estimator of the drift parameter in the fractional Ornstein–Uhlenbeck process observed over $[0, T]$ converges to the standard normal distribution with a uniform error rate of the order $O(T^{-1/2})$ for the case $H > 1/2$ where H is the Hurst exponent of the fractional Brownian motion driving the Ornstein–Uhlenbeck process. Then based on discrete observations, it introduces several approximate minimum contrast estimators and studies their rate of weak convergence to normal distribution.

The theory of Stein–Malliavin is very powerful. Nourdin and Peccati (2012) used this theory for Berry–Esseen theorem. Chaos expansion is used to prove the central limit theorem for multiple stochastic integral. For the rest of the chapter, we use the Stein–Malliavian theory. It uses the fourth moment. In this chapter, we use this theory to obtain the Berry–Esseen theorem for MLE for FOU process. We also study Berry–Esseen theorem for discretely observed FOU process. It does not use the stochastic integral. This gives the optimal Berry–Esseen bound.

The chapter introduces Berry–Esseen Stein–Malliavin calculus for maximum likelihood estimator in the fractional Ornstein–Uhlenbeck process. It is a new method to obtain optimal Berry–Esseen bound.

13.2 Exact Berry–Esseen Bounds

Stein's Heuristics

Let $Z \sim N(0,1)$. Let $\mu_p := E(Z^p),\ p \geq 1$. We know that $\mu_p = 0$ if p is odd and $\mu_p = (p-1)!! := \frac{p!}{2^{p/2}(p/2)!}$ if p is even. The sequence μ_p is completely determined by the recurrence relation $\mu_1 = 0, \mu_2 = 1$ and $\mu_p = (p-1) \times \mu_{p-2}$ for every $p \geq 3$. Now introduce $f_p(x) = x^p$.

$$E[Z \times f_{p-1}(Z)] = E(f'_{p-1}(Z)).$$

J. P. N. Bishwal, *Parameter Estimation in Stochastic Volatility Models*,
https://doi.org/10.1007/978-3-031-03861-7_13

This relation holds for f_p being replaced by any smooth (C^1 function with sub-polynomial derivative) function f.

From this follows, the *Stein's Lemma: A random variable* X *has* $N(0,1)$ *distribution iff* $E[Xf(X) - f'(X)] = 0$ *for every smooth function* f.

Thus one can measure the distance between X and Z by assessing the distance from zero of the quantity $E[Xf(X) - f'(X)]$.

In fact,

$$d(X,Z) \leq C \sup_{f\in\mathcal{F}} |E[Xf(X) - f'(X)]|.$$

When d is Kolmogorov distance, one can take $C = 1$ and take $\mathcal{F}$ equal to the collection of all bounded Lipschitz functions with Lipschitz constant less than or equal to 1.

$$d_{Kol}(X,Z) \leq \sup_{f\in\mathcal{F}} |E[Xf(X) - f'(X)]|.$$

Let $\{F_n, n \geq 1\}$ be a sequence of zero mean valued random variables, and consider a standard Gaussian random variable $Z \sim \mathcal{N}(0,1)$. Assume that each F_n is a functional of an infinite dimensional Gaussian field, and suppose that as $n \to \infty$, $F_n \to^D Z$.

Optimal Bound We say that a bound ϕ_n is optimal for the sequence $\{F_n\}$ and the distance d whenever there exists a constant $c \in (0,1)$ (independent of n) such that, for n sufficiently large

$$c < d(F_n, Z)/\phi_n \leq 1.$$

The *Kolmogorov distance* for any pair of random variables X and Y is defined as

$$d_{Kol}(X,Y) = \sup_{x\in\mathbb{R}} |P(X \leq x) - P(Y \leq x)|.$$

Stein equation is given by

$$1_{(-\infty,x)} - \Phi(x) = f'(x) - xf(x), \; x \in \mathbb{R},$$

where $\Phi(x) = P(Z \leq x), i.e., Z \sim \mathcal{N}(0,1)$, and 1_A stands for the indicator of the set A.

For every fixed x, Stein equation admits a solution $f_x(z) = e^{z^2/2} \int_{-\infty}^{z} [1_{(-\infty,x)}$ $(a) - \Phi(x)]e^{a^2/2}da$ such that $\|f_x\|_\infty \leq \sqrt{2\pi}/4$ and $\|f_x'\|_\infty \leq 1$.

The *Wasserstein distance* between the laws of two random variables X and Y is given by

$$d_W(X,Y) := \sup_{g\in\mathcal{L}, |g|_{Lip}\leq 1} |E[g(X) - g(Y)]|,$$

where $\mathcal{L}$ is the class of all Lipschitz functions.

Let $\mathcal{H}$ be a real separable Hilbert space, and let $\mathcal{H}^{\otimes q}$ (resp. $\mathcal{H}^{\odot q}$) be the second tensor product (resp. 2nd symmetric tensor product) of $\mathcal{H}$.

We write $X = \{X_h, h \in \mathcal{H}\}$ to indicate centered isonormal Gaussian process on $\mathcal{H}$. For every $q \geq 0$, we denote by I_q the isometry between $\mathcal{H}^{\odot q}$

(equipped with the norm $\sqrt{q!}\|\cdot\|_{\mathcal{H}^{\otimes q}}$) and the qth Wiener chaos of X. Note that if $\mathcal{H}$ is a σ-finite measure space with no atoms, each random variable $I_q(h), h \in \mathcal{H}^{\otimes q}$ has the form of a multiple Wiener–Itô integral of order q.

Let DF_n be the Malliavin derivative of F_n and L^{-1} be the pseudo-inverse of the Ornstein–Uhlenbeck generator. DF_n is a random element with values in a Hilbert space $\mathcal{H}$.

We denote by δ the adjoint of the operator D called the divergence operator. The *integration by parts* (IBP) formula says that

$$E(F\delta(u)) = E\langle DF, u\rangle_{\mathcal{H}}.$$

The *integration by parts* formula for Malliavin calculus is

$$E[Yf(Y)] = E[\langle DY, -DL^{-1}Y\rangle_{\mathcal{H}} f'(Y)].$$

This gives

$$P(F_n \le x) - \Phi(x) = E[f_x'(F_n) - F_n f_x(F_n)] = E[f_x'(F_n)(1 - \langle DF_n, -DL^{-1}F_n\rangle_{\mathcal{H}}]. \tag{1.6}$$

If $F = \sum_{q=0}^{\infty} I_q(f_q)$, then

$$DF = \sum_{q=1}^{\infty} qI_{q-1}(f_q(\cdot, a)),$$

where I_q is the multiple stochastic Wiener–Itô integral of order q.

The Ornstein–Uhlenbeck operator L satisfies $LL^{-1}(F) = F - E(F)$ for all $F \in L^2(\Omega, \sigma(X), P)$.

Mehler's Formula It gives a mixture type representation of the infinite dimensional Ornstein–Uhlenbeck semigroup.

Wiener Chaos For every $q \ge 1$, we write $\mathcal{H}_q$ in order to indicate the qth Wiener chaos of a Gaussian process X. $\mathcal{H}_q$ is the closed linear subspace of $L^2(\Omega, \mathcal{F}, P)$ generated by random variables of the type $H_q(X(h))$, where $h \in \mathcal{H}$ is such that $\|h\|_{\mathcal{H}} = 1$, where H_q stands for the qth Hermite polynomial.

The qth Hermite polynomial is defined as

$$H_q(x) = (-1)^q e^{\frac{x^2}{2}} \frac{d^q}{dx^q} e^{\frac{x^2}{2}}, x \in \mathbb{R}, q \ge 1.$$

For instance, $H_1(x) = x, H_2(x) = x^2 - 1$.

$$\frac{d}{dx} H_q(x) = qH_{q-1}(x).$$

$$\Phi^{(q)}(x) = (-1)^{q-1} H_{q-1}(x) \frac{e^{-x^2/2}}{\sqrt{2\pi}}.$$

We also use the convention $\mathcal{H}_0 = \mathbb{R}$. For any $q \geq 1$, the mapping

$$I_q(h^{\otimes q}) = q!H_q(X(h))$$

can be extended to a linear isometry. For $q = 0$, we write $I_0(c) = c,\ c \in \mathbb{R}$.

Hypercontractivity Random variables living in a fixed Wiener chaos are hypercontractive. More precisely, assume that Y belongs to the qth Wiener chaos $\mathcal{H}_q, q \geq 1$. Then Y has finite variance by construction, and for all $p \in [2, \infty)$, one has the estimate

$$E(|Y|^p) \leq (p-1)E(Y^2)^{p/2}.$$

Chaotic Decomposition The Wiener chaos decomposition says that the space $L^2(\Omega, \mathcal{F}, P)$ can be decomposed into the infinite orthogonal sum of the spaces $\mathcal{H}_q$. It follows that any square integrable random variable $Y \in L^2(\Omega, \mathcal{F}, P)$ admits the following chaotic expansion

$$Y = \sum_{q=0}^{\infty} I_q(f_q),$$

where $f_0 = E(Y)$ and the kernels $f_q \in \mathcal{H}^{\odot q}$ are uniquely determined.

Multiplication Rule If $f \in \mathcal{H}^{\odot q}$ and $g \in \mathcal{H}^{\odot q}$, then

$$I_p(f)I_q(g) = \sum_{r=0}^{p\wedge q} r!\binom{p}{r}\binom{q}{r} I_{p+q-2r}(f\tilde{\otimes}_r g).$$

$$I_2(f)^2 = \sum_{s=0}^{2} s!\binom{2}{s}^2 I_{4-2s}(f\tilde{\otimes}_s f).$$

If $F = I_2(f)$ the double Wiener–Itô integral and $f \in \mathcal{H}^{\odot 2}$, then

$$\langle DF, -DL^{-1}F\rangle_{\mathcal{H}} = \langle DI_2(f), -DL^{-1}I_2(f)\rangle_{\mathcal{H}} = \frac{1}{2}\|DI_2(f)\|^2_{\mathcal{H}}.$$

$$Var\left(\frac{1}{2}\|DI_2(f)\|^2_{\mathcal{H}}\right) \leq \frac{1}{6}\left(E(I_2(f))^2 - 3\right).$$

By using Cauchy–Schwarz inequality to the RHS of (1.6) and by using the fact that f'_x is bounded by 1, one obtains

$$d_{Kol}(F_n, Z) \leq \sqrt{1 - \langle DF_n, -DL^{-1}F_n\rangle_{\mathcal{H}}]} \leq \sqrt{\frac{1}{6}(E(I_2(f))^2 - 3)}.$$

We shall use the following theorem to obtain the optimal Berry–Esseen bound.

Theorem 2.1 *(Nourdin and Peccati 2009) Let $F_n, n \geq 1$ be a sequence of centered and square integrable functionals of some isonormal Gaussian process $X = \{X(h) : h \in \mathcal{H}\}$ such that $E(F_n^2) \to 1$ as $n \to \infty$. Suppose the following three conditions hold:*

(i) For every n, one has that $F_n \in \mathbb{D}^{1,2}$, and F_n has an absolutely continuous law with respect to Lebesgue measure.

(ii) $\phi_n = \sqrt{E\left(\frac{1}{2}\|DF_n\|^2 - 1\right)}$ is finite for every n as $n \to \infty, \phi_n \to 0$; there exists $m \geq 1$ such that $\phi_n > 0$ for $n \geq m$.

(iii) As $n \to \infty$, the two-dimensional vector $\left(F_n,\ \frac{1}{2}\|DF_n\|^2 - 1\right)$ converges in distribution to a centered two-dimensional Gaussian vector (N_1, N_2) such that $E(N_1^2) = E(N_2^2) = 1$ and $E(N_1 N_2) = \rho$.

Then the upper bound $d_{Kol}(F_n, N) \leq \phi_n$ holds.

Moreover, for every $x \in \mathbb{R}$

$$\phi_n^{-1}[P(F_n \leq x) - \Phi(x)] \to \frac{\rho}{3}(x^2 - 1)\frac{e^{-x^2/2}}{\sqrt{2\pi}} = \frac{\rho}{3}\Phi^{(3)}(x).$$

As a consequence, if $\rho \neq 0$, there exists a constant $c \in (0,1)$, as well as an integer $n_0 \geq 1$, such that (1.3) holds for $d = d_{Kol}$ and for every $n \geq n_0$.

Theorem 2.2 *(Edgeworth Expansion) Let $F_n, n \geq 1$ be a sequence of centered and square integrable functionals of some isonormal Gaussian process $X = \{X(h) : h \in \mathcal{H}\}$ such that $E(F_n^2) = 1$. Suppose that conditions (i)–(iii) of Theorem 1.1 are satisfied and also that:*

(a) For every n, one has $E|F_n|^3 < \infty$.

(b) There exists $\epsilon > 0$ such that $\sup_{n \geq 1} E|F_n|^{2+\epsilon} < \infty$.

Then as $n \to \infty$,

$$\frac{1}{2\phi_n}E(F_n^3) \to -\rho,$$

and for every $x \in \mathbb{R}$, one has the following one-term local Edgeworth expansion

$$P(F_n \leq x) - \Phi(x) + \frac{1}{3!}E(F_n^3)\Phi^{(3)}(x) = o(\phi_n).$$

The fractional Ornstein–Uhlenbeck process is an extension of Ornstein–Uhlenb eck process with fractional Brownian motion (FBM) driving term. In finance, it is known as fractional Vasicek model and is being extensively used these days as one-factor short-term interest rate model that takes into account the long-memory effect of the interest rate, Hog and Frederiksen (2006). The model parameter is usually unknown and must be estimated from data. Bishwal (2008a) studied the maximum likelihood estimation in partially observed fractional stochastic differential equations.

To estimate the drift parameter, we use an alternative to the maximum likelihood estimator and study the accuracy of distributional approximation of the resulting estimators from continuous data and high-frequency discrete data.

Bootstrap Confidence Interval
The parametric model is

$$dX_t = -\theta\ X_t dt + dW_t, \theta > 0.$$

MLE of θ based on $\{X_t, 0 \le t \le T\}$ is given by

$$\hat{\theta}_T = -\frac{\int_0^T X_t dX_t}{\int_0^T X_t^2 dt}.$$

The empirical nonparametric model is given by

$$dY_t = -\hat{\theta}_t\ Y_t dt + dB_t,$$

where B is a Brownian motion independent of W.
The bootstrap MLE based on $\{Y_t, 0 \le t \le T\}$ is given by

$$\hat{\hat{\theta}}_T = \frac{\int_0^T Y_t dY_t}{\int_0^T Y_t^2 dt}.$$

The α-th quantile is given by

$$\widetilde{\omega}_\alpha(T) := \inf_{y \in \mathbb{R}} \left\{ P\left(\sqrt{\frac{T}{2\hat{\theta}_T}} \left(\hat{\hat{\theta}}_T - \hat{\theta}_T \right) \le y \right) \ge \alpha \right\}.$$

In the next theorem, we obtain bootstrap confidence interval for the fractional OU process.

Theorem 2.3 *The bootstrap confidence interval is given by*

$$I_T^{boot}(\alpha) := \left[\hat{\theta}_T - \frac{\widetilde{\omega}_\alpha(T)}{\sqrt{\frac{T}{2\hat{\theta}_T}}}, \infty \right).$$

Conditional on $\hat{\theta}_T$, the quantile $\widetilde{\omega}$ can be expanded as

$$\widetilde{\omega}_\alpha(T) = q_\alpha + \frac{1}{\sqrt{T}} h_\theta(q_\alpha) + o\left(\frac{1}{\sqrt{T}}\right),$$

where q_α is α-th quantile of the normal distribution. Thus

$$P\left(\theta \in I_T^{boot}(\alpha)\right) - \alpha = o\left(\frac{1}{\sqrt{T}}\right).$$

Edgeworth Expansion for MCE

Theorem 2.4

$$P_\nu[\sqrt{T}(\hat{\theta}_T - \theta)/\sigma \le x] \le \Phi(x) + T^{-1/2}\phi(x)(a_1 + a_2(1 - x^2)) + O_P(T^{-1})$$

uniformly in $x \in \mathbb{R}$, *where*

$$\sigma^2 = \frac{E(\bar{F}_0)^2}{E(L_0)}, \bar{F}_0 = F_0 - \theta L_0, \alpha = E(L_0)$$

$$a_1 = E_\mu(\bar{F}_0) - E_\nu(\bar{\nu}), a_2 = \frac{\kappa - 3\rho\sigma^2}{6\alpha\sigma^3}, \kappa = E(\bar{F}_0^3), \rho = E(\bar{F}_0 L_0).$$

Regenerative Method

A cadlag process X is called a *regenerative process* if there exists an increasing sequence of finite random times $\{\tau_j\}_{j\ge 1}$ such that

$$\{X_t\}_{0\le t\le \tau_1}, \{X_t\}_{\tau_1\le t\le \tau_2}, \ldots, \{X_t\}_{\tau_j \le t\le \tau_{j+1}}, \ldots$$

are independent and

$$\{X_t\}_{\tau_1\le t\le \tau_2}, \ldots, \{X_t\}_{\tau_j \le t\le \tau_{j+1}}, \ldots$$

are identically distributed. The random times τ_j is called a $j-$th regenerative epoch.

$L_j := \tau_{j+1} - \tau_j, j \ge 1$ is an i.i.d. sequence. It holds that $\tau_j \to \infty$ a.s. as $j \to \infty$. By the law of large numbers for i.i.d. sequences,

$$\text{P-lim}_{T\to\infty}\hat{\theta}_T = \text{P-lim}_{T\to\infty}\frac{-T}{2\int_0^T X_t^2 dt} = \text{P-lim}_{N\to\infty}\frac{\tau_1 - \tau_N}{2\sum_{j=0}^{N-1}\int_{\tau_j}^{\tau_{j+1}} X_t^2 dt} = \theta.$$

Let

$$M_T := \max\{j : \tau_{j+1} \le T\}.$$

Define the studentized statistics

$$\breve{\theta}_T = \frac{-\sum_{j=1}^{M_T} L_j}{2\sum_{j=1}^{M_T} F_j}, \quad \hat{\sigma}_T^2 = \frac{\sum_{j=1}^{M_T} |L_j - \breve{\theta}_T F_j|^2}{2\sum_{j=1}^{M_T} F_j}.$$

Theorem 2.5

$$P_\mu[\sqrt{T}(\breve{\theta}_T - \theta)/\hat{\sigma}_T \le x] \le \Phi(x) + T^{-1/2}\phi(x)(\hat{a}_2(2x^2 + 1)) + O_P(T^{-1})$$

uniformly in $x \in \mathbb{R}$, where $\hat{a}_2$ is an estimator of a_2 with $\sqrt{T}(\hat{a}_2 - a_2) = O(1)$. We can use for instance

$$\hat{\sigma}^2 = \frac{1}{M_T}\sum_{j=1}^{M_T}\frac{\check{F}_j^2}{\hat{\alpha}},\ \check{F}_j = F_j - \check{\theta}L_j,\ \hat{\alpha} = \frac{1}{M_T}\sum_{j=1}^{M_T} L_j$$

$$\hat{a}_2 = \frac{\hat{\kappa} - 3\hat{\rho}\hat{\sigma}^2}{6\hat{\alpha}\hat{\sigma}^3},\ \hat{\kappa} = \frac{1}{M_T}\sum_{j=1}^{M_T}\check{F}_j^3,\ \hat{\rho} = \frac{1}{M_T}\sum_{j=1}^{M_T}\check{F}_j L_j.$$

Let $\mathcal{F}_T = \{(F_j, L_j)\}, j = 1, 2, \ldots, M_T$ be the observed regenerative blocks. Let $(F_j^*, L_j^*), j = 1, 2, \ldots, M_T$ be an i.i.d. sequence and each (F_j^*, L_j^*) be uniformly distributed on $\mathcal{F}_T$. Here, M_T and $\mathcal{F}_T$ are fixed conditionally on the observation $\{X_t, 0 \le t \le T\}$.

Define the bootstrap statistics:

$$\check{\theta}_T^* = \frac{-\sum_{j=1}^{M_T} L_j^*}{2\sum_{j=1}^{M_T} F_j^*} = \frac{-T^*}{2\sum_{j=1}^{M_T} F_j^*},$$

where

$$T^* = \sum_{j=1}^{M_T} L_j^*,\ \{\hat{\sigma}_T^*\}^2 = \frac{\sum_{j=1}^{M_T} |L_j^* - \check{\theta}_T F_j^*|^2}{2\sum_{j=1}^{M_T} F_j^*}.$$

Theorem 2.6

$$P_\mu^*[\sqrt{T^*}(\check{\theta}_T^* - \check{\theta}_T)/\sigma_T^* \le x] \le \Phi(x) + T^{-1/2}\phi(x)(\hat{b} + \hat{a}_2(2x^2 + 1)) + O_P(T^{-1})$$

uniformly in $x \in \mathbb{R}$, where P_μ^* is the conditional probability given $\{X_t, 0 \le t \le T\}$.

The rest of the chapter is organized as follows: In Sect. 3, we obtain Berry–Esseen bounds for continuous observation. In Sect. 4, we obtain Berry–Esseen bounds for discrete observations. Berry–Esseen bound for least squares estimator is proved. In Sect. 5, we obtain asymptotic results for nonergodic case.

Likelihood Flow

Probabilistic representation of likelihood: We apply the Malliavin calculus to obtain *the derivative of the likelihood function as a conditional expectation.* Then we obtain the LAN property. Use of this method to obtain asymptotic normality of the corresponding estimator does not need the RIED condition. MLE is given by

$$\hat{\theta} = -\frac{E(\int_0^T X_t dX_t | X^{(n)})}{E(\int_0^T X_t^2 dt | X^{(n)})}.$$

We use the property of double Wiener chaos expansion. The asymptotic Fisher information is given by $\frac{H\Gamma(2H)}{\theta^{2H}}$.

Local Asymptotic Normality

$$Z_n(\theta, \theta + \frac{u}{\sqrt{n\Delta_n}}) \to^{\mathcal{L}} u\mathcal{N}\left(0, \frac{H\Gamma(2H)}{\theta^{2H}}\right) - u^2 \frac{H\Gamma(2H)}{\theta^{2H}}.$$

The model satisfies the LAN as $\Delta_n \to 0$ and $n\Delta_n \to \infty$.

13.3 Continuous Sampling

Let $(\Omega, \mathcal{F}, \{\mathcal{F}_t\}_{t\geq 0}, P)$ be a stochastic basis on which is defined the fractional Ornstein–Uhlenbeck process X_t satisfying the Itô stochastic differential equation

$$dX_t = -\theta X_t dt + dW_t, \quad t \geq 0, \quad X_0 = 0, \tag{2.1}$$

where $\{W_t\}$ is a fractional Wiener process with the filtration $\{\mathcal{F}_t\}_{t\geq 0}$ and $\theta > 0$ is the unknown parameter to be estimated on the basis of continuous observation of the process $\{X_t\}$ on the time interval $[0, T]$.

Let the realization $\{X_t, 0 \leq t \leq T\}$ be denoted by X_0^T. Let P_θ^T be the measure generated on the space (C_T, B_T) of continuous functions on $[0, T]$ with the associated Borel σ-algebra B_T generated under the supremum norm by the process X_0^T and P_0^T be the standard Wiener measure. When θ is the true value of the parameter, P_θ^T is absolutely continuous with respect to P_0^T, and the Radon–Nikodym derivative (likelihood) of P_θ^T with respect to P_0^T based on X_0^T is given by

$$L_T(\theta) := \frac{dP_\theta^T}{dP_0^T}(X_0^T) = \exp\left\{-\theta \int_0^T X_t dX_t - \frac{\theta^2}{2} \int_0^T X_t^2 dt\right\}. \tag{2.2}$$

Consider the score function, the derivative of the log-likelihood function, which is given by

$$Y_T(\theta) := -\int_0^T X_t dX_t - \theta \int_0^T X_t^2 dt. \tag{2.3}$$

A solution of $Y_T(\theta) = 0$ provides the maximum likelihood estimate (MLE)

$$\hat{\theta}_T := -\frac{\int_0^T X_t dX_t}{\int_0^T X_t^2 dt}. \tag{2.4}$$

The MLE can be represented as

$$\hat{\theta}_T := -\frac{X_T^2 - X_0^2 -{}_1 F_1(2H-1, 2H+1, \theta T)T^{2H}}{\int_0^T X_t^2 dt},$$

where ${}_1F_1(a,b,c)$ is the Kummer's confluent hypergeometric function

$$ {}_1F_1(a,b,c) = \sum_{n=0}^{\infty} \frac{a^{(n)}c^n}{b^{(n)}n!}, $$

where

$$ a^{(n)} = a(a+1)(a+2)\dots(a+n-1). $$

It is an analytic function of complex variables a, b, c except for poles at $b = 1, 2, \dots$. From the series representation, we get that ${}_1F_1(0,b,c) = 1$. If $b > a > 0$, the function ${}_1F_1(a,b,c)$ can be represented as an integral

$$ {}_1F_1(a,b,c) = \frac{\Gamma(b)}{\Gamma(a)\Gamma(b-a)} \int_0^1 e^{cu}u^{a-1}(1-u)^{b-a-1}du. $$

We consider the estimator

$$ \breve{\theta} = \frac{{}_1F_1(2H-1, 2H+1, \theta T)T^{2H}}{\int_0^T X_t^2 dt}, $$

$$ \breve{\theta}_T - \theta = \frac{J_T}{G_T}, $$

where

$$ J_T ={}_1 F_1(2H-1, 2H+1, \theta T)T^{2H} - \theta G_T, $$

$$ \theta_T - \theta = \frac{F_T}{G_T} $$

where

$$ F_T = -\frac{1}{2} I_2\left(e^{-\theta|t-s|}\right), \quad G_T = \int_0^T X_t^2 dt, $$

$$ \left(\frac{T}{2\theta}\right)^{1/2} (\theta_T - \theta) = \frac{(\frac{2\theta}{T})^{1/2} F_T}{(\frac{2\theta}{T}) G_T} $$

$$ \left(\frac{2\theta}{T}\right)^{1/2} F_T = -\left(\frac{\theta}{2T}\right)^{1/2} I_2\left(e^{-\theta|t-s|}\right) =: M_T. $$

Note that M_T is in the second Wiener chaos of W. Hence by Nourdin and Peccati (2009),

$$ \| \mathcal{L}(M_T) - \mathcal{L}(Z) \|_{TV} = 2\sqrt{\frac{1}{6}|E(M_T^4) - 3| + \frac{3 + E(M_T^2)}{2}|E(M_T^2) - 1|}, $$

where Z is a generic standard normal random variable.

When $H = 1/2$, we have

$$E\left(I_2\left(e^{-\theta|t-s|}\right)\right)^2 = E\left(2\int_0^T X_t dW_t\right)^2 = 4E\int_0^T X_t^2 dt$$

$$= 4\int_0^T \frac{1-e^{-2\theta t}}{2\theta} dt = \frac{2}{\theta}\left(T - \frac{1-e^{-2\theta T}}{2\theta}\right).$$

Minimum Contrast Estimator

Using fractional Itô formula (Lemma D.2.3), the score function $Y_T(\theta)$ can be written as

$$Y_T(\theta) = -\frac{1}{2}X_T^2 - \int_0^T \left[\theta X_t^2 - H(2H-1)\int_0^t s^{2H-2}e^{-\theta s}ds\right]dt, \tag{2.5}$$

$$\frac{1}{T}\int_0^T\int_0^t s^{2H-2}e^{-\theta s}dsdt - \frac{1}{2T}X_T^2 \to H\Gamma(2H)\theta^{1-2H} \quad as \quad T \to \infty.$$

Consider the estimating function

$$M_T(\theta) := \theta\int_0^T X_t^2 dt - H\Gamma(2H)\theta^{1-2H} \tag{2.6}$$

and the minimum contrast estimate (MCE)

$$\tilde{\theta}_T := \left(\frac{TH\Gamma(2H)}{\int_0^T X_t^2 dt}\right)^{1/2H} = \left(\frac{T\Gamma(2H+1)}{2\int_0^T X_t^2 dt}\right)^{1/2H}$$

$$= \left(\frac{T}{\mu_H \int_0^T X_t^2 dt}\right)^{1/2H} = \left(\frac{T}{\mu_H G_T}\right)^{1/2H}, \tag{2.7}$$

where

$$\mu_H := \frac{2}{\Gamma(2H+1)} \text{ and } G_T := \int_0^T X_t^2 dt.$$

Note that the MCE does not involve stochastic integral unlike the MLE. It is well known that $\tilde{\theta}_T$ is strongly consistent and asymptotically $\mathcal{N}(0, \mathcal{I}_H(\theta))$ distributed as $T \to \infty$ in the case $H < 3/4$ (see Hu and Nualart (2010)). We develop asymptotic theory for the case $H = 3/4$ and $H > 3/4$. We also obtain Berry–Esseen bounds for all the three cases.

Observe that

$$\tilde{\theta}_T - \theta = \left(\frac{T}{\mu_H G_T}\right)^{1/2H} - \theta = \frac{\left(\frac{T}{\mu_H}\right)^{1/2H} - \theta(G_T)^{1/2H}}{(G_T)^{1/2H}}.$$

Hence,

$$\left(\frac{T}{\mathcal{I}_H(\theta)}\right)^{1/2} (\tilde{\theta}_T - \theta) = \frac{\left(\frac{\mathcal{I}_H(\theta)}{T}\right)^{1/2} N_T}{\left(\frac{\mathcal{I}_H(\theta)}{T}\right) G_T}, \tag{2.8}$$

where

$$-N_T := (\theta^{2H} G_T)^{1/2H} - \left(\frac{T}{\mu_H}\right)^{1/2H}.$$

Let

$$\mathcal{I}_H(\theta) := \frac{2\theta^{2H}}{\Gamma(2H+1)} = \mu_H \theta^{2H}.$$

Observe that $E(X_t^2) \to \mathcal{I}_H^{-1}(\theta)$.. By Taylor expansion, for any two random variables a and b, we have

$$a^{\frac{1}{2H}} - b^{\frac{1}{2H}} = \frac{1}{2H} b^{\frac{1}{2H}-1}(a-b) + \frac{1}{2}(a-b)^2 a^*,$$

where a^* is a random point between a and b. Hence,

$$\begin{aligned} -N_T &= (\theta^{2H} G_T)^{1/2H} - \left(\frac{T}{\mu_H}\right)^{1/2H} \\ &= \frac{1}{2H}\left(\frac{T}{\mu_H}\right)^{\frac{1}{2H}-1}\left(\theta^{2H} G_T - \frac{T}{\mu_H}\right) + \frac{1}{2}\left(\theta^{2H} G_T - \frac{T}{\mu_H}\right)^2 a^* \\ &=: Z_{1,T} + Z_{2,T}, \end{aligned}$$

where a^* is a random point between $\theta^{2H} I_T$ and $\frac{T}{\mu_H}$.

We show that

$$Z_{2,T} \to 0 \text{ in probability as } T \to \infty \text{ and } \left(\frac{\mathcal{I}_H(\theta)}{T}\right)^{1/2} Z_{1,T} \to^D \mathcal{N}(0,1).$$

First we obtain exact Berry–Esseen bound for $\left(\frac{\mathcal{I}_H(\theta)}{T}\right)^{1/2} Z_{1,T}$.

Define the covariance function as

$$r_H(t) = E(X_t X_0).$$

Let $\Phi(\cdot)$ denote the standard normal distribution function. Let $\mathcal{R}(\cdot)$ denote the Rosenblatt distribution. Throughout the chapter, C denotes a generic constant (which does not depend on T and x).

Let

$$(\mathcal{J}_H(\theta))^{1/2} := \frac{4\theta^2}{3}, \quad \text{and} \quad (\mathcal{K}_H(\theta))^{1/2} := \theta^2[H(4H-2)/(4H-3)]^{-1/2}.$$

We have:

Lemma 3.1
(a) For $H \in (0, 3/4)$

$$\sup_{x\in\mathbb{R}} \left| P\left\{ \left(\frac{\mathcal{I}_H(\theta)}{T} \right)^{1/2} Z_{1,T} \le x \right\} - \Phi(x) \right| \le CT^{-1/2}.$$

(b) For $H = 3/4$ (the singular case)

$$\sup_{x\in\mathbb{R}} \left| P\left\{ \left(\frac{\mathcal{J}_H(\theta)}{T\log T} \right)^{1/2} Z_{1,T} \le x \right\} - \Phi(x) \right| \le C(\log T)^{-1/2}.$$

(c) For $H \in (3/4, 1)$

$$\sup_{x\in\mathbb{R}} \left| P\left\{ \left(\frac{\mathcal{K}_H(\theta)}{T^{4H-2}} \right)^{1/2} Z_{1,T} \le x \right\} - \mathcal{R}(x) \right| \le CT^{3/4-H}.$$

Proof Note that

$$X_t = \int_{-\infty}^{t} e^{-\theta(t-s)} dW_s^H, \ t \ge 0, \ \theta \in \mathbb{R}_+,$$

where the integral converges pathwisely in the usual Riemann–Stieltjes sense. The process is asymptotically stationary.

If $H \ne 1/2$, $r_H(t) = Ct^{2H-2} + O(t^{2H-4})$ as $t \to \infty$. If $H = 1/2$, $r_{1/2}(t) = e^{-\theta|t|}$ for all $t \in \mathbb{R}$.

The tools of the proof are CLT for multiple stochastic integrals by Peccati and Tudor (2005).

Let

$$Y_t := \theta^H \mu_H^{1/2} X_t.$$

Let

$$A_T(v_1, v_2) := \frac{1}{\sqrt{2TL(T)}} \int_{v_1 T}^{v_2 T} (Y_u^2 - 1)du, \ v_2 \ge v_1 \ge 0,$$

where

$$L(T) := \int_0^T r_H^2(u)du = \frac{4\sin^2(\pi H)}{\pi\theta} \int_0^\infty \frac{s^{2-4H}}{(1+s^2)^2} ds$$

$$E\left(\int_0^T (Y_u^2 - 1)du \right)^{2(1+p)} \le C\left(2\int_0^T r_H^2(u)du \right)^{1+p}, 0 < p \le 1.$$

We want to show that as $T \to \infty$

$$A_T(v_1, v_2) \to^D \sqrt{2}(W_{v_2} - W_{v_1}). \tag{2.9}$$

By Proposition 1.1.4 in Nualart (1995) and a stochastic version of the stochastic Fubini's theorem, we express $Z_{1,T}$ as a second Wiener chaos:

$$\begin{aligned} Z_{1,T} &= \tfrac{1}{2H}\left(\tfrac{T}{\mu_H}\right)^{\frac{1}{2H}-1}\left(\theta^{2H} G_T - \tfrac{T}{\mu_H}\right) \\ &= \tfrac{1}{2H}\left(\tfrac{T}{\mu_H}\right)^{\frac{1}{2H}-1}\left(\theta^{2H}\textstyle\int_0^T X_t^2 dt - \tfrac{T}{\mu_H}\right) \\ &= \tfrac{1}{2H}\left(\tfrac{T}{\mu_H}\right)^{\frac{1}{2H}-1}\left(\theta^{2H}\textstyle\int_{\mathbb{R}^2}\int_0^T e^{-\theta(t-u_1)}e^{-\theta(t-u_2)}dW^{\otimes 2}(u_1,u_2) - \tfrac{T}{\mu_H}\right), \end{aligned} \tag{2.10}$$

where the right hand side is understood as a multiple Wiener integral (MWI). The covariance function $r : \mathbb{R} \to \mathbb{R}$ is defined as

$$E(X_t X_0) = r(t), \ t \geq 0.$$

Define

$$L(T) := \int_0^T r^2(u)du, \quad L_{|\cdot|}(T) := \int_0^T |r(u)|^2 du.$$

If we show that

$$E\left(A_T(v_1, v_2)\right)^2 \to 2E(W_{v_2} - W_{v_1})^2, \tag{2.11}$$

$$E\left(A_T(v_1, v_2)A_T(v_3, v_4)\right) \to 0, \tag{2.12}$$

$$E\left(A_T(v_1, v_2)\right)^4 \to 2E(W_{v_2} - W_{v_1})^4, \tag{2.13}$$

then by (2.10)

$$A_T(v_1, v_2) \to^D \sqrt{2}(W_{v_2} - W_{v_1}).$$

(i) Proof of (2.11): We show (2.11) by Wick's lemma 3.3.1 (d).

As r is symmetric and L is slowly varying function, we get from Fubini's theorem and Karamata's theorem that as $T \to \infty$

$$\begin{aligned} \int_{v_1T}^{v_2T}\int_{v_1T}^{v_2T} r^2(u_1-u_2)du_1du_2 &= \int_0^T (v_2 - v_1)L(u)du \\ &\sim 2(v_2 - v_1)TL(T). \end{aligned} \tag{2.14}$$

Thus by Wick's lemma as $T \to \infty$

$$E\left(A_T(v_1, v_2)\right)^2 = \frac{4}{\sqrt{2TL(T)}}\int_{v_1T}^{v_2T}\int_{uT}^{wT} r^2(u_1-u_2)du_1du_2 \to 2(v_2 - v_1).$$

This completes the proof of (2.11).

(ii) Proof of (2.12): Pick $v_4 > v_3 \geq v_2 \geq v_1 \geq 0$. We have

$$\int_{v_3T}^{v_4T}\int_{v_1T}^{v_2T} r^2(u_1-u_2)du_1du_2$$

$$= \left[\int_0^{(v_4-v_1)T} + \int_0^{(v_3-v_2)T} - \int_0^{(v_4-v_2)T} - \int_0^{(v_3-v_1)T}\right] L(u)du$$

for $T > 0$, and thus by Karamata's theorem as $T \to \infty$

$$\int_{v_3T}^{v_4T}\int_{v_1T}^{v_2T} r^2(u_1-u_2)du_1du_2 = o(TL(T)). \tag{2.15}$$

Equation (2.15) and Wick's lemma imply (2.12) since as $T \to \infty$

$$E\left(A_T(v_1,v_2)A_T(v_3,v_4)\right) = \frac{4}{\sqrt{2TL(T)}}\int_{v_3T}^{v_4T}\int_{v_1T}^{v_2T} r^2(u_1-u_2)du_1du_2 \to 0.$$

(iii) Proof of (2.13): Let $\epsilon > 0$.

$$\lim_{T\to\infty} L(T) = \int_0^\infty r^2(u)du = \infty.$$

By Hölder inequality and Karamata's theorem for $v > \epsilon > 0$,

$$\int_0^{vT} r^m(u)du \leq (\epsilon T)^{1-m/q}L^{m/q}(\epsilon T) + [(v-\epsilon)T]^{1-m/q}[L(vT)-L(\epsilon T)]^{m/q}$$

$$\sim (\epsilon T)^{1-m/q}L^{m/q}(T), T\to\infty, 1\leq m<q.$$

Thus

$$\int_0^{vT} r^m(u)du = o(T^{1-m/q})L^{m/q}(T),\ T\to\infty, 1\leq m<q, v>0. \tag{2.16}$$

On the other hand, from the inequality between the geometric and arithmetic mean, we have for all $m \in \mathbb{N}, v > 0$ and $0 \leq w_1, \ldots, w_m \leq vT$,

$$\int_0^{vT} \prod_{i=1}^m |r(u-w_i)|du \leq 2\int_0^{vT} |r^m(u)|du.$$

Fix $v_2 > v_1 \geq 0$.

$$\left|\int_{v_1T}^{v_2T}\int_{v_1T}^{v_2T}\int_{v_1T}^{v_2T}\int_{v_1T}^{v_2T} r(u_1-u_2)r(u_3-r_4)du_1du_2du_3du_4\right|$$

$$\leq \left| \int_{v_1 T}^{v_2 T} \int_{v_1 T}^{v_2 T} \int_{v_1 T}^{v_2 T} \int_{v_1 T}^{v_2 T} r(u_1 - u_2) r(u_2 - u_3) r(u_3 - u_4) du_1 du_2 du_3 du_4 \right|.$$

Thus,

$$\left| \int_{v_1 T}^{v_2 T} \int_{v_1 T}^{v_2 T} \int_{v_1 T}^{v_2 T} \int_{v_1 T}^{v_2 T} r(u_1 - u_2) r(u_3 - u_4) du_1 du_2 du_3 du_4 \right|$$

$$\leq 8 u_2 T L(v_2 T) \int_0^{v_2 T} |r(u)| du_1 \int_0^{v_2 T} |r(u)| du_1$$

$$= o(T^2 L^2(T)), \quad T \to \infty.$$

Consequently, by Wick's lemma and Fubini's theorem as $T \to \infty$,

$$E\left(A_T(v_1, v_2)\right)^4 = 6 \left(\int_{v_3 T}^{v_4 T} \int_{v_1 T}^{v_2 T} r^2(u_1 - u_2) du_1 du_2 \right)^2 + o(T^2 L^2(T)).$$

This completes the proof of (2.13) in view of (2.14). □

Second Wiener Chaos and Fourth Moment Condition

Let $F_T = I_2(f)$ with $E(F_T) = 0$ and $Var(F_T) = 1$. Since

$$L^{-1} F_T = -\frac{F_T}{2},$$

$\langle DF_T, -DL^{-1} F_T \rangle_{\mathcal{H}} = \frac{1}{2} \|F_T\|^2$. Thus

$$d_{Kol}(F_T, Z) \leq \sqrt{Var(\langle DF_T, -DL^{-1} F_T \rangle_{\mathcal{H}})} \leq \sqrt{\frac{1}{6}(E(F_T^4) - 3)}.$$

$$DF_T = 2 I_1(f), \; E(F_T^2) = 2\|f\|^2_{\mathcal{H}^{\otimes 2}}, \; DF_T^3 = 3 F_T^2 DF_T, \; E(F_T^4) = \frac{3}{2} E(F_T^2 \|DF_T\|).$$

For a sequence of chaotic random variables with unit variance, the speed of convergence to zero of the fourth cumulants $E(F_T^4) - 3$ also determines the speed of convergence in the Kolmogorov distance.

The speed of convergence to zero of the k-th cumulant of F_T is always dominated by the speed of convergence of the fourth cumulant $E(F_T^4) - 3$.

$$E(F_T^k) - E(z^k) \leq C E(F_T^4) - 3.$$

One could also discuss here the convergence of moments of estimators.

Using (Buchmann and Chan, 2009, Theorem 3.1), the following four theorem shows the limiting distribution of the MCE.

Theorem 3.1

(a) For $H = 3/4$

$$\sup_{x\in\mathbb{R}}\left|P\left\{(T\log T\mathcal{I}_H^{-1}(\theta))^{1/2}(\tilde{\theta}_T-\theta)\leq x\right\}-\Phi(x)\right|\to 0.$$

(b) For $H > 3/4$

$$\sup_{x\in\mathbb{R}}\left|P\left\{(T\mathcal{K}_H^{-1}(\theta))^{1-2H}(\tilde{\theta}_T-\theta)\leq x\right\}-\mathcal{R}(x)\right|\to 0.$$

(c) For $H = 3/4$

$$\sup_{x\in\mathbb{R}}\left|P\left\{(T\log T\mathcal{J}_H^{-1}(\theta))^{1/2}(\hat{\theta}_T-\theta)\leq x\right\}-\Phi(x)\right|\to 0.$$

(d) For $H > 3/4$

$$\sup_{x\in\mathbb{R}}\left|P\left\{(T\mathcal{K}_H^{-1}(\theta))^{1-2H}(\hat{\theta}_T-\theta)\leq x\right\}-\mathcal{R}(x)\right|\to 0.$$

We obtain the following Berry–Esseen theorems:

Theorem 3.2

(a) For $H < 3/4$

$$\sup_{x\in\mathbb{R}}\left|P\left\{T\mathcal{I}_H^{-1}(\theta))^{1/2}(\tilde{\theta}_T-\theta)\leq x\right\}-\Phi(x)\right|\leq CT^{-1/2}.$$

(b) For $H = 3/4$

$$\sup_{x\in\mathbb{R}}\left|P\left\{(T\log T\mathcal{J}_H^{-1}(\theta))^{1/2}(\tilde{\theta}_T-\theta)\leq x\right\}-\Phi(x)\right|\leq C(\log T)^{-1/2}.$$

(c) For $H > 3/4$

$$\sup_{x\in\mathbb{R}}\left|P\left\{(T\mathcal{K}_H^{-1}(\theta))^{1-2H}(\tilde{\theta}_T-\theta)\leq x\right\}-\mathcal{R}(x)\right|\leq CT^{3/4-H}.$$

We use the method of Breton and Nourdin (2008) by expressing the integral G_T as a second Wiener chaos I_2. We also use stochastic Fubini theorem.

We need the following Lemmas for the proof of the theorems.

Lemma 3.2

(a) For $H = 3/4$, for every $\delta > 0$,

$$P\left\{\left|(T\log T)^{-1}G_T-1\right|\geq\delta\right\}\to 0.$$

(b) For $H = 3/4$

$$\sup_{x\in\mathbb{R}}\left|P\left\{(T\log T)^{-1/2}N_T \le x\right\} - \Phi(x)\right| \to 0.$$

(c) For $H > 3/4$, for every $\delta > 0$,

$$P\left\{\left|(T)^{-2+4H}G_T - 1\right| \ge \delta\right\} \to 0.$$

(d) For $H > 3/4$

$$\sup_{x\in\mathbb{R}}\left|P\left\{(T)^{-1+2H}N_T \le x\right\} - \mathcal{R}(x)\right| \to 0.$$

Proof We use the stochastic Fubini theorem to convert the Lebesgue integral into a multiple Wiener integral. Then we use the central limit theorem for double Wiener chaos. See Buchmann and Chan (2009). First we obtain Berry–Esseen bound for $\left(\frac{\mathcal{I}_H(\theta)}{T}\right)^{1/2} Z_{1,T}$.

For $H \in (0, 3/4)$,

$$\sup_{x\in\mathbb{R}}\left|P\left\{\left(\frac{\mathcal{I}_H(\theta)}{T}\right)^{1/2} Z_{1,T} \le x\right\} - \Phi(x)\right| \le CT^{-1/2}.$$

Recall that

$$Z_{1,T} = \frac{1}{2H}\left(\frac{T}{\mu_H}\right)^{\frac{1}{2H}-1}\left(\theta^{2H}G_T - \frac{T}{\mu_H}\right)$$

$$= \frac{1}{2H}\left(\frac{T}{\mu_H}\right)^{\frac{1}{2H}-1}\left(\theta^{2H}\int_0^T X_t^2 dt - \frac{T}{\mu_H}\right)$$

$$= \frac{1}{2H}\left(\frac{T}{\mu_H}\right)^{\frac{1}{2H}-1}\left(\theta^{2H}\int_{\mathbb{R}^2}\int_0^T e^{-\theta(t-u_1)}e^{-\theta(t-u_2)}dW^{\otimes 2}(u_1,u_2) - \frac{T}{\mu_H}\right)$$

and

$$\sigma^2(T) = \mathrm{Var}(Z_{1,T}) = \mathrm{Var}\left(\frac{1}{2H}\left(\frac{T}{\mu_H}\right)^{\frac{1}{2H}-1}\left(\theta^{2H}\int_0^T X_t^2 dt - \frac{T}{\mu_H}\right)\right)$$

$$= \frac{2T^{\frac{1}{H}-2}\theta^{4H}}{4H^2\mu_H^2}\int_0^T\int_0^T r^2(u-v)dudv.$$

As $T \to \infty$,

$$\sigma^2(T) \to \sigma^2(\infty) = \frac{2\theta^{4H}}{4H^2\mu_H^2}\int_{-\infty}^{\infty} r^2(x)dx.$$

We know that $Z_{1,T} \to^D U \sim N(0,1)$.

$$DZ_{1,T} = \frac{2}{\sigma(T)T^{2-\frac{1}{H}}} \int_0^T I_1(1_{[u,u+1]})1_{[u,u+1]}du.$$

We first show that as $T \to \infty$

$$\sqrt{T}\left(\frac{1}{2}\|DZ_{1,T}\|^2 - 1\right) \to^D N(0, \hat{\sigma}^2(\infty)) \qquad (2.17),$$

where

$$\hat{\sigma}^2(\infty) := \frac{8}{\sigma^4(\infty)} \int_{\mathbb{R}}^3 r(x_1)r(x_2)r(x_3)r(x_2+x_3-x_1)dx_1dx_2dx_3$$

$$\begin{aligned}
\|DZ_{1,T}\|^2 &= \frac{4}{\sigma^2 T}\int_0^T\int_0^T r(u-v)I_1(1_{[u,u+1]})I_1(1_{[v,v+1]})dudv \\
&= \frac{4}{\sigma^2 T}\int_0^T\int_0^T I_2(1_{[u,u+1]})\otimes(1_{[v,v+1]})r(u-v)dudv \\
&\quad +\frac{4}{\sigma^2 T}\int_0^T\int_0^T I_0(1_{[u,u+1]}\otimes 1_{[v,v+1]})r^2(u-v)dudv \\
&= \frac{4}{\sigma^2 T}\int_0^T\int_0^T I_4(1^{\otimes 2}_{[u,u+1]})\otimes(1^{\otimes}2_{[v,v+1]})r(u-v)dudv \\
&\quad +\frac{4}{\sigma^2 T}\int_0^T\int_0^T I_2(1_{[u,u+1]}\otimes 1_{[v,v+1]})r^2(u-v)dudv
\end{aligned}$$

yielding

$$\|DZ_{1,T}\|^2 - 1 = \frac{2}{\sigma^2 T}\int_0^T\int_0^T I_2(1_{[u,u+1]}\otimes 1_{[v,v+1]})r(u-v)dudv.$$

We shall first show that as $T \to \infty$

$$\frac{1}{\sqrt{T}}\int_0^T\int_0^T I_2(1_{[u,u+1]}\otimes 1_{[v,v+1]})r(u-v)dudv \to^D N(0,\hat{\sigma}^2(\infty)). \qquad (2.18)$$

Observe first that

$$\begin{aligned}
\hat{\sigma}^2(T) &:= Var\left(\frac{1}{\sqrt{T}}\int_0^T\int_0^T I_2(1_{[u,u+1]}\otimes 1_{[v,v+1]})r(u-v)dudv\right) \\
&= \frac{24}{T}\int_0^T\int_0^T\int_0^T\int_0^T r(u-v)r(w-z)r(u-w)r(v-z)dudvdwdz \\
&\to \hat{\sigma}^2(\infty)
\end{aligned}$$

so that (2.9) will take place if and only if

$$Q_T := \frac{1}{\hat{\sigma}(T)\sqrt{T}} \int_0^T \int_0^T I_2(1_{[u,u+1]} \otimes 1_{[v,v+1]}) r(u-v) du dv \to^D \mathcal{N}(0,1). \tag{2.19}$$

We have

$$DQ_T = \frac{1}{\hat{\sigma}(T)\sqrt{T}} \int_0^T \int_0^T r(u-v) I_1(1_{[v,v+1]} \otimes 1_{[u,u+1]}) 1_{[u,u+1]} du dv.$$

$$\begin{aligned}
&\|DQ_1\|^2 \\
= \; & \frac{4}{\hat{\sigma}^2(T)T} \int_0^T \int_0^T \int_0^T \int_0^T r(u_1-u_2) r(u_3-u_4) r(u_1-u_3) \\
& I_1\left(1_{[u_1,u_1+1]} \otimes 1_{[u_2,u_2+1]}\right) I_1\left(1_{[u_3,u_3+1]} \otimes 1_{[u_4,u_4+1]}\right) du_1 du_2 du_3 du_4 \\
= \; & \frac{4}{\hat{\sigma}^2(T)T} \int_0^T \int_0^T \int_0^T \int_0^T r(u_1-u_2) r(u_3-u_4) r(u_1-u_3) \\
& I_1\left(1_{[u_1,u_1+1]} \otimes 1_{[u_3,u_3+1]}\right) I\left(1_{[u_2,u_2+1]} \otimes 1_{[u_4,u_4+1]}\right) du_1 du_2 du_3 du_4 \\
& + \frac{4}{\hat{\sigma}^2(T)T} \int_0^T \int_0^T \int_0^T \int_0^T r(u_1-u_2) r(u_3-u_4) r(u_1-u_3) \\
& I_1\left(1_{[u_1,u_1+1]} \otimes 1_{[u_3,u_3+1]}\right) I_1\left(1_{[u_2,u_2+1]} \otimes 1_{[u_4,u_4+1]}\right) du_1 du_2 du_3 du_4.
\end{aligned}$$

Consequently,

$$\begin{aligned}
&\frac{1}{2}\|DQ_1\|^2 - 1 \\
= \; & \frac{4}{\hat{\sigma}^2(T)T} \int_0^T \int_0^T \int_0^T \int_0^T r(u_1-u_2) r(u_3-u_4) r(u_1-u_3) \\
& I_1\left(1_{[u_1,u_1+1]} \otimes 1_{[u_2,u_2+1]} \otimes 1_{[u_3,u_3+1]} \otimes 1_{[u_4,u_4+1]}\right) du_1 du_2 du_3 du_4 \\
+ \; & \frac{4}{\hat{\sigma}^2(T)T} \int_0^T \int_0^T \int_0^T \int_0^T r(u_1-u_2) r(u_3-u_4) r(u_1-u_3) \\
& I_1\left(1_{[u_1,u_1+1]} \otimes 1_{[u_3,u_3+1]}\right) du_1 du_2 du_3 du_4.
\end{aligned}$$

By the orthogonality and isometric properties of multiple stochastic integrals

$$\begin{aligned}
& Var\left(\sqrt{T}\left(\|DZ_{1,T}\|^2 - 1\right)\right) \\
= \; & \frac{8}{\hat{\sigma}^4(T)T} \left\langle \int_0^T \int_0^T 1_{[u,u+1]} \otimes 1_{[v,v+1]} r(u-v) du dv, \right. \\
& \left. \int_0^T \int_0^T 1_{[w,w+1]} \otimes 1_{[z,z+1]} r(w-z) dw dz \right\rangle \\
= \; & \frac{8}{\hat{\sigma}^4(T)T} \int_0^T \int_0^T \int_0^T \int_0^T r(u-v) r(w-z) r(u-w) r(v-z) du dv dw dz
\end{aligned}$$

from which we easily deduce that

$$Var\left(\sqrt{T}\left(\|DZ_{1,T}\|^2-1\right)\right)\to\hat{\sigma}^2(\infty)$$

as $T\to\infty$. Let us prove that

$$\left(Z_{1,T},\ \sqrt{T}\left(\|DZ_{1,T}\|^2-1\right)\right)\to^D(U,V) \tag{2.20}$$

with (U,V) is a centered Gaussian vector such that $E(U^2)=1,\ E(V^2)=\hat{\sigma}^2(\infty)$ and

$$E(UV)=-\hat{\gamma}(\infty)=\frac{4}{\sigma^3(\infty)}\int_{\mathbb{R}}\int_{\mathbb{R}}r(x)r(y)r(x-y)dxdy.$$

We already know that $Z_{1,T}\to^D U$ and also that (2.17) is verified. We have proved (2.17) by first decomposing $\sqrt{T}\left(\|DZ_{1,T}\|^2-1\right)$ into a finite sum of multiple Wiener–Itô integrals, and each multiple integral satisfies an appropriate CLT. It is sufficient to show the convergence

$$E\left(Z_{1,T}\times\sqrt{T}\left(\|DZ_{1,T}\|^2-1\right)\right)\to\frac{4}{\sigma^3(\infty)}\int_{\mathbb{R}}\int_{\mathbb{R}}r(x)r(y)r(x-y)dxdy. \tag{2.21}$$

$$\begin{aligned}
&E\left(Z_{1,T}\times\sqrt{T}\left(\|DZ_{1,T}\|^2-1\right)\right)\\
=\ &\frac{2}{T\sigma^3(T)}\int_0^T\int_0^T\int_0^T r(u-v)E\\
&\times\left(I_2(1^{\otimes 2}_{[w,w+1]}))I_2(1_{[u,u+1]}\otimes 1_{[v,v+1]}\right)dudvdw\\
=\ &\frac{4}{T\sigma^3(T)}\int_0^T\int_0^T\int_0^T r(u-v)r(u-w)r(w-v)dudvdw\\
\to\ &\frac{4}{\sigma^3(\infty)}\int_{\mathbb{R}}\int_{\mathbb{R}}r(x)r(y)r(x-y)dxdy\\
=\ &\frac{4}{\sigma^3(\infty)}\int_{\mathbb{R}}\int_{\mathbb{R}}r(x)r(y)r(x-y)dxdy.
\end{aligned}$$

From (2.19) as $T\to\infty$, we have

$$\phi_T:=Var\left(\sqrt{T}\left(\|DZ_{1,T}\|^2-1\right)\right)\sim T^{-1/2}\hat{\sigma}(\infty)$$

and

$$\left(Z_{1,T},\ \sqrt{T}\left(\|DZ_{1,T}\|^2-1\right)\right)\to^D(U,\ \hat{\sigma}(\infty)^{-1}V). \tag{2.22}$$

As a consequence, with $\rho=\frac{\hat{\gamma}(\infty)}{\hat{\sigma}(\infty)}$, as $T\to\infty$, we have

$$\phi_T^{-1}[P(Z_{1,T}\le x)-\Phi(x)]\to\frac{\hat{\gamma}(\infty)}{3\hat{\sigma}(\infty)}(x^2-1)\Phi'(x)=\frac{\rho}{3}\Phi'''(x).$$

Optimal Rate

Suppose that

$$d_{Kol}(F_T; Z) \leq \psi_T \downarrow 0,$$

where $Z \sim \mathcal{N}(0,1)$.

Recall that ψ_T is said to provide an optimal rate if there exists a $c \in (0,1)$ such that

$$c < \frac{d_{Kol}(F_T; Z)}{\psi_T} \leq 1$$

for large enough T.

Reverse Berry–Esseen Inequality (Barbour and Hall (1984))

Barbour and Hall (1984) studied the lower bound on the convergence to Poisson distribution.

Here optimal rate is implied by the fact that, for some $x \in \mathbb{R}$,

$$\frac{P(F_T \leq x) - P(Z \leq x)}{\psi_T} \to \alpha(x) \neq 0.$$

Random Time Change of Double Stochastic Integral

This method is more satisfactory than the method of moments. But the implied rate by this method is sub-optimal.

Method of Cumulants

This method gives the optimal bound. The cumulant or semi-invariant of a random variable Y with finite moments of all orders and characteristic function $\psi_Y(t) = E[e^{itY}]$ is given by

$$\chi_n(Y) = (-i)^n \frac{d^n}{dt^n} \log \psi_Y(t)\big|_{t=0}, n \geq 1.$$

Square integrable functionals of Gaussian fields can always be represented in terms of a series of multiple Wiener–Itô integrals, and the moments and cumulants of multiple integrals can be computed by means of *diagram formula.*

Contraction

The connection between moments and contraction comes from the formula:

Asymptotic Independence

Multiple integrals of different orders are asymptotically independent. Peccati and Tudor (2005) showed that for a vector of multiple integrals whose covariance matrix is converging, the componentwise convergence to a Gaussian distribution always implies joint convergence.

13.3.1 Rate Convergence in the Non-CLT

Gamma Approximations (Chi-Square Approximations)

Proposition 3.1 Let $F_T = I_2(f_T)$. Suppose $E(F_T^2) \to 1$ and $\|f_T \otimes_1 f_T\|_{\mathcal{H}^{\otimes 2}} \to 0$ as $T \to \infty$. Then by defining

$$H_T = I_4(f_T \tilde{\otimes} f_T),$$

one has

$$E\left[\left(2 + 2H_T - \frac{1}{4}\|DH_T\|_{\mathcal{H}}^2\right)\right] \to 0$$

and

$$d_{\mathcal{H}_1}(F_T^2 - 1, Z^2 - 1) \leq 8\sqrt{2}\|f_T \otimes_1 f_T\|_{\mathcal{H}^{\otimes 2}} + \sqrt{2\pi E\left[\left(2 + 2H_T - \frac{1}{4}\|DH_T\|_{\mathcal{H}}^2\right)\right]}.$$

Example

$$d_{\mathcal{H}_1}(F_n,\ Z^2 - 1) \leq Cn^{-1/2},$$

where $F_n = \frac{1}{n}\sum_{k,l=0}^{n-1} a(k-l)(G_k G_l - \delta_{kl})$, where (G_k) is a sequence of centered i.i.d. standard Gaussian random variables with $a_0 = 1,\ a(r) = a(-r), r \in \mathbb{Z}$ and $\sum_{r\in\mathbb{Z}} |a(r) - 1| < \infty$.

Without loss of generality, we can assume that $G_k = B_{k+1} - B_k$, where B is the standard Brownian motion.

Lemma 3.3 Let $F = I_2(f)$. Then

$$E(\|DF\|_{\mathcal{H}}^4) = \frac{2}{3}E(F^4) + 2(E(F^2))^2.$$

Fractional Ornstein–Uhlenbeck Process

Let $(\Omega, \mathcal{F}, \{\mathcal{F}_t\}_{t\geq 0}, P)$ be a stochastic basis on which is defined the Ornstein–Uhlenbeck process $\{X_t\}$ satisfying the Itô stochastic differential equation

$$dX_t = \theta X_t dt + dW_t^H, \quad t \geq 0, \quad X_0 = 0, \tag{1.1}$$

where $\{W_t^H\}$ is a fractional Brownian motion with $H > 1/2$ with the filtration $\{\mathcal{F}_t\}_{t\geq 0}$ and $\theta > 0$ is the unknown parameter to be estimated on the basis of continuous observation of the process $\{X_t\}$ on the time interval $[0, T]$.

Define

$$\begin{aligned} \kappa_H &:= 2H\Gamma(3/2 - H)\Gamma(H + 1/2), \\ k_H(t, s) &:= \kappa_H^{-1}(s(t - s))^{\frac{1}{2} - H} \end{aligned}$$

$$\lambda_H := \frac{2H\Gamma(3-2H)\Gamma(H+\frac{1}{2})}{\Gamma(3/2-H)}$$

$$v_t \equiv v_t^H := \lambda_H^{-1} t^{2-2H}, \; M_t^H := \int_0^t k_H(t,s) dW_s^H.$$

From Norros et al. (1999), it is well known that M_t^H is a Gaussian martingale, called the fundamental martingale whose variance function $\langle M^H \rangle_t$ is v_t^H.

Moreover, the natural filtration of the martingale M^H coincides with the natural filtration of the FBM W^H since

$$W_t^H := \int_0^t K(t,s) dM_s^H$$

holds for $H \in (1/2, 1)$, where

$$K_H(t,s) := H(2H-1) \int_s^t r^{H-\frac{1}{2}} (r-s)^{H-\frac{3}{2}} dr, \quad 0 \leq s \leq t,$$

and for $H = 1/2$, the convention $K_{1/2} \equiv 1$ is used.

Define

$$Q_t := \frac{d}{dv_t} \int_0^t k_H(t,s) X_s ds.$$

It is easy to see that

$$Q_t = \frac{\lambda_H}{2(2-2H)} \left\{ t^{2H-1} Z_t + \int_0^t r^{2H-1} dZ_s \right\}.$$

Define the process $Z = (Z_t, t \in [0,T])$ by

$$Z_t := \int_0^t k_H(t,s) dX_s.$$

The following facts are known from Kleptsyna and Le Breton (2002):

(i) Z is the fundamental semimartingale associated with the process X.
(ii) Z is a $(\mathcal{F}_t)$-semimartingale with the decomposition

$$Z_t = \theta \int_0^t Q_s dv_s + M_t^H.$$

(iii) X admits the representation

$$X_t = \int_0^t K_H(t,s) dZ_s.$$

(iv) The natural filtrations $(\mathcal{Z}_t)$ of Z and $(\mathcal{X}_t)$ of X coincide.

Let the realization $\{X_t, 0 \le t \le T\}$ be denoted by X_0^T. Let P_θ^T be the measure generated on the space (C_T, B_T) of continuous functions on $[0, T]$ with the associated Borel σ-algebra B_T generated under the supremum norm by the process X_0^T and P_0^T be the standard Wiener measure. Applying Girsanov type formula for FBM, when θ is the true value of the parameter, P_θ^T is absolutely continuous with respect to P_0^T, and the Radon–Nikodym derivative (likelihood) of P_θ^T with respect to P_0^T based on X_0^T is given by

$$L_T(\theta) := \frac{dP_\theta^T}{dP_0^T}(X_0^T) = \exp\left\{\theta \int_0^T Q_t dZ_t - \frac{\theta^2}{2}\int_0^T Q_t^2 dv_t\right\}. \tag{1.2}$$

Consider the score function, the derivative of the log-likelihood function, which is given by

$$R_T(\theta) := \int_0^T Q_t dZ_t - \theta \int_0^T Q_t^2 dv_t. \tag{1.3}$$

A solution of $R_T(\theta) = 0$ provides the maximum likelihood estimate (MLE)

$$\hat{\theta}_T := \frac{\int_0^T Q_t dZ_t}{\int_0^T Q_t^2 dv_t}. \tag{1.4}$$

The LSE is given by

$$\theta_T := \frac{\int_0^T X_t dX_t}{\int_0^T X_t^2 d_t}. \tag{1.4}$$

$$\theta_T - \theta = \frac{F_T}{G_T},$$

where

$$F_T = \frac{1}{2} I_2\left(e^{-\theta|t-s|}\right), \quad G_T = \int_0^T X_t^2 dt.$$

Strictly speaking, $\hat{\theta}_T$ is not the maximum likelihood estimate of θ since $\hat{\theta}_T$ may take negative values where as the parameter θ is assumed to be strictly positive. For an exact definition of MLE, see Kutoyants (1994a,b). Nevertheless, we use this terminology.

Kleptsyna and Le Breton (2002) showed that $\hat{\theta}_T$ is a strongly consistent estimator of θ. Using the Fourier method, we prove a Berry–Esseen type theorem for the estimator $\tilde{\theta}_T$ that gives the asymptotic normality.

Using the fractional Itô formula, the score function $R_T(\theta)$ can be written as

$$R_T(\theta) = \frac{1}{2}\left[\frac{\lambda_H}{(2-2H)} Z_T \int_0^T t^{2H-1} dZ_t - T\right] - \theta \int_0^T Q_t^2 dv_t. \tag{1.5}$$

Consider the contrast function

$$c_T(\theta) = -\frac{T}{2} - \theta \int_0^T Q_t^2 dv_t \tag{1.6}$$

and the minimum contrast estimate (MCE)

$$\tilde{\theta}_T := \frac{-T}{2\int_0^T Q_t^2 dv_t}. \tag{1.7}$$

M-estimator is reduced to the minimum contrast estimator. Note that the MCE does not involve stochastic integral unlike the MLE.

Minimum contrast estimation in diffusion processes was studied by Lanksa (1979), and M-estimation in diffusion processes was studied by Yoshida (1988).

Note that using the Skorokhod embedding of martingale, which has, since long, been the basic tool for normal approximation of martingales, will not give a rate better than $O(T^{-1/4})$ (see Borokov 1973). To obtain the rate of normal approximation of the order $O(T^{-1/2})$, we adopt the Fourier method followed by the squeezing technique of Pfanzagl (1971) and Michel and Pfanzagl (1971) developed for the minimum contrast estimator in the i.i.d. case.

Observe that

$$\left(\frac{T}{-2\theta}\right)^{1/2}(\tilde{\theta}_T - \theta) = \frac{\left(\frac{-2\theta}{T}\right)^{1/2} J_T}{\left(\frac{2\theta}{T}\right) I_T}, \tag{1.8}$$

where

$$J_T := \theta I_T - \frac{T}{2} \quad \text{and} \quad I_T := \int_0^T Q_t^2 dv_t.$$

We study the large deviations of the estimator $\tilde{\theta}_T$. For the standard Ornstein–Uhlenbeck process, large deviations of $\hat{\theta}_T$ and $\tilde{\theta}_T$ were obtained in Florens-Landais and Pham (1999). In particular, it was shown that the large deviation probabilities of $\hat{\theta}_T$ are identical to those of $\tilde{\theta}_T$ for $\theta \leq \theta_0/3$ but weaker for $\theta > \theta_0/3$, where θ_0 is the true value of the parameter. We generalize this result to the fractional Ornstein–Uhlenbeck case.

Proposition 1.1 $\tilde{\theta}_T$ satisfies the large deviations principle with good rate function $I(\theta)$ given by

$$I(\theta) = \begin{cases} -\frac{(\theta-\theta_0)^2}{4\theta} & : \theta \leq \frac{\theta_0}{3} \\ 2\theta - \theta_0 & : \theta > \frac{\theta_0}{3}. \end{cases}$$

Bercu et al. (2009) showed the large deviation for the maximum likelihood estimator. By following Florens-Landais and Pham (1999), the above large deviations hold for the minimum contrast estimators. We omit the details.

Let $\Phi(\cdot)$ denote the standard normal distribution function. Throughout the chapter, C denotes a generic constant (which does not depend on T and x).

We start with some preliminary lemmas.

Lemma 1.1 For every $\delta > 0$,

$$P\left\{\left|\left(\frac{2\theta}{T}\right) I_T - 1\right| \geq \delta\right\} \leq CT^{-1}\delta^{-2}.$$

Lemma 1.2

$$\sup_{x\in\mathbb{R}}\left|P\left\{\left(\frac{2\theta}{T}\right)^{1/2} J_T \leq x\right\} - \Phi(x)\right| \leq C\,T^{-1/2}.$$

The following lemma gives the error rate on the difference of the characteristic function of the denominator of the MCE and the normal characteristic function.

Lemma 1.3

(a) Let $\phi_T(z_1) := E\exp(z_1 I_T),\ z_1 \in \mathbb{C}$. Then $\phi_T(z_1)$ exists for $|z_1| \leq \delta$, for some $\delta > 0$, and is given by

$$\phi_T(z_1) = \left\{\frac{4\sin\pi H\sqrt{\theta^2 - 2z_1}e^{-\theta T}}{\pi T D_T^H(\theta;\sqrt{\theta^2 - 2z_1}}\right\}^{1/2},$$

and we choose the principal branch of the square root, where

$$\begin{aligned} D_T^H(\theta;\alpha) &:= [\alpha\cosh(\frac{\alpha}{2}T) - \theta\sinh(\frac{\alpha}{2}T)]^2 J_{-H}(\frac{\alpha}{2}T)J_{H-1}(\frac{\alpha}{2}T) \\ &- [\alpha\sinh(\frac{\alpha}{2}T) - \theta\cosh(\frac{\alpha}{2}T)]^2 J_{1-H}(\frac{\alpha}{2}T)J_H(\frac{\alpha}{2}T) \end{aligned}$$

for $\alpha > 0$ and J_ν is the modified Bessel function of first kind of order ν.
(b) Let

$$\psi_{T,x} := \left(\frac{2\theta}{T}\right)^{1/2} J_T - \left((\frac{2\theta}{T})I_T - 1\right)x.$$

Then for $|x| \leq 2(\log T)^{1/2}$ and for $|u| \leq \epsilon T^{1/2}$, where ϵ is sufficiently small

$$\left|E\exp(iu\psi_{T,x}) - \exp(\frac{-u^2}{2})\right| \leq C\exp(\frac{-u^2}{4})(|u| + |u|^3)T^{-1/2}.$$

Proof
(a) Kleptsyna and Le Breton (2002) proved the following Cameron–Martin type formula for $a > 0$

$$E\exp(-aI_T) = \left\{\frac{4\sin\pi H\sqrt{\theta^2 + 2a}e^{-\theta T}}{\pi T D_T^H(\theta;\sqrt{\theta^2 + 2a}}\right\}^{1/2}.$$

For $H = 1/2$, this formula reduces to the Novikov's formula given in Liptser and Shiryayev (1978). By analytic continuation, this formula extends to the complex plane $z_1 \in \mathbb{C}$.

(b) Observe that

$$\begin{aligned}
& E\exp(iu\psi_{T,x}) \\
= \quad & E\exp\left[-iu\left(\frac{2\theta}{T}\right)^{1/2} J_T - iu\left((\frac{2\theta}{T})^{2-2H} I_T - 1\right)x\right] \\
= \quad & E\exp\left[-iu\left(\frac{2\theta}{T}\right)^{1/2}\left\{\theta I_T - \frac{T}{2}\right\} - iu\left((\frac{2\theta}{T})I_T - 1\right)x\right] \\
= \quad & E\exp(z_1 I_T + z_3) = \exp(z_3)\phi_T(z_1),
\end{aligned}$$

where $z_1 := -iu\theta\delta_{T,x}$ and $z_3 := \frac{iuT}{2}\delta_{T,x}$ with $\delta_{T,x} := \left(\frac{2\theta}{T}\right)^{1/2} + \frac{2x}{T}$. Note that $\phi_T(z_1)$ satisfies the conditions of (a) by choosing ϵ sufficiently small. Let $\alpha_{1,T}(u), \alpha_{2,T}(u), \alpha_{3,T}(u)$, and $\alpha_{4,T}(u)$ be functions that are $O(|u|T^{-1/2})$, $O(|u|^2T^{-1/2}), O(|u|^3T^{-3/2})$, and $O(|u|^3T^{-1/2})$, respectively. Note that for the given range of values of x and u, the conditions on z_1 for part (a) of Lemma are satisfied. Further, with

$$\beta_T(u) := 1 + iu\frac{\delta_{T,x}}{\theta} + \frac{u^2\delta_{T,x}^2}{2\theta^2},$$

we have

$$\begin{aligned}
& (\theta^2 - 2z_1)^{1/2} \\
= \quad & \theta\left[1 - \frac{z_1}{\theta^2} - \frac{z_1^2}{2\theta^4} + \frac{z_1^3}{2\theta^8} + \cdots\right] \\
= \quad & \theta\left[1 + iu\frac{\delta_{T,x}}{\theta} + \frac{u^2\delta_{T,x}^2}{2\theta^2} + \frac{iu^3\delta_{T,x}^3}{2\theta^3} + \cdots\right] \\
= \quad & \theta[1 + \alpha_{1,T}(u) + \alpha_{2,T}(u) + \alpha_{3,T}(u)] \\
= \quad & \theta\beta_T(u) + \alpha_{3,T}(u) \\
= \quad & \theta[1 + \alpha_{1,T}(u)].
\end{aligned}$$

Thus

$$\gamma - \theta = \alpha_{1,T}, \ \gamma + \theta = 2\theta + \alpha_{1,T}.$$

Hence, the above expectation equals

$$\begin{aligned}
& \exp\left(z_3 + \tfrac{\theta T}{2}\right) \times \\
& \left[\frac{2\theta\beta_T(u) + \alpha_{3,T}(u)}{\alpha_{1,T}\exp\{-\theta T\beta_T(u) + \alpha_{4,T}(u)\} + (2\theta + \alpha_{1,T}(u))\exp\{\theta T\beta_T(u) + \alpha_{4,T}(u)\}}\right]^{1/2} \\
= & \left[\frac{1 + \alpha_{1,T}(u)}{\alpha_{1,T}\exp(\chi_T(u)) + (1 + \alpha_{1,T}(u))\exp(\psi_T(t))}\right]^{1/2},
\end{aligned}$$

where

$$\begin{aligned}\chi_T(u) &= -\theta T\beta_T(u) + \alpha_{4,T}(u) - 2z_3 - \theta T\\ &= -2\theta T + \alpha_{1,T}(u) + t^2\alpha_{1,T}(u)\end{aligned}$$

and

$$\begin{aligned}\psi_T(u) &= \theta T\beta_T(u) + \alpha_{4,T}(u) - 2z_3 - \theta T\\ &= \theta T\left[1 + iu\frac{\delta_{T,x}}{\theta} + \frac{u^2\delta_{T,x}^2}{2\theta^2}\right] + \alpha_{4,T}(u) - iuT\delta_{T,x} - \theta T\\ &= \frac{u^2T}{2\theta}\left[\left(\frac{2\theta}{T}\right)^{1/2} + \frac{2x}{T}\right]^2\\ &= u^2 + u^2\alpha_{1,T}(u).\end{aligned}$$

Hence, for the given range of values of u, $\chi_T(t) - \psi_T(u) \leq -\theta T$.

Hence, the above expectation equals

$$\begin{aligned}&\exp(-\frac{u^2}{2})(1+\alpha_{1,T})^{1/2}\left[\alpha_{1,T}\exp\{-2\theta T + \alpha_{1,T} + t^2\alpha_{1,T}\}\right.\\ &\left.+(1+\alpha_{1,T}(t))\exp\{t^2\alpha_{1,T}(t)\}\right]^{-1/2}\\ = \;&\exp(-\frac{u^2}{2})\left[1+\alpha_{1,T})(1+\alpha_{1,T}(1+\alpha_{1,T})\exp\{-\theta T + \alpha_{1,T} + t^2\alpha_{1,T}\}\right]\\ &\times\exp(u^2\alpha_{1,T}(u)).\end{aligned}$$

This completes the proof of the lemma. □

Kleptsyna and Le Breton (2002) obtained the following fractional Cameron –Martin type formula:

Denote $V_T := \int_0^T Q_s^2 dv_s$. Let $\phi_T^H(u) := E\exp(-uV_T),\ u > 0$. Then $\phi_T^H(u)$ is given by

$$\phi_T^H(u) = \left\{\frac{4\sin\pi H\sqrt{\theta^2+2u}e^{-\theta T}}{\pi T D_T^H(\theta;\sqrt{\theta^2+2u})}\right\}^{1/2}, \tag{1.9}$$

where

$$\begin{aligned}D_T^H(\theta;\alpha) \;:=\; &[\alpha\cosh(\frac{\alpha}{2}T) - \theta\sinh(\frac{\alpha}{2}T)]^2 J_{-H}(\frac{\alpha}{2}T)J_{H-1}(\frac{\alpha}{2}T)\\ &-[\alpha\sinh(\frac{\alpha}{2}T) - \theta\cosh(\frac{\alpha}{2}T)]^2 J_{1-H}(\frac{\alpha}{2}T)J_H(\frac{\alpha}{2}T)\end{aligned}$$

for $\alpha > 0$ and J_ν is the modified Bessel function of first kind of order ν.

We generalize the above fractional Cameron–Martin type formula to obtain joint generating function of U_T and V_T.

Lemma 1.3
(c) Let $\Psi_T^H(z_1, z_2) := E\exp(z_1 V_T + z_2 U_T), z_1, z_2 \in \mathbb{C}$. Then $\Psi_T^H(z_1, z_2)$ exists for $|z_i| \le \delta$, 1 = 1,2 for some $\delta > 0$ and is given by

$$\Psi_T^H(z_1, z_2) = \exp\left(\frac{-\theta T}{2}\right)\left[\frac{4(\sin \pi H)\gamma}{\pi T \mathcal{D}_T^H(\gamma; z_2)}\right]^{1/2}, \tag{1.10}$$

where $\gamma := (\theta^2 + 2z_1)^{1/2}$ and we choose the principal branch of the square root and

$$\begin{aligned}\mathcal{D}_T^H(\gamma; z_2) \quad &:= \quad [\gamma \cosh(\frac{\gamma}{2}T) - (\theta - 2z_2)\sinh(\frac{\gamma}{2}T)]^2 J_{-H}(\frac{\gamma}{2}T) J_{H-1}(\frac{\gamma}{2}T) \\ &\quad -[\gamma \sinh(\frac{\gamma}{2}T) - (\theta - 2z_2)\cosh(\frac{\gamma}{2}T)]^2 J_{1-H}(\frac{\gamma}{2}T) J_H(\frac{\gamma}{2}T),\end{aligned}$$

and J_ν is the modified Bessel function of first kind of order ν.

Proof We use analytic continuation, fractional Itô formula (Lemma D.2.3) along with change of measure technique as in Liptser and Shiryayev (1978) and Kleptsyna and Le Breton (2002).

First assume $z_i = a_i \in \mathbb{R}, i = 1, 2$, and a_i are sufficiently small. Define $\gamma := (\theta^2 + 2z_1)^{1/2}$ and $dX_t^\gamma = \gamma X_t^\gamma + dW_t^H, X_0^\gamma = 0$. Also recall that $dX_t^\theta = \theta X_t^\theta + dW_t^H$.

Then

$$\frac{dP_\theta^T}{dP_\gamma^T}(Z_0^T) = \left\{(\theta - \gamma)\int_0^T Q_t^\gamma dZ_t^\gamma - \frac{(\theta^2 - \gamma^2)}{2}\int_0^T (Q_t^\gamma)^2 dv_t\right\}. \tag{1.11}$$

Note that

$$\Psi_T^H(a_1, a_2) := E\exp(a_1 V_T^\theta + a_2 U_T^\theta).$$

If we change the measure to that generated by Z_t^γ, then by (1.11),

$$\begin{aligned}&\Psi_T^H(a_1, a_2) \\ &= E\left[\exp(a_1 \int_0^T (Q_t^\gamma)^2 dv_t + a_2 (Z_T^\gamma)^2)\right. \\ &\quad \left. -(\gamma - \theta)\int_0^T Q_t^\gamma dZ_t^\gamma - a_1 \int_0^T (Q_t^\gamma)^2 dv_t\right] \\ &= E\exp\left[a_2 U_T^\gamma - (\gamma - \theta)\int_0^T Q_t^\gamma dZ_t^\gamma\right].\end{aligned} \tag{1.12}$$

By fractional Itô formula (Lemma D.2.3)

$$\int_0^T Q_s^\gamma dZ_s^\gamma = \frac{1}{2}\left[\frac{\lambda_H}{2 - 2H} Z_T^\gamma \int_0^T t^{2H-1} dZ_s^\gamma - T\right]. \tag{1.13}$$

Using this in (1.12), we have

$$\Psi_T^H(a_1,a_2) = E\exp\left[\left(a_2 - \frac{\gamma-\theta}{2}\right)\frac{\lambda_H}{2-2H} Z_T^\gamma \int_0^T t^{2H-1} dZ_s^\gamma - \left(\frac{\gamma-\theta}{2}\right)T\right]. \tag{1.14}$$

Let $\lambda_H^* := \frac{\lambda_H}{2-2H}$.

Note that $(Z_t^\gamma, \int_0^T t^{2H-1} dZ_s^\gamma)'$ is a Gaussian pair with covariance matrix denoted by $R^\gamma(T)$.

Thus,

$$\Psi_T^H(a_1,a_2) = \exp\left(\frac{(\gamma-\theta)T}{2}\right)\left\{\det[I + \frac{(\gamma-\theta)\lambda_H^*}{2} JR^\gamma(T)]\right\}^{-1/2}, \tag{1.15}$$

where the 2×2 matrices I and J are defined by

$$I = \begin{pmatrix} 1 & 0 \\ 0 & 1 \end{pmatrix}$$

and

$$J = \begin{pmatrix} 0 & 1 \\ 1 & 0 \end{pmatrix}.$$

$R^\gamma(T)$ satisfies the Lyapunov differential equation

$$\dot{R}^\gamma(T) = \frac{\gamma}{2}\left\{A_H(T)R^\gamma(T) + R^\gamma(T)A_H'(T)\right\} + \frac{1}{\lambda_H^*} B_H(T);\ T \geq 0;\ R^\gamma(0) = 0, \tag{1.16}$$

where

$$A_H(s) = \begin{pmatrix} 1 & s^{1-2H} \\ s^{2H-1} & 1 \end{pmatrix}$$

and

$$B_H(s) = \begin{pmatrix} s^{1-2H} & 1 \\ 1 & s^{2H-1} \end{pmatrix}.$$

Now solving for $R^\gamma(T)$ as in Kleptsyna and Le Breton (2002)

$$\begin{aligned} &\det[I + \frac{(\gamma-\theta)\lambda_H^*}{2} JR^\gamma(T)] \\ &= \frac{\gamma\pi T}{4\pi \sin \pi H}\left\{\left[1 + \frac{\gamma-\theta}{\gamma} e^{\frac{\gamma T}{2}} \sinh \frac{\gamma T}{2}\right]^2 J_{-H}(\tfrac{\gamma}{2}T) J_{H-1}(\tfrac{\gamma}{2}T)\right. \\ &\quad \left. - \left[1 - \frac{\gamma-\theta}{\gamma} e^{\frac{\gamma T}{2}} \cosh \frac{\gamma T}{2}\right]^2 J_{1-H}(\tfrac{\gamma}{2}T) J_H(\tfrac{\gamma}{2}T)\right\}, \end{aligned}$$

where J_ν is the modified Bessel function of first kind of order ν.

Substituting this into (1.15) yields (1.10) for z_i's real, around a neighborhood of zero. Thus there is no problem of existence of $\Psi_T^H(z_1, z_2)$ around zero in $\mathbb{C}$, and since we have shown that the moment generating function exists, $\Psi_T^H(z_1,z_2)$. On the other hand, (1.10) defines an analytic function in the relevant domain and agrees with $\Psi_T^H(z_1,z_2)$ for z_1, z_2 real. □

To obtain the rate of normal approximation for the MCE, we need the following estimate on the tail behavior of the MCE.

Lemma 1.4

$$P\left\{(\frac{T}{2\theta})^{1/2}|\tilde{\theta}_T-\theta|\geq 2(\log T)^{1/2}\right\}\leq CT^{-1/2}.$$

Proof Observe that

$$\begin{aligned}
& P\left\{(\frac{T}{2\theta})^{1/2}|\tilde{\theta}_T-\theta|\geq 2(\log T)^{1/2}\right\} \\
= \; & P\left\{\left|\frac{(\frac{2\theta}{T})^{1/2}J_T}{(\frac{2\theta}{T})^{2-2H}I_T}\right|\geq 2(\log T)^{1/2}\right\} \\
\leq \; & P\left\{\left|(\frac{2\theta}{T})^{1/2}J_T\right|\geq (\log T)^{1/2}\right\}+P\left\{\left|\frac{2\theta}{T}I_T\right|\leq\frac{1}{2}\right\} \\
\leq \; & \left|P\left\{(\frac{2\theta}{T})^{1/2}|J_T|\geq (\log T)^{1/2}\right\}-2\Phi(-(\log T)^{1/2})\right| \\
& +2\Phi(-(\log T)^{1/2})+P\left\{\left|\frac{2\theta}{T}I_T-1\right|\geq\frac{1}{2}\right\} \\
\leq \; & \sup_{x\in\mathbb{R}}\left|P\left\{(\frac{2\theta}{T})^{1/2}|J_T|\geq x\right\}-2\Phi(-x)\right| \\
\leq \; & \sup_{x\in\mathbb{R}}\left|P\left\{(\frac{2\theta}{T})^{1/2}|J_T|\geq x\right\}-2\Phi(-x)\right| \\
& +2\Phi(-(\log T)^{1/2})+P\left\{\left|(\frac{2\theta}{T})^{2-2H}I_T-1\right|\geq\frac{1}{2}\right\} \\
\leq \; & CT^{-1/2}+C(T\log T)^{-1/2}+CT^{-1} \\
\leq \; & CT^{-1/2}.
\end{aligned}$$

The bounds for the first and third terms come from Lemmas 1.2 and 1.1, respectively, and that for the middle term comes from (Feller, 1957, p. 166). □

Now we are ready to obtain the uniform rate of normal approximation of the distribution of the MCE.

Theorem 1.5

$$\sup_{x\in\mathbb{R}}\left|P\left\{(\frac{T}{2\theta})^{1/2}(\tilde{\theta}_T-\theta)\leq x\right\}-\Phi(x)\right|\leq C_\theta T^{-1/2}.$$

Proof We shall consider two possibilities (i) and (ii).
(i) $|x| > 2(\log T)^{1/2}$.
We shall give a proof for the case $x > 2(\log T)^{1/2}$. The proof for the case $x < -2(\log T)^{1/2}$ runs similarly. Note that

$$\left|P\left\{(\frac{T}{2\theta})^{1/2}(\tilde{\theta}_T - \theta) \le x\right\} - \Phi(x)\right| \le P\left\{(\frac{T}{2\theta})^{1/2}(\tilde{\theta}_T - \theta) \ge x\right\} + \Phi(-x).$$

But $\Phi(-x) \le \Phi(-2(\log T)^{1/2}) \le CT^{-2(1/2)}$. See (Feller, 1957, p. 166).
Moreover by Lemma 1.4, we have

$$P\left\{(\frac{T}{2\theta})^{1/2}(\tilde{\theta}_T - \theta) \ge 2(\log T)^{1/2}\right\} \le CT^{-1/2}.$$

Hence,

$$\left|P\left\{(\frac{T}{2\theta})^{1/2}(\tilde{\theta}_T - \theta) \le x\right\} - \Phi(x)\right| \le CT^{-1/2}.$$

(ii) $|x| \le 2(\log T)^{1/2}$.

$$\text{Let} \quad A_T := \left\{(\frac{T}{2\theta})^{1/2}|\tilde{\theta}_T - \theta| \le 2(\log T)^{1/2}\right\} \text{ and } B_T := \left\{\frac{I_T}{T} > c_0\right\},$$

where $0 < c_0 < \frac{1}{2\theta}$. By Lemma 1.4, we have

$$P(A_T^c) \le CT^{-1/2}. \tag{1.17}$$

By Lemma 1.1, we have

$$P(B_T^c) = P\{(\frac{2\theta}{T})^{2-2H} I_T - 1 < 2\theta c_0 - 1\} < P\{|\frac{2\theta}{T} I_T - 1| > 1 - 2\theta c_0\} \le CT^{-1}. \tag{1.18}$$

Let b_0 be some positive number. On the set $A_T \cap B_T$ for all $T > T_0$ with $4b_0(\log T_0)^{1/2}(\frac{2\theta}{T_0})^{1/2} \le c_0$, we have

$$\begin{aligned}
& (\frac{T}{2\theta})^{1/2}(\tilde{\theta}_T - \theta) \le x \\
\Rightarrow \quad & I_T + b_0 T(\tilde{\theta}_T - \theta) < I_T + (\frac{T}{2\theta})^{1/2} 2b_0\theta x \\
\Rightarrow \quad & (\frac{T}{2\theta})^{1/2}(\tilde{\theta}_T - \theta)[I_T + b_0 T(\theta_T - \theta)] < x[I_T + (\frac{T}{2\theta})^{1/2} 2b_0\theta x] \\
\Rightarrow \quad & (\tilde{\theta}_T - \theta)I_T + b_0 T(\theta_T - \theta)^2 < (\frac{2\theta}{T})^{1/2} I_T x + 2b_0\theta x^2 \\
\Rightarrow \quad & -J_T + (\tilde{\theta}_T - \theta)I_T + b_0 T(\tilde{\theta}_T - \theta)^2 < -J_T + (\frac{2\theta}{T})^{1/2} I_T x + 2b_0\theta x^2 \\
\Rightarrow \quad & 0 < -J_T + (\frac{2\theta}{T})^{1/2} I_T x + 2b_0\theta x^2
\end{aligned}$$

since

$$\begin{aligned} & I_T + b_0T(\theta_T - \theta) \\ > \ & Tc_0 + b_0T(\theta_T - \theta) \\ > \ & 4b_0(\log T)^{1/2}(\frac{2\theta}{T})^{1/2} - 2b_0(\log T)^{1/2}(\frac{2\theta}{T})^{1/2} \\ = \ & 2b_0(\log T)^{1/2}(\frac{2\theta}{T})^{1/2} > 0. \end{aligned}$$

On the other hand, on the set $A_T \cap B_T$ for all $T > T_0$ with $4b_0(\log T_0)^{1/2}(\frac{2\theta}{T_0})^{1/2} \leq c_0$, we have

$$\begin{aligned} & (\frac{T}{2\theta})^{1/2}(\theta_T - \theta) > x \\ \Rightarrow \ & I_T - b_0T(\tilde{\theta}_T - \theta) < I_T - (\frac{T}{2\theta})^{1/2}2b_0\theta x \\ \Rightarrow \ & (\frac{T}{2\theta})^{1/2}(\tilde{\theta}_T - \theta)[I_T - b_0T(\theta_T - \theta)] > x[I_T - (\frac{T}{2\theta})^{1/2}2b_0\theta x] \\ \Rightarrow \ & (\tilde{\theta}_T - \theta)I_T - b_0T(\tilde{\theta}_T - \theta)^2 > (\frac{2\theta}{T})^{1/2}I_Tx - 2b_0\theta x^2 \\ \Rightarrow \ & -J_T + (\tilde{\theta}_T - \theta)I_T - b_0T(\theta_T - \theta)^2 > -J_T + (\frac{2\theta}{T})^{1/2}I_Tx - 2b_0\theta x^2 \\ \Rightarrow \ & 0 > -J_T + (\frac{2\theta}{T})^{1/2}I_Tx - 2b_0\theta x^2 \end{aligned}$$

since

$$\begin{aligned} & I_T - b_0T(\tilde{\theta}_T - \theta) \\ > \ & Tc_0 - b_0T(\tilde{\theta}_T - \theta) \\ > \ & 4b_0(\log T)^{1/2}(\frac{2\theta}{T})^{1/2} - 2b_0(\log T)^{1/2}(\frac{2\theta}{T})^{1/2} \\ = \ & 2b_0(\log T)^{1/2}(\frac{2\theta}{T})^{1/2} > 0. \end{aligned}$$

Hence,

$$0 < -J_T + (\frac{2\theta}{T})^{1/2}I_Tx - 2b_0\theta x^2 \Rightarrow (\frac{T}{2\theta})^{1/2}(\theta_T - \theta) \leq x.$$

Letting $D^{\pm}_{T,x} := \left\{-J_T + (\frac{2\theta}{T})^{1/2}I_Tx \pm 2b_0\theta x^2 > 0\right\}$, we obtain

$$D^{-}_{T,x} \cap A_T \cap B_T \subseteq A_T \cap B_T \cap \left\{(\frac{T}{2\theta})^{1/2}(\theta_T - \theta) \leq x\right\} \subseteq D^{+}_{T,x} \cap A_T \cap B_T. \quad (1.19)$$

If it is shown that

$$\left|P\left\{D^{\pm}_{T,x}\right\} - \Phi(x)\right| \leq CT^{-1/2} \quad (1.20)$$

for all $T > T_0$ and $|x| \leq 2(\log T)^{1/2}$, then the theorem would follow from (1.17)–(1.19).

We shall prove (1.20) for $D^+_{T,x}$. The proof for $D^-_{T,x}$ is analogous. Observe that

$$\begin{aligned}
& \left|P\left\{D^+_{T,x}\right\} - \Phi(x)\right| \\
= & \left|P\left\{(\frac{2\theta}{T})^{1/2}J_T - \left((\frac{2\theta}{T})^{2-2H}I_T - 1\right)x < x + 2(\frac{2\theta}{T})^{1/2}b_0\theta x^2\right\} - \Phi(x)\right| \\
\leq & \sup_{y\in\mathbb{R}}\left|P\left\{(\frac{2\theta}{T})^{1/2}J_T - \left(\frac{2\theta}{T}I_T - 1\right)x \leq y\right\} - \Phi(y)\right| \\
& + \left|\Phi\left(x + (\tfrac{2\theta}{T})^{1/2}b_0\theta x^2\right) - \Phi(x)\right| \\
=: & \Delta_1 + \Delta_2.
\end{aligned} \tag{1.21}$$

Lemma 1.3(b) and Esseen's smoothing lemma immediately yield

$$\Delta_1 \leq CT^{-1/2}. \tag{1.22}$$

On the other hand, for all $T > T_0$,

$$\Delta_2 \leq 2(\frac{2\theta}{T})^{1/2}b_0\theta x^2(2\pi)^{-1/2}\exp(-\bar{x}^2/2),$$

where

$$|\bar{x} - x| \leq 2(\frac{2\theta}{T})^{1/2}b_0\theta x^2.$$

Since $|x| \leq 2(\log T)^{1/2}$, it follows that $|\bar{x}| > |x|/2$ for all $T > T_0$, and consequently,

$$\Delta_2 \;\leq\; 2(\frac{2\theta}{T})^{1/2}b_0\theta x^2(2\pi)^{-1/2}x^2\exp(-x^2/8) \leq CT^{-1/2}. \tag{1.23}$$

From (1.21)–(1.23), we obtain

$$\left|P\left\{D^+_{T,x}\right\} - \Phi(x)\right| \leq CT^{-1/2}.$$

This completes the proof of the theorem. □

13.4 Discrete Sampling

The discrete observations are $X_{t_1}, X_{t_2}, \ldots, X_{t_n}$ with $t_i - t_{i-1} = \Delta$ for $i = 1, 2, \ldots, n$. We have the following Berry–Esseen theorem for the quadratic transformation of O–U process.

Theorem 4.1 With $t_{i+1} - t_i = \frac{T}{n}$, $i = 1, 2, \ldots, n$ as $n \to \infty$

$$d_{Kol}(Q_n, Z) \leq c_H n^{-(\frac{1}{2}\wedge(\frac{3}{2}-2H))},$$

where

$$Q_n := \frac{1}{\sigma_H \sqrt{n}} \sum_{i=1}^{n} [n^{2H} (X_{t_{i+1}} - X_{t_i})^2].$$

When $H = 1/2$, the rate is $n^{-1/2}$. This rate is still valid for $H < 1/2$. For $H > 1/2$, the rate is $n^{2H-\frac{3}{2}}$.

The LSE is given by

$$\hat{\theta}_n = \frac{1}{\Delta} \log \frac{\sum_{i=1}^{n} X_{t_i} X_{t_{i-1}}}{\sum_{i=1}^{n} X_{t_{i-1}}^2}.$$

Parametric Euler model is given by

$$X_{t_i} - X_{t_{i-1}} = \theta b(X_{t_{i-1}})(t_i - t_{i-1}) + \sigma(X_{t_{i-1}})(W_{t_i} - W_{t_{i-1}}).$$

The approximate maximum likelihood estimator (AMLE) is given by

$$\hat{\theta}_{n,T} := \frac{\sum_{i=1}^{n} \frac{b(X_{t_{i-1}})}{\sigma^2(X_{t_{i-1}})} (X_{t_i} - X_{t_{i-1}})}{\sum_{i=1}^{n} \frac{b^2(X_{t_{i-1}})}{\sigma^2(X_{t_{i-1}})} (t_i - t_{i-1})}.$$

Empirical Euler model is given by

$$Y_{t_i} - Y_{t_{i-1}} = \hat{\theta}_{n,T} b(Y_{t_{i-1}})(t_i - t_{i-1}) + \sigma(Y_{t_{i-1}})(B_{t_i} - B_{t_{i-1}}),$$

where B is a Brownian motion independent of W.

Bootstrap AMLE is given by

$$\hat{\hat{\theta}}_{n,T} := \frac{\sum_{i=1}^{n} Y_{t_{i-1}} (Y_{t_i} - Y_{t_{i-1}})}{\sum_{i=1}^{n} Y_{t_{i-1}}^2 (t_i - t_{i-1})}.$$

The α-th quantile is defined as

$$\tilde{\omega}_\alpha(n,T) := \inf_{y \in \mathbb{R}} \left\{ P\left(\sqrt{\frac{T}{2\hat{\theta}_{n,T}}} \left(\hat{\hat{\theta}}_{n,T} - \hat{\theta}_{n,T} \right) \leq y \right) \geq \alpha \right\}.$$

Theorem 4.2 The bootstrap confidence interval is given by

$$I_{n,T}^{boot}(\alpha) := \left[\hat{\theta}_{n,T} - \frac{\tilde{\omega}_\alpha(T)}{\sqrt{\frac{T}{2\hat{\theta}_{n,T}}}}, \infty \right).$$

Conditional on $\hat{\theta}_{n,T}$, the quantile $\tilde{\omega}$ can be expanded as

$$\tilde{\omega}_\alpha(n,T) = q_\alpha + \frac{1}{\sqrt{T}} h_\theta(q_\alpha) + o\left(\frac{1}{\sqrt{T}} \right).$$

$$P\left(\theta \in I_{n,T}^{boot}(\alpha)\right) - \alpha = o\left(\frac{1}{\sqrt{T}}\right).$$

Theorem 4.3
(a) For $H < 3/4$

$$\sup_{x\in\mathbb{R}} \left| P\left\{ (n\mathcal{I}_H^{-1}(\theta))^{1/2}(\hat{\theta}_n - \theta) \le x \right\} - \Phi(x) \right| \to 0 \text{ as } n \to \infty.$$

(b) For $H = 3/4$

$$\sup_{x\in\mathbb{R}} \left| P\left\{ (n\log n\mathcal{J}_H^{-1}(\theta))^{1/2}(\hat{\theta}_n - \theta) \le x \right\} - \Phi(x) \right| \to 0 \text{ as } n \to \infty.$$

(c) For $H > 3/4$

$$\sup_{x\in\mathbb{R}} \left| P\left\{ (n\mathcal{K}_H^{-1}(\theta))^{1-2H}(\hat{\theta}_n - \theta) \le x \right\} - \mathcal{R}(x) \right| \to 0 \text{ as } n \to \infty.$$

Next we consider high-frequency data. For $i = 1, 2, \ldots, n,\ \ t_i - t_{i-1} := Tn^{-1} \to 0$ as $n \to \infty$.

The approximate minimum contrast estimator (AMCE) is given by

$$\tilde{\theta}_{n,T} := \left(\frac{n}{\mu_H \sum_{i=1}^{n} X_{t_{i-1}}^2} \right)^{1/2H}. \tag{3.1}$$

Another efficient estimator is given by

$$\hat{\theta}_{n,T} := \left(\frac{2n}{\mu_H (X_{t_0}^2 + X_{t_n}^2)} \right)^{1/2H}. \tag{3.2}$$

Theorem 4.4
(a) For $H < 3/4$

$$\sup_{x\in\mathbb{R}} \left| P\left\{ (T\mathcal{I}_H^{-1}(\theta))^{1/2}(\tilde{\theta}_{n,T} - \theta) \le x \right\} - \Phi(x) \right| \to 0$$

as $T \to \infty$ and $Tn^{-1/2} \to 0$.
(b) For $H = 3/4$

$$\sup_{x\in\mathbb{R}} \left| P\left\{ (T\log T\mathcal{J}_H^{-1}(\theta))^{1/2}(\tilde{\theta}_{n,T} - \theta) \le x \right\} - \Phi(x) \right| \to 0$$

as $T \to \infty$ and $Tn^{-1/2} \to 0$.
(c) For $H > 3/4$

$$\sup_{x\in\mathbb{R}} \left| P\left\{ (T\mathcal{K}_H^{-1}(\theta))^{1-2H}(\tilde{\theta}_{n,T} - \theta) \le x \right\} - \mathcal{R}(x) \right| \to 0$$

as $T \to \infty$ and $Tn^{-1/2} \to 0$.

(d) For $H < 3/4$

$$\sup_{x\in\mathbb{R}} \left| P\left\{ (T\mathcal{I}_H^{-1}(\theta))^{1/2}(\hat{\theta}_{n,T} - \theta) \le x \right\} - \Phi(x) \right| \to 0$$

as $T \to \infty$ and $Tn^{-1/2} \to 0$.

(e) For $H = 3/4$

$$\sup_{x\in\mathbb{R}} \left| P\left\{ (T \log T \mathcal{J}_H^{-1}(\theta))^{1/2}(\hat{\theta}_{n,T} - \theta) \le x \right\} - \Phi(x) \right| \to 0$$

as $T \to \infty$ and $Tn^{-1/2} \to 0$.

(f) For $H > 3/4$

$$\sup_{x\in\mathbb{R}} \left| P\left\{ (T\mathcal{K}_H^{-1}(\theta))^{1-2H}(\hat{\theta}_{n,T} - \theta) \le x \right\} - \mathcal{R}(x) \right| \to 0$$

as $T \to \infty$ and $Tn^{-1/2} \to 0$.

Proof

$$\tilde{\theta}_{n,T} - \theta = \left(\frac{n}{\mu_H \sum_{i=1}^n X_{t_{i-1}}^2} \right)^{1/2H} - \theta = (H(2H-1)\alpha_T)^{-1/2H} \hat{\theta}_{n,T}^{1/2H} - \theta,$$

where

$$\alpha_T := H(2H-1) \int_0^T \int_0^t e^{-\theta u} u^{2H-2} du dt.$$

The relationship between the Young integral (the pathwise Riemann–Stieltjes integral) and the Skorokhod integral is given by

$$\int_0^T X_s dW_s = \int_0^T X_s \delta W_s + \alpha_T,$$

$$\theta \int_0^T X_t^2 dt - \alpha_T = \int_0^T X_t \delta W_t - \frac{X_T^2}{2},$$

$$\int_0^T X_t dX_t = -\theta \int_0^T X_t^2 dt + \int_0^T X_t \delta W_t.$$

By the change of variable formula

$$\int_0^T X_t dX_t = \frac{X_T^2}{T},$$

$$\frac{X_T^2}{T} = -\theta \int_0^T X_t^2 dt + \int_0^T X_t \delta W_t,$$

$$\int_0^T X_t \delta W_t = \frac{X_T^2}{T} + \theta \int_0^T X_t^2$$

Let $\xi_t := \int_0^t e^{\theta s} dW_s$ and $F_T := \frac{1}{\sqrt{T}} \int_0^T X_t \delta W_t$. Then

$$X_{t_{i-1}} - X_t = \xi_{t_{i-1}}(e^{-\theta t_{i-1}} - e^{-\theta t}) + e^{-\theta t}(\xi_{t_{i-1}} - \xi_t),$$

$$\begin{aligned}
& P\left(\left|\frac{\theta}{\alpha_T} I_T - 1\right| \geq \rho\right) \\
\leq \; & P\left(\left|\frac{1}{\alpha_T}\int_0^T X_t \delta W_t - \frac{X_T^2}{2\alpha_T}\right| \geq \rho\right) \\
\leq \; & P\left(\left|\frac{1}{\alpha_T}\int_0^T X_t \delta W_t\right| \geq \frac{\rho}{2}\right) + P\left(\frac{X_T^2}{2\alpha_T} \geq \frac{\rho}{2}\right) \\
\leq \; & \frac{T^2 E(F_T^2)}{\rho^2 \alpha_T^2} + \frac{E(X_T^2)}{\rho \alpha_T} \\
\leq \; & \frac{1}{\rho^2 T} + \frac{1}{\rho T} \leq \frac{1}{\rho^2 T}.
\end{aligned}$$

Thus

$$\begin{aligned}
& P\left(\left|\frac{\theta I_{n,T}}{\alpha_T} - 1\right| > \rho\right) \\
\leq \; & P\left(\left|\frac{\theta}{\alpha_T}\int_{t_{i-1}}^{t_i} [X_{t_{i-1}}^2 - X_t^2] dt\right| > \rho/2\right) + P\left(\left|\frac{\theta I_T}{\alpha_T} - 1\right| > \rho/2\right) \\
\leq \; & C\frac{T^H}{\rho n^H} + C\frac{1}{\rho^2 T}
\end{aligned}$$

since

$$\begin{aligned}
& E\left|\frac{\theta}{\alpha_T}\sum_{i=1}^n \int_{t_{i-1}}^{t_i} [X_{t_{i-1}}^2 - X_t^2] dt\right| \\
\leq \; & \frac{\theta}{\alpha_T}\sum_{i=1}^n \int_{t_{i-1}}^{t_i} E\left|X_{t_{i-1}}^2 - X_t^2\right| dt \\
\leq \; & \frac{2\theta \sup_{t \geq 0}(E[X_t^2])^{1/2}}{\alpha_T} \sum_{i=1}^n \int_{t_{i-1}}^{t_i} E\left(\left|X_{t_{i-1}}^2 - X_t^2\right|^2\right)^{1/2} dt \\
\leq \; & C\frac{T^2/n + T^{H+1}/n^H}{T} \\
\leq \; & CT^H/n^H
\end{aligned}$$

since $\sup_{t \geq 0}(E[X_t^2]) < C < \infty$.

We know that

$$F_T = \frac{1}{2\sqrt{T}} I_2\left(e^{-\theta(t-s)} I_{[0,T]}^{\otimes 2}(t,s)\right), \quad E(\| DF_T \|^2) = 2E(F_T^2).$$

$$d_{Kol}\left(\frac{F_T}{\sqrt{E(F_T^2)}}, \; N\right) \leq T^{4H-3}.$$

$E(F_T^2) \to A_H(\theta)$ as $T \to \infty$, where

$$A_H(\theta) := \theta^{1-4H}\left(H^2(4H-1)\left[\Gamma(2H)^2 + \frac{\Gamma(2H)\Gamma(3-4H)\Gamma(4H-1)}{\Gamma(2-2H)}\right]\right).$$

13.5 Approximate Minimum Contrast Estimator

Let $\Phi(\cdot)$ denote the standard normal distribution function. Throughout the chapter, C denotes a generic constant (perhaps depending on θ, but not on anything else).

Let

$$I_T := \int_0^T X_t^2 dt, \quad I_{n,T} := \frac{T}{n}\sum_{i=1}^{n} X_{t_{i-1}}^2, \quad J_{n,T} := \frac{T}{n}\sum_{i=2}^{n+1} X_{t_{i-1}}^2.$$

$$I_{n,T} := \frac{T}{n}[X_{t_0}^2 + X_{t_1}^2 + \ldots + X_{t_{n-1}}^2], \; J_{n,T} := \frac{T}{n}[X_{t_1}^2 + X_{t_2}^2 + \ldots + X_{t_n}^2]$$

$$\frac{I_{n,T} + J_{n,T}}{2} = \frac{T}{n}\left[\frac{1}{2}(X_{t_0}^2 + X_{t_n}^2) + X_{t_1}^2 + \ldots + X_{t_{n-1}}^2\right].$$

We need the following lemmas in the sequel.

Lemma 5.1 For every $\delta > 0$,

$$P\left\{\left|\frac{B_H(\theta)}{T} I_T - 1\right| \geq \delta\right\} \leq CT^{-1}\delta^{-2}.$$

Lemma 5.2

$$\sup_{x\in\mathbb{R}}\left|P\left\{\left(\frac{B_H(\theta)}{T}\right)^{1/2}\left(\theta I_T - \frac{T}{2}\right) \leq x\right\} - \Phi(x)\right| \leq CT^{-1/2}.$$

Lemma 5.3

$$\text{(a)} \quad E|I_{n,T} - I_T|^2 = O\left(\frac{T^4}{n^2}\right).$$

$$\text{(b)} \quad E\left|\frac{I_{n,T} + J_{n,T}}{2} - I_T\right|^2 = O\left(\frac{T^4}{n^2}\right).$$

The following theorem gives the bound on the error of normal approximation of the AMCEs. Note that part (a) uses parameter dependent nonrandom norming. While this is useful for testing hypotheses about θ, it may not necessarily give a confidence interval. The normings in parts (b) and (c) are sample dependent, which can be used for obtaining a confidence interval. The following theorem shows that the asymptotic normality of the AMCEs needs $T \to \infty$ and $\frac{T}{\sqrt{n}} \to 0$.

Theorem 5.1 Denote $b_{n,T} := O(\max(T^{-1/2}(\log T)^{1/2}, (\frac{T^4}{n^2})(\log T)^{-1}))$ and $B_H(\theta) := \frac{A_H(\theta)\theta^{2H}}{H\Gamma(2H)}$.

(a) $$\sup_{x\in\mathbb{R}} \left| P\left\{ \left(\frac{T}{B_H(\theta)}\right)^{1/2} (\tilde{\theta}_{n,T,F} - \theta) \le x \right\} - \Phi(x) \right| = O(b_{n,T}),$$

(b) $$\sup_{x\in\mathbb{R}} \left| P\left\{ I_{n,T}^{1/2}(\tilde{\theta}_{n,T,F} - \theta) \le x \right\} - \Phi(x) \right| = O(b_{n,T}),$$

(c) $$\sup_{x\in\mathbb{R}} \left| P\left\{ \left(\frac{T}{B_H(|\tilde{\theta}_{n,T,F}|)}\right)^{1/2} (\tilde{\theta}_{n,T,F} - \theta) \le x \right\} - \Phi(x) \right| = O(b_{n,T}).$$

Proof

(a) Observe that with $Y_T := -\theta I_T - \frac{T}{2}$ and $I_T := \int_0^T X_t^2 dt$,

$$\left(\frac{T}{B_H(\theta)}\right)^{1/2} (\tilde{\theta}_T - \theta) = \frac{\left(\frac{B_H(\theta)}{T}\right)^{1/2} Y_T}{\left(-\frac{2\theta}{T}\right) I_T}. \tag{2.1}$$

Thus, we have $I_{n,T}\tilde{\theta}_{n,T,F} = -\dfrac{T}{2} = Y_T + \theta I_T$. Hence,

$$\begin{aligned}
&\left(\frac{T}{B_H(\theta)}\right)^{1/2} (\tilde{\theta}_{n,T,F} - \theta) \\
= &\ \frac{\left(\frac{T}{B_H(\theta)}\right)^{1/2} Y_T + \theta\left(\frac{T}{B_H(\theta)}\right)^{1/2} (I_T - I_{n,T})}{I_{n,T}} \\
= &\ \frac{\left(\frac{B_H(\theta)}{T}\right)^{1/2} Y_T + \left(-\frac{B_H(\theta)}{T}\right)^{1/2} (I_T - I_{n,T})}{\left(-\frac{B_H(\theta)}{T}\right) I_{n,T}},
\end{aligned} \tag{2.2}$$

$$\begin{aligned}
&P\left\{\left|\left(\frac{\theta}{\alpha_T}\right) I_{n,T} - 1\right| > \epsilon\right\} \\
= &\ \left\{\left|\left(\frac{\theta}{\alpha T}\right) (I_{n,T} - I_T + I_T) - 1\right| > \epsilon\right\} \\
\le &\ P\left\{\left|\left(\frac{\theta}{\alpha_T}\right) I_T - 1\right| > \frac{\epsilon}{2}\right\} + P\left\{\left(\frac{\theta}{\alpha_T}\right) |I_{n,T} - I_T| > \frac{\epsilon}{2}\right\} \\
\le &\ C\exp\left(\frac{T\theta}{16}\epsilon^2\right) + \frac{16\theta^2}{T^2}\frac{E|I_{n,T} - I_T|^2}{\epsilon^2} \\
\le &\ C\exp\left(\frac{T\theta}{16}\epsilon^2\right) + C\frac{T^2/n^2}{\epsilon^2}.
\end{aligned} \tag{2.3}$$

Further,

$$
\begin{aligned}
& \sup_{x\in\mathbb{R}} \left| P\left\{ \left(\tfrac{T}{B_H(\theta)}\right)^{1/2} (\tilde{\theta}_{n,T,F} - \theta) \le x \right\} - \Phi(x) \right| \\
= & \sup_{x\in\mathbb{R}} \left| P\left\{ \frac{\left(\frac{B_H(\theta)}{T}\right)^{1/2} Y_T + \left(-\frac{2\theta}{T}\right)^{1/2}(I_T - I_{n,T})}{\left(\frac{B_H(\theta)}{T}\right) I_{n,T}} \le x \right\} - \Phi(x) \right| \\
\le & \sup_{x\in\mathbb{R}} \left| P\left\{ \left(\tfrac{B_H(\theta)}{T}\right)^{1/2} Y_T \le x \right\} - \Phi(x) \right| \\
& + P\left\{ \left| \theta \left(\tfrac{B_H(\theta)}{T}\right)^{1/2} (I_{n,T} - I_T) \right| > \epsilon \right\} \\
& + P\left\{ \left| \left(\tfrac{B_H(\theta)}{T}\right) I_{n,T} - 1 \right| > \epsilon \right\} + 2\epsilon \\
\le & \; CT^{-1/2} + \theta^2 \frac{\left(\frac{B_H(\theta)}{T}\right) E|I_{n,T} - I_T|^2}{\epsilon^2} \\
& + C\exp\left(\tfrac{T\theta}{4}\epsilon^2\right) + C\tfrac{T^2}{n^2\epsilon^2} + 2\epsilon, \qquad (2.4) \\
& \text{(the bound for the 3rd term in the r.h.s. of (2.4)} \\
& \text{is obtained from (2.3))} \\
\le & \; CT^{-1/2} + C\tfrac{T^2}{n^2\epsilon^2} + C\exp\left(\tfrac{T\theta}{4}\epsilon^2\right) + C\tfrac{T}{n^2\epsilon^2} + \epsilon \qquad (2.5) \\
& \text{(by Lemma 2.3(a)).}
\end{aligned}
$$

Choosing $\epsilon = CT^{-1/2}(\log T)^{1/2}$, the terms in the right hand side of (2.5) are of the order
$O(\max(T^{-1/2}(\log T)^{1/2}, (\frac{T^4}{n^2})(\log T)^{-1}))$. □

(b) From (2.1), we have $I_{n,T}^{1/2}(\tilde{\theta}_{n,T,F} - \theta) = \frac{Y_T + \theta(I_T - I_{n,T})}{I_{n,T}^{1/2}}$. Then,

$$
\begin{aligned}
& \sup_{x\in\mathbb{R}} \left| P\left\{ I_{n,T}^{1/2}(\tilde{\theta}_{n,T,F} - \theta) \le x \right\} - \Phi(x) \right| \\
= & \sup_{x\in\mathbb{R}} \left| P\left\{ \frac{Y_T}{I_{n,T}^{1/2}} + \theta \frac{I_T - I_{n,T}}{I_{n,T}^{1/2}} \le x \right\} - \Phi(x) \right| \\
\le & \sup_{x\in\mathbb{R}} \left| P\left\{ \frac{Y_T}{I_{n,T}^{1/2}} \le x \right\} - \Phi(x) \right| + P\left\{ \left| \frac{\theta(I_T - I_{n,T})}{I_{n,T}^{1/2}} \right| > \epsilon \right\} + \epsilon \\
=: & \; U_1 + U_2 + \epsilon. \qquad (2.6)
\end{aligned}
$$

We have from (2.3)

$$
U_1 \le CT^{-1/2} + C\exp\left(\frac{T\theta}{16}\epsilon^2\right) + C\frac{T^2}{n^2\epsilon^2} + \epsilon. \qquad (2.7)
$$

Now,

$$
U_2 = P\left\{ |\theta| \left| \frac{I_{n,T} - I_T}{I_{n,T}^{1/2}} \right| > \epsilon \right\} = P\left\{ |\theta| \frac{\left| \left(-\frac{2\theta}{T}\right)^{1/2} (I_{n,T} - I_T) \right|}{\left| \left(-\frac{2\theta}{T}\right)^{1/2} I_{n,T}^{1/2} \right|} > \epsilon \right\}
$$

$$\begin{aligned}
&\leq P\left\{\left|\left(-\frac{2\theta}{T}\right)^{1/2}\right| |I_{n,T} - I_T| > \delta\right\} \\
&\quad + P\left\{\left|\left(-\frac{2\theta}{T}\right)^{1/2} I_{n,T}^{1/2} - 1\right| > \delta_1\right\} \qquad (2.8)\\
&\quad (\text{where } \delta = \epsilon - C\epsilon^2 \text{ and } \delta_1 = (\epsilon - \delta)/\epsilon > 0) \\
&\leq \left(-\frac{2\theta}{T}\right) \frac{E|I_{n,T} - I_T|^2}{\delta^2} + P\left\{\left|\left(-\frac{2\theta}{T}\right) I_{n,T} - 1\right| > \delta_1\right\} \\
&\leq C\frac{T^3}{n^2\delta^2} + C\exp\left(\frac{T\theta}{16}\delta_1^2\right) + C\frac{T^2}{n^2\delta_1^2}. \qquad (2.9)
\end{aligned}$$

Here, the bound for the first term in the right hand side of (2.7) comes from Lemma 2.2(c) and that for the second term is obtained from (2.3).

Now, using the bounds (2.7) and (2.9) in (2.6) with $\epsilon = CT^{-1/2}(\log T)^{1/2}$, we obtain that the terms in (2.6) are of the order $O(\max(T^{-1/2}(\log T)^{1/2}, (\frac{T^4}{n^2})(\log T)^{-1}))$. □

(c) Let

$$G_T := \left\{|\tilde{\theta}_{n,T,F} - \theta| \leq d_T\right\} \qquad \text{and} \qquad d_T := CT^{-1/2}(\log T)^{1/2}.$$

On the set G_T, expanding $(2|\tilde{\theta}_{n,T}|)^{1/2}$, we obtain $(-2\tilde{\theta}_{n,T})^{-1/2}$ $= (-2\theta)^{1/2}\left[1 - \frac{\theta - \tilde{\theta}_{n,T,F}}{\theta}\right]^{-1/2} = (-2\theta)^{1/2}\left[1 + \frac{1}{2}\left(\frac{\theta - \tilde{\theta}_{n,T,F}}{\theta}\right)\right] + O(d_T^2)$.
Then,

$$\begin{aligned}
&\sup_{x\in\mathbb{R}}\left|P\left\{\left(\frac{T}{2|\tilde{\theta}_{n,T,F}|}\right)^{1/2}(\tilde{\theta}_{n,T,F} - \theta) \leq x\right\} - \Phi(x)\right| \\
&\leq \sup_{x\in\mathbb{R}}\left\{P\left(\frac{T}{2|\tilde{\theta}_{n,T}|}\right)^{1/2}(\tilde{\theta}_{n,T,F} - \theta) \leq x, G_T\right\} + P(G_T^c).
\end{aligned}$$

Now,

$$\begin{aligned}
P(G_T^c) &= P\left\{|\tilde{\theta}_{n,T,F} - \theta| > CT^{-1/2}(\log T)^{1/2}\right\} \\
&= P\left\{\left(\frac{T}{B_H(\theta)}\right)^{1/2}|\tilde{\theta}_{n,T,F} - \theta| > C(\log T)^{1/2}(B_H(\theta))^{-1/2}\right\} \\
&\leq C\max\left(T^{-1/2}(\log T)^{1/2}, \frac{T^3}{n^2}(\log T)^{-1}\right) \\
&\quad + 2(1 - \Phi(\log T^{1/2}(B_H(\theta))^{-1/2})) \\
&\quad (\text{by Theorem 2.1(a)})
\end{aligned}$$

$$\leq C \max\left(T^{-1/2}(\log T)^{1/2}, \frac{T^3}{n^2}(\log T)^{-1}\right).$$

On the set G_T, $\left|\left(\frac{\tilde{\theta}_{n,T,F}}{\theta}\right)^{1/2} - 1\right| \leq CT^{-1/2}(\log T)^{1/2}$. Hence, upon choosing $\epsilon = CT^{-1/2}(\log T)^{1/2}, C$ large, we obtain

$$\begin{aligned}
&\left|P\left\{\left(\frac{T}{B_H(\tilde{\theta}_{n,T,F})}\right)^{1/2}(\tilde{\theta}_{n,T,F} - \theta) \leq x, G_T\right\} - \Phi(x)\right| \\
\leq\ & \left|P\left\{\left(\frac{T}{B_H(\theta)}\right)^{1/2}(\tilde{\theta}_{n,T,F} - \theta) \leq x, G_T\right\}\right| \\
& + P\left\{\left|\left(\frac{\tilde{\theta}_{n,T,F}}{\theta}\right)^{1/2} - 1\right| > \epsilon, G_T\right\} + \epsilon \\
& \text{(by Lemma 5.1(b))} \\
\leq\ & C \max\left(T^{-1/2}(\log T)^{1/2}, \frac{T^4}{n^2}(\log T)^{-1}\right) \\
& \text{(by Theorem 5.1(a)).} \qquad \square
\end{aligned}$$

In the following theorem, we improve the bound on the error of normal approximation using a mixture of random and nonrandom normings. Thus asymptotic normality of the AMCEs needs $T \to \infty$ and $\frac{T}{n^{2/3}} \to 0$ that are sharper than the bound in Theorem 5.1.

Theorem 5.2

$$\sup_{x\in\mathbb{R}} \left|P\left\{I_{n,T}\left(\frac{B_H(\theta)}{T}\right)^{1/2}(\tilde{\theta}_{n,T,F} - \theta) \leq x\right\} - \Phi(x)\right|$$
$$= O\left(\max\left(T^{-1/2}, \left(\frac{T^3}{n^2}\right)^{1/3}\right)\right).$$

Proof From (2.2), we have

$I_{n,T}\left(\frac{B_H(\theta)}{T}\right)^{1/2}(\tilde{\theta}_{n,T,F} - \theta) = \left(\frac{B_H(\theta)}{T}\right)^{1/2} Y_T + \theta\left(\frac{B_H(\theta)}{T}\right)^{1/2}(I_T - I_{n,T})$.

Hence, by Lemma 2.1–2.3,

$$\begin{aligned}
&\sup_{x\in\mathbb{R}} \left|P\left\{I_{n,T}\left(\frac{B_H(\theta)}{T}\right)^{1/2}(\tilde{\theta}_{n,T,F} - \theta) \leq x\right\} - \Phi(x)\right| \\
=\ & \sup_{x\in\mathbb{R}} \left|P\left\{\left(\frac{B_H(\theta)}{T}\right)^{1/2} Y_T + \theta\left(\frac{B_H(\theta)}{T}\right)^{1/2}(I_T - I_{n,T}) \leq x\right\} - \Phi(x)\right| \\
\leq\ & \sup_{x\in R} \left|P\left\{\left(\frac{B_H(\theta)}{T}\right)^{1/2} Y_T \leq x\right\} - \Phi(x)\right|
\end{aligned}$$

$$+P\left\{\left|\theta\left(\frac{B_H(\theta)}{T}\right)^{1/2}(I_T - I_{n,T})\right| > \epsilon\right\} + \epsilon$$

$$\leq \; CT^{-1/2} + C\frac{E|I_T - I_{n,T}|^2}{T\epsilon^2} + \epsilon \leq CT^{-1/2} + C\frac{T^3}{n^2\epsilon^2} + \epsilon.$$

Choosing $\epsilon = (\frac{T^3}{n^2})^{1/3}$, the theorem follows. □

The following theorem gives stochastic bound on the error of approximation of the continuous MCE by AMCEs.

Theorem 5.3

$$(a) \;\; |\tilde{\theta}_{n,T} - \tilde{\theta}_T| = O_P\left(\frac{T}{n}\right)^{1/2}.$$

$$(b) \;\; |\tilde{\theta}_{n,T,z} - \tilde{\theta}_T| = O_P\left(\frac{T^2}{n^2}\right)^{1/2}.$$

Proof From (1.9) and (1.14), we have

$$\tilde{\theta}_T = -\frac{T}{2I_T}, \quad \tilde{\theta}_{n,T} = -\frac{T}{2I_{n,T}}.$$

Hence, applying Lemma 1.2 with the aid of Lemma 2.3(a) and noting that $|\frac{I_{n,T}}{T}| = O_P(1)$ and $|\frac{I_T}{T}| = O_P(1)$, the part (a) of theorem follows.

From (1.9) and (1.16), we have

$$\tilde{\theta}_T = -\frac{T}{2I_T}, \quad \tilde{\theta}_{n,T,z} = -\frac{T}{I_{n,t} + J_{n,T}}.$$

Applying Lemma 1.2 with the aid of Lemma 2.3(b) and noting that $|\frac{J_{n,T}}{T}| = O_P(1)$ and $|\frac{I_T}{T}| = O_P(1)$, the part (b) of theorem follows. □

13.6 Higher Order Estimators

Let the continuous realization $\{X_t, 0 \leq t \leq T\}$ be denoted by X_0^T. Let P_θ^T be the measure generated on the space (C_T, B_T) of continuous functions on $[0, T]$ with the associated Borel σ-algebra B_T generated under the supremum norm by the process X_0^T, and let P_0^T be the standard Wiener measure. It is well known that when θ is the true value of the parameter, P_θ^T is absolutely continuous with respect to P_0^T, and the Radon–Nikodym derivative (likelihood) of P_θ^T with respect to P_0^T based on X_0^T is given by

$$L_T(\theta) := \frac{dP_\theta^T}{dP_0^T}(X_0^T) = \exp\left\{\theta\int_0^T X_t dX_t - \frac{\theta^2}{2}\int_0^T X_t^2 dt\right\}. \tag{1.2}$$

Consider the score function, the derivative of the log-likelihood function, which is given by

$$\gamma_T(\theta) := \int_0^T X_t dX_t - \theta \int_0^T X_t^2 dt. \tag{1.3}$$

A solution of the estimating equation $\gamma_T(\theta) = 0$ provides the conditional maximum likelihood estimate (MLE)

$$\hat{\theta}_T := \frac{\int_0^T X_t dX_t}{\int_0^T X_t^2 dt}. \tag{1.4}$$

Strictly speaking, $\hat{\theta}_T$ is not the maximum likelihood estimate of θ since $\hat{\theta}_T$ may take positive values, whereas the parameter θ is assumed to be strictly negative. For an exact definition of MLE, see Kutoyants (1994a,b). Nevertheless, this definition of MLE is widely used. It is well known that $\hat{\theta}_T$ is strongly consistent and $T^{1/2}(\hat{\theta}_T - \theta)$ asymptotically $\mathcal{N}(0, -2\theta)$ distributed as $T \to \infty$ (see Kutoyants 1984a,b). Bishwal (2000a) obtained weak convergence bound of the order $O(T^{-1/2})$ for the MLE. Bishwal (2000b) obtained weak convergence bound of the order $O(T^{-1/2})$ for the Bayes estimators, with smooth priors and loss functions. These two papers used parameter dependent nonrandom norming. Bishwal (1999a,b) obtained weak convergence bound of the order $O(T^{-1/2})$ for the MLE and Bayes estimators using two different random normings that are useful for computation of a confidence interval. Bishwal and Bose (2001) obtained weak convergence bound for two approximate maximum likelihood estimators (AMLEs).

Note that in the stationary case where X_0 has $\mathcal{N}(0, -1/2\theta)$ distribution, the exact log-likelihood function is given by

$$l_T(\theta) := \frac{1}{2}\log\left(-\frac{\theta}{\pi}\right) - \frac{\theta^2}{2}\int_0^T X_t^2 dt + \left(\frac{\theta}{2}\right)[X_T^2 + X_0^2] + \frac{\theta T}{2}.$$

In this case, the likelihood equation has two solutions:

$$\breve{\theta}_{T,1} = \frac{[X_T^2 + X_0^2 - T] + \{[X_T^2 + X_0^2]^2 - \pi \int_0^T X_t^2 dt\}^{1/2}}{4\int_0^T X_t^2 dt}, \tag{1.5}$$

$$\breve{\theta}_{T,2} = \frac{[X_T^2 + X_0^2 - T] - \{[X_T^2 + X_0^2]^2 - \pi \int_0^T X_t^2 dt\}^{1/2}}{4\int_0^T X_t^2 dt}. \tag{1.6}$$

As an alternative to maximum likelihood method and to obtain estimators with higher order accuracy, our aim is to use contrast functions. Using Itô formula, the score function $\gamma_T(\theta)$ can be written as

$$\gamma_T(\theta) = \frac{1}{2}X_T^2 - \int_0^T (\theta X_t^2 + \frac{1}{2})dt. \tag{1.7}$$

Consider the contrast function

$$M_T(\theta) = -\int_0^T (\theta X_t^2 + \frac{1}{2})dt \tag{1.8}$$

and the minimum contrast estimate (MCE)

$$\tilde{\theta}_T := -\frac{T}{2}\left\{\int_0^T X_t^2 dt\right\}^{-1}. \tag{1.9}$$

M-estimator is reduced to the minimum contrast estimator. It is well known that $\tilde{\theta}_T$ is strongly consistent and asymptotically $\mathcal{N}(0, \ -2\theta)$ distributed as $T \to \infty$ (see Lanksa 1979). The large deviations of $\hat{\theta}_T$ and $\tilde{\theta}_T$ were obtained in Florens-Landais and Pham (1999). In particular, it was shown the large deviation probabilities of $\hat{\theta}_T$ are identical to those of $\tilde{\theta}_T$ for $\theta \leq \theta_0/3$ but weaker for $\theta > \theta_0/3$, where θ_0 is the true value of the parameter. However, as far as the rate of convergence to normality is concerned, Bishwal (2002) showed that $\hat{\theta}_T$ and $\tilde{\theta}_T$ have the same Berry–Esseen bound of the order $O(T^{-1/2})$.

In this chapter, we obtain weak convergence bounds for several approximate minimum contrast estimators (AMCEs) that are simpler and robust. In order to define the approximate minimum contrast estimators (AMCEs), we use various discrete approximations of the integral in the definition (1.9) of MCE.

Define a weighted sum of squares

$$M_{n,T} := \frac{T}{n}\left\{\sum_{i=1}^{n} w_{t_i} X_{t_{i-1}}^2 + \sum_{i=2}^{n+1} w_{t_i} X_{t_{i-1}}^2\right\}, \tag{1.10}$$

where $w_{t_i} \geq 0$ is a weight function.

Denote the discrete increasing functions

$$I_{n,T} := \frac{T}{n}\sum_{i=1}^{n} X_{t_{i-1}}^2, \tag{1.11}$$

$$J_{n,T} := \frac{T}{n}\sum_{i=2}^{n+1} X_{t_{i-1}}^2 = \frac{T}{n}\sum_{i=1}^{n} X_{t_i}^2. \tag{1.12}$$

General weighted AMCE is defined as

$$\tilde{\theta}_{n,T} := -\left\{\frac{2}{n} M_{n,T}\right\}^{-1}. \tag{1.13}$$

With $w_{t_i} = 1$, we obtain the forward AMCE as

$$\tilde{\theta}_{n,T,F} := -\left\{\frac{2}{n} I_{n,T}\right\}^{-1}. \tag{1.14}$$

With $w_{t_i} = 0$, we obtain the backward AMCE as

$$\tilde{\theta}_{n,T,B} := -\left\{\frac{2}{n}J_{n,T}\right\}^{-1}. \tag{1.15}$$

Analogous to the estimators for the discrete AR (1) model, we define the simple symmetric and weighted symmetric estimators (see Fuller 1996):

With $w_{t_i} = 0.5$, the simple symmetric AMCE is defined as

$$\tilde{\theta}_{n,T,z} := -\left\{\frac{1}{n}[I_{n,T} + J_{n,T}]\right\}^{-1} = -\left\{\frac{2}{n}\sum_{i=2}^{n} X_{t_{i-1}}^2 + 0.5(X_{t_0}^2 + X_{t_n}^2)\right\}^{-1}. \tag{1.16}$$

With the weight function

$$w_{t_i} = \begin{cases} 0 & : \ i = 1 \\ \frac{i-1}{n} & : \ i = 2, 3, \cdots, n \\ 1 & : \ i = n+1 \end{cases}$$

the weighted symmetric AMCE is defined as

$$\tilde{\theta}_{n,T,w} := -\left\{\frac{2}{n}\sum_{i=2}^{n} X_{t_{i-1}}^2 + \frac{1}{n}\sum_{i=1}^{n} X_{t_{i-1}}^2\right\}^{-1}. \tag{1.17}$$

Note that estimator (1.13) is analogous to the trapezoidal rule in numerical analysis. One can instead use the midpoint rule to define another estimator

$$\tilde{\theta}_{n,T,A} := -\left\{\frac{2}{n}\sum_{i=1}^{n}\left(\frac{X_{t_{i-1}} + X_{t_i}}{2}\right)^2\right\}^{-1}. \tag{1.18}$$

One can use the Simpson's rule to define another estimator where the denominator is a convex combination of the denominators in (1.11) and (1.12)

$$\tilde{\theta}_{n,T,S} := -\left\{\frac{1}{3n}\sum_{i=1}^{n}\left\{X_{t_{i-1}}^2 + 4\left(\frac{X_{t_{i-1}} + X_{t_i}}{2}\right)^2 + X_{t_i}^2\right\}\right\}^{-1}. \tag{1.19}$$

In general, one can generalize Simpson's rule as

$$\tilde{\theta}_{n,T,GS} := -\left\{\frac{2}{n}\sum_{i=1}^{n}\left\{\alpha\frac{X_{t_{i-1}}^2 + X_{t_i}^2}{2} + (1-\alpha)\left(\frac{X_{t_{i-1}} + X_{t_i}}{2}\right)^2\right\}\right\}^{-1} \tag{1.20}$$

for any $0 \le \alpha \le 1$.

The case $\alpha = 0$ produces the estimator (1.18). The case $\alpha = 1$ produces the estimator (1.17). The case $\alpha = \frac{1}{3}$ produces the estimator (1.19).

I propose a very general form of the quadrature based estimator as

$$\tilde{\theta}_{n,T,w} := -\left\{\frac{2}{n}\sum_{i=1}^{n}\sum_{j=1}^{m}\left[(1-s_j)X_{t_{i-1}} + s_j X_{t_i}\right]^2 p_j\right\}^{-1}, \tag{1.21}$$

where $p_j,\ \ j \in \{1, 2, \cdots, m\}$, is a probability mass function of a discrete random variable S on $0 \le s_1 < s_2 < \cdots < s_m \le 1$ with $P(S = s_j) := p_j,\ j \in \{1, 2, \cdots, m\}$.

Denote the k-th moment of the random variable S as $\mu_k := \sum_{j=1}^{m} s_j^k p_j,\ k = 1, 2, \cdots$.

If one chooses the probability distribution as uniform distribution for which the moments are a harmonic sequence $(\mu_1, \mu_2, \mu_3, \mu_4, \mu_5, \mu_6, \cdots) = (\frac{1}{2}, \frac{1}{3}, \frac{1}{4}, \frac{1}{5}, \frac{1}{7}, \cdots)$, then there is no change in rate of convergence than second order. If one can construct a probability distribution for which the harmonic sequence is truncated at a point, then there is a rate of convergence improvement at the point of truncation.

Given a positive integer m, construct a probability mass function $p_j,\ \ j \in \{1, 2, \cdots, m\}$ on $0 \le s_1 < s_2 < \cdots < s_m \le 1$ such that

$$\sum_{j=1}^{m} s_j^r p_j = \frac{1}{r+1},\ r \in \{0, \cdots, m-2\} \tag{1.22}$$

$$\sum_{j=1}^{m} s_j^{N-1} p_j \neq \frac{1}{m}. \tag{1.23}$$

Neither the probabilities p_j nor the atoms, s_j, of the distribution are specified in advance.

This problem is related to the truncated Hausdorff moment problem. I obtain examples of such probability distributions and use them to get higher order accurate (up to sixth order) AMCEs.

The order of approximation error (rate of convergence) of an estimator is $n^{-\nu}$, where

$$\nu := \inf\left\{k : \mu_k \neq \frac{1}{1+k},\ \mu_j = \frac{1}{1+j}, j = 1, 2, \cdots, k-1\right\}. \tag{1.24}$$

We construct probability distributions satisfying these moment conditions and obtain estimators of the rate of convergence up to order 6.

Probability $p_1 = 1$ at the point $s_1 = 0$ gives the estimator (1.11) for which $\mu_1 = 0$. Note that $\mu_1 \neq \frac{1}{2}$. Thus $\nu = 1$.

Probability $p_1 = 1$ at the point $s_1 = 1$ gives the estimator (1.12) for which $\mu_1 = 1$. Note that $\mu_1 \neq \frac{1}{2}$. Thus $\nu = 1$.

Probabilities $(p_1, p_2) = (\frac{1}{2}, \frac{1}{2})$ at the respective points $(s_1, s_2) = (0, 1)$ produce the estimator $\tilde{\theta}_{n,T,Z}$ for which $(\mu_1, \mu_2) = (\frac{1}{2}, \frac{1}{4})$. Thus $\nu = 2$.

Probability $p_j = 1$ at the point $s_j = \frac{1}{2}$ produces the estimator $\tilde{\theta}_{n,T,A}$ for which $(\mu_1, \mu_2) = (\frac{1}{2}, \frac{1}{2})$. Thus $\nu = 2$.

Probabilities $(p_1, p_2) = (\frac{1}{4}, \frac{3}{4})$ at the respective points $(s_1, s_2) = (0, \frac{2}{3})$ produce the asymmetric estimator

$$\tilde{\theta}_{n,T,3} := -\left\{\frac{2}{n}\frac{1}{4}\sum_{i=1}^{n}\left[(X_{t_{i-1}})^2 + 3(\frac{X_{t_{i-1}}+2X_{t_i}}{3})^2\right]\right\}^{-1} \tag{1.25}$$

for which $(\mu_1, \mu_2, \mu_3) = (\frac{1}{2}, \frac{1}{3}, \frac{2}{9})$. Thus $\nu = 3$.

Probabilities $(p_1, p_2) = \frac{3}{4}, \frac{1}{4}$ at the respective points $(s_1, s_2) = (\frac{1}{3}, 1)$ produce asymmetric estimator

$$\tilde{\theta}_{n,T,4} := -\left\{\frac{2}{n}\frac{1}{4}\sum_{i=1}^{n}\left[3(\frac{2X_{t_{i-1}}+X_{t_i}}{3})^2 + (X_{t_i})^2\right]\right\}^{-1} \tag{1.26}$$

for which $(\mu_1, \mu_2, \mu_3) = (\frac{1}{2}, \frac{1}{3}, \frac{10}{36})$. Thus $\nu = 3$.

Probabilities $(p_1, p_2, p_3) = (\frac{1}{6}, \frac{2}{3}, \frac{1}{6})$ at the respective points $(s_1, s_2, s_3) = (0, \frac{1}{2}, 1)$ produce the estimator $\tilde{\theta}_{n,T,5}$ for which $(\mu_1, \mu_2, \mu_3, \mu_4) = (\frac{1}{2}, \frac{1}{3}, \frac{1}{4}, \frac{5}{25})$. Thus $\nu = 4$.

Probabilities $(p_1, p_2, p_3, p_4) = (\frac{1}{8}, \frac{3}{8}, \frac{3}{8}, \frac{1}{8})$ at the respective points $(s_1, s_2, s_3, s_4) = (0, \frac{1}{3}, \frac{2}{3}, 1)$ produce the symmetric estimator

$$\tilde{\theta}_{n,T,5} :=$$
$$-\left\{\frac{2}{n}\frac{1}{8}\sum_{i=1}^{n}\left[(X_{t_{i-1}})^2 + 3(\frac{2X_{t_{i-1}}+X_{t_i}}{3})^2 + 3(\frac{X_{t_{i-1}}+2X_{t_i}}{3})^2 + (X_{t_i})^2\right]\right\}^{-1} \tag{1.27}$$

for which $(\mu_1, \mu_2, \mu_3, \mu_4) = (\frac{1}{2}, \frac{1}{3}, \frac{1}{4}, \frac{11}{54})$. Thus $\nu = 4$.

Probabilities $(p_1, p_2, p_3, p_4, p_5) = (\frac{1471}{24192}, \frac{6925}{24192}, \frac{1475}{12096}, \frac{2725}{12096}, \frac{5675}{24192}, \frac{1721}{24192})$ at the respective points $(s_1, s_2, s_3, s_4, s_5) = (0, \frac{1}{5}, \frac{2}{5}, \frac{3}{5}, \frac{4}{5}, 1)$ produce the asymmetric estimator

$$\begin{aligned}\tilde{\theta}_{n,T,7} := -\Big\{\frac{2}{n}\frac{1}{24192}\sum_{i=1}^{n}\Big[&1471(X_{t_{i-1}})^2 + 6925(\frac{X_{t_{i-1}}+X_{t_i}}{5})^2\\ &+2950(\frac{2X_{t_{i-1}}+2X_{t_i}}{5})^2 + 5450(\frac{3X_{t_{i-1}}+3X_{t_i}}{5})^2\\ &+5675(\frac{4X_{t_{i-1}}+4X_{t_i}}{5})^2 + 1721(X_{t_i})^2\Big]\Big\}^{-1}\end{aligned} \tag{1.28}$$

for which $(\mu_1, \mu_2, \mu_3, \mu_4, \mu_5, \mu_6) = (\frac{1}{2}, \frac{1}{3}, \frac{1}{4}, \frac{1}{5}, \frac{841}{5040})$. Thus $\nu = 5$.

Probabilities $(p_1, p_2, p_3, p_4, p_5) = (\frac{7}{90}, \frac{16}{45}, \frac{2}{15}, \frac{16}{45}, \frac{7}{90})$ at the respective points $(s_1, s_2, s_3, s_4, s_5) = (0, \frac{1}{4}, \frac{1}{2}, \frac{3}{4}, 1)$ produce the symmetric estimator $\tilde{\theta}_{n,T,8}$ given by

$$\begin{aligned}\tilde{\theta}_{n,T,8} := -\Big\{\frac{2}{n}\frac{1}{90}\sum_{i=1}^{n}\Big[&7(X_{t_{i-1}})^2 + 32(\frac{3X_{t_{i-1}}+X_{t_i}}{4})^2 + 12(\frac{X_{t_{i-1}}+X_{t_i}}{2})^2\\ &+32(\frac{X_{t_{i-1}}+3X_{t_i}}{4})^2 + 7(t_i, X_{t_i})^2\Big]\Big\}^{-1}\end{aligned} \tag{1.29}$$

for which $(\mu_1, \mu_2, \mu_3, \mu_4, \mu_5, \mu_6) = (\frac{1}{2}, \frac{1}{3}, \frac{1}{4}, \frac{1}{5}, \frac{1}{6}, \frac{110}{768})$. Thus $\nu = 6$.

Probabilities $(p_1, p_2, p_3, p_4, p_5) = (\frac{19}{288}, \frac{75}{288}, \frac{50}{288}, \frac{50}{288}, \frac{75}{288}, \frac{19}{288})$ at the respective points
$(s_1, s_2, s_3, s_4, s_5) = (0, \frac{1}{5}, \frac{2}{5}, \frac{3}{5}, \frac{4}{5}, 1)$ produce symmetric estimator

$$\begin{aligned}\tilde{\theta}_{n,T,9} := -\Big\{\tfrac{2}{n}\tfrac{1}{288}\textstyle\sum_{i=1}^{n}\Big[19(X_{t_{i-1}})^2 + 75(\tfrac{4X_{t_{i-1}}+X_{t_i}}{5})^2 \\ +50(\tfrac{3X_{t_{i-1}}+2X_{t_i}}{5})^2 + 50(\tfrac{2X_{t_{i-1}}+3X_{t_i}}{5})^2 \\ +75(\tfrac{X_{t_{i-1}}+4X_{t_i}}{5})^2 + 19(X_{t_i})^2\Big]\Big\}^{-1} \qquad (1.30)\end{aligned}$$

for which $(\mu_1, \mu_2, \mu_3, \mu_4, \mu_5, \mu_6) = (\frac{1}{2}, \frac{1}{3}, \frac{1}{4}, \frac{1}{5}, \frac{1}{6}, \frac{3219}{22500})$. Thus $\nu = 6$.

The estimator $\tilde{\theta}_{n,T,S}$ is based on the arithmetic mean of $I_{n,T}$ and $J_{n,T}$. One can use geometric mean and harmonic mean instead. The geometric mean based symmetric AMCE (which is based on the ideas of partial autocorrelation) is defined as

$$\tilde{\theta}_{n,T,G} := \frac{-T/2}{\sqrt{I_{n,T}\ J_{n,T}}}. \qquad (1.31)$$

The harmonic mean based symmetric AMCE is defined as

$$\tilde{\theta}_{n,T,H} := \frac{-T}{\frac{1}{I_{n,T}} + \frac{1}{J_{n,T}}}. \qquad (1.32)$$

Note that

$$\tilde{\theta}_{n,T,H} \ \leq\ \tilde{\theta}_{n,T,G} \ \leq\ \tilde{\theta}_{n,T,S}. \qquad (1.33)$$

It is a desirable property of an estimator to lie inside the parameter space. The estimators $\tilde{\theta}_{n,T,H}$ and $\tilde{\theta}_{n,T,G}$ may lie inside the parameter space even if the estimator $\tilde{\theta}_{n,T,F}$ or $\tilde{\theta}_{n,T,B}$ may not.

Note that the symmetric estimators use both the end points of observations. These estimators are sufficient statistics. So one can take advantage of Rao–Blackwell theorem. If one excludes the end points of observations (as in the forward and backward estimators), then the estimators will not be sufficient statistics.

From a time series viewpoint, a discretely observed stationary Ornstein–Uhlenbeck model is a nonlinear (in the parameter) autoregressive model given by

$$X_{t_i} = \exp(\theta(t_i - t_{i-1})) + \epsilon_i, \quad i \geq 1, \qquad (1.34)$$

where

$$\epsilon_i := \int_{t_{i-1}}^{t_i} \exp\{\theta(t_i - u)\}dW_u \qquad (1.36),$$

which are independent $\mathcal{N}\left(0, \frac{e^{2\theta(t_i - t_{i-1})} - 1}{2\theta}\right)$ random variables. The errors are heteroscedastic if the data points are unequally spaced, but as we assume equally spaced data points, the errors are homoscedastic and i.i.d.

$\mathcal{N}(0, V_{n,T}(\theta))$, where

$$V_{n,T}(\theta) := \frac{e^{2\theta\frac{T}{n}} - 1}{2\theta}. \tag{1.37}$$

The log-likelihood is

$$l_{n,T}(\theta) := -\frac{n}{2}\log 2\pi - \frac{n}{2}V_{n,T}(\theta) - \frac{1}{2V_{n,T}(\theta)}\sum_{i=1}^{n}\left(X_{t_i} - e^{\frac{T}{n}\theta}X_{t_{i-1}}\right)^2. \tag{1.38}$$

The exact maximum likelihood estimator based on equally spaced discrete observations is explicit and is given by

$$\hat{\theta}_{n,T} := \frac{n}{T}\log\frac{\sum_{i=1}^{n} X_{t_{i-1}}^2}{\sum_{i=1}^{n} X_{t_i}X_{t_{i-1}}}. \tag{1.39}$$

Note that

$$\breve{\theta}_{n,T} = 1 - \frac{n}{T}\exp\{\frac{T}{n}\hat{\theta}_{n,T}\}.$$

As far as the rate of convergence of this estimator to the continuous conditional MLE is concerned, it is of first order.

The Euler estimator that is the conditional least squares estimator is given by

$$\breve{\theta}_{n,T} := \frac{n}{T}\frac{\sum_{i=1}^{n} X_{t_{i-1}}(X_{t_i} - X_{t_{i-1}})}{\sum_{i=1} X_{t_{i-1}}^2}. \tag{1.40}$$

Another contrast function, which is the least squares quadratic, is

$$\kappa_{n,T}(\theta) := \frac{2\theta}{e^{2\theta\frac{T}{n}} - 1}\sum_{i=1}^{n}\left(X_{t_i} - e^{\frac{T}{n}\theta}X_{t_{i-1}}\right)^2. \tag{1.41}$$

The solution of $\kappa'_{n,T}(\theta) = 0$ provides an AMCE but does not have an explicit form.

The simple symmetric approximate MLE was introduced and studied in Mishra and Bishwal (1995) for a more general diffusion process. Stochastic bound between the difference of the AMLE and the continuous MLE was obtained.

In this chapter, we obtain the rate of weak convergence to normal distribution of some AMCEs using different normings. We also obtain stochastic bound on the difference of the AMCEs and their continuous counterpart MCE when T is fixed.

Properties of Estimators

Let $\Phi(\cdot)$ denote the standard normal distribution function. Throughout the chapter, C denotes a generic constant (perhaps depending on θ, but not on anything else).

We need the following lemmas in the sequel.

Lemma 6.1 (Bishwal, 2000a) For every $\delta > 0$,

$$P\left\{\left|\frac{2\theta}{T}I_T - 1\right| \geq \delta\right\} \leq CT^{-1}\delta^{-2}.$$

Lemma 6.2 (Bishwal, 2000a)

$$\sup_{x\in\mathbb{R}}\left|P\left\{\left(-\frac{2\theta}{T}\right)^{1/2}\left(\theta I_T - \frac{T}{2}\right) \leq x\right\} - \Phi(x)\right| \leq CT^{-1/2}.$$

Lemma 6.3

$$(a)\ \ E|I_{n,T} - I_T|^2 = O\left(\frac{T^4}{n^2}\right).$$

$$(b)\ \ E|\frac{I_{n,T} + J_{n,T}}{2} - I_T|^2 = O\left(\frac{T^4}{n^2}\right).$$

Part (a) is from Bishwal and Bose (2001). The proof of part (b) is analogous to part (a). We omit the details.

The following theorem gives the bound on the error of normal approximation of the AMCEs. Note that part (a) uses parameter dependent nonrandom norming. While this is useful for testing hypotheses about θ, it may not necessarily give a confidence interval. The normings in parts (b) and (c) are sample dependent, which can be used for obtaining a confidence interval. The following theorem shows that asymptotic normality of the AMCEs needs $T \to \infty$ and $\frac{T}{\sqrt{n}} \to 0$.

Theorem 6.1 Denote $b_{n,T} := O(\max(T^{-1/2}(\log T)^{1/2}, (\frac{T^4}{n^2})(\log T)^{-1}))$.

(a) $\sup_{x\in\mathbb{R}}\left|P\left\{\left(-\frac{T}{2\theta}\right)^{1/2}(\tilde{\theta}_{n,T,F} - \theta) \leq x\right\} - \Phi(x)\right| = O(b_{n,T})$,

(b) $\sup_{x\in\mathbb{R}}\left|P\left\{I_{n,T}^{1/2}(\tilde{\theta}_{n,T,F} - \theta) \leq x\right\} - \Phi(x)\right| = O(b_{n,T})$,

(c) $\sup_{x\in\mathbb{R}}\left|P\left\{\left(\frac{T}{2|\tilde{\theta}_{n,T,F}|}\right)^{1/2}(\tilde{\theta}_{n,T,F} - \theta) \leq x\right\} - \Phi(x)\right| = O(b_{n,T})$.

Proof

(a) Observe that

$$\left(-\frac{T}{2\theta}\right)^{1/2}(\tilde{\theta}_T - \theta) = \frac{\left(-\frac{2\theta}{T}\right)^{1/2} Y_T}{\left(-\frac{2\theta}{T}\right) I_T} \tag{2.1}$$

where

$$Y_T := -\theta I_T - \frac{T}{2} \quad \text{and} \quad I_T := \int_0^T X_t^2 dt.$$

Thus, we have $I_{n,T}\tilde{\theta}_{n,T,F} = -\frac{T}{2} = Y_T + \theta I_T$. Hence,

$$
\begin{aligned}
&\left(-\frac{T}{2\theta}\right)^{1/2} (\tilde{\theta}_{n,T,F} - \theta) \\
= \ & \frac{\left(-\frac{T}{2\theta}\right)^{1/2} Y_T + \theta\left(-\frac{T}{2\theta}\right)^{1/2} (I_T - I_{n,T})}{I_{n,T}} \\
= \ & \frac{\left(-\frac{2\theta}{T}\right)^{1/2} Y_T + \left(-\frac{2\theta}{T}\right)^{1/2} (I_T - I_{n,T})}{\left(-\frac{2\theta}{T}\right) I_{n,T}}, \qquad (2.2) \\
&P\left\{\left|\left(\frac{-2\theta}{T}\right)(I_{n,T} - 1)\right| > \epsilon\right\} \\
= \ & \left\{\left|\left(\frac{-2\theta}{T}\right)(I_{n,T} - I_T + I_T) - 1\right| > \epsilon\right\} \\
\leq \ & P\left\{\left|\left(\frac{-2\theta}{T}\right) I_T - 1\right| > \frac{\epsilon}{2}\right\} + P\left\{\left(-\frac{2\theta}{T}\right) |I_{n,T} - I_T| > \frac{\epsilon}{2}\right\} \\
\leq \ & C \exp\left(\frac{T\theta}{16}\epsilon^2\right) + \frac{16\theta^2}{T^2} \frac{E|I_{n,T} - I_T|^2}{\epsilon^2} \\
\leq \ & C \exp\left(\frac{T\theta}{16}\epsilon^2\right) + C\frac{T^2/n^2}{\epsilon^2}. \qquad (2.3)
\end{aligned}
$$

Next, observe that $\sup_{x\in\mathbb{R}} \left|P\left\{\left(-\frac{T}{2\theta}\right)^{1/2} (\tilde{\theta}_{n,T,F} - \theta) \leq x\right\} - \Phi(x)\right|$

$$
\begin{aligned}
= \ & \sup_{x\in\mathbb{R}} \left|P\left\{\frac{\left(-\frac{2\theta}{T}\right)^{1/2} Y_T + \left(-\frac{2\theta}{T}\right)^{1/2} (I_T - I_{n,T})}{\left(-\frac{2\theta}{T}\right) I_{n,T}} \leq x\right\} - \Phi(x)\right| \\
\leq \ & \sup_{x\in\mathbb{R}} \left|P\left\{\left(-\frac{2\theta}{T}\right)^{1/2} Y_T \leq x\right\} - \Phi(x)\right| \\
& + P\left\{\left|\theta\left(-\frac{2\theta}{T}\right)^{1/2} (I_{n,T} - I_T)\right| > \epsilon\right\} \\
& + P\left\{\left|\left(-\frac{2\theta}{T}\right) I_{n,T} - 1\right| > \epsilon\right\} + 2\epsilon \\
\leq \ & CT^{-1/2} + \theta^2 \frac{\left(-\frac{2\theta}{T}\right) E|I_{n,T} - I_T|^2}{\epsilon^2} \\
& + C\exp\left(\frac{T\theta}{4}\epsilon^2\right) + C\frac{T^2}{n^2\epsilon^2} + 2\epsilon, \qquad (2.4)
\end{aligned}
$$

(the bound for the 3rd term in the right hand side of (2.4) is obtained from (2.3))

$$
\leq \ CT^{-1/2} + C\frac{T^2}{n^2\epsilon^2} + C\exp\left(\frac{T\theta}{4}\epsilon^2\right) + C\frac{T}{n^2\epsilon^2} + \epsilon \qquad (2.5)
$$

(by Lemma 2.3(a)).

Choosing $\epsilon = CT^{-1/2}(\log T)^{1/2}$, the terms in the right hand side of (2.5) are of the order
$O(\max(T^{-1/2}(\log T)^{1/2}, (\frac{T^4}{n^2})(\log T)^{-1}))$. □

(b) From (2.1), we have

$$
I_{n,T}^{1/2}(\tilde{\theta}_{n,T,F} - \theta) = \frac{Y_T + \theta(I_T - I_{n,T})}{I_{n,T}^{1/2}}.
$$

Then,

$$
\sup_{x\in\mathbb{R}} \left|P\left\{I_{n,T}^{1/2}(\tilde{\theta}_{n,T,F} - \theta) \leq x\right\} - \Phi(x)\right|
$$

$$
\begin{aligned}
&= \sup_{x\in\mathbb{R}} \left| P\left\{ \frac{Y_T}{I_{n,T}^{1/2}} + \theta \frac{I_T - I_{n,T}}{I_{n,T}^{1/2}} \le x \right\} - \Phi(x) \right| \\
&\le \sup_{x\in\mathbb{R}} \left| P\left\{ \frac{Y_T}{I_{n,T}^{1/2}} \le x \right\} - \Phi(x) \right| + P\left\{ \left| \frac{\theta(I_T - I_{n,T})}{I_{n,T}^{1/2}} \right| > \epsilon \right\} + \epsilon \\
&=: U_1 + U_2 + \epsilon. \qquad (2.6)
\end{aligned}
$$

We have from (2.3),

$$
U_1 \le CT^{-1/2} + C\exp\left(\frac{T\theta}{16}\epsilon^2\right) + C\frac{T^2}{n^2\epsilon^2} + \epsilon. \qquad (2.7)
$$

Now,

$$
\begin{aligned}
U_2 &= P\left\{ |\theta| \left| \frac{I_{n,T} - I_T}{I_{n,T}^{1/2}} \right| > \epsilon \right\} = P\left\{ |\theta| \frac{\left| \left(-\frac{2\theta}{T}\right)^{1/2} (I_{n,T} - I_T) \right|}{\left| \left(-\frac{2\theta}{T}\right)^{1/2} I_{n,T}^{1/2} \right|} > \epsilon \right\} \\
&\le P\left\{ \left| \left(-\frac{2\theta}{T}\right)^{1/2} \right| |I_{n,T} - I_T| > \delta \right\} \\
&\quad + P\left\{ \left| \left(-\frac{2\theta}{T}\right)^{1/2} I_{n,T}^{1/2} - 1 \right| > \delta_1 \right\} \qquad (2.8) \\
&\quad (\text{where } \delta = \epsilon - C\epsilon^2 \text{ and } \delta_1 = (\epsilon - \delta)/\epsilon > 0) \\
&\le \left(-\frac{2\theta}{T}\right) \frac{E|I_{n,T} - I_T|^2}{\delta^2} + P\left\{ \left| \left(-\frac{2\theta}{T}\right) I_{n,T} - 1 \right| > \delta_1 \right\} \\
&\le C\frac{T^3}{n^2\delta^2} + C\exp\left(\frac{T\theta}{16}\delta_1^2\right) + C\frac{T^2}{n^2\delta_1^2}. \qquad (2.9)
\end{aligned}
$$

Here, the bound for the first term in the right hand side of (2.7) comes from Lemma 2.2(c) and that for the second term is obtained from (2.3).

Now, using the bounds (2.7) and (2.9) in (2.6) with $\epsilon = CT^{-1/2}(\log T)^{1/2}$, we obtain that the terms in (2.6) are of the order $O(\max(T^{-1/2}(\log T)^{1/2}, (\frac{T^4}{n^2})(\log T)^{-1}))$. □

(c) Let

$$
G_T := \left\{ |\tilde{\theta}_{n,T,F} - \theta| \le d_T \right\} \quad \text{and} \quad d_T := CT^{-1/2}(\log T)^{1/2}.
$$

On the set G_T, expanding $(2|\tilde{\theta}_{n,T}|)^{1/2}$, we obtain

$$
(-2\tilde{\theta}_{n,T})^{-1/2} = (-2\theta)^{1/2} \left[1 - \frac{\theta - \tilde{\theta}_{n,T,F}}{\theta} \right]^{-1/2}
$$

$$= (-2\theta)^{1/2}\left[1 + \frac{1}{2}\left(\frac{\theta - \tilde{\theta}_{n,T,F}}{\theta}\right)\right] + O(d_T^2).$$

Then,

$$\begin{aligned}
&\sup_{x\in\mathbb{R}}\left|P\left\{\left(\frac{T}{2|\tilde{\theta}_{n,T,F}|}\right)^{1/2}(\tilde{\theta}_{n,T,F} - \theta) \le x\right\} - \Phi(x)\right| \\
\le\ &\sup_{x\in\mathbb{R}}\left\{P\left(\frac{T}{2|\tilde{\theta}_{n,T}|}\right)^{1/2}(\tilde{\theta}_{n,T,F} - \theta) \le x, G_T\right\} + P(G_T^c).
\end{aligned}$$

Now,

$$\begin{aligned}
P(G_T^c) &= P\left\{|\tilde{\theta}_{n,T,F} - \theta| > CT^{-1/2}(\log T)^{1/2}\right\} \\
&= P\left\{\left(-\frac{T}{2\theta}\right)^{1/2}|\tilde{\theta}_{n,T,F} - \theta| > C(\log T)^{1/2}(-2\theta)^{-1/2}\right\} \\
&\le C\max\left(T^{-1/2}(\log T)^{1/2}, \frac{T^3}{n^2}(\log T)^{-1}\right) \\
&\quad +2(1 - \Phi\log T^{1/2}(-2\theta)^{-1/2}) \\
&\quad \text{(by Theorem 2.1(a))} \\
&\le C\max\left(T^{-1/2}(\log T)^{1/2}, \frac{T^3}{n^2}(\log T)^{-1}\right).
\end{aligned}$$

On the set G_T, $\left|\left(\frac{\tilde{\theta}_{n,T,F}}{\theta}\right)^{1/2} - 1\right| \le CT^{-1/2}(\log T)^{1/2}$. Hence, upon choosing $\epsilon = CT^{-1/2}(\log T)^{1/2}$, C large, we obtain

$$\begin{aligned}
&\left|P\left\{\left(\frac{T}{-2\tilde{\theta}_{n,T,F}}\right)^{1/2}(\tilde{\theta}_{n,T,F} - \theta) \le x, G_T\right\} - \Phi(x)\right| \\
\le\ &\left|P\left\{\left(\frac{T}{-2\theta}\right)^{1/2}(\tilde{\theta}_{n,T,F} - \theta) \le x, G_T\right\}\right| \\
&+P\left\{\left|\left(\frac{\tilde{\theta}_{n,T,F}}{\theta}\right)^{1/2} - 1\right| > \epsilon, G_T\right\} + \epsilon \\
&\text{(by Lemma 1.1(b))} \\
\le\ &C\max\left(T^{-1/2}(\log T)^{1/2}, \frac{T^4}{n^2}(\log T)^{-1}\right) \\
&\text{(by Theorem 2.1(a)).}
\end{aligned}$$

□

In the following theorem, we improve the bound on the error of normal approximation using a mixture of random and nonrandom normings. Thus asymptotic normality of the AMCEs needs $T \to \infty$ and $\frac{T}{n^{2/3}} \to 0$ that are sharper than the bound in Theorem 2.1.

Theorem 6.2
$\sup_{x\in\mathbb{R}} \left|P\left\{I_{n,T}\left(-\frac{2\theta}{T}\right)^{1/2}(\tilde{\theta}_{n,T,F}-\theta) \le x\right\} - \Phi(x)\right|$
$= O\left(\max\left(T^{-1/2}, \left(\frac{T^3}{n^2}\right)^{1/3}\right)\right).$

Proof From (2.2), we have

$$I_{n,T}\left(-\frac{2\theta}{T}\right)^{1/2}(\tilde{\theta}_{n,T,F}-\theta) = \left(-\frac{2\theta}{T}\right)^{1/2} Y_T + \theta\left(-\frac{2\theta}{T}\right)^{1/2}(I_T - I_{n,T}).$$

Hence, by Lemma 2.1–2.3

$$\begin{aligned}
&\sup_{x\in\mathbb{R}}\left|P\left\{I_{n,T}\left(-\frac{2\theta}{T}\right)^{1/2}(\tilde{\theta}_{n,T,F}-\theta) \le x\right\} - \Phi(x)\right| \\
= &\sup_{x\in\mathbb{R}}\left|P\left\{\left(-\frac{2\theta}{T}\right)^{1/2} Y_T + \theta\left(-\frac{2\theta}{T}\right)^{1/2}(I_T - I_{n,T}) \le x\right\} - \Phi(x)\right| \\
\le &\sup_{x\in R}\left|P\left\{\left(-\frac{2\theta}{T}\right)^{1/2} Y_T \le x\right\} - \Phi(x)\right| \\
&+P\left\{\left|\theta\left(-\frac{2\theta}{T}\right)^{1/2}(I_T - I_{n,T})\right| > \epsilon\right\} + \epsilon \\
\le &\ CT^{-1/2} + C\frac{E|I_T - I_{n,T}|^2}{T\epsilon^2} + \epsilon \le CT^{-1/2} + C\frac{T^3}{n^2\epsilon^2} + \epsilon.
\end{aligned}$$

Choosing $\epsilon = (\frac{T^3}{n^2})^{1/3}$, the theorem follows. □

The following theorem gives stochastic bound on the error of approximation of the continuous MCE by AMCEs.

Theorem 6.3

$$(a)\quad |\tilde{\theta}_{n,T} - \tilde{\theta}_T| = O_P\left(\frac{T}{n}\right)^{1/2}.$$

$$(b)\quad |\tilde{\theta}_{n,T,z} - \tilde{\theta}_T| = O_P\left(\frac{T^2}{n^2}\right)^{1/2}.$$

Proof From (1.9) and (1.14), we have

$$\tilde{\theta}_T = -\frac{T}{2I_T}, \quad \tilde{\theta}_{n,T} = -\frac{T}{2I_{n,T}}.$$

Hence, applying Lemma 1.2 with the aid of Lemma 2.3(a) and noting that $|\frac{I_{n,T}}{T}| = O_P(1)$ and $|\frac{I_T}{T}| = O_P(1)$, the part (a) of theorem follows.

From (1.9) and (1.16), we have

$$\tilde{\theta}_T = -\frac{T}{2I_T}, \quad \tilde{\theta}_{n,T,z} = -\frac{T}{I_{n,t} + J_{n,T}}.$$

Applying Lemma 1.2 with the aid of Lemma 2.3(b) and noting that $|\frac{J_{n,T}}{T}| = O_P(1)$ and $|\frac{I_T}{T}| = O_P(1)$, the part (b) of theorem follows. □

Remarks

(1) The bounds in Theorems 6.1 and 6.2 are uniform over compact subsets of the parameter space.

(2) It remains to investigate the nonuniform rates of convergence to normality that are more useful.

(3) Weak convergence rate for AMCE remains open when $T \to \infty$ and $\frac{T}{n} \to 0$.

(4) Weighted estimators converge faster than the unweighted estimators if the weights are chosen according to (1.22) and (1.23).

13.7 Fractional Discrete Sampling

We introduce some new approximate minimum contrast estimators of the drift parameter in the fractional Ornstein–Uhlenbeck process based on discretely sampled data and obtain rates of weak convergence of the distributions of the estimators to the standard normal distribution using random, nonrandom, and mixed normings.

Let $(\Omega, \mathcal{F}, \{\mathcal{F}_t\}_{t\geq 0}, P)$ be a stochastic basis on which is defined the Ornstein–Uhlenbeck process $\{X_t\}$ satisfying the Itô stochastic differential equation

$$dX_t = \theta\, X_t\, dt + \, dW_t^H, \; t \geq 0, \;\; X_0 = 0, \tag{2.1}$$

where $\{W_t\}$ is a standard Wiener process with the filtration $\{\mathcal{F}_t\}_{t\geq 0}$ and $\theta < 0$ is the unknown parameter to be estimated on the basis of discrete observations of the process $\{X_t\}$ at times $0 = t_0 < t_1 < \cdots < t_n = T$ with $t_i - t_{i-1} = \frac{T}{n}, i = 1, 2 \cdots, n$. We assume two types of high-frequency data with long observation time:

(1) $T \to \infty, n \to \infty, \frac{T}{\sqrt{n}} \to 0$, (2) $T \to \infty, n \to \infty, \frac{T}{n^{2/3}} \to 0$.

Let the continuous realization $\{Z_t, 0 \leq t \leq T\}$ be denoted by X_0^T. Let P_θ^T be the measure generated on the space (C_T, B_T) of continuous functions on $[0, T]$ with the associated Borel σ-algebra B_T generated under the supremum norm by the process Z_0^T, and let P_0^T be the standard Wiener measure. It is well known that when θ is the true value of the parameter, P_θ^T is absolutely continuous with respect to P_0^T and the Radon–Nikodym derivative (likelihood) of P_θ^T with respect to P_0^T based on Z_0^T is given by

$$L_T(\theta) := \frac{dP_\theta^T}{dP_0^T}(Z_0^T) = \exp\left\{\theta \int_0^T Q_t dZ_t - \frac{\theta^2}{2}\int_0^T Q_t^2 dv_t\right\}. \tag{2.2}$$

Consider the score function, the derivative of the log-likelihood function, which is given by

$$\gamma_T(\theta) := \int_0^T Q_t dZ_t - \theta \int_0^T Q_t^2 dv_t. \tag{2.3}$$

A solution of the estimating equation $\gamma_T(\theta) = 0$ provides the conditional maximum likelihood estimate (MLE)

$$\hat{\theta}_T := \frac{\int_0^T Q_t dZ_t}{\int_0^T Q_t^2 dv_t}. \tag{2.4}$$

Strictly speaking, $\hat{\theta}_T$ is not the maximum likelihood estimate of θ since $\hat{\theta}_T$ may take positive values, whereas the parameter θ is assumed to be strictly negative. For an exact definition of MLE, nevertheless, this definition of MLE is widely used. It is well known that $\hat{\theta}_T$ is strongly consistent and $T^{1/2}(\hat{\theta}_T - \theta)$ asymptotically $\mathcal{N}(0, -2\theta)$ distributed as $T \to \infty$ (see Bercu et al. (2009)).

Note that in the stationary case where X_0 has $\mathcal{N}(0, -1/2\theta)$ distribution, the exact log-likelihood function is given by

$$l_T(\theta) := \frac{1}{2}\log\left(-\frac{\theta}{\pi}\right) - \frac{\theta^2}{2}\int_0^T Q_t^2 dv_t + \left(\frac{\theta}{2}\right)[Q_T^2 + Q_0^2] + \frac{\theta T}{2}.$$

In this case, the likelihood equation has two solutions:

$$\breve{\theta}_{T,1} = \frac{[Q_T^2 + Q_0^2 - T] + \{[Q_T^2 + Q_0^2]^2 - \pi \int_0^T Q_t^2 dt\}^{1/2}}{4\int_0^T Q_t^2 dt}, \tag{2.5}$$

$$\breve{\theta}_{T,2} = \frac{[Q_T^2 + Q_0^2 - T] - \{[Q_T^2 + Q_0^2]^2 - \pi \int_0^T X_t^2 dt\}^{1/2}}{4\int_0^T Q_t^2 dt}. \tag{2.6}$$

As an alternative to maximum likelihood method and to obtain estimators with higher order accuracy, our aim is to use contrast functions. Using Itô formula, the score function $\gamma_T(\theta)$ can be written as

$$\gamma_T(\theta) = \frac{1}{2}Q_T^2 - \int_0^T (\theta Q_t^2 + \frac{1}{2})dt. \tag{2.7}$$

Consider the contrast function

$$M_T(\theta) = -\int_0^T (\theta Q_t^2 + \frac{1}{2})dt \tag{2.8}$$

and the minimum contrast estimate (MCE)

$$\tilde{\theta}_T := -\frac{T}{2}\left\{\int_0^T Q_t^2 dt\right\}^{-1}. \tag{2.9}$$

M-estimator is reduced to the minimum contrast estimator. It is well known that $\tilde{\theta}_T$ is strongly consistent and asymptotically $\mathcal{N}(0, -2\theta)$ distributed as $T \to \infty$. The large deviations of $\hat{\theta}_T$ and $\tilde{\theta}_T$ were obtained in Florens-Landais and Pham (1999). In particular, it was shown the large deviation probabilities of $\hat{\theta}_T$ are identical to those of $\tilde{\theta}_T$ for $\theta \leq \theta_0/3$ but weaker for $\theta > \theta_0/3$, where θ_0 is the true value of the parameter. In this chapter, we obtain weak convergence bounds for several approximate minimum contrast estimators (AMCEs) that are simpler and robust. In order to define the approximate minimum contrast estimators (AMCEs), we use various discrete approximations of the integral in the definition (2.9) of MCE.

Note that for equally spaced data

$$v_{t_i} - v_{t_{i-1}} = \lambda_H^{-1}\left(\frac{T}{n}\right)^{2-2H}[i^{2-2H} - (i-1)^{2-2H}].$$

We have

$$\begin{aligned}
Q_t &= \frac{d}{dv_t}\int_0^t k_H(t,s)X_s ds\\
&= \kappa_H^{-1}\frac{d}{dv_t}\int_0^t s^{1/2-H}(t-s)^{1/2-H}X_s ds\\
&= \kappa_H^{-1}\lambda_H t^{2H-1}\frac{d}{dt}\int_0^t s^{1/2-H}(t-s)^{1/2-H}X_s ds\\
&= \kappa_H^{-1}\lambda_H t^{2H-1}\int_0^t \frac{d}{dt}s^{1/2-H}(t-s)^{1/2-H}X_s ds\\
&= \kappa_H^{-1}\lambda_H t^{2H-1}\int_0^t s^{1/2-H}(t-s)^{-1/2-H}X_s ds.
\end{aligned}$$

The process Q depends continuously on X, and the discrete observations of X do not produce the discrete observations of Q. The process Q can be approximated by

$$\tilde{Q}_n = \kappa_H^{-1}\lambda_H n^{2H-1}\sum_{j=0}^{n-1} j^{1/2-H}(n-j)^{-1/2-H}X_j.$$

It is easy to show that $\tilde{Q}_n \to Q_t$ almost surely as $n \to \infty$.

Define a partition $0 \leq u_1 < u_2 < u_3 < \cdots < u_{m_i} = t_i, \quad i = 1, 2, \cdots, n$.

$$\tilde{Q}_{t_i} = \kappa_H^{-1}\lambda_H t_i^{2H-1}\sum_{j=1}^{m_i} u_j^{1/2-H}(u_{m_i} - u_j)^{-1/2-H}X_{u_j}(u_j - u_{j-1}).$$

It is easy to show that $\tilde{Q}_{t_i} \to Q_t$ almost surely as $m_i \to \infty$ for each $i = 1, 2, \cdots, n$.

Now we define the weighted approximate minimum contrast estimators (AMCEs).

Define a weighted sum of squares

$$G_{n,T} := \left\{ \sum_{i=1}^{n} w_{t_i} \tilde{Q}^2_{t_{i-1}} (v_{t_i} - v_{t_{i-1}}) + \sum_{i=2}^{n+1} w_{t_i} \tilde{Q}^2_{t_{i-1}} (v_{t_i} - v_{t_{i-1}}) \right\}, \tag{2.10}$$

where $w_{t_i} \geq 0$ is a weight function.

Denote the discrete increasing functions

$$I_{n,T} := \sum_{i=1}^{n} \tilde{Q}^2_{t_{i-1}} (v_{t_i} - v_{t_{i-1}}), \tag{2.11}$$

$$J_{n,T} := \sum_{i=2}^{n+1} \tilde{Q}^2_{t_{i-1}} (v_{t_i} - v_{t_{i-1}}) = \sum_{i=1}^{n} \tilde{Q}^2_{t_i} (v_{t_i} - v_{t_{i-1}}). \tag{2.12}$$

General weighted AMCE is defined as

$$\tilde{\theta}_{n,T} := -\left\{ \frac{2}{T} G_{n,T} \right\}^{-1}. \tag{2.13}$$

With $w_{t_i} = 1$, we obtain the forward AMCE as

$$\tilde{\theta}_{n,T,F} := -\left\{ \frac{2}{T} I_{n,T} \right\}^{-1}. \tag{2.14}$$

With $w_{t_i} = 0$, we obtain the backward AMCE as

$$\tilde{\theta}_{n,T,B} := -\left\{ \frac{2}{T} J_{n,T} \right\}^{-1}. \tag{2.15}$$

Analogous to the estimators for the discrete AR(1) model, we define the simple symmetric and weighted symmetric estimators (see Fuller 1996):

With $w_{t_i} = 0.5$, the simple symmetric AMCE is defined as

$$\tilde{\theta}_{n,T,z} := -\left\{ \frac{1}{T} [I_{n,T} + J_{n,T}] \right\}^{-1}. \tag{2.16}$$

Note that

$$\tilde{\theta}_{n,T,z} = -\left\{ \frac{2}{T} \sum_{i=2}^{n} \tilde{Q}^2_{t_{i-1}} (v_{t_i} - v_{t_{i-1}}) + 0.5 (\tilde{Q}^2_{t_0} + \tilde{Q}^2_{t_n}) (v_{t_i} - v_{t_{i-1}}) \right\}^{-1}.$$

With the weight function

$$w_{t_i} = \begin{cases} 0 & : \quad i = 1 \\ \frac{i-1}{n} & : \quad i = 2, 3, \cdots, n \\ 1 & : \quad i = n+1 \end{cases}$$

the weighted symmetric AMCE is defined as

$$\tilde{\theta}_{n,T,w} := -\left\{\frac{2}{T}\sum_{i=2}^{n} \tilde{Q}^2_{t_{i-1}}(v_{t_i} - v_{t_{i-1}}) + \frac{1}{n}\sum_{i=1}^{n} \tilde{Q}^2_{t_{i-1}}(v_{t_i} - v_{t_{i-1}})\right\}^{-1}. \quad (2.17)$$

Note that estimator (2.13) is analogous to the trapezoidal rule in numerical analysis. One can instead use the midpoint rule to define another estimator

$$\tilde{\theta}_{n,T,A} := -\left\{\frac{2}{T}\sum_{i=1}^{n} \left(\frac{\tilde{Q}_{t_{i-1}} + \tilde{Q}_{t_i}}{2}\right)^2 (v_{t_i} - v_{t_{i-1}})\right\}^{-1}. \quad (2.18)$$

One can use the Simpson's rule to define another estimator where the denominator is a convex combination of the denominators in (2.11) and (2.12)

$$\tilde{\theta}_{n,T,S} := -\left\{\frac{1}{3T}\sum_{i=1}^{n} \left\{\tilde{Q}^2_{t_{i-1}} + 4\left(\frac{\tilde{Q}_{t_{i-1}} + \tilde{Q}_{t_i}}{2}\right)^2 + \tilde{Q}^2_{t_i}\right\}(v_{t_i} - v_{t_{i-1}})\right\}^{-1}. \quad (2.19)$$

In general, one can generalize Simpson's rule as

$$\tilde{\theta}_{n,T,GS} :=$$

$$-\left\{\frac{2}{T}\sum_{i=1}^{n} \left\{\alpha \frac{\tilde{Q}^2_{t_{i-1}} + \tilde{Q}^2_{t_i}}{2} + (1-\alpha)\left(\frac{\tilde{Q}_{t_{i-1}} + \tilde{Q}_{t_i}}{2}\right)^2\right\}(v_{t_i} - v_{t_{i-1}})\right\}^{-1} \quad (2.20)$$

for any $0 \leq \alpha \leq 1$.

The case $\alpha = 0$ produces the estimator (2.18). The case $\alpha = 1$ produces the estimator (2.17). The case $\alpha = \frac{1}{3}$ produces the estimator (2.19).

We propose a very general form of the quadrature based estimator as

$$\tilde{\theta}_{n,T,w} := -\left\{\frac{2}{T}\sum_{i=1}^{n}\sum_{j=1}^{m} \left[(1-s_j)\tilde{Q}_{t_{i-1}} + s_j\tilde{Q}_{t_i}\right]^2 p_j(v_{t_i} - v_{t_{i-1}})\right\}^{-1}, \quad (2.21)$$

where p_j, $j \in \{1, 2, \cdots, m\}$, is a probability mass function of a discrete random variable S on $0 \leq s_1 < s_2 < \cdots < s_m \leq 1$ with $P(S = s_j) := p_j$, $j \in \{1, 2, \cdots, m\}$. Denote the k-th moment of the random variable S as $\mu_k := \sum_{j=1}^{m} s_j^k p_j$, $k = 1, 2, \cdots$.

If one chooses the probability distribution as uniform distribution for which the moments are a harmonic sequence $(\mu_1, \mu_2, \mu_3, \mu_4, \mu_5, \mu_6, \cdots) = (\frac{1}{2}, \frac{1}{3}, \frac{1}{4}, \frac{1}{5}, \frac{1}{7}, \cdots)$, then there is no change in the rate of convergence than second order. If one can construct a probability distribution for which the harmonic sequence is truncated at a point, then there is a rate of convergence improvement at the point of truncation.

Given a positive integer m, construct a probability mass function p_j, $j \in \{1, 2, \cdots, m\}$ on $0 \le s_1 < s_2 < \cdots < s_m \le 1$ such that

$$\sum_{j=1}^{m} s_j^r p_j = \frac{1}{r+1}, \quad r \in \{0, \cdots, m-2\} \tag{2.22}$$

$$\sum_{j=1}^{m} s_j^{N-1} p_j \neq \frac{1}{m}. \tag{2.23}$$

Neither the probabilities p_j nor the atoms, s_j, of the distribution are specified in advance.

This problem is related to the truncated Hausdorff moment problem. I obtain examples of such probability distributions and use them to get higher order accurate (up to sixth order) AMCEs.

The order of approximation error (the rate of convergence) of an estimator is $n^{-\nu}$, where

$$\nu := \inf\left\{k : \mu_k \neq \frac{1}{1+k}, \ \mu_j = \frac{1}{1+j}, j = 1, 2, \cdots, k-1\right\}. \tag{2.24}$$

I construct probability distributions satisfying these moment conditions and obtain estimators of the rate of convergence up to order 6.

Probability $p_1 = 1$ at the point $s_1 = 0$ gives the estimator (1.11) for which $\mu_1 = 0$. Note that $\mu_1 \neq \frac{1}{2}$. Thus $\nu = 1$.

Probability $p_1 = 1$ at the point $s_1 = 1$ gives the estimator (1.12) for which $\mu_1 = 1$. Note that $\mu_1 \neq \frac{1}{2}$. Thus $\nu = 1$.

Probabilities $(p_1, p_2) = (\frac{1}{2}, \frac{1}{2})$ at the respective points $(s_1, s_2) = (0, 1)$ produce the estimator $\tilde{\theta}_{n,T,Z}$ for which $(\mu_1, \mu_2) = (\frac{1}{2}, \frac{1}{4})$. Thus $\nu = 2$.

Probability $p_j = 1$ at the point $s_j = \frac{1}{2}$ produces the estimator $\tilde{\theta}_{n,T,A}$ for which $(\mu_1, \mu_2) = (\frac{1}{2}, \frac{1}{2})$. Thus $\nu = 2$.

Probabilities $(p_1, p_2) = (\frac{1}{4}, \frac{3}{4})$ at the respective points $(s_1, s_2) = (0, \frac{2}{3})$ produce the asymmetric estimator

$$\tilde{\theta}_{n,T,3} := -\left\{\frac{2}{T}\frac{1}{4}\sum_{i=1}^{n}\left[(\tilde{Q}_{t_{i-1}})^2 + 3(\frac{\tilde{Q}_{t_{i-1}} + 2\tilde{Q}_{t_i}}{3})^2\right]\right\}^{-1} \tag{2.25}$$

for which $(\mu_1, \mu_2, \mu_3) = (\frac{1}{2}, \frac{1}{3}, \frac{2}{9})$. Thus $\nu = 3$.

Probabilities $(p_1, p_2) = \frac{3}{4}, \frac{1}{4}$ at the respective points $(s_1, s_2) = (\frac{1}{3}, 1)$ produce asymmetric estimator

$$\tilde{\theta}_{n,T,4} := -\left\{\frac{2}{T}\frac{1}{4}\sum_{i=1}^{n}\left[3(\frac{2\tilde{Q}_{t_{i-1}} + \tilde{Q}_{t_i}}{3})^2 + (\tilde{Q}_{t_i})^2\right](v_{t_i} - v_{t_{i-1}})\right\}^{-1} \quad (2.26)$$

for which $(\mu_1, \mu_2, \mu_3) = (\frac{1}{2}, \frac{1}{3}, \frac{10}{36})$. Thus $\nu = 3$.

Probabilities $(p_1, p_2, p_3) = (\frac{1}{6}, \frac{2}{3}, \frac{1}{6})$ at the respective points $(s_1, s_2, s_3) = (0, \frac{1}{2}, 1)$ produce the estimator $\tilde{\theta}_{n,T,5}$ for which $(\mu_1, \mu_2, \mu_3, \mu_4) = (\frac{1}{2}, \frac{1}{3}, \frac{1}{4}, \frac{5}{25})$. Thus $\nu = 4$.

Probabilities $(p_1, p_2, p_3, p_4) = (\frac{1}{8}, \frac{3}{8}, \frac{3}{8}, \frac{1}{8})$ at the respective points $(s_1, s_2, s_3, s_4) = (0, \frac{1}{3}, \frac{2}{3}, 1)$ produce the symmetric estimator

$$\begin{aligned}&\tilde{\theta}_{n,T,5} := \\ &-\left\{\frac{2}{T}\frac{1}{8}\sum_{i=1}^{n}\left[(\tilde{Q}_{t_{i-1}})^2 + 3(\frac{2\tilde{Q}_{t_{i-1}} + \tilde{Q}_{t_i}}{3})^2 + 3(\frac{\tilde{Q}_{t_{i-1}} + 2\tilde{Q}_{t_i}}{3})^2 + (\tilde{Q}_{t_i})^2\right]\right. \\ &\left.(v_{t_i} - v_{t_{i-1}})\right\}^{-1} \qquad (2.27)\end{aligned}$$

for which $(\mu_1, \mu_2, \mu_3, \mu_4) = (\frac{1}{2}, \frac{1}{3}, \frac{1}{4}, \frac{11}{54})$. Thus $\nu = 4$.

Probabilities $(p_1, p_2, p_3, p_4, p_5) = (\frac{1471}{24192}, \frac{6925}{24192}, \frac{1475}{12096}, \frac{2725}{12096}, \frac{5675}{24192}, \frac{1721}{24192})$ at the respective points $(s_1, s_2, s_3, s_4, s_5) = (0, \frac{1}{5}, \frac{2}{5}, \frac{3}{5}, \frac{4}{5}, 1)$ produce the asymmetric estimator

$$\begin{aligned}&\tilde{\theta}_{n,T,7} := \\ &-\left\{\tfrac{2}{T}\tfrac{1}{24192}\textstyle\sum_{i=1}^{n}\left[1471(\tilde{Q}_{t_{i-1}})^2 + 6925(\frac{\tilde{Q}_{t_{i-1}}+\tilde{Q}_{t_i}}{5})^2 + 2950(\frac{2\tilde{Q}_{t_{i-1}}+2\tilde{Q}_{t_i}}{5})^2\right.\right. \\ &\left.+5450(\frac{3\tilde{Q}_{t_{i-1}}+3\tilde{Q}_{t_i}}{5})^2 + 5675(\frac{4\tilde{Q}_{t_{i-1}}+4\tilde{Q}_{t_i}}{5})^2 + 1721(\tilde{Q}_{t_i})^2\right] \\ &\left.(v_{t_i} - v_{t_{i-1}})\right\}^{-1} \qquad (2.28)\end{aligned}$$

for which $(\mu_1, \mu_2, \mu_3, \mu_4, \mu_5, \mu_6) = (\frac{1}{2}, \frac{1}{3}, \frac{1}{4}, \frac{1}{5}, \frac{841}{5040})$. Thus $\nu = 5$.

Probabilities $(p_1, p_2, p_3, p_4, p_5) = (\frac{7}{90}, \frac{16}{45}, \frac{2}{15}, \frac{16}{45}, \frac{7}{90})$ at the respective points $(s_1, s_2, s_3, s_4, s_5) = (0, \frac{1}{4}, \frac{1}{2}, \frac{3}{4}, 1)$ produce the symmetric estimator $\tilde{\theta}_{n,T,8}$ given by

$$\begin{aligned}&\tilde{\theta}_{n,T,8} := \\ &-\left\{\frac{2}{T}\frac{1}{90}\sum_{i=1}^{n}\left[7(\tilde{Q}_{t_{i-1}})^2 + 32(\frac{3\tilde{Q}_{t_{i-1}} + \tilde{Q}_{t_i}}{4})^2 + 12(\frac{\tilde{Q}_{t_{i-1}} + \tilde{Q}_{t_i}}{2})^2\right.\right. \\ &\left.\left.+32(\frac{\tilde{Q}_{t_{i-1}} + 3\tilde{Q}_{t_i}}{4})^2 + 7(t_i, \tilde{Q}_{t_i})^2\right](v_{t_i} - v_{t_{i-1}})\right\}^{-1} \qquad (2.29)\end{aligned}$$

for which $(\mu_1, \mu_2, \mu_3, \mu_4, \mu_5, \mu_6) = (\frac{1}{2}, \frac{1}{3}, \frac{1}{4}, \frac{1}{5}, \frac{1}{6}, \frac{110}{768})$. Thus $\nu = 6$.

Probabilities $(p_1, p_2, p_3, p_4, p_5) = (\frac{19}{288}, \frac{75}{288}, \frac{50}{288}, \frac{50}{288}, \frac{75}{288}, \frac{19}{288})$ at the respective points
$(s_1, s_2, s_3, s_4, s_5) = (0, \frac{1}{5}, \frac{2}{5}, \frac{3}{5}, \frac{4}{5}, 1)$ produce symmetric estimator

$$\begin{aligned}
&\tilde{\theta}_{n,T,9} := \\
&-\left\{\frac{2}{T}\frac{1}{288}\sum_{i=1}^{n}\left[19(\tilde{Q}_{t_{i-1}})^2 + 75(\frac{4\tilde{Q}_{t_{i-1}} + \tilde{Q}_{t_i}}{5})^2 + 50(\frac{3\tilde{Q}_{t_{i-1}} + 2\tilde{Q}_{t_i}}{5})^2\right.\right. \\
&\left.+50(\frac{2\tilde{Q}_{t_{i-1}} + 3\tilde{Q}_{t_i}}{5})^2 + 75(\frac{\tilde{Q}_{t_{i-1}} + 4\tilde{Q}_{t_i}}{5})^2 + 19(\tilde{Q}_{t_i})^2\right] \\
&\left.(v_{t_i} - v_{t_{i-1}})\right\}^{-1} \qquad (2.30)
\end{aligned}$$

for which $(\mu_1, \mu_2, \mu_3, \mu_4, \mu_5, \mu_6) = (\frac{1}{2}, \frac{1}{3}, \frac{1}{4}, \frac{1}{5}, \frac{1}{6}, \frac{3219}{22500})$. Thus $\nu = 6$.

The estimator $\tilde{\theta}_{n,T,S}$ is based on the arithmetic mean of $I_{n,T}$ and $J_{n,T}$. One can use geometric mean and harmonic mean instead. The geometric mean based symmetric AMCE (which is based on the ideas of partial autocorrelation) is defined as

$$\tilde{\theta}_{n,T,G} := \frac{-T/2}{\sqrt{I_{n,T}\, J_{n,T}}}. \qquad (2.31)$$

The harmonic mean based symmetric AMCE is defined as

$$\tilde{\theta}_{n,T,H} := \frac{-T}{\frac{1}{I_{n,T}} + \frac{1}{J_{n,T}}}. \qquad (2.32)$$

Note that

$$\tilde{\theta}_{n,T,H} \;\le\; \tilde{\theta}_{n,T,G} \;\le\; \tilde{\theta}_{n,T,S}. \qquad (2.33)$$

It is a desirable property of an estimator to lie inside the parameter space. The estimators $\tilde{\theta}_{n,T,H}$ and $\tilde{\theta}_{n,T,G}$ may lie inside the parameter space even if the estimator $\tilde{\theta}_{n,T,F}$ or $\tilde{\theta}_{n,T,B}$ may not.

Note that the symmetric estimators use both the end points of observations. These estimators are sufficient statistics. So one can take advantage of Rao–Blackwell theorem. If one excludes the end points of observations (as in the forward and backward estimators), then the estimators will not be sufficient statistics.

From a time series viewpoint, a discretely observed stationary fractional Ornstein–Uhlenbeck model is a nonlinear (in the parameter) autoregressive model given by

$$\tilde{Q}_{t_i} = \exp(\theta(v_{t_i} - v_{t_{i-1}})) + \epsilon_i, \quad i \ge 1 \qquad (2.34),$$

where

$$\epsilon_i := \int_{v_{t_{i-1}}}^{v_{t_i}} \exp\{\theta(v_{t_i} - u)\} dW_u^H, \qquad (2.36)$$

which are independent $\mathcal{N}\left(0, \frac{e^{2\theta(v_{t_i}-v_{t_{i-1}})}-1}{2\theta}\right)$ random variables. The errors are heteroscedastic if the data points are unequally spaced, but as we assume equally spaced data points, the errors are homoscedastic and i.i.d. $\mathcal{N}(0, V_{n,T}(\theta))$, where

$$V_{n,T}(\theta) := \frac{e^{2\theta\frac{T}{n}}-1}{2\theta}. \tag{2.37}$$

The log-likelihood is

$$l_{n,T}(\theta) := -\frac{n}{2}\log 2\pi - \frac{n}{2}V_{n,T}(\theta) - \frac{1}{2V_{n,T}(\theta)}\sum_{i=1}^{n}\left(\tilde{Q}_{t_i} - e^{\frac{T}{n}\theta}\tilde{Q}_{t_{i-1}}\right)^2. \tag{2.38}$$

The exact maximum likelihood estimator based on equally spaced discrete observations is explicit and is given by

$$\hat{\theta}_{n,T} := \frac{n}{T}\log\frac{\sum_{i=1}^{n}\tilde{Q}^2_{t_{i-1}}}{\sum_{i=1}^{n}\tilde{Q}_{t_i}\tilde{Q}_{t_{i-1}}}. \tag{2.39}$$

Note that

$$\breve{\theta}_{n,T} = 1 - \frac{n}{T}\exp\{\frac{T}{n}\hat{\theta}_{n,T}\}.$$

As far as the rate of convergence of this estimator to the continuous conditional MLE is concerned, it is of first order.

The Euler estimator that is the conditional least squares estimator is given by

$$\breve{\theta}_{n,T} := \frac{n}{T}\frac{\sum_{i=1}^{n}\tilde{Q}_{t_{i-1}}(\tilde{Q}_{t_i} - \tilde{Q}_{t_{i-1}})}{\sum_{i=1}\tilde{Q}^2_{t_{i-1}}}. \tag{2.40}$$

Another contrast function, which is the least squares quadratic, is

$$\kappa_{n,T}(\theta) := \frac{2\theta}{e^{2\theta\frac{T}{n}}-1}\sum_{i=1}^{n}\left(\tilde{Q}_{t_i} - e^{\frac{T}{n}\theta}\tilde{Q}_{t_{i-1}}\right)^2. \tag{2.41}$$

The solution of $\kappa'_{n,T}(\theta) = 0$ provides an AMCE but does not have an explicit form.

The simple symmetric approximate MLE was introduced and studied in Mishra and Bishwal (1995) for a more general diffusion process. Stochastic bound between the difference of the AMLE and the continuous MLE was obtained.

In this chapter, we obtain the rate of weak convergence to normal distribution of some AMCEs using different normings. We also obtain stochastic bound on the difference of the AMCEs and their continuous counterpart MCE when T is fixed.

Lemma 7.1

$$(a)\quad E|I_{n,T}-I_T|^2 = O\left(\tfrac{T^4}{n^2}\right)^{2-2H}$$
$$(b)\quad E|\tfrac{I_{n,T}+J_{n,T}}{2}-I_T|^2 = O\left(\tfrac{T^4}{n^2}\right)^{2-2H}.$$

Proof Let $g(t) := \tilde{Q}_{t_{i-1}} - \tilde{Q}_t$ for $t_{i-1} \le t < t_i, \quad i = 1, 2, \dots, n$.
It is easy to show that
$E|\tilde{Q}_{t_{i-1}} - \tilde{Q}_t|^{2k} \le C(t_{i-1}-t)^{2kH}, k = 1, 2, \dots$.
Hence,

$$\begin{aligned}
& E|Y_{n,T} - Y_T|^2 \\
= \; & E|\sum_{i=1}^{n} \tilde{Q}_{t_{i-1}}[W_{t_i} - W_{t_{i-1}}] - \int_0^T \tilde{Q}_t dW_t|^2 \\
= \; & E|\int_0^T g(t)dW_t|^2 = \int_0^T E(g^2(t))dt \\
\le \; & C\sum_{i=1}^{n}\int_{t_{i-1}}^{t_i} |t_{i-1}-t|dt \\
= \; & Cn\frac{(t_i - t_{i-1})^2}{2} = C\frac{T^2}{n}.
\end{aligned}$$

$$\tilde{Q}_{t_i} - \tilde{Q}_{t_{i-1}} = \int_{t_{i-1}}^{t_i} \theta\tilde{Q}_t dt + [W_{t_i} - W_{t_{i-1}}].$$

We obtain

$$\begin{aligned}
& E|Z_{n,T} - Z_T|^2 \\
= \; & E|\sum_{i=1}^{n} \tilde{Q}_{t_{i-1}}[\tilde{Q}_{t_i} - \tilde{Q}_{t_{i-1}}] - \int_0^T \tilde{Q}_t dW_t|^2 \\
= \; & E|\sum_{i=1}^{n}\int_{t_{i-1}}^{t_i} \theta\tilde{Q}_t\tilde{Q}_{t_{i-1}}dt + \sum_{i=1}^{n}\tilde{Q}_{t_{i-1}}[W_{t_i} - W_{t_{i-1}}] \\
& - \int_0^T \theta\tilde{Q}_t^2 dt - \int_0^T \tilde{Q}_t dW_t|^2 \\
\le \; & 2E|\sum_{i=1}^{n}\tilde{Q}_{t_{i-1}}[W_{t_i} - W_{t_{i-1}}] - \int_0^T \tilde{Q}_t dW_t|^2 \\
& +2\theta^2 E|\sum_{i=1}^{n}\int_{t_{i-1}}^{t_i} \tilde{Q}_t[\tilde{Q}_{t_{i-1}} - \tilde{Q}_t]dt|^2. \\
=: \; & N_1 + N_2.
\end{aligned}$$

N_1 is $O(\frac{T^2}{n})$ by Lemma 7.2(a). To estimate N_2 let $f_i(t) = \tilde{Q}_t[\tilde{Q}_{t_{i-1}} - \tilde{Q}_t]$ for $t_{i-1} \le t < t_i$, $i = 1, 2, \dots, n$. Then

$$\begin{aligned} & E|\sum_{i=1}^{n}\int_{t_{i-1}}^{t_i} f(t)dt|^2 \\ = & \sum_{i=1}^{n} E|\int_{t_{i-1}}^{t_i} f_i(t)dt|^2 + 2\sum_{i,j=1,i<j}^{n} E\left[\int_{t_{i-1}}^{t_i} f_i(t)dt\int_{t_{i-1}}^{t_i} f_j(s)ds\right] \\ =: & M_1 + M_2. \end{aligned}$$

By the boundedness of $E(\tilde{Q}_t^4)$, we have

$$\begin{aligned} & E(f^2(t)) \\ = & E\{\tilde{Q}_t^2[\tilde{Q}_{t_{i-1}} - \tilde{Q}_t]^2\} \\ \le & \{E(\tilde{Q}_t^4)\}^{1/2}\{E[\tilde{Q}_{t_{i-1}} - \tilde{Q}_t]^4\}^{1/2} \\ \le & C(t_{i-1} - t). \end{aligned}$$

$$\begin{aligned} M_1 &= \sum_{i=1}^{n} E|\int_{t_{i-1}}^{t_i} f(t)dt|^2 \\ &\le \sum_{i=1}^{n}(t_i - t_{i-1})\int_{t_{i-1}}^{t_i} E(f^2(t))dt \\ &\le C\frac{T}{n}\sum_{i=1}^{n}\int_{t_{i-1}}^{t_i}(t - t_{i-1})dt \\ &\le C\frac{T}{n}\sum_{i=1}^{n}(t_i - t_{i-1})^2 = C\frac{T^3}{n^2}. \end{aligned}$$

$$\begin{aligned} M_2 &= 2\sum_{i,j=1,i<j}^{n} E\int_{t_{i-1}}^{t_i}\int_{t_{j-1}}^{t_i}[f_i(t)f_j(s)]dtds \\ &= 2\sum_{i,j=1,i<j}^{n}\int_{t_{i-1}}^{t_i}\int_{t_{j-1}}^{t_j} E[f_i(t)f_j(s)]dtds. \end{aligned}$$

Now by Lemma 7.3

$$\begin{aligned} & E[f_i(t)f_j(s)] \\ = & E[\tilde{Q}_t(\tilde{Q}_{t_{i-1}} - \tilde{Q}_t)\tilde{Q}_s(X(t_{j-1}) - \tilde{Q}_s)] \\ = & E[\tilde{Q}_t(\tilde{Q}_{t_{i-1}} - \tilde{Q}_t)]E[\tilde{Q}_s(X(t_{j-1}) - \tilde{Q}_s)] \\ + & E[\tilde{Q}_t\tilde{Q}_s]E[(\tilde{Q}_{t_{i-1}} - \tilde{Q}_t)(X(t_{j-1}) - \tilde{Q}_s)] \\ + & E[\tilde{Q}_t(X(t_{j-1}) - \tilde{Q}_s)]E[\tilde{Q}_s(\tilde{Q}_{t_{i-1}} - \tilde{Q}_t)] \\ =: & A_1 + A_2 + A_3. \end{aligned}$$

Note that

$$\tilde{Q}_t = \int_0^t e^{\theta(t-u)} dW_u, t \geq 0.$$

Let $a = e^\theta$. For $s \geq t$, we have

$$\begin{aligned}
& E(\tilde{Q}_t \tilde{Q}_s) \\
= \ & E(\int_0^t e^{\theta(t-u)} dW_u)(\int_0^s e^{\theta(s-u)} dW_u) \\
= \ & \int_0^t e^{\theta(t+s-2u)} du \\
= \ & \frac{1}{2\theta}[a^{s+t} - a^{s-t}].
\end{aligned}$$

Now

$$\begin{aligned}
& E(\tilde{Q}_t - \tilde{Q}_{t_{i-1}})(\tilde{Q}_s - X(t_{j-1})) \\
= \ & E(\tilde{Q}_t \tilde{Q}_s) - E(\tilde{Q}_t X(t_{j-1})) - E(\tilde{Q}_{t_{i-1}} \tilde{Q}_s) + E(X(t_{i-1} X(t_{j-1})) \\
= \ & \frac{1}{2\theta}(a^s - a^{t_{j-1}})[(a^t - a^{t_{i-1}}) + (a^{-t_{i-1}} - a^{-t})] \\
= \ & \frac{1}{2\theta}(s - t_{j-1}) a^{t^*} [(t - t_{i-1} a^{t^{**}} + (t - t_{i-1}) a^{-t^{***}}] \\
& (\text{where} t_{j-1} < t^* < s, t_{i-1} < t^{**}, t^{***} < t) \\
\leq \ & \frac{1}{2\theta}(s - t_{j-1}) a^t (t - t_{i-1}) a^{t_{i-1}} + (s - t_{j-1}) a^t (t - t_{i-1}) a^{-t}] \\
\leq \ & C(s - t_{j-1})(t - t_{i-1}).
\end{aligned}$$

Thus $A_2 \leq C(s - t_{j-1})(t - t_{i-1})$ since $|E(\tilde{Q}_t \tilde{Q}_s)|$ is bounded.

Next

$$\begin{aligned}
& |E[\tilde{Q}_t(X(t_{i-1} - \tilde{Q}_t)]| \\
= \ & \frac{1}{2|\theta|}[a^{t+t_{i-1}} - a^{t-t_{i-1}} - a^{2t} + 1] \\
= \ & \frac{1}{2|\theta|} a^t [a^{t_{i-1}} - a^{-t_{i-1}} - a^t + a^{-t}] \\
\leq \ & \frac{1}{2|\theta|} a^t (t - t_{i-1})[a^{t_{i-1}} + a^{-t}] \\
\leq \ & C(t - t_{i-1})
\end{aligned}$$

and

$$\begin{aligned}
& |E[\tilde{Q}_s(\tilde{Q}_s - X(t_{j-1})]| \\
= \ & \frac{1}{2|\theta|}[a^{2s} - 1 - a^{s+t_{j-1}} + a^{s+t_{j-1}}]
\end{aligned}$$

$$
\begin{aligned}
&= \frac{1}{2|\theta|}a^s[a^s - a^{-s} - a^{t_{j-1}} + a^{-t_{j-1}}] \\
&\le \frac{1}{2|\theta|}a^s(s - t_{j-1})[a^{t_{j-1}} + a^{-s}] \\
&\le C(s - t_{j-1}).
\end{aligned}
$$

Thus $A_1 \le C(s - t_{j-1})(t - t_{i-1})$.

Next

$$
\begin{aligned}
&|E[\tilde{Q}_t(\tilde{Q}_s - X(t_{j-1})]| \\
&= \frac{1}{2|\theta|}[a^{s+t} - a^{s-t} - a^{t+t_{j-1}} + a^{t_{j-1}-t}] \\
&= \frac{1}{2|\theta|}a^t(a^s - a^{t_{j-1}}) \\
&\le \frac{1}{2|\theta|}a^t(1 - a^{-2t})(s - t_{j-1})a^t \\
&\le (a^{2t} - 1)(s - t_{j-1}) \\
&\le C(s - t_{j-1})
\end{aligned}
$$

and

$$
\begin{aligned}
&|E[\tilde{Q}_s(\tilde{Q}_t - \tilde{Q}_{t_{i-1}}]| \\
&= \frac{1}{2|\theta|}[a^{t+s} - a^{s-t} - a^{s+t_{i-1}} + a^{s-t_{i-1}}] \\
&= \frac{1}{2|\theta|}a^s[a^t - a^{-t} - a^{t_{i-1}} + a^{-t_{i-1}}] \\
&\le \frac{1}{2|\theta|}a^s(t - t_{i-1})[a^{t_{i-1}} + a^{-t}] \\
&\le C(t - t_{i-1}).
\end{aligned}
$$

Thus $A_3 \le C(s - t_{j-1})(t - t_{i-1})$.
Hence, $E[f_i(t)f_j(s)] \le C(s - t_{j-1})(t - t_{i-1})$.

Thus,

$$
\begin{aligned}
M_2 &= 2\sum_{i,j=1,i<j}^{n}\int_{t_{i-1}}^{t_i}\int_{t_{j-1}}^{t_j} E[f_i(t)f_j(s)]dtds \\
&\le C\sum_{i,j=1,i<j}^{n}\int_{t_{i-1}}^{t_i}\int_{t_{j-1}}^{t_j}(t - t_{i-1})(s - t_{j-1})dtds \\
&= C\sum_{i,j=1,i<j}^{n}(t_{i-1} - t_i)^2(t_{j-1} - t_i)^2 \\
&= Cn^2(\frac{T}{n})^4 = C\frac{T^4}{n^2}.
\end{aligned}
$$

Hence, N_2 is $O(\frac{T^3}{n^2})$.

Let $h_i(t) := \tilde{Q}^2_{t_{i-1}} - \tilde{Q}^2_t$.

$$
\begin{aligned}
& E|I_{n,T} - I_t|^2 \\
= & E|\sum_{i=1}^n \tilde{Q}^2_{t_{i-1}}(t_i - t_{i-1}) - \int_0^T \tilde{Q}^2_t dt|^2 \\
= & E|\sum_{i=1}^n \int_{t_{i-1}}^{t_i} [\tilde{Q}^2_{t_{i-1}} - \tilde{Q}^2_t]dt|^2 \\
= & E|\sum_{i=1}^n \int_{t_{i-1}}^{t_i} h_i(t)dt|^2 \\
= & \sum_{i=1}^n E|\int_{t_{i-1}}^{t_i} h_i(t)dt|^2 + 2 \sum_{i,j=1,i<j}^n E \int_{t_{i-1}}^{t_i} \int_{t_{j-1}}^{t_j} h_i(t)h_j(s)dtds \\
=: & B_1 + B_2.
\end{aligned}
$$

$$
\begin{aligned}
Eh^2(t) &= E[\tilde{Q}^2_{t_{i-1}} - \tilde{Q}^2_t]^2 \\
&= E[\tilde{Q}_{t_{i-1}} - \tilde{Q}_t]^2[\tilde{Q}_{t_{i-1}} + \tilde{Q}_t]^2 \\
&\le \{E[\tilde{Q}_{t_{i-1}} - \tilde{Q}_t]^4\}^{1/2}\{\{E[\tilde{Q}_{t_{i-1}} + \tilde{Q}_t]^4\}^{1/2} \\
&\le C(t - t_{i-1})
\end{aligned}
$$

(by (2.1) and the boundedness of the second term)

$$
\begin{aligned}
B_1 &= \sum_{i=1}^n E|\int_{t_{i-1}}^{t_i} h_i(t)dt|^2 \\
&\le \sum_{i=1}^n (t_i - t_{i-1}) \int_{t_{i-1}}^{t_i} E(h_i^2(t))dt \\
&\le C\frac{T}{n} \sum_{i=1}^n \int_{t_{i-1}}^{t_i} (t - t_{i-1})dt \\
&= C\frac{T^3}{n^2}.
\end{aligned}
$$

Note that

$$
\begin{aligned}
& E[h_i(t)h_j(s)] \\
= & E(\tilde{Q}^2_{t_{i-1}} - \tilde{Q}^2_t)(\tilde{Q}^2_{t_{j-1}} - \tilde{Q}^2_s) \\
= & E(\tilde{Q}_{t_{i-1}} - \tilde{Q}_t)(\tilde{Q}_{t_{i-1}} + \tilde{Q}_t)(\tilde{Q}_{t_{j-1}} - \tilde{Q}_s)(\tilde{Q}_{t_{j-1}} + \tilde{Q}_s).
\end{aligned}
$$

Now using Lemma 2.1 and proceeding similar to the estimation of M_2, it is easy to see that

$$
B_2 \le C\frac{T^4}{n^2}.
$$

Combining B_1 and B_2, (c) follows. □

The following theorem gives the bound on the error of normal approximation of the AMCEs. Note that part (a) uses parameter dependent nonrandom norming. While this is useful for testing hypotheses about θ, it may not necessarily give a confidence interval. The normings in parts (b) and (c) are sample dependent, which can be used for obtaining a confidence interval. The following theorem shows that asymptotic normality of the AMCEs needs $T \to \infty$ and $\frac{T}{\sqrt{n}} \to 0$.

Theorem 7.1 Denote $b_{n,T} := O(\max(T^{-1/2}(\log T)^{1/2}, (\frac{T^4}{n^2})(\log T)^{-1}))$.

(a) $$\sup_{x\in\mathbb{R}}\left|P\left\{\left(-\frac{T}{2\theta}\right)^{1/2}(\tilde{\theta}_{n,T,F}-\theta)\le x\right\}-\Phi(x)\right| = O(b_{n,T}),$$

(b) $$\sup_{x\in\mathbb{R}}\left|P\left\{I_{n,T}^{1/2}(\tilde{\theta}_{n,T,F}-\theta)\le x\right\}-\Phi(x)\right| = O(b_{n,T}),$$

(c) $$\sup_{x\in\mathbb{R}}\left|P\left\{\left(\frac{T}{2|\tilde{\theta}_{n,T,F}|}\right)^{1/2}(\tilde{\theta}_{n,T,F}-\theta)\le x\right\}-\Phi(x)\right| = O(b_{n,T}).$$

Proof

(a) Observe that

$$\left(-\frac{T}{2\theta}\right)^{1/2}(\tilde{\theta}_T-\theta) = \frac{\left(-\frac{2\theta}{T}\right)^{1/2}Y_T}{\left(-\frac{2\theta}{T}\right)I_T} \tag{2.42}$$

where

$$Y_T := -\theta I_T - \frac{T}{2} \quad\text{and}\quad I_T := \int_0^T \tilde{Q}_t^2 dt.$$

Thus, we have

$$I_{n,T}\tilde{\theta}_{n,T,F} = -\frac{T}{2} = Y_T + \theta I_T.$$

Hence,

$$\left(-\frac{T}{2\theta}\right)^{1/2}(\tilde{\theta}_{n,T,F}-\theta) = \frac{\left(-\frac{T}{2\theta}\right)^{1/2}Y_T + \theta\left(-\frac{T}{2\theta}\right)^{1/2}(I_T - I_{n,T})}{I_{n,T}}$$

$$= \frac{\left(-\frac{2\theta}{T}\right)^{1/2}Y_T + \left(-\frac{2\theta}{T}\right)^{1/2}(I_T - I_{n,T})}{\left(-\frac{2\theta}{T}\right)I_{n,T}}. \tag{2.43}$$

Further,

$$P\left\{\left|\left(\frac{-2\theta}{T}\right)(I_{n,T}-1)\right| > \epsilon\right\} = \left\{\left|\left(\frac{-2\theta}{T}\right)(I_{n,T}-I_T+I_T)-1\right| > \epsilon\right\}$$

$$\le P\left\{\left|\left(\frac{-2\theta}{T}\right)I_T - 1\right| > \frac{\epsilon}{2}\right\} + P\left\{\left(-\frac{2\theta}{T}\right)|I_{n,T}-I_T| > \frac{\epsilon}{2}\right\}$$

$$
\begin{aligned}
&\le C\exp\left(\frac{T\theta}{16}\epsilon^2\right) + \frac{16\theta^2}{T^2}\,\frac{E|I_{n,T}-I_T|^2}{\epsilon^2} \\
&\le C\exp\left(\frac{T\theta}{16}\epsilon^2\right) + C\frac{T^2/n^2}{\epsilon^2}. \qquad (2.44)
\end{aligned}
$$

Next, observe that

$$
\begin{aligned}
&\sup_{x\in\mathbb{R}}\left|P\left\{\left(-\frac{T}{2\theta}\right)^{1/2}(\tilde{\theta}_{n,T,F}-\theta)\le x\right\}-\Phi(x)\right| \\
=\ &\sup_{x\in\mathbb{R}}\left|P\left\{\frac{\left(-\frac{2\theta}{T}\right)^{1/2}Y_T+\left(-\frac{2\theta}{T}\right)^{1/2}(I_T-I_{n,T})}{\left(-\frac{2\theta}{T}\right)I_{n,T}}\le x\right\}-\Phi(x)\right| \\
\le\ &\sup_{x\in\mathbb{R}}\left|P\left\{\left(-\frac{2\theta}{T}\right)^{1/2}Y_T\le x\right\}-\Phi(x)\right| \\
&+P\left\{\left|\theta\left(-\frac{2\theta}{T}\right)^{1/2}(I_{n,T}-I_T)\right|>\epsilon\right\} \\
&+P\left\{\left|\left(-\frac{2\theta}{T}\right)I_{n,T}-1\right|>\epsilon\right\}+2\epsilon \\
\le\ &CT^{-1/2}+\theta^2\frac{\left(-\frac{2\theta}{T}\right)E|I_{n,T}-I_T|^2}{\epsilon^2}+C\exp\left(\frac{T\theta}{4}\epsilon^2\right) \\
&+C\frac{T^2}{n^2\epsilon^2}+2\epsilon, \qquad (2.45)
\end{aligned}
$$

(the bound for the 3rd term in the right hand side of (2.45) is obtained from (2.44))

$$
\le\ CT^{-1/2}+C\frac{T^2}{n^2\epsilon^2}+C\exp\left(\frac{T\theta}{4}\epsilon^2\right)+C\frac{T}{n^2\epsilon^2}+\epsilon \qquad (2.46)
$$

(by Lemma 7.3(a)).

Choosing $\epsilon = CT^{-1/2}(\log T)^{1/2}$, the terms in the right hand side of (2.5) are of the order $O(\max(T^{-1/2}(\log T)^{1/2}, (\frac{T^4}{n^2})(\log T)^{-1}))$. □

(b) From (2.1), we have

$$
I_{n,T}^{1/2}(\tilde{\theta}_{n,T,F}-\theta)=\frac{Y_T+\theta(I_T-I_{n,T})}{I_{n,T}^{1/2}}.
$$

Then,

$$
\begin{aligned}
&\sup_{x\in\mathbb{R}}\left|P\left\{I_{n,T}^{1/2}(\tilde{\theta}_{n,T,F}-\theta)\le x\right\}-\Phi(x)\right| \\
=\ &\sup_{x\in\mathbb{R}}\left|P\left\{\frac{Y_T}{I_{n,T}^{1/2}}+\theta\frac{I_T-I_{n,T}}{I_{n,T}^{1/2}}\le x\right\}-\Phi(x)\right|
\end{aligned}
$$

$$\begin{aligned}
&\leq \sup_{x\in\mathbb{R}}\left|P\left\{\frac{Y_T}{I_{n,T}^{1/2}}\leq x\right\}-\Phi(x)\right|+P\left\{\left|\frac{\theta(I_T-I_{n,T})}{I_{n,T}^{1/2}}\right|>\epsilon\right\}+\epsilon\\
&=: U_1+U_2+\epsilon. \qquad (2.47)
\end{aligned}$$

We have from (2.44),

$$U_1\leq CT^{-1/2}+C\exp\left(\frac{T\theta}{16}\epsilon^2\right)+C\frac{T^2}{n^2\epsilon^2}+\epsilon. \tag{2.48}$$

Now,

$$\begin{aligned}
U_2 &= P\left\{|\theta|\left|\frac{I_{n,T}-I_T}{I_{n,T}^{1/2}}\right|>\epsilon\right\}=P\left\{|\theta|\frac{\left|\left(-\frac{2\theta}{T}\right)^{1/2}(I_{n,T}-I_T)\right|}{\left|\left(-\frac{2\theta}{T}\right)^{1/2}I_{n,T}^{1/2}\right|}>\epsilon\right\}\\
&\leq P\left\{\left|\left(-\frac{2\theta}{T}\right)^{1/2}\right||I_{n,T}-I_T|>\delta\right\}\\
&\quad +P\left\{\left|\left(-\frac{2\theta}{T}\right)^{1/2}I_{n,T}^{1/2}-1\right|>\delta_1\right\} \qquad (2.49)\\
&\quad (\text{where } \delta=\epsilon-C\epsilon^2 \text{ and } \delta_1=(\epsilon-\delta)/\epsilon>0)\\
&\leq \left(-\frac{2\theta}{T}\right)\frac{E|I_{n,T}-I_T|^2}{\delta^2}+P\left\{\left|\left(-\frac{2\theta}{T}\right)I_{n,T}-1\right|>\delta_1\right\}\\
&\leq C\frac{T^3}{n^2\delta^2}+C\exp\left(\frac{T\theta}{16}\delta_1^2\right)+C\frac{T^2}{n^2\delta_1^2}. \qquad (2.50)
\end{aligned}$$

Here, the bound for the first term in the right hand side of (2.7) comes from Lemma 2.2(c) and that for the second term is obtained from (2.44).

Now, using the bounds (2.48) and (2.50) in (2.47) with $\epsilon = CT^{-1/2}$ $(\log T)^{1/2}$, we obtain that the terms in (2.47) are of the order $O(\max(T^{-1/2}(\log T)^{1/2}, (\frac{T^4}{n^2})(\log T)^{-1}))$. □

(c) Let

$$G_T:=\left\{|\tilde{\theta}_{n,T,F}-\theta|\leq d_T\right\}, \qquad \text{and} \qquad d_T:=CT^{-1/2}(\log T)^{1/2}.$$

On the set G_T, expanding $(2|\tilde{\theta}_{n,T}|)^{1/2}$, we obtain

$$\begin{aligned}
(-2\tilde{\theta}_{n,T})^{-1/2} &= (-2\theta)^{1/2}\left[1-\frac{\theta-\tilde{\theta}_{n,T,F}}{\theta}\right]^{-1/2}\\
&= (-2\theta)^{1/2}\left[1+\frac{1}{2}\left(\frac{\theta-\tilde{\theta}_{n,T,F}}{\theta}\right)\right]+O(d_T^2).
\end{aligned}$$

Then,

$$
\begin{aligned}
&\sup_{x\in\mathbb{R}}\left|P\left\{\left(\frac{T}{2|\tilde{\theta}_{n,T,F}|}\right)^{1/2}(\tilde{\theta}_{n,T,F}-\theta)\le x\right\}-\Phi(x)\right| \\
\le\ &\sup_{x\in\mathbb{R}}\left\{P\left(\frac{T}{2|\tilde{\theta}_{n,T}|}\right)^{1/2}(\tilde{\theta}_{n,T,F}-\theta)\le x, G_T\right\}+P(G_T^c).
\end{aligned}
$$

Now,

$$
\begin{aligned}
P(G_T^c) &= P\left\{|\tilde{\theta}_{n,T,F}-\theta|>CT^{-1/2}(\log T)^{1/2}\right\} \\
&= P\left\{\left(-\frac{T}{2\theta}\right)^{1/2}|\tilde{\theta}_{n,T,F}-\theta|>C(\log T)^{1/2}(-2\theta)^{-1/2}\right\} \\
&\le C\max\left(T^{-1/2}(\log T)^{1/2},\frac{T^3}{n^2}(\log T)^{-1}\right) \\
&\quad +2(1-\Phi\log T^{1/2}(-2\theta)^{-1/2}) \\
&\quad \text{(by Theorem 2.1(a))} \\
&\le C\max\left(T^{-1/2}(\log T)^{1/2},\frac{T^3}{n^2}(\log T)^{-1}\right).
\end{aligned}
$$

On the set G_T,

$$
\left|\left(\frac{\tilde{\theta}_{n,T,F}}{\theta}\right)^{1/2}-1\right|\le CT^{-1/2}(\log T)^{1/2}.
$$

Hence, upon choosing $\epsilon=CT^{-1/2}(\log T)^{1/2}, C$ large, we obtain

$$
\begin{aligned}
&\left|P\left\{\left(\frac{T}{-2\tilde{\theta}_{n,T,F}}\right)^{1/2}(\tilde{\theta}_{n,T,F}-\theta)\le x, G_T\right\}-\Phi(x)\right| \\
\le\ &\left|P\left\{\left(\frac{T}{-2\theta}\right)^{1/2}(\tilde{\theta}_{n,T,F}-\theta)\le x, G_T\right\}\right| \\
&+P\left\{\left|\left(\frac{\tilde{\theta}_{n,T,F}}{\theta}\right)^{1/2}-1\right|>\epsilon, G_T\right\}+\epsilon \\
&\text{(by Lemma 1.1(b))} \\
\le\ &C\max\left(T^{-1/2}(\log T)^{1/2},\frac{T^4}{n^2}(\log T)^{-1}\right) \\
&\text{(by Theorem 2.1(a)).}
\end{aligned}
$$

□

In the following theorem, we improve the bound on the error of normal approximation using a mixture of random and nonrandom normings. Thus asymptotic normality of the AMCEs needs $T \to \infty$ and $\frac{T}{n^{2/3}} \to 0$ that are sharper than the bound in Theorem 2.1.

Theorem 7.2

$$\sup_{x\in\mathbb{R}} \left|P\left\{I_{n,T}\left(-\tfrac{2\theta}{T}\right)^{1/2}(\tilde{\theta}_{n,T,F}-\theta) \le x\right\} - \Phi(x)\right| \\ = O\left(\max\left(T^{-1/2}, \left(\tfrac{T^3}{n^2}\right)^{1/3}\right)\right).$$

Proof From (2.2), we have

$$I_{n,T}\left(-\frac{2\theta}{T}\right)^{1/2}(\tilde{\theta}_{n,T,F}-\theta) = \left(-\frac{2\theta}{T}\right)^{1/2} Y_T + \theta\left(-\frac{2\theta}{T}\right)^{1/2}(I_T - I_{n,T}).$$

Hence, by Lemma 2.1–2.3

$$\begin{aligned} & \sup_{x\in\mathbb{R}} \left|P\left\{I_{n,T}\left(-\frac{2\theta}{T}\right)^{1/2}(\tilde{\theta}_{n,T,F}-\theta) \le x\right\} - \Phi(x)\right| \\ = & \sup_{x\in\mathbb{R}} \left|P\left\{\left(-\frac{2\theta}{T}\right)^{1/2} Y_T + \theta\left(-\frac{2\theta}{T}\right)^{1/2}(I_T - I_{n,T}) \le x\right\} - \Phi(x)\right| \\ \le & \sup_{x\in R} \left|P\left\{\left(-\frac{2\theta}{T}\right)^{1/2} Y_T \le x\right\} - \Phi(x)\right| \\ & +P\left\{\left|\theta\left(-\frac{2\theta}{T}\right)^{1/2}(I_T - I_{n,T})\right| > \epsilon\right\} + \epsilon \\ \le & \; CT^{-1/2} + C\frac{E|I_T - I_{n,T}|^2}{T\epsilon^2} + \epsilon \le CT^{-1/2} + C\frac{T^3}{n^2\epsilon^2} + \epsilon. \end{aligned}$$

Choosing $\epsilon = (\frac{T^3}{n^2})^{1/3}$, the theorem follows. □

The following theorem gives stochastic bound on the error of approximation of the continuous MCE by AMCEs.

Theorem 7.3

$$(a)\;\; |\tilde{\theta}_{n,T} - \tilde{\theta}_T| = O_P\left(\frac{T}{n}\right)^{1/2}.$$

$$(b)\;\; |\tilde{\theta}_{n,T,z} - \tilde{\theta}_T| = O_P\left(\frac{T^2}{n^2}\right)^{1/2}.$$

Proof From (1.9) and (1.14), we have

$$\tilde{\theta}_T = -\frac{T}{2I_T}, \quad \tilde{\theta}_{n,T} = -\frac{T}{2I_{n,T}}.$$

Hence, applying Lemma 1.2 with the aid of Lemma 2.4(a) and noting that $|\frac{I_{n,T}}{T}| = O_P(1)$ and $|\frac{I_T}{T}| = O_P(1)$, the part (a) of theorem follows.

From (1.9) and (1.16), we have

$$\tilde{\theta}_T = -\frac{T}{2I_T}, \quad \tilde{\theta}_{n,T,z} = -\frac{T}{I_{n,t} + J_{n,T}}.$$

Applying Lemma 1.2 with the aid of Lemma 2.4(b) and noting that $|\frac{J_{n,T}}{T}|$ $= O_P(1)$ and $|\frac{I_T}{T}| = O_P(1)$, the part (b) of theorem follows. □

Chapter 14

Approximate Maximum Likelihood Estimation in Sub-fractional Hybrid Stochastic Volatility Model

14.1 Introduction

We consider the hybrid stochastic volatility model under the risk-neutral measure given by

$$
\begin{aligned}
dS_t &= X_t dt + \sqrt{V_{t-}} S_t dW_t + \rho_{\lambda t} dL_{\tau_{\lambda t}}, \\
dV_t &= -\lambda V_t dt + \upsilon_{\lambda t-} dL_{\tau_{\lambda t}}, \\
dX_t &= \alpha(\beta - X_t)dt + \sigma X_t^{\gamma_t} dW_t^H, \\
d\rho_t &= ((2\zeta - \eta) - \eta\rho_t)dt + \theta\sqrt{(1+\rho_t)(1-\rho_t)}dZ_t, \\
d\xi_t &= \kappa(\mu - \xi_t)dt + \varsigma\sqrt{\xi_t}dB_t, \\
d\gamma_t &= \varpi(\psi - \delta))dt + \sqrt{\chi}dM_t, \\
d\tau_t &= \xi_{t-}dt,
\end{aligned}
$$

where L_t is a Levy process, W^H is a sub-fractional Brownian motion, and B_t, Z_t, and M_t are the standard Brownian motions. Here, S_t is the asset price that is a geometric jump-diffusion, V_t is the stochastic volatility that is a Levy O–U process, X_t is the stochastic interest rate that is a sub-fractional CKLS process, ρ_t is the stochastic leverage Jacobi (Beta) process, ξ_t is a volatility modulation of the driving Levy subordinator that is a CIR process, τ_t is the stochastic time change of the driving Levy subordinator that is a time integral of the CIR process, γ_t is the stochastic elasticity model that is a CIR process, υ is the volatility that is independent of L, and all the 14 parameters in the model are positive.

This chapter introduces several approximate maximum likelihood estimators of the parameters of the sub-fractional Chan–Karolyi–Longstaff–Sanders (CKLS) interest rate model and obtains their rates of convergence. A new algorithm inspired by Newton–Cotes formula is presented to improve the accuracy of estimation. The estimators are useful for simulation of interest rates. The proposed new algorithm could be useful for other stochastic computation. It also proposes a generalization of the CKLS interest rate model with sub-fractional Brownian motion drivers that preserves medium range memory.

J. P. N. Bishwal, *Parameter Estimation in Stochastic Volatility Models*,
https://doi.org/10.1007/978-3-031-03861-7_14

14.2 Term Structure Models and Derivative Pricing

High-frequency and ultra-high-frequency data analysis in finance is the recent trend of investigation, see Ait-Sahalia and Jacod (2014). We introduce some new approximate maximum likelihood estimators of the drift parameters in the Chan–Karolyi–Longstaff–Sanders model (CKLS model hereafter), introduced by Chan et al. (1992), based on discretely sampled interest rate data, and propose a new algorithm to obtain the rates of convergence of the estimators to the corresponding continuous maximum likelihood estimators.

Recently, Filipovic et al. (2017) studied linear-rational term structure models with martingale error terms that ensure non-negative interest rates, accommodate unspanned factors affecting volatility and risk premiums, and admit semi-analytical solutions to swaptions—an important class of interest rate derivatives—that underlie the pricing and hedging of mortgage-backed securities, callable agency securities, life insurance products, and a wide variety of structured products. A parsimonious model specification within the linear-rational class has a very good fit to both interest rate swaps and swaptions from 1997 to 2003 and captures many features of term structure, volatility, and risk premium dynamics—including when interest rates are close to the zero lower bound (ZLB). The term structure is assumed to be driven by three factors.

Polynomial diffusions, studied in Filipovic and Larsson (2016), represent an extension of affine class that has linear drift and quadratic diffusion functions. A state price density is a positive semimartingale $\zeta_t = e^{-\alpha t} p(X_t)$ where p is a positive polynomial of a factor diffusion process X defined on a filtered probability space. The cash flow $C_T = q(X_T)$ for some polynomial q. The fact that pq itself being a polynomial implies that the model price $\Pi(t,T) = \frac{1}{\zeta_t} E[\zeta_T C_T | \mathcal{F}_t]$ becomes a rational function of X_t with coefficients given in terms of matrix exponential. Polynomial diffusion models thus yield closed form expressions for any security with cash flows specified as a polynomial function of X, which makes them universally applicable in finance. This includes financial market models for interest rates (with $C_T = 1$), credit risk in a doubly stochastic framework (with C_T the conditional survival probability), stochastic volatility (with C_T the spot variance), and commodities and electricity (with C_T the spot price). They showed uniqueness of polynomial diffusions via moment determinacy in the classical theory of moment problems in combination with pathwise uniqueness. Mixed moments of all finite dimensional marginal distributions of polynomial diffusions are uniquely determined by its generator, and uniqueness follows whenever these moments determine the underlying distribution. This is often true, for instance in the affine case or when the state space is compact, or more generally if exponential moments exist. There are however situations exist where the moment problem fails. In that case, Yamada–Watanabe type arguments (see Ikeda and Watanabe (1989)) are used, which give uniqueness in one dimensional case as well as when the process dynamics exhibits a certain hierarchical structure. These uniqueness results do not depend on the geometry of the state space. Existence boils down to a stochastic invariance problem that is solved for semialgebraic state spaces.

Zhou (2003) used one dimensional polynomial jump-diffusions to build short rate models that were estimated by data using generalized method of moments approach. Examples of non-affine polynomial processes include multidimensional Jacobi or Fisher–Wright processes (Ethier, 1976), Pearson diffusions (Forman and Sørensen, 2008), and Dunkl processes (Dunkl, 1992; Gallardo and Yor, 2006). A short rate model based on the Jacobi process was presented by Delbaen and Shirakawa (2002), and Larsen and Sørensen (2007) used the same process for exchange rate modeling. Forman and Sørensen (2008) studied parameter estimation in non-affine polynomial diffusions.

Polynomial jump-diffusion models that nest affine jump-diffusions are studied in Filipovic and Larsson (2017). A jump-diffusion model is polynomial if its extended generator maps any polynomial to a polynomial of equal or lower degree. As a consequence, its conditional moments can be computed in closed form. This property renders polynomial jump-diffusions computationally tractable and perfectly suited for financial asset pricing models. Many commonly occurring jump-diffusions are polynomial, for example, Ornstein–Uhlenbeck processes, square-root diffusions, Jacobi or Wright–Fisher diffusions, Levy processes, and geometric Levy processes. Modulo integrability conditions on the jumps, affine jump-diffusions are polynomials but the converse is not true so that polynomial jump-diffusions truly extend the class of affine jump-diffusions. Affine property of jump-diffusion is not invariant under exponentiation or subordination in general. Beyond the affine class, polynomial jump-diffusions include Dunkl processes and univariate diffusion volatility models, e.g., extended Stein–Stein model and extended Hull–White model.

Another example of polynomial diffusion is GARCH diffusion

$$dX_t = \alpha(\beta - X_t)dt + \sqrt{2\beta}X_t dW_t$$

with $\alpha, \beta > 0$, whose invariant distribution is inverse gamma distribution with shape parameter 2 and scale parameter $1/\beta$. When in the stationary case when X_0 has the invariant distribution, we have $E(X_t) = \beta$ and $E(X_t^2) = \infty$.

Affine property is not invariant under exponentiation. Non-affine polynomial diffusions: consider the square-root process

$$dX_t = \alpha(\beta - X_t)dt + \sigma\sqrt{X_t}dW_t,$$

which is an affine diffusion. The augmented process $\bar{X}_t = (X_t, X_t^2)$ satisfies

$$d\bar{X}_{1t} = (b + \gamma\bar{X}_{1t})dt + \gamma\sqrt{\bar{X}_{1t}}dW_t$$

$$d\bar{X}_{2t} = ((2b + \sigma^2 \bar{X}_{1t}) + 2\beta \bar{X}_{2t})dt + \sqrt{\bar{X}_{1t}\bar{X}_{2t}}dW_t.$$

This shows that $\bar{X}_t = (X_t, X_t^2)$ is not affine, while it is a polynomial.

One can extend the GARCH diffusion to have jumps driven by Poisson random measure that is a special case of the Levy driven linear SDE.

Consider the jump-diffusion operator of the form

$$\mathcal{G}f(x) = \frac{1}{2}a(x)f''(x) + b(x)f'(x) + \int_{\mathbb{R}} (f(x+\xi) - f(x) - \xi f'(x))\nu(x, d\xi)$$

for some measurable maps a and b and transition kernel $\nu(x, d\xi)$ satisfying $\nu(x, \{0\}) = 0$ and $\int_{\mathbb{R}} |\xi| \wedge |\xi|^2 \nu(x, d\xi) < \infty$.

Let X_t be a jump-diffusion with the extended generator $\mathcal{G}$. That is, X_t is a special semimartingale, and

$$f(X_t) - f(X_0) - \int_0^t \mathcal{G}f(X_s)ds$$

is a local martingale for any bounded C^2 function $f(x)$.

Polynomial asset price models are given by financial market with m primary assets with price processes given by

$$S_{i,t} = S_{i,0} \exp\left\{ \int_0^t r_s ds + Y_{i,t} \right\}, 1 \le i \le m, t \ge 0,$$

where r_t is the risk-free rate and $Y_t = (Y_{1,t}, Y_{2,t}, \ldots, Y_{m,t})$ are logarithmic excess return processes with $Y_0 = 0$.

Let X_t be a polynomial jump-diffusion such that $Z_t = (X_t, Y_t)$ is a jump-diffusion with extended generator $\mathcal{G}$. We assume that

$$\int_{\mathbb{R}} (e^{\eta_i} - 1 - \eta_i)\nu^Y(x, d\eta) < \infty.$$

The price processes are special semimartingales with the decomposition

$$\frac{dS_{i,t}}{S_{i,t-}} = (r_t + \epsilon_i(X_t))dt + dY_{i,t}^c + \int_{\mathbb{R}} (e^{\eta_i} - 1)(\mu^Y(d\eta, dt) - \nu^Y(X_{t-}, d\eta)), \ i \ge 1, \ t \ge 0,$$

where Y_t^c denotes the continuous martingale part of Y_t, $\mu^Y(d\eta, dt)$ is the integer-valued random measure associated with the jumps of Y_t, and the simple excess rates of return are given by

$$\epsilon_i(x) = b_i^Y(x) + \frac{1}{2}a_{ii}^Y(x) + \int_{\mathbb{R}} (e^{\eta_i} - 1 - \eta_i)\nu^Y(x, d\eta).$$

The measure P is the risk-neutral measure so that the discounted price process $\exp(-\int_0^t r_s ds)S_{i,t}$ is a local martingale. This is achieved by setting $\epsilon_i(x) = 0$. But $\epsilon_i(x)$ cannot be zero if $Z_t = (X_t, Y_t)$ is a polynomial jump-diffusion, other than affine in general. But Z_t can be embedded into a higher dimensional polynomial jump-diffusion such that $\epsilon_i(x) = 0$.

European call option at time $t = 0$ written on assets $S_{i,t}$ with strike price K and maturity T is given by

$$E[\exp\left(-\int_0^T r_s ds)S_{i,T} - K\right)^+ |\mathcal{F}_0] = E[S_{i,0}\exp(Y_{i,T}) - K\exp(-\int_0^T r_s ds))^+|\mathcal{F}_0].$$

Linear-Rational Framework

A sufficient condition for the absence of arbitrage opportunities (no free lunch with vanishing risk) in a model of a financial market is the existence of a state price density that is a positive and adapted process ζ_t on a filtered probability space $(\Omega, \mathcal{F}, \{\mathcal{F}_t\}_{t\geq 0}, P)$ such that the price $\Pi(t,T)$ at time t of any time-T cash flow C_T is given by

$$\Pi(t,T) = \frac{1}{\zeta_t} E[\zeta_t C_T|\mathcal{F}_t] = \frac{1}{\zeta_t} E_t[\zeta_t C_T].$$

We may interpret P as the historical measure.

Filipovic et al. (2017) modeled the term structure models directly specifying the state price density. They assume a multivariate factor process Z_t, which has a linear drift, and a state price density, which is a linear function of Z_t. In this case, the bond prices and short rate become linear-rational functions, that is the ratios of linear functions of Z_t, which is why the framework is referred as *linear-rational.* They distinguish between factors that are spanned by the term structure and that are unspanned and provide conditions such that all of the factors in Z_t are spanned. A key feature of the framework is that the term structure depends on the drift of Z_t. This leaves freedom to specify exogeneous factors feeding into the martingale part of Z_t. Such factors give rise to unspanned fractional stochastic volatility (UFSV) and can be recovered from bond derivative prices. The UFSV factors that directly affect the instantaneous bond return covariances and those that affect the expected future bond return covariances are distinguished.

The joint factor process is affine, and swaptions can be priced semi-analytically.

A linear-rational (LR) term structure models consist of two components: A factor process Z_t with linear drift and a state price density ζ_t given a linear function of Z_t. We assume that Z_t has the dynamics of the form

$$dZ_t = \alpha(\beta - Z_t)dt + dM_t^H,$$

where $\alpha > 0, \beta > 0$, and M_t^H is a sub-fractional martingale. The state price density is assumed to be given by

$$\zeta_t = e^{-\delta t}(\phi + \psi Z_t),$$

where $\phi \in \mathbb{R}$ and $\psi \in \mathbb{R}$ such that $\phi + \psi x > 0$ for all $x \in \mathbb{R}$ and for some $\delta \in \mathbb{R}$. The role of the parameter δ is to keep the short rate non-negative. The linear drift of Z_t implies that the conditional expectations are of the linear form:

$$E_t[Z_t] = \beta + e^{-\alpha(T-t)}(Z_t - \beta),\ t \leq T.$$

An immediate consequence is that the zero coupon bond prices and the short rate become linear functions of Z_t.

The basic pricing formula with $C_T = 1$ shows that zero coupon bond prices are given by

$$P(t,T) = F(T-t, Z_t),$$

where

$$F(\tau, z) = e^{-\delta\tau}\frac{\phi + \psi\beta + \psi e^{-\tau\alpha}(z-\beta)}{\phi + \psi Z_t}.$$

The short rate is obtained by the formula

$$r_t = -\partial_T \log P(t,T)|_{T=t}$$

and is given by

$$r_t = \delta - \frac{\psi\alpha(\beta - Z_t)}{\phi + \psi Z_t}.$$

Provided the short rate is bounded from below, short rate is guaranteed to be non-negative by choosing δ large enough. This leads to an intrinsic choice of δ as the smallest value that yields a non-negative short rate:

$$\delta^* = \sup_{z\in\mathbb{R}} \frac{\psi\alpha(\beta - z)}{\phi + \psi z}, \quad \delta_* = \inf_{z\in\mathbb{R}} \frac{\psi\alpha(\beta - z)}{\phi + \psi z},$$

and we set $\delta = \delta^*$ provided this is finite. The short rate then satisfies

$$r_t \in [0, \delta^* - \delta_*].$$

Notice that δ^* and δ_* depend on drift parameters of Z_t that are estimated from data. Therefore, a crucial step in the model validation process is to verify that the range of possible short rates is sufficiently wide. Finally, when $\alpha > 0$, we can verify that $(-1/\tau)\log F(\tau, z)$ converges to δ when τ goes to infinity. Hence, δ can be interpreted as the *infinite maturity zero coupon bond yield.*

Every LR process $Z_t = (\zeta_t; \zeta_t X_t)$ can be represented as a linearity-generating (LG) process (ζ_t, X_t) introduced by Gabaix (2007) and vice versa.

Unspanned Fractional Stochastic Volatility Factors

Let Z_t have a fractional diffusive dynamics of the form

$$dZ_t = \alpha(\beta - Z_t)dt + \sigma(Z_t, U_t)dW_t^H,$$

where U_t is a UFSV factor process given by

$$dU_t = \alpha(\beta - U_t)dt + \sigma\sqrt{U_t}dB_t^H,$$

where B_t^H is another sub-FBM correlated with W_t^H.

Using fractional Itô's formula (Lemma D.2.3), the state price can be written as

$$d\zeta_t/\zeta_t = -r_t dt - \lambda_t dW_t^H,$$

where

$$\lambda_t = -\frac{\sigma(Z_t, U_t)\psi}{\phi + \psi Z_t}$$

is the market price of risk. Then the dynamics of $P(t,T)$ are given by

$$\frac{dP(t,T)}{P(t,T)} = (r_t + \nu(t,T)\lambda_t)dt + \nu(t,T)dW_t^H,$$

where

$$\nu(t,T) = \frac{\sigma(Z_t, U_t)\partial_z F(T-t, Z_t)}{F(T-t, Z_t)}.$$

To see that the term structure is uneffected by UFSV factors, consider the following bond price decomposition. The time t-price of a zero coupon bond with maturity T equals its discounted expected price at a nearby future date $t + dt$ plus the risk premium,

$$P(t,T) = P(t, t+dt)E_t[P(t+dt, T)]P(t,T) + \mathrm{Cov}_t\left[\frac{\zeta_{t+dt}}{\zeta_t}, P(t+dt, T)\right].$$

While $P(t, t+dt) = 1 - r_t dt$ does not depend on U_t, the expected time-$(t+dt)$ price does depend on U_t due to the nonlinear dependence on Z_{t+dt},

$$E_t[P(t+dt,T)] = P(t,T) + (-\frac{\partial}{\partial\tau}F(T-t, Z_t)) + \partial_z F(T-t, Z_t)\alpha(\beta - Z_t)$$

$$+\frac{1}{2}\partial_z^2 F(T-t, Z_t)a(U_t, Z_t)dt.$$

But this dependence is offset by the risk premium on the right hand side of the price equation

$$\mathrm{Cov}_t\left[\frac{\zeta_{t+dt}}{\zeta_t}, P(t+dt, T)\right] = -P(t,T)\nu(t,T)\lambda_t dt.$$

It is straightforward to verify that

$$\frac{1}{2}\partial_z^2 F(T-t, Z_t)a(U_t, Z_t)dt - P(t,T)\nu(t,T)\lambda_t = 0.$$

Z and U are highly correlated across specifications.

Pricing and Estimation for Linear-Rational Fractional Square-Root (LRFSQ) Model

Swaps and Swaptions

An interest rate swap is a contract between two parties who agree to exchange cash flows over a pre-specified time grid. The holder of a payer swap contract pays a fixed rate and receives a floating rate on a notional amount.

The holder of a receiver swap contract pays a fixed rate and receives a fixed rate on a notional amount.

The floating rate is usually tied to the London Interbank Offered Rate (LIBOR), a daily fixed benchmark for interest rates in various currencies.

A payer swaption gives the holder the right but not the obligation to enter a payer swap at a pre-specified fixed strike rate.

A receiver swaption gives the holder the right but not the obligation to enter a receiver swap at a pre-specified fixed strike rate.

Linear-rational term structure models have the important advantage of allowing for tractable swaption pricing.

Swaptions form an important class of derivatives that allow to price and hedge interest rate risk. They underlie callable mortgage-backed securities, life insurance products, and a wide variety of structured products. The outstanding notional amount in the swap market is in the order of hundreds of trillions of US dollars.

Swaptions can be divided into three categories according to their exercise timing rights. European swaptions can be exercised only at the expiration date. American swaptions allow the holder to enter the swap on any dates that falls within the range of two dates. Bermudan swaptions constitute a simplified variant of American swaptions where exercise is only possible on a finite time grid. The standard approach is to approximate American option by the Bermudan swaptions and price the later using simple tree models or Monte Carlo simulation based methods. Recently, Filipovic and Kitapbeyev (2018) priced American swaptions in the one-factor linear rations term structure model analytically in continuous time that boils down to an undiscounted optimal stopping problem for a scalar diffusion process. The later problem is reduced to a free-boundary problem that they tackle by the local time-space calculus. They characterize the optimal stopping boundary as the unique solution to a nonlinear integral equation that can be readily solved analytically. Using these boundaries, they obtain the American swaption price and the optimal exercise strategies in terms of swap rates for both payer and receiver swaptions.

A fixed versus floating interest rate swap (IRS) on $[T_0, T_n]$ is specified by a tenor structure of recent and payment dates $T_0 = t_0 < t_1 < \ldots < t_N = T_n$ for the floating leg and tenor structure $T_0 < T_1 < \ldots < T_n$ for the fixed leg. We let $\delta_1 := t_i - t_{i-1}$ and $\Delta := T_i - T_{i-1}$ denote the length between the tenor dates.

We make the simplifying assumption that the discount factors implied by the state price density reflect the same credit and liquidity characteristics of LIBOR. The value of the payer swap (paying fixed and receiving floating) at time $t \leq T_0$ is given by

$$\Pi_t^{swap} = P(t, T_0) - P(t, T_n) - \Delta K \sum_{j=1}^{n} P(t, T_j),$$

where K is fixed annualized interest rate. This valuation formula was market standard until the financial crisis.

The time-t forward rate swap rate $S_t^{T_0, T_n}$ is the rate K that makes the value of the swap rate equal to zero. It is given by

$$S_t^{T_0, T_n} = \frac{P(t, T_0) - P(t, T_n)}{\Delta \sum_{i=1}^{n} P(t, T_i)},$$

which is linear-rational in Z_t. The forward rate swap becomes the spot rate swap at time T_0.

A payer swaption is an option to enter the IRS (interest rate swap) paying the fixed leg at a predetermined rate and receiving the floating leg.

A European payer swaption expiring at T_0 on a swap with the characteristics described above has value at expiration of

$$C_{T_0} = (\Pi_{T_0}^{swap})^+ = \left(\sum_{i=0}^{n} c_i P(T_0, T_i) \right)^+ = \frac{1}{\zeta_{T_0}} \left(\sum_{i=1}^{n} c_i E_{T_0}[\zeta_{T_i}] \right)^+,$$

where the coefficients c_i can easily be read off from the expression of the payer swap.

Swap In a linear-rational term structure model, the conditional expectations $E_{T_0}[\zeta_{T_i}]$ are explicit functions of Z_{T_0}. Specifically, we have

$$C_{T_0} = p_{swap}(Z_{T_0})^+ / \zeta_{T_0},$$

where $p_{swap}(z)$ is the explicit linear function

$$p_{swap}(z) = \sum_{i=1}^{n} c_i e^{-\delta T_i} (\phi + \psi\beta + \psi e^{-\alpha(T_i - T_0)}(z - \beta)).$$

Swaption The swaption price at time $t \leq T_0$ is then obtained by the fundamental pricing formula

$$\Pi_t^{swaption} = \frac{1}{\zeta_t} E_t[\zeta_{T_0} C_{T_0}] = \frac{1}{\zeta_t} E_t[p_{swap}(Z_{T_0})^+].$$

To compute the price, one has to evaluate the conditional expectation on the right side of the above equation. If the conditional distribution of Z_{T_0} is known, this can be done via a direct numerical integration over $\mathbb{R}$.

When U is constant and $\sigma(x)\sqrt{x}$, the transition density $p(\hat{x}; u, x, t)$ is noncentral chi-squared.

There is an alternative efficient approach via Fourier transform methods.

The probability density function p of Z_u is recovered by the Levy inversion formula:

$$p(\hat{x}; u, x, t) = \frac{1}{2\pi} \int_{\mathbb{R}} e^{-iw\hat{x}} \hat{q}(iw; u, x, t) dw$$

for $\hat{x}, x > 0$ and $u \geq t \geq 0$.

Theorem 2.1 The swaption price is given by

$$\Pi_t^{swaption} = \frac{1}{\zeta_t \pi} \int_0^\infty \text{Re}\left[\frac{\hat{q}(\mu + i\lambda)}{(\mu + i\lambda)^2}\right] d\lambda,$$

where $\hat{q}(x) = E_t[\exp(x\ p_{swap}(Z_{T_0})]$ for $x \in \mathbb{C}$ and $\hat{q}(\mu) < \infty$ for $\mu > 0$.

Lemma 2.1

$$E_\rho[Z_{\rho+\tau}] = \beta + e^{-k\tau}(Z_\rho - \beta)$$

for any bounded stopping time ρ and any deterministic $\tau \geq 0$.

Swap Risk Premium

These are risk premiums in swap contracts.

Sharpe Ratio

Unconditional mean and volatility of excess returns increase with bond maturity, and unconditional Sharp ratio decreases with bond maturity. Implied volatility is a robust predictor of excess returns.

We have the following *Exponential Affine Transform Formula*

Lemma 2.2

$$E_t[e^{u+vX_T}] = e^{\Phi(T-t)+\Psi(T-t)X_t}.$$

The exponents here satisfy complex valued ODEs inspired by Lemma D.2.3. See Duffie et al. (2000) in the nonfractional case.

Pricing Collateralized Interest Rate Derivatives

Swap contracts are always collateralized. The interest rate derivative price can be calculated using Theorem 2.1.

Parameter Estimation

The model is estimated by quasi-maximum likelihood estimation (QMLE) method (by approximating true noncentral transition density by Gaussian density) along with nonlinear unscented Kalman filter.

Let X_t be the vector containing the factors, and Y_t be the vector consisting of term structure of swap rates and NIVs observed at time t.

The measurement equation is given by

$$Y_t = h(X_t, \Theta) + u_t,$$

where h is the pricing function, Θ is the vector of model parameters, and u_t is a vector of i.i.d. Gaussian pricing errors with covariance matric Σ. X_t is a square-root fractional diffusion process whose transition density is unknown. We approximate the transition equation by

$$X_t = \Phi_0 + \Phi_x X_{t-1} + w_t,$$

where $w_t \sim N(0, Q_t)$ and Q_t is a linear function of X_{t-1}.

The Kalman filer produces one-step ahead forecasts for $Y_t, \hat{Y}_{t|t-1}$, and the corresponding error covariance matrix $F_{t|t-1}$ from which we construct the log-likelihood function is given by

$$L(\Theta) = -\frac{1}{2}\sum_{t=1}^{T}\left(n_t \log(2\pi) + \log|F_{t|t-1}| + (Y_t - \hat{Y}_{t|t-1})^{\mathsf{T}} F_{t|t-1}^{-1}(Y_t - \hat{Y}_{t|t-1})\right),$$

where T is the number of observation dates and n_t is the number of observations in Y_t. The QMLE is given by

$$\widehat{\Theta} = \arg\max_{\Theta} L(\Theta).$$

The QMLE is not consistent, since the conditional variance matrix Q_t in the recursions depends on the Kalman filter estimates of the factors instead of the true but unobservable values.

Sub-fractional CKLS Model

The CKLS model is a very popular one-factor short rate model in term structure of interest rates. This general one-factor interest rate model comprises a linear drift with constant elasticity of variance. Hence, this is also called the constant elasticity of variance (CEV) model or Cox–Ross model. For time-varying elasticity of variance model, see Fan et al. (2003). We generalize the CKLS model to sub-fractional Brownian motion noise that will have long-memory nonstationary increments. For the model to be useful and for calibration purposes, it becomes necessary to estimate the unknown parameters in the model from discrete interest rate data. See the monograph Bishwal (2008a) for asymptotic results on approximate likelihood estimators and approximate Bayes estimators for drift parameter estimation of discretely observed diffusions based on high-frequency data. To estimate the drift parameter, we use approximate maximum likelihood estimators and study the accuracy of approximation in terms of rate of convergence of the resulting estimators from high- frequency data.

Let $(\Omega, \mathcal{F}, \{\mathcal{F}_t\}_{t\geq 0}, P)$ be a stochastic basis on which is defined the sub-fractional Brownian motion (sub-FBM) W_t^H which is a centered Gaussian process with covariance function

$$C_H(s,t) = s^{2H} + t^{2H} - \frac{1}{2}\left[(s+t)^{2H} + |s-t|^{2H}\right], \quad s,t > 0$$

for $0 < H < 1$ introduced by Bojdecki et al. (2004). The interesting feature of sub-FBM is that this process has some of the main properties of FBM, but the increments of the process are nonstationary, more weakly correlated on non-overlapping time intervals than that of FBM, and its covariance decays polynomially at a higher rate as the distance between the intervals tends to infinity. The parameter H governs the long-memory behavior of the model. For $H = 0.5$, this process reduces to the standard Brownian motion.

Recall that a fractional Brownian motion has the covariance

$$\tilde{C}_H(s,t) = \frac{1}{2}\left[s^{2H} + t^{2H} - |s-t|^{2H}\right], \quad s,t > 0.$$

On the same stochastic basis, we define the fractional CKLS process $\{X_t\}$ satisfying the fractional Itô stochastic differential equation

$$dX_t = \alpha(\beta - X_t)dt + \sigma X_t^{\gamma} dW_t^H, \ t \geq 0, \ \ X_0 = x_0,$$

where $\{W_t, t \geq 0\}$ is a standard sub-fractional Wiener process with the filtration $\{\mathcal{F}_t\}_{t\geq 0}$ and H is the Hurst parameter that describes the medium range memory of the process lying in $[0.5, 1]$ that is assumed to be known. The unknown parameters are α that is the mean reversion parameter, β is the central tendency parameter, σ is the standard deviation or volatility, and γ is the elasticity of variance. The unknown parameters are to be estimated from discrete observations of the interest rate process $\{X_t\}$. The special case $\gamma = 0$ gives sub-fractional Vasicek model, $\gamma = 1$ gives the sub-fractional Rendleman–Bartter model, and $\gamma = 1/2$ gives sub-fractional Cox–Ingersoll–Ross model that is a popular one-factor equilibrium model used for short-term interest rate modeling, see Hull (2015). Sub-fractional Ho–Lee model and sub-fractional Hull–White models are nonhomegeneous generalizations of the sub-fractional Vasicek model, see Brigo and Mercurio (2001) for the short rate dependent case $H = 0.5$. Using Itô's formula, it is easy to see that the process $X_t^{1-\gamma}$ is a sub-fractional square-root diffusion process. Thus once an estimate of γ is obtained, one can simulate the process using simulation methods of square-root diffusion, see Glasserman (2004).

Model parsimony is an important problem in finance. We propose the following generalization of Ait-Sahalia model (see Ait-Sahalia (1996, 1999)) that we call the *Generalized Ait-Sahalia Model* that is given by

$$\begin{aligned} dX_t &= (\alpha_0 + \alpha_1 X_t^{v_1} + \alpha_2 X_t^{v_2} + \alpha_3 X_t^{v_3} + \alpha_4 X_t^{v_4} (\log X_t)^{v_5} + \alpha_5 \sin X_t)\ dt \\ &\quad + \sqrt{\beta_0 + \beta_1 X_t^{\gamma_1} + \ \beta_2 X_t^{\gamma_2} + \beta_3 X_t^{\gamma_3} \exp(\gamma_4 X_t)}\ dW_t^H, \end{aligned}$$

where W_t^H is the sub-fractional Brownian motion (sub-fBm).

The sine term in the model comes from exchange rate modeling by Larsen and Sørensen (2007). This is a model with 19 parameters. Special cases of this model are sub-fractional generalizations of Ait-Sahalia diffusions, Pearson diffusions, Jacobi diffusions, CIR diffusion, CKLS diffusion, CHLS diffusion (Conley et al. (1997)), OU diffusion, Vasicek diffusion, radial OU diffusion, Linetsky cube-root diffusions (see Linetsky (2004)), Lewis diffusion, Ahn–Gao diffusion, Duffie–Kan diffusion, Black–Scholes diffusion, Black–Karasinski diffusion, Schwartz diffusion, and Periodic diffusion.

Thus we obtain diffusions with invariant distributions as normal, log-normal, gamma, inverse gamma, noncentral chi-square, skewed Student-t, scaled F, Beta, logistic, and Gompertz and hyperbolic distributions. Typical value of H based on Canadian interest rate data is around 0.63.

We recall some facts from sub-fractional Pearson diffusion processes. Fractional Pearson diffusion is a stationary solution to the fractional stochastic differential equation of the form

$$dX_t = -\theta(X_t - \mu)dt + \sqrt{2\theta(aX_t^2 + bX_t + c)}\ dW_t^H,$$

where $\theta > 0$ and a, b and c are such that the square root is well defined when X_t is in the state space.

This is a special case of the following equation

$$dX_t = -\theta(X_t - \mu)dt + \sigma(X_t)\ dW_t^H,$$

where $\sigma^2(X_t) = 2\theta(aX_t^2 + bX_t + c)$.

For $H = 0.5$, this is a diffusion process with given invariant measure ν that admits a density p with respect to Lebesgue measure. Assume that the density p is continuous, bounded, has finite variance, and p is strictly positive in the interval $(l, u)(-\infty \le l \le u \le \infty)$ and is zero outside (l, u). Let μ be the expectation of the measure ν and

$$\sigma^2(x) = \frac{2\int_l^x (\mu - y)p(y)dy}{p(x)} = \frac{2\mu F(x) - 2\int_l^x yp(y)dy}{p(x)},\ x \in (l, u),$$

where $F(x) = \int_{-\infty}^x p(y)dy$ is the distribution function associated with the density p.

The SDE with the above diffusion coefficient has a Markovian weak solution. We have $E(\sigma^2(X)) < \infty$ when $X \sim p$. The solution is ergodic with invariant density p.

Sub-fractional Student Diffusion The sub-fractional Student diffusion is given by

$$dX_t = -\theta X_t dt + \sqrt{\frac{2\theta}{\nu - 1}(\nu + X_t^2)}\, dW_t^H.$$

For $H = 0.5$, the invariant density is given by

$$f(x) = \frac{\Gamma((\nu+1)/2)}{\sqrt{\nu\pi}\Gamma(\nu/2)}\left(1 + \frac{1}{\nu}x^2\right)^{-(\nu+1)/2}, \quad x \in \mathbb{R},\ \nu > 0.$$

$$= \frac{1}{\sqrt{\nu}B(\nu/2, 1/2))[1 + (x/\nu)^2]^{(\nu+1)/2]}},$$

where $B(\cdot, \cdot)$ is the beta function.

Sub-fractional Skew Student Diffusion

$$dX_t = -\theta X_t dt + \sqrt{\frac{2\theta}{\nu - 1}\left[X_t^2 + 2\varpi\nu^{1/2}X_t + (1 + \varpi^2)\nu\right]}\, dW_t^H.$$

The parameter ϖ is the skewness parameter. The case $\varpi = 0$ gives the sub-fractional symmetric Student diffusion.

Sub-fractional Gamma Diffusion

$$dX_t = -\theta(X_t - \mu)dt + \sqrt{2\theta X_t}\, dW_t^H.$$

Sub-fractional Snedecor Diffusion

$$dX_t = -\theta(X_t - \mu)dt + \sqrt{2\theta a X_t(X_t + 1)}\, dW_t^H.$$

Sub-fractional Jacobi Diffusion

$$dX_t = -\theta(X_t - \mu)dt + \sqrt{2\theta a X_t(X_t - 1)}\, dW_t^H.$$

Sub-fractional Uniform Diffusion

$$dX_t = -\theta(X_t - \frac{1}{2})dt + \sqrt{\theta X_t(1 - X_t)}\, dW_t^H.$$

Sub-fractional Logistic Diffusion

$$dX_t = -\theta\left[1 - 2\mu + (1 - \mu)e^{X_t} - \mu e^{-1} - 8a\cosh^4\left(\frac{X_t}{2}\right)\right] dt$$

$$+2\sqrt{-a\theta}\cosh\left(\frac{X_t}{2}\right) dW_t^H.$$

Sub-fractional Gompertz Diffusion

$$dX_t = (\alpha X_t - \beta X_t \ln X_t)\, dt + \sigma X_t\, dW_t^H.$$

A crucial feature of the CIR model is that both the infinitesimal drift $\alpha(\beta - x)$ and the variance $\sigma^2 x$ are affine functions of the state variable. However, empirical studies contradict the CIR specification. Using the sub-fractional Bessel process with constant drift, we find some non-affine specifications that are analytically tractable.

Sub-fractional Non-affine Models

Sub-fractional Bessel Process

$$dY_t = ((\nu + 1/2)Y_t^{-1} + \mu)dt + \ dW_t^H.$$

Applying Itô's lemma with $X_t = 4(\sigma Y_t)^{-2}$, we obtain

$$dX_t = -\theta(X_t^{1/2} - \mu)X_t^{3/2}dt + \sigma X_t^{3/2}\ dW_t^H.$$

This diffusion with nonlinear drift and infinitesimal variance $\sigma^2 x^3$.

Applying Itô's lemma with $X_t = (\sigma Y_t)^{-1}$, we obtain

$$dX_t = -\theta(X_t - \mu)X_t^2 dt + \sigma X_t^2\ dW_t^H.$$

This diffusion with nonlinear drift and infinitesimal variance $\sigma^2 x^4$.

Sub-fractional Bessel Process with Linear Drift

$$dY_t = ((\nu + 1/2)Y_t^{-1} + \mu Y_t)dt + \ dW_t^H$$

Sub-fractional 3/2 Model

By fractional Itô's lemma (Lemma D.2.3), the *sub-fractional reciprocal squared process*, i.e., taking $X_t = (4/\sigma^2)Y_t^{-2}$, we obtain the sub-fractional 3/2 model

$$dX_t = -\theta(X_t - \mu)X_t dt + \sigma X_t^{3/2}\ dW_t^H.$$

This diffusion with nonlinear drift and infinitesimal variance $\sigma^2 x^3$ is the reciprocal of the square-root CIR process. This process was proposed by Cox et al. (1985) as a model for the inflation rate in their three-factor inflation rate model.

Any European style interest rate derivative with some payoff function $f(X_t)$ at anytime $t > 0$ can be valued by integrating the payoff function against the state price density:

$$E\left[\exp(-\int_0^t X_u du)f(X_t)|X_0 = x\right] = \int_0^\infty f(y)\pi(t; x, y)dy.$$

The state price density $\pi(t; x, y)$ can be interpreted as the transition density of the diffusion *killed at the linear rate*. The state price density of the short rate of the above non-affine models can be easily obtained from the transition density of the Bessel process with constant drift.

14.3 Newton–Cotes Distribution and Drift Estimators

Hereafter, we focus on the CKLS diffusion. The initial value of the process $x_0 \geq 0$ for nonnegativity of the solution. For simplicity of exposition, we assume $x_0 = 0$. Let the continuous realization $\{X_t, 0 \leq t \leq T\}$ be denoted by X_0^T. Let $P_{(\alpha,\beta)}^T$ be the measure generated on the space (C_T, B_T) of continuous functions on $[0, T]$ with the associated Borel σ-algebra B_T generated under the supremum norm by the process X_0^T, and let P_0^T be the standard Wiener measure. It is well known that when (α, β) is the true value of the parameter vector, $P_{(\alpha,\beta)}^T$ is absolutely continuous with respect to P_0^T, and the Radon–Nikodym derivative (likelihood) of $P_{(\alpha,\beta)}^T$ with respect to P_0^T based on observation X_0^T is given by

$$\begin{aligned} L_T(\alpha, \beta) &:= \frac{dP_{(\alpha,\beta)}^T}{dP_0^T}(X_0^T) \\ &= \exp\left\{\int_0^T \alpha(\beta - X_t)\sigma^{-2}X_t^{1-2\gamma}dX_t - \frac{1}{2}\int_0^T \alpha(\beta - X_t)^2\sigma^{-2}X_t^{-2\gamma}dt\right\}. \end{aligned}$$

Consider the score function, the derivative of the log-likelihood function, which is given by

$$l_T(\alpha, \beta) := \sigma^{-2}\left\{\int_0^T \alpha(\beta - X_t)X_t^{1-2\gamma}dX_t - \frac{1}{2}\int_0^T \alpha(\beta - X_t)^2 X_t^{-2\gamma}dt\right\}.$$

A solution of the estimating equation $l_T(\alpha, \beta) = 0$ provides the conditional maximum likelihood estimates (MLEs)

$$\hat{\alpha}_T := \frac{\int_0^T X_t^{-2\gamma}dX_t \int_0^T X_t^{1-2\gamma}dt - \int_0^T X_t^{1-2\gamma}dX_t \int_0^T X_t^{-2\gamma}dt}{\int_0^T X_t^{-2\gamma}dt \int_0^T X_t^{2-2\gamma}dt - \int_0^T X_t^{1-2\gamma}dt \int_0^T X_t^{1-2\gamma}dt}$$

and

$$\hat{\beta}_T := \frac{\int_0^T X_t^{-2\gamma}dX_t \int_0^T X_t^{2-2\gamma}dt - \int_0^T X_t^{1-2\gamma}dX_t \int_0^T X_t^{1-2\gamma}dt}{\int_0^T X_t^{-2\gamma}dX_t \int_0^T X_t^{1-2\gamma}dt - \int_0^T X_t^{1-2\gamma}dX_t \int_0^T X_t^{-2\gamma}dt}.$$

We transform the Itô integrals $\int$ into the Stratonovich integrals $\oint$ as follows:
For a smooth function $f(\cdot)$, we have

$$\int_0^T f(X_t)dX_t = \oint_0^T f(X_t)dX_t - \frac{\sigma^2}{2}\int_0^T f'(X_t)X_t^{2\gamma}dt.$$

For simplicity, we assume that $\sigma = 1$. Thus

$$\int_0^T X_t^{-2\gamma}dX_t = \oint_0^T X_t^{-2\gamma}dX_t + \gamma\int_0^T X_t^{-1}dt,$$

$$\int_0^T X_t^{1-2\gamma} dX_t = \oint_0^T X_t^{1-2\gamma} dX_t - \frac{1-2\gamma}{2} T,$$

$$\int_0^T X_t^{2-2\gamma} dX_t = \oint_0^T X_t^{2-2\gamma} dX_t - (1-\gamma) \int_0^T X_t dt.$$

Thus Stratonovich integral based MLEs are given by

$$\hat{\alpha}_T := \frac{\left(\oint_0^T X_t^{-2\gamma} dX_t + \gamma \int_0^T X_t^{-1} dt\right) \int_0^T X_t^{1-2\gamma} dt - \left(\oint_0^T X_t^{1-2\gamma} dX_t - \frac{1-2\gamma}{2} T\right) \int_0^T X_t^{-2\gamma} dt}{\int_0^T X_t^{-2\gamma} dt \int_0^T X_t^{2-2\gamma} dt - \int_0^T X_t^{1-2\gamma} dt \int_0^T X_t^{1-2\gamma} dt},$$

$$\hat{\beta}_T := \frac{\left(\oint_0^T X_t^{-2\gamma} dX_t + \gamma \int_0^T X_t^{-1} dt\right) \int_0^T X_t^{2-2\gamma} dt - \left(\oint_0^T X_t^{1-2\gamma} dX_t - \frac{1-2\gamma}{2} T\right) \int_0^T X_t^{1-2\gamma} dt}{\left(\oint_0^T X_t^{-2\gamma} dX_t + \gamma \int_0^T X_t^{-1} dt\right) \int_0^T X_t^{1-2\gamma} dt - \left(\oint_0^T X_t^{1-2\gamma} dX_t - \frac{1-2\gamma}{2} T\right) \int_0^T X_t^{-2\gamma} dt}.$$

Laurini and Hotta (2013) studied estimation in CIR fractional short-term interest rate models driven by fractional Brownian motion using simulation based *indirect inference method.*

In this chapter, we obtain rates of convergence for several approximate maximum likelihood estimators (AMLEs) based on discrete observations of the interest rate model. We assume that the process $\{X_t\}$ is observed at discrete time points $0 = t_0 < t_1 < \cdots < t_n = T$ with $t_i - t_{i-1} = \frac{T}{n}, i = 1, 2 \cdots, n$. We assume two types of high-frequency discrete data with long observation time: (1) $T \to \infty, n \to \infty, \frac{T}{\sqrt{n}} \to 0$, 2) $T \to \infty, n \to \infty, \frac{T}{n^{2/3}} \to 0$.

The volatility parameter σ can be estimated as

$$\hat{\sigma}^2_{n,T} := \frac{\sum_{i=1}^n (X_{t_i} - X_{t_{i-1}})^2}{\sum_{i=1}^n X^2_{t_{i-1}} (t_i - t_{i-1})}.$$

For a weight function $w_{t_i} \geq 0$, define the weighted AMLEs

$$\begin{aligned}
\tilde{\alpha}_{n,T} \quad := \quad & \Bigg(\Bigg[\Big\{ \textstyle\sum_{i=1}^n w_{t_i} X_{t_{i-1}}^{-2\gamma} + \sum_{i=2}^{n+1} w_{t_i} X_{t_{i-1}}^{-2\gamma} \Big\} (X_{t_i} - X_{t_{i-1}}) \\
& + \gamma \Big\{ \textstyle\sum_{i=1}^n w_{t_i} X_{t_{i-1}}^{-1} + \sum_{i=2}^{n+1} w_{t_i} X_{t_{i-1}}^{-1} \Big\} (t_i - t_{i-1}) \Bigg] \\
& \Bigg[\Big\{ \textstyle\sum_{i=1}^n w_{t_i} X_{t_{i-1}}^{1-2\gamma} + \sum_{i=2}^{n+1} w_{t_i} X_{t_{i-1}}^{1-2\gamma} \Big\} (t_i - t_{i-1}) \\
& - \Big\{ \textstyle\sum_{i=1}^n w_{t_i} X_{t_{i-1}}^{1-2\gamma} + \sum_{i=2}^{n+1} w_{t_i} X_{t_{i-1}}^{1-2\gamma} \Big\} (X_{t_i} - X_{t_{i-1}}) - \frac{1-2\gamma}{2} T \Bigg] \\
& \Bigg[\Big\{ \textstyle\sum_{i=1}^n w_{t_i} X_{t_{i-1}}^{2-2\gamma} + \sum_{i=2}^{n+1} w_{t_i} X_{t_{i-1}}^{2-2\gamma} \Big\} (t_i - t_{i-1}) \Bigg] \Bigg) \\
& \Bigg(\Bigg[\Big\{ \textstyle\sum_{i=1}^n w_{t_i} X_{t_{i-1}}^{-2\gamma} + \sum_{i=2}^{n+1} w_{t_i} X_{t_{i-1}}^{-2\gamma} \Big\} (t_i - t_{i-1}) \Bigg] \\
& \Bigg[\Big\{ \textstyle\sum_{i=1}^n w_{t_i} X_{t_{i-1}}^{2-2\gamma} + \sum_{i=2}^{n+1} w_{t_i} X_{t_{i-1}}^{2-2\gamma} \Big\} (t_i - t_{i-1}) \\
& - \Big\{ \textstyle\sum_{i=1}^n w_{t_i} X_{t_{i-1}}^{1-2\gamma} + \sum_{i=2}^{n+1} w_{t_i} X_{t_{i-1}}^{1-2\gamma} \Big\} (t_i - t_{i-1}) \Bigg] \\
& \Bigg[\Big\{ \textstyle\sum_{i=1}^n w_{t_i} X_{t_{i-1}}^{1-2\gamma} + \sum_{i=2}^{n+1} w_{t_i} X_{t_{i-1}}^{1-2\gamma} \Big\} (t_i - t_{i-1}) \Bigg] \Bigg)^{-1},
\end{aligned}$$

$$
\begin{aligned}
\tilde{\beta}_{n,T} := & \Big(\Big[\Big\{\textstyle\sum_{i=1}^{n} w_{t_i} X_{t_{i-1}}^{-2\gamma} + \sum_{i=2}^{n+1} w_{t_i} X_{t_{i-1}}^{-2\gamma}\Big\}(X_{t_i} - X_{t_{i-1}}) \\
& + \gamma\Big\{\textstyle\sum_{i=1}^{n} w_{t_i} X_{t_{i-1}}^{-1} + \sum_{i=2}^{n+1} w_{t_i} X_{t_{i-1}}^{-1}\Big\}(t_i - t_{i-1})\Big] \\
& \Big[\Big\{\textstyle\sum_{i=1}^{n} w_{t_i} X_{t_{i-1}}^{2-2\gamma} + \sum_{i=2}^{n+1} w_{t_i} X_{t_{i-1}}^{2-2\gamma}\Big\}(t_i - t_{i-1}) \\
& - \Big\{\textstyle\sum_{i=1}^{n} w_{t_i} X_{t_{i-1}}^{1-2\gamma} + \sum_{i=2}^{n+1} w_{t_i} X_{t_{i-1}}^{1-2\gamma}\Big\}(X_{t_i} - X_{t_{i-1}}) - \frac{1-2\gamma}{2} T\Big] \\
& \Big[\Big\{\textstyle\sum_{i=1}^{n} w_{t_i} X_{t_{i-1}}^{1-2\gamma} + \sum_{i=2}^{n+1} w_{t_i} X_{t_{i-1}}^{1-2\gamma}\Big\}(t_i - t_{i-1})\Big]\Big) \\
& \Big(\Big[\Big\{\textstyle\sum_{i=1}^{n} w_{t_i} X_{t_{i-1}}^{-2\gamma} + \sum_{i=2}^{n+1} w_{t_i} X_{t_{i-1}}^{-2\gamma}\Big\}(X_{t_i} - X_{t_{i-1}})\Big] \\
& \Big[\Big\{\textstyle\sum_{i=1}^{n} w_{t_i} X_{t_{i-1}}^{1-2\gamma} + \sum_{i=2}^{n+1} w_{t_i} X_{t_{i-1}}^{1-2\gamma}\Big\}(t_i - t_{i-1}) \\
& - \Big\{\textstyle\sum_{i=1}^{n} w_{t_i} X_{t_{i-1}}^{1-2\gamma} + \sum_{i=2}^{n+1} w_{t_i} X_{t_{i-1}}^{1-2\gamma}\Big\}(X_{t_i} - X_{t_{i-1}})\Big] \\
& \Big[\Big\{\textstyle\sum_{i=1}^{n} w_{t_i} X_{t_{i-1}}^{-2\gamma} + \sum_{i=2}^{n+1} w_{t_i} X_{t_{i-1}}^{-2\gamma}\Big\}(t_i - t_{i-1})\Big]\Big)^{-1}.
\end{aligned}
$$

With $w_{t_i} = 1$, we obtain the forward AMLEs

$$
\begin{aligned}
\tilde{\alpha}_{n,T,F} := & \Big[\textstyle\sum_{i=1}^{n} X_{t_{i-1}}^{1-2\gamma}(X_{t_i} - X_{t_{i-1}}) - \frac{1-\gamma}{2}\sum_{i=1}^{n} X_{t_{i-1}}^{\gamma}(t_i - t_{i-1})\Big] \\
& \Big[\textstyle\sum_{i=1}^{n} X_{t_{i-1}}^{2-2\gamma}(t_i - t_{i-1})\Big]^{-1},
\end{aligned}
$$

$$
\begin{aligned}
\tilde{\beta}_{n,T,F} := & \Big[\textstyle\sum_{i=1}^{n} X_{t_{i-1}}^{1-2\gamma}(X_{t_i} - X_{t_{i-1}}) - \frac{1-\gamma}{2}\sum_{i=1}^{n} X_{t_{i-1}}^{\gamma}(t_i - t_{i-1})\Big] \\
& \Big[\textstyle\sum_{i=1}^{n} X_{t_{i-1}}^{2-2\gamma}(t_i - t_{i-1})\Big]^{-1}.
\end{aligned}
$$

With $w_{t_i} = 0$, we obtain the backward AMLEs

$$
\begin{aligned}
\tilde{\alpha}_{n,T,B} := & \Big[\textstyle\sum_{i=1}^{n} X_{t_i}^{1-2\gamma}(X_{t_i} - X_{t_{i-1}}) - \frac{1-\gamma}{2}\sum_{i=1}^{n} X_{t_i}^{\gamma}(t_i - t_{i-1})\Big] \\
& \Big[\textstyle\sum_{i=1}^{n} X_{t_i}^{2-2\gamma}(t_i - t_{i-1})\Big]^{-1},
\end{aligned}
$$

$$
\begin{aligned}
\tilde{\beta}_{n,T,B} := & \Big[\textstyle\sum_{i=1}^{n} X_{t_i}^{1-2\gamma}(X_{t_i} - X_{t_{i-1}}) - \frac{1-\gamma}{2}\sum_{i=1}^{n} X_{t_i}^{\gamma}(t_i - t_{i-1})\Big] \\
& \Big[\textstyle\sum_{i=1}^{n} X_{t_i}^{2-2\gamma}(t_i - t_{i-1})\Big]^{-1}.
\end{aligned}
$$

Analogous to the estimators for the discrete first order autoregressive model AR(1), we define the simple symmetric and weighted symmetric estimators as follows:

With $w_{t_i} = 1/2$, the simple symmetric AMLE is defined as

$$
\begin{aligned}
\tilde{\alpha}_{n,T,M} := & \Big[\Big\{\textstyle\sum_{i=2}^{n} X_{t_{i-1}}^{1-2\gamma} + \frac{1}{2}(X_{t_0}^{1-2\gamma} + X_{t_n}^{1-2\gamma})\Big\}(X_{t_i} - X_{t_{i-1}}) \\
& - \frac{1-\gamma}{2}\Big\{\textstyle\sum_{i=2}^{n} X_{t_{i-1}}^{\gamma} + 0.5(X_{t_0}^{\gamma} + X_{t_n}^{\gamma})\Big\}(t_i - t_{i-1})\Big] \\
& \Big[\Big\{\textstyle\sum_{i=2}^{n} X_{t_{i-1}}^{2-2\gamma} + 0.5(X_{t_0}^{2-2\gamma} + X_{t_n}^{2-2\gamma})\Big\}(t_i - t_{i-1})\Big]^{-1},
\end{aligned}
$$

$$
\begin{aligned}
\tilde{\beta}_{n,T,M} \quad := \quad & \Big[\Big\{\textstyle\sum_{i=2}^{n} X_{t_{i-1}}^{1-2\gamma} + \frac{1}{2}(X_{t_0}^{1-2\gamma} + X_{t_n}^{1-2\gamma})\Big\}(X_{t_i} - X_{t_{i-1}}) \\
& -\tfrac{1-\gamma}{2}\Big\{\textstyle\sum_{i=2}^{n} X_{t_{i-1}}^{\gamma} + 0.5(X_{t_0}^{\gamma} + X_{t_n}^{\gamma})\Big\}(t_i - t_{i-1})\Big] \\
& \Big[\Big\{\textstyle\sum_{i=2}^{n} X_{t_{i-1}}^{2-2\gamma} + 0.5(X_{t_0}^{2-2\gamma} + X_{t_n}^{2-2\gamma})\Big\}(t_i - t_{i-1})\Big]^{-1}.
\end{aligned}
$$

With the weight function

$$
w_{t_i} = \begin{cases} 0 & : \quad i = 1 \\ \frac{i-1}{n} & : \quad i = 2, 3, \cdots, n \\ 1 & : \quad i = n+1 \end{cases}
$$

the weighted symmetric AMLE is defined as

$$
\begin{aligned}
\tilde{\alpha}_{n,T,D} \quad := \quad & \Big[\Big\{\textstyle\sum_{i=2}^{n} X_{t_{i-1}}^{1-2\gamma} + \sum_{i=1}^{n} X_{t_{i-1}}^{1-2\gamma}\Big\}(X_{t_i} - X_{t_{i-1}}) \\
& -\tfrac{1-\gamma}{2}\Big\{\textstyle\sum_{i=2}^{n} X_{t_{i-1}}^{\gamma} + \sum_{i=1}^{n} X_{t_{i-1}}^{\gamma}\Big\}(t_i - t_{i-1})\Big] \\
& \Big[\Big\{\textstyle\sum_{i=2}^{n} X_{t_{i-1}}^{2-2\gamma} + \sum_{i=1}^{n} X_{t_{i-1}}^{2-2\gamma}\Big\}(t_i - t_{i-1})\Big]^{-1}.
\end{aligned}
$$

$$
\begin{aligned}
\tilde{\beta}_{n,T,D} \quad := \quad & \Big[\Big\{\textstyle\sum_{i=2}^{n} X_{t_{i-1}}^{1-2\gamma} + \sum_{i=1}^{n} X_{t_{i-1}}^{1-2\gamma}\Big\}(X_{t_i} - X_{t_{i-1}}) \\
& -\tfrac{1-\gamma}{2}\Big\{\textstyle\sum_{i=2}^{n} X_{t_{i-1}}^{\gamma} + \sum_{i=1}^{n} X_{t_{i-1}}^{\gamma}\Big\}(t_i - t_{i-1})\Big] \\
& \Big[\Big\{\textstyle\sum_{i=2}^{n} X_{t_{i-1}}^{2-2\gamma} + \sum_{i=1}^{n} X_{t_{i-1}}^{2-2\gamma}\Big\}(t_i - t_{i-1})\Big]^{-1}.
\end{aligned}
$$

Note the above two estimators are analogous to the trapezoidal rule in numerical analysis. One can instead use the midpoint rule to define the estimators

$$
\begin{aligned}
\tilde{\alpha}_{n,T,A} \quad := \quad & \Big[\textstyle\sum_{i=1}^{n}\left(\frac{X_{t_{i-1}}+X_{t_i}}{2}\right)^{1-2\gamma}(X_{t_i} - X_{t_{i-1}}) \\
& -\tfrac{1-\gamma}{2}\textstyle\sum_{i=1}^{n}\left(\frac{X_{t_{i-1}}+X_{t_i}}{2}\right)^{\gamma}(t_i - t_{i-1})\Big] \\
& \Big[\textstyle\sum_{i=1}^{n}\left(\frac{X_{t_{i-1}}+X_{t_i}}{2}\right)^{2-2\gamma}(t_i - t_{i-1})\Big]^{-1},
\end{aligned}
$$

$$
\begin{aligned}
\tilde{\beta}_{n,T,A} \quad := \quad & \Big[\textstyle\sum_{i=1}^{n}\left(\frac{X_{t_{i-1}}+X_{t_i}}{2}\right)^{1-2\gamma}(X_{t_i} - X_{t_{i-1}}) \\
& -\tfrac{1-\gamma}{2}\textstyle\sum_{i=1}^{n}\left(\frac{X_{t_{i-1}}+X_{t_i}}{2}\right)^{\gamma}(t_i - t_{i-1})\Big] \\
& \Big[\textstyle\sum_{i=1}^{n}\left(\frac{X_{t_{i-1}}+X_{t_i}}{2}\right)^{2-2\gamma}(t_i - t_{i-1})\Big]^{-1}.
\end{aligned}
$$

One can use the Simpson's rule to define the estimators that are a convex combination of trapezoidal and midpoint estimators:

$$
\begin{aligned}
\tilde{\alpha}_{n,T,S} \quad := \quad & \Big[\tfrac{1}{3}\textstyle\sum_{i=1}^{n}\Big\{X_{t_{i-1}}^{1-2\gamma} + 4\left(\frac{X_{t_{i-1}}+X_{t_i}}{2}\right)^{1-2\gamma} + X_{t_i}^{1-2\gamma}\Big\}(X_{t_i} - X_{t_{i-1}}) \\
& -\tfrac{1-\gamma}{6}\textstyle\sum_{i=1}^{n}\Big\{X_{t_{i-1}}^{\gamma} + 4\left(\frac{X_{t_{i-1}}+X_{t_i}}{2}\right)^{\gamma} + X_{t_i}^{-1}\Big\}(t_i - t_{i-1})\Big] \\
& \Big[\tfrac{1}{3}\textstyle\sum_{i=1}^{n}\Big\{X_{t_{i-1}}^{2-2\gamma} + 4\left(\frac{X_{t_{i-1}}+X_{t_i}}{2}\right)^{2-2\gamma} + X_{t_i}^{2-2\gamma}\Big\}(t_i - t_{i-1})\Big]^{-1},
\end{aligned}
$$

$$\begin{aligned}\tilde{\beta}_{n,T,S} &:= \Big[\tfrac{1}{3}\textstyle\sum_{i=1}^{n}\Big\{X_{t_{i-1}}^{1-2\gamma}+4\left(\frac{X_{t_{i-1}}+X_{t_i}}{2}\right)^{1-2\gamma}+X_{t_i}^{1-2\gamma}\Big\}(X_{t_i}-X_{t_{i-1}})\\ &\quad -\tfrac{1-\gamma}{6}\textstyle\sum_{i=1}^{n}\Big\{X_{t_{i-1}}^{\gamma}+4\left(\frac{X_{t_{i-1}}+X_{t_i}}{2}\right)^{\gamma}+X_{t_i}^{-1}\Big\}(t_i-t_{i-1})\Big]\\ &\quad \Big[\tfrac{1}{3}\textstyle\sum_{i=1}^{n}\Big\{X_{t_{i-1}}^{2-2\gamma}+4\left(\frac{X_{t_{i-1}}+X_{t_i}}{2}\right)^{2-2\gamma}+X_{t_i}^{2-2\gamma}\Big\}(t_i-t_{i-1})\Big]^{-1}.\end{aligned}$$

The estimator $\tilde{\beta}_{n,T,D}$ is based on the arithmetic mean of the forward and the backward sum. One can use geometric mean and harmonic mean instead for positive diffusions. Note that the symmetric estimators use both the end points of observations. These estimators are sufficient statistics. So one can take advantage of Rao–Blackwell theorem. If one excludes the end points of observations (as in the forward and backward estimators), then the estimators will not be sufficient statistics. The simple symmetric approximate MLE was introduced and studied first in Mishra and Bishwal (1995) for a more general diffusion process. Stochastic bound on the difference between the AMLE and the continuous MLE was obtained.

In general, one can generalize Simpson's rule for any $0 \le \theta \le 1$ as

$$\begin{aligned}\tilde{\alpha}_{n,T,GS} &:= \Big[\textstyle\sum_{i=1}^{n}\Big\{\theta\frac{X_{t_{i-1}}^{1-2\gamma}+X_{t_i}^{1-2\gamma}}{2}+(1-\theta)\left(\frac{X_{t_{i-1}}+X_{t_i}}{2}\right)^{1-2\gamma}\Big\}(X_{t_i}-X_{t_{i-1}})\\ &\quad -\tfrac{1-\gamma}{2}\textstyle\sum_{i=1}^{n}\Big\{\theta\frac{X_{t_{i-1}}^{-1}+X_{t_i}^{\gamma}}{2}+(1-\theta)\left(\frac{X_{t_{i-1}}+X_{t_i}}{2}\right)^{\gamma}\Big\}(t_i-t_{i-1})\Big]\\ &\quad \Big[\textstyle\sum_{i=1}^{n}\Big\{\theta\frac{X_{t_{i-1}}^{2-2\gamma}+X_{t_i}^{2-2\gamma}}{2}+(1-\theta)\left(\frac{X_{t_{i-1}}+X_{t_i}}{2}\right)^{2-2\gamma}\Big\}(t_i-t_{i-1})\Big]^{-1}\end{aligned}$$

$$\begin{aligned}\tilde{\beta}_{n,T,GS} &:= \Big[\textstyle\sum_{i=1}^{n}\Big\{\theta\frac{X_{t_{i-1}}^{1-2\gamma}+X_{t_i}^{1-2\gamma}}{2}+(1-\theta)\left(\frac{X_{t_{i-1}}+X_{t_i}}{2}\right)^{1-2\gamma}\Big\}(X_{t_i}-X_{t_{i-1}})\\ &\quad -\tfrac{1-\gamma}{2}\textstyle\sum_{i=1}^{n}\Big\{\theta\frac{X_{t_{i-1}}^{\gamma}+X_{t_i}^{\gamma}}{2}+(1-\theta)\left(\frac{X_{t_{i-1}}+X_{t_i}}{2}\right)^{\gamma}\Big\}(t_i-t_{i-1})\Big]\\ &\quad \Big[\textstyle\sum_{i=1}^{n}\Big\{\theta\frac{X_{t_{i-1}}^{2-2\gamma}+X_{t_i}^{2-2\gamma}}{2}+(1-\theta)\left(\frac{X_{t_{i-1}}+X_{t_i}}{2}\right)^{2-2\gamma}\Big\}(t_i-t_{i-1})\Big]^{-1}.\end{aligned}$$

The case $\theta = 0$ produces the A-estimators. The case $\theta = 1$ produces the M-estimators. The case $\theta = \frac{1}{3}$ produces the S-estimators.

We propose a very general form of the quadrature based estimator as

$$\begin{aligned}\tilde{\beta}_{n,T,Q} &:= \Big[\textstyle\sum_{i=1}^{n}\sum_{j=1}^{m}\left[(1-s_j)X_{t_{i-1}}+s_jX_{t_i}\right]^{1-2\gamma}p_j\,(X_{t_i}-X_{t_{i-1}})\\ &\quad -\tfrac{1-\gamma}{2}\Big\{\textstyle\sum_{i=1}^{n}\sum_{j=1}^{m}\left[(1-s_j)X_{t_{i-1}}+s_jX_{t_i}\right]^{\gamma}p_j\Big\}(t_i-t_{i-1})\Big]\\ &\quad \Big[\Big\{\textstyle\sum_{i=1}^{n}\sum_{j=1}^{m}\left[(1-s_j)X_{t_{i-1}}+s_jX_{t_i}\right]^{1-2\gamma}p_j\Big\}(t_i-t_{i-1})\Big]^{-1}\end{aligned}$$

where p_j, $j \in \{1, 2, \cdots, m\}$, is a probability mass function of a discrete random variable S on $0 \le s_1 < s_2 < \cdots < s_m \le 1$ with $P(S = s_j) := p_j$, $j \in \{1, 2, \cdots, m\}$. Denote the k-th moment of the random variable S as $\mu_k := \sum_{j=1}^{m} s_j^k p_j$, $k = 1, 2, \cdots$.

If one chooses the probability distribution as uniform distribution for which the moments are a harmonic sequence $(\mu_1, \mu_2, \mu_3, \mu_4, \mu_5, \mu_6, \cdots) =$

$(\frac{1}{2}, \frac{1}{3}, \frac{1}{4}, \frac{1}{5}, \frac{1}{6}, \frac{1}{7}, \cdots)$, then there is no change in rate of convergence than second order. If one can construct a probability distribution for which the harmonic sequence is truncated at a point, then there is an improvement in the rate of convergence at the point of truncation. We conjecture the following construction:

Given a positive integer m, construct a probability mass function p_j, $j \in \{1, 2, \cdots, m\}$ on $0 \le s_1 < s_2 < \cdots < s_m \le 1$ such that

$$\sum_{j=1}^{m} s_j^r p_j = \frac{1}{r+1}, \quad r \in \{0, \cdots, m-2\}, \quad \sum_{j=1}^{m} s_j^{N-1} p_j \neq \frac{1}{m}.$$

Neither the probabilities p_j nor the atoms, s_j, of the distribution are specified in advance.

This problem is related to the truncated Hausdorff moment problem. We obtain examples of such probability distributions and use them to get higher order accurate (up to sixth order) AMLEs.

The order of approximation error (rate of convergence) of an estimator is $n^{-\nu}$, where

$$\nu := \inf\left\{k : \mu_k \neq \frac{1}{1+k}, \ \mu_j = \frac{1}{1+j}, j = 1, 2, \cdots, k-1\right\}.$$

In order to obtain higher order approximations, one needs to concentrate on higher order moments of the weighting probability distribution.

Newton–Cotes Probability Distribution

Let δ_a be a Dirac measure at the point a. For $N \ge 2$, the Newton–Cotes distribution is given by

$$\nu_N = \sum_{j=0}^{2N-2} \gamma_j \delta_{j/(2N-2)} \quad \text{where } \gamma_j = \int_0^1 \prod_{k \neq j} \frac{2(N-1)u - k}{j-k} du.$$

For $N = 0$, $\nu_0 = \delta_0$. For $N = 1$, $\nu_1 = (\delta_0 + \delta_1)/2$, the trapezoidal distribution. For $N = 2$, the Simpson's distribution is given by

$$\nu_2 = \sum_{j=0}^{2} \gamma_j \delta_{j/2}, \quad \text{where } \gamma_j = \int_0^1 \prod_{k \neq j} \frac{2u - k}{j-k} du.$$

Simply

$$\nu_2 = \gamma_0 \delta_0 + \gamma_1 \delta_{1/2} + \gamma_2 \delta_1 \quad \text{where } \gamma_0 = \frac{1}{6}, \ \gamma_1 = \frac{2}{6}, \ \gamma_2 = \frac{1}{6}.$$

Boole's Probability Distribution

The Boole distribution is defined as

$$\nu_3 = \sum_{j=0}^{4} \gamma_j \delta_{j/4} \quad \text{where} \quad \gamma_j = \int_0^1 \underset{k\neq j}{\Pi} \frac{4u-k}{j-k} du.$$

Simply

$$\nu_3 = \gamma_0\delta_0 + \gamma_1\delta_{1/4} + \gamma_2\delta_{1/2} + \gamma_3\delta_{3/4} + \gamma_4\delta_1,$$

where

$$\gamma_0 = \frac{7}{90}, \gamma_1 = \frac{32}{90}, \gamma_2 = \frac{12}{90}, \gamma_3 = \frac{32}{90}, \gamma_4 = \frac{7}{90}.$$

The Boole probability distribution is a symmetric probability distribution with support: $(0, \frac{1}{4}, \frac{1}{2}, \frac{3}{4}, 1)$, probability mass: $(\frac{7}{90}, \frac{32}{90}, \frac{12}{90}, \frac{32}{90}, \frac{7}{90})$, and moments: $\mu_1 = \frac{1}{2}, \mu_2 = \frac{1}{3}, \mu_3 = \frac{1}{4}, \mu_4 = \frac{1}{5}, \mu_5 = \frac{1}{6}, \mu_6 = \frac{110}{768}$. Here $\mu_6 \neq \frac{1}{7}$. Thus rate is $\nu = 6$.

Define the Boole's approximation of the stochastic integral $B_T := \int_0^T f(X_t)dW_t$ as

$$\begin{aligned} B_{n,T} \quad := \quad & \tfrac{1}{90}\textstyle\sum_{i=1}^{n} \Big[7f(t_{i-1}, X_{t_{i-1}}) + 32f\left(\tfrac{3t_{i-1}+t_i}{4}, \tfrac{3X_{t_{i-1}}+X_{t_i}}{4}\right) \\ & +12f\left(\tfrac{t_{i-1}+t_i}{2}, \tfrac{X_{t_{i-1}}+X_{t_i}}{2}\right) + 32f\left(\tfrac{t_{i-1}+3t_i}{4}, \tfrac{X_{t_{i-1}}+3X_{t_i}}{4}\right) \\ & +7f(t_i, X_{t_i})\Big](W_{t_i} - W_{t_{i-1}}). \end{aligned}$$

For fixed T, $B_{n,T} \xrightarrow{L_2} B_T$ as $n \to \infty$. The rate of convergence of this discrete approximation is $O(n^{-6})$.

We construct probability distributions satisfying these moment conditions and obtain estimators of the rate of convergence up to order 6. Probability $p_1 = 1$ at the point $s_1 = 0$ gives the F-estimators for which $\mu_1 = 0$. Note that $\mu_1 \neq \frac{1}{2}$. Thus $\nu = 1$. Probability $p_1 = 1$ at the point $s_1 = 1$ gives the B-estimators for which $\mu_1 = 1$. Note that $\mu_1 \neq \frac{1}{2}$. Thus $\nu = 1$. Probabilities $(p_1, p_2) = (\frac{1}{2}, \frac{1}{2})$ at the respective points $(s_1, s_2) = (0, 1)$ produce the estimator $\tilde{\beta}_{n,T,Z}$ for which $(\mu_1, \mu_2) = (\frac{1}{2}, \frac{1}{4})$. Thus $\nu = 2$. Probability $p_j = 1$ at the point $s_j = \frac{1}{2}$ produces the estimator $\tilde{\beta}_{n,T,A}$ for which $(\mu_1, \mu_2) = (\frac{1}{2}, \frac{1}{2})$. Thus $\nu = 2$. Probabilities $(p_1, p_2) = (\frac{1}{4}, \frac{3}{4})$ at the respective points $(s_1, s_2) = (0, \frac{2}{3})$ produce the asymmetric estimators $\tilde{\alpha}_{n,T,3}$ and $\tilde{\beta}_{n,T,3}$ for which $(\mu_1, \mu_2, \mu_3) = (\frac{1}{2}, \frac{1}{3}, \frac{2}{9})$. Thus $\nu = 3$. Probabilities $(p_1, p_2) = (\frac{3}{4}, \frac{1}{4})$ at the respective points $(s_1, s_2) = (\frac{1}{3}, 1)$ produce asymmetric estimators $\tilde{\alpha}_{n,T,4}$ and $\tilde{\alpha}_{n,T,4}$ for which $(\mu_1, \mu_2, \mu_3) = (\frac{1}{2}, \frac{1}{3}, \frac{10}{36})$. Thus $\nu = 3$. Probabilities $(p_1, p_2, p_3) = (\frac{1}{6}, \frac{2}{3}, \frac{1}{6})$ at the respective points $(s_1, s_2, s_3) = (0, \frac{1}{2}, 1)$ produce the estimators $\tilde{\alpha}_{n,T,5}$ and $\tilde{\beta}_{n,T,5}$ for which $(\mu_1, \mu_2, \mu_3, \mu_4) = (\frac{1}{2}, \frac{1}{3}, \frac{1}{4}, \frac{5}{25})$. Thus $\nu = 4$. Probabilities $(p_1, p_2, p_3, p_4) = (\frac{1}{8}, \frac{3}{8}, \frac{3}{8}, \frac{1}{8})$ at the respective points $(s_1, s_2, s_3, s_4) = (0, \frac{1}{3}, \frac{2}{3}, 1)$ produce the symmetric estimators $\tilde{\alpha}_{n,T,6}$ and $\tilde{\alpha}_{n,T,6}$ for which $(\mu_1, \mu_2, \mu_3, \mu_4) = (\frac{1}{2}, \frac{1}{3}, \frac{1}{4}, \frac{11}{54})$. Thus $\nu = 4$. Probabilities $(p_1, p_2, p_3, p_4, p_5) = (\frac{1471}{24192}, \frac{6925}{24192}, \frac{1475}{12096}, \frac{2725}{12096}, \frac{5675}{24192}, \frac{1721}{24192})$ at the respective

points $(s_1, s_2, s_3, s_4, s_5) = (0, \frac{1}{5}, \frac{2}{5}, \frac{3}{5}, \frac{4}{5}, 1)$ produce the asymmetric estimators $\tilde{\alpha}_{n,T,7}$ and $\tilde{\alpha}_{n,T,7}$ for which $(\mu_1, \mu_2, \mu_3, \mu_4, \mu_5, \mu_6) = (\frac{1}{2}, \frac{1}{3}, \frac{1}{4}, \frac{1}{5}, \frac{841}{5040})$. Thus $\nu = 5$. Probabilities $(p_1, p_2, p_3, p_4, p_5) = (\frac{7}{90}, \frac{16}{45}, \frac{2}{15}, \frac{16}{45}, \frac{7}{90})$ at the respective points $(s_1, s_2, s_3, s_4, s_5) = (0, \frac{1}{4}, \frac{1}{2}, \frac{3}{4}, 1)$ produce the symmetric estimators $\tilde{\alpha}_{n,T,8}$ and $\tilde{\beta}_{n,T,8}$ for which $(\mu_1, \mu_2, \mu_3, \mu_4, \mu_5, \mu_6) = (\frac{1}{2}, \frac{1}{3}, \frac{1}{4}, \frac{1}{5}, \frac{1}{6}, \frac{110}{768})$. Thus $\nu = 6$. Probabilities $(p_1, p_2, p_3, p_4, p_5) = (\frac{19}{288}, \frac{75}{288}, \frac{50}{288}, \frac{50}{288}, \frac{75}{288}, \frac{19}{288})$ at the respective points $(s_1, s_2, s_3, s_4, s_5)$
$= (0, \frac{1}{5}, \frac{2}{5}, \frac{3}{5}, \frac{4}{5}, 1)$ produce symmetric estimators $\tilde{\alpha}_{n,T,9}$ and $\tilde{\beta}_{n,T,9}$ for which $(\mu_1, \mu_2, \mu_3,$
$\mu_4, \mu_5, \mu_6) = (\frac{1}{2}, \frac{1}{3}, \frac{1}{4}, \frac{1}{5}, \frac{1}{6}, \frac{3219}{22500})$. Thus $\nu = 6$.

$$
\begin{aligned}
\tilde{\alpha}_{n,T,3} := \Bigg(& \Bigg[\Bigg\{\frac{1}{4}\sum_{i=1}^{n}\Big[(X_{t_{i-1}})^{1-2\gamma} + 3(\frac{X_{t_{i-1}} + 2X_{t_i}}{3})^{1-2\gamma}\Big]\Bigg\}(X_{t_i} - X_{t_{i-1}}) \\
& +\gamma\Bigg\{\frac{1}{4}\sum_{i=1}^{n}\Big[(X_{t_{i-1}})^{-1} + 3(\frac{X_{t_{i-1}} + 2X_{t_i}}{3})^{-1}\Big]\Bigg\}(t_i - t_{i-1})\Bigg] \\
& \Bigg[\Bigg\{\frac{1}{4}\sum_{i=1}^{n}\Big[(X_{t_{i-1}})^{1-2\gamma} + 3(\frac{X_{t_{i-1}} + 2X_{t_i}}{3})^{1-2\gamma}\Big]\Bigg\}(X_{t_i} - X_{t_{i-1}})\Bigg] \\
& - \Bigg\{\frac{1}{4}\sum_{i=1}^{n}\Big[(X_{t_{i-1}})^{1-2\gamma} + 3(\frac{X_{t_{i-1}} + 2X_{t_i}}{3})^{1-2\gamma}\Big]\Bigg\}(X_{t_i} - X_{t_{i-1}}) - \frac{1-2\gamma}{2}T\Bigg] \\
& \Bigg[\Bigg\{\frac{1}{4}\sum_{i=1}^{n}\Big[(X_{t_{i-1}})^{-2\gamma} + 3(\frac{X_{t_{i-1}} + 2X_{t_i}}{3})^{-2\gamma}\Big]\Bigg\}(t_i - t_{i-1})\Bigg]\Bigg) \\
& \Bigg(\Bigg[\Bigg\{\frac{1}{4}\sum_{i=1}^{n}\Big[(X_{t_{i-1}})^{-2\gamma} + 3(\frac{X_{t_{i-1}} + 2X_{t_i}}{3})^{-2\gamma}\Big]\Bigg\}(t_i - t_{i-1})\Bigg] \\
& \Bigg[\Bigg\{\frac{1}{4}\sum_{i=1}^{n}\Big[(X_{t_{i-1}})^{2-2\gamma} + 3(\frac{X_{t_{i-1}} + 2X_{t_i}}{3})^{2-2\gamma}\Big]\Bigg\}(t_i - t_{i-1})\Bigg] \\
& - \Bigg[\Bigg\{\frac{1}{4}\sum_{i=1}^{n}\Big[(X_{t_{i-1}})^{1-2\gamma} + 3(\frac{X_{t_{i-1}} + 2X_{t_i}}{3})^{1-2\gamma}\Big]\Bigg\}(t_i - t_{i-1})\Bigg]^2\Bigg)^{-1},
\end{aligned}
$$

$$
\begin{aligned}
\tilde{\beta}_{n,T,3} := \Bigg(& \Bigg[\Bigg\{\frac{1}{4}\sum_{i=1}^{n}\Big[(X_{t_{i-1}})^{-2\gamma} + 3(\frac{X_{t_{i-1}} + 2X_{t_i}}{3})^{-2\gamma}\Big]\Bigg\}(X_{t_i} - X_{t_{i-1}}) \\
& +\gamma\Bigg\{\frac{1}{4}\sum_{i=1}^{n}\Big[(X_{t_{i-1}})^{-1} + 3(\frac{X_{t_{i-1}} + 2X_{t_i}}{3})^{-1}\Big]\Bigg\}(t_i - t_{i-1})\Bigg] \\
& \Bigg[\Bigg\{\frac{1}{4}\sum_{i=1}^{n}\Big[(X_{t_{i-1}})^{2-2\gamma} + 3(\frac{X_{t_{i-1}} + 2X_{t_i}}{3})^{2-2\gamma}\Big]\Bigg\}(t_i - t_{i-1})\Bigg] \\
& - \Bigg\{\frac{1}{4}\sum_{i=1}^{n}\Big[(X_{t_{i-1}})^{1-2\gamma} + 3(\frac{X_{t_{i-1}} + 2X_{t_i}}{3})^{1-2\gamma}\Big]\Bigg\}(X_{t_i} - X_{t_{i-1}}) - \frac{1-2\gamma}{2}T\Bigg] \\
& \Bigg[\Bigg\{\frac{1}{4}\sum_{i=1}^{n}\Big[(X_{t_{i-1}})^{-2\gamma} + 3(\frac{X_{t_{i-1}} + 2X_{t_i}}{3})^{-2\gamma}\Big]\Bigg\}(t_i - t_{i-1})\Bigg]\Bigg) \\
& \Bigg(\Bigg[\Bigg\{\frac{1}{4}\sum_{i=1}^{n}\Big[(X_{t_{i-1}})^{-2\gamma} + 3(\frac{X_{t_{i-1}} + 2X_{t_i}}{3})^{-2\gamma}\Big]\Bigg\}(X_{t_i} - X_{t_{i-1}})\Bigg]
\end{aligned}
$$

$$+\gamma\left\{\frac{1}{4}\sum_{i=1}^{n}\left[(X_{t_{i-1}})^{-1}+3(\frac{X_{t_{i-1}}+2X_{t_i}}{3})^{-1}\right]\right\}(t_i-t_{i-1})\Bigg]$$

$$\left[\left\{\frac{1}{4}\sum_{i=1}^{n}\left[(X_{t_{i-1}})^{1-2\gamma}+3(\frac{X_{t_{i-1}}+2X_{t_i}}{3})^{1-2\gamma}\right]\right\}(t_i-t_{i-1})\right]$$

$$-\left(\left\{\frac{1}{4}\sum_{i=1}^{n}\left[(X_{t_{i-1}})^{1-2\gamma}+3(\frac{X_{t_{i-1}}+2X_{t_i}}{3})^{1-2\gamma}\right]\right\}(X_{t_i}-X_{t_{i-1}})-\frac{1-2\gamma}{2}T\right]$$

$$\left[\left\{\frac{1}{4}\sum_{i=1}^{n}\left[(X_{t_{i-1}})^{-2\gamma}+3(\frac{X_{t_{i-1}}+2X_{t_i}}{3})^{-2\gamma}\right]\right\}(t_i-t_{i-1})\right]\Bigg)^{-1}$$

$$\tilde{\alpha}_{n,T,4}:=\Bigg(\Bigg[\left\{\frac{1}{4}\sum_{i=1}^{n}\left[3(\frac{2X_{t_{i-1}}+X_{t_i}}{3})^{-2\gamma}+(X_{t_i})^{-2\gamma}\right]\right\}(X_{t_i}-X_{t_{i-1}})$$

$$+\gamma\left\{\frac{1}{4}\sum_{i=1}^{n}\left[3(\frac{2X_{t_{i-1}}+X_{t_i}}{3})^{-1}+(X_{t_i})^{-1}\right]\right\}(t_i-t_{i-1})\Bigg]$$

$$\left[\left\{\frac{1}{4}\sum_{i=1}^{n}\left[3(\frac{2X_{t_{i-1}}+X_{t_i}}{3})^{1-2\gamma}+(X_{t_i})^{1-2\gamma}\right]\right\}(t_i-t_{i-1})\right]$$

$$-\left[\left\{\frac{1}{4}\sum_{i=1}^{n}\left[3(\frac{2X_{t_{i-1}}+X_{t_i}}{3})^{1-2\gamma}+(X_{t_i})^{1-2\gamma}\right]\right\}(X_{t_i}-X_{t_{i-1}})-\frac{1-2\gamma}{2}T\right]$$

$$\left[\left\{\frac{1}{4}\sum_{i=1}^{n}\left[3(\frac{2X_{t_{i-1}}+X_{t_i}}{3})^{-2\gamma}+(X_{t_i})^{-2\gamma}\right]\right\}(t_i-t_{i-1})\right]\Bigg)$$

$$\Bigg(\Bigg[\left\{\frac{1}{4}\sum_{i=1}^{n}\left[3(\frac{2X_{t_{i-1}}+X_{t_i}}{3})^{-2\gamma}+(X_{t_i})^{-2\gamma}\right]\right\}(t_i-t_{i-1})\Bigg]$$

$$\left[\left\{\frac{1}{4}\sum_{i=1}^{n}\left[3(\frac{2X_{t_{i-1}}+X_{t_i}}{3})^{2-2\gamma}+(X_{t_i})^{2-2\gamma}\right]\right\}(t_i-t_{i-1})\right]$$

$$-\left[\left\{\frac{1}{4}\sum_{i=1}^{n}\left[3(\frac{2X_{t_{i-1}}+X_{t_i}}{3})^{1-2\gamma}+(X_{t_i})^{1-2\gamma}\right]\right\}(t_i-t_{i-1})\right]^2\Bigg)^{-1},$$

$$\tilde{\beta}_{n,T,4}:=\Bigg(\Bigg[\left\{\frac{1}{4}\sum_{i=1}^{n}\left[3(\frac{2X_{t_{i-1}}+X_{t_i}}{3})^{-2\gamma}+(X_{t_i})^{-2\gamma}\right]\right\}(X_{t_i}-X_{t_{i-1}})$$

$$+\gamma\left\{\frac{1}{4}\sum_{i=1}^{n}\left[3(\frac{2X_{t_{i-1}}+X_{t_i}}{3})^{-1}+(X_{t_i})^{-1}\right]\right\}(t_i-t_{i-1})\Bigg]$$

$$\left[\left\{\frac{1}{4}\sum_{i=1}^{n}\left[3(\frac{2X_{t_{i-1}}+X_{t_i}}{3})^{2-2\gamma}+(X_{t_i})^{2-2\gamma}\right]\right\}(t_i-t_{i-1})\right]$$

$$-\left[\left\{\frac{1}{4}\sum_{i=1}^{n}\left[3(\frac{2X_{t_{i-1}}+X_{t_i}}{3})^{1-2\gamma}+(X_{t_i})^{1-2\gamma}\right]\right\}(X_{t_i}-X_{t_{i-1}})-\frac{1-2\gamma}{2}T\right]$$

$$\left[\left\{\frac{1}{4}\sum_{i=1}^{n}\left[3(\frac{2X_{t_{i-1}}+X_{t_i}}{3})^{1-2\gamma}+(X_{t_i})^{1-2\gamma}\right]\right\}(t_i-t_{i-1})\right]\Bigg)$$

$$\Bigg(\Bigg[\left\{\frac{1}{4}\sum_{i=1}^{n}\left[3(\frac{2X_{t_{i-1}}+X_{t_i}}{3})^{-2\gamma}+(X_{t_i})^{-2\gamma}\right]\right\}(t_i-t_{i-1})\Bigg]$$

$$+\gamma\left[\left\{\frac{1}{4}\sum_{i=1}^{n}\left[3(\frac{2X_{t_{i-1}}+X_{t_i}}{3})^{-1}+(X_{t_i})^{-1}\right]\right\}(t_i-t_{i-1})\right]$$
$$\left[\left\{\frac{1}{4}\sum_{i=1}^{n}\left[3(\frac{2X_{t_{i-1}}+X_{t_i}}{3})^{1-2\gamma}+(X_{t_i})^{1-2\gamma}\right]\right\}(t_i-t_{i-1})\right]$$
$$-\left[\left\{\frac{1}{4}\sum_{i=1}^{n}\left[3(\frac{2X_{t_{i-1}}+X_{t_i}}{3})^{1-2\gamma}+(X_{t_i})^{1-2\gamma}\right]\right\}(t_i-t_{i-1})-\frac{1-2\gamma}{2}T\right]$$
$$\left.\left[\left\{\frac{1}{4}\sum_{i=1}^{n}\left[3(\frac{2X_{t_{i-1}}+X_{t_i}}{3})^{-2\gamma}+(X_{t_i})^{-2\gamma}\right]\right\}(t_i-t_{i-1})\right]^2\right)^{-1}$$

$$\tilde{\alpha}_{n,T,6}:=$$
$$\left(\left[\left\{\frac{1}{8}\sum_{i=1}^{n}\left[(X_{t_{i-1}})^{-2\gamma}+3(\frac{2X_{t_{i-1}}+X_{t_i}}{3})^{-2\gamma}+3(\frac{X_{t_{i-1}}+2X_{t_i}}{3})^{-2\gamma}+(X_{t_i})^{-2\gamma}\right]\right\}(X_{t_i}-X_{t_{i-1}})\right.\right.$$
$$\left.+\gamma\left\{\frac{1}{8}\sum_{i=1}^{n}\left[(X_{t_{i-1}})^{-1}+3(\frac{2X_{t_{i-1}}+X_{t_i}}{3})^{-1}+3(\frac{X_{t_{i-1}}+2X_{t_i}}{3})^{-1}+(X_{t_i})^{-1}\right]\right\}(t_i-t_{i-1})\right]$$
$$\left[\left\{\frac{1}{8}\sum_{i=1}^{n}\left[(X_{t_{i-1}})^{2-2\gamma}+3(\frac{2X_{t_{i-1}}+X_{t_i}}{3})^{2-2\gamma}+3(\frac{X_{t_{i-1}}+2X_{t_i}}{3})^{2-2\gamma}+(X_{t_i})^{2-2\gamma}\right]\right\}(t_i-t_{i-1})\right]$$
$$-\left[\frac{1}{8}\sum_{i=1}^{n}\left[(X_{t_{i-1}})^{1-2\gamma}+3(\frac{2X_{t_{i-1}}+X_{t_i}}{3})^{1-2\gamma}+3(\frac{X_{t_{i-1}}+2X_{t_i}}{3})^{1-2\gamma}+(X_{t_i})^{1-2\gamma}\right](X_{t_i}-X_{t_{i-1}})\right.$$
$$\left.-\frac{1-2\gamma}{2}T\right]$$
$$\left.\left[\left\{\frac{1}{8}\sum_{i=1}^{n}\left[(X_{t_{i-1}})^{1-2\gamma}+3(\frac{2X_{t_{i-1}}+X_{t_i}}{3})^{1-2\gamma}+3(\frac{X_{t_{i-1}}+2X_{t_i}}{3})^{1-2\gamma}+(X_{t_i})^{1-2\gamma}\right]\right\}(t_i-t_{i-1})\right]\right)$$
$$\left(\left[\left\{\frac{1}{8}\sum_{i=1}^{n}\left[(X_{t_{i-1}})^{-2\gamma}+3(\frac{2X_{t_{i-1}}+X_{t_i}}{3})^{-2\gamma}+3(\frac{X_{t_{i-1}}+2X_{t_i}}{3})^{-2\gamma}+(X_{t_i})^{-2\gamma}\right]\right\}(X_{t_i}-X_{t_{i-1}})\right.\right.$$
$$\left.+\gamma\left\{\frac{1}{8}\sum_{i=1}^{n}\left[(X_{t_{i-1}})^{-1}+3(\frac{2X_{t_{i-1}}+X_{t_i}}{3})^{-1}+3(\frac{X_{t_{i-1}}+2X_{t_i}}{3})^{-1}+(X_{t_i})^{-1}\right]\right\}(t_i-t_{i-1})\right]$$
$$\left[\left\{\frac{1}{8}\sum_{i=1}^{n}\left[(X_{t_{i-1}})^{1-2\gamma}+3(\frac{2X_{t_{i-1}}+X_{t_i}}{3})^{1-2\gamma}+3(\frac{X_{t_{i-1}}+2X_{t_i}}{3})^{1-2\gamma}+(X_{t_i})^{1-2\gamma}\right]\right\}(t_i-t_{i-1})\right]$$
$$-\left[\frac{1}{8}\sum_{i=1}^{n}\left[(X_{t_{i-1}})^{1-2\gamma}+3(\frac{2X_{t_{i-1}}+X_{t_i}}{3})^{1-2\gamma}+3(\frac{X_{t_{i-1}}+2X_{t_i}}{3})^{1-2\gamma}+(X_{t_i})^{1-2\gamma}\right]\right.$$
$$\left.\left.(t_i-t_{i-1})\right]^2\right)^{-1},$$

$$\tilde{\beta}_{n,T,6}:=$$
$$\left(\left[\left\{\frac{1}{8}\sum_{i=1}^{n}\left[(X_{t_{i-1}})^{-2\gamma}+3(\frac{2X_{t_{i-1}}+X_{t_i}}{3})^{-2\gamma}+3(\frac{X_{t_{i-1}}+2X_{t_i}}{3})^{-2\gamma}+(X_{t_i})^{-2\gamma}\right]\right\}(X_{t_i}-X_{t_{i-1}})\right.\right.$$
$$\left.+\gamma\left\{\frac{1}{8}\sum_{i=1}^{n}\left[(X_{t_{i-1}})^{-1}+3(\frac{2X_{t_{i-1}}+X_{t_i}}{3})^{-1}+3(\frac{X_{t_{i-1}}+2X_{t_i}}{3})^{-1}+(X_{t_i})^{-1}\right]\right\}(t_i-t_{i-1})\right]$$
$$\left[\left\{\frac{1}{8}\sum_{i=1}^{n}\left[(X_{t_{i-1}})^{1-2\gamma}+3(\frac{2X_{t_{i-1}}+X_{t_i}}{3})^{1-2\gamma}+3(\frac{X_{t_{i-1}}+2X_{t_i}}{3})^{1-2\gamma}+(X_{t_i})^{1-2\gamma}\right]\right\}(t_i-t_{i-1})\right]$$
$$-\left[\frac{1}{8}\sum_{i=1}^{n}\left[(X_{t_{i-1}})^{1-2\gamma}+3(\frac{2X_{t_{i-1}}+X_{t_i}}{3})^{1-2\gamma}+3(\frac{X_{t_{i-1}}+2X_{t_i}}{3})^{1-2\gamma}+(X_{t_i})^{1-2\gamma}\right](X_{t_i}-X_{t_{i-1}})\right.$$
$$\left.-\frac{1-2\gamma}{2}T\right]$$
$$\left.\left[\left\{\frac{1}{8}\sum_{i=1}^{n}\left[(X_{t_{i-1}})^{-2\gamma}+3(\frac{2X_{t_{i-1}}+X_{t_i}}{3})^{-2\gamma}+3(\frac{X_{t_{i-1}}+2X_{t_i}}{3})^{-2\gamma}+(X_{t_i})^{-2\gamma}\right]\right\}(t_i-t_{i-1})\right]\right)$$
$$\left(\left[\left\{\frac{1}{8}\sum_{i=1}^{n}\left[(X_{t_{i-1}})^{-2\gamma}+3(\frac{2X_{t_{i-1}}+X_{t_i}}{3})^{-2\gamma}+3(\frac{X_{t_{i-1}}+2X_{t_i}}{3})^{-2\gamma}+(X_{t_i})^{-2\gamma}\right]\right\}(X_{t_i}-X_{t_{i-1}})\right.\right.$$

$$+\gamma\left\{\frac{1}{8}\sum_{i=1}^{n}\left[(X_{t_{i-1}})^{-1}+3(\frac{2X_{t_{i-1}}+X_{t_i}}{3})^{-1}+3(\frac{X_{t_{i-1}}+2X_{t_i}}{3})^{-1}+(X_{t_i})^{-1}\right]\right\}(t_i-t_{i-1})\Bigg]$$

$$\left[\left\{\frac{1}{8}\sum_{i=1}^{n}\left[(X_{t_{i-1}})^{1-2\gamma}+3(\frac{2X_{t_{i-1}}+X_{t_i}}{3})^{1-2\gamma}+3(\frac{X_{t_{i-1}}+2X_{t_i}}{3})^{1-2\gamma}+(X_{t_i})^{1-2\gamma}\right]\right\}(t_i-t_{i-1})\right]$$

$$-\left[\frac{1}{8}\sum_{i=1}^{n}\left[(X_{t_{i-1}})^{1-2\gamma}+3(\frac{2X_{t_{i-1}}+X_{t_i}}{3})^{1-2\gamma}+3(\frac{X_{t_{i-1}}+2X_{t_i}}{3})^{1-2\gamma}+(X_{t_i})^{1-2\gamma}\right](X_{t_i}-X_{t_{i-1}})\right.$$

$$\left.-\frac{1-2\gamma}{2}T\right]$$

$$\left[\left\{\frac{1}{8}\sum_{i=1}^{n}\left[(X_{t_{i-1}})^{-2\gamma}+3(\frac{2X_{t_{i-1}}+X_{t_i}}{3})^{-2\gamma}+3(\frac{X_{t_{i-1}}+2X_{t_i}}{3})^{-2\gamma}+(X_{t_i})^{-2\gamma}\right]\right\}(t_i-t_{i-1})\right]\Bigg)^{-1}$$

$\tilde{\alpha}_{n,T,7} :=$

$$\Bigg(\Bigg[\left\{\frac{1}{24192}\sum_{i=1}^{n}\left[1471(X_{t_{i-1}})^{-2\gamma}+6925(\frac{X_{t_{i-1}}+X_{t_i}}{5})^{-2\gamma}+2950(\frac{2X_{t_{i-1}}+2X_{t_i}}{5})^{-2\gamma}\right.\right.$$

$$\left.\left.+5450(\frac{3X_{t_{i-1}}+3X_{t_i}}{5})-2\gamma+5675(\frac{4X_{t_{i-1}}+4X_{t_i}}{5})^{-2\gamma}+1721(X_{t_i})^{-2\gamma}\right]\right\}(X_{t_i}-X_{t_{i-1}})$$

$$+\gamma\left\{\frac{1}{24192}\sum_{i=1}^{n}\left[1471(X_{t_{i-1}})^{-1}+6925(\frac{X_{t_{i-1}}+X_{t_i}}{5})^{-1}+2950(\frac{2X_{t_{i-1}}+2X_{t_i}}{5})^{-1}\right.\right.$$

$$\left.\left.+5450(\frac{3X_{t_{i-1}}+3X_{t_i}}{5})^{-1}+5675(\frac{4X_{t_{i-1}}+4X_{t_i}}{5})^{-1}+1721(X_{t_i})^{-1}\right]\right\}(t_i-t_{i-1})\Bigg]$$

$$-\Bigg[\left\{\frac{1}{24192}\sum_{i=1}^{n}\left[1471(X_{t_{i-1}})^{1-2\gamma}+6925(\frac{X_{t_{i-1}}+X_{t_i}}{5})^{1-2\gamma}+2950(\frac{2X_{t_{i-1}}+2X_{t_i}}{5})^{1-2\gamma}\right.\right.$$

$$\left.\left.+5450(\frac{3X_{t_{i-1}}+3X_{t_i}}{5})^{1-2\gamma}+5675(\frac{4X_{t_{i-1}}+4X_{t_i}}{5})^{1-2\gamma}+1721(X_{t_i})^{2}\right]\right\}(X_{t_i}-X_{t_{i-1}})$$

$$-\frac{1-2\gamma}{2}T\Bigg]$$

$$\Bigg[\left\{\frac{1}{24192}\sum_{i=1}^{n}\left[1471(X_{t_{i-1}})^{-2\gamma}+6925(\frac{X_{t_{i-1}}+X_{t_i}}{5})^{-2\gamma}+2950(\frac{2X_{t_{i-1}}+2X_{t_i}}{5})^{-2\gamma}\right.\right.$$

$$\left.\left.+5450(\frac{3X_{t_{i-1}}+3X_{t_i}}{5})^{-2\gamma}+5675(\frac{4X_{t_{i-1}}+4X_{t_i}}{5})^{-2\gamma}+1721(X_{t_i})^{2}\right]\right\}(t_i-t_{i-1})\Bigg]\Bigg)$$

$$\Bigg(\Bigg[\left\{\frac{1}{24192}\sum_{i=1}^{n}\left[1471(X_{t_{i-1}})^{-2\gamma}+6925(\frac{X_{t_{i-1}}+X_{t_i}}{5})^{-2\gamma}+2950(\frac{2X_{t_{i-1}}+2X_{t_i}}{5})^{-2\gamma}\right.\right.$$

$$\left.\left.+5450(\frac{3X_{t_{i-1}}+3X_{t_i}}{5})^{2}+5675(\frac{4X_{t_{i-1}}+4X_{t_i}}{5})^{-2\gamma}+1721(X_{t_i})^{-2\gamma}\right]\right\}(t_i-t_{i-1})\Bigg]$$

$$\Bigg[\left\{\frac{1}{24192}\sum_{i=1}^{n}\left[1471(X_{t_{i-1}})^{2-2\gamma}+6925(\frac{X_{t_{i-1}}+X_{t_i}}{5})^{2-2\gamma}+2950(\frac{2X_{t_{i-1}}+2X_{t_i}}{5})^{2-2\gamma}\right.\right.$$

$$\left.\left.+5450(\frac{3X_{t_{i-1}}+3X_{t_i}}{5})^{2-2\gamma}+5675(\frac{4X_{t_{i-1}}+4X_{t_i}}{5})^{2-2\gamma}+1721(X_{t_i})^{2-2\gamma}\right]\right\}(t_i-t_{i-1})\Bigg]$$

$$-\Bigg[\left\{\frac{1}{24192}\sum_{i=1}^{n}\left[1471(X_{t_{i-1}})^{1-2\gamma}+6925(\frac{X_{t_{i-1}}+X_{t_i}}{5})^{1-2\gamma}+2950(\frac{2X_{t_{i-1}}+2X_{t_i}}{5})^{1-2\gamma}\right.\right.$$

$$\left.\left.+5450(\frac{3X_{t_{i-1}}+3X_{t_i}}{5})^{1-2\gamma}+5675(\frac{4X_{t_{i-1}}+4X_{t_i}}{5})^{1-2\gamma}+1721(X_{t_i})^{2}\right]\right\}(t_i-t_{i-1})\Bigg]$$

$$\Bigg[\left\{\frac{1}{24192}\sum_{i=1}^{n}\left[1471(X_{t_{i-1}})^{1-2\gamma}+6925(\frac{X_{t_{i-1}}+X_{t_i}}{5})^{1-2\gamma}+2950(\frac{2X_{t_{i-1}}+2X_{t_i}}{5})^{1-2\gamma}\right.\right.$$

$$+5450(\frac{3X_{t_{i-1}}+3X_{t_i}}{5})^{1-2\gamma}+5675(\frac{4X_{t_{i-1}}+4X_{t_i}}{5})^{1-2\gamma}+1721(X_{t_i})^{1-2\gamma}\Big]\Big\}$$
$$(t_i-t_{i-1})\Big]\Big)^{-1},$$

$$\tilde{\beta}_{n,T,7}:=$$
$$\Bigg(\Bigg[\Bigg\{\frac{1}{24192}\sum_{i=1}^{n}\Big[1471(X_{t_{i-1}})^{-2\gamma}+6925(\frac{X_{t_{i-1}}+X_{t_i}}{5})^{-2\gamma}+2950(\frac{2X_{t_{i-1}}+2X_{t_i}}{5})^{-2\gamma}$$
$$+5450(\frac{3X_{t_{i-1}}+3X_{t_i}}{5})^{-2\gamma}+5675(\frac{4X_{t_{i-1}}+4X_{t_i}}{5})^{-2\gamma}+1721(X_{t_i})^{-2\gamma}\Big]\Bigg\}(X_{t_i}-X_{t_{i-1}})$$
$$+\gamma\Bigg\{\frac{1}{24192}\sum_{i=1}^{n}\Big[1471(X_{t_{i-1}})^{-1}+6925(\frac{X_{t_{i-1}}+X_{t_i}}{5})^{-1}+2950(\frac{2X_{t_{i-1}}+2X_{t_i}}{5})^{-1}$$
$$+5450(\frac{3X_{t_{i-1}}+3X_{t_i}}{5})^{-1}+5675(\frac{4X_{t_{i-1}}+4X_{t_i}}{5})^{-1}+1721(X_{t_i})^{-1}\Big]\Bigg\}(t_i-t_{i-1})\Bigg]$$
$$\Bigg[\Bigg\{\frac{1}{24192}\sum_{i=1}^{n}\Big[1471(X_{t_{i-1}})^{2-2\gamma}+6925(\frac{X_{t_{i-1}}+X_{t_i}}{5})^{2-2\gamma}+2950(\frac{2X_{t_{i-1}}+2X_{t_i}}{5})^{2-2\gamma}$$
$$+5450(\frac{3X_{t_{i-1}}+3X_{t_i}}{5})^{2-2\gamma}+5675(\frac{4X_{t_{i-1}}+4X_{t_i}}{5})^{2-2\gamma}+1721(X_{t_i})^{2-2\gamma}\Big]\Bigg\}(t_i-t_{i-1})\Bigg]$$
$$-\Bigg[\Bigg\{\frac{1}{24192}\sum_{i=1}^{n}\Big[1471(X_{t_{i-1}})^{1-2\gamma}+6925(\frac{X_{t_{i-1}}+X_{t_i}}{5})^{1-2\gamma}+2950(\frac{2X_{t_{i-1}}+2X_{t_i}}{5})^{1-2\gamma}$$
$$+5450(\frac{3X_{t_{i-1}}+3X_{t_i}}{5})^{1-2\gamma}+5675(\frac{4X_{t_{i-1}}+4X_{t_i}}{5})^{1-2\gamma}+1721(X_{t_i})^{1-2\gamma}\Big]\Bigg\}$$
$$(X_{t_i}-X_{t_{i-1}})-\frac{1-2\gamma}{2}T\Bigg]$$
$$\Bigg[\Bigg\{\frac{1}{24192}\sum_{i=1}^{n}\Big[1471(X_{t_{i-1}})^{1-2\gamma}+6925(\frac{X_{t_{i-1}}+X_{t_i}}{5})^{1-2\gamma}+2950(\frac{2X_{t_{i-1}}+2X_{t_i}}{5})^{1-2\gamma}$$
$$+5450(\frac{3X_{t_{i-1}}+3X_{t_i}}{5})^{1-2\gamma}+5675(\frac{4X_{t_{i-1}}+4X_{t_i}}{5})^{1-2\gamma}+1721(X_{t_i})^{1-2\gamma}\Big]\Bigg\}$$
$$(t_i-t_{i-1})\Bigg]\Bigg)$$
$$\Bigg(\Bigg[\Bigg\{\frac{1}{24192}\sum_{i=1}^{n}\Big[1471(X_{t_{i-1}})^{-2\gamma}+6925(\frac{X_{t_{i-1}}+X_{t_i}}{5})^{-2\gamma}+2950(\frac{2X_{t_{i-1}}+2X_{t_i}}{5})^{-2\gamma}$$
$$+5450(\frac{3X_{t_{i-1}}+3X_{t_i}}{5})^{-2\gamma}+5675(\frac{4X_{t_{i-1}}+4X_{t_i}}{5})^{-2\gamma}+1721(X_{t_i})^{2}\Big]\Bigg\}(t_i-t_{i-1})\Bigg]$$
$$+\gamma\Bigg[\Bigg\{\frac{1}{24192}\sum_{i=1}^{n}\Big[1471(X_{t_{i-1}})^{-1}+6925(\frac{X_{t_{i-1}}+X_{t_i}}{5})^{-1}+2950(\frac{2X_{t_{i-1}}+2X_{t_i}}{5})^{-1}$$
$$+5450(\frac{3X_{t_{i-1}}+3X_{t_i}}{5})^{2}+5675(\frac{4X_{t_{i-1}}+4X_{t_i}}{5})^{-1}+1721(X_{t_i})^{-1}\Big]\Bigg\}(t_i-t_{i-1})\Bigg]$$
$$-\Bigg[\Bigg\{\frac{1}{24192}\sum_{i=1}^{n}\Big[1471(X_{t_{i-1}})^{1-2\gamma}+6925(\frac{X_{t_{i-1}}+X_{t_i}}{5})^{1-2\gamma}+2950(\frac{2X_{t_{i-1}}+2X_{t_i}}{5})^{1-2\gamma}$$
$$+5450(\frac{3X_{t_{i-1}}+3X_{t_i}}{5})^{1-2\gamma}+5675(\frac{4X_{t_{i-1}}+4X_{t_i}}{5})^{1-2\gamma}+1721(X_{t_i})^{1-2\gamma}\Big]\Bigg\}$$
$$(X_{t_i}-X_{t_{i-1}})-\frac{1-2\gamma}{2}T\Bigg]$$

$$\Big[\Big\{\frac{1}{24192}\sum_{i=1}^{n}\Big[1471(X_{t_{i-1}})^{-2\gamma}+6925(\frac{X_{t_{i-1}}+X_{t_i}}{5})^{-2\gamma}+2950(\frac{2X_{t_{i-1}}+2X_{t_i}}{5})^{-2\gamma}$$
$$+5450(\frac{3X_{t_{i-1}}+3X_{t_i}}{5})^{-2\gamma}+5675(\frac{4X_{t_{i-1}}+4X_{t_i}}{5})^{-2\gamma}+1721(X_{t_i})^{-2\gamma}\Big]\Big\}$$
$$(t_i-t_{i-1})]\Big)^{-1}$$

$$\tilde{\alpha}_{n,T,8}:=$$
$$\Big(\Big[\Big\{\frac{1}{90}\sum_{i=1}^{n}\Big[7(X_{t_{i-1}})^{-2\gamma}+32(\frac{3X_{t_{i-1}}+X_{t_i}}{4})^{-2\gamma}+12(\frac{X_{t_{i-1}}+X_{t_i}}{2})^{-2\gamma}$$
$$+32(\frac{X_{t_{i-1}}+3X_{t_i}}{4})^{-2\gamma}+7(X_{t_i})^{-2\gamma}\Big]\Big\}(X_{t_i}-X_{t_{i-1}})$$
$$+\gamma\Big\{\frac{1}{90}\sum_{i=1}^{n}\Big[7(X_{t_{i-1}})^{-1}+32(\frac{3X_{t_{i-1}}+X_{t_i}}{4})^{-1}+12(\frac{X_{t_{i-1}}+X_{t_i}}{2})^{-1}$$
$$+32(\frac{X_{t_{i-1}}+3X_{t_i}}{4})^{-1}+7(X_{t_i})^{\gamma}\Big]\Big\}(t_i-t_{i-1})\Big]$$
$$\Big[\Big\{\frac{1}{90}\sum_{i=1}^{n}\Big[7(X_{t_{i-1}})^{1-2\gamma}+32(\frac{3X_{t_{i-1}}+X_{t_i}}{4})^{1-2\gamma}+12(\frac{X_{t_{i-1}}+X_{t_i}}{2})^{1-2\gamma}$$
$$+32(\frac{X_{t_{i-1}}+3X_{t_i}}{4})^{1-2\gamma}+7(X_{t_i})^{1-2\gamma}\Big]\Big\}(t_i-t_{i-1})\Big]$$
$$-\Big[\Big\{\frac{1}{90}\sum_{i=1}^{n}\Big[7(X_{t_{i-1}})^{1-2\gamma}+32(\frac{3X_{t_{i-1}}+X_{t_i}}{4})^{1-2\gamma}+12(\frac{X_{t_{i-1}}+X_{t_i}}{2})^{1-2\gamma}$$
$$+32(\frac{X_{t_{i-1}}+3X_{t_i}}{4})^{1-2\gamma}+7(t_i,X_{t_i})^{1-2\gamma}\Big]\Big\}(X_{t_i}-X_{t_{i-1}})-\frac{1-2\gamma}{2}T\Big]$$
$$\Big[\Big\{\frac{1}{90}\sum_{i=1}^{n}\Big[7(X_{t_{i-1}})^{-2\gamma}+32(\frac{3X_{t_{i-1}}+X_{t_i}}{4})^{-2\gamma}+12(\frac{X_{t_{i-1}}+X_{t_i}}{2})^{-2\gamma}$$
$$+32(\frac{X_{t_{i-1}}+3X_{t_i}}{4})^{-2\gamma}+7(X_{t_i})^{-2\gamma}\Big]\Big\}(t_i-t_{i-1})\Big]\Big)$$
$$\Big(\Big[\Big\{\frac{1}{90}\sum_{i=1}^{n}\Big[7(X_{t_{i-1}})^{-2\gamma}+32(\frac{3X_{t_{i-1}}+X_{t_i}}{4})^{-2\gamma}+12(\frac{X_{t_{i-1}}+X_{t_i}}{2})^{-2\gamma}$$
$$+32(\frac{X_{t_{i-1}}+3X_{t_i}}{4})^{-2\gamma}+7(X_{t_i})^{-2\gamma}\Big]\Big\}(t_i-t_{i-1})\Big]$$
$$\Big[\Big\{\frac{1}{90}\sum_{i=1}^{n}\Big[7(X_{t_{i-1}})^{2-2\gamma}+32(\frac{3X_{t_{i-1}}+X_{t_i}}{4})^{2-2\gamma}+12(\frac{X_{t_{i-1}}+X_{t_i}}{2})^{2-2\gamma}$$
$$+32(\frac{X_{t_{i-1}}+3X_{t_i}}{4})^{2-2\gamma}+7(X_{t_i})^{-1}\Big]\Big\}(t_i-t_{i-1})\Big]$$
$$-\Big[\Big\{\frac{1}{90}\sum_{i=1}^{n}\Big[7(X_{t_{i-1}})^{1-2\gamma}+32(\frac{3X_{t_{i-1}}+X_{t_i}}{4})^{1-2\gamma}+12(\frac{X_{t_{i-1}}+X_{t_i}}{2})^{1-2\gamma}$$
$$+32(\frac{X_{t_{i-1}}+3X_{t_i}}{4})^{1-2\gamma}+7(X_{t_i})^{1-2\gamma}\Big]\Big\}(t_i-t_{i-1})\Big]$$
$$\Big[\Big\{\frac{1}{90}\sum_{i=1}^{n}\Big[7(X_{t_{i-1}})^{1-2\gamma}+32(\frac{3X_{t_{i-1}}+X_{t_i}}{4})^{1-2\gamma}+12(\frac{X_{t_{i-1}}+X_{t_i}}{2})^{1-2\gamma}$$

$$+32(\frac{X_{t_{i-1}}+3X_{t_i}}{4})^{1-2\gamma}+7(X_{t_i})^{1-2\gamma}\Big]\Big\}(t_i-t_{i-1})\Big]\Big)^{-1},$$

$$\tilde{\beta}_{n,T,8}:=$$

$$\Bigg(\Bigg[\Bigg\{\frac{1}{90}\sum_{i=1}^{n}\Big[7(X_{t_{i-1}})^{-2\gamma}+32(\frac{3X_{t_{i-1}}+X_{t_i}}{4})^{-2\gamma}+12(\frac{X_{t_{i-1}}+X_{t_i}}{2})^{-2\gamma}$$

$$+32(\frac{X_{t_{i-1}}+3X_{t_i}}{4})^{-2\gamma}+7(X_{t_i})^{-2\gamma}\Big]\Bigg\}(X_{t_i}-X_{t_{i-1}})$$

$$+\gamma\Bigg\{\frac{1}{90}\sum_{i=1}^{n}\Big[7(X_{t_{i-1}})^{-1}+32(\frac{3X_{t_{i-1}}+X_{t_i}}{4})^{-1}+12(\frac{X_{t_{i-1}}+X_{t_i}}{2})^{-1}$$

$$+32(\frac{X_{t_{i-1}}+3X_{t_i}}{4})^{-1}+7(X_{t_i})^{\gamma}\Big]\Bigg\}(t_i-t_{i-1})\Bigg]$$

$$\Bigg[\Bigg\{\frac{1}{90}\sum_{i=1}^{n}\Big[7(X_{t_{i-1}})^{2-2\gamma}+32(\frac{3X_{t_{i-1}}+X_{t_i}}{4})^{2-2\gamma}+12(\frac{X_{t_{i-1}}+X_{t_i}}{2})^{2-2\gamma}$$

$$+32(\frac{X_{t_{i-1}}+3X_{t_i}}{4})^{2-2\gamma}+7(X_{t_i})^{2-2\gamma}\Big]\Bigg\}(t_i-t_{i-1})\Bigg]$$

$$-\Bigg[\Bigg\{\frac{1}{90}\sum_{i=1}^{n}\Big[7(X_{t_{i-1}})^{1-2\gamma}+32(\frac{3X_{t_{i-1}}+X_{t_i}}{4})^{1-2\gamma}+12(\frac{X_{t_{i-1}}+X_{t_i}}{2})^{1-2\gamma}$$

$$+32(\frac{X_{t_{i-1}}+3X_{t_i}}{4})^{1-2\gamma}+7(t_i,X_{t_i})^{1-2\gamma}\Big]\Bigg\}(X_{t_i}-X_{t_{i-1}})-\frac{1-2\gamma}{2}T\Bigg]$$

$$\Bigg[\Bigg\{\frac{1}{90}\sum_{i=1}^{n}\Big[7(X_{t_{i-1}})^{1-2\gamma}+32(\frac{3X_{t_{i-1}}+X_{t_i}}{4})^{1-2\gamma}+12(\frac{X_{t_{i-1}}+X_{t_i}}{2})^{1-2\gamma}$$

$$+32(\frac{X_{t_{i-1}}+3X_{t_i}}{4})^{1-2\gamma}+7(X_{t_i})^{1-2\gamma}\Big]\Bigg\}(t_i-t_{i-1})\Bigg]\Bigg)$$

$$\Bigg(\Bigg[\Bigg\{\frac{1}{90}\sum_{i=1}^{n}\Big[7(X_{t_{i-1}})^{-2\gamma}+32(\frac{3X_{t_{i-1}}+X_{t_i}}{4})^{-2\gamma}+12(\frac{X_{t_{i-1}}+X_{t_i}}{2})^{-2\gamma}$$

$$+32(\frac{X_{t_{i-1}}+3X_{t_i}}{4})^{-2\gamma}+7(X_{t_i})^{-2\gamma}\Big]\Bigg\}(t_i-t_{i-1})$$

$$+\gamma\Bigg\{\frac{1}{90}\sum_{i=1}^{n}\Big[7(X_{t_{i-1}})^{-1}+32(\frac{3X_{t_{i-1}}+X_{t_i}}{4})^{-1}+12(\frac{X_{t_{i-1}}+X_{t_i}}{2})^{-1}$$

$$+32(\frac{X_{t_{i-1}}+3X_{t_i}}{4})^{-1}+7(X_{t_i})^{-1}\Big]\Bigg\}(t_i-t_{i-1})\Bigg]$$

$$\Bigg[\Bigg\{\frac{1}{90}\sum_{i=1}^{n}\Big[7(X_{t_{i-1}})^{1-2\gamma}+32(\frac{3X_{t_{i-1}}+X_{t_i}}{4})^{1-2\gamma}+12(\frac{X_{t_{i-1}}+X_{t_i}}{2})^{1-2\gamma}$$

$$+32(\frac{X_{t_{i-1}}+3X_{t_i}}{4})^{1-2\gamma}+7(X_{t_i})^{1-2\gamma}\Big]\Bigg\}(t_i-t_{i-1})\Bigg]$$

$$-\Bigg[\Bigg\{\frac{1}{90}\sum_{i=1}^{n}\Big[7(X_{t_{i-1}})^{1-2\gamma}+32(\frac{3X_{t_{i-1}}+X_{t_i}}{4})^{1-2\gamma}+12(\frac{X_{t_{i-1}}+X_{t_i}}{2})^{1-2\gamma}$$

$$+32(\frac{X_{t_{i-1}}+3X_{t_i}}{4})^{1-2\gamma}+7(X_{t_i})^{1-2\gamma}\Big]\Bigg\}(t_i-t_{i-1})-\frac{1-2\gamma}{2}T\Bigg]$$

$$\Bigg[\Bigg\{\frac{1}{90}\sum_{i=1}^{n}\Big[7(X_{t_{i-1}})^{-2\gamma}+32(\frac{3X_{t_{i-1}}+X_{t_i}}{4})^{-2\gamma}+12(\frac{X_{t_{i-1}}+X_{t_i}}{2})^{-2\gamma}$$

$$+32(\frac{X_{t_{i-1}}+3X_{t_i}}{4})^{-2\gamma}+7(X_{t_i})^{-2\gamma}\Big]\Big\}(t_i-t_{i-1})\Big]\Big)^{-1},$$

$\tilde{\alpha}_{n,T,9} :=$

$$\Big(\Big[\Big\{\frac{1}{288}\sum_{i=1}^{n}\Big[19(X_{t_{i-1}})^{-2\gamma}+75(\frac{4X_{t_{i-1}}+X_{t_i}}{5})^{-2\gamma}+50(\frac{3X_{t_{i-1}}+2X_{t_i}}{5})^{-2\gamma}$$

$$+50(\frac{2X_{t_{i-1}}+3X_{t_i}}{5})^{-2\gamma}+75(\frac{X_{t_{i-1}}+4X_{t_i}}{5})^{1-2\gamma}+19(X_{t_i})^{-2\gamma}\Big]\Big\}(X_{t_i}-X_{t_{i-1}})$$

$$+\gamma\Big[\Big\{\frac{1}{288}\sum_{i=1}^{n}\Big[19(X_{t_{i-1}})^{-1}+75(\frac{4X_{t_{i-1}}+X_{t_i}}{5})^{-1}+50(\frac{3X_{t_{i-1}}+2X_{t_i}}{5})^{-1}$$

$$+50(\frac{2X_{t_{i-1}}+3X_{t_i}}{5})^{-1}+75(\frac{X_{t_{i-1}}+4X_{t_i}}{5})^{-1}+19(X_{t_i})^{-1}\Big]\Big\}(t_i-t_{i-1})\Big]$$

$$\Big[\Big[\Big\{\frac{1}{288}\sum_{i=1}^{n}\Big[19(X_{t_{i-1}})^{1-2\gamma}+75(\frac{4X_{t_{i-1}}+X_{t_i}}{5})^{1-2\gamma}+50(\frac{3X_{t_{i-1}}+2X_{t_i}}{5})^{1-2\gamma}$$

$$+50(\frac{2X_{t_{i-1}}+3X_{t_i}}{5})^{1-2\gamma}+75(\frac{X_{t_{i-1}}+4X_{t_i}}{5})^{2-2\gamma}+19(X_{t_i})^{1-2\gamma}\Big]\Big\}(t_i-t_{i-1})\Big]$$

$$-\Big[\Big[\Big\{\frac{1}{288}\sum_{i=1}^{n}\Big[19(X_{t_{i-1}})^{1-2\gamma}+75(\frac{4X_{t_{i-1}}+X_{t_i}}{5})^{1-2\gamma}+50(\frac{3X_{t_{i-1}}+2X_{t_i}}{5})^{1-2\gamma}$$

$$+50(\frac{2X_{t_{i-1}}+3X_{t_i}}{5})^{1-2\gamma}+75(\frac{X_{t_{i-1}}+4X_{t_i}}{5})^{2-2\gamma}+19(X_{t_i})^{1-2\gamma}\Big]\Big\}(X_{t_i}-X_{t_{i-1}})$$

$$-\frac{1-2\gamma}{2}T\Big]$$

$$\Big[\Big\{\frac{1}{288}\sum_{i=1}^{n}\Big[19(X_{t_{i-1}})^{-2\gamma}+75(\frac{4X_{t_{i-1}}+X_{t_i}}{5})^{-2\gamma}+50(\frac{3X_{t_{i-1}}+2X_{t_i}}{5})^{-2\gamma}$$

$$+50(\frac{2X_{t_{i-1}}+3X_{t_i}}{5})^{-2\gamma}+75(\frac{X_{t_{i-1}}+4X_{t_i}}{5})^{-2\gamma}+19(X_{t_i})^{-2\gamma}\Big]\Big\}(X_{t_i}-X_{t_{i-1}})\Big)$$

$$\Big(\Big[\Big\{\frac{1}{288}\sum_{i=1}^{n}\Big[19(X_{t_{i-1}})^{-2\gamma}+75(\frac{4X_{t_{i-1}}+X_{t_i}}{5})^{-2\gamma}+50(\frac{3X_{t_{i-1}}+2X_{t_i}}{5})^{-2\gamma}$$

$$+50(\frac{2X_{t_{i-1}}+3X_{t_i}}{5})^{-2\gamma}+75(\frac{X_{t_{i-1}}+4X_{t_i}}{5})^{\gamma}+19(X_{t_i})^{-2\gamma}\Big]\Big\}(t_i-t_{i-1})\Big]$$

$$\Big[\Big[\Big\{\frac{1}{288}\sum_{i=1}^{n}\Big[19(X_{t_{i-1}})^{2-2\gamma}+75(\frac{4X_{t_{i-1}}+X_{t_i}}{5})^{2-2\gamma}+50(\frac{3X_{t_{i-1}}+2X_{t_i}}{5})^{2-2\gamma}$$

$$+50(\frac{2X_{t_{i-1}}+3X_{t_i}}{5})^{2-2\gamma}+75(\frac{X_{t_{i-1}}+4X_{t_i}}{5})^{2-2\gamma}+19(X_{t_i})^{2-2\gamma}\Big]\Big\}(t_i-t_{i-1})\Big]$$

$$-\Big[\Big[\Big\{\frac{1}{288}\sum_{i=1}^{n}\Big[19(X_{t_{i-1}})^{1-2\gamma}+75(\frac{4X_{t_{i-1}}+X_{t_i}}{5})^{1-2\gamma}+50(\frac{3X_{t_{i-1}}+2X_{t_i}}{5})^{1-2\gamma}$$

$$+50(\frac{2X_{t_{i-1}}+3X_{t_i}}{5})^{1-2\gamma}+75(\frac{X_{t_{i-1}}+4X_{t_i}}{5})^{1-2\gamma}+19(X_{t_i})^{1-2\gamma}\Big]\Big\}(t_i-t_{i-1})\Big]^2\Big)^{-1},$$

$\tilde{\beta}_{n,T,9} :=$

$$\Big(\Big[\Big\{\frac{1}{288}\sum_{i=1}^{n}\Big[19(X_{t_{i-1}})^{-2\gamma}+75(\frac{4X_{t_{i-1}}+X_{t_i}}{5})^{-2\gamma}+50(\frac{3X_{t_{i-1}}+2X_{t_i}}{5})^{-2\gamma}$$

$$+50(\frac{2X_{t_{i-1}}+3X_{t_i}}{5})^{-2\gamma}+75(\frac{X_{t_{i-1}}+4X_{t_i}}{5})^{1-2\gamma}+19(X_{t_i})^{-2\gamma}\Big]\Big\}(X_{t_i}-X_{t_{i-1}})$$

$$+\gamma\left[\left\{\frac{1}{288}\sum_{i=1}^{n}\left[19(X_{t_{i-1}})^{-1}+75(\frac{4X_{t_{i-1}}+X_{t_i}}{5})^{-1}+50(\frac{3X_{t_{i-1}}+2X_{t_i}}{5})^{-1}\right.\right.\right.$$
$$\left.\left.\left.+50(\frac{2X_{t_{i-1}}+3X_{t_i}}{5})^{-1}+75(\frac{X_{t_{i-1}}+4X_{t_i}}{5})^{-1}+19(X_{t_i})^{-1}\right]\right\}(t_i-t_{i-1})\right]$$
$$\left[\left[\left\{\frac{1}{288}\sum_{i=1}^{n}\left[19(X_{t_{i-1}})^{2-2\gamma}+75(\frac{4X_{t_{i-1}}+X_{t_i}}{5})^{2-2\gamma}+50(\frac{3X_{t_{i-1}}+2X_{t_i}}{5})^{2-2\gamma}\right.\right.\right.\right.$$
$$\left.\left.\left.+50(\frac{2X_{t_{i-1}}+3X_{t_i}}{5})^{2-2\gamma}+75(\frac{X_{t_{i-1}}+4X_{t_i}}{5})^{2-2\gamma}+19(X_{t_i})^{2-2\gamma}\right]\right\}(t_i-t_{i-1})\right]$$
$$-\left[\left[\left\{\frac{1}{288}\sum_{i=1}^{n}\left[19(X_{t_{i-1}})^{1-2\gamma}+75(\frac{4X_{t_{i-1}}+X_{t_i}}{5})^{1-2\gamma}+50(\frac{3X_{t_{i-1}}+2X_{t_i}}{5})^{1-2\gamma}\right.\right.\right.\right.$$
$$\left.\left.+50(\frac{2X_{t_{i-1}}+3X_{t_i}}{5})^{1-2\gamma}+75(\frac{X_{t_{i-1}}+4X_{t_i}}{5})^{2-2\gamma}+19(X_{t_i})^{1-2\gamma}\right]\right\}(X_{t_i}-X_{t_{i-1}})$$
$$\left.-\frac{1-2\gamma}{2}T\right]$$
$$\left[\left\{\frac{1}{288}\sum_{i=1}^{n}\left[19(X_{t_{i-1}})^{-2\gamma}+75(\frac{4X_{t_{i-1}}+X_{t_i}}{5})^{1-2\gamma}+50(\frac{3X_{t_{i-1}}+2X_{t_i}}{5})^{1-2\gamma}\right.\right.\right.$$
$$\left.\left.+50(\frac{2X_{t_{i-1}}+3X_{t_i}}{5})^{1-2\gamma}+75(\frac{X_{t_{i-1}}+4X_{t_i}}{5})^{1-2\gamma}+19(X_{t_i})^{1-2\gamma}\right]\right\}(X_{t_i}-X_{t_{i-1}})\Bigg)$$
$$\Bigg(\left[\left[\left\{\frac{1}{288}\sum_{i=1}^{n}\left[19(X_{t_{i-1}})^{-2\gamma}+75(\frac{4X_{t_{i-1}}+X_{t_i}}{5})^{-2\gamma}+50(\frac{3X_{t_{i-1}}+2X_{t_i}}{5})^{-2\gamma}\right.\right.\right.\right.$$
$$\left.\left.\left.+50(\frac{2X_{t_{i-1}}+3X_{t_i}}{5})^{-2\gamma}+75(\frac{X_{t_{i-1}}+4X_{t_i}}{5})^{\gamma}+19(X_{t_i})^{-2\gamma}\right]\right\}(X_{t_i}-X_{t_{i-1}})\right]$$
$$+\gamma\left[\left\{\frac{1}{288}\sum_{i=1}^{n}\left[19(X_{t_{i-1}})^{-1}+75(\frac{4X_{t_{i-1}}+X_{t_i}}{5})^{-1}+50(\frac{3X_{t_{i-1}}+2X_{t_i}}{5})^{-1}\right.\right.\right.$$
$$\left.\left.\left.+50(\frac{2X_{t_{i-1}}+3X_{t_i}}{5})^{-1}+75(\frac{X_{t_{i-1}}+4X_{t_i}}{5})^{-1}+19(X_{t_i})^{-1}\right]\right\}(t_i-t_{i-1})\right]$$
$$\left[\left[\left\{\frac{1}{288}\sum_{i=1}^{n}\left[19(X_{t_{i-1}})^{2-2\gamma}+75(\frac{4X_{t_{i-1}}+X_{t_i}}{5})^{1-2\gamma}+50(\frac{3X_{t_{i-1}}+2X_{t_i}}{5})^{1-2\gamma}\right.\right.\right.\right.$$
$$\left.\left.\left.+50(\frac{2X_{t_{i-1}}+3X_{t_i}}{5})^{1-2\gamma}+75(\frac{X_{t_{i-1}}+4X_{t_i}}{5})^{2-2\gamma}+19(X_{t_i})^{1-2\gamma}\right]\right\}(t_i-t_{i-1})\right]$$
$$-\left[\left[\left\{\frac{1}{288}\sum_{i=1}^{n}\left[19(X_{t_{i-1}})^{1-2\gamma}+75(\frac{4X_{t_{i-1}}+X_{t_i}}{5})^{1-2\gamma}+50(\frac{3X_{t_{i-1}}+2X_{t_i}}{5})^{1-2\gamma}\right.\right.\right.\right.$$
$$\left.\left.+50(\frac{2X_{t_{i-1}}+3X_{t_i}}{5})^{1-2\gamma}+75(\frac{X_{t_{i-1}}+4X_{t_i}}{5})^{1-2\gamma}+19(X_{t_i})^{1-2\gamma}\right]\right\}(X_{t_i}-X_{t_{i-1}})$$
$$\left.-\frac{1-2\gamma}{2}T\right]$$

$$\left[\left[\left\{\frac{1}{288}\sum_{i=1}^{n}\left[19(X_{t_{i-1}})^{1-2\gamma}+75(\frac{4X_{t_{i-1}}+X_{t_i}}{5})^{1-2\gamma}+50(\frac{3X_{t_{i-1}}+2X_{t_i}}{5})^{1-2\gamma}\right.\right.\right.\right.$$
$$\left.\left.\left.\left.+50(\frac{2X_{t_{i-1}}+3X_{t_i}}{5})^{1-2\gamma}+75(\frac{X_{t_{i-1}}+4X_{t_i}}{5})^{1-2\gamma}+19(X_{t_i})^{1-2\gamma}\right]\right\}(t_i-t_{i-1})\right]\right)^{-1}.$$

14.3.1 Summary of Newton–Cotes Distributions

Here we summarize the different distributions used to construct the estimators.

Itô Distribution
Support: $s_1 = 0$
Probability: $p_1 = 1$
Moment: $\mu_1 = 0$
Rate: $\nu = 1$

McKean Distribution
Support: $s_1 = 1$
Probability: $p_1 = 1$
Moment: $\mu_1 = 1$
Rate: $\nu = 1$

Distribution 1 (Fisk Distribution)
Support: $(s_1, s_2) = (0, 1)$
Probability: $(p_1, p_2) = (\frac{1}{2}, \frac{1}{2})$
Moment: $(\mu_1, \mu_2) = (\frac{1}{2}, \frac{1}{4})$
Rate: $\nu = 2$

Distribution 2 (Stratonovich Distribution)
Support: $s_1 = \frac{1}{2}$
Probability: $p_1 = 1$
Moment: $(\mu_1, \mu_2) = (\frac{1}{2}, \frac{1}{2})$
Rate: $\nu = 2$

Distribution 3
Support: $(s_1, s_2) = (0, \frac{2}{3})$
Probability: $(p_1, p_2) = (\frac{1}{4}, \frac{3}{4})$
Moment: $(\mu_1, \mu_2, \mu_3) = (\frac{1}{2}, \frac{1}{3}, \frac{2}{9})$
Rate: $\nu = 3$

Distribution 4
Support: $(s_1, s_2) = (\frac{1}{3}, 1)$
Probability: $(p_1, p_2) = \frac{3}{4}, \frac{1}{4}$

Moment: $(\mu_1, \mu_2, \mu_3) = (\frac{1}{2}, \frac{1}{3}, \frac{10}{36})$
Rate: $\nu = 3$

Distribution 5 (Simpson's 2/3 Distribution)
Support: $(s_1, s_2, s_3) = (0, \frac{1}{2}, 1)$
Probability: $(p_1, p_2, p_3) = (\frac{1}{6}, \frac{2}{3}, \frac{1}{6})$
Moment: $(\mu_1, \mu_2, \mu_3, \mu_4) = (\frac{1}{2}, \frac{1}{3}, \frac{1}{4}, \frac{5}{24})$
Rate: $\nu = 4$

Distribution 6 (Simpson's 3/8 Distribution)
Support: $(s_1, s_2, s_3, s_4) = (0, \frac{1}{3}, \frac{2}{3}, 1)$
Probability: $(p_1, p_2, p_3, p_4) = (\frac{1}{8}, \frac{3}{8}, \frac{3}{8}, \frac{1}{8})$
Moment: $(\mu_1, \mu_2, \mu_3, \mu_4) = (\frac{1}{2}, \frac{1}{3}, \frac{1}{4}, \frac{11}{54})$
Rate: $\nu = 4$

Distribution 7
Support: $(s_1, s_2, s_3, s_4, s_5) = (0, \frac{1}{5}, \frac{2}{5}, \frac{3}{5}, \frac{4}{5}, 1)$
Probability: $(p_1, p_2, p_3, p_4, p_5) = (\frac{1471}{24192}, \frac{6925}{24192}, \frac{1475}{12096}, \frac{2725}{12096}, \frac{5675}{24192}, \frac{1721}{24192})$
Moment: $(\mu_1, \mu_2, \mu_3, \mu_4, \mu_5, \mu_6) = (\frac{1}{2}, \frac{1}{3}, \frac{1}{4}, \frac{1}{5}, \frac{841}{5040})$.
Rate: $\nu = 5$

Distribution 8 (Boole's Distribution)
Support: $(s_1, s_2, s_3, s_4, s_5) = (0, \frac{1}{4}, \frac{1}{2}, \frac{3}{4}, 1)$
Probability: $(p_1, p_2, p_3, p_4, p_5) = (\frac{7}{90}, \frac{32}{90}, \frac{2}{15}, \frac{32}{90}, \frac{7}{90})$
Moment: $(\mu_1, \mu_2, \mu_3, \mu_4, \mu_5, \mu_6) = (\frac{1}{2}, \frac{1}{3}, \frac{1}{4}, \frac{1}{5}, \frac{1}{6}, \frac{110}{768})$
Rate: $\nu = 6$

Distribution 9
Support: $(s_1, s_2, s_3, s_4, s_5) = (0, \frac{1}{5}, \frac{2}{5}, \frac{3}{5}, \frac{4}{5}, 1)$
Probability: $(p_1, p_2, p_3, p_4, p_5) = (\frac{19}{288}, \frac{75}{288}, \frac{50}{288}, \frac{50}{288}, \frac{75}{288}, \frac{19}{288})$
Moment: $(\mu_1, \mu_2, \mu_3, \mu_4, \mu_5, \mu_6) = (\frac{1}{2}, \frac{1}{3}, \frac{1}{4}, \frac{1}{5}, \frac{1}{6}, \frac{3219}{22500})$
Rate: $\nu = 6$

14.4 Test Function Estimator of Elasticity of Volatility and Stochastic Elasticity Model

One way to estimate the elasticity parameter γ is to minimize a generalized method of moments (GMM) criterion function based on a combined set of moment conditions constructed from the level and difference test functions, whereas the elasticity is treated as an unknown parameter to be estimated along with the drift parameters. To facilitate the interpretation of the GMM test statistics, we can estimate the elasticity parameter γ by the two-step GMM estimation procedure proposed in Conley et al. (1997). In the first step, we use our estimators of the drift parameters of the previous section as a function of the variance elasticity γ and plug them into the moment conditions formed from test functions of the first differences to estimate γ. Then we estimate γ by GMM method. Based on empirical data fitting to the model, the value of the elasticity parameter is known to be near 0.75.

Consider a test function φ in the domain of the generator $\mathcal{A}$ of the sub-fractional diffusion process X satisfying the sub-fractional SDE

$$dX_t = \mu(X_t)dt + \sigma(X_t)dW_t^H.$$

Since $E[\varphi(X_t)]$ is constant over time, it has zero derivative. We have

$$E[\mathcal{A}\varphi(X_t)] = E[\mu(X_t)\varphi'(X_t) + \frac{1}{2}\sigma^2(X_t)\varphi''(X_t)] = 0.$$

An efficient test function would be l_T, the derivative of log-likelihood. The resulting test function estimator using $E[\mathcal{A}l_T] = 0$ will be efficient. It will be more efficient than the quasi-maximum likelihood estimator (QMLE) that uses the moment condition $E[l_T] = 0$. In effect, the application of the generator to the score function adjusts the moment conditions optimally for the presence of temporal dependence.

One can use localized test functions by multiplying the first derivative by a smooth kernel K.

Nonhomogeneous time-dependent elasticity of volatility has been proposed in Fan et al. (2003). We propose a sub-fractional stochastic elasticity of volatility model

$$d\gamma_t = (\alpha_1 + \beta_1\gamma_t)dt + \sigma_1\sqrt{\gamma_t}(\rho W_t^H + \sqrt{1-\rho^2}B_t^H),$$

where W_t^H and B_t^H are two independent sub-fractional Brownian motions and ρ is the correlation between the interest rate and the elasticity.

14.5 Conclusion

We proposed a new financial model called the sub-fractional CKLS model by generalizing the CKLS model that preserves medium range memory and whose increments are nonstationary. We proposed a new interest rate model called the Generalized Ait-Sahalia Model (GAM) being driven by sub-fractional Brownian motion that preserves medium range memory and whose increments are nonstationary. We proposed a sub-fractional stochastic elasticity of volatility model. We obtained several new estimators of the CKLS interest rate model using a new algorithm that would be useful for simulation, estimation, and bond pricing.

Appendix

A. Time Series Regression and Discrete Financial Models

1. Autoregressive Moving Average (ARMA) Models

1.1 Introduction

Any array of random variables indexed by integers that are associated can be classified as a time series. However, times series is normally thought of as an ordered sequence of data points. Time series is a sequence of observations collected over an interval of time. The daily closing prices of stocks from ten different companies are recorded. These prices are then transformed into log returns using the formula for continuously compounded returns. This log return at time t is denoted by r_t and is defined as

$$r_t = \log(1 + R_t) = \log(P_t/P_{(t-1)}) = p_t - p_{(t-1)}, \tag{1.1}$$

where P_t is the price of an asset at time t, $1 + R_t$ is the gross return, and $p_t = \log(P_t)$. Log returns are approximately equal to returns because if x is approximately small, then $\log(1+x) \approx x$. The advantage to using log returns is simplicity of multiperiod returns. A k period log return is just the sum of single log returns, rather than the product of returns. A stochastic process is a sequence of random variables and can be thought of as the population analogue of a time series. Equivalently, a times series can be viewed as a sample from a stochastic process. A time series model is just a statistical model for a stochastic process.

1.2 *Behavior of Returns*

Returns cannot be predicted because they are random. However, if certain assumptions are made regarding the data, then we can classify the data into stationary and nonstationary. A process is said to be stationary if all aspects of its behavior are unchanged by shifts in time. The assumption that future returns will be similar to past returns is classified as stationary or near stationary data. If no such assumption can be made, then the returns are classified as nonstationary.

If returns are classified as nonstationary, this process is either explosive or a random walk. If $R_1, R_2, \ldots$ are returns from a single asset, then a common model is that they are i.i.d. $N(\mu, \sigma^2)$ for some unknown μ and σ^2. However, two flaws exist with this model. First, no bounds exist on the values that a normally distributed random variable can take. This implies that loss can be unlimited, but the liability is limited, because you can lose no

J. P. N. Bishwal, *Parameter Estimation in Stochastic Volatility Models*,
https://doi.org/10.1007/978-3-031-03861-7

more than the amount you initially invested. Second, multiperiod returns are not normally distributed since $1 + R_t(k)$ is a product and not a sum. The product of a normally distributed variable is not normal, but the sum is still normally distributed. However, this can be resolved by using the log-normal model. The lognormal model assumes the log returns are i.i.d. and $r_t = \log(1 + R_t) \sim N(\mu, \sigma^2)$. Therefore, the gross returns $1 + R_t$ are exponentially distributed and, therefore, positive. Also, $\log(1 + R_t(k))$ is a sum of logarithms and since a sum of normally distributed random variables is also normal, now both issues have been resolved. The normality of single period returns implies the normality of multiperiod returns. During this analysis, two questions of interest will be addressed. The first question is whether or not the returns behave like a geometric random walk model. A geometric random walk is the exponential of a random walk. The second issue is model selection. If the returns do in fact behave like a geometric random walk model, we are finished. However, if the returns behave differently, other models need to be considered and tested. Several different models can be used to represent stochastic processes. These models include autoregressive (AR), integrated (I), and moving average (MA). These models are linear and depend on previous data points. Combinations of these models include the autoregressive moving average (ARMA) model and the autoregressive integrated moving average (ARIMA) model.

1.3 *Normal Probability Plots*

Normal probability plots are used to determine if the returns do indeed come from a normal distribution. Normality is easy to test. If the QQ plot shows a relatively straight line, then the data is considered normal; however, if the data shows deviations from a straight line, then a different model should be considered.

1.4 *Autocorrelation and Autocovariance Function*

Throughout this paper, the autocorrelation function is denoted as $\rho(\cdot)$. Also, $\rho(h) = \rho(-h)$. This is because the correlation between Y_t and $Y_(t+h)$ is the same as the correlation between Y_{t+h} and Y_t. The autocorrelation function of the same variable is $\rho(0) = 1$. The autocovariance function is denoted as $\gamma(\cdot)$ and has the following properties:

$$\gamma(h) = \sigma^2 \rho(h) \tag{1.2}$$

$$\gamma(0) = \sigma^2 \tag{1.3}$$

$$\gamma(h) = \sigma^2 \rho(h). \tag{1.4}$$

Therefore,

$$\rho(h) = (\gamma(h))/\sigma^2 = (\gamma(h))/(\gamma(0)). \tag{1.5}$$

Equation (1.5) is just the covariance of Y_t and Y_{t+h} divided by the variance of Y_t.

1.5 Weak White Noise

White noise is the simplest example of a stationary process. A sequence $Y_1, Y_2, \ldots$ is a weak white noise with a mean of μ and a variance of σ^2, where σ is a constant, denoted as WhiteNoise(μ, σ^2), if for all i

$$E(Y_i) = \mu \tag{1.6}$$

$$Var(Y_i) = \sigma^2 \tag{1.7}$$

$$Corr(Y_i, Y_j) = 0 \ \forall \ i \neq j. \tag{1.8}$$

In addition, if the sequence is independent and from a normal distribution, then the process can also be called Gaussian white noise process. A weak white noise process is weakly stationary when

$$\rho(0) = 1 \tag{1.9}$$

$$\rho(h) = 0 \text{ if } h \neq 0 \tag{1.10}$$

$$\gamma(0) = \sigma^2 \tag{1.11}$$

$$\gamma(h) = 0 \text{ if } h \neq 0. \tag{1.12}$$

The term weakly refers to the assumption that only the mean, variance, and covariance are stationary. No other distributional characteristics such as kurtosis, quantile, or skewness are assumed stationary. Because of the lack of correlation, any future white noise cannot be predicted from past values. $E(Y_t|Y_1, Y_2, , Y_{t-1}) = \mu$. This expression states that any future values are completely independent of the past or present values. In addition, the best predictor of any value is simply the mean even if any of the previous values have been unobserved. The estimation of the parameters μ and σ^2 are best represented by the mean of the sample $\bar{Y}$ and sample standard deviation s^2. To estimate the autocovariance function, use the sample autocovariance function

$$\hat{\gamma}(h) = \frac{1}{n}\sum_{j=1}^{n-h}(Y_{j+h} - \bar{Y})(Y_j - \bar{Y}). \tag{1.13}$$

To estimate the autocorrelation function (ACF), use the sample autocorrelation function (SACF)

$$\hat{\rho}(h) = \frac{\hat{\gamma}(h)}{\hat{\gamma}(0)}. \tag{1.14}$$

Although a stationary process can be classified as parsimonious, the process is still not sufficiently parsimonious because an infinite amount of parameters still exist. A class of stationary time series models with only a finite number of parameters is needed. The ARIMA models are precisely this, and the simplest of these models are the autoregressive models.

1.6 *Autoregressive (AR) Process*

1.6.1 *AR(1) Process*

Although white noise is not an appropriate model for time series with correlation, times series models with correlation can be constructed out of white noise. The simplest stationary processes with correlation are autoregressive processes. In these stochastic processes, Y_t is modeled as a weighted sum of past observations plus a white noise error term. The simplest of these is the first order process, AR(1). Only the immediately previous value has a direct effect on the current value. The AR(1) equation is a standard linear difference equation where

$$Y_t - \mu = \phi(Y_{t-1} - \mu) + \epsilon_t \tag{1.15}$$

for all t. A difference equation is an equation that relates an observation to its past values. The parameter is known as the mean of the process. The term $\phi(Y_{t-1} - \mu)$ represents how much feedback or memory of the past is included in the present value of the process. The error term ϵ_t makes up for the variability in the system. This variability occurs when the system moves from one time period to the next. The solution desired for this equation is unique and independent of time, a stationary process. If the solution is independent of time, then an initial condition can be avoided, which can be difficult to determine. Also, in order to forecast, having a solution as a function of past errors is necessary.

1.6.2 *Properties of a Stationary AR(1) Process*

When the AR(1) process is stationary, $|phi| < 1$ and has the following properties:

$$Y_t = \mu + \epsilon_t + \phi\epsilon_{t-1} + \ldots = \mu + \sum_{h=0}^{\infty} \phi^h \epsilon_{t-h} \tag{1.16}$$

$$E(Y_t) = \mu \forall\, t \tag{1.17}$$

$$Cov\left(\sum_{i=0}^{\infty} \epsilon_{t-i}\phi^i, \sum_{j=1}^{\infty} \epsilon_{t+h-j}\phi^j\right) = (\sigma_\epsilon^2 \phi^{|h|})/(1 - \phi^2) \tag{1.18}$$

$$\gamma(0) = Var(Y_t) = (\sigma_\epsilon^2)/(1 - \phi^2) \;\; \forall t \tag{1.19}$$

$$\gamma(h) = Cov(Y_t, Y_{t+h}) = (\sigma_\epsilon^2 \phi^{|h|})/(1 - \phi^2) \; \forall t \text{ and } \forall h \tag{1.20}$$

$$\gamma(h) = Corr(Y_t, Y_{t+h}) = \phi^{|h|} \;\; \forall t \text{ and } \; \forall h. \tag{1.21}$$

These formulas only hold when $|\phi| < 1$ and only for an AR(1) process. In order for Y_t to be stationary, Y_0 must begin in the stationary process such that $E(Y_0) = \mu$ and $Var(Y_0) = (\sigma_\epsilon^2)/(1-\phi^2)$. Otherwise, the process must have started in the infinite past so that (1.1) holds t. The autocorrelation function (ACF) of an AR(1) process depends only on the parameter ϕ. While this is excellent for parsimony, the ACF of an AR(1) process only has a limited range of shapes.

1.6.3 *Nonstationary AR(1) Process*

A nonstationary process occurs either when $\phi = 1$, a random walk, or $|\phi| > 1$, explosive behavior. For $\phi = 1$, the model is reduced into the following random walk process

$$Y_t = Y_{t-1} + \epsilon_t. \tag{1.22}$$

If we started the above model at an arbitrary point, then

$$Y_t = Y_0 + \epsilon_1 + \epsilon_2 + \ldots + \epsilon_t. \tag{1.23}$$

Then $E(Y_t|Y_0) = Y_0 \forall t$. This value is constant and depends entirely on the arbitrary starting point. In addition, $Var(Y_t|Y_0) = t\sigma_\epsilon^2$. This value is not stationary and increases linearly as t increases. Therefore, the increasing variance accounts for the random walking the graph walk in the process. Here Y_t taking increasingly longer trips away from its mean. The model for $|\phi| > 1$ when the process starts at Y_0 and $\mu = 0$ is

$$Y_t = \phi Y_{t-1} + \epsilon_t = \phi(\phi Y_{t-2} + \epsilon_{t-1}) + \epsilon_t = \phi^2 Y_{t-2} + \phi\epsilon_{t-1} + \epsilon_t =$$

$$\ldots = \epsilon_t + \phi\epsilon_{t-1} + \phi^2\epsilon_{t-2} + \ldots + \phi^{t-1}\epsilon_1 + \phi^t Y_0. \tag{1.24}$$

Then, $E(Y_t) = Y_0$ and $Var(Y_t) = \sigma^2(1 + \phi^2 + \phi^4 + \ldots + \phi^{2(t-1)})$ $= \sigma^2(\phi^{2t-1}/(\phi^2 - 1)$. Because the absolute value of ϕ is greater than 1, the variance of this process increases geometrically fast as t approaches infinity. The following graphs are provided to illustrate examples of time series for different values of ϕ.

1.6.4 *Estimation of an AR(1) Process*

In finance applications, asset prices are normally given, but analysis is done on the log returns. Therefore, a transformation must first be made. One can take the logarithm and difference of the closing prices. The least squares estimates of ϕ and μ can be found by minimizing the sum of squared error term

$$S = \sum_{t=2}^{n+1}[Y_t - \mu - \phi(Y_{t-1} - \mu)]^2. \tag{1.25}$$

By solving $\partial S/\partial\phi = 0$ and $\partial S/\partial\mu = 0$. This gives the results of

$$\hat{\phi} = \left(\sum_{t=2}^{n+1}(Y_t - \bar{Y}_n)(Y_{t-1} - (\bar{Y}'_n))\right)\left(\sum_{t=2}^{n+1}(Y_{t-1} - \bar{Y}'_n)^2\right)^{-1} \tag{1.26}$$

$$\hat{\mu} = \bar{Y}_n - \hat{\phi}\bar{Y}'_n \tag{1.27}$$

$$\bar{Y}_n = \frac{1}{n}\sum_{t=2}^{n+1} Y_t \tag{1.28}$$

$$\bar{Y}'_n = \frac{1}{n}\sum_{t=2}^{n+1} Y_{t-1}. \tag{1.29}$$

If the error terms are also Gaussian white noise, then the least squares estimates are also the maximum likelihood estimates. When ϕ and μ have been estimated, an estimate for the white noise process can now be found. Using the equation

$$\epsilon_t = (Y_t - \mu) - \phi(Y_{t-1} - \mu) \tag{1.30}$$

the residuals $\hat{\epsilon}_1, \hat{\epsilon}_2, \ldots, \hat{\epsilon}_t$ are used to estimate $\epsilon_1, \epsilon_2, \ldots, \epsilon_t$ and are defined as

$$\hat{\epsilon}_t = (Y_t - \hat{\mu}) - \hat{\phi}(Y_(t-1) - \hat{\mu}). \tag{1.31}$$

These estimates are used to check the assumption that $Y_1, Y_2, \ldots, Y_t$ are an AR(1) process. Any autocorrelation among the residuals supports the assumption that the data is not an autoregressive process.

1.6.5 *AR(p) Models*

The stochastic process Y_t is an AR(p) process if

$$Y_t - \mu = \phi_1(Y_{t-1} - \mu) + \phi_2(Y_{t-2} - \mu) + \ldots + \phi_p(Y_{t-p} - \mu) + \epsilon_t, \tag{1.32}$$

where $\epsilon_1, \ldots, \epsilon_n$ is white noise with a distribution $N(0, \sigma_\epsilon^2)$. In this model, the last p values of the process, $Y_t, \ldots, Y_{t-p}$, feed back into the current value Y_t. This is a linear model, which can be rewritten with lagged values of the time series, and expressed as

$$Y_t = \beta_0 + \phi_1 Y_{t-1} + \phi_2 Y_{t-2} + \ldots + \phi_p Y_{t-p}, \tag{1.33}$$

where the constant term is $\beta_0 = \{1 - (\phi_1 + \phi_2 + \ldots + \phi_p)\}\mu$. Most of the concepts that were discussed for the AR(1) process can also be applied to the AR(p) process.

1.7 *Moving Average (MA) Process*

1.7.1 *MA(1) Process*

A process Y_t is a moving average process if Y_t can be represented as a weighted average or moving average of the past values of the white noise process ϵ_t rather than the past values of Y_t. The moving average process of order 1 is the expressed as the following

$$Y_t - \mu = \epsilon_t - \theta\epsilon_{t-1}, \tag{1.34}$$

where, as shown before, the ϵ_t's are white noise where $\epsilon \sim N(0, \sigma_\epsilon^2)$.

1.7.2 *Properties of an MA(1) Process*

The following is a list of general properties for the MA(1) process:

$$E(Y_t) = \mu \tag{1.35}$$

$$Var(Y_t) = \sigma_\epsilon^2(1 + \theta^2) \tag{1.36}$$

$$\gamma(1) = -\theta\sigma_\epsilon^2 \tag{1.37}$$

$$\gamma(h) = 0 \text{ if } |h| > 1 \tag{1.38}$$

$$\rho = \rho(1) = -\theta/(1 + \theta^2) \tag{1.39}$$

$$\rho(h) = 0 \text{ if } |h| > 1. \tag{1.40}$$

Looking at these properties, one notices that an MA(1) process has zero correlation at every lag except 0 and 1. Equation (1.35) gives $\rho\theta^2 + \theta + \rho = 0$. Solving this quadratic equation for θ gives the estimator of θ as follows

$$\hat{\theta} = (-1 \pm \sqrt{(1 - 4\hat{\rho}^2)})/(2\hat{\rho}), \tag{1.41}$$

where

$$\hat{\rho} = (\sum_{t=2}^{n+1}(Y_t - \bar{Y}_n)(Y_{t-1} - (\bar{Y}_n')))/(\sum_{t=2}^{n+1}(Y_{t-1} - \bar{Y}_n')^2), \tag{1.42}$$

where $\bar{Y}_n'$ and $\bar{Y}_n$ are defined in Eqs. (1.28) and (1.29).

1.7.3 *General MA Processes*

The more general formula for an MA process of order q is the following equation

$$Y_t = \mu + \epsilon_t - \theta_1\epsilon_{t-1} - \theta_2\epsilon_{t-2} - \ldots - \theta_q\epsilon_{t-q}. \tag{1.43}$$

1.7.4 *Autoregressive Moving Average (ARMA) Process*

The ARMA process combines both the AR and MA processes, which can be represented as

$$Y_t - \mu = \phi_1(Y_{t-1} - \mu) + \ + \phi_p(Y_{t-p} - \mu) + \epsilon_t - \theta_1\epsilon_{t-1} - \ldots - \theta_q\epsilon_{t-q}. \quad (1.44)$$

This equation shows how Y_t is dependent on the lagged values of itself and the lagged values of the white noise. The model has two parameters, p which is the AR order and q which is the MA order. Thus the ARMA model has $p + q + 1$ parameters. If a process is ARMA(0, 0), then $p = q = 0$ and the equation reduces to the white noise

$$Y_t - \mu = \epsilon_t. \quad (1.45)$$

1.8 *Autoregressive Integrated Moving Average (ARIMA) Process*

1.8.1 *The Backwards Operator*

The backwards operator is defined as $BY_t = Y_{t-1}$. More generally, the backwards operator can be expressed as $B^kY_t = Y_{t-k}$. Therefore, the backwards operator B backs up one unit of time while the operator B^k repeatedly backs up repeatedly so time is backed up k time units. However, $Bc = c$ for any constant since constants do not change with time. The ARMA process is expressed using the backwards operator as the following equation

$$(1 - \phi_1 B - \ldots - \phi_p B^p)(Y_t - \mu) = (1 - \theta_1 B - \ldots - \theta_q B^q)\epsilon_t. \quad (1.46)$$

This expression can be rewritten without the backwards operator as

$$(Y_t - \mu) = \phi_1(Y_{t-1} - \mu) + \ldots + \phi_p(Y_{t-p} - \mu) + \epsilon_t - \theta_1\epsilon_{t-1} - \ldots - \theta_q\epsilon_{t-q}. \quad (1.47)$$

This equation shows how Y_t is dependent on the lagged values of itself and the lagged values of the white noise.

1.8.2 *The Difference Operator*

The difference operator is defined as $\Delta = 1 - B$, where B is the backwards operator. Therefore,

$$\Delta Y_t = Y_t - BY_t = Y_t - Y_{t-1}. \quad (1.48)$$

Thus, differencing a time series produces a new time series consisting of the changes in the original series. This can be applied to log price, then the log return is defined as

$$r_t = \Delta p_t. \quad (1.49)$$

Recall, r_t is the log return at time t and $p_t = log(P_t)$ where P_t is the price of an asset at time t. Differencing can be iterated. This can be shown with the following example

$$\Delta^2 Y_t = \Delta(\Delta Y_t) = \Delta(Y_t - Y_{t-1}) = (Y_t - Y_{t-1}) - (Y_{t-1} - Y_{t-2}) = Y_t - 2Y_{t-1} + Y_{t-2}. \tag{1.50}$$

A time series Y_t is said to be ARIMA(p, d, q) model if $\Delta^d Y_t$ is ARMA(p, q) model. Here d is the order of differencing, thus ARIMA(p, d, q) model has $p + q + d + 1$ parameters. The parameter d is also known as the memory parameter. The Y_t process is ARIMA(p, d, q) when $\Delta^d Y_t$ satisfies the following equation.

$$(\Delta^d Y)_t - \mu = \phi_1((\Delta^d Y)_{t-1} - \mu) + \ldots + \phi_p((\Delta^d Y)_{t-p} - \mu) + \epsilon_t - \theta_1 \epsilon_{t-1} - \ldots - \theta_q \epsilon_{t-q}. \tag{1.51}$$

Note that ARIMA(p,0,q)=ARMA(p,q). Also, note that ARIMA(p,0,0)= ARMA(p,0)=AR(p) and ARIMA(0,0,q)=ARMA(0,q)=MA(q). A random walk is modeled as ARIMA(0,1,0).

1.9 *Model Selection*

1.9.1 *AIC and SBC*

AIC and SBC are model selection criteria based on the log likelihood and can be used to find the appropriate value for p and q given that d has already been selected. The parameter d can be found by looking at the SACF of Y_t, ΔY_t,and$\Delta^2 Y_t$. If the SACF decays to zero very slowly and is consistently positive and is well above the upper test bound for 30 or more lags, then the process is nonstationary. If this is true for Y_t then Y_t should be differenced at least once. If the SACF of ΔY_t looks stationary, then use $d = 1$. Otherwise, if the SACF of $\Delta^2 Y_t$ looks stationary then use $d = 2$. If a time series is encountered where $\Delta^2 Y_t$ does not look stationary, then $d > 2$ should be considered. AIC (Akaike's Information Criteria) is defined as

$$-2\log(L) + 2(p + q), \tag{1.52}$$

where L is the likelihood evaluated at the MLE. SBC (Schwarz's Bayesian Criterion) also known as BIC (Bayesian Information Criteria) is defined as

$$-2\log(L) + \log(n)(p + q), \tag{1.53}$$

where n is the length of the time series. The model selection process states that the best model is the model that minimizes the criterion. Both criteria will tend to select models with large values of the likelihood, since large values of L mean that the observed data are likely under the model. The terms

$2(p+q)$ and $\log(n)(p+q)$ are penalties for lack of parsimony. Both AIC and SBC try to trade-off a good fit to the data measured by L with the objective to use as few parameters as possible. Note that $\log(n) > 2$ if $n \geq 8$. Since most time series are larger than 8, then SBC penalizes more than AIC. Therefore, AIC will choose models with more parameters than SBC. Compared to SBC, with AIC, the trade-off is more in favor of a large value of L than a small value of $p+q$. AIC is designed to select the model that will predict best and is less concerned than SBC with having a few too many parameters. SBC is designed to select the true values of p and q exactly. The best AIC model is normally close to the best SBC model, so they are often the same model.

2. GARCH Models

Suppose we have noticed that recent daily returns have been unusually volatile. We might expect that tomorrow's return is also more variable than usual. At this time, many models are not applied to this situation which based on the assumption that error terms are i.i.d. normally distributed. In this part ,we will study GARCH(p,q) model which is widely used in economics and finance because it is especially suitable for volatility analysis and prediction. Specifically, ARCH(q) model is a special case of GARCH(p,q) model when $p=0$. As before, let ϵ_t be Gaussian white noise with unit variance. Then a_t is an ARCH(q) process if

$$a_t = \sigma_t \epsilon_t, \tag{1.54}$$

where

$$\sigma_t = ((\alpha_0 + \sum_{i=1}^{q} \alpha_i a_{t-i}^2))^{1/2} \tag{1.55}$$

is the conditional standard deviation of a_t given the past values $a_{t-1}, a_{t-2}, \dots$ of this process.

A drawback of ARCH(q) model is that conditional standard deviation process has high volatility in short bursts. GARCH models permit a wider persistent volatility. The GARCH(p,q) model is

$$a_t = \sigma_t \epsilon_t, \tag{1.56}$$

where

$$\sigma_t = ((\alpha_0 + \sum_{i=1}^{q} \alpha_i a_{t-i}^2 + \sum_{i=1}^{p} \beta_i \sigma_{t-i}^2))^{1/2} \tag{1.57}$$

and ϵ_t is a Gaussian white noise with unit variance. The $\alpha_0 \alpha_i, \beta_i$ in models are constants.

A GARCH(p,q) process is stationary with a finite variance if

$$\sum_{i=1}^{q} \alpha_i + \sum_{i=1}^{p} \beta_i < 1. \tag{1.58}$$

A deficiency of ARCH(q) models is that conditional standard deviation process has high-frequency oscillation with high volatility coming in short burst. Compared to ARCH(q) models, GARCH models keep a wide range of behavior, particularly, persistent high variability and low variability.

2.1 I-GARCH Models

A GARCH(p,q) process is called an I-GARCH process if

$$\sum_{i=1}^{q} \alpha_i + \sum_{i=1}^{p} \beta_i = 1. \tag{1.58}$$

I-GARCH processes are either nonstationary or if they are stationary they have an infinite variance. Notice that the conditional standard deviation plot is much like the plot of conditional standard deviation in GARCH(1,1). As $\alpha_1 \to 1, \beta_1 \to 0$,the I-GARCH(1,1) behaves more like an ARCH(1) process, in which the volatility comes in sharp bursts.

2.2 GARCH-M Models

In some cases, it is useful to take conditional standard deviation as one of the regression variables. For example, investors expect higher conditional variability may causes higher returns, because market demands a higher risk premium for higher risk. Therefore, we can use GARCH-in-mean, or GARCH-M models to model this behavior. They have the form

$$Y_t = X_t^\top \gamma + \delta \sigma_t + a_t, \tag{1.59}$$

where a_t is a GARCH model with conditional standard deviation process σ_t. The process σ_t and the components of X_t are the predictor variables. δ and the components of are regression coefficients. The main difference between predictor variables X_t and σ_t is that X_t contains observable variable whereas σ_t must be estimated.

2.3 E-GARCH Models

In finance, the leverage effect predicts that an asset's returns may become more volatile when its price decreases. We can use E-GARCH model to estimate financial volatility of daily returns. The exponential GARCH, or E-GARCH, model is

$$\log(\sigma_t) = \alpha_0 + \sum_{i=1}^{q} \alpha_1 g(\epsilon_{t-i}) + \sum_{i=1}^{p} \beta_i \log(\sigma_{t-i}), \tag{1.60}$$

where

$$g(\epsilon_t) = \theta \epsilon_t + \gamma |\epsilon_t| - E(|\epsilon_t|) \tag{1.61}$$

and $\epsilon_t = a_t/\sigma_t$. Since $\log(\sigma_t)$ can be negative, there is no constrains for α_1, β_1 can only be positive. The leverage effect occurs when an asset's return become more volatile as its price decreases, which can be seen in an E-GARCH process when $\theta < 0$.

2.4 The GARCH Zoo

There are many other types of GARCH models. Such as, M-GARCH (multivariate GARCH), A-GARCH(asymmetric GARCH), NGARCH (nonlinear asymmetric GARCH), and so on. For this chapter, we only introduce the most widely used models.

2.5 Model Selection

The Akaike information criterion (AIC) is a measure of the relative quality of statistical models for a given set of data. Given a collection of models for the data, usually we choose the model with the lowest AIC. Hence, AIC provides a means for model selection. In general case, AIC can be expressed as: $AIC = 2k - 2\ln(L)$, where k is number of parameters, L is the maximum value of the likelihood function for the model.

2.6 GARCH Estimation

Consider the AR(1) model $Y_t = \phi Y_{t-1} + \epsilon_t, \quad t = 1, 2, \ldots, n$ where the initial random variable Y_0 is fixed, $\{\epsilon_t\}$ is a Gaussian white noise sequence with variance 1 and $0 < \phi < 1$. It is easy to derive the ACF of Y_t and the least squares estimator of ϕ.

For an ARCH(1) model, $Y_t = \epsilon_t(\alpha_0 + \alpha_1 Y_{t-1}^2)^{1/2}, \quad t = 1, 2, \ldots, n$ where $\alpha_0 > 0$ and $0 \le \alpha_1 < 1$ and ϵ_t is a Gaussian white noise sequence with variance 1, it is easy to derive the ACF of Y_t process and the Y_t^2 process and the conditional least squares estimators of (α_0, α_1).

For the AR(1)/ARCH(1) model

$$u_t = \alpha + \beta u_{t-1} + a_t$$

$$a_t = \epsilon_t(1 + \gamma a_{t-1}^2)^{1/2}$$

with ϵ_t being WhiteNoise(0,1). It is easy to calculate $E(u_2|u_1,\ u_0) =$, Var $(u_2|u_1,\ u_0)$, the mean of u_t. the variance and ACF of u_t. and the ACF of a_t^2.

Autoregressive Gamma Process: Finally, we have the ARG (autoregressive gamma) representation of the Heston model:

$$y_i = \sqrt{x_i}\xi_i, \quad x_i = \sum_{i=1}^{N_t} G_i + \epsilon_i,$$

where x_i is ARG process. N_t is Poisson, G is Gamma, and ϵ is exponential.

3. Market Security Valuation

3.1 Portfolio Theory

Portfolio optimization stems from a theory based on two principles:
(1) Investors want to maximize the expected return of their portfolio.
(2) Investors want to minimize the risk, which is defined as the standard deviation of the return.

These principles are constantly battling each other, because riskier assets tend to have higher expected returns. This is because investors demand a reward for bearing risk in their investment. The incentive to invest in risky assets is called the *risk premium*. The risk premium is simply the difference between the expected return of a risky asset and the risk-free rate of return. There are optimal compromises between expected return and risk, though. The goal of this section is to show how we can maximize expected return while minimizing risk.

3.2 Estimating $E(R)$ and σ_R

Since Treasury bill rates are published online and in most newspapers, the value of the risk-free rate, μ_f, is known. If returns on the asset can be assumed to be stationary, we can take a time series of past returns and use the sample mean and standard deviation to estimate $E(R)$ and σ_R. Additionally, if we assume that the asset is not stationary, we could adjust these estimates upward or downward based on our opinion of the asset. An important consideration is the length of the time series and how that affects the estimation of $E(R)$ and σ_R. A long time series of 10 to 20 years will produce less variable estimates, but if the series has slowly drifting parameters (i.e., not stationary), a shorter series of 1 to 2 years will give a more realistic representation of the future. Since stationary is a relative term, virtually every time series can be considered stationary over short enough time periods.

3.3 Risk Versus Expected Return

The mathematics of mixing risky assets can be explained most easily when there are only two risky assets. We will start here and eventually add a risk-free asset to the portfolio of two risky assets.

Suppose we have two risky assets with returns R_1 and R_2 in proportions w and $1-w$, respectively. The return is $R = wR_1 + (1-w)R_2$. The expected return is $E(R) = w\mu_1 + (1-w)\mu_2$. If we let ρ_{12} be the correlation between the two risky asset's returns, then the variance of the portfolio's return is:

$$\sigma_R^2 = w^2\sigma_1^2 + (1-w)^2\sigma_2^2 + 2w(1-w)\rho_{12}\sigma_1\sigma_2.$$

The leftmost point in the reward-risk parabola has the minimum value of risk and is called the *minimum variance portfolio*. The points on this curve that have an expected return of at least the minimum variance portfolio are called the *efficient frontier*. The portfolios along the efficient frontier are called *mean-variance-efficient portfolios.*

3.4 Estimating Means, Standard Deviations, and Covariances

Using a time series of past returns on the first risky asset, we can estimate μ_1 and σ_1. Denote this time series by $R_{1,1}, \ldots, R_{1,n}$ where the first subscript indicates the asset and the second subscript indicates the time. Denote the sample mean and standard deviation of this series by $\bar{R}_1$ and s_1. Denote the second time series by $R_{2,1}, \ldots, R_{2,n}$. Estimates for μ_2 and σ_2 can be found similarly with the sample mean and standard deviation of this series by $\bar{R}_2$ and s_2. The covariance σ_{12} can be estimated by the sample covariance

$$\hat{\sigma}_{12} = \frac{1}{n} \sum_{t=1}^{n} (R_{1,t} - \bar{R}_1)(R_{2,t} - \bar{R}_2).$$

The correlation ρ_{12} can be estimated by the sample correlation

$$\hat{\rho}_{12} = \frac{\hat{\sigma}_{12}}{s_1 s_2}.$$

The estimator $\hat{\rho}_{12}$, sometimes denoted by r_{12}, is called the sample cross-correlation coefficient between R_1 and R_2 at lag 0, since we are correlating the return on the first risky asset with the return on the second during the same time period (i.e., no lag time).

3.5 Combining Two Risky Assets with a Risk-Free Asset

The ultimate goal is to optimize portfolios with many risky assets and a risk-free asset, but many concepts needed for that can be understood when there are only two risky assets combined with a risk-free asset.

3.5.1 Tangency Portfolio with Two Risky Assets

As mentioned previously, each point on the efficient frontier is $(\sigma_R, E(R))$ for some value of w between 0 and 1. If w is fixed, then we have a fixed portfolio of two risky assets. Now we can mix the portfolio of risky assets with the risk-free asset. The point F near the origin is labeled Risk-Free Asset and gives the values of $(\sigma_R, E(R))$ for the risk-free asset. Intuitively, $\sigma_R = 0$ at F. The possible values of $(\sigma_R, E(R))$ for a portfolio consisting of the fixed portfolio of risky assets and the risk-free asset are a line connecting point F with a point on the efficient frontier. The slope of each line connecting X to each portfolio is called its *Sharpe's Ratio*, named after William Sharpe whom we will discuss later regarding the Capital Asset Pricing Model. Sharpe's ratio can be thought of as a reward-to-risk ratio. It is the ratio of the reward quantified by the excess expected return to the risk as measured by standard deviation. Hence, a line with a larger slope gives a higher expected return for a given level of risk. Therefore, the larger the Sharpe's ratio, the better regardless of the risk one is willing to accept. The point labeled *Tangency Portfolio* represents the portfolio with the largest Sharpe's ratio. It is called the tangency portfolio because its line is tangent to the efficient frontier. The optimal or efficient portfolios mix the tangency portfolio of two risky assets with the risk-free asset and have two properties:

1. It has a higher expected return than any other portfolio with the same (or smaller) risk.
2. It has a smaller risk than any other portfolio with the same (or smaller) expected return.

Thus we can only improve (increase) the expected return of an efficient portfolio by accepting higher risk, and we can only improve (reduce) the risk of an efficient portfolio by accepting a smaller expected return.

Also, note that every efficient portfolio uses the same mix of the two risky assets, namely the tangency portfolio. Only the proportion allocated to the tangency portfolio and the proportion allocated to the risk-free asset change. In other words, we do not move along the efficient frontier once we have the tangency portfolio; we only move along the line connecting the risk-free asset to the tangency portfolio. The tangency portfolio clearly plays a very key role in portfolio optimization, so finding it is an important task. Let μ_1, μ_2, and μ_f be the expected returns on the two risky assets and the risk-free asset again. Let σ_1 and σ_2 be the standard deviations of the returns on the two risky assets and let ρ_{12} be the correlation between the returns on the risky assets. Define $V_1 = \mu_1 - \mu_f$ and $V_2 = \mu_2 - \mu_f$, where V_1 and V_2 are the excess returns. The tangency portfolio then uses weight

$$w_T = \frac{V_1\sigma_2^2 - V_2\rho_{12}\sigma_1\sigma_2}{V_1\sigma_2^2 + V_2\sigma_1^2 - (V_1 + V_2)\rho_{12}\sigma_1\sigma_2}$$

for the first risky asset and weight $(1 - w_T)$ for the second risky asset.

Let R_T be the return, $E(R_T)$ be the expected return, and σ_T be the standard deviation of the return on the tangency portfolio. $E(R_T)$ and σ_T can be found by first using the above formula to calculate w_T and then using the formulas

$$E(R_T) = w_T\mu_1 + (1 - w_T)\mu_2$$

and

$$\sigma_T = \sqrt{w_T^2\sigma_1^2 + (1 - w_T)^2\sigma_2^2 + 2\rho_{12}\sigma_1\sigma_2}.$$

3.5.2 Combining the Tangency Portfolio with the Risk-Free Asset

Now we can finally add the risk-free asset to the tangency portfolio. Let R be the return on the portfolio that allocates a fraction ω of the investment to the tangency portfolio and $(1-\omega)$ to the risk-free asset. Then $R = \omega R_T + (1-\omega)\mu_f = \mu_f + \omega(R_T - R_f)$ such that

$$E(R) = \mu_f + \omega\{(E(R_T) - \mu_f)\} \text{ and } \sigma_R = \omega\sigma_T.$$

5.3 Effect of ρ_{12}

Positive correlation between the two risky assets is bad because it increases risk. The two assets tend to move together which increases the volatility of the portfolio. This means when the assets are doing well, they will do extremely well since both will increase in value, but if both assets decline in value simultaneously, large losses will be realized. Conversely, negative correlation is good. This is the idea of diversifying or hedging your risk. If the assets are negatively correlated, a negative return for one tends to occur with a positive return for the other, so the volatility of the portfolio decreases. As correlation decreases, the Sharpe's ratio of the tangency portfolio returns increases (i.e., higher return with lower risk). This means that when ρ_{12} is small, the efficient portfolios have less risk for a given expected return compared to when ρ_{12} is large.

3.5.3 Risk-Efficient Portfolios with N Risky Assets

When we have more than two risky assets, the mathematics becomes much more complex. For this reason, one can use software to help model risk-efficient portfolios with N risky assets. In this case, we show how a greater quantity of risky assets compares to our model with only three risky assets and a risk-free asset. An important concept to grasp in portfolio optimization is that diversification helps to lower risk and maximize expect return.

4. The Capital Asset Pricing Model

The Capital Asset Pricing Model (CAPM) was originally introduced in 1961 by William Sharpe and Jack Treynor based on the earlier work of Harry Markowitz on portfolio optimization theory. Sharpe and Markowitz received Nobel Prize in Economics in 1990 for their contribution. The most widely recognized model is the Sharpe-Lintner CAPM, which was the result of William Sharpe and John Lintner working independently on the model in the early 1960s. Interestingly, the two economists were studying the exact same model but from different points of view. The CAPM is used to estimate the cost of capital for a firm (Lintner's perspective) and to evaluate the performance of managed portfolios (Sharpe's perspective). They eventually realized they were solving and saying much of the same, which is that the CAPM helps to make predictions about how to measure risk and the relation between expected return and risk. This model has been the most used asset pricing model (possibly the only one) in the finance industry for calculating the cost of equity or portfolio performance, but it is not perfect.

4.1 Building the CAPM from the Markowitz Model

The CAPM builds on the Markowitz model of portfolio choice, which we covered informally above. To review, an investor selects a portfolio at time $t-1$ that produces a stochastic return at time t. The model assumes that investors are risk-averse and they care only about the mean and variance of their one-period investment return when choosing among portfolios. This means that investors will choose mean-variance-efficient portfolios to maximize expected return or minimize the variance of the portfolio return, given the other. The portfolio model provides an algebraic condition on asset weights in mean-variance-efficient portfolios. The CAPM turns this algebraic statement into a testable prediction about the relation between risk and expected return by identifying a portfolio that must be efficient if asset prices are to clear the market of all assets. Sharpe and Lintner added two key assumptions to the Markowitz model.

The first assumption was complete agreement: given market clearing asset prices at time $t-1$, investors agree on the joint distribution of asset returns from $t-1$ to t. The second assumption is that there is borrowing and lending at a risk-free rate. Without risk-free borrowing and lending, only portfolios along the efficient frontier are mean-variance-efficient since these portfolios also maximize expected return, given their return variances. To obtain the mean-variance-portfolios with risk-free borrowing and lending, we again draw a line from the risk-free rate to the tangency portfolio. This implies that all efficient portfolios are combinations of the risk-free asset and a single risky tangency portfolio. This is the idea of the Capital Asset Pricing Model; all investors see the same opportunity set and they combine the same risky tangency portfolio with risk-free borrowing or lending. Since all investors hold the same tangency portfolio, it must be the value-weight market portfolio of risky assets. We now call the tangency portfolio the market portfolio, and each risky asset's weight must be the total market value of all outstanding units of the asset divided by the total market value of all risky assets. Also, the risk-free rate must be set to clear the market for risk-free borrowing and lending. The assumptions made in the CAPM imply that the market portfolio must be on the efficient frontier if the asset market is to clear. Thus, the algebraic relation that holds for any minimum variance portfolio must hold for the market portfolio. Then if there are N risky assets, the Minimum Variance Condition for M is

$$E(R_i) = E(R_{ZM}) + [E(R_M) - E(R_{ZM})]\beta_{iM}, i = 1, 2, \ldots, N,$$

where $E(R_i)$ is the expected return on asset i, and β_{iM}, the market beta of asset i is the covariance of its return with the market return divided by the variance of the market return.

$$\beta_{iM} = \frac{cov(R_i, R_M)}{\sigma^2(R_M)}.$$

$E(R_{ZM})$ is the expected return on assets that have market betas equal to zero, which means their returns are uncorrelated with the market return. $[E(R_M) - E(R_{ZM})]\beta_{iM}$ is a risk premium—the market beta of asset i times the premium per unit of beta, which is the expected market return minus $E(R_{ZM})$. Since β_{iM} is also the slope in the regression of its return on the market return, a common interpretation of beta is that it measures the sensitivity of the asset's return to variation in the market return. This is correct but another interpretation of beta, with the spirit of the Markowitz model that underlies the CAPM, is that the risk (denominator of β_{iM}) of the market portfolio is a weighted average of the covariance risks of the assets in the market portfolio (numerators of β_{iM} for different assets). In economic terms, β_{iM} is proportional to the risk each dollar invested in asset i contributes to the market portfolio. The final step in development of the Sharpe-Lintner model is to use the assumption of risk-free borrowing and lending to substitute $E(R_{ZM})$. A risky asset's return is uncorrelated with the market return when the average of the asset's covariances with the returns on other assets just offsets the variance of the asset's return. Such a risky asset is considered riskless in the market portfolio, because it contributes nothing to the variance of the market return. The expected return on assets that are uncorrelated with the market returns must equal the risk-free rate, R_f . The relation between expected return and beta then becomes the Sharpe-Lintner CAPM equation:

$$E(R_i) = R_f + [E(R_M) - R_f]\beta_{iM}, i = 1, \ldots, N.$$

In words, the expected return on any asset i is the risk-free interest rate plus a risk premium.

4.2 The Capital Asset Pricing Model Assumptions

The CAPM uses various assumptions about the markets and investors' behavior to establish a set of equilibrium conditions that allow investors to predict the return of an asset for its level of systematic risk. Systematic risk is also known as the aggregate, non-diversifiable, or market risk. The CAPM uses a measure of systematic risk that can be compared with other assets in the market. This measure of risk allows investors to improve their portfolios and to find their required rate of return.

The Capital Asset Pricing Model is based on the following assumptions:
(1) The Market prices are in equilibrium (i.e., supply equals demand).
(2) All investors are rational and risk-averse.
(3) All investors have the same forecast of expected returns and risk.

(4) All investors are price takers, i.e., they cannot influence the prices.
(5) Markets are frictionless (the borrowing rate is equal to the lending rate).
(6) All information is available at the same time to all investors.
(7) Trade without transaction or taxation costs.
(8) All assets are perfectly divisible and liquid.
(9) All investors are broadly diversified across a range of investments.

Under an efficient market hypothesis (EMH), which states that financial markets are *informationally efficient*, all the previously mentioned assumptions are reasonable and valid. The validity of the CAPM can only be guaranteed if all of these assumptions are true. The Capital Market Line The Capital Market Line (CML) is the tangent line drawn from the risk-free point to the feasible region for risky assets. CML relates the excess expected return on an efficient portfolio to its risk. Excess expected return is the difference between the expected return from the risk-free rate of the return. It is also called the risk premium. Let R be the return on the given efficient portfolio, i.e., the mixture of the Market portfolio and the risk-free asset. The equation for the CML is given by:

$$\mu_R = \mu_f + (\mu_M - \mu_f)\frac{\sigma_M}{\sigma_R}. \tag{4.1}$$

Here μ_R is the expected return of the portfolio, $\mu_(f)$is the risk-free rate of return and μ_M is the expected return of the market portfolio, σ_M is the standard deviation of the return on the market portfolio, and σ_R is the standard deviation of the return on the portfolio. The excess expected return of a portfolio R is $\mu_R - \mu_f$ and the expected return of the market portfolio is $\mu_M - \mu_f$. In (1.1) μ_f, μ_M, and σ_M are constants. What varies are σ_R and μ_R. These vary as we change the efficient portfolio R.

The slope of the CML is given by $(\mu_M - \mu_f)/\sigma_M$ which can be interpreted as the ratio of the risk premium to the standard deviation of the market portfolio. This is also known as Sharpe's reward to-risk ratio. The Eq. (1.1) can be rewritten as

$$(\mu_R - \mu_f)/\sigma_R = (\mu_M - \mu_f)/\sigma_M$$

which says that the reward-to-risk ratio for any efficient portfolio equals to the reward-to-risk ratio for the market portfolio. The CML is easy to derive. Consider an efficient portfolio that assigns a proportion weight (ω) of its assets to the market portfolio and $(1 - \omega)$ to the risk-free assets. Then

$$R = \omega R_M + (1 - \omega)\mu_f = \mu_f + \omega(R_M - \mu_f). \tag{4.2}$$

Therefore, taking expectation both sides in (4.2),

$$\mu_R = \mu_f + \omega(\mu_M - \mu_f). \tag{4.3}$$

Also from (1.2)

$$\sigma_R = \omega\sigma_M \tag{4.4}$$

which gives

$$\omega = \sigma_R/\sigma_M. \tag{4.5}$$

Substituting (4.5) into (4.3) gives the CML. One can view $\omega = \sigma_R/\sigma_M$ as an index of the risk aversion of the investor. The smaller the value of ω, the more risk-averse the investor. If an investor places zero weight ($\omega = 0$) on an asset, then that investor is 100% in risk-free assets. Similarly, if an investor places 100% weight ($\omega = 1$) on an asset, then that investor is totally invested in the tangency portfolio of risky assets. The plot of reward as quantified by expected return versus the risk as measured by standard deviation of the return is a parabola. The left most point on the parabola achieves the minimum value of the risk and is called the minimum variance portfolio. The Sharpe ratio is the ratio of the reward to the risk. A line joining the risk-free rate to the parabola of reward versus risk with a large slope gives a higher expected return for a given level of risk. Sharpe ratio is the slope of the line. So larger the Sharpe ratio, the better is the expected return regardless of what level of risk one is willing to accept. The portfolio with highest Sharpe ratio is called Tangency Portfolio. The efficient frontier is the part of the parabola that has an expected return at least as larger as the minimum variance portfolio. This is the optimal or efficient portfolio for the purpose of mixing the tangency portfolio of two risky assets with the risk-free asset. Each efficient portfolio has two properties:

(1) It has a higher expected return than any other portfolio with the same (or smaller) risk.

(2) It has a smaller risk than any other portfolio with the same (or smaller) expected return.

Therefore, we could only improve (reduce) the risk of an efficient portfolio by accepting a worse (smaller) expected return, and we can only improve (increase) the expected return of an efficient portfolio by accepting worse (higher) risk.

4.3 The Security Market Line

The Security Market Line (SML) relates the excess return on an asset to the slope of its regression on the market portfolio. Suppose that there are many securities indexed by j define

$$\beta_j = \sigma_{jM}/\sigma_M^2, \tag{4.6}$$

where σ_{jM} is the covariance between the returns on the j-th security and the market portfolio. By using returns of the market portfolio as the predictor variable, β_j is the slope of the best linear predictor of the j-th security's return. Another way to interpret the significance of β_j is based on linear regression. Linear regression is a method for estimating the coefficients of the best linear predictor based upon data. To apply linear regression, suppose that we have a bivariate time series $(R_{j,t}, R_{M,t})_{t=1}^{n}$ of returns on the j-th asset and the market portfolio. Then the estimated slope of the linear regression of $R_{j,t}$ on $R_{M,t}$ is

$$\hat{\beta}_j = (\sum_{t=1}^{n}(R_{j,t} - \bar{R}_j)((R_{M,t} - \bar{R}_M))/\sum_{t=1}^{n}(R_{M,t} - \bar{R}_M)^2. \qquad (4.7)$$

After multiplying the numerator and the denominator by n^{-1}, $\hat{\beta}_j$ becomes an estimate of σ_{jM} divided by an estimate of σ_M^2 and, therefore, the expression (4.6) is an estimate of β_j. The Security Market Line (SML) is given by:

$$\mu_j - \mu_f = \beta_j(\mu_M - \mu_f). \qquad (4.8)$$

Let μ_j be the expected return on the j-th security, then $\mu_j - \mu_f$ is the risk premium for that security. Here, β_j is the independent variable in the linear equation, not the slope; more precisely, μ_j is a linear function of β_j with slope of $(\mu_M - \mu_f)$. The SML says that the risk premium of the j-th asset is the product of its beta and the risk premium of the market portfolio. Therefore, β_j measures both the riskiness of the j-th asset and the reward for assuming that risk. When $\beta_j = 0$, a risky asset that is uncorrelated with the market portfolio will have an expected rate of return equal to the risk-free rate. There is no expected excess return over risk-free even the investor bears some risk in holding a risky asset with zero beta. When $\beta_j = 1$, a risky asset which is perfectly correlated with the market portfolio has the same expected rate of return as the market portfolio. When $\beta_j > 1$, the expected excess rate of return is higher than the market portfolio, also known as an aggressive asset. When $\beta_j < 1$, the asset is said to be non-aggressive or risk-averse. When $\beta_j < 0$, a risky asset with negative beta reduces the variance of the portfolio. This risk reduction potential of asset with negative β_j is something like paying premium to reduce risk.

4.4 Comparison of the CML with SML

The primary difference between the SML and the CML is that the SML applies to any asset but the CML applies only to the return of an efficient portfolio. It can be arranged so as to relate the excess expected return of that portfolio to be excess expected return of the market portfolio:

$$\mu_R - \mu_f = (\sigma_R/\sigma_M)(\mu_M - \mu_f). \qquad (4.9)$$

The SML applies to any asset like the CML relates its excess expected return to the excess expected return of the market portfolio:

$$\mu_j - \mu_f = \beta_j(\mu_M - \mu_f). \qquad (4.10)$$

If we take an efficient portfolio and consider it as an asset, then μ_R and μ_j both denote the expected return on that portfolio/asset. Both (4.9) and (4.10) hold so that

$$\sigma_R/\sigma_M = \beta_R. \tag{4.11}$$

4.5 The Security Characteristic Line

The Security Characteristic Line is a representation of the SML in the form of a regression model. The characteristic line takes the form of a straight line with the y-axis intercept representing the return of a security in excess of the risk-free return and the x-axis representing that for a portfolio made up of all assets in the market. The values that make up the characteristic line are obtained by performing a statistical regression analysis as shown in Figure 3.

The Security Characteristic Line (SCL) is given as

$$R_{j,t} = \mu_{f,t} + \beta_j(R_{M,t} - \mu_{f,t}) + \epsilon_{j,t}, \tag{4.12}$$

where $\epsilon_{j,t}$ is $N(0, \sigma^2_{e,j})$. It is often assumed that the $\epsilon_{j,t}$ s are uncorrelated across assets, that is, that $\epsilon_{j,t}$ is uncorrelated with $_{j',t}$, for $j \neq j'$. This assumption has important ramifications for risk reduction by diversification. The SML gives us information about expected returns, but not about the variance of the returns. The characteristic line gives us a probability model of the returns, not just a model of their expected values. Hence, the characteristic line is said to be a return generating process. The similarity between the SML and characteristic line is that the regression line $E(Y|X) = \beta_0 + \beta_1 X$ gives the expected value of Y given X but not the conditional probability distribution of Y given X. The regression model $Y_t = \beta_0 + \beta_1 X_t + \epsilon_t$, and $\epsilon_t \sim N(0, \sigma^2)$ gives us this conditional probability distribution. The characteristic line implies that

$$\sigma_j^2 = \beta_j^2 \sigma_M^2 + \sigma^2_{\epsilon,j}. \tag{4.13}$$

The total risk of the j-th asset is

$$\sigma_j = (\beta_j^2 \sigma_M^2 + \sigma^2_{\epsilon,j})^{1/2}. \tag{4.14}$$

The risk has two components: $\beta_j^2 \sigma_M^2$ is called the market or systematic component of risk and $\sigma^2_{\epsilon,j}$ is called non-market, unsystematic, or unique component of risk.

4.6 Reducing Unique Risk by Diversification

The systematic component of risk cannot be reduced by diversification, but the unsystematic component of risk can be reduced or even eliminated by sufficient diversification. Suppose that there are N assets with returns $R_{1,t}, \ldots, R_{N,t}$ for holding period t. If we form a portfolio with weights $\omega_1, \ldots,$ ω_N then the return of the portfolio is

$$R_{P,t} = \omega_1 R_{1,t} + \ldots + \omega_N R_{N,t}. \tag{4.15}$$

Let $R_{M,t}$ be the return on the market portfolio. According to the characteristic line model

$$R_{j,t} = \mu_{f,t} + \beta_j(R_{M,t} - \mu_{f,t}) + \epsilon_{j,t}$$

so that

$$R_{P,t} = \mu_{f,t} + (\sum_{j=1}^{N} \beta_j \omega_j)(R_{M,t} - \mu_{f,t}) + \sum_{j=1}^{N} \omega_j \epsilon_{j,t}. \tag{4.16}$$

Therefore, the portfolio beta is

$$\beta_P = \sum_{j=1}^{N} \beta_j \omega_j \tag{4.17}$$

and the epsilon for the portfolio is

$$\epsilon_{P,t} = \sum_{j=1}^{N} \omega_j \epsilon_{j,t} \tag{4.18}$$

assume that $\epsilon_{1,t}, \ldots, \epsilon_{N,t}$ are uncorrelated, then

$$\sigma^2_{\epsilon,P} = \sum_{j=1}^{N} \omega_j^2 \sigma^2_{\epsilon,j}. \tag{4.19}$$

4.7 Possible Negative Beta

Beta can be negative. Beta measures the risk added by an investment to a well-diversified portfolio. By that definition, any asset added to a portfolio, makes the overall risk of the portfolio go down, has a negative beta. A negative beta investment represents insurance against some macroeconomic risk that affects the rest of the portfolio adversely. In fact, adding an asset with a negative beta to a portfolio can reduce the portfolio's risk more than adding a risk-free asset. However, according to the SML, the expected return on an asset with a negative beta is lower than the risk-free rate. A standard example of an asset that often has a negative beta is gold or mining companies, which acts as a hedge against higher inflation.

4.8 Stationarity Test

Before we could estimate the betas of the security, we have to ensure that the Time Series model is in fact stationary by using the Time Series regression tests. This step is very important in our analysis for two important reasons: 1. The stationarity of a time series can strongly influence its behavior and properties. For example, persistence of stocks will be infinite for nonstationary time series.

4.9 Beta Function

The assumption that the beta of a security is constant over time is rather strong assumption to make, especially when the time series chosen contain a large time space. It only becomes natural to assume that beta could vary over time. Beta is not directly observable, in the sense that an asset's beta for a particular holding period cannot be observed. Remember that β_j, the beta of the j-th asset, is defined by

$$R^*_{j,t} = \alpha_j + \beta_j R^*_{M,t} + \epsilon_{j,t}. \tag{4.20}$$

We simplify this equation by assuming that the asset is priced correctly according to the CAPM so that $\alpha_j = 0$. Since both β_j and $\epsilon_{j,t}$ are unknown in this equation, we cannot solve for them. However, we could estimate β_j by regression using a time series of data. However, this gives us a β_j that is constant over the time interval during which the data were collected. Consider a more complex model with time-varying beta:

$$R^*_{j,t} = \beta_{j,t} R^*_{M,t} + \epsilon_{j,t}, \; t = 1, \ldots, n. \tag{4.21}$$

At this juncture, the values $\beta_{j,1}, \ldots, \beta_{j,n}$ cannot be estimated; hence we make a reasonable assumption that $\beta_{j,t}$ varies with time. Here, we assume that $\beta_{j,t}$ is linear in t so that

$$\beta_{j,t} = b_0 + b_1 t, \tag{4.22}$$

where b_0 and b_1 represent the intercept and the slope of the time dependent beta, respectively. To make the model more symmetrical with the time series we use

$$\beta_{j,t} = \beta_j + \gamma_j((t - ((n+1))/2)/(n-1)) \tag{4.23}$$

where $(n+1)/2$ is the midpoint of the time variable $t = 1, \ldots, n$. In this model, the constant term is not the slope, but is the value of beta when t is halfway between the beginning and the end of the time space. The denominator is the length of the discrete time interval $[1, n]$. When $t = 1$, $\gamma_j = -1/2$ and when $t = n$, $\gamma_j = 1/2$. Now, substituting (4.24) in (4.22) we get the model,

$$R^*_{j,t} = \beta_j(R^*_{M,t}) + \gamma_j[R^*_{M,t}((t - ((n+1))/2)/(n-1))] + \epsilon_{j,t}. \tag{4.24}$$

4.10 Multi-Factor Models

The security characteristic line is a regression model (4.12)

$$R_{j,t} = \mu_{f,t} + \beta_j(R_{M,t} - \mu_{f,t}) + \epsilon_{j,t}. \tag{4.25}$$

The variable $(R_{M,t} - \mu_{f,t})$ is called a factor, which is the excess return on the market. In CAPM, the market factor is the only source of risk besides the unsystematic risk of each asset. Factor models generalize CAPM by allowing more factors than simply the market risk and the unique risk of each asset. A multi-factor model is

$$R_{j,t} = \mu_{f,t} + \beta_{0,j} + \beta_{1,j} F_{1,t} \ldots + \beta_{p,j} F_{p,t} + \epsilon_{j,t}, \tag{4.26}$$

where $F_{1,t}, \ldots, F_{p,t}$ are the values of p factors at time t. An example of factors are the stock market average, dividend yield of S&P 500, a measure of the risk of corporate bonds, interest rate variables, employment rate, inflation rate, monthly growth rate in industrial production, etc. The factors are to be correlated among themselves and are meant to simplify and reduce the amount of randomness required in an analysis of the assets. The factors are chosen by the modeler and depend on the type of assets being considered. With enough factors all commonalities between assets should be accounted for in the models. Then the error term should represent factors truly unique to the individual assets and, therefore, should be uncorrelated across j, ($\epsilon_{j,t} = 0$). The multi-factor model is a multiple linear regression model. If a factor model includes the returns on the market portfolio and other factors, then according to the CAPM, the betas on the other factors should be zero, so testing that they are zero also tests the CAPM.

With $p = 2$, one has the two-factor model:

$$R_{j,t} = \mu_{f,t} + \beta_{0,j} + \beta_{1,j}(R_{M,t} - \mu_{f,t}) + \beta_{2,j} F_{2,t} + \epsilon_{j,t}.$$

Fama and French (1995) introduced the Three-Factor Model which is given by

$$R_{j,t} = \mu_{f,t} + \beta_{0,j} + \beta_{1,j}(R_{M,t} - \mu_{f,t}) + \beta_{2,j} F_{2,t} + \beta_{3,j} F_{3,t} + \epsilon_{j,t}, \tag{4.27}$$

where the factors F_2 and F_3 are SML and HML, respectively. The factor SML (small minus large) is the difference in returns of portfolios of small stocks and portfolio of large stocks and the factor HML (high minus low) is the difference in returns on a portfolio of high book-to-market value (BE/ME) stocks and a portfolio of low BE/ME stocks. Multi-factor models refer to model parsimony. A statistical model should have as few parameters as possible like the CAPM model. One reasonable way of having few parameters is good is that each unknown parameter is another quantity that must be estimated and each estimate is a source of estimation error. On the other hand, a model must have enough parameters to adequately describe the behavior of data. A model with too few parameters can create bias as the model does not fit the data well. A model without excess parameters is parsimonious. Large models have less bias but more variability. Models with too few parameters and sizable bias underfit, while model with too many parameters overfit. Thus CAPM models underfit and Fama-French model is parsimonious.

For more details in this area see Ruppert (2004) and Ruppert (2011).

B. Stochastic Calculus

Consider the stochastic basis $(\Omega, \mathcal{F}, \{\mathcal{F}\}_{t\geq 0}, P)$. Let $\{W_t, t \geq 0\}$ be a standard Brownian motion on this stochastic basis.

Itô Integral

Let $\{\gamma_t, 0 \leq t \leq T\}$ be an $\mathbb{R}$- valued adapted process for which

$$P\left(\int_0^T \gamma_t^2 dt < \infty\right) = 1$$

for some fixed T, then the Itô integral $\int_0^T \gamma_t dW_t$ is well defined.

Itô Formula

For the SDE

$$dX_t = a(t, X_t)dt + b(t, X_t)dW_t,$$

for a smooth function $f(t,x)$ the Itô formula is given by

$$\begin{aligned} df(t, X_t) &= [f_t(t, X_t) + f_x(t, X_t)a(t, X_t) + \frac{1}{2} f_{xx}(t, X_t)b^2(t, X_t)]dt \\ &\quad + f_x(t, X_t)b(t, X_t)dW_t \\ &= \mathcal{L}^0 f(t, X_t)dt + \mathcal{L}^1 f(t, X_t)dW_t, \end{aligned}$$

where the Itô operators are given by

$$\mathcal{L}^0 = \frac{\partial}{\partial t} + a\frac{\partial}{\partial x} + \frac{1}{2}b^2\frac{\partial^2}{\partial x^2}, \quad \mathcal{L}^1 = b\frac{\partial}{\partial x}.$$

B.1 *Stochastic Differential Equations*

$$dX_t = a(t, X_t)dt + b(t, X_t)dW_t, 0 \leq t \leq T.$$

Existence and Uniqueness of Solutions:
(a) Lipschitz condition
(b) Linear Growth condition

B.2 *Markov Property*

$$E[f(X_t)|\mathcal{F}_s] = E[f(X_t)|X_s] \ \forall\ 0 \leq s < t < T$$

and all bounded Borel functions $f : \mathbb{R} \to \mathbb{R}$. The process is strong Markov process if this condition continues to hold with s replaced by any stopping time taking values in $[0, T]$.

B.3 *Martingales*

$$E[|X_t|] < \infty \ \forall \ t \geq 0,$$

$$E[X_t|\mathcal{F}_s] = E[X_t|X_s] \ \forall \ 0 \leq s < t < \infty.$$

B.4 *Change of Measure*

If P and Q are equivalent and $E_Q[|X|] < \infty$, then

$$E[X|\mathcal{F}_t] = \left(\frac{dQ}{dP}\right)_t^{-1} E\left[X\frac{dQ}{dP}|\mathcal{F}_t\right].$$

B.5 *Girsanov Formula*

Let

$$X_t = \exp(-\frac{1}{2}\int_0^t \gamma_s^2 ds + \int_0^t \gamma_s dW_s).$$

If $E_P[X_T] = 1$ is a martingale and the measure Q on $(\Omega, \mathcal{F}_T^W)$ defined by

$$\frac{dQ}{dP} = X_T$$

is equivalent to P. Under Q, the process $W_t^Q := W_t - \int_0^t \gamma(u)du$ is a standard Brownian motion with respect to $\mathcal{F}_T^W$.

B.5 *Novikov Condition*

If

$$E[\exp(\frac{1}{2}\int_0^T \gamma_t^2 dt)] < \infty,$$

then X_t in the Girsanov formula is a martingale.

B.6 *Levy Area Formula*

Clearly

$$E\exp(uW_1^2) = \frac{1}{\sqrt{1-2u}}.$$

From Chapter 17 in Liptser and Shiryayev (1977) we have

$$E\exp\left(-\frac{1}{2}\int_0^T W_t^2 dt\right) = \frac{2}{e^T + e^{-T}} = \frac{1}{\sqrt{\cosh T}}.$$

For the Ornstein–Uhlenbeck process

$$dX_t = -\theta X_t dt + dW_t, \ \theta > 0$$

with $I_t := \int_0^T X_t^2 dt$, from Chapter 17 in Liptser and Shiryayev (1977), we have

$$E \exp\left(uT^{-1}2\theta I_T\right) = \exp\left(\frac{\theta T}{2}\right)\left[\frac{2\gamma}{(\gamma-\theta)e^{-\gamma T}+(\gamma+\theta)e^{\gamma T}}\right]^{1/2}$$

$$= \exp\left(\frac{\theta T}{2}\right)\left[\frac{2\gamma}{(\gamma-\theta)e^{-\gamma T}+(\gamma+\theta)e^{\gamma T}}\right]^{1/2},$$

where

$$\gamma := \left(\theta^2 - 2uT^{-1}2\theta\right)^{1/2}.$$

B.7 *Realized Measures*

Denote the squared difference of returns by

$$\Delta s_{t_{i-1}}^2 := (s_{t_i} - s_{t_{i-1}})^2.$$

The realized volatility is defined as

$$R_{n,T} := \sum_{i=1}^{n} \Delta s_{t_{i-1}}^2.$$

It is well known that

$$\operatorname*{P\text{-}lim}_{n\to\infty} R_{n,T} = I_T.$$

See Barndorff-Nielsen and Shephard (2001).

C. Stein's Method

See Chen, Goldstein, and Shao (2011) and Nourdin and Peccati (2012).

D. Fractional Stochastic Calculus

1. Fractional Versus Stable Process

Let $\{W_t^H, t \geq 0\}$ be a fractional Brownian motion with Hurst parameter $H \in (0,1)$. For $H \neq 1/2$, the process is neither a Markov process nor a martingale. For $H > 1/2$, the process displays long memory and for $H < 1/2$, the process displays short memory.

Instead of using fractional Brownian motion, to capture discontinuities and auto correlation, one can use stable densities. Associated with them is an exponent α measuring the frequency of discontinuous jumps, or heavy-tailedness. α ranges from 0 to 2 inclusive, 2 being Gaussian. Moments greater than order α do not exist, so a process with α equal to 1.65, for example, will not have a standard deviation!

(1) The relation with the Hurst exponent is $\alpha = 1/H$. So antipersistent data which fold up on themselves will have α greater than 2, discontinuous and long-memory data will have α less than 2.

(2) In a loose way α can be taken to be the dimension of the time series. Thus a curve is needed to cover a surface, so needs to be as dense as possible thus continuity for Brownian motion ($\alpha = 2$), but need not cover entire surface, as thing being covered has smaller dimension, so discontinuities for α less than 2.

(3) A process with Brownian motion noise will have a derivative price derived by a PDE with time derivative equal to second space derivative. Solving BS amounts to changing to a heat equation: $dU/dt = A*U$, or where U is heat operator (F-K semigroup). The problem is to formulate a Feynman-Kac solution for stable process underlying involves: $dU/dt = ([A^{0.5}]^{\alpha}) * U$: no longer the heat operator, but the α-th power of its square root, or the α-th power of the space derivative.

(4) When α is a fraction less than 2, then one will have fractional operators, and fractional derivatives, so arbitrage is not possible. However, Merton pointed out that most of the time stable or jump-diff processes are better behaved than Brownian motion: one can hedge most of the time, except at rare discontinuities.

(5) Any stable process can be decomposed as a sum of a drift, diffusion, and a compound Poisson process (familiar as trend plus noise plus jumps model of Merton). As a result one can avoid fractional operators and derivatives, and obtain option pricing formula as infinite sums of Brownian motion (B-S) formulae. Another way is to use Zolotarev's infinite sum expansion of the density directly (see Feller (1971), p. 581).

(6) If one watches the dispersion, which is the equivalent of volatility for a stable process, it differs in size, not in kind. Hence using a stable model does not get rid of stochastic volatility, nor smiles. The two are similar topologically (have similar shape), but as α decreases to above 1, the dispersion becomes a greater multiple of volatility. A stable VaR model gives 99% 1-day limits about twice the Gaussian case for α around 1.65, but for 95% 1-day VaR gives slight smaller limits. The tail is heavier, the rump is thinner and has higher peak.

(7) The fundamental question of "Which period one should use to measure historic volatility?" becomes "which periods one should use to measure historic α and dispersion?", doubling the problem by making the problem harder.

(8) However, one can get explicit results on objects like jump probability, so one can calculate the likelihood that market sails past the stop point without hitting it, and gauge liquidity.

(9) Rather than complementing existing models for equity/rates/FX, value added will probably come from new markets like weather, electricity and network traffic.

(10) One can visualize α as the degree to which an asset is marked to market to its driving fundamental. α near 2 is frequent marking to market, near one means pegged currency unpegged or policy instrument (Fed rate). For convergence and fundamental traders one can filter indicators by measuring their α vs the inherited α of the market traded, with acceptance/usage if α decreases, thus filtering out jumps. One can also choose markets/hedge funds this way.

(11) Parsimony: A stable process can be completely specified by 4 parameters measuring discontinuity frequency (α), skewness (β), dispersion (γ), and location (δ). The last two correspond to the diffusion and drift for diffusions. In addition since they are not martingales, a Girsanov type theorem to choose arbitrary drift is not immediately available (discontinuities disallow arbitrage argument replacing drift by riskless rate). So the fifth element, a market price of risk, must be introduced as in pre-martingale bond models. So there are six parameters. A B-S type model will need in addition to spot, strike, rate, dividend, maturity, volatility data, the α and the skewness and market price of risk. Fudges assuming zero skew, zero market price of risk mean just one extra parameter α, and replacing the riskless rate by the location (drift) parameter do the same, but then we have pre-B-S, Samuelson type models which require drift estimation. So for stable processes even using Monte Carlo needs caution.

(12) If one is into martingales judging, there is a mention of stable subordinators in Revuz and Yor (1991). Changing the time parameter in Brownian motion to a new time scale with stable exponent $\alpha/2$, then the time-changed Brownian motion will be stable with exponent α. This has been rediscovered by Geman, in her articles on time-changed BM (clock change). To get equations one notes that one is dealing with subordinated semigroups, one of which is BM with heat kernel, and the other (the time) is a compound Poisson type. So to get the operator for the time-changed BM one has to integrate heat operators w.r.t. the directing process τ_t. Then the F-K type formula will be $dU/dt = Au$, where A is the operator obtained by the former integration. This is analogous to Merton's sum of B-S formulae for jump-diffusions.

(13) A way to get intuition is by noting the forward equation for a compound Poisson process will be given by time derivative equal to integral over jumps, and tacking on diffusion terms (second space derivative), drift terms (first space derivative), and killing terms (first order term in u as in Feynman-Kac). One can do this because of the Levy-Khintchine theorem allows one to represent any Levy process as a sum of a diffusion, drift, compound Poisson integral, and killing term.

2. Fractional Brownian Calculus

2.1 Fractional Brownian Motion

The fractional Brownian motion (fBm, in short), which provides a suitable generalization of the Brownian motion, is one of the simplest stochastic processes exhibiting long range dependence. It was introduced by Kolmogorov (1940) in a Hilbert space framework and later on studied by Levy (1948) and in detail by Mandelbrot and Van Ness (1968).

Consider a probability space $(\Omega, \mathcal{F}, \mathbb{P})$ on which all random variables and processes below are defined.

A fractional Brownian motion $\{W_t^H, t \geq 0\}$ with Hurst parameter $H \in (0,1)$ is a centered Gaussian process with continuous sample paths whose covariance kernel is given by

$$E(W_t^H W_s^H) = \frac{V_H}{2}(s^{2H} + t^{2H} - |t-s|^{2H}), \quad s,t \geq 0,$$

where

$$V_H := var(W_1^H) = \frac{1}{[\Gamma(H+\frac{1}{2})]^2}\left\{\frac{1}{2H} + \int_1^\infty \left[u^{H-\frac{1}{2}} - (u-1)^{H-\frac{1}{2}}\right]^2 du\right\}.$$

With $V_H = 1$, fBm is called a *normalized fBm.*

Properties

(1) It has stationary increments: $E(W_t^H - W_s^H)^2 = |t-s|^{2H}, t,s \geq 0$.

(2) $W_0^H = 0, E(W_t^H) = 0, E(W_t)^2 = |t|^{2H}, t \geq 0$.

(3) When $H = \frac{1}{2}, W_t^{\frac{1}{2}}$ is the standard Brownian motion. The increments are independent.

(4) The process is self-similar or scale invariant, i.e.,

$$(W_{\alpha t}^H, t \geq 0) =^d (\alpha^H W_t^H, t \geq 0), \alpha > 0.$$

H is also called the self-similarity parameter.

(5) The increments of the fBm are negatively correlated for $H < \frac{1}{2}$ and positively correlated for $H > \frac{1}{2}$.

(6) For $H > \frac{1}{2}$, fBm is a long-memory process since the covariance between far apart increments decrease to zero as a power law:

$$r(n) := E[W_1^H(W_{1+n}^H - W_n^H)] \sim C_H n^{2H-2}, \quad \sum_{n=1}^\infty r(n) = \infty.$$

This property is also called long-range dependence. The parameter H measures the intensity of the long-range dependence.

(7) The sample paths of W^H are almost surely Holder continuous of any order less than H, but not Holder continuous of any order greater than H, hence continuous but nowhere differentiable.

(8) For any H, it has a finite $\frac{1}{H}$ variation, i.e.,

$$0 < \sup_{\Pi} E \sum_{t_i \in \Pi} \left[\left| W^H_{t_{i+1}} - W^H_{t_i} \right|^{\frac{1}{H}} \right] < \infty.$$

(9) Law of the Iterated Logarithm (Arcones (1995)):

$$P\left(\overline{\lim}_{t \to 0+} \frac{W^H_t}{t^H (\log \log t^{-1})^{\frac{1}{2}}} = \sqrt{V_H} \right) = 1.$$

Self-similarity of fBm leads to

$$P\left(\overline{\lim}_{t \to 0+} \frac{W^H_{\frac{1}{t}}}{(\log \log t^{-1})^{\frac{1}{2}}} = \sqrt{V_H} \right) = 1.$$

Setting $u = \frac{1}{t}$,

$$P\left(\overline{\lim}_{u \to \infty} \frac{W^H_u}{u^H (\log \log u^{-1})^{\frac{1}{2}}} = \sqrt{V_H} \right) = 1.$$

Strong Law of Large Numbers:

$$\lim_{u \to \infty} \frac{W^H_u}{u} = 0 \ \ a.s.$$

(10) fBm can be represented as a stochastic integral with respect to standard Brownian motion B (Mandelbrot and Van Ness, 1968). For $H > \frac{1}{2}$,

$$W^H_t = \frac{1}{\Gamma(H + \frac{1}{2})} \left\{ \int_{-\infty}^{0} [(t-s)^{H-\frac{1}{2}} - (-s)^{H-\frac{1}{2}}] dB_s + \int_0^t (t-s)^{H-\frac{1}{2}} dB_s \right\}.$$

Standard Brownian motion can be written as a stochastic integral w.r.t W^H_t (see, Igloi and Terdik (1999)):

$$B_t = \frac{1}{\Gamma(\frac{3}{2} - H)} \left\{ \int_{-\infty}^{0} [(t-s)^{-H+\frac{1}{2}} - (-s)^{-H+\frac{1}{2}}] dW^H_s + \int_0^t (t-s)^{-H+\frac{1}{2}} dW^H_s \right\}.$$

(11) With topological dimension n, the fractal dimension of fBm is $n + 1 - H$. Hausdorff dimension of one dimensional fBm is $2 - H$.

(12) *Existence of fBm*:

(i) It can be defined by a stochastic integral w.r.t. Brownian motion.

(ii) It can be constructed by Kolmogorov extension theorem (see, Samorodnitsky and Taqqu (1994)).

(iii) It can be defined as the weak limit of some random walks with strong correlations (see, Taqqu (1975)).

(13) For $H \neq \frac{1}{2}$, the fBm is not a semimartingale and not a Markov process, but a Dirichlet process.

(14) *Dirichlet Process*: A process is called a Dirichlet process if it can be decomposed as the sum of a local martingale and an adapted process of zero quadratic variation (zero energy). Obviously it is a larger class of processes than semimartingales.

For $H < \frac{1}{2}$, the quadratic variation of W^H is infinite. For $H > \frac{1}{2}$, the quadratic variation of W^H is zero. Hence for $H > \frac{1}{2}$, W^H is a Dirichlet process.

Note that the estimation of the parameter H based on observation of fractional Brownian motion has already been paid some attention, see, e.g., Peltier and Levy Vehel (1994) and the references therein. However, we assume H to be known.

2.2 Stochastic Integral w.r.t. fBm

For $H \neq \frac{1}{2}$, the classical theory of stochastic integration with respect to semimartingales is not applicable to stochastic integration with respect to fBm. Since fBm is a Gaussian process, stochastic integration with respect to Gaussian process is applicable.

For integration questions related to fractional Brownian motion, see Pipiras and Taqqu (2000). Now there exist several approaches to stochastic integration with respect to fBm.

(i) Classical Riemann sum approach: Lin (1995), Dai and Heyde (1996), and Kleptsyna et al. (1998c)

(ii) Malliavin calculus approach: Decreusefond and Ustunel (1998, 1999), Coutin and Decreusefond (1999), and Alos et al. (2000, 2001)

(iii) Wick calculus approach: Duncan et al. (2000)

(iv) Pathwise calculus: Zahle (1998, 1999), Ruzmaikina (2000)

(v) Dirichlet calculus: Lyons and Zhang (1994)

(vi) Rough path analysis: Lyons (1998)

Lin (1995) introduced the stochastic integral as follows: Let ϕ be a left continuous bounded Lebesgue measurable function with right limits, called *sure processes.* Then

$$\int_0^1 \psi(t)dW_t^H = \text{l.i.m.}_{|\pi|\to\infty} \sum_{t_i\in\pi} \psi(t_{i-1})(W_{t_i}^H - W_{t_{i-1}}^H).$$

The indefinite integral is defined as

$$\int_0^t \psi(s)dW_s^H = \int_0^1 \psi(t)I_{[0,t]}dW_t^H.$$

This integral has a continuous version and a Gaussian process. However,

$$E\left(\int_0^t \psi(s)dW_s^H\right) \neq 0.$$

To overcome this situation, Duncan, Hu, and Pasik-Duncan (2000) introduced an integral using *Wick calculus* for which

$$E\left(\int_0^t f(s)dW_s^H\right) = 0.$$

They defined integrals of both Itô and Stratonovich type.

We shall discuss the Wick calculus approach here. Wiener integral for deterministic kernel was defined by Gripenberg and Norros (1996).

Let $\phi : \mathbb{R}_+ \times \mathbb{R} \to \mathbb{R}$ be a Borel measurable deterministic function. Let

$$L^2_\phi(\mathbb{R}_+) := \left\{ f : |f|^2_\phi : \int_0^\infty \int_0^\infty f(s)f(t)\phi(s,t)dsdt < \infty \right\}.$$

The inner product in the Hilbert space L^2_ϕ is denoted by $\langle \cdot, \cdot \rangle_\phi$.

If $f, g \in L^2_\phi$, then $\int_0^\infty f_s dW_s^H$ and $\int_0^\infty g_s dW_s^H$ are well defined zero mean, Gaussian random variables with variances $|f|^2_\phi$ and $|g|^2_\phi$, respectively, and covariance

$$E\left(\int_0^\infty f_s dW_s^H \int_0^\infty g_s dW_s^H\right) = \int_0^\infty \int_0^\infty f_s g_s \phi(s,t)dsdt =: \langle f, g\rangle_\phi.$$

Let $(\Omega, \mathcal{F}, P)$ be the probability space on which W^H is defined. For $f \in L^2_\phi$, define $\epsilon : L^2_\phi \to L^1(\Omega, \mathcal{F}, P)$ as

$$\begin{aligned}\epsilon(f) &:= \exp\left\{\int_0^\infty f_t dW_t^H - \frac{1}{2}\int_0^\infty \int_0^\infty f_s f_t \phi(s,t)dsdt\right\}\\ &= \exp\left\{\int_0^\infty f_t dW_t^H - \frac{1}{2}\int_0^\infty |f|^2_\phi\right\}\end{aligned}$$

which is called an exponential function.

Let $\mathcal{E}$ be the linear span of exponentials, i.e.,

$$\mathcal{E} = \{\sum_{k=1}^n a_k \epsilon(f_k) : n \in \mathbb{N}, a_k \in \mathbb{R}, f_k \in L^2_\phi(\mathbb{R}_+), k = \overline{1,n}\}.$$

The Wick product of two exponentials is defined as

$$\epsilon(f) \diamond \epsilon(g) = \epsilon(f+g).$$

For distinct $f_1, f_2, \cdots, f_n \in L^2_\phi, \epsilon(f_1), \epsilon(f_2), \cdots, \epsilon(f_n)$ are independent. It can be extended to define the Wick product of two functionals F and G in $\mathcal{E}$.

An Analogue of Malliavin Derivative: Wick Derivative

The ϕ-derivative of a random variable $F \in L^p$ in the direction of Φg where $g \in L^2_\phi$ is defined as

$$D_{\Phi g}F(\omega) = \lim_{\delta \to 0} \frac{1}{\delta}\left[F\left(\omega + \delta \int_0^{\cdot} (\Phi g)(u)du\right) - F(\omega)\right]$$

if the limit exists in $L^p(\Omega, \mathcal{F}, P)$.

If there is a process $(D^\phi F_s, s \geq 0)$ such that

$$D_{\Phi g}F = \int_0^\infty D^\phi F_s g_s ds \quad a.s.$$

for all $g \in L^2_\phi$, then F is said to be ϕ-differentiable. Let $F : [0, T] \times \Omega \to \mathbb{R}$ be a stochastic process. The process is said to be ϕ-differentiable if for each $t \in [0, T], F(t, \cdot)$ is ϕ-differentiable and $D^\phi_s F_t$ is jointly measurable.

Chain Rule If $f : \mathbb{R} \to \mathbb{R}$ is smooth and $F : \Omega \to \mathbb{R}$ is $\phi-$ differentiable, then $f(F)$ is also $\phi-$ differentiable and

$$D_{\Phi g}f(F) = f'(F)D_{\phi g}F$$

and

$$D^\phi_s f(F) = f'(F)D^\phi_s(F).$$

(1) If $g \in L^2_\phi, F \in L^2(\Omega, \mathcal{F}, P)$ and $D_{\Phi g}F \in L^2(\Omega, \mathcal{F}, P)$, then

$$F \diamond \int_0^\infty g_s dW_s^H = F \int_0^\infty g_s dW_s^H - D_{\Phi g}F.$$

(2) If $g, h \in L^2_\phi$ and $F, G \in \mathcal{E}$, then

$$E\left(F \diamond \int_0^\infty g_s dW_s^H \;\; G \diamond \int_0^\infty h_s dW_s^H\right) = E\left[D_{\Phi g}F D_{\Phi h}G + FG\langle g, h\rangle_\phi\right].$$

Let $\pi_n : 0 < t_1 < t_2 < \cdots < t_n = T$.

Let $\mathcal{L}[0, T]$ be the family of stochastic processes on F on $[0, T]$ such that $E|F|^2_\phi < \infty, F$ is $\phi-$ differentiable, the trace of $(D^\phi_s F_t, 0 \leq s \leq T, 0 \leq t \leq T)$ exists and $E\int_0^T (D^\phi_s F_s)^2 ds < \infty$ and for each sequence of partitions $\{\pi_n, n \in \mathbb{N}\}$ such that $|\pi_n| \to 0$

$$\sum_{i=0}^{n-1} E\left\{\int_{t_i^{(n)}}^{t_{i+1}^{(n)}} |D^\phi_s F^\pi_{t_i^{(n)}} - D^\phi_s F_s| ds\right\}^2 \to 0$$

and $E|F^\pi - F|^2_\phi \to 0$ as $n \to \infty$.

For $F \in \mathcal{L}[0,T]$, define

$$\int_0^T F_s dW_s^H = l.i.m._{|\pi|\to 0} \sum_{i=0}^{n-1} F_{t_i} \diamond (W_{t_{i+1}}^H - W_{t_i}^H).$$

Proposition 2.2.1 Let $F, G, \in \mathcal{L}[0,T]$. Then
(i) $E\left(\int_0^T F_s dW_s^H\right) = 0.$
(ii) $E\left(\int_0^T F_s dW_s^H\right)^2 = E\left\{\left(D_s^\phi F_s ds\right)^2 + \left|I_{[0,T]}F\right|_\phi^2\right\}.$
(iii) $\int_o^t (aF_s + bG_s)dW_s^H = a\int_0^t F_s dW_s^H + b\int_0^t G_s dW_s^H a.s.$
(iv) If $E\left[\sup_{0\le s\le T} F_s\right]^2 < \infty$ and $\sup_{0\le s\le T} E|D_s^\phi F_s|^2 < \infty$, then $\{\int_0^t F_s dW_s^H,$ $0 \le t \le T\}$ has a continuous version.

Here it is not assumed that $(F_s, s \in [0,T])$ is adapted to the fBm. Assume that $D_s^\phi F_s = 0, s \in [0,T]$. Then
(v) $E\left(\int_0^T F_s dW_s^H\right)^2 = \left|I_{[0,T]}F\right|_\phi^2 = E\int_0^T\int_0^T F_u F_v \phi(u,v)dudv.$

Fractional version of *Stratonovich Integral* is defined as

$$\int_0^t F_s \delta W_s^H := \int_0^t F_s dW_s^H + \int_0^t D_s^\phi F_s ds \quad a.s.$$

2.3 Fractional Itô Formula

Lemma 2.3

(a) If $f : \mathbb{R} \to \mathbb{R}$ is a twice continuously differentiable function with bounded derivatives of order two, then

$$f(W_T^H) - f(W_0^H) = \int_0^T f'(W_s^H)dW_s^H + H\int_0^T s^{2H-2} f''(W_s^H)ds \quad a.s.$$

For $H = \frac{1}{2}$, it gives the classical Itô formula for standard Brownian motion.

General Itô Formula

(b) Let $\{F_u, 0 \le u \le T\}$ and $\{G_u, 0 \le u \le T\}$ be stochastic processes in $\mathcal{L}[0,T]$. Assume that there exists an $\alpha > 1 - H$ such that

$$E|F_u - F_v|^2 \le C|u-v|^{2\alpha},$$

$$\lim_{|u-v|\to 0} E\{|D_u^\phi(F_u - F_v)|^2\} = 0$$

and

$$E \sup_{0\le s\le T} |G_s| < \infty.$$

Let

$$dX_t = G_t dt + F_t dW_t^H, X_0 = \xi \in \mathbb{R}, 0 \le t \le T,$$

i.e.,

$$X_t = \xi + \int_0^t G_s ds + \int_0^t F_s dW_s^H.$$

Let $f : \mathbb{R} \to \mathbb{R}$ be C_b^1 in the first variable and C_b^2 in the second variable and let $\left(\frac{\partial f}{\partial x}(s, X_s), s \in [0,T]\right) \in \mathcal{L}[0,T]$. Then

$$\begin{aligned} & f(t, X_t) \\ = \; & f(0,\xi) + \int_0^t \frac{\partial f}{\partial s}(s, X_s)ds + \int_0^t \frac{\partial f}{\partial x}(s, X_s)G_s ds + \int_0^t \frac{\partial f}{\partial x}(s, X_s)F_s dW_s^H \\ & + \int_0^t \frac{\partial^2 f}{\partial x^2}(s, X_s)F_s D_s^\phi X_s ds. \end{aligned}$$

Itô formula for Stratonovich Type Integral

(c) Let $\{F_t, 0 \le t \le T\}$ satisfy the above assumptions. Let

$$\eta_t = \int_0^t F_s \delta W_s^H.$$

Let $g \in C_b^2$ and $\left(\frac{\partial g}{\partial x}(s, \eta_s)F_s, s \in [0,T]\right) \in \mathcal{L}[0,T]$.

Then for $t \in [0,T]$,

$$g(t, \eta_t) = g(0,0) + \int_0^t \frac{\partial g}{\partial s}(s, \eta_s)ds + \int_0^t \frac{\partial g}{\partial x}(s, \eta_s)F_s dW_s^H,$$

i.e.,

$$\delta g(t, \eta_t) = g_t(t, \eta_t)dt + g_x(t, \eta_t)d\eta_t.$$

E. Statistics of Diffusions

E.1 Frequentist Methods

See Liptser and Shiryayev (1978), Kutoyants (1984a,b), Kutoyants (1994a,b), Kutoyants (2004), Bishwal (2008a), Iacus (2008), and Iacus and Yoshida (2018).

E.2 Bayesian Methods

See Bishwal (2008a).

E.3 Jacobi Beta Process

The Jacobi process $(V_t, t \geq 0)$ satisfies the SDE

$$dV_t = (\alpha - \beta V_t)dt + \sigma\sqrt{V_t(1 - V_t)}dW_t,$$

where α, β, and σ are positive constants and W is a standard Brownian motion. This process takes values in $[0, 1]$ and is mean reverting to $\alpha/beta$ at speed β. The invariant distribution of the process is beta distribution. For proof regarding existence of a solution to the equation, see Warren and Yor (1998). See Veraart and Veraart (2012) for more details on this model.

Bibliography

Abergel, F., Anane, M., Chakraborti, A., Jedidi, I.M. (2016): *Limit Order Books*, Cambridge University Press, Delhi.

Adler, R., Muller, P. and Rozovskii, B.L. (1996): *Stochastic Modeling in Physical Oceanography*, Birkhauser, Boston - Berlin.

Aihara, S.I. (1992) Regularized maximum likelihood estimate for an infinite dimensional parameter in stochastic parabolic systems, *SIAM J. Cont. Optim.* **30**, 745–764.

Aihara, S.I. (1994): Maximum likelihood estimate for discontinuous parameter in stochastic hyperbolic systems, *Acta Applicandae Mathematicae* **35**, 131–151.

Aihara, S.I. (1998): Identification of a discontinuous parameter in stochastic parabolic systems, *Applied Math. Optimization* **37**, 43–69.

Aihara, S.I. and Bagchi, A. (1988): Parameter identification for stochastic diffusion equations with unknown boundary conditions, *App. Math. Optimization* **17**, 15–36.

Aihara, S.I. and Bagchi, A. (1989): Infinite dimensional parameter identification for stochastic parabolic systems, *Statist. Probab. Letters* **8**, 279–287.

Aihara, S.I. and Bagchi, A. (1991): Parameter identification for hyperbolic stochastic systems, *J. Math. Anal. Appl.* **160**, 485–499.

Ait-Sahalia, Y. (1996): Testing continuous time models of the spot interest rates, *Review of Financial Studies* **9**, 385–426.

Ait-Sahalia, Y. (1999): Transition densities for interest rates and other nonlinear diffusions, *Journal of Finance* **54**, 1361–1395.

Ait-Sahalia, Y. (2002): Maximum likelihood estimation of discretely sampled diffusions: a closed form approximation approach, *Econometrica* **70**, 223–262.

J. P. N. Bishwal, *Parameter Estimation in Stochastic Volatility Models*,
https://doi.org/10.1007/978-3-031-03861-7

Ait-Sahalia, Y. and Jacod, J. (2014): *High-Frequency Financial Econometrics*, Princeton University Press, Princeton and Oxford.

Ait-Sahalia, Y. and Jacod, J. (2018): Semimartingale: Itô or not?, *Stochastic Processes and their Applications* **128**, 233–254.

Ait-Sahalia, Y. and Hansen, L. (2009) *Handbook of Financial Econometrics, Volume I*, Elsevier Science.

Ait-Sahalia, Y. and Hansen, L. (2010): *Handbook of Financial Econometrics, Volume II*, Elsevier Science.

Ait-Sahalia, Y., Mykland, P.A. and Zhang, L. (2011): Ultra high frequency volatility estimation with dependent microstructure noise, *Journal of Econometrics* **160**, 160–175.

Ait-Sahalia, Y., Mykland, P.A. and Zhang, L. (2005): How often to sample a continuous time process in the presence of market microstructure noise, *Review of Financial Studies* **18**, 351–416.

Aldous, D.J. and Eagleson, G.K. (1978): On mixing and stability of limit theorems, *Ann. Probab.* **6**, 325–331.

Alfonsi, A. (2005): On the discretization schemes for the CIR (and Bessel squared) processes, *Monte Carlo Methods and Applications* **11**, 355–384.

Alfonsi, A. (2010): High order discretization schemes for the CIR process: Application to affine term structure and Heston models, *Mathematics of Computation* **79**, 209–237.

Alfonsi, A. (2013): Strong order one convergence of a drift implicit Euler scheme: Application to CIR process, *Statistics and Probability Letters* **83** (2), 602–607.

Al Gerbi, A., Jourdain, B. and Clement, E. (2018): Asymptotics for the normalized error of the Ninomiya-Victoir scheme, *Stochastic Processes and their Applications* **128**, 1889–1928.

Alos, E., Mazet, O. and Nualart, D. (2000): Stochastic calculus with respect to fractional Brownian motion with Hurst parameter less than $\frac{1}{2}$, *Stoch. Proc. Appl.* **86** (1), 121–139.

Alos, E., Mazet, O. and Nualart, D. (2001): Stochastic calculus with respect to Gaussian processes, *Annals of Probability* **29** (2), 766–801.

Anscombe, F.J. (1952): Large sample theory of sequential estimation, *Proc. Camb. Phil. Soc.* **48**, 600–607.

Applebaum, D. (2009): *Levy Processes and Stochastic Calculus*, Second Edition, Cambridge University Press, New York.

Arato, M. (1978): On the statistical examination of continuous state Markov processes, I, II, III, IV, *Selected Transl. in Math. Statist. and Probability* **14**, 203–287, American Mathematical Society, Providence.

Arato, M. (1982): *Linear Stochastic Systems with Constant Coefficients: A Statistical Approach*, Lecture Notes in Control and Information Sciences **45**, Springer-Verlag, New York.

Arato, M. Kolmogorov, A.N. and Sinai, Ya. G. (1962): On estimation of parameters of complex stationary Gaussian processes, *Doklady of Academy of Sciences* **146**(4), 747–750.

Arcones, M. (1995): On the law of logarithm for Gaussian processes, *Journal of Theoretical Probability*, **8**, 877–903.

Arnold, L. (1974): *Stochastic Differential Equations*, Wiley, New York.

Bagchi, A. and Borkar, V. (1984): Parameter identification in infinite dimensional linear systems, *Stochastics* **12**, 201–213.

Bahadur, R.R., Zabell, S.L. and Gupta, J.C. (1980): Large deviation, tests and estimates, In: *Asymptotic Theory of Statistical Tests* (Ed.) I.M. Chakravarti, Academic Press, New York, 33–64.

Bain, A. and Crisan, D. (2009): *Fundamentals of Stochastic Filtering*, Springer, New York.

Bally, V., Caramellino, L. and Cont, R. (2016): *Stochastic Integration by Parts and Functional Itô Calculus*, Springer, Switzerland.

Bally, V. and Talay, D. (1996): The law of the Euler scheme for stochastic differential equations I: Convergence rate of the distribution function, *Probability Theory and Related Fields* **104**, 43–60.

Bandi, F. and Russel, J. (2008): Microstructure noise, realized variance and optimal sampling, *Review of Economic Studies* **75**, 339–369.

Barbour, A.D. and Hall, P. (1984): Stein's method and the Berry-Esseen theorem, *Australian Journal of Statistics* **26** (1), 8–15.

Barczy, M., Ben Alaya, M., Kebaier, A. and Pap, G. (2018): Asymptotic properties of maximum likelihood estimator for the growth rate for a jump-type CIR process based on continuous time observations, *Stochastic Processes and their Applications* **128**, 1135–1164.

Barndorff-Nielsen, O.E. and Schmiegel, J. (2009): Brownian semistationary processes and volatility/intermittency, *Radon Series Comp. Appl. Math.* **8**, 1–26.

Barndorff-Nielsen, O.E. and Shephard, N. (2002): Normal modified stable processes, *Theory of Probability and Mathematical Statistics*, **65**,7–20.

Barndorff-Nielsen, O.E. and Sørensen, M. (1994): A review of some aspects of asymptotic likelihood theory for stochastic processes, *Int. Statist. Review* **62**, 133–165.

Barndorff-Nielsen, O.E. and Shephard, N. (2001): Non-Gaussian Ornstein-Uhlenbeck-based models and some of their uses in financial economics (with discussion), *Journal of the Royal Statistical Society,* Series B, **63**, 167–241.

Barndorff-Nielsen, O.E. and Shephard, N. (2002a): Econometric analysis of realized volatility and its use in estimating stochastic volatility models, *J. Royal. Statist. Society*, Series B, **64**, 253–280.

Barndorff-Nielsen, O.E. and Shephard, N. (2002b): How accurate is the asymtotic approximation to the distribution of realised variance? In: *Identification and Inference for Econometric Models* (Festschrift for Thomas J. Rothenberg) , Eds. D.W.F. Andrews, J.L. Powell, P.A. Ruud, J.H. Stock, Cambridge University Press.

Barndorff-Nielsen, O.E. and Shephard, N. (2002c): Estimating quadratic variance using realised variance, *Journal of Applied Econometrics* **17**, 457–477.

Barndorff-Nielsen, O.E. and Shephard, N. (2002d) :Econometric analysis of realized volatility and its use in estimating stochastic volatility models, *Journal of the Royal Statistical Society,* Series B **64**, 253–280.

Barndorff-Nielsen, O.E. and Shephard, N. (2003): Realised power variation and stochastic volatility, *Bernoulli* **9**, 243–265.

Barndorff-Nielsen, O.E. and Shephard, N. (2004a): Econometric analysis of realised covariation: high frequency based covariance, regression and correlation in financial economics, *Econometrica* **72**, 885–925.

Barndorff-Nielsen, O.E. and Shephard, N. (2004b): Power and bipower variation with stochastic volatility and jumps (with discussion), *Journal of Financial Econometrics* **2**, 1–48.

Barndorff-Nielsen, O.E. and Shephard, N. (2005a): Impact of jumps on returns and realised variances: econometric analysis of time-deformed Lvy processes, *Journal of Econometrics*, **131**, 217–252.

Barndorff-Nielsen, O.E. and Shephard, N. (2005b): Power variation and time change, *Theory of Probability and Its Applications* **50**, 1–15.

Barndorff-Nielsen, O.E. and Shephard, N. (2005c): Multipower variation and stochastic volatility, In: M. do Rosario Grossinho, A.N. Shiryayev, M.L. Esquivel and P.E. Oliveira, *Stochastic Finance*, Springer, New York, 73-82.

Barndorff-Nielsen, O.E. and Shephard, N. (2006): Econometrics of testing for jumps in financial economics using bipower variation, *Journal of Financial Econometrics* **4**, 1–30.

Barndorff-Nielsen, O.E. and Shephard, N. (2006a): Multipower variation and stochastic volatility, In *Stochastic Finance,* Eds. A.N. Shiryayev, M.R. Grossinho, P.E. Olieviera, M.L. Esquivel, Springer, Boston, MA, 73–82.

Barndorff-Nielsen, O.E. and Shephard, N. (2006b): Econometrics of testing for jumps in financial economics using bipower variation, *Journal of Financial Econometrics* **4** (1), 1–30.

Barndorff-Nielsen, O.E. and Shephard, N. (2006c): Power variation and time change, *Theory of Probability and Its Applications* **50** (1), 1–15.

Barndorff-Nielsen, O.E. and Shephard, N. (2006d): Impact of jumps on returns and realised variances: econometric analysis of time-deformed Lvy processes, *Journal of Econometrics.* **131**, (1–2), 217–252.

Barndorff-Nielsen, O.E. and Shephard, N. (2007): Variations, jumps, market frictions and high frequency data in financial econometrics, In: *Advances in Economics and Econometrics*, Eds. R. Bundell, W. Newey, T. Persson., Oxford University Press, 328–372.

Barndorff-Nielsen, O.E., Gravensen, S.E., Jacod, J., and Podolskij, M., Shephard, N. (2005): A central limit theorem for realised power and bipower variations of continuous semimartingales, In: (Eds.) Y. Kabanov and R. Lipster, *From Stochastic Analysis to Mathematical Finance, Festschrift for Albert Shiryaev*, Springer., 33–68.

Barndorff-Nielsen, O.E., Graversen, S.E., Shephard, N. (2004): Power variation and stochastic volatility: A review and some new results, *Journal of Applied Probability* **41**, 133–143.

Barndorff-Nielsen, O.E., Graversen, S.E., Jacod, J., Shephard, N. (2006): Limit theorems for bipower variation in financial econometrics, *Econometric Theory* **22**, 677–719.

Barndorff-Nielsen, O.E. and Schmiegel, J. (2009): Brownian semistationary processes and volatility/intermittency, *Radon Series Comp. Appl. Math.* **8**, 1–26.

Barndorff-Nielsen, O.E. and Shiryayev, A. (2015): *Change of Time and Change of Measure*, World Scientific, Singapore.

Barrieu, P., Rouault, A. and Yor, M. (2004): A study of the Hartman-Watson distribution motivated by numerical provlems related to the pricing of Asian options, *Journal of Applied Probability* **41**, 1049–1058.

Basawa, I.V. and Prakasa Rao, B.L.S. (1980): *Statistical Inference for Stochastic Processes*, Academic Press, New York-London.

Basawa, I.V. and Scott, D.J. (1983): *Asymptotic Optimal Inference for Non-ergodic Models*, Lecture Notes in Statistics **17**, Springer-Verlag, Berlin.

Basu, A. (1983a): Asymptotic theory of estimation in non-linear stochastic differential equations for the multiparameter case, *Sankhyā Ser. A* **45**, 56–65.

Basu, A. (1983b): The Bernstein-von Mises theorem for a certain class of diffusion processes, *Sankhyā Ser. A* **45**, 150–160.

Bauer, K. (1980): On asymptotic properties of estimates of the drift coefficient parameter of a diffusion process, *Theor. Probab. Appl.* **25**, 429–431.

Bender, C. and Marquardt, T. (2007): Stochastic calculus for convoluted Levy processes, *Bernoulli* **14**, 499–518.

Bellach, B. (1980): Consistency, asymptotic normality and asymptotic efficiency of the maximum likelihood estimator in linear stochastic differential equations, *Math. Operationsforch. Statist., Ser. Statistics* **11**, 227–266.

Bellach, B. (1983): Parameter estimation in linear stochastic differential equations and their asymptotic properties, *Math. Operationsforch. Statist. Ser. Statistics* **14**, 141–191.

Bender, C. and Marquardt, T. (2008): Stochastic calculus for convoluted Levy processes, *Bernoulli* **14**, 499–518.

Bentkus, V., Pap, G. and Yor, M. (2003): Optimal bounds for Cauchy approximations for the winding distribution of planar Brownian motion. *Journal of Theoretical Probability* **16** (2), 345–361.

Beran, R. (1994): *Statistics for Long-Memory Processes*, Chapman & Hall: New York.

Bercu, B., Coutin, L. and Savy, N. (2009): Sharp large deviations for the fractional Ornetsin-Uhlenbeck process, *Theory Probab. Appl.* **55** (4), 575–610.

Bertin, K., Torres, S. and Tudor, C.A. (2011): Drift parameter estimation in fractional diffusions driven by perturbed random walks, *Statistics and Probability Letters* **81**, 243–249.

Berger, J.O. (1985): *Statistical Decision Theory and Bayesian Analysis*, Springer-Verlag, New York.

Bergstrom, A.R. (ed.) (1976): *Statistical Inference in Continuous Time Economic Models*, North-Holland, Amsterdam.

Bergstrom, A.R. (1988): The history of continuous time econometric models, *Econometric Theory* **4**, 365–383.

Bertoin, J. (1996): *Levy Processes*, Cambridge University Press, Cambridge.

Bertoin, J. (2019): Ergodic aspects of some Ornstein-Uhlenbeck type processes related to Levy processes, *Stochastic Processes and their Applications* **129**, 1443–1454.

Beskos, A. and Roberts, G.O. (2005): Exact simulation of diffusions, *Annals of Applied Probabability* **15** (4), 2422–2444.

Benth, F.E. (2003): On arbitrage-free pricing of weather derivatives based on fractional Brownian motion, *Applied Mathematical Finance* **10**, 303–324.

Bibby, B.M. and Sørensen, M. (1995a): Martingale estimation functions for discretely observed diffusion processes, *Bernoulli* **1**, 17–39.

Biagini, F, Hu, Y., Oksendal, B. and Zhang, T. (2008): *Stochastic Calculus for Fractional Brownian Motion and Applications*, Springer-Verlag, London.

Bibby, B.M. and Sørensen, M. (1995b): On estimation for discretely observed diffusions: A review, *Theory of Stochastic Processes* **2** (18), 49–56.

Bibinger, M., Jirac, M. and Reiss, M. (2015): Volatility estimation under one-sided errors with applications to limit order books, *Annals of Applied Probability* **26** (5), 2754–2790.

Bilecki, T.R. and Rutkowski, M. (2004): *Credit Risk: Modeling, Valuation and Hedging*, Springer-Verlag, Berlin.

Billingsley, P. (1961): *Statistical Inference for Markov Processes*, Chicago University Press, Chicago.

Bishwal, J.P.N. (1999a): Large deviations inequalities for the maximum likelihood estimator and the Bayes estimators in the nonlinear stochastic differential equations, *Statistics and Probability Letters* **43** (2), 207–215.

Bishwal, J.P.N. (1999b): Bayes and sequential estimation in Hilbert space valued stochastic differential equations, *Journal of the Korean Statistical Society* **28** (1), 96–108.

Bishwal, J.P.N. (2000a): Sharp Berry-Esseen bound for the maximum likelihood estimator in the Ornstein-Uhlenbeck process, *Sankhya,* Series A **62** (1), 1–10.

Bishwal, J.P.N. (2000b): Rates of convergence of the posterior distributions and the Bayes estimators in the Ornstein- Uhlenbeck process, *Random Operators and Stochastic Equations* **8** (1), 51–70.

Bishwal, J.P.N. (2001): Accuracy of normal approximation for the maximum likelihood and the Bayes estimators in the Ornstein-Uhlenbeck process using random norming, *Statistics and Probability Letters* **52**, 427–439.

Bishwal, J.P.N. (2002): The Bernstein-von Mises theorem and spectral asymptotics of Bayes estimators for parabolic SPDEs, *Journal of the Australian Mathematical Society* **72** (2), 289–300.

Bishwal, J.P.N. (2003): Maximum likelihood estimation in partially observed stochastic differential system driven by fractional Brownian motion, *Stochastic Analysis and Applications* **21** (5), 995–1007.

Bishwal, J.P.N. (2006a): Rates of weak convergence of the approximate minimum contrast estimators for the discretely observed Ornstein-Uhlenbeck process, *Statistics and Probability Letters* **76** (13), 1397–1409.

Bishwal, J.P.N. (2006b): Sequential maximum likelihood estimation in semimartingales, *Journal of Statistics and Applications* **1** (2-4), 15–25.

Bishwal, J.P.N. (2007a): A new estimating function for discretely sampled diffusions, *Random Operators and Stochastic Equations* **15** (1), 65–88.

Bishwal, J.P.N. (2007b): Large deviations and Berry-Esseen inequalities for estimators in nonlinear nonhomogeneous diffusions, *RevStat-Statistical Journal* **5** (3), 249–267.

Bishwal, J.P.N. (2008a): *Parameter Estimation in Stochastic Differential Equations*, Lecture Notes in Mathematics, **1923**, Springer-Verlag.

Bishwal, J.P.N. (2008b): Large deviations in testing fractional Ornstein-Uhlenbeck models, *Statistics and Probability Letters* **78** (8), 953–962

Bishwal, J.P.N. (2009a): Berry-Esseen inequalities for discretely observed diffusions, *Monte Carlo Methods and Applications* **15** (3), 229–239.

Bishwal, J.P.N. (2009b): M-Estimation for discretely sampled diffusions, *Theory of Stochastic Processes* **15** (31)(2), 62–83.

Bishwal, J.P.N. (2010a): Uniform rate of weak convergence for the minimum contrast estimator in the Ornstein-Uhlenbeck process, *Methodology and Computing in Applied Probability* **12** (3), 323–334.

Bishwal, J.P.N. (2010b): Maximum likelihood estimation in Skorohod stochastic differential equations, *Proceedings of the American Mathematical Society* **138** (4), 1471–1478.

Bishwal, J.P.N. (2010c): Sequential Monte Carlo methods for stochastic volatility models: a review, *Journal of Interdisciplinary Mathematics* **13** (6), 619–635.

Bishwal, J.P.N. (2010d): Conditional least squares estimation in diffusion processes based on Poisson sampling, *Journal of Applied Probability and Statistics* **5** (2), 169–180.

Bishwal, J.P.N. (2011a): Berry-Esseen inequalities for the discretely observed Ornstein-Uhlenbeck-Gamma process, *Markov Processes and Related Fields* **17** (1), 119–150.

Bishwal, J.P.N. (2011b): Minimum contrast estimation in fractional Ornstein-Uhlenbeck process: continuous and discrete sampling, *Fractional Calculus and Applied Analysis* **14** (3), 375–410.

Bishwal, J.P.N. (2011c): Sieve estimator for fractional stochastic partial differential equation, *Annals of Constantin Brancusi-Engineering Series* **5** (1), 9–18.

Bishwal, J.P.N. (2011d): Milstein approximation of posterior density of diffusions, *International Journal of Pure and Applied Mathematics* **68** (4), 403–414.

Bishwal, J.P.N. (2011e): Sufficiency and Rao-Blackwellization of Vasicek model, *Theory of Stochastic Processes* **17** (1), 12–15.

Bishwal, J.P.N. (2011f): Some new estimators of integrated volatility, *Open Journal of Statistics* **1** (2), 74–80.

Bishwal, J.P.N. (2011g): Estimation in interacting diffusions: continuous and discrete sampling, *Applied Mathematics* **2** (9), 1154–1158.

Bishwal, J.P.N. (2011h): Stochastic moment problems and hedging of generalized Black-Scholes options, *Applied Numerical Mathematics* **61** (12), 1271–1280.

Bishwal, J.P.N. (2011i): Maximum quasi-likelihood estimation in fractional Levy stochastic volatility model, *Journal of Mathematical Finance* **1** (3), 12–15.

Bishwal, J.P.N. (2011j): Financial extremes: a short review, *Advances and Applications in Statistics* **25** (1), 1–14.

Bishwal, J.P.N. (2017a): Conditional least squares estimation for discretely sampled nonergodic diffusions, *Asian Research Journal of Mathematics* **7** (4), 1–18.

Bishwal, J.P.N. (2017b): Valuation of real options under persistent shocks, *Journal of Statistics and Management Systems* **20** (5), 801–815.

Bishwal, J.P.N. (2017c): Hypothesis testing for fractional stochastic partial differential equations with applications to neurophysiology and finance, *Asian Research Journal of Mathematics* **4** (1), 1–24.

Bishwal, J.P.N. (2018a): Bernstein-von Mises theorem and small noise Bayesian asymptotics for parabolic stochastic partial differential equations, *Theory of Stochastic Processes* **23** (1), 6–17.

Bishwal, J.P.N. (2018b): Sequential maximum likelihood estimation in nonlinear non-Markov diffusion type processes, *Dynamic Systems and Applications* **27** (1), 107–124.

Bishwal, J.P.N. (2021a): A new algorithm for approximate maximum likelihood estimation in sub-fractional Chan-Karolyi-Longstaff-Sanders model, *Asian Journal of Probability and Statistics* **13** (3), 62–88.

Bishwal, J.P.N. (2021b): Statistics of SPDEs: From linear to nonlinear, *European Journal of Statistics* **1** (1), 1–57.

Bishwal, J.P.N. (2021c): Berry-Esseen bounds of approximate Bayes estimators for the discretely observed Ornstein-Uhlenbeck process, *Asian Journal of Statistical Sciences* **1** (2), 83–122.

Bishwal, J.P.N. (2022a): Mixingale estimation function for SPDEs with random sampling, *European Journal of Statistics* **2** (1), 1–13.

Bishwal, J.P.N. (2022b): Berry-Esseen bounds of the quasi maximum likelihood estimators for the discretely observed diffusions, *Applied Math* 2 (2), 39–53.

Bishwal, J.P.N. (2022c): On the Stratonovich estimator for the Ito diffusion, *European Journal of Mathematical Analysis* **2** (1), 1–13.

Bishwal, J.P.N. (2022d): Parameter estimation in Levy driven stochastic volatility models, *Journal of Financial Economics and Statistics* **1** (1), 1–31.

Bishwal, J.P.N. and Bose, A. (1995): Speed of convergence of the maximum likelihood estimator in the Ornstein-Uhlenbeck process, *Calcutta Statistical Association Bulletin* **45**, 245–251.

Bishwal, J.P.N. and Bose, A. (2001): Rates of convergence of approximate maximum likelihood estimators in the Ornstein-Uhlenbeck process, *Computers and Mathematics with Applications* **42** (1–2), 23–48.

Bjork, T. (1998): *Arbitrage in Continuous Time*, Oxford University Press, Oxford.

Black, F. and Scholes, M. (1973): The pricing of options and corporate liabilities, *Journal of Political Economy* **81**, 637–659.

Blumenthal, R.M. and Getoor, R.K. (1961): Sample functions of stochastic processes with stationary independent increments, *J. Math. Mech.* **10**, 493–516.

Bojdecki, T., Gorostiza, L.G. and Talarczyk, A. (2004): Sub-fractional Brownian motion and its relation to occupation times, *Statistics and Probability Letters*, **69**, 405–419.

Borkar, V. and Bagchi, A. (1982): Parameter estimation in continuous time stochastic processes, *Stochastics* **8**, 193–212.

Borokov, A.A. (1973): On the rate of convergence for the invariance principle, *Theory Probab. Appl.* **18**, 217–234.

Borkovec, M. and Kluppelberg, C. (2007): Extremal behavior of diffusion models in finance, *Extremes* **1** (1), 47–80.

Borovkov, K. and Decrouez, G. (2013): Ornstein-Uhlenbeck type processes with heavy distribution tails, *Theory of Probabability and its Applications* **57** (3), 396–418.

Borup, D. and Jakobsen, J.S. (2019): Capturing volatility persistence: a dynamically realized EGARCH-MIDAS model **19**, 1839–1855.

Borwanker, J.D., Kallianpur, G. and Prakasa Rao, B.L.S. (1971): The Bernstein-von Mises theorem for Markov processes, *Ann. Math. Statist.* **42** (1971), 1241–1253.

Bose, A. (1985): Rate of convergence of the maximum likelihood estimator in the Ornstein-Uhlenbeck process, *Technical Report No. 4/85*, Stat-Math Unit, Indian Statistical Institute, Calcutta.

Bose, A. (1986a): Berry-Esseen bound for the maximum likelihood estimator in the Ornstein-Uhlenbeck process, *Sankhyā Ser. A* **48**, 181–187.

Bose, A. (1986b): *Asymptotic study of estimators in some discrete and continuous time models*, Ph.D. Thesis, Indian Statistical Institute, Calcutta.

Bossy, M. and Olivero, H. (2018): Strong convergence of the symmetrized Milstein scheme for some CEV-like SDEs, *Bernoulli* **24**, 1995–2042.

Brigo, D. and Mercurio, F. (2001): *Interest Rate Models: Theory and Practice (with smile, inflation and credit)*, Second Edition, Springer, Hiedelberg.

Brandt, M.W. and Santa-Clara, P. (2002): Simulated likelihood estimation of diffusions with an application to exchange rate dynamics in incomplete markets, *Journal of Financial Economics* **63** (2), 161–210.

Brassesco, S. and Garcia-Pire, S.C. (2014): On the density of the winding number of planar Brownian motion, *Journal of Theoretical Probability* **27** (3), 899–914.

Breton, J.C. and Nourdin, I. (2008): Error bounds on the non-normal approximation of Hermite power variations of fractional Brownian motion, *Electronic Communications in Probability* **13**, 482–493.

Brockwell, P.J., Davis, R.A. and Yang, Y. (2007): Estimation for non-negative Levy driven Ornstein-Uhlenbeck processes, *Journal of Applied Probability* **44**, 977–989.

Brown, B.M. and Hewitt, J.T. (1975a): Asymptotic likelihood theory for diffusion processes, *J. Appl. Prob.* **12**, 228–238.

Brown, B.M. and Hewitt, J.T. (1975b): Inference for the diffusion branching process, *J. Appl. Prob.* **12**, 588–594.

Broze, L., Scaillet, O. and Zakoian, J.M. (1998): Quasi indirect inference for diffusion processes, *Econometric Theory* **14**, 161–186.

Buchmann, B. and Chan, N.H. (2007): Asymptotic theory of least squares estimators for nearly unstable processes under strong dependence, *Annals of Ststistics* **35**, 2001–2017.

Buchmann, B. and Chan, N.H. (2009): Integrated functionals of normal and fractional processes *Annals of Applied Probability* **19**, 49–70.

Buchmann, B. and Kluppelberg, C. (2005): Maxima of stochastic processes driven by fractional Brownian motion, *Advances in Applied Probability* **37**, 743–764.

Buchmann, B. and Kluppelberg, C. (2006): Fractional integral equations and state space transform *Bernoulli* **12**, 431–456.

Campbell, J.Y., Lo, A.W. and MacKinlay, A.C. (1997): *The Ecnometrics of Financial Markets*, Princeton University Press, Princeton.

Campillo, F. and Le Gland, F. (1989): MLE for partially observed diffusions: direct maximization vs. the EM algorithm, *Stoch. Process. Appl.* **33**, 245–274.

Cappe, O., Moulines, E. and Ryden, T.(2005): *Inference in Hidden Markov Models*, Springer, New York.

Caravenna, F. and Corbetta, J. (2018): The asymptotic smile of a multiscaling stochastic volatility model, *Stochastic Processes and their Applications*, 1034–1071.

Carmona, R. and Tehranchi, M. (2006): *Interest Rate Model: an Infinite Dimensional Stochastic Analysis Perspective*, Springer-Verlag, Berlin.

Carr, P., Gabaix, X. and Wu, L. (2011): Linearity-generating processes, unspanned stochastic volatility and interest-rate options, *Econometrics: Mathematical Methods and Programming*, 1–51.

Carr, P., Geman, H., Madan, D. and Yor, M. (2002): The fine structure of asset returns: An empirical investigation, *Journal of Business*, **75** (2), 305–332.

Carr, P., Geman, H., Madan, D., and Yor, M. (2003): Stochastic volatility for Levy processes, *Mathematical Finance* **13**, 345–382.

Cartea, A., Jaimungal, S. and Penalva, J. (2015): *Algorithmic and High-Frequency Trading*, Cambridge University Press, Cambridge.

Chambers, R.L. and Heathcote, C.R. (1981): On the estimation of slope and identification of outliers in linear regression, *Biometrica* **68**, 21–33.

Chan, K.C., Karloyi, G.A., Longstaff, F.A. Sanders, A.B. (1992): An empirical comparison of the short term interest rates, *Journal of Finance* **47**, 1209–1227.

Chang, J. and Chen, S.X. (2011): On the approximate maximum likelihood estimation for diffusion processes, *Annals of Statistics* **39**, 2820–2851.

Chatterjee, S. (1997): Another look at the jackknife: further examples of generalized bootstrap, *Statist. Probab. Letters* **40**, 307–319.

Chatterjee, S. and A. Bose (2005): Generalized bootstrap for estimating equations, *Annals of Statistics* **33**, 414–436.

Chen, L.H.Y., Goldstein, L. and Shao, Q.M. (2011): *Normal Approximation by Stein's Method*, Springer-Verlag, Berlin.

Chen, N. and Glasserman, P. (2007): Malliavin Greeks without Malliavin calculus, *Stochastic Processes and their Applications* **117**, 1689–1723.

Chib, S., Nardari, F. and Shephard, N. (2006): Analysis of high dimensional multivariate stochastic volatility models, *Journal of Econometrics* **134**, 341–371.

Chow, Y.S. and Teicher, H. (1997): *Probability Theory: Independence, Interchangebility and Martingales*, Third Edition, Springer-Verlag, New York.

Chronopoulou, A. and Spiliopoulos, K. (2018): Sequential Monte Carlo for stochastic volatility models, *Quantitative Finance*, **18**, 507–517.

Clement, E. (1993): Inference statistique des processes de diffusion, *Research Report No. 9404*, CREST-INSEE, Universite Paris VI.

Clement, E. (1995): Estimation de processus de diffusion par methodes simulees, *Research Report No. 9507*, CREST-INSEE, Universite Paris VI.

Clement, E. (1997a): Estimation of diffusion processes by simulated moments methods, *Scand. J. Statist.* **24**(3), 353–369.

Clement, E. (1997b): Bias correction for the estimation of discretized diffusion processes from an approximated likelihood function, *Theor. Probab. Appl.* **42**, 283–288.

Clement, E. and Gloter, A. (2019): Estimating functions for SDEs driven by stable Levy processes, *Annals of the Institute of Henri Poincare-Probability and Statistics*, **55**, 1316–1348.

Clement, E., Gloter, A. and Nguyen, H. (2018): Asymptotics in small time for the density of stochastic differential equation driven by a stable Levy process, *ESAIM: Probability and Statistics*, **22**, 58–95.

Clement, E., Gloter, A. and Nguyen, H. (2019): LAMN property for the drift and volatility parameters of a SDE driven by a stable Levy process, *ESAIM: Probability and Statistics* **23**, 136–175.

Clinet, S. and Yoshida, N. (2017): Statistical inference for ergodic point processes and application to limit order book, *Stochastic Processes and their Applications* **127**, 1800–1839.

Cohen, S.N. and Elliott, R.J. (2015): *Stochastic Calculus and Applications*, Second Edition, Birkhauser, Heidelberg.

Comte, F., Coutin, L. and Renault, E. (2012): Affine fractional stochastic volatility models with application to option pricing, *Annals of Finance*, **8**, 337–378.

Comte, F. and Renault, E. (1996): Long memory continuous time models, *J. Econometrics*, *73*, 101–149.

Comte, F. and E. Renault, E. (1998): Long memory in continuous- time stochastic volatility models, *Mathematical Finance*, **8**, 291–323.

Conley, T.G., Hansen, L.P., Luttmer, E.G.J. and Scheinkman, J.A. (1997): Short term interest rates as subordinated diffusions, *Review of Financial Studies* **10**, 525–578.

Cont, R. and Tankov, P. (2004): *Financial Modeling with Jump Processes*, Chapman and Hall/CRC, Boca Raton.

Corradi, V. and Swanson, N.R. (2005): Bootstrap specification test for diffusion processes, *Journal of Econometrics* **124**, 117–148.

Coutin, L. and Decreusefond, L. (1997): Stochastic differential equations driven by a fractional Brownian motion, *Tech. Report* 97C004, ENST Paris.

Coutin, L. and Decreusefond, L. (1999): Abstract nonlinear filtering in the presence of fractional Brownian motion, *Ann. Applied Probab.* **9** (4), 1058–1090.

Cox, J.C., Ingersoll, J.E. and Ross, S.A. (1985): The theory of term structure of interest rates, *Econometrica* **53**, 363–384.

Crepey, S. (2013): *Financial Modeling: A Backward Stochastic Differential Equation Perspective*, Springer-Verlag, Berlin.

Curtain, R. and Prichard, A. (1978): *Infinite Dimensional Linear Systems Theory*, Springer-Verlag, Berlin.

Czado, C. and S. Haug, S. (2010): An ACD-ECOGARCH(1,1) model, *Journal of Financial Econometrics* **8**, 335–344.

Dacunha-Castelle, D. and Duflo, M. (1986): *Probability and Statistics II*, Springer-Verlag, Berlin.

Dacunha-Castelle, D. and Florens-Zmirou, D. (1986): Estimation of the coefficients of a diffusion from discrete observations, *Stochastics* **19**, 263–284.

Dai, W. and Heyde, C.C. (1996): Itô formula with respect to fractional Brownian motion and its applications, *J. Appl. Math. Stoch. Anal.*, **9**, 439–448.

Dambis, K.E. (1965): On the decomposition of continuous submartingales, *Theor. Probab. Appl* **10**, 401–410.

Davis, R. A. (1982): Maximum and minimum of one-dimensional diffusions, *Stochastic Processes and their Applications* **13**, 1–9.

David, R.A. and Dunsmuir, W.T.M. (1997): Least absolute deviation estimation for regression with ARMA errors, *Journal of Theoretical Probability*, **10**, 481–497.

Davis, R.A., Knight, K. and Liu, J. (1992): M-estimation for autoregressions with infinite variance, *Stochastic Processes and their Applications*, **40**, 145–180.

Davis, R.A. and McCormick, W.P. (1989): Estimation for first order autoregressive processes with positive or bounded innovations *Stochastic Process. Appl.* **31**, 237–250.

Decreusefond, L. and Ustunel, A.S. (1998): Fractional Brownian motion: theory and applications, In: *Fractional Differential Systems: Models, Methods and Applications, ESAIM Proceedings* **5**, 75–86.

Decreusefond, L. and Ustunel, A.S. (1999): Stochastic analysis of the fractional Brownian motion, *Potential Analysis* **10**, 177–214.

Del Moral, P (2004): *Feynman-Kac Formulae*, Springer-Verlag, New York.

Delbaen, F. and Shirakawa, H. (2002): An interest rate model with upper and lower bounds, *Asia Pacific Financial Markets* **9**, 191–209.

Delbaen, F and Schachermayer, W (2006): *The Mathematics of Arbitrage*, Springer-Verlag, Berlin.

Dembo, A. and Zeitouni, O. (2010): *Large Deviations Techniques and Applications*, Second Edition, Springer-Verlag, Berlin.

Dereich, S., Neuenkirch, A. and Szpruch, L. (2012): An Euler type method for the strong approximation of the Cox-Ingersoll-Ross process, *Proceedings of the Royal Society, Series A* **468**, 1105–1115.

Dieterich, W., Fulde, P. and Peschel, I. (1980): Theoretical models for superionic conductors, *Adv. Physics* **29**, 527–605.

Dietz, H.M. (1989): Asymptotic properties of maximum likelihood estimators in diffusion type models, Preprint No. 228, Humboldt-Universitat Zu Berlin, Sektion Mathematik.

Dietz, H.M. (1992): A non-Markovian relative to the Ornstein-Uhlenbeck process and some of its local statistical properties, *Scand. J. Statist.* **19**, 363–379.

Dietz, H.M. and Kutoyants, Yu. A. (1997): A class of minimum-distance estimators for diffusion processes with with ergodic properties, *Statist. Decisions* **15**, 211–227.

Dohnal, G. (1987): On estimating the diffusion coefficient, *J. Appl. Prob.* **24**, 105–114.

Doob, J.L. (1953): *Stochastic Processes*, John Wiley, New York.

Dorogovcev, A. Ja. (1976): The consistency of an estimate of a parameter of a stochastic differential equation, *Theory Prob. Math. Statist.* **10**, 73–82.

Doucet, A., de Feitas, J.F.G. and Gordon, N.J. (eds.) (2001): *Sequential Monte Carlo Methods in Practice*, Springer-Verlag, New York.

Dubins, L.E. and Schwaz, G. (1965): On continuous martingales, *Proceedings of the National Academy of Sciences* **53** (3), 913–916.

Duffie, D. (1995): *Dynamic Asset Pricing Model*, Second Edition, Princeton University Press, Princeton.

Duffie, D. and Glynn, P. (2004): Estimation of Continuous time Markov processes sampled at random time intervals, *Econometrics*, **72**, 1773–1808.

Duffie, D. and Kan, R. (1996): A yield-factor model of interest rates, *Mathematical Finance* **6**, 379–406.

Duffie, D., Pan, J. and Singleton, K.J. (2000): Transform analysis and asset pricing for affine jump-diffusions, *Econometrica* **68**, 1343–1376.

Duffie, D. and Singleton, K.J. (1997): An econometric model of the term structure of interest rate swap yields, *Journal of Finance* **52**, 1287–1323.

Duncan, T.E., Hu, Y. and Pasik-Duncan, B. (2000): Stochastic calculus for fractional Brownian motion I, *SIAM J. Control. Optim.* **38**, 582–612.

Dunkl, C.F. (1992): Hankel transforms associated with finite reflection groups, *Contemporary Mathematics*, **138**, 123–138.

Dupuis, P. and Kushner, H.J. (1989): Stochastic approximations and large deviations: upper bounds and w.p.1 convergence, *SIAM J. Control Optim.* **27**, 1108–1135.

Durbin, J. and S.J. Koopman (1997): Monte Carlo maximum likelihood estimation for non-Gaussian state space models, *Biometrika*, **84**, 669–684.

Durbin, J. and S.J. Koopman (2000): Time series analysis of non-Gaussian observations based on state space models from both classical and Bayesian perspectives, *Journal of the Royal Statistical Society*, Series B, **62**, 3–56.

Durrett, R. (2019): *Probability: Theory and Examples*, Fourth Edition, Cambridge University Press, Cambridge, England.

Duvernet, L., Robert, C.Y. and Rosenbaum, M. (2010): Testing the type of a semi-martingale: Itô against multifractal, *Electronic Journal of Statistics* **4**, 1300–1323.

Efron, B. (1979): Bootstrap methods: Another look at the jackknife, *Ann. Statist.* **7** (1), 1–26.

Elerian, O. (1998): A note on the existence of a closed form conditional transition density for the Milstein scheme, Economics Discussion Paper, Nuffield College, Oxford.

Elerian, O., Chib, S. and Shephard, N. (2001): Likelihood inference for discretely observed nonlinear diffusions, *Econometrica*, **69**, 959–993.

Elliot, R.J. (1982): *Stochastic Calculus and Applications*, Springer-Verlag, New York.

Embrechts, P. and Maejima, M. (2002): *Selfsimilar Processes*, Princeton University Press, Princeton and Oxford.

Embrechts, P., Klupelberg, C. and Mikosh, T. (1997): *Modeling Extremal Events for Insurance and Finance*, Springer-Verlag, Berlin.

Engle, R.F. (1982): Autoregressive conditional heteroscedasticity with estimates of the variance of United Kingdom inflation, *Econometrica*, **50**, 987–1007.

Engle, R. and Russell, J. (1998): The autoregressive conditional duration model: a new model for irregularly spaced data, *Econometrica* **66**, 1127–1162.

Es-Sebaiy, K. and Viens, F.G. (2019): Optimal rates for parameter estimation of stationary Gaussian processes, *Stochastic Processes and their Applications* **129**, 3018–3054.

Ethier, S.N. (1976): A class of degenerate diffusion processes arising in population genetics, *Communications in Pure and Applied Mathematics* **29**, 483–493.

Fama, E.F. and French, K.R. (1995): Size and book-to-market factors in earnings and returns, *Journal of Finance* **50** (1), 131–155.

Fan J., Jiang J., Zhang, C., Zhou Z. (2003): Time-dependent diffusion models for term structure dynamics, *Statistica Sinica* **13**, 965–992.

Fasen, V., Kluppelberg, C. and Lindner, A. (2006): Extremal behavior of stochastic volatility models, In: *Stochastic Finance* (Eds.) M.R. Grossinho, A.N. Shiryayev, M.L. Esquivel, P.E. Oliviera, 107–155, Springer.

Fasen, V. and Kluppelberg, C. (2007): Extremes of supOU processes, In: *Stochastic Analysis and Applications: The Abel Symposium* (Eds.) F.E. Benth, G. DiNunno, T. Lindstrom, B. Oksendal, T. Zhang, 340–359, Springer.

Feigin, P.D. (1976): Maximum likelihood estimation for continuous time stochastic processes, *Adv. Appl. Prob.* **8**, 712–736.

Fearnhead, P. (2002): Markov chain Monte Carlo, sufficient statistics and particle filters *Journal of Computational and Graphical Statistics*, **11**, 848–862.

Fearnhead, P. (2005): Using random quasi-Monte Carlo within pa rticle filters with application to financial time series, *Journal of Computational and Graphical Statistics*, **14**, 751–769.

Feigin, P.D. (1979): Some comments concerning curious singularity, *J. Appl. Prob.* **16**, 440–444.

Feller, W. (1957): *An Introduction to Probability Theory and its Applications*, Vol. I, Wiley, New York.

Feller, W. (1971): *An Introduction to Probability Theory and its Applications*, Vol. II, Wiley, New York.

Filipovic, D. (2009): *Term Structure Models: A Graduate Course*, Springer, Berlin.

Filipovic, D. and Larsson, M. (2016): Polynomial diffusions and applications in finance, *Finance and Stochastics* **20**, 931–972.

Filipovic, D. and Larsson, M. (2017): Polynomial jump-diffusion models, *Stochastic Systems* **10**, 71–97.

Filipovic, D., Larsson, M. and Statti, F. (2019): Unspanned stochastic volatility for the multi-factor CIR model, *Mathematical Finance* **29**, 827–836.

Filipovic, D., Larsson, M., Trolle, A.B. (2017): Linear rational term structure models, *Journal of Finance* **72** (2), 655–704.

Filipovic, D. and Kitapbeyev, Y. (2018): On the American swaption in the linear-rational framework, *Quantitative Finance*, **18**, 1865–1876.

Fisk, D.L. (1963): Quasimartingales and stochastic integrals, *Technical Report No. 1*, Department of Mathematics, Michigan State University, East Lansing.

Friedman, A. (1975): *Stochastic Differential Equations*, Vol. I, Academic Press, New York.

Florens-Landais, D. and Pham, H. (1999): Large deviations in estimation of an Ornstein-Uhlenbeck model, *J. Appl. Probab.* **36**, 60–77.

Florens-Zmirou, D. (1989): Approximate discrete-time schemes for statistics of diffusion processes, *Statistics* **20**, 547–557.

Florens-Zmirou, D. (1991): Statistics on crossings of discretized diffusions and local time, *Stoch. Proc. Appl.* **39**, 139–151.

Florens-Zmirou, D. (1993): On estimating the diffusion coefficient from discrete observations, *J. Appl. Probab.* **30**, 790–804.

Forman, J.L. and Sørensen, M. (2008): The Pearson diffusions: a class of statistically tractable diffusion processes, *Scandinavian Journal of Statistics* **35**, 438–465.

Fournie, E., Lasry, J.M., Lebuchoux, J, Lions, P.L. and Touzi, N. (1999): Applications of Monte Carlo methods to Malliavin calculus in finance, *Finance and Stochastics* **3**, 391–412.

Fournie, E., Lasry, J.M., Lebuchoux, J, Lions, P.L. and Touzi, N. (2001): Applications of Monte Carlo methods to Malliavin calculus in finance II, *Finance and Stochastics* **5**, 201–236.

Fournier, N. (2011): Simulation and approximation of Levy-driven stochastic differential equations, *ESAIM: Probability and Statistics* **15**, 233-248.

Friz, P.K., Gatheral, J., Gulisashvili, A., Jacquier, A. and Tiechmann, J. (2015): *Large Deviations and Asymptotic Methods in Finance*, Springer-New York.

Fuchs, C. (2013): *Inference for Diffusion Processes with Applications in Life Sciences*, Springer, New York.

Fuller, W.A. (1996): *Introduction to Statistical Time Series*, Second Edition, Wiley, New York.

Gabaix, X. (2007): Linearity-generating processes: A modelling tool yielding closed forms for asset prices, *Working Paper, New York University.*

Gallant, A.R. and Long, J.R. (1997): Estimating stochastic differential equations efficiently by minimum chi-squared, *Biometrika* **84**, 125–141.

Gallardo, L. and Yor, M. (2006): A chaotic representation property of the multidimensional Dunkl processes, *Annals of Probability* **34**, 1530–1549.

Garbaczewski, P. and Olkiewicz, R. (2000): Orstein-Uhlenbeck-Cauchy process, *Journal of Mathematical Physics* **41**, 6843–6860.

Gard, T.C. (1988): *An Introduction to Stochastic Differential Equations*, Marcel Dekker.

Gaver and Lewis (1980): First order autoregressive gamma sequences and point processes, *Adv. Appl. Probab.* **12**, 727–745.

Gelb, A. (ed.) (1974): *Applied Optimal Estimation*, MIT Press, Cambridge, MA.

Genon-Catalot, V. (1987): *Observations Partielles de Diffusions: Traitement Statistique dans L' Asymptotique de la variance*, Thesis, Universite Paris Sud, Orsay, Cedex.

Genon-Catalot, V. (1990): Maximum contrast estimation for diffusion processes from discrete observations, *Statistics* **21**, 99–116.

Genon-Catalot, V. and Jacod, J. (1993): On the estimation of the diffusion coefficient for multidimensional diffusion processes, *Ann. Inst. Henri Poincaré, Probabilités et Statistiques* **29**, 119–151.

Genon-Catalot, V. and Jacod, J. (1994): On the estimation of diffusion coefficient for diffusion processes, *Scand. J. Statist.* **21**, 193–221.

Genon-Catalot, V., Jeantheau, T. and Laredo, C. (1998a): Parameter estimation for discretely observed stochastic volatility models, *Bernoulli* **4**(1), 1–18.

Genon-Catalot, V., Jeantheau, T. and Laredo, C. (1998b): Limit theorems for for discretely observed stochastic volatility models, *Bernoulli* **4**(3), 283–303.

Ghosal, S. (1994): *Asymptotic Properties of the posterior distributions and study of some nonregular cases,* Ph.D. Thesis, Indian Statistical Institute, Calcutta.

Ghosal, S. , Ghosh, J.K. and Samanta, T. (1995): On convergence of posterior distributions, *Ann. Statist.* **23**, 2145–2152.

Ghosh, J.K., Ghosal, S. and Samanta, T. (1994): Stability and convergence of the posterior in some nonregular problems, In: *Statistical Decision Theory and Related Topics V*, (Eds.) Gupta, S.S. and Berger, J.O., Proceedings of the Fifth Purdue International Symposium, Springer-Verlag, New York, 183–200.

Giesecke, K. and Smelov, D. (2013): Exact sampling for jump-diffusions, *Operations Research* **61** (4), 894–907.

Giesecke, K., Sirigano, J. and Schwenkler, G. (2020): Inference in large financial systems, *Mathematical Finance* **30** (1), 3–46.

Gikhman, I.I. and Skorohod, A.V. (1972): *Stochastic Differential Equations*, Springer-Verlag, Berlin.

Glasserman, P. (2004): *Monte Carlo Methods in Financial Engineering,* Springer, New York.

Goncalves, S. and Meddahi, N. (2009): Bootstrapping the realised volatility, *Econometrica* **77**, 283–306.

Gourieroux, C. and Jasiak, J. (2001): *Financial Econometrics: Problems, Models and Methods*, Princeton University Press, Princeton.

Gourieroux, C. and Jasiak, J. (2006): Autoregressive gamma process, *Journal of Forecasting* **25** (2), 129–152.

Gouriéroux, C., Monfort, A. and Renault, E. (1993): Indirect inference, *Journal of Applied Econometrics* **8**, 585–618.

Gouriéroux, C. and Monfort, A. (1994): Indirect inference for stochastic differential equations, *Discussion Paper 9443*, CREST-INSEE, University of Paris VI.

Grambsch, P. (1983): Sequential sampling based on the observed Fisher information to guarantee the accuracy of the maximum likelihood estimator, *Ann. Statist.* **11**, 68–77.

Greenwood, P.E. and Shiryayev, A.N. (1992): Asymptotic minimaxity of a sequential estimator of a first order autoregressive model, *Stoch. Stoch. Reports* **38**, 49–65.

Greenwood, P.E. and Wefelmeyer, W. (1993): Asymptotic minimax results for stochastic process families with critical points, *Stoch. Proc. Appl.* **44**, 107–116.

Grigelionis, B. (2012): *Student's t-distribution and Related Stochastic Processes*, Springer.

Gripenberg, G. and Norros, I. (1996): On the prediction of fractional Brownian motion, *Journal of Applied Probability* **33** (2), 400–410.

Gulisashvili, A. (2012): *Analytically Tractable Stochastic Stock Price Models*, Springer-Verlag, Berlin.

Gushchin, A.A. (1995): On asymptotic optimality of estimators of parameters under the LAQ condition, *Theory. Probab. Appl.* **40**, 261–272.

Gushchin, A.A. and Küchler, U. (1999): Asymptotic inference in linear stochastic differential equation with time delay, *Bernoulli* **5**, 1059–1098.

Guttorp, P. and Kulperger, R. (1984): Statistical inference for some Volterra population processes in a random environment, *Canadian J. Statist.* **12**, 289–302.

Hall, P. (1992): *The Bootstrap and Edgeworth Expansion*, Springer-Verlag.

Hall, P. and Heyde, C.C. (1980): *Martingale Limit Theory and its Applications*, Academic Press, New York.

Hall, P. and Presnell, B. (1999): Intentionally bised bootstrap methods, *Journal of the Royal Statistical Society,* Series B, **61** (1), 143–158.

Hansen, P.R. and Lunde, A. (2006): Realized variance and market microstructure, *Journal of Business and Economic Statistics* **24** (2) 127–161.

Hansen, L.P. and Scheinkman, J.A. (2009): Long-term risk: An Operator approach, *Econometrica* **77**, 177–234.

Harison, V. (1992): The Bernstein-von Mises theorem for a certain class of Gaussian diffusion processes, *Publications du Service de Mathematiques, Universite d' Antananarivo* **6**, 1–7.

Harison, V. (1996): Drift estimation of a certain class of diffusion processes from discrete observations, *Computers Math. Applic.* **31**, 121–133.

Hartman, P. and Watson, G.S. (1974): Normal distribution functions on the spheres and the modified Bessel functions, *Annals of Probability* **4**, 593–607.

Harvey A.C., Ruiz, E. and Shephard, N. (1994): Multivariate Stochastic Volatility Models, *Review of Economic Studies* **61,** 247–264. Also reprinted in Engle, R.F. (Ed.) ARCH models, Oxford: Oxford University Press.

Haug, S. and Czado, C. (2006a): Quasi maximum likelihood estimation and prediction in the compound Poisson ECOGARCH(1,1) model, *Sonderforschungsbereich 386,* No. 516.

Haug, S. and Czado, C. (2006b): A fractionally integrated ECOGARCH process, *Sonderforschungsbereich 386,* No. 484.

Haug, S. and Czado, C. (2007): An exponential continuous time GARCH process, *Journal of Applied Probability* **44**, 960–976.

Haug, S., Kluppelberg, C., Lindner, A. and Zapp, M. (2007): Method of moment estimation in the COGARCH(1,1) model, *Econometrics Journal* **10**, 320–341.

Heston, S.L. (1993): A closed-form solution for options with stochastic volatility, *Review of Financial Studies* **6**(2), 327–343.

Hinkley, D.V. (1977): Jackknifing in unbalanced situations, *Technometrics* **19** (3), 285–292.

Heyde, C.C. (1998): *Quasi-likelihood and its Applications*, Springer-Verlag, Berlin.

Hipp, C. and Michel,R. (1976): On the Bernstein-von Mises approximation of posterior distributions, *Ann. Statist.* **4**, 972–980.

Hefter, M. and Jentzen, A. (2019): On the arbitrarily slow convergence rates for strong numerical approximations of Cox-Ingersoll-Ross processes and squared Bessel processes, *Finance and Stochastics* **23**, 139–172.

Heyde (1997): *Quasi-Likelihood and Its Application: A General Approach to Optimal Parameter Estimation*, Springer, New-York.

Hoffmann, M., Vetter, M. and Dette, H. (2018): Nonparametric inference of gradual changes in the jump behavior of time-continuous processes, *Stochastic Processes and their Applications* **128**, 3679–3723.

Hog, E.P. and Frederiksen, P.H. (2006): The fractional Ornstein-Uhlenbeck process: Term structure theory and applications, *Scandinavian Working Papers in Business Administration*, No. F-2006-01.

Hopfner, R. (2014): *Asymptotic Statistics: With a view to Stochastic Processes*, De Gruyter, Berlin.

Hol, E.M.J.H. (2003a): Forecasting the variability of stock index returns with stochastic volatility models and implied volatility, In: *Empirical Studies on Volatility in International Stock Markets*, Dynamic Modeling and Econometrics in Economics and Finance, **6**, Springer, Boston, 71–97.

Hol, E.M.J.H. (2003b): Stock index volatility forecasting with high frequency data, In: *Empirical Studies on Volatility in International Stock Markets*, Dynamic Modeling and Econometrics in Economics and Finance, **6**, Springer, Boston, 99–127.

Horst, U. and Wenzelburger, J. (2008): On non-ergodic asset prices, *Economic Theory* **34**, 207–234.

Hu, Y. and Long, H. (2009): Least squares estimation of Ornstein-Uhlenbeck process driven by α-stable motions, *Stochastic Processes and their Applications* **119**, 2465–2480.

Hu, Y. and Long, H. (2010): Parameter estimation for a class of stochastic differential equations driven by small stable noises from discrete observations, *Acta Mathematical Scientia* **30**, 645–663.

Hu, Y. and Nualart, D. (2010): Parameter estimation for fractional Ornstein-Uhlenbeck process, *Statistics and Probability Letters* **80** (11–12), 1030–1038.

Huang, J., Ma, C. and Zhu, C. (2011): Estimation for discretely observed continuous stet branching process with immigration, *Statistics and Probability Letters* **81**, 1104–1111.

Huebner, M. (1993): *Parameter Estimation in Stochastic Differential Equations,* Ph.D. Thesis, University of Southern California, Los Angeles.

Huebner, M., Khasminskii, R.Z. and Rozovskii, B.L. (1992): Two examples of parameter estimation for stochastic partial differential equations, In: S. Cambanis, J.K. Ghosh, R.L. Karandikar, P.K. Sen (Eds.) *Stochastic Processes, Freschrift in honour of G. Kallianpur*, Springer-Verlag, Berlin, 149–160.

Huebner, M. and Rozovskii, B.L. (1995): On the asymptotic properties of maximum likelihood estimators for parabolic stochastic PDEs, *Prob. Theor. Rel. Fields* **103**, 143–163.

Hull, J.C. (2015): *Options, Futures and Other Derivatives*, Ninth Edition, Prentice Hall.

Hutton, J.E. and Nelson, P.I. (1984): Interchanging the order of differentiation and stochastic integration, *Stoch. Process. Appl.* **18**, 371–377.

Iacus, S.M. (2008): *Simulation and Inference for Stochastic Differential Equations*, Springer, New York.

Iacus, S.M. and Yoshida, N. (2018): *Simulation and Inference for Stochastic Processes with YUIMA*, Springer, New York.

Ibragimov, I.A. and Khasminskii, R.Z. (1981): *Statistical Estimation: Asymptotic Theory*, Springer-Verlag, Berlin.

Igloi, E. and Terdik, G. (1999): Long-range dependence through Gamma mixed Ornstein-Uhlenbeck processes, *Electronic Journal of Probability* **4**, 1–33.

Ikeda, N. and Watanabe, S. (1989): *Stochastic Differential Equations and Diffusion Processes*, Second Edition, North-Holland, Amsterdam (Kodansha Ltd., Tokyo).

Imkeller, P. (1994): Regularity of Skorohod integral processes based on integrands in a finite Wiener chaos, *Probability Theory and Related Fields* **98**, 137–142.

Itô, K. and McKean, H.P. Jr. (1974): *Diffusion Processes and their Sample Paths*, Springer-Verlag, Berlin.

Itô, K. (1984): *Foundations of Stochastic Differential Equations in Infinite Dimensional Spaces*, CBMS-NSF Reginal Conference Series in Applied Mathematics, Vol. 49, SIAM, Philadeplphia, Pennsylvania.

Ivenenko, D.O., Kulik, A.M. and Masuda, H. (2015): Uniform LAN property of locally stable Levy processes observed at high frequency, *ALEA: Latin American Journal of Probability and Mathematical Statistics*, **12**, 835–862.

Ivanov, A.V. (1976): The Berry-Esseen inequality for the distribution of the least squares estimate, *Mathematical Notes* **20**, 721–727.

Jacobsen, M. (1998): Discretely observed diffusions: classes of estimating functions and small Δ-optimality, *Preprint No. 11*, Department of Theoretical Statistics, University of Copenhagen.

Jacobsen, N.M. and Sørensen, M. (2019): Estimating functions for jump-diffusions, *Stochastic Processes and their Applications* **129**, 3282–3318.

Jacod, J. (2018): Limit of random measures associated with the increments of a Brownian semimartingale, *Journal of Financial Econometrics* **16**, 526–569.

Jacod, J. (2010): Statistics and high-frequency data, *Preprint, Paris-VI Universite,* 100 pages.

Jacod, J. and Shiryayev, A.N. (1987): *Limit Theorems for Stochastic Processes*, Springer-Verlag, Berlin.

Jacod, J. (1993): Random sampling in estimation problems for continuous Gaussian processes with independent increments, *Stoch. Proc. Appl.* **44**, 181–204.

Jacod, J and Protter, P. (2012): *Discretization of Processes*, Springer-Verlag, Berlin.

Jacod, J. and Shiryev, A.N. (1987): *Limit Theorems for Stochastic processes*, Springer-Verlag, Berlin.

James, M. and Le Gland, F. (1995): Consistent parameter estimation for partially observed diffusions with small noise, *Applied Math. Optim.* **32**, 47–72.

Janicki, A. and Weron, A. (1994): *Simulation and Chaotic Behavior of α-Stable Stochastic Processes*, Marcel Dekker, New York.

Jankunas, A. and Khasminskii, R.Z. (1997): Estimation of the parameters of linear homogeneous stochastic differential equations, *Stoch. Proc. Appl.* **72**, 205–219.

Jasra, A., Kamatani, K. and Masuda, H. (2019): Bayesian inference for stable Levy-driven stochastic differential equations with high freqnency data, *Scandinavian Journal of Statistics* **46**, 545–574.

Jazaerli, S. and Saporito, Y.F. (2017): Functional Itô calculus, path-dependence and the computation of Greeks, *Stochastic Processes and their Applications* **127**, 3997–4028.

Jeanblanc, M., Yor, M. and Chesney, M. (2009): *Mathematical Methods for Financial Markets*, Springer-Verlag, London.

Jeganathan, P. (1982): On the asymptotic theory of statistical inference when the limit of the log-likelihood ratios is mixed normal, *Sankhyā Ser. A* **44**, 173–212.

Jeganathan, P. (1995): Some aspects of asymptotic theory with applications to time series models, *Econometric Theory* **11**, 818–887.

Jennrich, R.I. and Bright, P.B. (1976): Fitting systems of linear differential equations using computer generated exact derivatives. *Technometrics* **18**, 385–392.

Jensen, B. and Poulsen, R. (1999): A comparison of approximate techniques for transition densities of diffusion processes, *Working Paper Series No. 38*, Center for Analytical Finance, University of Aaarhus - Aarhus School of Business.

Jensen, B. and Poulsen, R. (2002): Transition densities of diffusion processes: Numerical comparison and approximate techniques, *Journal of Derivatives* **9**, 18–32.

Jiang, G.J. and Knight, J.L. (2002): Estimation of continuous-time processes via the empirical characteristic function, *Journal of Business and Economic Statistics* **20**, 198–212.

Jiao, Y., Ma, C. and Scotti, S. (2017): Alpha-CIR model with branching processes in sovereign interest rate modeling, *Finance and Stochastics* **21**, 789–813.

Jondeau, E., Poon, S.H. and Rockinger, M. (2007): *Financial Modeling Under Non-Gaussian Distributions*, Springer-Verlag, London.

Jongbloed, G., van der Meulen, F.H. van der Vaart, A.W. (2005): Nonparametric inference for Levy driven Ornstein-Uhlenbeck processes, *Bernoulli* **11**, 759–791.

Jones, R.H. (1984): Fitting multivariate models to unequally spaced data, In: Time Series Analysis of Irregularity Observed Data (Ed. E. Parzen), *Lecture Notes in Statistics* **25** Springer-Verlag, New York, 155–188.

Kallianpur, G. (1980): *Stochastic Filtering Theory*, Springer-Verlag: New York.

Kallianpur, G. and Karandikar, R.L. (2000): *Introduction to Option Pricing Theory*, Birkhauser, Boston

Kallianpur, G. and Selukar, R.S. (1991): Parameter estimation in linear filtering, *J. Multivariate Anal.* **39**, 284–304.

Kallianpur, G. and Selukar, R.S. (1993): Estimation of Hilbert space valued parameter by the method of sieves, In :*Statistics and Probability, A R.R.Bahadur Festschrift* (Eds.) J.K.Ghosh, S.K.Mitra, K.R.Parthasarathy and B.L.S.Prakasa Rao, Wiley Eastern, 325–347.

Kallianpur, G. and Xiong, X. (1995): *Infinite Dimensional Stochastic Differential Equations*, Lecture Notes-Monograph Series, **26**, Institute of Mathematical Statistics, California.

Karandikar, R.L. (1983): Interchanging the order of stochastic integration and ordinary differentiation, *Sankhyā Ser. A* **45**, 120–124.

Karatzas, I. and Shreve, S.E. (1987): *Brownian Motion and Stochastic Calculus*, Springer-Verlag, New York.

Karlin, S. and Taylor, H.M. (1975): *A First Course in Stochastic Processes*, Second Edition, Academic Press, New York.

Karlin, S. and Taylor, H.M. (1981): *A Second Course in Stochastic Processes*, Academic Press, New York.

Kasonga, R.A. (1988): The consistency of a non-linear least squares estimator from diffusion processes, *Stoch. Proc. Appl.* **30**, 263–275.

Kasonga, R.A. (1990): Parameter estimation by deterministic approximation of a solution of a stochastic differential equation, *Commun. Statist. - Stoch. Models* **6**, 59–67.

Kawai, R. and Masuda, H. (2011a): On simulation of tempered stable random variables, *Journal of Computational and Applied Mathematics* **235**, 2873–2887.

Kawai, R. and Masuda, H. (2011b): On local asymptotic behavior of the likelihood function for Meixner Levy processes under high-frequency sampling, *Statistics and Probability Letters* **81**, 460–469.

Kawai, R. and Masuda, H. (2012): Infinite variation tempered stable Ornstein-Uhlenbeck processes with discrete observation, *Communications in Statistics-Simulation and Computation*, 125–139.

Kawai, R. and Masuda, H. (2013): Local asymptotic normality for normal inverse Gaussian Levy processes with high-frequency sampling,*ESAIM: Probability and Statistics* **17**, 13–32.

Kessler, M. (1997): Estimation of an ergodic diffusion from discrete observations, *Scand. J. Statist.* **24**(2), 211–229.

Kessler, M. (2000): Simple and explicit estimating functions for a discretely observed diffusion process, *Scand. J. Statist.* **27**, 65–82.

Kessler, M. and Sørensen, M. (1999): Estimating equations based on eigen function for a discretely observed diffusion process, *Bernoulli* **5**, 299–314.

Khasminskii, R.Z., Krylov, N. and Moshchuk, N. (1999): On estimation of parameters of linear stochastic differential equations, *Prob. Theor. Rel. Fields* **113** (3), 443–472.

Kim, Y.T. (1996): Parameter estimation for an infinite dimensional stochastic differential equations, *J. Korean Statistical Society* **25**, No. 2, 161–173.

Kim, Y.S., Rachev, S.T., Chung, D.M., Bianichi, M.L. (2008): A modified tempered stable distribution with volatility clustering, In: J.O. Soares, J. Pina, M. Catalao-Lopes, *New Developments in Financial Modelling*, Cambridge Scholars Publishing, Newcastle upon Tyne, UK.

Kim, Y.S., Rachev, S.T., Chung, D.M. and Bianichi, M.L. (2009): The modified tempered stable distribution, GARCH models and option pricing, *Probability and Mathematical Statistics* **29** (1), 91–117.

Kim S., N. Shephard and S. Chib (1998): Stochastic volatility: likelihood inference and comparison with ARCH models, *Review of Economic Studies* **65**, 361–393.

Kitagawa, G. (1987): Non-Gaussian state space modelling of nonstationary time series (with discussion), *Journal of the American Statistical Association* **82**, 1032–1063.

Klebaner, F.C. (2012): *Introduction to Stochastic Calculus with Applications*, Third Edition, Imperial College Press, London.

Kleptsyna, M.L. Kloeden, P.E. and Anh, V.V. (1998a): Linear filtering with fractional Brownian motion, *Stoch. Anal. Appl.* **16**, 907–914.

Kleptsyna, M.L., Kloeden, P.E. and Anh, V.V. (1998b): Nonlinear filtering with fractional Brownian motion, *Problems of Information Transmission* **34** (2) 65–76.

Kleptsyna, M.L., Kloeden, P.E. , Anh, V.V. (1998c): Existence and uniqueness theorem for fBm stochastic differential equations, *Problems of Information Transmission* **34** (4), 51–61.

Kleptsyna, M.L. , Le Breton, A. and Roubaud, M.C. (1998): An elementary approach to filtering in systems with fractional Brownian motion observation noise, In: *Probability Theory and Mathematical Statistics*, B. Grigelionis (Eds.), VSP/TEV, 373–392.

Kleptsyna, M.L. and Le Breton, A. (2002): Statistical analysis of the fractional Ornstein-Uhlenbeck type process, *Statistical Inference for Stochastic Processes* **5** (3), 229–248.

Kleptsyna, M.L. , Le Breton, A. , Roubaud, M.C. (2000): General approach to filtering with fractional Brownian noises- application to linear systems, *Stochastics and Stochastics Reports* **71**, 119–140.

Kloeden, P.E., Platen, E. and Schurz, H. (1994): *Numerical Solution of SDE Through Computer Experiments*, Springer-Verlag, Berlin.

Kloeden, P.E. and Platen, E. (1995): *Numerical Solution of Stochastic Differential Equations*, Second Print, Springer-Verlag, Berlin.

Kloeden, P.E., Schurz, H., Platen, H. and Sørensen, M. (1996): On the effects of discretization of estimators of drift parameters for diffusion processes, *J. Appl. Prob.* **33**, 1061–1076.

Knight, K. (1998): Limit distributions for L_1 regression estimators under general conditions, *Annals of Statistics*, **26**, 755–770.

Koenker, R. (2005): *Quantile Regression*, Cambridge University Press, Cambridge.

Kolmogorov, A.N. (1940): Wiener skewline and other interesting curves in Hilbert space, *Doklady Akad. Nauk* **26**, 115–118.

Konecky, F. (1990): Maximum likelihood estimator from a partially observed diffusions in the case of small measurement noise, *Statist. Decisions* **8**, 115–130.

Koopman S.J. and Uspensky, E.H. (2002): The stochastic volatility in mean model: Empirical evidence from international stock markets, *Journal of Applied Econometrics*, **17** (6), 667–689.

Koski, T. and Loges, W. (1985): Asymptotic statistical inference for a stochastic heat flow problem, *Statist. Probab. Letters* **3**, 185–189.

Koski, T. and Loges, W. (1986): On minimum contrast estimation for Hilbert space valued stochastic differential equations, *Stochastics* **16**, 217–225. functions, *Scand. J. Statist.* **21**, 421–431.

Kruhner, P. and Schnurr, A. (2018): Time change equations for Levy processes, *Stochastic Processes and their Applications* **128**, 963–978.

Krylov, N.V. (1995): *An Introduction to the Theory of Diffusion Processes*, American Mathematical Society, Providence.

Küchler, U. and Sørensen, M. (1994a): Exponential families of stochastic processes and Levy processes, *J. Statist. Planning Inf.* **39**, 211–237.

Küchler, U. and Sørensen, M. (1994b): Exponential families of stochastic processes with time continuous likelihood functions, *Scand. J. Statist.* **21**, 421–431.

Küchler, U. and Sørensen, M. (1997): *Exponential Family of Stochastic Process*, Springer-Verlag, New York.

Küchler, U. and Tappe, S. (2008): Bilateral gamma distributions and processes in financial mathematics, *Stochastic Processes and Their Applications* **118**, 261–283.

Küchler, U. and Tappe, S. (2013): Tempered stable distributions and processes **123**, 4256–4293.

Kulinich, G.L. (1975): On estimation of the drift parameter of a stochastic diffusion equation, *Theory Prob. Appl.* **20**, 384–387.

Kunita, H. (1990): *Stochastic Flows and Stochastic Differential Equations*, Cambridge Univ. Press, New York.

Kusuoka, S. (1982): The nonlinear transformation of Gaussian measure on Banach space and its absolute continuity I, *J. Fac. Sci. Univ. Tokyo, Sect. IA Math.* **7**, 567–597.

Kutoyants, Yu. A. (1977): Estimation of the trend parameter of a diffusion process in the smooth case, *Theory Probab. Appl.* **22**, 399–406.

Kutoyants, Yu. A. (1978): Estimation of a parameter of a diffusion process, *Theory Probab. Appl.* **23**, 641–649.

Kutoyants, Yu. A. (1984a): *Parameter Estimation for Stochastic Processes* (Translated and edited by B.L.S. Prakasa Rao), Research and Expositions in Mathematics, Vol. 6, Heldermann-Verlag, Berlin.

Kutoyants, Yu. A. (1984b): Parameter estimation for diffusion type processes of observations, *Math. Operationsforch. U. Statist., Ser. Statist.* **15**, 541–551.

Kutoyants, Yu. A. (1994a): *Identification of Dynamical Systems with Small Noise*, Kluwer, Dordrecht.

Kutoyants, Yu. A. (1994b): Nonconsistent estimation by diffusion type observations, *Statist. Probab. Letters.* **20**, 1–7.

Kutoyants, Yu. A. and Pohlman, H. (1994): Parameter estimation for Kalman-Bucy filter with small noise, *Statistics* **25**, 307–323.

Kutoyants, Yu. A. (2004): *Statistical Inference for Ergodic Diffusion Processes*, Springer, Berlin-New York.

Lai, T.L. and Xing, H. (2008): *Statistical Methods for Financial Markets*, Springer, New York.

Lanksa, V. (1979): Minimum contrast estimation in diffusion processes, *J. Appl. Prob.* **16**, 65–75.

Larsen, K.S. and Sørensen, M. (2007): Diffusion models for exchange rates in a target zone, *Mathematical Finance* **17**, 285–306.

Laredo, C. (1990): Sufficient condition for asymptotic sufficiency of incomplete observations of a diffusion process, *Ann. Statist.* **18**, 1158–1171.

Laurini, M.P. and Hotta, L.K. (2013): Indirect inference in fractional shoort-term interest rate diffusions, *Mathematics and Computers in Simulation* **94**, 109–126.

Lebovits, J. and Levy Vehel, J. (2014): White noise-based stochastic calculus with respect to multifractional Brownian motion, *Stochastics*, **86**, 87–124.

Le Breton, A. (1976): On continuous and discrete sampling for parameter estimation in diffusion type processes, *Mathematical Programming Study* **5**,124–144.

Le Breton, A. (1977): Parameter estimation in a linear stochastic differential equation, *Transactions of the Seventh Prague Conference on Information on Theory, Statistical Decision Functions and Random Processes and of the 1974 EMS, Vol. A*, Academica, Prague, 353–366.

Le Breton, A. (1998a): Filtering and parameter estimation in a simple linear system driven by fractional Brownian motion, *Statist. Probab. Letters* **38**, 263–274.

Le Breton, A. (1998b): A Girsanov type approach to filtering in a simple linear system driven by fractional Brownian motion, *C. R. Acad. Sci. Paris, Series I*, **326**, 997–1002.

Le Breton, A. and Musiela, M. (1984): A study of a one dimensional bilinear differential model for stochastic processes, *Prob. Math. Statist.* **4**, 91–107.

Le Cam, L. (1953): On some asymptotic properties of maximum likelihood estimates and related Bayes estimates, *Univ. Calif. Publ. Statist.* **1**, 277–330.

Le Cam, L. (1973): Convergence of estimates under dimensionality restrictions, *Ann. Statist* **1**, 38–53.

Le Cam, L. (1986): *Asymptotic Methods in Statistical Decision Theory*, Springer-Verlag, Berlin-New York.

Le Cam, L. and Yang, G.L. (1990): *Asymptotics in Statistics*, Springer-Verlag, New York.

Lee, C., Bishwal, J.P.N. and Lee, M. (2012): Sequential maximum likelihood estimation for reflected Ornstein-Uhlenbeck process, *Journal of Statistical Planning and Inference* **142** (5), 1234–1242.

Lee, T.S. and Kozin, F. (1977): Almost sure asymptotic likelihood theory for diffusion processes, *J. Appl. Prob.* **14**, 527–537.

Leonenko, N.N., Papic, I., Sikorskii, A. and Suvak, N. (2020): Approximation of heavy-tailed fractional Pearson diffusions in Skorokhod topology, *Journal of Mathematical Analysis and Applications*, **486**, 1–22.

Leonenko, N.N. and Suvak, N (2010): Statistical inference for student diffusion process, *Stoch. Anal. Appl.* **28**, 972–1002.

Leonenko, N.N. and Suvak, N (2010): Statistical inference for reciprocal gamma diffusion process, *Journal of Statistical Planning and Inference* **140**, 30–51.

Levanony, D. (1994): Conditional tail probabilities in continuous-time martingale LLN with application to parameter estimation in diffusions, *Stoch. Proc. Appl.* **51**, 117–134.

Levanony, D., Shwartz, A. and Zeitouni, O. (1993): Uniform decay and equicontinuity for normalized parameter dependent Itô integrals, *Stoch. Stoch. Reports* **43**, 9–28.

Levanony, D., Shwartz, A. and Zeitouni, O. (1994): Recursive identification in continuous-time stochastic processes, *Stoch. Proc. Appl.* **49**, 245–275.

Levy, P. (1948): Processus stochastiques et movement Brownien, Paris.

Lin, S. J. (1995): Stochastic analysis of fractional Brownian motions, *Stochastics and Stochastics Reports* **55**, 121–140.

Linetsky, V. (2004): The spectral representation of Bessel processes with constant drift: application in queueing and finance, *Journal of Applied Probability* **41**, 327–344.

Ling, S. (2005): Self-weighted least absolute deviation estimation for infinite variance autoregressive models, *Journal of Royal Statistical Society, Series B, Statistical Methodology* **67**, 381–393.

Linkov, Yu. N. (1990): Local asymptotic mixed normality for distributions of processes of diffusion type, *Theory Prob. Math. Statist.* **40**, 67–73.

Linkov, Yu. N. (1993): *Asymptotic Methods of Statistics for Stochastic Processes*, Naukova, Dumka (In Russian).

Liptser, R.S. and Shiryayev, A.N. (1977): *Statistics of Random Processes I: General Theory* Springer-Verlag, Berlin.

Liptser, R.S. and Shiryayev, A.N. (1978): *Statistics of Random Processes II: Applications* Springer-Verlag, Berlin.

Liptser, R.S. and Shiryayev, A.N. (1982): Rate of convergence in the central limit theorem for semimartingales, *Theor. Probab. Appl.* **27**, 1–13.

Liptser, R.S. and Shiryayev, A.N. (1989): *Theory of Martingales*, Kluwer, Dordrecht.

Liu, R.Y. (1988): Bootstrap procedure under some non-i.i.d. models, *Annals of Statistics* **16**, 1696–1708.

Liu, J.S. and Chen, R. (1995): Blind deconvolutions via sequential imputations, *Journal of the American Statistical Association* **90**, 567–576.

Liu, J.S. and Chen, R. (1995): Sequential Monte Carlo methods for dynamical systems, *Journal of the American Statistical Association*, **93**, 1032–1044.

Liu, Y., Nualart, E. and Tindel, S. (2019): LAN property for stochastic differential equations with additive fractional noise and continuous time observation, *Stochastic Processes and their Applications* **129**, 2880–2902.

Liu, Z. (2017): Jump-robust estimation of volatility with simultaneous presence of microstructure noise and multiple observations, *Finance and Stochastics* **21**, 427–469.

Liu, R.Y. and Singh, K. (1992): Efficiency and robustness in resampling, *Annals of Statistics* **20**, 370–384.

Lo, A.W. (1988): Maximum likelihood estimation for generalized Itô processes with discretely sampled data, *Econometric Theory* **4**, 231–247.

Lo, A.W. and MacKinlay, A.C. (1999): *A Non-Random Walk Down Wall Street*, Princeton University Press, Princeton.

Loges, W. (1984): Girsanov theorem in Hilbert space and an application to the statistics of Hilbert space-valued stochastic differential equations, *Stoch. Proc. Appl.* **17**, 243–263.

Long, H., Ma, C. and Shimizu, Y. (2017): Least squares estimators for stochastic differential equations driven by small Levy noises, *Stochastic Processes and their Applications*, **127**, 1475–1495.

Luschgy, H. (1992): Local asymptotic mixed normality for semimartingale experiments, *Probability Theory and Related Fields* **92** 151–176.

Luschgy, H. (1993): Second order behavior of Neyman-Pearson tests for stochastic processes, *Statistics and Decisions* **11**, 133–150.

Luschgy, H. (1994a): Asymptotic expansions for error probabilities for tests, In: Mandl, P and Huskova, M. (Eds.) *Asymptotic Statistics: Proceedings of the Fifth Prague Symposium*, Springer, 369–377.

Luschgy, H. (1994b): Asymptotic behavior of Neayman-Pearson tests for autoregressive processes, *Scandinavian Journal of Statistics* **21**, 461–473.

Luschgy, H. (1994c): Asymptotic inference for semimartingale models with singular parameter points, *Journal of Statistical Planning and Inference* **39**, 155–186.

Luschgy, H. (1995): Local asymptotic quadraticity of stochastic process based on stopping times, *Stochastic Processes and their Applications* **57**, 305–317.

Lyons, T.J. (1998): Differential equations driven by rough signals, *Rev. Mat. Iberoamericana* **14** (2), 215–310.

Lyons, T.J. and Zhang, T.S. (1994): Decomposition of Dirichlet processes and its application, *Ann. Probab.* **22**, 494–524.

Ma, R. (2015): Lamperti transformation for continuous-state branching process with competition and applications, *Statistics and Probability Letters* **107**, 11–17.

Malliavin, P. and Thalmaier, A. (2006): *Stochastic Calculus of Variation in Mathematical Finance,* Springer-Verlag, Berlin.

Mammen, E. (1993): Bootstrap and wild bootstrap for high dimensional linear models *Ann. Statist.* bf 21 (1), 255–285.

Mandelbrot, B. and Van Ness, J.W. (1968): Fractional Brownian motions, fractional noises and applications, *SIAM Review* **10**, 422–437.

Mandl, P. (1968): *Analytical Treatment of One-Dimensional Markov Processes*, Springer-Verlag, Berlin.

Mao, X. (1995): *Stochastic Differential Equations and Applications* Horwood Publishing, Chichester.

Marquardt, T. (2006a): Fractional Levy processes with application to long memory moving average processes, *Bernoulli* **12** (6), 1009–1126.

Marquardt, T. (2006b): Multivariate fractionally integrated CARMA processes, *Journal of Multivariate Analysis* **98**, 1705–1725.

Masuda, H. (2005): Classical method of moments for partially and discretely observed ergodic models, *Statistical Inference for Stochastic Processes* **8**, 25–50.

Masuda, H. (2009a): Joint estimation of discretely observed stable Levy processes with symmetric Levy density, *Journal of the Japan Statistical Society* **39**, 49–75.

Masuda, H. (2009b): On statistical aspects in calibrating geometric skewed stable asset price model, In: (Eds) Kijima, M., Hara, C., Tanaka, K. and Muromachi, *Recent Advances in Financial Engineering*, World Scientific, Singapore. 181–202.

Masuda, H. (2010a): Approximate quadratic estimating function for discretely observed Levy driven SDEs with application to a noise normality test, *Kyushu University Preprint*, MI 2010-21.

Masuda, H. (2010b): Asymptotic mixed normality in the self-weighted LAD estimation of Levy OU process, *Electronic Journal of Statistics* **4**, 525–565.

Masuda, H. (2011): On quasi-likelihood analysis for stochastic differential equations with jumps, *Proceedings of the International Statistical Institute, 55th World Statistical Congress*, 83–91.

Masuda, H. (2013a): Convergence of gaussian quasi-lilkelihood random fields for ergodic Levy driven SDE observed art high frequency, *Annals of Statistics* **41**, 1593–1641.

Masuda, H. (2013b): On optimal estimation of stable Ornstein-Uhlenbeck processes. *ArXiv.*

Masuda, H. (2013c): Asymptotics for functionals of self-normalized residuals of discretely observed stochastic processes, *Stochastic Processes and their Applications* **12**, 2752–2778.

Masuda, H. (2013d): Multi-step estimation procedure for stable Ornstein-Uhlenbeck processes, *Stochastic Processes and their Statistics in Finance*, Okinawa.

Masuda, H. (2015): Parametric estimation of levy processes, *Levy Matters IV*, Lecture Notes in Mathematics **2128**, Springer.

Masuda, H. (2019): Non-Gaussian quasi-likelihood estimation of SDE driven by locally stable Levy process, *Stochastic Processes and their Applications* **129**, 1013–1059.

Masuda, H. and Uehara, Y. (2017): Two-step estimation of ergodic Levy driven SDE, *Statistical Inference for Stochastic Processes* **20**, 105–127.

Masuda, H. and Uehara, Y. (2018): Estimating diffusion with compound Poisson jumps based on self-normalized residuals. *ArXiv:1802.03945*

Masuda, H. and Yoshida, N. (2005): On asymptotic expansion for Barndorff-Nielsen and Shephard's stochastic volatility models, *Stochastic Processes and their Applications* **115**, 1167–1186.

Masuda, H. and Yoshida, N. (2013): Edgeworth expansion for the integrated Levy driven Ornstein-Uhlenbeck process, *Electronic Communications in Probability* **18**, 1–10.

McKean, Jr. H.P. (1969): *Stochastic Integrals*, Academic Press, New York.

McLeish, D.L. and Kolkiewicz, A.W. (1997): Fitting diffusion models in finance, In: *Selected Proceedings of the Symposium on Estimating Functions* (Eds.) Basawa, I.V., Godambe, V.G. and Taylor, R.L., IMS Lecture Notes - Monograph Series **32**, 327–350.

McNeil, A.J., Frey, R. and Embrechts, P. (2005): *Quantitative Risk Management: Concepts, Techniques, Tools*, Princeton University Press, Princeton.

McShane, E.J. (1974): *Stochastic Calculus and Stochastic Models*, Academic Press, New York.

Meddahi, N. (2002): A theoretical comparison between integrated and realized volatility, *Journal of Applied Econometrics*, **17**, 479–508.

Merton, R.C. (1990): *Continuous-Time Finance*, Blackwell Publishing, Oxford.

Metivier, M. and Pellaumail, J. (1980): *Stochastic Integration*, Academic Press, New York.

Michel, R. and Pfanzagl, J. (1971): The accuracy of the normal approximation for minimum contrast estimate, *Zeit. Wahr. Verw. Gebiete* **18**, 73–84.

Mider, M., Jenkins, P.A., Pollcok, M., Roberts, G.O., Sørensen, M. (2019): Simulating bridges using confluent diffusions, *ArXiv*.

Milshtein, G.N. (1995): *Numerical Integration of Stochastic Differential Equations*, Kluwer, Norwell, MA.

Mishra, M.N. (1989): The Bernstein- von Mises theorem for a class of nonhomogeneous diffusion processes, *Statist. Decisions* **7**, 153–165.

Mishra, M.N. and Bishwal, J.P.N. (1995): Approximate maximum likelihood estimation for diffusion processes from discrete observations, *Stochastics and Stochastics Reports* **52**, 1–13.

Mishra, M.N. and Prakasa Rao, B.L.S. (1985a): On the Berry-Esseen theorem for maximum likelihood estimator for linear homogeneous diffusion processes, *Sankhyā, Ser A* **47**, 392–398.

Mishra, M.N. and Prakasa Rao, B.L.S. (1985b): Asymptotic study of maximum likelihood estmation for nonhomogeneous diffusion processes, *Statist. Decisions* **3**, 193–203.

Mishra, M.N. and Prakasa Rao, B.L.S. (1987): Rate of convergence in the Bernstein-von Mises theorem for a class of diffusion processes, *Stochastics* **22**, 59–75.

Mishra, M.N. and Prakasa Rao, B.L.S. (1991): Bounds on the equivalence of Bayes and maximum likelihood estimators for a class of diffusion processes, *Statistics* **22**, 613–625.

Mishura, Y. (2008): *Stochastic Calculus for Fractional Brownian Motion and Related Processes*, Lecture Notes in Mathematics **1929**, Springer-Verlag, Berlin.

Mizera, I. and Muller, C.H. (1999): Breakdown points and variation exponents of robust M-estimators in linear models, *Annals of Statistics* **27**, 1164–1177.

Mizera, I. and Muller, C.H. (2002): Breakdown points of Cauchy regression-scale estimators, *Statistics and Probability Letters* **57**, 79–89.

Mohapl, J. (1994): Maximum likelihood estimation in linear infinite dimensional models, *Commun. Statist. - Stoch. Models* **10**, 781–794.

Mohapl, J. (1997): On estimation in planar Ornstein-Uhlenbeck process, *Communications in Statistics - Stochastic Models* **13**, 435–455.

Molchan, S (2000): On the maximum of a fractional Brownian motion, *Theory of Probability and its Applications* **44** (1), 97–102.

Musiela, M. (1976): Risks of maximum likelihood estimators of parameters of an Itô type processes, *Bulletin de L' Academic Polonaise des Sciences, Serie de Sci. Math.* **24**, 1035–1039.

Musiela, M. (1984): *Processes de Diffusions: Aspects Probabilistics et Statistiques*, These l' Universite' Scientifique et Medicale de Grenoble, France.

B. Nielsen, and N. Shephard, (2003): Likelihood analysis of a first order autoregressive model with exponential innovations, *Journal of Time Series Analysis*, **24** (3), 337–344.

Musiela, M and Rutkowski, M. (1997): *Martingale Methods in Financial Modeling*, Springer-Verlag, Berlin.

Nicolay, D. (2014): *Asymptotic Chaos Expansions in Finance: Theory and Practice*, Springer-Verlag, London.

Nielsen, L.T. (1999): *Pricing and Hedging of Derivative Securities*, Oxford University Press, Oxford.

Ninomiya, S. and Victoir, N. (2008): Weak approximation of stochastic differential equations and application to derivative pricing, *Applied Mathematical Finance* **15**, 107–121.

Norros, I., Valkeila, E. and Virtamo, J. (1999): An elementary approach to a Girsanov formula and other analytical results on fractional Brownian motion, *Bernoulli* **5**, 571–587.

Nourdin, I and Peccati, G. (2009): Stein's method on Wiener chaos, *Probability Theory and Related Fields* **145**, 75–118.

Nourdin, I and Peccati, G. (2012): *Normal Approximation with Malliavin Calculus: From Stein's Method to Universality*, Cambridge University Press, Cambridge.

Nourdin, I. and Tran, T.T.D. (2019): Statistical inference for Vasicek-type model driven by Hermite processes, *Stochastic Processes and their Applications* **129**, 3774–3719.

Novikov, A. (1972): Sequential estimation of the parameters of diffusion type processes, *Mathematical Notes* **12**, 812–818.

Nunno, G.D, Oksendal, B and Proske, F. (2009): *Malliavin Calculus for Levy Processes with Applications to Finance*, Springer-Verlag, Berlin.

Oblakova, T.V. (1989): On the rate of convergence in the central limit theorem for stochastic integrals, *Russian Math. Surveys* **44**(2), 289–290.

Nualart, D. (1995): *Malliavin Calculus and Related Topics,* Springer-Verlag, Berlin.

Nualart, D. and Xu, F. (2019): Asymptotic behavior for an additive functional of two independent self-similar Gaussian processes, *Stochastic Processes and their Applications* **129**, 3981–4008.

Øksendal, O. (2003): *Stochastic Differential Equations*, 6th edition, Springer-Verlag, New York-Berlin.

Øksendal, O. and Sulem, A. (2007): *Applied Stochastic Control of Jump Diffusions*, Springer, New York.

El Omari, M., El Maroufy, H., Fuchs, C. (2020): Nonparametric estimation for fractional diffusion processes with random effects, *Statistics: A Journal of Theoretical and Applied Statistics* **53**, 753–769.

Overbeck, L. (1998): Estimation for continuous branching processes, *Scandinavian Journal of Statistics* **25**, 111–126.

Overbeck, L. and Ryden, T. (1997): Estimation in the Cox-Ingersoll-Ross model, *Econometric Theory* **13**, 430–461.

Pap, G and Yor, M. (2000): The accuracy of Cauchy approximation for the windings of planar Brownian motion, *Periodica Mathematica Hungarica* **41**, 213–226.

Pascucci, A. (2011): *PDE and Martingale Methods in Option Pricing*, Springer-Verlag Italia.

Peccati, G. and Taqqu, M.S. (2011): *Wiener Chaos: Moments, Cumulants and Diagrams*, Springer-Verlag Italia.

Peccati, G. and Tudor, C. (2005): Gaussian limits for vector-valued multiple stochastic integrals. In: Emery, M., Lodoux, M. and Yor, M. (eds) Seminaire de Probabilites XXXVIII, *Lecture Notes in Mathematics* **1857**, 247–262, Springer, Berlin-Hiedelberg.

Pages, G. and Panloup, F. (2012): Ergodic approximation of the distribution of a stationary diffusion: Rate of convergence, *Allans of Applied Probability*, **22**, 1059–1100.

Pardoux, E. and Veretennikov, A.Y. (2001): On Poisson equation and diffusion approximation 1, *Annals of Probability* **29** (3), 1061–1085.

Pardoux, E. and Veretennikov, A.Y. (2003): On Poisson equation and diffusion approximation 3, *Annals of Probability* **31** (3), 1166–1192.

Pedersen, A.R. (1994): Quasi likelihood inference for discretely observed diffusion processes, *Research Rep. No. 295*, Department of Theoretical Statistics, University of Åarhus.

Pedersen, A.R. (1995a): A new approach to maximum likelihood estimation for stochastic differential equations based on discrete observations, *Scand. J. Statist.* **22**, 55–71.

Pedersen, A.R. (1995b): Consistency and asymptotic normality of an approximate maximum likelihood estimator for discretely observed diffusion processes, *Bernoulli* **1**, 257–279.

Peltier, R.F., Levy Vehel, J. (1994): A new method for estimating the parameter of fractional Brownian motion, *INRIA Research Report No. 2396.*

Penev, S.I. (1985): Parametric statistical inference for multivariate diffusion processes using discrete observations, *Mathematics and Education in Mathematics*, Proc. Fourteenth Spring Conf. Bulg. Math., Sunny Beach, Sophia, 501–510.

Perman, M. and Wellner, J.A. (1996): On the distribution of Brownian areas, *Annals of Applied Probability* **6** (4), 1091–1111.

Petrov, V.V. (1995): *Limit Theorems of Probability Theory*, Oxford University Press, Oxford.

Petrov, V.V. (1975): *Sums of Independent Random Variables* Springer-Verlag, Berlin-New York.

Pfanzagl, J. (1971): The Berry-Esseen bound for minimum contrast estimators, *Metrika* **17**, 82–91.

Pipiras, V. and Taqqu, M.S. (2000): Integration questions related to fractional Brownian motion, *Probab. Theory Related Fields* **118** (2), 251–291.

Piterbarg, L. and Rozovskii, B.L. (1995): Maximum likelihood estimators in the equations of physical oceanography, In: *Stochastic Modelling in Physical Oceanography*, (Eds.) R.Adler, P.Muller, and B.Rozovskii, Birkhauser, Boston.

Piterbarg, L. and Rozovskii, B.L. (1996): On asymptotic problems of parameter estimation in stochastic PDEs: discrete time sampling, *Math. Methods Statist.* **6** (2), 200–223.

Pitt, M.K. and Shephard, N. (1999): Filtering via simulation: auxiliary particle filters, *Journal of the American Statistical Association* **94**, 590–599.

Platen, E. and Bruti-Liberati, N. (2010): *Numerical Solution of Stochastic Differential Equations with Jumps in Finance*, Springer, Berlin.

Portnoy, S. and Koenker, R. (1997): The Gaussian hare and Laplacian tortoise: computability of squared-error versus absolute-error estimators, *Statistical Science* **12**, 279–300.

Poulsen, R. (1999): Approximate maximum likelihood estimation of discretely observed diffusion processes, *Working Paper No. 29*, Centre for Analytical Finance, University of Aarhus - Aarhus School of Business.

Prakasa Rao, B.L.S. (1973): On the rate of convergence of estimators for Markov Processes, *Zeit. Wahr. Verw. Gebiete* **26**, 141–152.

Prakasa Rao, B.L.S. (1978): On the rate of convergence of the Bernstein-von Mises approximation for Markov processes, *Serdica Bulg. Math. Publ.* **4**, 36–42.

Prakasa Rao, B.L.S. (1980): The Bernstein - von Mises theorem for a class of diffusion processes, *Theory of Random Processes* **9**, 95–101 (in Russian).

Prakasa Rao, B.L.S. (1981): On Bayes estimation for diffusion fields, *In: Statistics: Applications and New Directions, Proc. ISI Golden Jubilee Conferences*, (Eds.) J.K. Ghosh and J. Roy, 504–511.

Prakasa Rao, B.L.S. (1982): Maximum probability estimation for diffusion processes, *In: Statistics and Probability, Essays in honour of C.R. Rao* (Eds.) G. Kallianpur, P.R. Krishnaiah, J.K. Ghosh, North-Holland, Amsterdam, 557–590.

Prakasa Rao, B.L.S. (1983): Asymptotic theory for non-linear least squares estimator for diffusion processes, *Math. Operationsforch Statist. Ser. Statistics* **14**, 195–209.

Prakasa Rao, B.L.S. (1984): On the exponential rate of convergence of the least squares estimator in the nonlinear regression model with Gaussian errors, *Statist. Probab. Letters* **2**, 139–142.

Prakasa Rao, B.L.S. (1985): Estimation of the drift for diffusion processes, *Statistics* **16**, 263–275.

Prakasa Rao, B.L.S. (1986a): Weak convergence of the least squares random process in the smooth case, *Statistics* **17**, 505–516.

Prakasa Rao, B.L.S. (1986b): The weak convergence of the least squares random fields and its applications, *Ann. Inst. Stat. Math.* **38** 475–483.

Prakasa Rao, B.L.S. (1988a): Law of the iterated logarithm for fluctuation of posterior distributions for a class of diffusion processes and a sequential test of power one, *Theor. Prob. Appl.* **33**, 269–275.

Prakasa Rao, B.L.S. (1988b): Statistical inference from sampled data for stochastic processes, In: *Statistical Inference from Stochastic Processes*, (Ed.) N.U.Prabhu *Contemporary Mathematics* **80**, 249–284, American Mathematical Society, Providence, Rhode-Island.

Prakasa Rao, B.L.S. (1999): *Statistical Inference for Diffusion Type Processes*, Arnold, London.

Prakasa Rao, B.L.S. (1999b): *Semimartingales and Their Statistical Inference*, CRC Press, Boca Raton, Florida.

Prakasa Rao, B.L.S. (2010): *Statistical Inference for Fractional Diffusion Processes*, Wiley, UK.

Prakasa Rao, B.L.S. and Rubin, H. (1981): Asymptotic theory of estimation in nonlinear stochastic differential equations, *Sankhyā Ser. A* **43**, 170–189.

Prakasa Rao, B.L.S. and Rubin, H. (1998): Uniform approximation for families of stochastic integrals, *Sankhyā, Ser. A* **60**, 57–73.

Prigent, J.L. (2003): *Weak Convergence of Financial Markets*, Springer-Verlag, Berlin.

Privault, N. (2012): *Am Elementary Introduction to Stochastic Interest Rate Modeling*, World Scientific, Singapore.

Privault, N. (2014): *Stochastic Finance - An Introduction with Market Examples*, Chapman and Hall/CRC, Boca Raton.

Profeta, C., Roynette, B. and Yor, M. (2010): *Option Prices as Probabilities: A New Look at Generalised Black-Scholes Formulae*, Springer-Verlag, Berlin.

Protter, P. (1990): *Stochastic Integration and Differential Equations: A New Approach*, Springer-Verlag, Berlin.

Quenoville, M.H. (1949): Approximate tests of correlation in time series, *J. Royal Statist. Soc. B* **11**, 68–84.

Rachev, S. and Mittnik, S. (2000): *Stable Paretian Models in Finance*, Wiley, New York.

Rachev, S.T., Menn, C. and Fabozzi, F.J. (2005): *Fat-Tailed and Skewed Asset Return Distributions*, Wiley, Hoboken.

Rebonato, R. (2003): *Interest-Rate Option Models*, Second Edition, John Wiley, New York.

Renyi, A. (1963): On stable sequences of events, *Sankhyā Ser. A* **25**, 293–302.

Resnick, S.I. (2007): *Heavy-Tail Phenomena: Probabilistic and Statistical Modeling*, Springer, New York.

Resnick, S.I. (2008): *Extreme Values, Regular variation and Point Processes*, Springer, New York.

Revuz, D. and Yor, M. (1991): *Continuous Martingales and Brownian Motion*, Springer-Verlag.

Rey, C. (2017): Convergence in total variation distance for a third order scheme for one dimensional diffusion process, *Monte Carlo Methods and Applications* **23**, 1–12.

Rogers, L.C.G. and Williams, D. (1987): *Diffusions, Markov Processes and Martingales Vol. 2: Itô Calculus*, John Wiley, New York.

Rogers, L.C.G. and Shi, Z. (1995): The value of an Asian option, *Journal of Applied Probability* **32**, 1077–1088.

Rootzen, H. (1980): Limit distributions for the error in approximation of stochastic integrals, *Ann. Proab.* **8**, 241–251.

Roussas, G.G. (1972): *Contiguity of Probability Measures*, Cambridge University Press.

Robinson, P.M. (1977): Estimation of a time series model from unequally spaced data, *Stoch. Proc. Appl.* **6**, 9–24.

Schmetterer, L. (1974): *Introduction to Mathematical Statistics*, Springer-Verlag, Berlin.

Ruppert, D. (2004): *Statistics and Finance: An Introduction*, Springer, New York.

Ruppert, D. (2011): *Statistics and Data Analysis for Financial Engineering*, Springer, New York.

Ruzmaikina, A.A. (2000): Stochastic calculus for fractional Brownian motion, *J. Statistical Physics*, **100**, 1049–1069.

Sabzikar, F. and Surgailis, D. (2018): Invarinace principles for tempered fractionally integrated processes, *Stochastic Processes and their Applications* **128**, 3419–3438.

Samorodnitsky, G. and Taqqu, M.S. (1994): *Stable Non-Gaussian Random Processes-Stochastic Models with Infinite Variance*, Chapman and Hall, London.

Sandmann, G. and S.J. Koopman (1998): Estimation of Stochastic Volatility Models via Monte Carlo Maximum Likelihood, *Journal of Econometrics*, **87**(2) 271–301.

Sato, K.I. (1999): *Levy Processes and Infinitely Divisible Distributions*, Cambridge University Press, Cambridge.

Schwartz, L. (1965): On Bayes procedures, *Z. Wahrscheinlichkeitstheorie* **4**, 10–26.

Schonkwiler, R.W. (2013): *Finance with Monte Carlo*, Springer, New York.

Shephard, N. and Bos, C.S. (2006): Inference for adaptive time series models: stochastic volatility and conditionally Gaussian state space form, *Econometric Reviews.* **25**, 219–244.

Shephard, N., Doornik, J.A. and Hendry, D.F. (2005): Parallel Computation in Econometrics: A Simplified Approach in *Handbook of Parallel Computing and Statistics*, Marcel Dekker.

Shephard, N. (2006): Stochastic Volatility, In: it New Palgrave Dictionary of Economics, 2nd Edition, (Eds.) Steven Durlauf and Lawrence Blume, 2006.

Shephard, N. and M.K. Pitt (1997): Likelihood analysis of non-Gaussian measurement time series, *Biometrika* **84**, 653–667.

Shevchenko, R. and Tudor, C.A. (2020): Parameter estimation for the Rosenblatt Ornstein-Uhlenbeck process with periodic mean, *Statistical Inference for Stochastic Processes*, **23**, 227–247.

Shimizu, Y. (2006): M-estimation for discretely observed diffusion processes with infinitely many jumps, *Statistical Inference for Stochastic Processes*, **9**, 179–225.

Shimizu, Y. and Yoshida, N. (2006): Estimation of parameters of diffusion processes with jumps from discrete observations, *Statistical Inference for Stochastic Processes* **9**, 227–277.

Shiryaev, A.N. (1999): *Essentials of Stochastic Finance: facts, Models, Theory*, World Scientific, Singapore.

Shiryayev, A.N. and Spokoiny, V.G. (1998): *Statistical Experiments and Decisions*, Math. Sci. Publishers.

Shoji, I. (1997): A note on asymptotic properties of estimator derived from the Euler method for diffusion processes at discrete times, *Statist. Probab. Letters* **36**, 153–159.

Shreve, S.E. (2004): *Stochastic Calculus for Finance I: The Binomial Asset Pricing Model*, Springer-Verlag, New York.

Shreve, S.E. (2004): *Stochastic Calculus for Finance II: Continuous-Time Models*, Springer-Verlag, New York.

Silvestrov, D.S. (2004): *Limit Theorems for Randomly Stopped Stochastic Processes*, Springer-Verlag, London.

Sim, C. (1990): First order autoregressive model for gamma and exponential processes *Journal of Applied Probability*, **27**, 325–332.

Simon, B. (1979): *Trace Ideals and Their Applications*, London Mathematical Society, Lecture Notes Series **35**, Cambridge University Press, Cambridge-New York.

Singer, H. (1993): Continuous time dynamical systems with sampled data, errors of measurement and unobserved components, *J. Time Series Analysis* **14**, 527–544.

Sirignano, J. and Spiliopoulos, K. (2017): Stochastic gradient descent in continuous time, *SIAM Journal of Financial Mathematics*, **8**, 933–961.

Sirignano, J. and Spiliopoulos, K. (2018): DGM: A deep learning algorithm for solving partial differential equations, *Journal of Computational Physics* **375** (15), 1339–1364.

Sirignano, J. and Spiliopoulos, K. (2020): Stochastic gradient descent in continuous time: A central limit theorem, *Stochastic Systems*, **10** (2), 124–151.

Sørensen, B.E. (1992): Continuous record asymptotics in systems of stochastic differential equations, *Econometric Theory* **8**, 28–51.

Sørensen, H. (1997): Parameter Estimation for Discretely Observed Diffusion Processes (In Danish), Master's Thesis, Department of Theoretical Statistics, University of Copenhagen.

Sørensen, H. (2001): Discretely observed diffusions: approximation of the continuous time score function, *Scandinavian Journal of Statistics* **28** (1), 113–121.

Sørensen, M. (1983): On maximum likelihood estimation in randomly stopped diffusion type processes, *Int. Statist. Review* **51**, 93–110.

Sørensen, M. (1997): Estimating functions for discretely observed diffusions: A review, In: *Selected Proceedings of the Symposium on Estimating Functions* (Eds.) Basawa, I.V., Godambe, V.G. and Taylor, R.L., IMS Lecture Notes - Monograph Series **32**, 305–325.

Sørensen, M. (1999): On asymptotics of estimating functions, *Brazilian Journal of Probability and Statistics* **13**, 111–136.

Spitzer, F. (1958): Some theorems concerning 2-dimensional Brownian motion, *Transactions of the American Mathematical Society* **87**, 187–197.

Spokoiny, V.G. (1992): Lower bounds of minimax risk in estimation problems without LAN condition, *Stoch. Stoch. Reports* **38**, 67–93.

Stockmarr, and Jacobsen, M. (1994): Gaussian diffusions and autoregressive processes: weak convergence and statistical inference, *Scand. J. Statistics* **21**, 403–419.

Stoyanov, J. (1984): Problems of estimation in continuous-discrete stochastic models, *Proc. Seventh Conf. on Prob. Theory, Statistical Decision Functions and Random Processes*, Brasov, Romania, 363–373.

Stoyanov, J. and Vladeva, D. (1982): Estimation of unknown parameter of continuous time stochastic processes by observations at random moments, *C.R. Acad. Bulgar. Sci.* **35**, 153–156.

Strasser, H. (1975): The asymptotic equivalence of Bayes and maximum likelihood estimation, *J. Multivariate Anal.* **5**, 206–226.

Strasser, H. (1976): Asymptotic properties of posterior distributions, *Zeit. Wahr. Verw. Gebiete* **35**, 269–282.

Strasser, H. (1977): Improved bounds for Bayes and maximum likelihood estimation, *Theory Prob. Appl.* **22**, 349–361.

Strasser, H. (1985): *Mathematical Theory of Statistics*, Walter de Gruyter, Berlin-New York.

Stratonovich, R.L. (1964): Novaya forma zapisi stokhasticheskikh integralov i uravenni, *Vestnik Moskov Univ. Ser I Mat. Mekh.*, **1**, 3–12. Translated as: A new representation for stochastic integrals and equations, *SIAM J. Control* (1966), 362–371.

Strauch, C. (2015): Sharp adaptive drift estimation for ergodic diffusions: The multivariate case, **125**, 2562–2602.

Straumann, D. (2005): Estimation in Conditionally Heteroscedastic Time Series Models, *Lecture Notes in Statistics*, **181**, Springer-Verlag, Berlin.

Straumann, D. and Mikosch, T. (2006): Quasi-maximum likelihood estimation in conditionally heteroscedastic time series: a stochastic recurrence equations approach, *Annals of Statistics* **34**, 2449–2495.

Stroock, D.W. and Varadhan, S.R.S. (1979): *Multidimensional Diffusion Process*, Springer-Verlag, Berlin.

Su, Y. and Wang, Y. (2019): Parameter estimation for fractional diffusions processes with discrete observations, *Journal of Function Spaces*, Article ID 9036285, 6 pages.

Surace, S. and Pfister, J. (2016): On maximum lilkelihood estimation of partially observed diffusion process, *ArXiv.*

Taniguchi, M. and Kakizawa, Y. (2000): *Asymptotic Theory of Statistical Inference for Time Series*, Springer-Verlag, New York.

Sweeting, T.J. and Adekola, A.O. (1987): Asymptotic posterior normality for stochastic processes revisited, *J. Royal Statist. Soc. B 49*, 215–222.

Tanaka, K. (2013): Distributions of the maximum likelihood and minimum contrast estimators associated with the fractional Ornstein-Uhlenbeck process, *Statistical Inference for Stochastic Processes* **36**, 173–192.

Taraskin, A.F. (1974): On the asymptotic normality of vector valued stochastic integrals and estimates of a multidimensional diffusion process, *Theory Prob. Math. Statist.* **2**, 209–224.

Taraskin, A.F. (1980): The asymptotic statistical problems for fields of diffusion type, *In: Stochastic Differential Systems* (Ed. B. Grigelionis) *Lecture Notes in Control and Information Sciences* **25**, Springer-Verlag, Berlin.

Taqqu, M.S. (1975): Weak convergence to fractional Brownian motion and to the Rosenblatt process, *Z. Wahr. verw. Gebiete* **31**, 287–302.

Tikhov, M.S. (1978): On optimal sequential estimation plans for nonquadratic loss functions, *Theory Prob. Appl.*, **23**, 132–138.

Tikhov, M.S. (1980) : On the sequential estimation of the drift coefficient of a diffusion process with quadratic and nonquadratic losses, *Theory Prob. Appl.* **25**, 607–614.

Thorisson, H. (2000): *Coupling, Stationary and Regeneration*, Springer-Verlag, New York.

Touti, A. (1993): Two theorems on convergence in distribution for stochastic integrals and statistical applications, *Theory Prob. Applic.* **38**, 95–117.

Trofimov, E.I. (1982): A stochastic equation for the maximum likelihood estimate of the parameter of a process of diffusion type, *Russian Math. Surveys* **37**(4), 221–222.

Tsitovich, I.I. (1977): On estimating the drift of a diffusion process from observations in a domain, *Theory Prob. Appl.* **22**, 851–858.

Tudor, C.A. (2013): *Analysis of Variations for Self-similar Processes: A Stochastic Calculus Approach*, Springer, New York.

Tudor, C.A. and Viens, F. (2007): Statistical aspects of the fractional stochastic calculus, *Annals of Statistics*, **35** (3), 1183–1212.

Uehara, Y. (2019): Statistical inference for misspecified ergodic Levy driven stochastic differential equation models, *Stochastic Processes and their Applications* **129** , 4051–4081.

Van der Vaart, A.W. (2000): *Asymptotic Statistics*, Cambridge University Press, Cambridge.

Van Kampen, N.G. (1981): *Stochastic Processes in Physics and Chemistry*, North Holland, Amsterdam.

Van der Meulen, F. and Schauer, M. (2018): Bayesian estimation of incompletely observed diffusions, *Stochastics* **90**, 641–662.

Vasicek, O. (1985): An equilibrium characterization of term structure dynamics, *Journal of Financial Economics* **5**, 177–188.

Veraart, A.E.D. and Veraart, L.A.M. (2012): Stochastic volatility and stochastic leverage, *Annals of Finance* **8**, 205–233.

Vetter, M. (2015): Estimation of integrated volatility of volatility with applications to goodness-of-fit testing, *Bernoulli* **21**, 2393–2418.

Walsh, J.B. (1986): An introduction to stochastic partial differential equations, *Lecture Notes in Mathematics* **1180**, 265–439, Springer, Berlin.

Warren. J. and Yor, M. (1998): The Brownian burglar: conditioning Brownian motion by its local time processes, *Seminaire de Probabilites* **XXXII**, 328–342.

Wilmott, P., Howison, S. and Dewynne, J. (1995): *The Mathematics of Financial Derivatives*, Cambridge University Press, Cambridge.

Woerner, J.H.C. (2005): Estimation of integrated volatility in stochastic volatility models, *Applied Stochastic Models in Business and Industry* **21**, 27–44.

Woerner, J.H.C. (2007): Inference in Levy-type stochastic volatility models, *Advances in Applied Probability* **39**, 531–549.

Wolfe, A.J. (1982): On a continuous analogue of the stochastic difference equation $X_n = \rho X_{n-1} + B_n$, *Stochastic Processes and their Applications* **12**, 301–312.

Wolfowitz, J. (1975): Maximum probability estimators in the classical and in the 'almost smooth' case, *Theory Probab. Appl.* **20**, 363–371.

Wu, C.F.J. (1986): Jackknife, bootstrap and other resampling methods in regression analysis, *Annals of Statistics* **14** (4), 1261–1295.

Yen, J.Y. and Yor, M. (2013): *Local Times and Excursion Theory for Brownian motion: A Tale of Wiener and Itô Measures*, Lecture Notes in Mathematics **2088**, Springer.

Yor, M. (1976): Sur quelques approximations d'integrals stochastiques, Seminaire de Probabilites XI, Lecture Notes in Mathematics **581**, Springer-Verlag.

Yor, M. (1997): The speed of convergence towards Cauchy distribution for the asymptotic windings of planar Brownian motion, *Paris VI Preprint.*

Yor, M (2001): *Exponential Functionals of Brownian Motion and Related Processes*, Springer, Berlin.

Yoshida, N. (1988): Robust M-estimators in diffusion processes, *Ann. Inst. Statist. Math.* **40**, 799–820.

Yoshida, N. (1990): Asymptotic behaviour of M-estimator and related random field for diffusion process, *Ann. Inst. Statist. Math.* **42**, 221–251.

Yoshida, N. (1992): Estimation for diffusion processes from discrete observations, *J. Multivariate Anal.* **41**, 220–242.

Yoshida, M. (1997): Malliavin calculus and asymptotic expansion for martingales, *Probab. Theory Relat. Fields* **109**, 301–342.

Yu, J. and Phillips, P.C.B. (2001): Gaussian estimation of continuous time model of the short term interest rate models, *Cowles Foundation Discussion Paper No. 1309*, Yale University.

Zagst, R. (2002): *Interest Rate Management*, Springer-Verlag, Berlin.

Zahle, M. (1998): Integration with respect to fractal functions and stochastic calculus, *Prob. Theory. Rel. Fields* **111**, 333–374.

Zhang, L. (2006): Efficient estimation of stochastic volatility models using noisy observations: a multi-scale approach, *Bernoulli* **12**, 1019–1043.

Zhang, L., Mykland, P. and Ait-Sahalia, Y. (2005): A tale of two time scales: determining integrated volatility with noisy high-frequency data, *Journal of American Statistical Association* **100**, 1394–1411.

Zhang, L., Mykland, P. and Ait-Sahalia, Y. (2011): Edgeworth expansions of realized volatility and related estimators, *Journal of Econometrics* **160** (1), 190–203.

Zhou, B. (1996): High-frequency data and volatility in foreign-exchange rates, *Journal of Business and Economic Statistics*, **14**, 45–52.

Zhou, H. (2003): Itô conditional moment generator and estimation of short rate processes, *Journal of Financial Econometrics* **1**, 250–271.

Index

J. P. N. Bishwal, *Parameter Estimation in Stochastic Volatility Models*, https://doi.org/10.1007/978-3-031-03861-7